Lecture Notes in Computer Science 12907

More information about this subseries at http://www.springer.com/series/7412

Marleen de Bruijne · Philippe C. Cattin ·
Stéphane Cotin · Nicolas Padoy ·
Stefanie Speidel · Yefeng Zheng ·
Caroline Essert (Eds.)

Medical Image Computing and Computer Assisted Intervention – MICCAI 2021

24th International Conference
Strasbourg, France, September 27 – October 1, 2021
Proceedings, Part VII

 Springer

Editors
Marleen de Bruijne 🆔
Erasmus MC - University Medical Center
Rotterdam
Rotterdam, The Netherlands

University of Copenhagen
Copenhagen, Denmark

Stéphane Cotin 🆔
Inria Nancy Grand Est
Villers-lès-Nancy, France

Stefanie Speidel 🆔
National Center for Tumor Diseases
(NCT/UCC)
Dresden, Germany

Caroline Essert 🆔
ICube, Université de Strasbourg, CNRS
Strasbourg, France

Philippe C. Cattin 🆔
University of Basel
Allschwil, Switzerland

Nicolas Padoy 🆔
ICube, Université de Strasbourg, CNRS
Strasbourg, France

Yefeng Zheng 🆔
Tencent Jarvis Lab
Shenzhen, China

ISSN 0302-9743   ISSN 1611-3349 (electronic)
Lecture Notes in Computer Science
ISBN 978-3-030-87233-5   ISBN 978-3-030-87234-2 (eBook)
https://doi.org/10.1007/978-3-030-87234-2

LNCS Sublibrary: SL6 – Image Processing, Computer Vision, Pattern Recognition, and Graphics

This Springer imprint is published by the registered company Springer Nature Switzerland AG
The registered company address is: Gewerbestrasse 11, 6330 Cham, Switzerland

Preface

The 24th edition of the International Conference on Medical Image Computing and Computer Assisted Intervention (MICCAI 2021) has for the second time been placed under the shadow of COVID-19. Complicated situations due to the pandemic and multiple lockdowns have affected our lives during the past year, sometimes perturbing the researchers work, but also motivating an extraordinary dedication from many of our colleagues, and significant scientific advances in the fight against the virus. After another difficult year, most of us were hoping to be able to travel and finally meet in person at MICCAI 2021, which was supposed to be held in Strasbourg, France. Unfortunately, due to the uncertainty of the global situation, MICCAI 2021 had to be moved again to a virtual event that was held over five days from September 27 to October 1, 2021. Taking advantage of the experience gained last year and of the fast-evolving platforms, the organizers of MICCAI 2021 redesigned the schedule and the format. To offer the attendees both a strong scientific content and an engaging experience, two virtual platforms were used: Pathable for the oral and plenary sessions and SpatialChat for lively poster sessions, industrial booths, and networking events in the form of interactive group video chats.

These proceedings of MICCAI 2021 showcase all 531 papers that were presented at the main conference, organized into eight volumes in the Lecture Notes in Computer Science (LNCS) series as follows:

- Part I, LNCS Volume 12901: Image Segmentation
- Part II, LNCS Volume 12902: Machine Learning 1
- Part III, LNCS Volume 12903: Machine Learning 2
- Part IV, LNCS Volume 12904: Image Registration and Computer Assisted Intervention
- Part V, LNCS Volume 12905: Computer Aided Diagnosis
- Part VI, LNCS Volume 12906: Image Reconstruction and Cardiovascular Imaging
- Part VII, LNCS Volume 12907: Clinical Applications
- Part VIII, LNCS Volume 12908: Microscopic, Ophthalmic, and Ultrasound Imaging

These papers were selected after a thorough double-blind peer review process. We followed the example set by past MICCAI meetings, using Microsoft's Conference Managing Toolkit (CMT) for paper submission and peer reviews, with support from the Toronto Paper Matching System (TPMS), to partially automate paper assignment to area chairs and reviewers, and from iThenticate to detect possible cases of plagiarism.

Following a broad call to the community we received 270 applications to become an area chair for MICCAI 2021. From this group, the program chairs selected a total of 96 area chairs, aiming for diversity — MIC versus CAI, gender, geographical region, and a mix of experienced and new area chairs. Reviewers were recruited also via an open call for volunteers from the community (288 applications, of which 149 were selected by the program chairs) as well as by re-inviting past reviewers, leading to a total of 1340 registered reviewers.

We received 1630 full paper submissions after an original 2667 intentions to submit. Four papers were rejected without review because of concerns of (self-)plagiarism and dual submission and one additional paper was rejected for not adhering to the MICCAI page restrictions; two further cases of dual submission were discovered and rejected during the review process. Five papers were withdrawn by the authors during review and after acceptance.

The review process kicked off with a reviewer tutorial and an area chair meeting to discuss the review process, criteria for MICCAI acceptance, how to write a good (meta-)review, and expectations for reviewers and area chairs. Each area chair was assigned 16–18 manuscripts for which they suggested potential reviewers using TPMS scores, self-declared research area(s), and the area chair's knowledge of the reviewers' expertise in relation to the paper, while conflicts of interest were automatically avoided by CMT. Reviewers were invited to bid for the papers for which they had been suggested by an area chair or which were close to their expertise according to TPMS. Final reviewer allocations via CMT took account of reviewer bidding, prioritization of area chairs, and TPMS scores, leading to on average four reviews performed per person by a total of 1217 reviewers.

Following the initial double-blind review phase, area chairs provided a meta-review summarizing key points of reviews and a recommendation for each paper. The program chairs then evaluated the reviews and their scores, along with the recommendation from the area chairs, to directly accept 208 papers (13%) and reject 793 papers (49%); the remainder of the papers were sent for rebuttal by the authors. During the rebuttal phase, two additional area chairs were assigned to each paper. The three area chairs then independently ranked their papers, wrote meta-reviews, and voted to accept or reject the paper, based on the reviews, rebuttal, and manuscript. The program chairs checked all meta-reviews, and in some cases where the difference between rankings was high or comments were conflicting, they also assessed the original reviews, rebuttal, and submission. In all other cases a majority voting scheme was used to make the final decision. This process resulted in the acceptance of a further 325 papers for an overall acceptance rate of 33%.

Acceptance rates were the same between medical image computing (MIC) and computer assisted interventions (CAI) papers, and slightly lower where authors classified their paper as both MIC and CAI. Distribution of the geographical region of the first author as indicated in the optional demographic survey was similar among submitted and accepted papers.

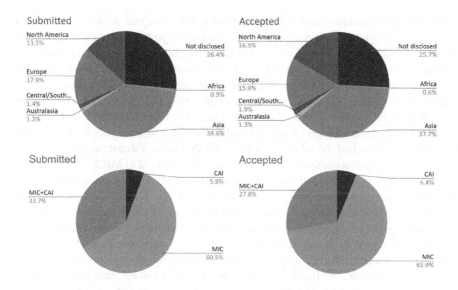

New this year, was the requirement to fill out a reproducibility checklist when submitting an intention to submit to MICCAI, in order to stimulate authors to think about what aspects of their method and experiments they should include to allow others to reproduce their results. Papers that included an anonymous code repository and/or indicated that the code would be made available were more likely to be accepted. From all accepted papers, 273 (51%) included a link to a code repository with the camera-ready submission.

Another novelty this year is that we decided to make the reviews, meta-reviews, and author responses for accepted papers available on the website. We hope the community will find this a useful resource.

The outstanding program of MICCAI 2021 was enriched by four exceptional keynote talks given by Alyson McGregor, Richard Satava, Fei-Fei Li, and Pierre Jannin, on hot topics such as gender bias in medical research, clinical translation to industry, intelligent medicine, and sustainable research. This year, as in previous years, high-quality satellite events completed the program of the main conference: 28 workshops, 23 challenges, and 14 tutorials; without forgetting the increasingly successful plenary events, such as the Women in MICCAI (WiM) meeting, the MICCAI Student Board (MSB) events, the 2nd Startup Village, the MICCAI-RSNA panel, and the first "Reinforcing Inclusiveness & diverSity and Empowering MICCAI" (or RISE-MICCAI) event.

MICCAI 2021 has also seen the first edition of CLINICCAI, the clinical day of MICCAI. Organized by Nicolas Padoy and Lee Swanstrom, this new event will hopefully help bring the scientific and clinical communities closer together, and foster collaborations and interaction. A common keynote connected the two events. We hope this effort will be pursued in the next editions.

We would like to thank everyone who has contributed to making MICCAI 2021 a success. First of all, we sincerely thank the authors, area chairs, reviewers, and session chairs for their dedication and for offering the participants and readers of these proceedings content of exceptional quality. Special thanks go to our fantastic submission platform manager Kitty Wong, who has been a tremendous help in the entire process from reviewer and area chair selection, paper submission, and the review process to the preparation of these proceedings. We also thank our very efficient team of satellite events chairs and coordinators, led by Cristian Linte and Matthieu Chabanas: the workshop chairs, Amber Simpson, Denis Fortun, Marta Kersten-Oertel, and Sandrine Voros; the challenges chairs, Annika Reinke, Spyridon Bakas, Nicolas Passat, and Ingerid Reinersten; and the tutorial chairs, Sonia Pujol and Vincent Noblet, as well as all the satellite event organizers for the valuable content added to MICCAI. Our special thanks also go to John Baxter and his team who worked hard on setting up and populating the virtual platforms, to Alejandro Granados for his valuable help and efficient communication on social media, and to Shelley Wallace and Anna Van Vliet for marketing and communication. We are also very grateful to Anirban Mukhopadhay for his management of the sponsorship, and of course many thanks to the numerous sponsors who supported the conference, often with continuous engagement over many years. This year again, our thanks go to Marius Linguraru and his team who supervised a range of actions to help, and promote, career development, among which were the mentorship program and the Startup Village. And last but not least, our wholehearted thanks go to Mehmet and the wonderful team at Dekon Congress and Tourism for their great professionalism and reactivity in the management of all logistical aspects of the event.

Finally, we thank the MICCAI society and the Board of Directors for their support throughout the years, starting with the first discussions about bringing MICCAI to Strasbourg in 2017.

We look forward to seeing you at MICCAI 2022.

September 2021

Marleen de Bruijne
Philippe Cattin
Stéphane Cotin
Nicolas Padoy
Stefanie Speidel
Yefeng Zheng
Caroline Essert

Organization

General Chair

Caroline Essert — Université de Strasbourg, CNRS, ICube, France

Program Chairs

Marleen de Bruijne — Erasmus MC Rotterdam, The Netherlands, and University of Copenhagen, Denmark

Philippe C. Cattin — University of Basel, Switzerland

Stéphane Cotin — Inria, France

Nicolas Padoy — Université de Strasbourg, CNRS, ICube, IHU, France

Stefanie Speidel — National Center for Tumor Diseases, Dresden, Germany

Yefeng Zheng — Tencent Jarvis Lab, China

Satellite Events Coordinators

Cristian Linte — Rochester Institute of Technology, USA

Matthieu Chabanas — Université Grenoble Alpes, France

Workshop Team

Amber Simpson — Queen's University, Canada

Denis Fortun — Université de Strasbourg, CNRS, ICube, France

Marta Kersten-Oertel — Concordia University, Canada

Sandrine Voros — TIMC-IMAG, INSERM, France

Challenges Team

Annika Reinke — German Cancer Research Center, Germany

Spyridon Bakas — University of Pennsylvania, USA

Nicolas Passat — Université de Reims Champagne-Ardenne, France

Ingerid Reinersten — SINTEF, NTNU, Norway

Tutorial Team

Vincent Noblet — Université de Strasbourg, CNRS, ICube, France

Sonia Pujol — Harvard Medical School, Brigham and Women's Hospital, USA

Clinical Day Chairs

Nicolas Padoy	Université de Strasbourg, CNRS, ICube, IHU, France
Lee Swanström	IHU Strasbourg, France

Sponsorship Chairs

Anirban Mukhopadhyay	Technische Universität Darmstadt, Germany
Yanwu Xu	Baidu Inc., China

Young Investigators and Early Career Development Program Chairs

Marius Linguraru	Children's National Institute, USA
Antonio Porras	Children's National Institute, USA
Daniel Racoceanu	Sorbonne Université/Brain Institute, France
Nicola Rieke	NVIDIA, Germany
Renee Yao	NVIDIA, USA

Social Media Chairs

Alejandro Granados Martinez	King's College London, UK
Shuwei Xing	Robarts Research Institute, Canada
Maxence Boels	King's College London, UK

Green Team

Pierre Jannin	INSERM, Université de Rennes 1, France
Étienne Baudrier	Université de Strasbourg, CNRS, ICube, France

Student Board Liaison

Éléonore Dufresne	Université de Strasbourg, CNRS, ICube, France
Étienne Le Quentrec	Université de Strasbourg, CNRS, ICube, France
Vinkle Srivastav	Université de Strasbourg, CNRS, ICube, France

Submission Platform Manager

Kitty Wong	The MICCAI Society, Canada

Virtual Platform Manager

John Baxter	INSERM, Université de Rennes 1, France

Program Committee

Ehsan Adeli	Stanford University, USA
Iman Aganj	Massachusetts General Hospital, Harvard Medical School, USA
Pablo Arbelaez	Universidad de los Andes, Colombia
John Ashburner	University College London, UK
Meritxell Bach Cuadra	University of Lausanne, Switzerland
Sophia Bano	University College London, UK
Adrien Bartoli	Université Clermont Auvergne, France
Christian Baumgartner	ETH Zürich, Switzerland
Hrvoje Bogunovic	Medical University of Vienna, Austria
Weidong Cai	University of Sydney, Australia
Gustavo Carneiro	University of Adelaide, Australia
Chao Chen	Stony Brook University, USA
Elvis Chen	Robarts Research Institute, Canada
Hao Chen	Hong Kong University of Science and Technology, Hong Kong SAR
Albert Chung	Hong Kong University of Science and Technology, Hong Kong SAR
Adrian Dalca	Massachusetts Institute of Technology, USA
Adrien Depeursinge	HES-SO Valais-Wallis, Switzerland
Jose Dolz	ÉTS Montréal, Canada
Ruogu Fang	University of Florida, USA
Dagan Feng	University of Sydney, Australia
Huazhu Fu	Inception Institute of Artificial Intelligence, United Arab Emirates
Mingchen Gao	University at Buffalo, The State University of New York, USA
Guido Gerig	New York University, USA
Orcun Goksel	Uppsala University, Sweden
Alberto Gomez	King's College London, UK
Ilker Hacihaliloglu	Rutgers University, USA
Adam Harrison	PAII Inc., USA
Mattias Heinrich	University of Lübeck, Germany
Yi Hong	Shanghai Jiao Tong University, China
Yipeng Hu	University College London, UK
Junzhou Huang	University of Texas at Arlington, USA
Xiaolei Huang	The Pennsylvania State University, USA
Jana Hutter	King's College London, UK
Madhura Ingalhalikar	Symbiosis Center for Medical Image Analysis, India
Shantanu Joshi	University of California, Los Angeles, USA
Samuel Kadoury	Polytechnique Montréal, Canada
Fahmi Khalifa	Mansoura University, Egypt
Hosung Kim	University of Southern California, USA
Minjeong Kim	University of North Carolina at Greensboro, USA

Ender Konukoglu	ETH Zürich, Switzerland
Bennett Landman	Vanderbilt University, USA
Ignacio Larrabide	CONICET, Argentina
Baiying Lei	Shenzhen University, China
Gang Li	University of North Carolina at Chapel Hill, USA
Mingxia Liu	University of North Carolina at Chapel Hill, USA
Herve Lombaert	ÉTS Montréal, Canada, and Inria, France
Marco Lorenzi	Inria, France
Le Lu	PAII Inc., USA
Xiongbiao Luo	Xiamen University, China
Dwarikanath Mahapatra	Inception Institute of Artificial Intelligence, United Arab Emirates
Andreas Maier	FAU Erlangen-Nuremberg, Germany
Erik Meijering	University of New South Wales, Australia
Hien Nguyen	University of Houston, USA
Marc Niethammer	University of North Carolina at Chapel Hill, USA
Tingying Peng	Technische Universität München, Germany
Caroline Petitjean	Université de Rouen, France
Dzung Pham	Henry M. Jackson Foundation, USA
Hedyeh Rafii-Tari	Auris Health Inc, USA
Islem Rekik	Istanbul Technical University, Turkey
Nicola Rieke	NVIDIA, Germany
Su Ruan	Laboratoire LITIS, France
Thomas Schultz	University of Bonn, Germany
Sharmishtaa Seshamani	Allen Institute, USA
Yonggang Shi	University of Southern California, USA
Darko Stern	Technical University of Graz, Austria
Carole Sudre	King's College London, UK
Heung-Il Suk	Korea University, South Korea
Jian Sun	Xi'an Jiaotong University, China
Raphael Sznitman	University of Bern, Switzerland
Amir Tahmasebi	Enlitic, USA
Qian Tao	Delft University of Technology, The Netherlands
Tolga Tasdizen	University of Utah, USA
Martin Urschler	University of Auckland, New Zealand
Archana Venkataraman	Johns Hopkins University, USA
Guotai Wang	University of Electronic Science and Technology of China, China
Hongzhi Wang	IBM Almaden Research Center, USA
Hua Wang	Colorado School of Mines, USA
Qian Wang	Shanghai Jiao Tong University, China
Yalin Wang	Arizona State University, USA
Fuyong Xing	University of Colorado Denver, USA
Daguang Xu	NVIDIA, USA
Yanwu Xu	Baidu, China
Ziyue Xu	NVIDIA, USA

Zhong Xue	Shanghai United Imaging Intelligence, China
Xin Yang	Huazhong University of Science and Technology, China
Jianhua Yao	National Institutes of Health, USA
Zhaozheng Yin	Stony Brook University, USA
Yixuan Yuan	City University of Hong Kong, Hong Kong SAR
Liang Zhan	University of Pittsburgh, USA
Tuo Zhang	Northwestern Polytechnical University, China
Yitian Zhao	Chinese Academy of Sciences, China
Luping Zhou	University of Sydney, Australia
S. Kevin Zhou	Chinese Academy of Sciences, China
Dajiang Zhu	University of Texas at Arlington, USA
Xiahai Zhuang	Fudan University, China
Maria A. Zuluaga	EURECOM, France

Reviewers

Alaa Eldin Abdelaal
Khalid Abdul Jabbar
Purang Abolmaesumi
Mazdak Abulnaga
Maryam Afzali
Priya Aggarwal
Ola Ahmad
Sahar Ahmad
Euijoon Ahn
Alireza Akhondi-Asl
Saad Ullah Akram
Dawood Al Chanti
Daniel Alexander
Sharib Ali
Lejla Alic
Omar Al-Kadi
Maximilian Allan
Pierre Ambrosini
Sameer Antani
Michela Antonelli
Jacob Antunes
Syed Anwar
Ignacio Arganda-Carreras
Mohammad Ali Armin
Md Ashikuzzaman
Mehdi Astaraki
Angélica Atehortúa
Gowtham Atluri

Chloé Audigier
Kamran Avanaki
Angelica Aviles-Rivero
Suyash Awate
Dogu Baran Aydogan
Qinle Ba
Morteza Babaie
Hyeon-Min Bae
Woong Bae
Junjie Bai
Wenjia Bai
Ujjwal Baid
Spyridon Bakas
Yaël Balbastre
Marcin Balicki
Fabian Balsiger
Abhirup Banerjee
Sreya Banerjee
Shunxing Bao
Adrian Barbu
Sumana Basu
Mathilde Bateson
Deepti Bathula
John Baxter
Bahareh Behboodi
Delaram Behnami
Mikhail Belyaev
Aicha BenTaieb

Camilo Bermudez
Gabriel Bernardino
Hadrien Bertrand
Alaa Bessadok
Michael Beyeler
Indrani Bhattacharya
Chetan Bhole
Lei Bi
Gui-Bin Bian
Ryoma Bise
Stefano B. Blumberg
Ester Bonmati
Bhushan Borotikar
Jiri Borovec
Ilaria Boscolo Galazzo
Alexandre Bousse
Nicolas Boutry
Behzad Bozorgtabar
Nathaniel Braman
Nadia Brancati
Katharina Breininger
Christopher Bridge
Esther Bron
Rupert Brooks
Qirong Bu
Duc Toan Bui
Ninon Burgos
Nikolay Burlutskiy
Hendrik Burwinkel
Russell Butler
Michał Byra
Ryan Cabeen
Mariano Cabezas
Hongmin Cai
Jinzheng Cai
Yunliang Cai
Sema Candemir
Bing Cao
Qing Cao
Shilei Cao
Tian Cao
Weiguo Cao
Aaron Carass
M. Jorge Cardoso
Adrià Casamitjana
Matthieu Chabanas

Ahmad Chaddad
Jayasree Chakraborty
Sylvie Chambon
Yi Hao Chan
Ming-Ching Chang
Peng Chang
Violeta Chang
Sudhanya Chatterjee
Christos Chatzichristos
Antong Chen
Chang Chen
Cheng Chen
Dongdong Chen
Geng Chen
Hanbo Chen
Jianan Chen
Jianxu Chen
Jie Chen
Junxiang Chen
Lei Chen
Li Chen
Liangjun Chen
Min Chen
Pingjun Chen
Qiang Chen
Shuai Chen
Tianhua Chen
Tingting Chen
Xi Chen
Xiaoran Chen
Xin Chen
Xuejin Chen
Yuhua Chen
Yukun Chen
Zhaolin Chen
Zhineng Chen
Zhixiang Chen
Erkang Cheng
Jun Cheng
Li Cheng
Yuan Cheng
Farida Cheriet
Minqi Chong
Jaegul Choo
Aritra Chowdhury
Gary Christensen

Daan Christiaens
Stergios Christodoulidis
Ai Wern Chung
Pietro Antonio Cicalese
Özgün Çiçek
Celia Cintas
Matthew Clarkson
Jaume Coll-Font
Toby Collins
Olivier Commowick
Pierre-Henri Conze
Timothy Cootes
Luca Corinzia
Teresa Correia
Hadrien Courtecuisse
Jeffrey Craley
Hui Cui
Jianan Cui
Zhiming Cui
Kathleen Curran
Claire Cury
Tobias Czempiel
Vedrana Dahl
Haixing Dai
Rafat Damseh
Bilel Daoud
Neda Davoudi
Laura Daza
Sandro De Zanet
Charles Delahunt
Yang Deng
Cem Deniz
Felix Denzinger
Hrishikesh Deshpande
Christian Desrosiers
Blake Dewey
Neel Dey
Raunak Dey
Jwala Dhamala
Yashin Dicente Cid
Li Ding
Xinghao Ding
Zhipeng Ding
Konstantin Dmitriev
Ines Domingues
Liang Dong

Mengjin Dong
Nanqing Dong
Reuben Dorent
Sven Dorkenwald
Qi Dou
Simon Drouin
Niharika D'Souza
Lei Du
Hongyi Duanmu
Nicolas Duchateau
James Duncan
Luc Duong
Nicha Dvornek
Dmitry V. Dylov
Oleh Dzyubachyk
Roy Eagleson
Mehran Ebrahimi
Jan Egger
Alma Eguizabal
Gudmundur Einarsson
Ahmed Elazab
Mohammed S. M. Elbaz
Shireen Elhabian
Mohammed Elmogy
Amr Elsawy
Ahmed Eltanboly
Sandy Engelhardt
Ertunc Erdil
Marius Erdt
Floris Ernst
Boris Escalante-Ramírez
Maria Escobar
Mohammad Eslami
Nazila Esmaeili
Marco Esposito
Oscar Esteban
Théo Estienne
Ivan Ezhov
Deng-Ping Fan
Jingfan Fan
Xin Fan
Yonghui Fan
Xi Fang
Zhenghan Fang
Aly Farag
Mohsen Farzi

Lina Felsner
Jun Feng
Ruibin Feng
Xinyang Feng
Yuan Feng
Aaron Fenster
Aasa Feragen
Henrique Fernandes
Enzo Ferrante
Jean Feydy
Lukas Fischer
Peter Fischer
Antonio Foncubierta-Rodríguez
Germain Forestier
Nils Daniel Forkert
Jean-Rassaire Fouefack
Moti Freiman
Wolfgang Freysinger
Xueyang Fu
Yunguan Fu
Wolfgang Fuhl
Isabel Funke
Philipp Fürnstahl
Pedro Furtado
Ryo Furukawa
Jin Kyu Gahm
Laurent Gajny
Adrian Galdran
Yu Gan
Melanie Ganz
Cong Gao
Dongxu Gao
Linlin Gao
Siyuan Gao
Yixin Gao
Yue Gao
Zhifan Gao
Alfonso Gastelum-Strozzi
Srishti Gautam
Bao Ge
Rongjun Ge
Zongyuan Ge
Sairam Geethanath
Shiv Gehlot
Nils Gessert
Olivier Gevaert

Sandesh Ghimire
Ali Gholipour
Sayan Ghosal
Andrea Giovannini
Gabriel Girard
Ben Glocker
Arnold Gomez
Mingming Gong
Cristina González
German Gonzalez
Sharath Gopal
Karthik Gopinath
Pietro Gori
Michael Götz
Shuiping Gou
Maged Goubran
Sobhan Goudarzi
Dushyant Goyal
Mark Graham
Bertrand Granado
Alejandro Granados
Vicente Grau
Lin Gu
Shi Gu
Xianfeng Gu
Yun Gu
Zaiwang Gu
Hao Guan
Ricardo Guerrero
Houssem-Eddine Gueziri
Dazhou Guo
Hengtao Guo
Jixiang Guo
Pengfei Guo
Xiaoqing Guo
Yi Guo
Yulan Guo
Yuyu Guo
Krati Gupta
Vikash Gupta
Praveen Gurunath Bharathi
Boris Gutman
Prashnna Gyawali
Stathis Hadjidemetriou
Mohammad Hamghalam
Hu Han

Liang Han
Xiaoguang Han
Xu Han
Zhi Han
Zhongyi Han
Jonny Hancox
Xiaoke Hao
Nandinee Haq
Ali Hatamizadeh
Charles Hatt
Andreas Hauptmann
Mohammad Havaei
Kelei He
Nanjun He
Tiancheng He
Xuming He
Yuting He
Nicholas Heller
Alessa Hering
Monica Hernandez
Carlos Hernandez-Matas
Kilian Hett
Jacob Hinkle
David Ho
Nico Hoffmann
Matthew Holden
Sungmin Hong
Yoonmi Hong
Antal Horváth
Md Belayat Hossain
Benjamin Hou
William Hsu
Tai-Chiu Hsung
Kai Hu
Shi Hu
Shunbo Hu
Wenxing Hu
Xiaoling Hu
Xiaowei Hu
Yan Hu
Zhenhong Hu
Heng Huang
Qiaoying Huang
Yi-Jie Huang
Yixing Huang
Yongxiang Huang

Yue Huang
Yufang Huang
Arnaud Huaulmé
Henkjan Huisman
Yuankai Huo
Andreas Husch
Mohammad Hussain
Raabid Hussain
Sarfaraz Hussein
Khoi Huynh
Seong Jae Hwang
Emmanuel Iarussi
Kay Igwe
Abdullah-Al-Zubaer Imran
Ismail Irmakci
Mobarakol Islam
Mohammad Shafkat Islam
Vamsi Ithapu
Koichi Ito
Hayato Itoh
Oleksandra Ivashchenko
Yuji Iwahori
Shruti Jadon
Mohammad Jafari
Mostafa Jahanifar
Amir Jamaludin
Mirek Janatka
Won-Dong Jang
Uditha Jarayathne
Ronnachai Jaroensri
Golara Javadi
Rohit Jena
Rachid Jennane
Todd Jensen
Won-Ki Jeong
Yuanfeng Ji
Zhanghexuan Ji
Haozhe Jia
Jue Jiang
Tingting Jiang
Xiang Jiang
Jianbo Jiao
Zhicheng Jiao
Amelia Jiménez-Sánchez
Dakai Jin
Yueming Jin

Bin Jing
Anand Joshi
Yohan Jun
Kyu-Hwan Jung
Alain Jungo
Manjunath K N
Ali Kafaei Zad Tehrani
Bernhard Kainz
John Kalafut
Michael C. Kampffmeyer
Qingbo Kang
Po-Yu Kao
Neerav Karani
Turkay Kart
Satyananda Kashyap
Amin Katouzian
Alexander Katzmann
Prabhjot Kaur
Erwan Kerrien
Hoel Kervadec
Ashkan Khakzar
Nadieh Khalili
Siavash Khallaghi
Farzad Khalvati
Bishesh Khanal
Pulkit Khandelwal
Maksim Kholiavchenko
Naji Khosravan
Seyed Mostafa Kia
Daeseung Kim
Hak Gu Kim
Hyo-Eun Kim
Jae-Hun Kim
Jaeil Kim
Jinman Kim
Mansu Kim
Namkug Kim
Seong Tae Kim
Won Hwa Kim
Andrew King
Atilla Kiraly
Yoshiro Kitamura
Tobias Klinder
Bin Kong
Jun Kong
Tomasz Konopczynski

Bongjin Koo
Ivica Kopriva
Kivanc Kose
Mateusz Kozinski
Anna Kreshuk
Anithapriya Krishnan
Pavitra Krishnaswamy
Egor Krivov
Frithjof Kruggel
Alexander Krull
Elizabeth Krupinski
Serife Kucur
David Kügler
Hugo Kuijf
Abhay Kumar
Ashnil Kumar
Kuldeep Kumar
Nitin Kumar
Holger Kunze
Tahsin Kurc
Anvar Kurmukov
Yoshihiro Kuroda
Jin Tae Kwak
Yongchan Kwon
Francesco La Rosa
Aymen Laadhari
Dmitrii Lachinov
Alain Lalande
Tryphon Lambrou
Carole Lartizien
Bianca Lassen-Schmidt
Ngan Le
Leo Lebrat
Christian Ledig
Eung-Joo Lee
Hyekyoung Lee
Jong-Hwan Lee
Matthew Lee
Sangmin Lee
Soochahn Lee
Étienne Léger
Stefan Leger
Andreas Leibetseder
Rogers Jeffrey Leo John
Juan Leon
Bo Li

Chongyi Li
Fuhai Li
Hongming Li
Hongwei Li
Jian Li
Jianning Li
Jiayun Li
Junhua Li
Kang Li
Mengzhang Li
Ming Li
Qing Li
Shaohua Li
Shuyu Li
Weijian Li
Weikai Li
Wenqi Li
Wenyuan Li
Xiang Li
Xiaomeng Li
Xiaoxiao Li
Xin Li
Xiuli Li
Yang Li
Yi Li
Yuexiang Li
Zeju Li
Zhang Li
Zhiyuan Li
Zhjin Li
Gongbo Liang
Jianming Liang
Libin Liang
Yuan Liang
Haofu Liao
Ruizhi Liao
Wei Liao
Xiangyun Liao
Roxane Licandro
Gilbert Lim
Baihan Lin
Hongxiang Lin
Jianyu Lin
Yi Lin
Claudia Lindner
Geert Litjens

Bin Liu
Chi Liu
Daochang Liu
Dong Liu
Dongnan Liu
Feng Liu
Hangfan Liu
Hong Liu
Huafeng Liu
Jianfei Liu
Jingya Liu
Kai Liu
Kefei Liu
Lihao Liu
Mengting Liu
Peng Liu
Qin Liu
Quande Liu
Shengfeng Liu
Shenghua Liu
Shuangjun Liu
Sidong Liu
Siqi Liu
Tianrui Liu
Xiao Liu
Xinyang Liu
Xinyu Liu
Yan Liu
Yikang Liu
Yong Liu
Yuan Liu
Yue Liu
Yuhang Liu
Andrea Loddo
Nicolas Loménie
Daniel Lopes
Bin Lou
Jian Lou
Nicolas Loy Rodas
Donghuan Lu
Huanxiang Lu
Weijia Lu
Xiankai Lu
Yongyi Lu
Yueh-Hsun Lu
Yuhang Lu

Imanol Luengo
Jie Luo
Jiebo Luo
Luyang Luo
Ma Luo
Bin Lv
Jinglei Lv
Junyan Lyu
Qing Lyu
Yuanyuan Lyu
Andy J. Ma
Chunwei Ma
Da Ma
Hua Ma
Kai Ma
Lei Ma
Anderson Maciel
Amirreza Mahbod
S. Sara Mahdavi
Mohammed Mahmoud
Saïd Mahmoudi
Klaus H. Maier-Hein
Bilal Malik
Ilja Manakov
Matteo Mancini
Tommaso Mansi
Yunxiang Mao
Brett Marinelli
Pablo Márquez Neila
Carsten Marr
Yassine Marrakchi
Fabio Martinez
Andre Mastmeyer
Tejas Sudharshan Mathai
Dimitrios Mavroeidis
Jamie McClelland
Pau Medrano-Gracia
Raghav Mehta
Sachin Mehta
Raphael Meier
Qier Meng
Qingjie Meng
Yanda Meng
Martin Menten
Odyssée Merveille
Islem Mhiri

Liang Mi
Stijn Michielse
Abhishek Midya
Fausto Milletari
Hyun-Seok Min
Zhe Min
Tadashi Miyamoto
Sara Moccia
Hassan Mohy-ud-Din
Tony C. W. Mok
Rafael Molina
Mehdi Moradi
Rodrigo Moreno
Kensaku Mori
Lia Morra
Linda Moy
Mohammad Hamed Mozaffari
Sovanlal Mukherjee
Anirban Mukhopadhyay
Henning Müller
Balamurali Murugesan
Cosmas Mwikirize
Andriy Myronenko
Saad Nadeem
Vishwesh Nath
Rodrigo Nava
Fernando Navarro
Amin Nejatbakhsh
Dong Ni
Hannes Nickisch
Dong Nie
Jingxin Nie
Aditya Nigam
Lipeng Ning
Xia Ning
Tianye Niu
Jack Noble
Vincent Noblet
Alexey Novikov
Jorge Novo
Mohammad Obeid
Masahiro Oda
Benjamin Odry
Steffen Oeltze-Jafra
Hugo Oliveira
Sara Oliveira

Arnau Oliver
Emanuele Olivetti
Jimena Olveres
John Onofrey
Felipe Orihuela-Espina
José Orlando
Marcos Ortega
Yoshito Otake
Sebastian Otálora
Cheng Ouyang
Jiahong Ouyang
Xi Ouyang
Michal Ozery-Flato
Danielle Pace
Krittin Pachtrachai
J. Blas Pagador
Akshay Pai
Viswanath Pamulakanty Sudarshan
Jin Pan
Yongsheng Pan
Pankaj Pandey
Prashant Pandey
Egor Panfilov
Shumao Pang
Joao Papa
Constantin Pape
Bartlomiej Papiez
Hyunjin Park
Jongchan Park
Sanghyun Park
Seung-Jong Park
Seyoun Park
Magdalini Paschali
Diego Patiño Cortés
Angshuman Paul
Christian Payer
Yuru Pei
Chengtao Peng
Yige Peng
Antonio Pepe
Oscar Perdomo
Sérgio Pereira
Jose-Antonio Pérez-Carrasco
Fernando Pérez-García
Jorge Perez-Gonzalez
Skand Peri

Matthias Perkonigg
Mehran Pesteie
Jorg Peters
Jens Petersen
Kersten Petersen
Renzo Phellan Aro
Ashish Phophalia
Tomasz Pieciak
Antonio Pinheiro
Pramod Pisharady
Kilian Pohl
Sebastian Pölsterl
Iulia A. Popescu
Alison Pouch
Prateek Prasanna
Raphael Prevost
Juan Prieto
Sergi Pujades
Elodie Puybareau
Esther Puyol-Antón
Haikun Qi
Huan Qi
Buyue Qian
Yan Qiang
Yuchuan Qiao
Chen Qin
Wenjian Qin
Yulei Qin
Wu Qiu
Hui Qu
Liangqiong Qu
Kha Gia Quach
Prashanth R.
Pradeep Reddy Raamana
Mehdi Rahim
Jagath Rajapakse
Kashif Rajpoot
Jhonata Ramos
Lingyan Ran
Hatem Rashwan
Daniele Ravì
Keerthi Sravan Ravi
Nishant Ravikumar
Harish RaviPrakash
Samuel Remedios
Yinhao Ren

Yudan Ren
Mauricio Reyes
Constantino Reyes-Aldasoro
Jonas Richiardi
David Richmond
Anne-Marie Rickmann
Leticia Rittner
Dominik Rivoir
Emma Robinson
Jessica Rodgers
Rafael Rodrigues
Robert Rohling
Michal Rosen-Zvi
Lukasz Roszkowiak
Karsten Roth
José Rouco
Daniel Rueckert
Jaime S. Cardoso
Mohammad Sabokrou
Ario Sadafi
Monjoy Saha
Pramit Saha
Dushyant Sahoo
Pranjal Sahu
Maria Sainz de Cea
Olivier Salvado
Robin Sandkuehler
Gianmarco Santini
Duygu Sarikaya
Imari Sato
Olivier Saut
Dustin Scheinost
Nico Scherf
Markus Schirmer
Alexander Schlaefer
Jerome Schmid
Julia Schnabel
Klaus Schoeffmann
Andreas Schuh
Ernst Schwartz
Christina Schwarz-Gsaxner
Michaël Sdika
Suman Sedai
Anjany Sekuboyina
Raghavendra Selvan
Sourya Sengupta

Youngho Seo
Lama Seoud
Ana Sequeira
Maxime Sermesant
Carmen Serrano
Muhammad Shaban
Ahmed Shaffie
Sobhan Shafiei
Mohammad Abuzar Shaikh
Reuben Shamir
Shayan Shams
Hongming Shan
Harshita Sharma
Gregory Sharp
Mohamed Shehata
Haocheng Shen
Li Shen
Liyue Shen
Mali Shen
Yiqing Shen
Yiqiu Shen
Zhengyang Shen
Kuangyu Shi
Luyao Shi
Xiaoshuang Shi
Xueying Shi
Yemin Shi
Yiyu Shi
Yonghong Shi
Jitae Shin
Boris Shirokikh
Suprosanna Shit
Suzanne Shontz
Yucheng Shu
Alberto Signoroni
Wilson Silva
Margarida Silveira
Matthew Sinclair
Rohit Singla
Sumedha Singla
Ayushi Sinha
Kevin Smith
Rajath Soans
Ahmed Soliman
Stefan Sommer
Yang Song

Youyi Song
Aristeidis Sotiras
Arcot Sowmya
Rachel Sparks
William Speier
Ziga Spiclin
Dominik Spinczyk
Jon Sporring
Chetan Srinidhi
Anuroop Sriram
Vinkle Srivastav
Lawrence Staib
Marius Staring
Johannes Stegmaier
Joshua Stough
Robin Strand
Martin Styner
Hai Su
Yun-Hsuan Su
Vaishnavi Subramanian
Gérard Subsol
Yao Sui
Avan Suinesiaputra
Jeremias Sulam
Shipra Suman
Li Sun
Wenqing Sun
Chiranjib Sur
Yannick Suter
Tanveer Syeda-Mahmood
Fatemeh Taheri Dezaki
Roger Tam
José Tamez-Peña
Chaowei Tan
Hao Tang
Thomas Tang
Yucheng Tang
Zihao Tang
Mickael Tardy
Giacomo Tarroni
Jonas Teuwen
Paul Thienphrapa
Stephen Thompson
Jiang Tian
Yu Tian
Yun Tian

Aleksei Tiulpin
Hamid Tizhoosh
Matthew Toews
Oguzhan Topsakal
Antonio Torteya
Sylvie Treuillet
Jocelyne Troccaz
Roger Trullo
Chialing Tsai
Sudhakar Tummala
Verena Uslar
Hristina Uzunova
Régis Vaillant
Maria Vakalopoulou
Jeya Maria Jose Valanarasu
Tom van Sonsbeek
Gijs van Tulder
Marta Varela
Thomas Varsavsky
Francisco Vasconcelos
Liset Vazquez Romaguera
S. Swaroop Vedula
Sanketh Vedula
Harini Veeraraghavan
Miguel Vega
Gonzalo Vegas Sanchez-Ferrero
Anant Vemuri
Gopalkrishna Veni
Mitko Veta
Thomas Vetter
Pedro Vieira
Juan Pedro Vigueras Guillén
Barbara Villarini
Satish Viswanath
Athanasios Vlontzos
Wolf-Dieter Vogl
Bo Wang
Cheng Wang
Chengjia Wang
Chunliang Wang
Clinton Wang
Congcong Wang
Dadong Wang
Dongang Wang
Haifeng Wang
Hongyu Wang

Hu Wang
Huan Wang
Kun Wang
Li Wang
Liansheng Wang
Linwei Wang
Manning Wang
Renzhen Wang
Ruixuan Wang
Sheng Wang
Shujun Wang
Shuo Wang
Tianchen Wang
Tongxin Wang
Wenzhe Wang
Xi Wang
Xiaosong Wang
Yan Wang
Yaping Wang
Yi Wang
Yirui Wang
Zeyi Wang
Zhangyang Wang
Zihao Wang
Zuhui Wang
Simon Warfield
Jonathan Weber
Jürgen Weese
Dong Wei
Donglai Wei
Dongming Wei
Martin Weigert
Wolfgang Wein
Michael Wels
Cédric Wemmert
Junhao Wen
Travis Williams
Matthias Wilms
Stefan Winzeck
James Wiskin
Adam Wittek
Marek Wodzinski
Jelmer Wolterink
Ken C. L. Wong
Chongruo Wu
Guoqing Wu

Ji Wu
Jian Wu
Jie Ying Wu
Pengxiang Wu
Xiyin Wu
Ye Wu
Yicheng Wu
Yifan Wu
Tobias Wuerfl
Pengcheng Xi
James Xia
Siyu Xia
Wenfeng Xia
Yingda Xia
Yong Xia
Lei Xiang
Deqiang Xiao
Li Xiao
Yiming Xiao
Hongtao Xie
Lingxi Xie
Long Xie
Weidi Xie
Yiting Xie
Yutong Xie
Xiaohan Xing
Chang Xu
Chenchu Xu
Hongming Xu
Kele Xu
Min Xu
Rui Xu
Xiaowei Xu
Xuanang Xu
Yongchao Xu
Zhenghua Xu
Zhoubing Xu
Kai Xuan
Cheng Xue
Jie Xue
Wufeng Xue
Yuan Xue
Faridah Yahya
Ke Yan
Yuguang Yan
Zhennan Yan

Changchun Yang
Chao-Han Huck Yang
Dong Yang
Erkun Yang
Fan Yang
Ge Yang
Guang Yang
Guanyu Yang
Heran Yang
Hongxu Yang
Huijuan Yang
Jiancheng Yang
Jie Yang
Junlin Yang
Lin Yang
Peng Yang
Xin Yang
Yan Yang
Yujiu Yang
Dongren Yao
Jiawen Yao
Li Yao
Qingsong Yao
Chuyang Ye
Dong Hye Ye
Menglong Ye
Xujiong Ye
Jingru Yi
Jirong Yi
Xin Yi
Youngjin Yoo
Chenyu You
Haichao Yu
Hanchao Yu
Lequan Yu
Qi Yu
Yang Yu
Pengyu Yuan
Fatemeh Zabihollahy
Ghada Zamzmi
Marco Zenati
Guodong Zeng
Rui Zeng
Oliver Zettinig
Zhiwei Zhai
Chaoyi Zhang

Daoqiang Zhang
Fan Zhang
Guangming Zhang
Hang Zhang
Huahong Zhang
Jianpeng Zhang
Jiong Zhang
Jun Zhang
Lei Zhang
Lichi Zhang
Lin Zhang
Ling Zhang
Lu Zhang
Miaomiao Zhang
Ning Zhang
Qiang Zhang
Rongzhao Zhang
Ru-Yuan Zhang
Shihao Zhang
Shu Zhang
Tong Zhang
Wei Zhang
Weiwei Zhang
Wen Zhang
Wenlu Zhang
Xin Zhang
Ya Zhang
Yanbo Zhang
Yanfu Zhang
Yi Zhang
Yishuo Zhang
Yong Zhang
Yongqin Zhang
You Zhang
Youshan Zhang
Yu Zhang
Yue Zhang
Yueyi Zhang
Yulun Zhang
Yunyan Zhang
Yuyao Zhang
Can Zhao
Changchen Zhao
Chongyue Zhao
Fenqiang Zhao
Gangming Zhao

He Zhao
Jun Zhao
Li Zhao
Qingyu Zhao
Rongchang Zhao
Shen Zhao
Shijie Zhao
Tengda Zhao
Tianyi Zhao
Wei Zhao
Xuandong Zhao
Yiyuan Zhao
Yuan-Xing Zhao
Yue Zhao
Zixu Zhao
Ziyuan Zhao
Xingjian Zhen
Guoyan Zheng
Hao Zheng
Jiannan Zheng
Kang Zheng
Shenhai Zheng
Yalin Zheng
Yinqiang Zheng
Yushan Zheng
Jia-Xing Zhong
Zichun Zhong

Bo Zhou
Haoyin Zhou
Hong-Yu Zhou
Kang Zhou
Sanping Zhou
Sihang Zhou
Tao Zhou
Xiao-Yun Zhou
Yanning Zhou
Yuyin Zhou
Zongwei Zhou
Dongxiao Zhu
Hancan Zhu
Lei Zhu
Qikui Zhu
Xinliang Zhu
Yuemin Zhu
Zhe Zhu
Zhuotun Zhu
Aneeq Zia
Veronika Zimmer
David Zimmerer
Lilla Zöllei
Yukai Zou
Lianrui Zuo
Gerald Zwettler
Reyer Zwiggelaar

Outstanding Reviewers

Neel Dey   New York University, USA
Monica Hernandez   University of Zaragoza, Spain
Ivica Kopriva   Rudjer Boskovich Institute, Croatia
Sebastian Otálora   University of Applied Sciences and Arts Western
                        Switzerland, Switzerland
Danielle Pace   Massachusetts General Hospital, USA
Sérgio Pereira   Lunit Inc., South Korea
David Richmond   IBM Watson Health, USA
Rohit Singla   University of British Columbia, Canada
Yan Wang   Sichuan University, China

Honorable Mentions (Reviewers)

Mazdak Abulnaga	Massachusetts Institute of Technology, USA
Pierre Ambrosini	Erasmus University Medical Center, The Netherlands
Hyeon-Min Bae	Korea Advanced Institute of Science and Technology, South Korea
Mikhail Belyaev	Skolkovo Institute of Science and Technology, Russia
Bhushan Borotikar	Symbiosis International University, India
Katharina Breininger	Friedrich-Alexander-Universität Erlangen-Nürnberg, Germany
Ninon Burgos	CNRS, Paris Brain Institute, France
Mariano Cabezas	The University of Sydney, Australia
Aaron Carass	Johns Hopkins University, USA
Pierre-Henri Conze	IMT Atlantique, France
Christian Desrosiers	École de technologie supérieure, Canada
Reuben Dorent	King's College London, UK
Nicha Dvornek	Yale University, USA
Dmitry V. Dylov	Skolkovo Institute of Science and Technology, Russia
Marius Erdt	Fraunhofer Singapore, Singapore
Ruibin Feng	Stanford University, USA
Enzo Ferrante	CONICET/Universidad Nacional del Litoral, Argentina
Antonio Foncubierta-Rodríguez	IBM Research, Switzerland
Isabel Funke	National Center for Tumor Diseases Dresden, Germany
Adrian Galdran	University of Bournemouth, UK
Ben Glocker	Imperial College London, UK
Cristina González	Universidad de los Andes, Colombia
Maged Goubran	Sunnybrook Research Institute, Canada
Sobhan Goudarzi	Concordia University, Canada
Vicente Grau	University of Oxford, UK
Andreas Hauptmann	University of Oulu, Finland
Nico Hoffmann	Technische Universität Dresden, Germany
Sungmin Hong	Massachusetts General Hospital, Harvard Medical School, USA
Won-Dong Jang	Harvard University, USA
Zhanghexuan Ji	University at Buffalo, SUNY, USA
Neerav Karani	ETH Zurich, Switzerland
Alexander Katzmann	Siemens Healthineers, Germany
Erwan Kerrien	Inria, France
Anitha Priya Krishnan	Genentech, USA
Tahsin Kurc	Stony Brook University, USA
Francesco La Rosa	École polytechnique fédérale de Lausanne, Switzerland
Dmitrii Lachinov	Medical University of Vienna, Austria
Mengzhang Li	Peking University, China
Gilbert Lim	National University of Singapore, Singapore
Dongnan Liu	University of Sydney, Australia

Bin Lou	Siemens Healthineers, USA
Kai Ma	Tencent, China
Klaus H. Maier-Hein	German Cancer Research Center (DKFZ), Germany
Raphael Meier	University Hospital Bern, Switzerland
Tony C. W. Mok	Hong Kong University of Science and Technology, Hong Kong SAR
Lia Morra	Politecnico di Torino, Italy
Cosmas Mwikirize	Rutgers University, USA
Felipe Orihuela-Espina	Instituto Nacional de Astrofísica, Óptica y Electrónica, Mexico
Egor Panfilov	University of Oulu, Finland
Christian Payer	Graz University of Technology, Austria
Sebastian Pölsterl	Ludwig-Maximilians Universität, Germany
José Rouco	University of A Coruña, Spain
Daniel Rueckert	Imperial College London, UK
Julia Schnabel	King's College London, UK
Christina Schwarz-Gsaxner	Graz University of Technology, Austria
Boris Shirokikh	Skolkovo Institute of Science and Technology, Russia
Yang Song	University of New South Wales, Australia
Gérard Subsol	Université de Montpellier, France
Tanveer Syeda-Mahmood	IBM Research, USA
Mickael Tardy	Hera-MI, France
Paul Thienphrapa	Atlas5D, USA
Gijs van Tulder	Radboud University, The Netherlands
Tongxin Wang	Indiana University, USA
Yirui Wang	PAII Inc., USA
Jelmer Wolterink	University of Twente, The Netherlands
Lei Xiang	Subtle Medical Inc., USA
Fatemeh Zabihollahy	Johns Hopkins University, USA
Wei Zhang	University of Georgia, USA
Ya Zhang	Shanghai Jiao Tong University, China
Qingyu Zhao	Stanford University, China
Yushan Zheng	Beihang University, China

Mentorship Program (Mentors)

Shadi Albarqouni	Helmholtz AI, Helmholtz Center Munich, Germany
Hao Chen	Hong Kong University of Science and Technology, Hong Kong SAR
Nadim Daher	NVIDIA, France
Marleen de Bruijne	Erasmus MC/University of Copenhagen, The Netherlands
Qi Dou	The Chinese University of Hong Kong, Hong Kong SAR
Gabor Fichtinger	Queen's University, Canada
Jonny Hancox	NVIDIA, UK

Nobuhiko Hata   Harvard Medical School, USA
Sharon Xiaolei Huang   Pennsylvania State University, USA
Jana Hutter   King's College London, UK
Dakai Jin   PAII Inc., China
Samuel Kadoury   Polytechnique Montréal, Canada
Minjeong Kim   University of North Carolina at Greensboro, USA
Hans Lamecker   1000shapes GmbH, Germany
Andrea Lara   Galileo University, Guatemala
Ngan Le   University of Arkansas, USA
Baiying Lei   Shenzhen University, China
Karim Lekadir   Universitat de Barcelona, Spain
Marius George Linguraru   Children's National Health System/George
                              Washington University, USA
Herve Lombaert   ETS Montreal, Canada
Marco Lorenzi   Inria, France
Le Lu   PAII Inc., China
Xiongbiao Luo   Xiamen University, China
Dzung Pham   Henry M. Jackson Foundation/Uniformed Services
                              University/National Institutes of Health/Johns
                              Hopkins University, USA
Josien Pluim   Eindhoven University of Technology/University
                              Medical Center Utrecht, The Netherlands
Antonio Porras   University of Colorado Anschutz Medical
                              Campus/Children's Hospital Colorado, USA
Islem Rekik   Istanbul Technical University, Turkey
Nicola Rieke   NVIDIA, Germany
Julia Schnabel   TU Munich/Helmholtz Center Munich, Germany,
                              and King's College London, UK
Debdoot Sheet   Indian Institute of Technology Kharagpur, India
Pallavi Tiwari   Case Western Reserve University, USA
Jocelyne Troccaz   CNRS, TIMC, Grenoble Alpes University, France
Sandrine Voros   TIMC-IMAG, INSERM, France
Linwei Wang   Rochester Institute of Technology, USA
Yalin Wang   Arizona State University, USA
Zhong Xue   United Imaging Intelligence Co. Ltd, USA
Renee Yao   NVIDIA, USA
Mohammad Yaqub   Mohamed Bin Zayed University of Artificial
                              Intelligence, United Arab Emirates, and University
                              of Oxford, UK
S. Kevin Zhou   University of Science and Technology of China, China
Lilla Zollei   Massachusetts General Hospital, Harvard Medical
                              School, USA
Maria A. Zuluaga   EURECOM, France

Contents – Part VII

Clinical Applications - Dermatology

Clinical Applications - Fetal Imaging

Clinical Applications - Lung

Clinical Applications - Neuroimaging - Brain Development

Clinical Applications - Neuroimaging - DWI and Tractography

Clinical Applications - Neuroimaging - Functional Brain Networks

Clinical Applications - Abdomen

Learning More for Free - A Multi Task Learning Approach for Improved Pathology Classification in Capsule Endoscopy

Anuja Vats[1(✉)], Marius Pedersen[1], Ahmed Mohammed[1,2], and Øistein Hovde[1,3,4]

[1] NTNU, Gjøvik, Norway
{anuja.vats,marius.pedersen,mohammed.kedir}@ntnu.no
[2] SINTEF Digital, Smart Sensor Systems, Oslo, Norway
[3] University of Oslo, Oslo, Norway
[4] Innlandet Hospital Trust, Gjøvik, Norway
oistein.hovde@sykehuset-innlandet.no

Abstract. The progress in Computer Aided Diagnosis (CADx) of Wireless Capsule Endoscopy (WCE) is thwarted by the lack of data. The inadequacy in richly representative healthy and abnormal conditions results in isolated analyses of pathologies, that can not handle realistic multi-pathology scenarios. In this work, we explore how to learn more for free, from limited data through solving a WCE multicentric, multi-pathology classification problem. Learning more implies to learning more than full supervision would allow with the same data. This is done by combining self supervision with full supervision, under multi task learning. Additionally, we draw inspiration from the Human Visual System (HVS) in designing self supervision tasks and investigate if seemingly ineffectual signals within the data itself can be exploited to gain performance, if so, which signals would be better than others. Further, we present our analysis of the high level features as a stepping stone towards more robust multi-pathology CADx in WCE. Code accompanying this work will be made available on github.

Keywords: Capsule endoscopy · Multi task learning · Self supervision

1 Introduction

WCE is a diagnostic procedure for screening the lining (mucosa) of the Gastrointestinal Tract (GIT) for abnormalities. It has succeeded in being the first line of diagnosis for the middle part of the GIT [26]. Apart from being widely

Electronic supplementary material The online version of this chapter (https://doi.org/10.1007/978-3-030-87234-2_1) contains supplementary material, which is available to authorized users.

M. de Bruijne et al. (Eds.): MICCAI 2021, LNCS 12907, pp. 3–13, 2021.
https://doi.org/10.1007/978-3-030-87234-2_1

preferred for examining the small bowel, due to its non-invasive nature, it is also a preferred alternative to upper endoscopy and colonoscopy [19,21].

The procedure involves a pill-sized imaging device shaped as a capsule, to be swallowed by the patient. It traverses through the GIT taking pictures of the mucosa and lumen, transmitting them wirelessly to an external data recorder. The data from each capsule is a video stream that is 8 to 16 h long, depending on the capsule model [2,13,16] amounting to frames upwards from 50,000 to be investigated by experts. Moreover, a significant number of these frames contribute from little to no diagnostic information due to either debris and bubbles completely or partially occluding the mucosa or undesirable capsule orientations for long duration during its traversal [22].

Owing to its success in recognition tasks, deep learning has led to major breakthroughs in anomaly detection across medical domains. Soffer et al. [23] present a comprehensive review including deep learning based studies in WCE. Most recent approaches apply a Convolutional Neural Net(CNN) based feature extractor for Computer Aided Detection (CADe) of pathologies like polyps, ulcers, bleeding etc. [9,20,23,26]. A multicentric CADe system trained potentially from millions of images as in [8] can achieve an even higher detection rate than human-readers for certain anomalies. They perform the first of such a large scale multicentric study by rigorously fine-tuning a pre-trained CNN on more than 100 million images. While their method achieves high sensitivity in filtering out suspected anomalies, further attempt to classify the detected anomaly is not as successful. Considering the large amount of multicentre data used in this study, the inability to do so, reveals the complexity involved in reliably representing features that can classify multiple abnormalities beyond obvious confounders. Infact, this happens to be one of the key challenges in CADx arising from the lack of morphological descriptions of pathological and healthy conditions at disposal (the essential difference lies in CADx focusing on features for the purpose of characterizing an abnormality as opposed to localizing an abnormality within a frame (CADe)). Stidham et al. [24] discuss the consequences of such a challenge, with regards to tissue histology. Some of the complicating factors can be understood by considering the response of the GIT under abnormality:

1. It may exhibit similarity in visual characteristics for different conditions and different severity of these conditions.
2. It may exhibit visible differences in pathology appearances that may get overshadowed by other (medically irrelevant but visually pervasive) morphological similarities pertaining to the endoluminal scene (e.g. mucosal surface structures, occlusions, turbidity, illumination etc.)

Our strategy to combat this challenge is to learn more efficiently discriminative features with limited data for free. We achieve this by taking advantage of additional (seemingly ineffectual) teaching signals within data through two ways - self supervision and Multi Task Learning (MTL). The inspiration behind MTL comes from the observation that humans greatly benefit, in their learning, by performing multiple tasks as opposed to learning the same tasks in isolation [7]. By the same logic, neural networks benefit by sharing information across several related tasks, by means of a common underlying representation. Even when the primary focus is only on one of the tasks, a good choice of additional tasks can lead to better predictive performance by supplementing domain-specific information not necessarily captured by the main task [4,6,7,15,18,27].

Notwithstanding the challenges of WCE, gastroenterologists are usually able to find pathological areas within the image with relative ease. Most lesions and inflammations are obscure and vary considerably in scale considering the global context within images. In addition, the inherent presence of distortions causes further corruption of diagnostically relevant details. Despite this, gastroenterologists can not only identify and localize abnormality but also classify its type. That is, aside from the most difficult of cases [28] the doctor's ability to identify and classify pathologies remains relatively unhindered by distortions. This applies generally to the HVS. Continuing from this argument, if distortions are unavoidable in WCE and doctors can diagnose in spite of them, can they be exploited to provide a useful inductive bias for classification? To this end, we propose a novel MTL framework as a combination of two types of tasks - supervised pathology classification task (SPT) and self-supervised (SS) distortion level classification task (SSDT). The SSDT task is SS as the labels are freely obtained from distorting the images in a pre-calculated way to enable supervised learning from it. Further, we also present feature analysis for WCE CAD-CAP data [14]. To the best of our knowledge, we are the first ones to perform multicentric pathology classification on WCE using MTL and SS. Besides ours, the same classification has been presented in only one other work [25], however, their objectives and approach differ from ours significantly. The main contributions of this paper are

1. We are the first to apply MTL for WCE classification. We combine teaching signals from two different levels of supervision - full supervision (SPT) and self supervision (SSDT) to gain in performance.
2. We increase the pathology classification accuracy by 7% from a Single Task (ST) based pathology classification.
3. We explore the influence of SSDTs and discuss their role in performance.
4. We discover that the clusters of similarity that automatically emerge in a high level feature space correspond to those manually identified in other works [13, 22]. This understanding of the similarity explains the cause for the bottleneck of CADx typical in WCE.

2   Method

Major distortions in WCE are similar but aggravated from those identified in clinical endoscopy [1], owing to the uncontrolled motion of the capsule inside the GIT. In this work, three distortions have been selected for auxiliary SSDT - brightness (B), contrast (C) and motion blur (MB). One MTL experiment is a combination of SPT with one (or two) SSDT. The SSDT comprises of adding a known level of the selected distortion to be predicted onto a centrally cropped, resized (400×400) and randomly rotated image. Each level of distortion (numerical value between (0,3]) corresponds to an output category for SSDT B and C (for more details refer suppl. Task Configuraions). For SSDT MB, we choose to

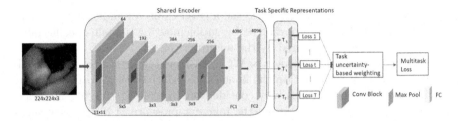

Fig. 1. Overview of our proposed MTL architecture. The encoder holds a shared representation that is simultaneously discriminative for multiple tasks, whereas the task specific layer specializes in a particular task.

add linear motion blur, with output categories corresponding to the direction of added motion [0,45,90,180]. The network at any time does two simultaneous-3 way-pathology and 4 way-distortion level classifications. The training scheme employed is random network initialization with hard parameter sharing [7], under which all layers upto the penultimate task-specific layers share the same representation. This is especially useful when the training data is limited. In having to generalize for multiple tasks using one representation, the network is less prone to overfitting [3]. Figure 1 illustrates the architecture, which has been adapted from AlexNet [12]. Five Convolutional (Conv) and two Fully Connected layers (FC1, FC2) are shared across the tasks whereas, the final FC layers are task specific. A single input image contains signal for both classification tasks at any time.

Objective Function - Every MTL is a joint optimization of two or more losses. In this problem, one of the losses corresponds to SPT while other to SSDT. The total objective conventionally is a weighted sum of individual objectives, in this case a cross-entropy objective for each task. Let $D^t = (x_i, y_i^t), i \in 1, 2, ..., N$ represent the dataset for a task t obtained from a set of tasks given by $t \in 1, 2, ..., T$. (x_i, y_i^t) represent the input output pair for task t such that, in our scheme, each image x_i contains the training signal for all the tasks, and y_i^t is the corresponding task label vector. If $\mathcal{L}_{cet}$ is the cross-entropy loss for task t, then a conventional formulation can be given by

$$\underset{\theta}{\text{argmin}} \sum_{t=1}^{T} \lambda_t \mathcal{L}_{cet}(\theta, f(X^t), Y^t) + \alpha \Omega(\theta) \qquad (1)$$

where $f(X^t)$ is the prediction for task t, θ are the network parameters and λ is the task-specific weight. The term $\alpha \Omega(\theta)$ corresponds to regularization concerning network weights with $\alpha \in [0, inf)$. This formulation assumes that optimal weighting of tasks λ_t is either known or can be found. In practice it requires manual tuning which is unintuitive and expensive. We adopt the approach proposed by Kendall et al.[10], where the contribution towards the total loss by a task is learned on the basis of homoscedastic uncertainty. This extends previous formulation for multiple classification tasks as

$$\underset{(\theta,\sigma)}{\text{argmin}} \sum_{t=1}^{T} \left(\frac{1}{\sigma_t^2} \mathcal{L}_{cet}(\theta, f(X^t), Y^t) + log\sigma_t \right) + \alpha\Omega(\theta) \qquad (2)$$

where $\frac{1}{\sigma_t^2}$, the relative contribution of a task to the loss, is an indirect estimate of the uncertainty of task t and learning σ_t can be interpreted as learning the temperature of a Boltzmann distribution [10]. As a task with high uncertainty begins contributing more towards the total loss compared to a task with lower uncertainty, the relative contribution and eventually task uncertainties themselves are optimized as training progresses, that is to say, the optimization in Eq.(2) with respect to both network parameters θ along with σ_t, allows the relative contribution of each task towards total loss to change during the course of training till a reasonably low ratio is reached for all tasks involved. The additional term $log\sigma_t$ is an additional regularizer, penalizing the tendency of σ_t towards larger values.

3   Experiments and Results

The data used for our multicentric pathology classification is from the Computer Assisted Diagnosis for Capsule Endoscopy Database CAD-CAP (GIANA Endoscopic Vision Challenge 2018) comprising of 20,000 normal and 5000 images of varying abnormalities [14]. Out of these, a balanced classification dataset with 1812 images belonging to three classes "normal", "inflammatory lesion" and "vascular lesion" (600, 607, 605 respectively) has been used. 80:10:10 split has been used for train, validation and test sets. In the challenge, the dataset has been used for CADe, while in this paper we use it for CADx. The five MTL configurations studied in this paper are - STL : fully supervised pathology classification (only SPT), MTL (P+X) : refers to (SPT + SSDT of X), where, SSDT of X is the SS auxiliary distortion task of distortion X. X could be Motion Blur (MB), Brightness (B) or Contrast (C). MTL (P+X+Y) is a three task configuration with SPT and a chosen combination of two SSDT of X. We train using Adam optimizer for 300 epochs with a batch size of 64, an initial learning rate of 0.01 and decay rate of 0.1 every 50 epochs on Nvidia Twin Titan RTX.

Results and Analysis - Table 1 shows the results of our experiments. Overall MTL (P+B) achieves the highest accuracy, but on account of maximizing the correct classification of normals, with no improvement (in fact worse discrimination as compared to STL) for the pathology classes. MTL (P+C) is the best with respect to pathology sensitivity, i.e. while the classification of normals is only slightly better than STL, change in image contrast enables pathologies to be better discriminated. MTL (P+MB) is marginally better than STL.

We observed that combining more than one SSDT with SPT (MTL P+B+C) does not further improve the overall performance, (Table 1, suppl. All MTL configurations). This result is not unexpected even when combining fully supervised

Table 1. Performance of STL and different MTL configurations. P_i and P_v refer to the pathology class of inflammatory and vascular lesions respectively. N refers to the class of normals.

Config	Accuracy (%)		Sensitivity (SPT)			Specificity (SPT)		
	Pathology	Distortion	P_i	N	P_v	P_i	N	P_v
STL	51.9	–	0.44	0.72	0.44	0.41	0.68	0.47
P+MB	52.1	40	0.45	0.70	0.38	0.49	0.57	0.44
P+C	57.1	80.95	**0.56**	0.75	**0.48**	0.55	0.68	0.53
P+B	**58.7**	85.7	0.42	**0.88**	0.35	0.46	0.74	0.39
P+B+C	57	73.5, 88.8	0.55	0.79	0.47	0.59	0.55	0.39

tasks [11]. However, compared to STL the gain by combining two tasks comes both in terms of overall accuracy (from correctly classifying normals as in the brightness task) and discriminating between pathologies (higher sensitivity from the contrast task). That is, this task combination can be seen as a way to gain a specific advantage from each of the two auxiliary SSDTs.

Further, we analyse how different MTL configurations impact the learning of the pathology task (SPT). Figure 2(a, b) show the training accuracy of SPT with and without MTL, (c) the plots for $\frac{1}{\sigma_t}$.

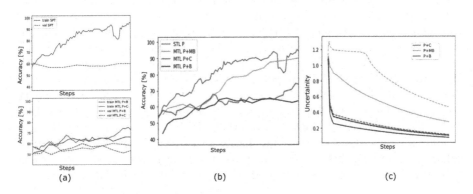

(a)   (b)   (c)

Fig. 2. Impact of SSDTs on the training performance of SPT. (a-top) Overfitting in SPT when trained as a supervised task. (a-bottom, b) Training of SPT under MTL, reduced overfitting with longer training time (c) Square-root of task-uncertainities $\frac{1}{\sigma_p^2}$, $\frac{1}{\sigma_d^2}$ given by $\frac{1}{\sigma_p}$(*solid*), $\frac{1}{\sigma_d}$(*dashed*) is shown in plot.

It can be seen that, in isolation SPT has a high tendency towards overfitting with quickly escalating training accuracy but plateauing validation accuracy (Fig. 2(a)). The addition of an auxiliary SSDT is seen to reduce this effect (Fig. 2(a, b)). This is a well known advantage of MTL [3] as an overall regularizer. This effect is directly linked to performance seen before, with brightness, contrast and motion blur in order of decreasing overall accuracy. This is also correlated with the trend in uncertainties. The nature of helpfulness of a task towards pathology can be inferred from how the uncertainties behave during training. From Fig. 2(c), for a helpful SSDT, both uncertainties for SPT ($\frac{1}{\sigma_p}$) and SSDT ($\frac{1}{\sigma_d}$) reduce in tandem, over the course of iterations as more and more confidence is gained from the task. In a relatively unhelpful task (MTL P+MB) however, higher uncertainties indicate this lack of confidence in the task, towards outputs. More specifically, choosing an auxiliary task more complicated (MB) than the main task (SPT) (as seen from the low distortion accuracy for MB in Table 1) can lead from little to no benefit from MTL. This complexity arising in the blur task may be due to the varying amount of inherent linear, rotational and motion blur in the images, that may make prediction of the exact level, especially difficult.

We further investigate the design consideration for an already beneficial task to remain beneficial. Figure 3 shows the training accuracy for two different design choices for SSDT C and B. Config 1 which results in better performance for both tasks is a task of higher complexity (smaller difference between levels of distortion, for this experiment this difference was found to be 0.1 or less for both SSDT B and C) compared to Config 2 (larger difference between levels, 0.3 or more) that results in poorer performance (no gain from STL). This may seem to be in contradiction to the observation before, however it is not. For an MTL to be beneficial the auxiliary task must neither be too easy nor too hard, as in either case, a plateauing effect is observed. MTL P+MB is a difficult task even for simpler designs (larger kernel size being simpler design as the added blur becomes increasingly easier to detect with larger kernels, and vice versa), however, for SSDT C and SSDT B which are tasks of lower complexity with regards to pathology, the plateauing occurs when they are made simple by design. This is done by increasing discrimination between the levels that correspond to output categories (refer Fig. 3, suppl. Task Configurations), In other words, since the auxiliary task is too easy, the distortion task has very high accuracy from the very beginning of the training process. Coupled with a difficult task (SPT), an easy SSDT loses its primary advantage of helping the pathology task out of early local minimas. These findings regarding the nature of beneficial tasks coincide with the observation from [5]. Figure 3(c) shows the task uncertainties during training.

A Note on Similarity - In this section, our focus is to answer the question about the correspondence between *representational similarity* and *medically-perceived similarity*. Medical similarity can be simply described as - two images having the same pathology are similar and different from all images with different pathology. Representational similarity on the other hand is the similarity as understood by the network on the basis of extracted features. A disagreement

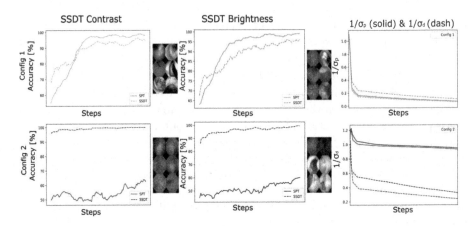

Fig. 3. Comparison of two different SSDT designs. Config 1. improves SPT beyond STL performance and vice versa for Config 2. Pre-processed images, adjacent to the graphs illustrate the distortion levels for the two configs.

between these two reveals the cause for confusion in the network. Although MTL is less prone to this than STL, it still is a major bottleneck in performance. We projected the features from FC2 (Fig. 1) on 3D after dimension reduction using Uniform Manifold Approximation and Projection for Dimension Reduction (UMAP) [17]. UMAP helps us interpret not only, what features dictate which images are most similar with respect to each other (local), but also what two features are more similar compared to others (global).

As seen in Fig. 4, MTL reveals the inherent variations within the dataset. The extracted features are such that while images with the same pathology lie close (even when in different clusters), pathology does not dictate similarity everywhere. The images within a cluster have representational similarity, that may be medically unimportant. In this work, we saw five such dominant factors emerge out of the feature space repeatedly - bubbles and turbidity, folds in mucosa, lumenal view, clear normal, clear pathology (these were verified by a medical expert to be similarities typical to endoluminal structure or scene, except for few clusters of clear pathology where pathology was the dominant feature). Moreover, at global scale, the feature "bubbles and turbidity" appears more dissimilar from all other features in all experiments. Images in between clusters show mingled characteristics. Similar dominant features have been used by other authors in works before [13,22]. We stress at this point that, unlike other works, these features emerge automatically for us as a result of the learning strategy applied and training under STL does not provide this advantage (refer suppl. UMAP for STL for more details).

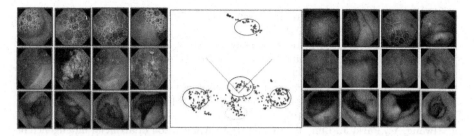

Fig. 4. UMAP of features from FC2 in MTL P+C. Each set of neighbors reveal a dominant similarity. Bubbles and turbid (blue), mucosal folds (red), clear normal (yellow), lumen (green), clear pathology (violet) (Color figure online)

4   Conclusion

In this paper, we demonstrate an MTL approach to gain performance over supervised WCE classification with no added data and a simple CNN architecture, purely by exploiting signals inherent in data under self supervision, additionally to supervision. Further, forcing the network into sharing features between multiple tasks reveals a hierarchy of features in WCE whereby factors that dictate similarity are not inspired only by the presence or absence of pathology, rather are typical of the endoluminal scene. Hence, images with different pathology can be easily perceived to be similar in the presence of such a factor. Our method can be used as a filter for these irrelevant salient patterns, allowing a better subsequent characterization of pathology in CADx. We also experimentally correlate task uncertainties with the nature (beneficial and non beneficial) of tasks [5] which can be used as a way to choose good auxiliary tasks for MTL. This framework is easily scalable to more complex architectures and auxiliary tasks.

References

1. Ali, S., et al.: Endoscopy artifact detection (EAD 2019) challenge dataset. arXiv preprint arXiv:1905.03209 (2019)
2. Atsawarungruangkit, A., Elfanagely, Y., Asombang, A.W., Rupawala, A., Rich, H.G.: Understanding deep learning in capsule endoscopy: can artificial intelligence enhance clinical practice? Artif. Intell. Gastrointest. Endosc. **1**(2), 33–43 (2020)
3. Baxter, J.: A bayesian/information theoretic model of learning to learn via multiple task sampling. Mach. Learn. **28**(1), 7–39 (1997)
4. Benton, A., Mitchell, M., Hovy, D.: Multitask learning for mental health conditions with limited social media data. In: Proceedings of the 15th Conference of the European Chapter of the Association for Computational Linguistics, vol. 1, Long Papers, pp. 152–162 (2017)
5. Bingel, J., Søgaard, A.: Identifying beneficial task relations for multi-task learning in deep neural networks. In: Proceedings of the 15th Conference of the European Chapter of the Association for Computational Linguistics, vol. 2, Short Papers, pp. 164–169. Association for Computational Linguistics, Valencia, Spain, April 2017. https://www.aclweb.org/anthology/E17-2026

6. Caruana, R.: Multitask learning. Mach. Learn. **28**(1), 41–75 (1997). https://doi.org/10.1023/A:1007379606734
7. Caruana, R.: Multitask learning: a knowledge-based source of inductive bias. In: Proceedings of the Tenth International Conference on Machine Learning, pp. 41–48. Morgan Kaufmann (1993)
8. Ding, Z., et al.: Gastroenterologist-level identification of small-bowel diseases and normal variants by capsule endoscopy using a deep-learning model. Gastroenterology **157**(4), 1044–1054 (2019)
9. Hwang, Y., Park, J., Lim, Y.J., Chun, H.J.: Application of artificial intelligence in capsule endoscopy: where are we now? Clin. Endosc. **51**(6), 547–551 (2018)
10. Kendall, A., Gal, Y., Cipolla, R.: Multi-task learning using uncertainty to weigh losses for scene geometry and semantics. In: Proceedings of the IEEE Conference on Computer Vision and Pattern Recognition, pp. 7482–7491 (2018)
11. Kokkinos, I.: Ubernet: training a universal convolutional neural network for low-, mid-, and high-level vision using diverse datasets and limited memory. In: Proceedings of the IEEE Conference on Computer Vision and Pattern Recognition, pp. 6129–6138 (2017)
12. Krizhevsky, A., Sutskever, I., Hinton, G.E.: ImageNet classification with deep convolutional neural networks. Commun. ACM **60**(6), 84–90 (2017)
13. Laiz, P., Vitria, J., Seguí, S.: Using the triplet loss for domain adaptation in WCE. In: Proceedings of the IEEE International Conference on Computer Vision Workshops, pp. 399–405 (2019)
14. Leenhardt, R., Li, C., Le Mouel, J.P., Rahmi, G., Saurin, J.C., Cholet, F., Boureille, A., Amiot, X., Delvaux, M., Duburque, C., et al.: CAD-CAP: a 25,000-image database serving the development of artificial intelligence for capsule endoscopy. Endosc. Int. Open **8**(3), E415 (2020)
15. Liu, X., Gao, J., He, X., Deng, L., Duh, K., Wang, Y.y.: Representation learning using multi-task deep neural networks for semantic classification and information retrieval. In: Proceedings of the 2015 Conference of the North American Chapter of the Association for Computational Linguistics: Human Language Technologies, pp. 912–921. Association for Computational Linguistics, Denver, Colorado, May–June 2015. https://doi.org/10.3115/v1/N15-1092, https://www.aclweb.org/anthology/N15-1092
16. McAlindon, M.E., Ching, H.L., Yung, D., Sidhu, R., Koulaouzidis, A.: Capsule endoscopy of the small bowel. Ann. Transl. Med. **4**(19), 369 (2016)
17. McInnes, L., Healy, J., Melville, J.: UMAP: uniform manifold approximation and projection for dimension reduction. arXiv preprint arXiv:1802.03426 (2018)
18. Misra, I., Shrivastava, A., Gupta, A., Hebert, M.: Cross-stitch networks for multi-task learning. In: Proceedings of the IEEE Conference on Computer Vision and Pattern Recognition, pp. 3994–4003 (2016)
19. Mohammed, A., Farup, I., Pedersen, M., Hovde, Ø., Yildirim Yayilgan, S.: Stochastic capsule endoscopy image enhancement. J. Imaging **4**(6), 75 (2018)
20. Muhammad, K., Khan, S., Kumar, N., Del Ser, J., Mirjalili, S.: Vision-based personalized wireless capsule endoscopy for smart healthcare: taxonomy, literature review, opportunities and challenges. Future Gener. Comput. Syst. **113**, 266–280 (2020)
21. Park, J., Cho, Y.K., Kim, J.H.: Current and future use of esophageal capsule endoscopy. Clin. Endosc. **51**(4), 317–322 (2018)
22. Seguí, S., Drozdzal, M., Pascual, G., Radeva, P., Malagelada, C., Azpiroz, F., Vitrià, J.: Generic feature learning for wireless capsule endoscopy analysis. Comput. Biol. Med. **79**, 163–172 (2016)

23. Soffer, S., Klang, E., Shimon, O., Nachmias, N., Eliakim, R., Ben-Horin, S., Kopylov, U., Barash, Y.: Deep learning for wireless capsule endoscopy: a systematic review and meta-analysis. Gastrointest. Endosc. **92**(4), 831–839 (2020)
24. Syed, S., Stidham, R.W.: Potential for standardization and automation for pathology and endoscopy in inflammatory bowel disease. Inflamm. Bowel Dis. **26**(10), 1490–1497 (2020)
25. Valério, M.T., Gomes, S., Salgado, M., Oliveira, H.P., Cunha, A.: Lesions multiclass classification in endoscopic capsule frames. Procedia Comput. Sci. **164**, 637–645 (2019)
26. Yang, Y.J.: The future of capsule endoscopy: the role of artificial intelligence and other technical advancements. Clin. Endosc. **53**(4), 387–394 (2020)
27. Zhang, Z., Luo, P., Loy, C.C., Tang, X.: Facial landmark detection by deep multitask learning. In: Fleet, D., Pajdla, T., Schiele, B., Tuytelaars, T. (eds.) ECCV 2014. LNCS, vol. 8694, pp. 94–108. Springer, Cham (2014). https://doi.org/10.1007/978-3-319-10599-4_7
28. Zheng, Y., Hawkins, L., Wolff, J., Goloubeva, O., Goldberg, E.: Detection of lesions during capsule endoscopy: physician performance is disappointing. Am. J. Gastroenterol. **107**(4), 554–560 (2012)

Learning-Based Attenuation Quantification in Abdominal Ultrasound

Myeong-Gee Kim[1], SeokHwan Oh[1], Youngmin Kim[1], Hyuksool Kwon[2], and Hyeon-Min Bae[1(✉)]

[1] School of Electrical Engineering , Korea Advanced Institute of Science and Technology, Daejeon, South Korea
{myeonggee.kim,joseph9337,hmbae}@kaist.ac.kr
[2] Department of Emergency Medicine, Seoul National University Bundang Hospital, Seong-nam, South Korea

Abstract. The attenuation coefficient (AC) of tissue in medical ultrasound has great potential as a quantitative biomarker due to its high sensitivity to pathological properties. In particular, AC is emerging as a new quantitative biomarker for diagnosing and quantifying hepatic steatosis. In this paper, a learning-based technique to quantify AC from pulse-echo data obtained through a single convex probe is presented. In the proposed method, ROI adaptive transmit beam focusing (TxBF) and envelope detection schemes are employed to increase the estimation accuracy and noise resilience, respectively. In addition, the proposed network is designed to extract accurate AC of the target region considering attenuation/sound speed/scattering of the propagating waves in the vicinities of the target region. The accuracy of the proposed method is verified through simulation and phantom tests. In addition, clinical pilot studies show that the estimated liver AC values using the proposed method are correlated strongly with the fat fraction obtained from magnetic resonance imaging ($R^2 = 0.89$, $p < 0.001$). Such results indicate the clinical validity of the proposed learning-based AC estimation method for diagnosing hepatic steatosis.

Keywords: Deep neural network · Attenuation coefficient · Hepatic steatosis · Medical ultrasound · Abdominal ultrasound

1 Introduction

Hepatic steatosis is a common liver disease and is the main cause of liver-related morbidity and mortality [1]. Liver biopsy is a gold standard for the diagnosis of

---

M.-G. Kim and S. Oh—Contributed equally.

Electronic supplementary material The online version of this chapter (https://doi.org/10.1007/978-3-030-87234-2_2) contains supplementary material, which is available to authorized users.

M. de Bruijne et al. (Eds.): MICCAI 2021, LNCS 12907, pp. 14–23, 2021.
https://doi.org/10.1007/978-3-030-87234-2_2

hepatic steatosis. However, the risk of infection and sampling errors are inevitable drawbacks of such an invasive method. Magnetic resonance imaging (MRI) can quantify hepatic steatosis non-invasively [2] but high cost and limited accessibility prevent wide-spread use. On the other hand, medical ultrasound (US) is widely used for the diagnosis of liver steatosis. However, conventional grayscale US compromises accuracy since it provides only operator-dependent qualitative information. The attenuation coefficient (AC) of the tissue has great potential as a quantitative biomarker due to its high sensitivity to pathological properties [3]. In particular, AC is emerging as a new quantitative biomarker for diagnosing and quantifying hepatic steatosis [4–7].

In this paper, a learning-based technique extracting the attenuation coefficient (AC) of the target region-of-interest (ROI) using a single ultrasound probe is proposed. To increase the accuracy and noise resilience, ROI adaptive beam focusing (TxBF) and envelope detection schemes are employed. Since the attenuation of the sound waves are functions of the propagation distance as well as the attenuation properties of the penetrating medium, the proposed network adaptively normalizes the intensity of the received data according to the location of the ROI to maintain estimation accuracy.

The proposed network is trained using simulation data and validated through simulation tests, phantom tests, and in-vivo tests. The clinical validity of the proposed method is verified through the correlation between the proton density fat fraction (PDFF) acquired using magnetic resonance imaging (MRI) and the extracted AC.

2   Dataset

Simulation data is generated using k-wave ultrasound simulation tool in MATLAB [8]. The simulation environment is modeled based on Vantage 64LE, Verasonics Inc., WA, USA, capable of transmitting on 128 channels and receiving on 64 channels simultaneously, together with a convex array probe (Humanscan, South Korea) featuring 128 elements with 0.475 mm pitch, 59.17° field of view, and 60 mm radius of curvature (ROC). The transmit frequency and sampling rate of the analog-to-digital converter are 3.5 MHz and 13.89 MHz, respectively. In the k-wave simulation, the sampling frequency is set to 41.67 MHz and the pulse-echo data is received for 210 μs. The simulation is accelerated by using a GPU (NVIDIA RTX2080Ti).

2.1   Details on the Simulation Phantom

The total grid dimension is 120 mm by 120 mm or 3000 by 3000 elements, with 100 perfectly matching layers on each side to absorb waves reaching computational boundaries. The simulation phantoms representing organs and lesions in the abdomen are created by placing 1 to 5 ellipses with a radius of 10 mm to 100 mm in arbitrary locations. In addition, 5 ellipses with an axial radius of 1mm to 2 mm and a lateral radius of 10 mm to 100 mm are placed at a depth of 30 mm to represent skin and subcutaneous fat. To model the deformation of the superficial tissues due to the shape of the convex array probe, the axial grid was distorted according to the ROC of the probe. The wave propagation properties of the human liver are modeled as follows: the sound speed (SS) is set from 1500 m/s to 1600 m/s, the AC from 0 dB/MHz/cm to 1.5 dB/MHz/cm, and the density of background from 900 kg/m^3 to 1100 kg/m^3. The SS of the objects representing subcutaneous fat is in the range of 1400 m/s and 1700 m/s. To implement scatters, 0 to 10 scatterers per unit area (wavelength by wavelength area) are uniformly distributed in the density range of 850 kg/m^3 and 1150 kg/m^3 (see Fig. 1).

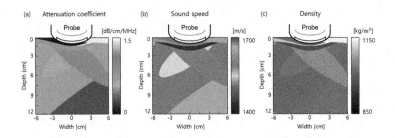

Fig. 1. Characteristic of the simulation phantom.

2.2   ROI Adaptive Tx Beam Focusing

The pulse-echo radio frequency (RF) data is obtained by transmitting multiple steered waves sequentially [9–11]. The transmission beam patterns are adjusted to concentrate the waves at a designated location to increase the SNR with the intention of monitoring the deep abdominal area. The transmission time delay, D_T, of each transmission sensor with respect to the focal depth and steering angle of the beam pattern is

$$D_T(Tx, \theta, F) = (\delta \cdot Tx)\sin\theta/c_0 + (\sqrt{F^2 + (\delta \cdot Tx)^2} - F)/c_0, \qquad (1)$$

where Tx is the order of the transmitting elements of an ultrasound probe, δ is the pitch of the transducer, c_0 is the average SoS in the soft tissue, and θ and F are the steering angle and focal depth of the transmission beam pattern, respectively. The nominal value of c_0 is 1540 m/s. In this study, the steering angles are $-12°$, $-8°$, $-4°$, $0°$, $4°$, $8°$, $12°$, and the focal depths are made adjustable from 40 mm to 90 mm with the granularity of 10 mm (see Fig. 2).

3   Deep Neural Network for AC Quantification

The attenuation characteristic of the tissue can be extracted by analyzing the time-varying intensity of reflected waves that have accumulated the attenuation profile of the tissues. A deep neural network featuring an ROI adaptive

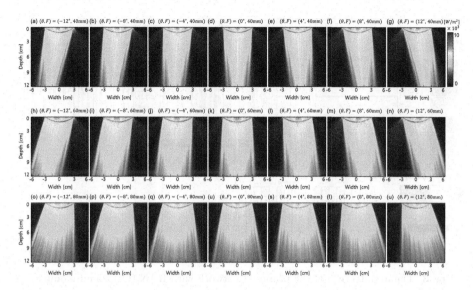

Fig. 2. The acoustic intensity of the ROI adaptive TxBF. (a)–(g), (h)–(n), and (o)–(u) are beam-formed Tx signals with Eq. 1 when F is 40 mm, 60 mm, and 80 mm, respectively.

normalization scheme is designed to extract accurate AC of the target from envelope-detected pulse-echo data.

3.1 Network Architecture

The architecture of the proposed network extracting the AC value from 7-angle beam-formed envelope-detected pulse-echo data, $E_{1:7}$, is illustrated in Fig. 3. The received pulse-echo data is cropped as a function of the target position prior to represent the observation region. The network is composed of two components: (1) downsampling encoders extracting the derivation of the individual envelop-detected pulse-echo data E_n through sequential convolutional computation and (2) a regression network extracting the AC of the target region from the correlation of the encoded derivations of $E_{1:7}$ considering geometric AC distribution surrounding the target.

Encoder. To encode attenuation information of the target embedded in $E_{(1:7)}$, the individual encoding path starts with filtering of 1×2 strided 3×3 two-dimensional convolution followed by a leaky rectified linear unit (ReLU) [12], which is equivalent to the mathematical differentiation after the separation of reflections. The encoding layer is designed as a residual network to manage the vanishing/exploding gradient problem [13].

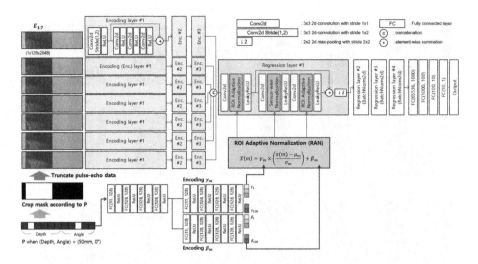

Fig. 3. The architecture of the proposed network with RAN (Color figure online)

AC Regression. To analyze the correlation of the encoded derivation profiles obtained from $E_{1:7}$, each output of the encoding path is channel-wise concatenated and then fed to a series of unit blocks employing skip connections (3×3 two-dimensional convolution – Normalization – Leaky ReLU). Since the cropped pulse-echo data is highly correlated with the observation area, the encoded $E_{1:7}$ is processed considering the location of the ROI. Inspired by the adaptive instance normalization [14], the encoded profile is adaptively normalized across the sensor dimension and subsequently scaled and shifted using ROI-dependent parameter vectors γ_m and β_m in the first regression layer. The ROI adaptive normalization (RAN) is given by

$$X(m) = \gamma_m \times \frac{x(m) - \mu_m}{\sigma_m} + \beta_m, \tag{2}$$

where $x(m)$ is the encoded data of the m-th sensor, $X(m)$ is the normalized feature of the m-th sensor, μ_m and σ_m are the mean and standard deviation of data of the m-th sensor, respectively. The ROI-dependent parameter vectors γ_m and β_m are adaptively formulated by processing one-hot vector, P, containing the depth and angle of the ROI via 4 common unit blocks (Fully-connected layer-ReLU) and 4 different unit blocks (see blue line in Fig. 3). Subsequent to the regression layer 1, a series of batch normalization is employed to manage the internal covariate shift problem [15]. Finally, the AC value is extracted via fully-connected layers.

3.2 Implementation Details

Time gain compensation is applied at a fixed rate of 0.75 dB/cm/MHz, and white Gaussian noise and quantization noise are added to the RF-data. The location of the ROI is adjustable between 40 mm and 90 mm on the depth-axis and $\pm 8°$

on the angular-axis and the size of ROI is set to ± 5 mm on the depth-axis and $\pm 2°$ on the angular-axis from the center. ROIs with heterogeneous tissues are excluded to maintain the homogeneity of the target tissues. The pulse-echo data is downsampled by a factor of 2, which provides sufficient resolution for the envelope change and cropped to cover ± 35 mm additional vertical region around the ROI. A total of 4,000 simulation phantoms were generated, where 3200 phantoms are used for the training, 400 phantoms are used for the validation, and 400 phantoms are used for the testing. The training objective of the network is to minimize the estimation loss, L1, between the ground truth AC value and the output of the network. L2 regularization is applied to enhance the convergence of the network [16]. The mathematical description of the objective function is

$$J = \underset{G}{\operatorname{argmin}} \, \mathbb{E}_{Y,E}[\|Y - G(E_{1:7}, P)\|_1] + \lambda \sum_i \|w_i\|^2. \tag{3}$$

where Y is the ground truth, $G(E_{1:7}, P)$ is the output of the proposed network, w_i is the weight of the network, and λ is the regularization parameter. The adam optimization algorithm [17] with an initial learning rate of 5×10^{-6} and a decay rate of 5×10^{-8} is applied. Dropout [18] with a retention probability of 0.5 is applied to reduce overfitting. Training stops at 250 epochs, which is determined based on the convergence of the loss in the validation set. The network is trained and tested by using PyTorch accelerated with an NVIDIA RTX2080Ti GPU.

4 Experiments and Results

The proposed method is evaluated through simulation, phantom, and in-vivo tests. For the phantom and in-vivo tests, an ultrasound system (Vantage 64LE, Verasonics Inc., WA, USA) with a convex probe (Humanscan, South Korea) and an MR system (Magnetom Verio 3T, Siemens, Germany) are used to obtain AC and MRI-PDFF, respectively. The reconstruction accuracy is evaluated by using the mean absolute error (MAE) and mean normalized absolute error (MNAE).

4.1 Simulation Test

The performance of the proposed method (RAN-ENV-AB) is validated by performing comprehensive ablation studies using 7 different variants (see Table 1). A fully convolutional network (FCN) [12] is employed as a baseline network. The methods with RAN network achieves 0.0665 in MAE and 8.87% in MNAE, which corresponds to an improvement of 0.0322 dB/cm/MHz in MAE and 4.29% in MNAE against methods with FCN, respectively. This indicates that RAN is effective for the accurate extraction of the AC. In addition, as compared to the FCN-based methods that suffer from 0.116 dB/cm/MHz drop in MAE (see Fig. 4 (a)), RAN-based networks experience only 0.0525 dB/cm/MHz drop in MAE (see Fig. 4 (b)) depending on the ROI depth. The RAN network with ROI adaptive TxBF achieves 0.0505 dB/cm/MHz in MAE and 6.73% in MNAE, which is a significant improvement of at least 0.0249 dB/cm/MHz in MAE and 3.32%

Table 1. Ablation study on ROI adaptive TxBF (AB), envelop detection (ENV), and network with RAN.

Method	TxBF AB	Preprocessing ENV	Network RAN	MAE [dB/cm/MHz]	MNAE [%]	PAR
FCN-RF w/o AB				0.1023	13.64	258M
RAN-RF w/o AB			✓	0.0829	11.05	260M
FCN-ENV w/o AB		✓		0.1229	16.39	**97M**
RAN-ENV w/o AB		✓	✓	0.0822	10.96	**99M**
FCN-RF-AB	✓			0.0852	11.36	258M
RAN-RF-AB	✓		✓	**0.0509**	**6.79**	260M
FCN-ENV-AB	✓	✓		0.0845	11.27	**97M**
RAN-ENV-AB	✓	✓	✓	**0.0501**	**6.68**	**99M**

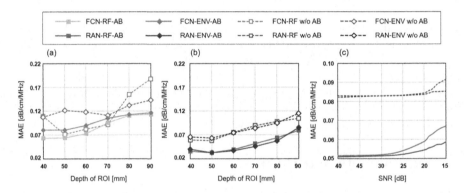

Fig. 4. MAE by depth of ROI of (a) methods using FCN and (b) methods featuring RAN network. (c) is MAE of the RAN-based methods under diverse input SNR.

in MNAE against the RAN network without ROI adaptive TxBF (see Fig. 4 (b)). In Fig. 4 (c), when the SNR of the received pulse-echo signal is decreased from 30dB to 15dB, the performance of RAN networks with and without envelope detection are degraded by 4.80×10^{-3} dB/cm/MHz and 11.65×10^{-3} dB/cm/MHz in MAE, respectively. The performance degradation of the RAN networks is reduced by 58.80% when envelope detection is used, which confirms the noise resiliency of the envelope detection scheme. Moreover, the number of network parameters (PAR) is reduced by 38% without performance degradation when envelope detection scheme is applied (see Table 1).

4.2 Phantom Test

The AC value (ground truth is 0.46 dB/cm/MHz) of the liver phantom (Model 057A, CIRS, USA) is extracted by using the proposed network (RAN-ENV-AB) and the baseline networks (RAN-ENV w/o AB, FCN-ENV-AB, and FCN-ENV w/o AB). In Fig. 5, the proposed RAN-ENV-AB outperforms all other variants

(0.0198 dB/cm/MHz in MAE and 4.31% in MNAE) and shows that the proposed method is valid in the real-world measurements. The proposed method featuring RAN outperforms those with FCN consistently irrespective of the ROI. The networks assisted by adaptive TxBF clearly demonstrate improved accuracy in AC value extraction. This suggests that adaptive TxBF and RAN are effective for the accurate extraction of the AC.

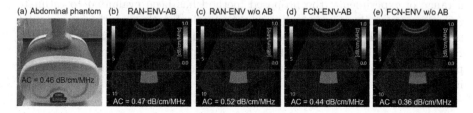

Fig. 5. Measurement setup and corresponding b-mode images with AC in the phantom test. (a) is the abdominal phantom and (b)-(e) are b-mode images with AC extracted by RAN-ENV-AB, RAN-ENV w/o AB, FB, FCN-ENV-AB, and FCN-ENV w/o AB, respectively.

4.3   In-Vivo Test

In-vivo liver data is acquired from 30 subjects (ranging from 21–39 years). PDFF is obtained from MRI and compared with the AC extracted by the proposed network (RAN-ENV-AB). The AC is extracted in the homogeneous liver parenchyma, excluding blood vessels is shown in Fig. 6(a)–(c). In Fig. 6(d), the AC values extracted by the proposed method demonstrate a strong correlation with those acquired with MRI-PDFF with a determination coefficient of $R^2 = 0.89, p < 0.001$.

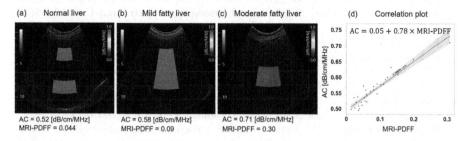

Fig. 6. B-mode images with AC in (a) normal, (b) mild, and (c) moderate fatty liver in the in-vivo test. (d) is correlation plot of the extracted AC by the proposed method (RAN-ENV-AB) versus MRI-PDFF.

5   Discussion and Conclusion

In this study, a learning-based AC quantification technique using a single convex probe is presented. The proposed method employs adaptive TxBF and envelope detection to increase the estimation accuracy and noise resilience. In addition, the proposed network adaptively normalizes the intensity of the received data according to the location of the ROI to maintain estimation accuracy. Through a range of experiments, the proposed method has been evaluated to estimate AC values correctly even in deep areas. In particular, through the in-vivo pilot study, we have shown that the proposed method is highly correlated with MRI-PDFF ($R^2 = 0.89$, $p < 0.001$), demonstrating the clinical potential of the proposed technique as a quantitative diagnosis modality in abdominal ultrasound including hepatic steatosis.

References

1. Browning, J.D., et al.: Prevalence of hepatic steatosis in an urban population in the United States: impact of ethnicity. Hepatology **40**(6), 1387–1395 (2004)
2. Kinner, S., Reeder, S.B., Yokoo, T.: Quantitative imaging biomarkers of NAFLD. Digest. Dis. Sci. **61**(5), 1337–1347 (2016). https://doi.org/10.1007/s10620-016-4037-1
3. McFarlin, B.L., Bigelow, T.A., Laybed, Y., O'Brien, W.D., Oelze, M.L., Abramowicz, J.S.: Ultrasonic attenuation estimation of the pregnant cervix: a preliminary report. Ultrasound Obstet. Gynecol. **36**(2), 218–225 (2010)
4. Karlas, T., et al.: Individual patient data me-ta-analysis of controlled attenuation parameter (CAP) technology for assessing steatosis. J. Hepatol. **66**(5), 1022–1030 (2017)
5. Samimi, K., Varghese, T.: Optimum diffraction-corrected frequency-shift estimator of the ultrasonic attenuation coefficient. IEEE Trans. Ultrason. Ferroelectr. Freq. Control. **63**(5), 691–702 (2016)
6. Coila, A.L., Lavarello, R.: Regularized spectral log difference technique for ultrasonic attenuation imaging. IEEE Trans. Ultrason. Ferroelectr. Freq. Control. **65**(3), 378–389 (2018)
7. Rau, R., Unal, O., Schweizer, D., Vishnevskiy, V., Goksel, O.: Attenuation imaging with pulse-echo ultrasound based on an acoustic reflector. In: Shen, D., et al. (eds.) MICCAI 2019. LNCS, vol. 11768, pp. 601–609. Springer, Cham (2019). https://doi.org/10.1007/978-3-030-32254-0_67
8. Treeby, B.E., Cox, B.T.: K-wave: MATLAB toolbox for the simulation and reconstruction of photoacoustic wave fields. J. Biomed. Opt. **15**(2), 021314 (2010)
9. Feigin, M., Freedman, D., Anthony, B.W.: A deep learning framework for single-sided sound speed inversion in medical ultrasound. IEEE Trans. Biomed. Eng. **67**(4), 1142–1151 (2019)
10. Oh, S., Kim, M.-G., Kim, Y., Bae, H.-M.: A learned representation for multi variable ultrasound lesion quantification. In: ISBI, pp. 1177–1181. IEEE (2021)
11. Kim, M.-G., Oh, S., Kim, Y., Kwon, H., Bae, H.-M.: Robust single-probe quantitative ultrasonic imaging system with a target-aware deep neural network. IEEE Trans. Biomed. Eng. (2021). https://doi.org/10.1109/TBME.2021.3086856

12. Xu, B., Wang, N., Chen, T., Li, M.: Empirical evaluation of rectified activations in convolutional network. arXiv preprint arXiv:1505.00853 (2015)
13. He, K., Zhang, X., Ren, S., Sun, J.: Deep residual learning for image recognition. In: CVPR, pp. 770–778 (2016)
14. Huang, X., Belongie, S.: Arbitrary style transfer in real-time with adaptive instance normalization. In: ICCV, pp. 1501–1510. IEEE (2017)
15. Ioffe, S., Szegedy, C.: Batch normalization: accelerating deep network training by reducing internal covariate shift. In: PMLR, pp. 448–456 (2015)
16. Girosi, F., Jones, M., Poggio, T.: Regularization theory and neural networks architectures. Neural Comput. **7**(2), 219–269 (1995)
17. Kingma, D., Ba, J.: Adam: a method for stochastic optimization. In: ICLR (2014)
18. Srivastava, N., Hinton, G., Krizhevsky, A., Sutskever, I., Salakhutdinov, R.: Dropout: a simple way to prevent neural networks from overfitting. J. Mach. Learn. Res. **15**(1), 1929–1958 (2014)

Colorectal Polyp Classification from White-Light Colonoscopy Images via Domain Alignment

Qin Wang[1,2], Hui Che[1,2], Weizhen Ding[1,2], Li Xiang[3], Guanbin Li[2,4],
Zhen Li[1,2(✉)], and Shuguang Cui[1,2]

[1] The Chinese University of Hong Kong, Shenzhen, China
`qinwang1@link.cuhk.edu.cn, lizhen@cuhk.edu.cn`
[2] Shenzhen Research Institute of Big Data, Shenzhen, China
[3] Longgang District People's Hospital of Shenzhen, Shenzhen, China
[4] Sun Yat-sen University, Guangzhou, China

Abstract. Differentiation of colorectal polyps is an important clinical examination. A computer-aided diagnosis system is required to assist accurate diagnosis from colonoscopy images. Most previous studies attempt to develop models for polyp differentiation using Narrow-Band Imaging (NBI) or other enhanced images. However, the wide range of these models' applications for clinical work has been limited by the lagging of imaging techniques. Thus, we propose a novel framework based on a teacher-student architecture for the accurate colorectal polyp classification (CPC) through directly using white-light (WL) colonoscopy images in the examination. In practice, during training, the auxiliary NBI images are utilized to train a teacher network and guide the student network to acquire richer feature representation from WL images. The feature transfer is realized by domain alignment and contrastive learning. Eventually the final student network has the ability to extract aligned features from only WL images to facilitate the CPC task. Besides, we release the first public-available paired CPC dataset containing WL-NBI pairs for the alignment training. Quantitative and qualitative evaluation indicates that the proposed method outperforms the previous methods in CPC, improving the accuracy by **5.6%** with very fast speed.

1 Introduction

Colorectal cancer (CRC) is one of the most common malignancies with a high mortality rate around the world [1]. Colorectal polyps are recognized as indicators of CRC, and they are roughly classified into two categories: hyperplastic and adenomatous [2]. Hyperplastic polyps are benign while adenomatous polyps have a high possibility of malignant transformation. Considering only the latter ones are required for surgical resection, precise differentiation is important to decrease unnecessary resection and unsuitable treatment. Colonoscopy is

---

Q. Wang and H. Che—Equal first authorship.

ⓒ Springer Nature Switzerland AG 2021
M. de Bruijne et al. (Eds.): MICCAI 2021, LNCS 12907, pp. 24–32, 2021.
https://doi.org/10.1007/978-3-030-87234-2_3

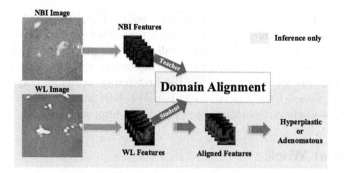

Fig. 1. The proposed teacher-student approach for polyp classification only utilizing WL images during inference. To improve the WL image based polyp differentiation accuracy, we adopt domain alignment to shift the distribution from WL features to NBI features during training, with the assistance of corresponding paired NBI images.

the preferred detection and diagnostic tool for colorectal polyps. However, due to varying illumination conditions, similar tissue representation, and occlusion, it is usually difficult to discriminate between benign and pre-cancerous polyps by conventional white-light (WL) observation, even for well-experienced endoscopists [3]. Therefore, an accurate and objective computer-aided classification system is demanded to assist clinical work.

Recent studies have achieved promising performance in colorectal polyp classification (CPC) by employing deep learning-based methods. Most works prefer to use datasets containing Narrow-Band Imaging (NBI) or Blue Light Imaging (BLI) images, owing to the enhanced visibility and superior performance [7]. For example, Usami *et al.* [11] proposed to distinguish benign/malignant polyps using WL, dye, and NBI images. In [2], authors achieved the highest accuracy of 95% by combining WL, BLI, and Linked Color Imaging (LCI) modalities.

Nevertheless, the widely used colonoscopy devices only have WL and NBI modes. Moreover, the acquisition of those advanced images is required to switch manually when polyps have been detected, while it usually suffers from missing detection in real clinic scenarios. Thus, the colorectal polyp detection using only WL endoscopy images is important but has not drawn sufficient attentions. Recently, Yang *et al.* [12] reported the classification results using WL images with the accuracy 79.5%. As shown, there is a large gap for the classification accuracy between using WL endoscopy images and enhanced images.

In this paper, we propose a novel framework as illustrated in Fig. 1 to facilitate the CPC task from WL colonoscopy images. To enhance low representative WL features, we adopt domain alignment to minimize the distance between WL and NBI feature distributions. Better feature representation in NBI images is transferred to the student network through domain alignment using adversarial learning and contrastive learning. Our main contributions are summarized in three-fold: (1) Through experiments, we prove that the CPC accuracy using

WL images is nearly 10% lower than that of using NBI images (88.9%) as input. Based on this observation, we define a new scheme that exploits NBI features to improve the WL-based classification results. (2) We propose a teacher-student model with GAN-based domain alignment and contrastive learning strategies to improve CPC. (3) We further release the first public-available polyp classification dataset named CPC-Paired, including WL-NBI image pairs. Our method achieves state-of-the-art performance (*i.e.*, $\sim 6\%$ improvement).

2   Related Work

Domain Alignment (DA). DA methods aim to align feature distributions between the source and target domains. Deep CORAL [9] defines a loss function to constrain the distance between the source and target domains in deep layer activations. In [6], correlation alignment is connected with entropy minimization to provide a solid performance. The above methods are applicable when target labels cannot be accessed. Other methods turn to minimize the difference between the source and target distributions in a shared feature space. The joint maximum mean discrepancy (JMMD) [4] is introduced to learn a transfer network by aligning the joint distributions of the network activations in domain-specific layers. Adversarial learning is adopted in domain adaptation to learn representations that the discriminator cannot distinguish between domains [10,13]. In this paper, we adopt this concept to align the features in different domains.

3   Method

3.1   Adversarial Learning for Domain Alignment

Inspired by [8], we adopt generative adversarial networks (GAN) to align the WL features with NBI features. As shown in Fig. 2, a teacher-student scheme is designed for the feature alignment. More specifically, we first pretrain a teacher feature extractor by only utilizing NBI images for CPC, where rich features can be extracted from NBI images to classify polyps. Then, we fix the teacher extractor to output NBI features X_p for aligning features X_a from student extractor. Particularly, the student extractor aims to extract features from WL images for polyp classification. However, the WL features X_a extracted from WL image are unsatisfactory for accurate polyp classification, rather than the features X_p from NBI images. Hence, to improve the classification accuracy of WL images, a discriminator D is introduced to align the WL features X_a with the rich NBI features X_p. The discriminator is optimized to distinguish between aligned WL features X_a and NBI features X_p (*i.e.*, NBI features are real and WL features are fake). As same with the GAN training manner, the discriminator D and student extractor are optimized alternatively. Therefore, the adversarial loss $\mathcal{L}_a$ supervises the student extractor to align its output with the teacher's (*i.e.*, NBI features X_p), which is shown in Eq. 1 where CE is the cross-entropy loss, Y_{nbi} indicates real label and takes 1 in practice.

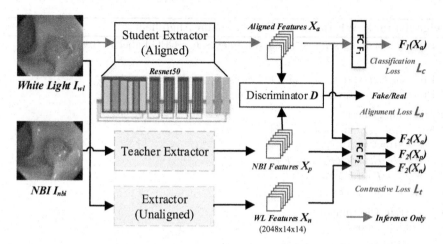

Fig. 2. The overview pipeline of the proposed method. First, we pretrain teacher and unaligned extractors by utilizing NBI images and WL images separately, which is shown in grey parts. To align the WL features X_a with more representative NBI features, an alignment loss $\mathcal{L}_a$ is designed to optimize the student extractor by introducing a discriminator for adversarial learning. Particularly, the discriminator aims to distinguish WL and NBI features. Section 3.1 illustrates details about alignment loss. Finally, the aligned accurate features X_a are fed in fully connected layer F_1 for polyp classification. Moreover, we exploit the contrastive learning loss $\mathcal{L}_t$ to shift the aligned WL features X_a much closer to NBI features X_p and far from the unaligned WL features X_n, which is introduced in details in Sect. 3.2. The blue arrow path indicates the inference phase. (Color figure online)

$$\mathcal{L}_a = CE(D(X_a), Y_{nbi}) = -log(D(X_a))Y_{nbi} \tag{1}$$

3.2 Contrastive Learning on CPC

As shown in Fig. 2, to further facilitate the model convergence and boost the performance, we design a novel contrastive loss $\mathcal{L}_t$ to take advantage of contrastive learning. More specifically, a naive unaligned feature extractor is pretrained to extract WL features X_n for CPC by only utilizing WL images as input. Then, the contrastive loss $\mathcal{L}_t$ can be formulated to supervise the student extractor to generate more representative features which are more similar to NBI features X_p and dissimilar to WL features X_n. Particularly, we take NBI features X_p as positive samples and unaligned WL features X_n as negative samples. To optimize aligned features X_a, the Kullback-Leibler (KL) divergence is adopted to constrain the distribution distance from X_a to WL features X_n (*i.e.*, negative samples) and NBI features X_p (*i.e.*, positive samples) in high-level semantic space, which is shown in Eq. 2. And F_2 is the fully connected (FC) layer to take feature maps for probability vectors generation, which is pretrained with the teacher extractor for classifying NBI images.

Algorithm 1: WL Image CPC via Domain Alignment

Input: NBI Images I_{nbi}; Paired WL Images I_{wl}; CPC Label Y; Test WL
Images I_s

1 Pretrain $Extractor_{teacher}$ and FC F_2 by I_{nbi} only;
2 Pretrain $Extractor_{unaligned}$ by I_{wl} only;

3 //Training Phase
4 For I_{nbi}^i, I_{wl}^i, Y^i in $\{I_{nbi}, I_{wl}, Y\}$
5   //Extract Features
6      $X_p \longleftarrow Extractor_{teacher}(I_{nbi}^i)$
7      $X_a \longleftarrow Extractor_{student}(I_{wl}^i)$
8      $X_n \longleftarrow Extractor_{unaligned}(I_{wl}^i)$

9   //Train Discriminator D
10   Minimize Loss $CE(D(X_p), 1) + CE(D(X_a), 0)$

11   //Train Student Extractor
12   Minimize CPC Loss $\mathcal{L}_c = CE(F_1(X_a), Y^i)$
13   Fix D and Minimize Alignment Loss $\mathcal{L}_a = CE(D(X_a), 1)$
14   Minimize Contrastive Loss $\mathcal{L}_t \longleftarrow Triplet(X_p, X_a, X_n)$
15 End

16 //Inference Phase
17 $\hat{Y} \longleftarrow F_1(Extractor_{student}(I_s))$
18 **Output**: CPC Prediction $\hat{Y}$

$$KL_{margin}(F_2(X_a), F_2(X_p)) \leq KL_{margin}(F_2(X_a), F_2(X_n)) \tag{2}$$

Finally, the triplet loss $\mathcal{L}_t$ is defined for contrastive learning in Eq. 3, where μ is a hyper-parameter we set as 0.85 in practice.

$$\mathcal{L}_t = \max(KL\,(F_2(X_a), F_2(X_p)) - KL\,(F_2(X_a), F_2(X_n)) + \mu, 0) \tag{3}$$

3.3 Loss Function

The overall training loss $\mathcal{L} = \mathcal{L}_c + \mathcal{L}_a + \mathcal{L}_t$ contains three parts. First, a conventional cross-entropy loss $\mathcal{L}_c = CE(F_1(X_a), Y)$ is applied to supervise student extractor and FC layer F_1 for binary classification (*i.e.*, hyperplastic or adenomatous), where Y is the ground truth. Then, the alignment loss $\mathcal{L}_a$ and contrastive loss $\mathcal{L}_t$ are utilized to align the WL features with NBI features, which make use of GAN and contrastive learning separately. Three loss functions are optimized jointly with equal weights. The Algorithm 1 illustrates the whole training and inference procedures.

4   Experiments and Results

4.1   Implementation Details

We implement our work by PyTorch. All models were trained for 500 epochs by Adam optimizer with learning rate 10^{-3} and weight decay 10^{-8} on single Nvidia V100 GPU. We randomly split the dataset into training and validation set by ratio 8:2 for training and evaluation. The training batch size is 16. We adopt random flipping and rotation for data augmentation. Additionally, we apply 5-fold cross-validation for all experiments, which randomly generates 5-fold train-valid settings.

4.2   Dataset and Preprocessing

We conduct the experiments on our CPC-Paired dataset[1]. The paired data means each WL image has a corresponding NBI image with the same polyp label. For each modal, a total of 307 adenomatous and 116 hyperplastic images are included. Our dataset consists of two parts: collated data from ISIT-UMR Colonoscopy Dataset [5] and clinical data collected from the hospital. ISIT-UMR Colonoscopy Dataset contains 76 short video sequences with category information. For our CPC task, we choose 21 hyperplastic lesions and 40 adenomas sequences. Each lesion in the video is recorded using both NBI and WL. We extract paired frames from videos to build an available dataset. The eventual collated data covers 102 adenomatous and 63 hyperplastic images in each modal. In addition, we collected 258 WL-NBI image pairs from 123 patients consisting of 205 adenoma images and 53 hyperplastic polyp images. We further annotate the bounding box of polyps to crop the corresponding area and scale it to 448×448 as input for the CPC task.

4.3   Network Architecture

In our framework, three extractors share the same backbone design. The backbone can be popular network architectures (*e.g.*, VGG, ResNet50, Inception-V3). More specifically, each extractor is utilized to mapping the original NBI I_{nbi} or WL images I_{wl} to a high-level feature space with the shape $2048 \times 14 \times 14$ (*e.g.*, ResNet50 backbone). Finally, each extractor is followed with a single FC layer to predict the final polyp class. In the pretrain stage, extractors and FC layers are optimized jointly(*e.g.*, teacher extractor and FC layer F_2) which will be fixed during the alignment training phase. The discriminator D consists of two convolution layers and two fully connected layers which aims to distinguish aligned WL features X_a and NBI features X_p.

---

[1] https://drive.google.com/drive/folders/1e2t5HhQf08sTAE_CPRNVgpi6YUKgQSHn? usp=sharing.

Table 1. The comparison between our approach and the previous best method in [12]. From the comparison, we can clearly notice that our approach surpasses the previous approach with a large margin(*e.g.*, $\sim 6\%$) among all backbones. 'FOLD X' indicates the different cross-validation split settings. 'Speed' indicates the inference time per image in millisecond. The 'Mean' averages the accuracy among all split settings, which gains 5.6% improvement by our approach and exactly proves the superior performance of the proposed alignment method.

	Backbone	Speed	FOLD1	FOLD2	FOLD3	FOLD4	FOLD5	Mean
Yang [12]	VGG	20.62 ms	78.2%	79.5%	77.3%	78.0%	77.9%	78.2%
Our			79.4%	81.1%	78.5%	79.1%	80.1%	79.7%
Yang [12]	InceptionV3	30.72 ms	81.1%	82.1%	80.5%	82.0%	81.3%	81.5%
Our			84.6%	85.7%	83.1%	85.3%	84.4%	84.6%
Yang [12]	ResNet50	17.07 ms	79.5%	80.5%	78.0%	80.3%	78.8%	79.7%
Our			**85.9%**	**86.1%**	**84.2%**	**85.8%**	**85.2%**	**85.3%**
Our w/o DA	ResNet50	17.07 ms	82.6%	83.3%	81.1%	83.7%	83.3%	82.9%
Our w/o CL			83.0%	84.1%	83.5%	84.4%	84.0%	84.0%

4.4 Results

Extensive experiments are conducted to demonstrate the superior performance of the proposed approach. Particularly, by 5-fold cross comparison in Table 1, we can observe that our method outperforms previous state-of-the-art approach [12] among all folds. Specifically, we improve the accuracy on all backbones including VGG, Inception-V3 and ResNet50, which proves the generality of the proposed model. We obtain the best performance of our approach with ResNet50 backbone (*i.e.*, the highest classification accuracy 85.3%, and 5.6% improvement on the mean score compared to the previous method). The ablation study further examines the gains of each component within the proposed approach as shown in Table 1. 'Our w/o DA' indicates removal of alignment loss and 'Our w/o CL' indicates ablation of contrastive loss. The CPC accuracy degradation of the ablation study exactly proves the effectiveness of each proposed component.

The qualitative analysis is shown in Fig. 3. Particularly, we extract aligned WL features X_a, unaligned WL features X_n and NBI features X_p for comparison. Obviously, the aligned WL features are more similar to NBI features than unaligned ones, which further demonstrates the superiority of aligned features and improvement of the proposed alignment approach.

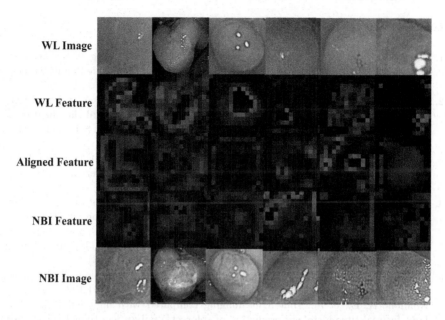

Fig. 3. The visualization comparison between WL feature, aligned feature, and NBI feature. From the comparison, we can obviously notice that the aligned feature (third row) is more similar to the NBI feature (fourth row) and less similar to the WL feature (second row), which exactly prove the effectiveness of the proposed domain alignment and contrastive learning approaches for domain shifting from WL to NBI. The aligned WL feature not only contains the original WL information but also provides more essential NBI domain representation for polyp classification.

5   Conclusion

For the purpose of investigating CPC, we release a polyp classification dataset CPC-Paired. To the best of our knowledge, this is the first public-available dataset including WL-NBI image pairs for this task. To improve the CPC accuracy of white-light (WL) images, we propose a teacher-student model for shifting the feature domain of WL images to NBI images which will be more representative for the CPC. Particularly, the novel alignment loss and contrastive loss are constructed to supervise the student model to generate more satisfactory features for the CPC. Extensive experiments consist of comparison, ablation study, and qualitative visualization, which sufficiently illustrate the effectiveness and superiority of our approach (*i.e.*, 5.6% accuracy improvement beyond the previous state-of-the-art approach on average).

Acknowledgement. The work was supported in part by Key Area R&D Program of Guangdong Province with grant No.2018B030338001, by the National Key R&D Program of China with grant No. 2018YFB1800800, by Shenzhen Outstanding Talents Training Fund, by Guangdong Research Project No. 2017ZT07X152, by NSFC-Youth 61902335, by Guangdong Regional Joint Fund-Key Projects 2019B1515120039, by The National Natural Science Foundation Fund of China (61931024), by helixon biotechnology company Fund and CCF-Tencent Open Fund.

References

1. Chen, P.J., Lin, M.C., Lai, M.J., Lin, J.C., Lu, H.H.S., Tseng, V.S.: Accurate classification of diminutive colorectal polyps using computer-aided analysis. Gastroenterology **154**(3), 568–575 (2018)
2. Fonollà, R., van der Zander, Q.E., Schreuder, R.M., Masclee, A.A., Schoon, E.J., van der Sommen, F., et al.: A CNN CADx system for multimodal classification of colorectal polyps combining WL, BLI, and LCI modalities. Appl. Sci. **10**(15), 5040 (2020)
3. Komeda, Y., et al.: Computer-aided diagnosis based on convolutional neural network system for colorectal polyp classification: preliminary experience. Oncology **93**(Suppl. 1), 30–34 (2017)
4. Long, M., Zhu, H., Wang, J., Jordan, M.I.: Deep transfer learning with joint adaptation networks. In: International Conference on Machine Learning, pp. 2208–2217. PMLR (2017)
5. Mesejo, P., et al.: Computer-aided classification of gastrointestinal lesions in regular colonoscopy. IEEE Trans. Med. Imaging **35**(9), 2051–2063 (2016)
6. Morerio, P., Cavazza, J., Murino, V.: Minimal-entropy correlation alignment for unsupervised deep domain adaptation. arXiv preprint arXiv:1711.10288 (2017)
7. Rondonotti, E., et al.: Blue-light imaging compared with high-definition white light for real-time histology prediction of colorectal polyps less than 1 centimeter: a prospective randomized study. Gastrointest. Endosc. **89**(3), 554–564 (2019)
8. Sankaranarayanan, S., Balaji, Y., Castillo, C.D., Chellappa, R.: Generate to adapt: Aligning domains using generative adversarial networks. In: Proceedings of the IEEE Conference on Computer Vision and Pattern Recognition, pp. 8503–8512 (2018)
9. Sun, B., Saenko, K.: Deep CORAL: correlation alignment for deep domain adaptation. In: Hua, G., Jégou, H. (eds.) ECCV 2016. LNCS, vol. 9915, pp. 443–450. Springer, Cham (2016). https://doi.org/10.1007/978-3-319-49409-8_35
10. Tzeng, E., Hoffman, J., Saenko, K., Darrell, T.: Adversarial discriminative domain adaptation. In: Proceedings of the IEEE Conference on Computer Vision and Pattern Recognition, pp. 7167–7176 (2017)
11. Usami, H., et al.: Colorectal polyp classification based on latent sharing features domain from multiple endoscopy images. Procedia Comput. Sci. **176**, 2507–2514 (2020)
12. Yang, Y.J., et al.: Automated classification of colorectal neoplasms in white-light colonoscopy images via deep learning. J. Clin. Med. **9**(5), 1593 (2020)
13. Zhang, W., Ouyang, W., Li, W., Xu, D.: Collaborative and adversarial network for unsupervised domain adaptation. In: Proceedings of the IEEE Conference on Computer Vision and Pattern Recognition, pp. 3801–3809 (2018)

Non-invasive Assessment of Hepatic Venous Pressure Gradient (HVPG) Based on MR Flow Imaging and Computational Fluid Dynamics

Kexin Wang[1], Shuo Wang[2,3], Minghua Xiong[4], Chengyan Wang[5(✉)],
and He Wang[5,6,7]

[1] Department of Physics, Fudan University, Shanghai, China
[2] Digital Medical Research Center, School of Basic Medical Sciences,
Fudan University, Shanghai, China
[3] Shanghai Key Laboratory of Medical Image Computing and Computer Assisted
Intervention, Shanghai, China
[4] Shanghai Zhiyu Software Information Co., Ltd, Shanghai, China
[5] Human Phenome Institute, Fudan University, Shanghai, China
{wangcy,hewang}@fudan.edu.cn
[6] Institute of Science and Technology for Brain-Inspired Intelligence,
Fudan University, Shanghai, China
[7] Key Laboratory of Computational Neuroscience and Brain-Inspired Intelligence
(Fudan University), Ministry of Education, Shanghai, China
https://faculty.fudan.edu.cn/wanghe/zh_CN/index.htm

Abstract. Clinically significant portal hypertension (CSPH) is a severe complication of chronic liver disease associated with cirrhosis, which is diagnosed by the measurement of hepatic venous pressure gradient (HVPG). However, HVPG measurement is invasive and therefore difficult to be widely applied in clinical routines. There is no currently available technique to measure HVPG noninvasively. Computational fluid dynamics (CFD) has been used for noninvasive measurement of vascular pressure gradient in the intracranial and coronary arteries. However, it has been scarcely employed in the hepatic vessel system due to the difficulties in reconstructing precise vascular anatomies and setting appropriate boundary conditions. Several computer tomography and ultrasound based studies have verified the effectiveness of virtual HVPG (vHVPG) by directly connecting the portal veins and hepatic veins before CFD simulations [12,16]. We apply the latest techniques of phase-contrast magnetic resonance imaging (PC-MRI) and DIXON to obtain the velocity and vessel anatomies at the same time. Besides, we improve the CFD pipeline in regards to the construction of vessel connections and reduction of calculation time. The proposed method shows high accuracy in the CSPH diagnosis in a study containing ten healthy volunteers and five patients. The MRI-based noninvasive HVPG measurement is promising in the clinical application of CSPH diagnosis.

Keywords: Computational fluid dynamics · Hepatic venous pressure gradient · Phase-contrast MRI

© Springer Nature Switzerland AG 2021
M. de Bruijne et al. (Eds.): MICCAI 2021, LNCS 12907, pp. 33–42, 2021.
https://doi.org/10.1007/978-3-030-87234-2_4

1   Introduction

Portal hypertension is a severe complication among patients with chronic liver disease. Currently, clinically significant portal hypertension (CSPH) is diagnosed with the measurement of hepatic venous pressure gradient (HVPG) larger than 10 mmHg. This HVPG measurement is conducted by putting a balloon catheter into the right hepatic vein to get both the free hepatic venous pressure (FHVP) and the wedged hepatic venous pressure (WHVP). The HVPG is defined as the difference between WHVP and FHVP. Although the gold standard of clinical diagnosis of CSPH is HVPG, it is invasive, probably leading to infection [6,7] and other complications [4], thus inaccessible to some patients.

Recently, several non-invasive measurements have been proposed to obtain HVPG, transient elastography (TE) for example [1,14]. However, TE is often unreliable for patients with obesity or intrahepatic inflammatory activities [3]. Some serum biomarkers including prothrombin index (PI) and aspartate amino-transferase (AST) to alanine aminotransferase (ALT) ratio (AAR) are more widely available in hospitals, but their accuracy is lower than TE.

Computational fluid dynamics (CFD) simulation has grown rapidly in the past few years for the diagnosis of cardiovascular diseases [2,13,18], carotid artery diseases [10] and intracranial artery occlusive diseases [9]. The virtual blood pressure gradient is measured by solving the Navier-Stokes equation with certain boundary conditions. CFD-based measurement of blood pressure gradient has achieved remarkable success in assessing intracranial and coronary arteries (e.g., cardiac fractional flow reserve (FFR)), while it is less explored in the hepatic vascular system due to two possible reasons. First, accurate CFD simulation is highly dependant on the precise reconstruction of vascular anatomy and the measurement of inflow velocity. However, the anatomies of hepatic vein and portal vein are difficult to be obtained. Currently available ultrasound (US) based velocity measurement is not robust and repeatable. Second, the computational efficiency is another limitation for its application in clinical routines due to the complex structure of liver vessels.

This paper proposes a one-stop solution for the noninvasive measurement of HVPG in both the healthy volunteers and patients with liver cirrhosis by using multi-contrast magnetic resonance imaging (MRI) and an efficient CFD model.

2   Methods and Experiments

2.1   Multi-contrast MRI

To avoid the use of contrast agents, multi-echo DIXON (mDIXON) imaging is conducted to obtain the anatomy of hepatic vessels in this study. A 3D gradient-echo (GRE) sequence with multiple echoes is used. All the subjects are scanned on a 3.0 T MR scanner (Ingenia, Philips, the Netherlands). The imaging parameters are: repetition time (TR) = 3.4 ms; echo time (TE) = 0.95 ms; field-of-view (FOV) = $350 \times 350 \times 120 \, \text{mm}^3$; flip angle (FA) = $10°$; spatial resolution = 0.87

$\times\ 0.87\ \times\ 2.0\,\mathrm{mm}^3$, which sets the limit for the smallest vessel reconstructed; SENSitivity Encoding (SENSE) factor = 2; scan time = 12.5 ms.

Blood flow velocity is measured using phase-contrast (PC) GRE sequence. Imaging is performed during free breathing with cardiac and respiratory gatings. The imaging parameters for PC-MRI are: TR = 2.9 ms; TE = 1.45 ms; FOV = $300\ \times\ 300\ \times\ 300\,\mathrm{mm}^3$; FA = 6°; spatial resolution = $0.87\ \times\ 0.87\ \times\ 2.0\,\mathrm{mm}^3$; SENSE factor = 2; encoding velocity = 40 cm/s. The acquisition time is nearly 2.5 min depending on the subject's heart rate. The blood flow acquisition plane is selected perpendicular to the portal vein on the morphological scout sequences.

2.2 Image Analysis

MRI data are transferred in Digital Imaging and Communications in Medicine (DICOM) format to a workstation for analysis. The vessels are automatically extracted from the mDIXON images using an pre-trained deep neural networks (DNN) [17]. Two separate pre-trained deep neural networks (DNNs) are applied to obtain the portal veins and hepatic veins automatically. The DNNs are trained on an additional dataset including 30 manually labelled cases in our previous work.

Extraction of the temporal flow velocity curve during a cardiac cycle is from the cross section of portal vein manually labeled by an experienced radiologist on the PC-MRI images. The flow velocity is calculated as the mean value of the cross section obtained during the 30 phases of several repeated cardiac cycles (see the process in Fig. 1. $u(t)$).

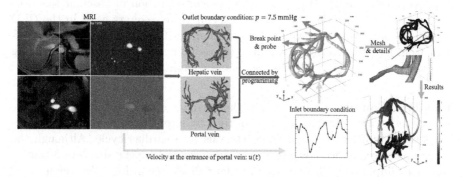

Fig. 1. The schematic diagram of the MRI based CFD model. From the MRI images on the left to the pressure distribution at the bottom right. Firstly, geometry of the hepatic vein and the portal vein are segmented individually, together with the attained velocity of flow at the entrance of the portal vein. The combination of the yellow and blue arrow is the green arrow, which represents applying the connected model and boundary conditions, then with meshing based on both practice, CFD calculation achieves the pressure distribution eventually. (Color figure online)

2.3   CFD Simulation

CFD simulation can provide accurate pressure measurement based on the geometry of blood vessels and the boundary conditions, e.g., the velocity and the reference pressure at the inlet and the outlet of the blood vessels. Finite element method (FEM) is applied to solve the Navier-Stokes (NS) equations. The main components of the simulation can be divided into three parts: a) dominant physics of the laminar fluid model, b) settings of the boundary conditions, and c) development of a solver in space-time.

Laminar Physics. A three-dimensional NS equations with the assumption of stationary incompressible fluid are applied in our study as the following:

$$\rho(\boldsymbol{u} \cdot \nabla)\boldsymbol{u} = \nabla \left(-p\mathbb{I} + \mu[\nabla \boldsymbol{u} + (\nabla \boldsymbol{u})^T]\right) + \boldsymbol{F} \tag{1}$$

$$\rho \nabla \cdot \boldsymbol{u} = 0 \tag{2}$$

where $\boldsymbol{u}$ is the flow velocity, p is the pressure, μ is the dynamic viscosity and ρ the density of the blood, which is assumed to be Newtonian fluid with $\mu = 0.005$ Pa·s and $\rho{=}1050$ kg/m^3 [16]. $\boldsymbol{F}$ is the volume force vector and is set to be zero as we ignore the gravity of blood. The Eq. 1 shows the balance of momentum from Newton's Second Law, and the Eq. 2 represents the constraint of continuity, which is incompressible.

We use only laminar physics here rather than the turbulence and non-Newtonian modelling, or Fluid-Structure Interaction (FSI) model, due to the small Reynolds number (Re = 27) and our consideration of vascular sclerosis. On the other hand, although the geometry of vessels counts a lot in the CFD model and admittedly, it will be more accurate if we use FSI, patients suffered from CSPH experience significantly reduced wall compliance, rendering our assumption of rigid wall reasonable and computationally efficient.

Boundary Condition. The inlet is the entrance of the portal vein, with the velocity set as a continuous function of time by interpolating cubic spline function into 30 uniformly-spaced time intervals in a cardiac cycle. Although the velocity has already been averaged over time, we prolong the simulation time of interest by duplicating the cardiac cycle four times (see in Fig. 2(b)), guaranteeing that the flow is fully developed. A smooth transition from zero is added at the beginning 100 ms to accelerate the convergence, as enlarged in Fig. 2(a).

The outlet is naturally the exit of the hepatic vein, and a reference pressure is set here. As vHVPG is calculated as the pressure difference before and after the balloon is inflated, it is irrelevant to the absolute value of the reference pressure. For the sake of similarity to clinical measurement and following the physiological condition, the constant value of 7.5 mmHg is set as the outlet pressure condition.

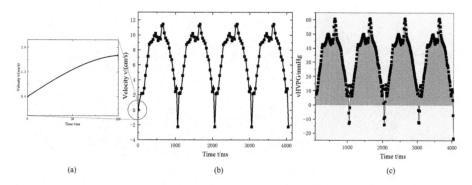

Fig. 2. Velocity settings in an example of a patient with liver cirrhosis. (a) The illustration of how to smooth the abrupt transition at the beginning 100 ms of a flow velocity period. (b) The exact interpolated function of velocity consists of four duplicate periods. (c) Corresponding vHVPG as a function of time. The grey denotes the section enclosed by the function and the x-axis, which is the integration of vHVPG during the calculation time considering the plus and minus sign. Then the integration by the time is the average vHVPG.

Solver. We use the Laminar Fluid module of COMSOL Multiphysics for finite element simulation. We use a fully coupled direct solver for CFD models. To avoid run-out-of-memory error, we cut the iteration to be ten thousand times. Two dependent variables are coupled in the equation, i.e., the pressure and velocity field. Moreover, algebraic multigrid (AMG) solvers are applied to obtain these two values at each node [5].

2.4   Experiments

This retrospective study includes a total of ten healthy volunteers and five patients with liver cirrhosis. The study was approved by the ethics committee of the local hospital. Informed consent was obtained from all the subjects prior to the examinations. The cirrhosis was diagnosis by means of liver biopsy.

Building Geometry. Portal veins and hepatic veins are extracted from PC-MRI separately and therefore the connection of capillary is needed. Since CFD largely depends on the geometry, the strategy to rebuild it is fundamental and original here in our work. As can be viewed in Fig. 3, in principle, the connection of portal vein and hepatic vein should get close to the reality, where both one-to-one and one-to-many connections exist. Therefore, the decisions to carry out which kind of the link and how to bridge the neighbour ports are important, and we demonstrate our procedure as below.

First of all, find two ends of the axis in two vessels, say P_0, P_1, P_2, P_3, from which the direction vector v_1 and v_2 will be obtained. Then defining the center point P_c by the midpoint of $P_1'P_3'$, where the spot P_1' is translated from P_1 along the vector v_1 by the distance of $|P_1P_3|/2$ and the same for P_3' but from P_3 along

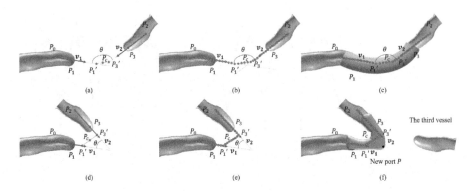

Fig. 3. Procedure to connect vessels by interpolation. The top line is the case when the angle between two vessels $\theta > 90°$, and the bottom line is $\theta < 90°$. $\overrightarrow{P_0 P_1}$ defines the directional vector v_1 of the vessel on the left and $\overrightarrow{P_2 P_3}$ works for v_2 on the right. P_1' (or P_3') is the translation of P_1 (or P_3) along v_1 (or v_2) at the distance of $|P_1 P_3|/2$. The midpoint P_c is in the middle of P_1' and P_3'. In (b) and (e), interpolation is demonstrated as green dots along $P_0, P_1, P_1', P_c, P_3', P_3$. As in (c), large θ results in a smooth bridge and no more connections are allowed for them, while in (f) the curve will provide a new point P, served as a new port to be connected with a third vessel. (Color figure online)

v_2. Next, interpolate enough points between the point set of $P_0, P_1, P_1', P_c, P_3', P_3$ and extrude the circle surface with the diameter of initial diameter on both sides along the interpolated curve. Finally, with some polishing, the connection is built up. However, if the angle between two vectors v_1 and v_2 is acute, the bridge will end up with a rather sharp curve, generating a new point P to be connected with other vessels.

After the connection, a break point is created for the modeling of WHVP, representing a strict block of the blood flow along the vessel. We ring cut the right-most secondary branch at the point of about 1 cm away from the fork. Afterwards, both sides are healed by the smooth smallest surface and then follows the remesh process. Remeshing includes uniforming the surface mesh and creating the volume mesh from the surface, providing an exact calculation domain for CFD with FEM. See in Fig. 4, pressure is evaluated near the break point from the side closer to the portal vein. The virtual FHVP (vFHVP) is from the model of complete blood vessels, while the virtual WHVP (vWHVP) is achieved in the cut-off model. Therefore, similar to the clinical measurement in practice, the virtual hepatic venous pressure gradient (vHVPG) is the difference between the above two pressures: vHVPG = vWHVP - vFHVP.

3   Results and Discussion

In this section we present the simulation result and prove the accuracy by anal-yse the data. In Fig. 4, typical examples of average pressure distribution in a

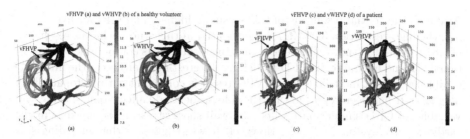

Fig. 4. Average pressure distribution of typical CFD model. The first two pictures show the pseudo-color images of flow pressure for a healthy volunteer and the other two are for a patient with liver cirrhosis. Arrows show measurement location. As for virtual free hepatic venous pressure (vFHVP), (a) = 9.8 mmHg, (c) = 9.6 mmHg; for virtual wedged hepatic venous pressure (vWHVP), (b) = 14.5 mmHg, (d) = 19.8 mmHg. vHVPG = vWHVP − vFHVP, and it is 4.7 mmHg for the healthy person, but 10.2 mmHg for the patient with liver disease. As CSPH is classified as HVPG ≥ 10 mmHg, the findings of these two cases fulfill this standard, therefore verifying our method of CFD model based on PC-MRI.

healthy volunteer and a patient with liver cirrhosis can be compared with each other. Three main properties verify the CFD model primarily: 1) Flow pressure decreases gradually from the inlet at the portal vein to the outlet at the hepatic vein. 2) Averagely, vWHVP is larger than vFHVP in both cases. 3) Obviously, the flow pressure of the patient is larger than that of the healthy volunteer. The simulation result of vHVPG is concluded in Table 1.

HVPG larger than 10 mmHg is defined as CSPH for the statistical analysis. Although the number of cases is limited, it still shows a significant increase of HVPG for the patients and 80 percents of them are found to be CSPH (HVPG ≤ 10 mmHg) according to our studies. Based on this confusing matrix, the kappa test has shown substantial consistency (k=0.7) between the diagnosis based on histopathology and our virtual HVPG measurements. Besides, the normal HVPG is suggested to be in the range of 3–5 mmHg for healthy adults [3], which is also fulfilled in our studies.

Table 1. Summary of the vHVPG for different groups

Group	Average vHVPG (mmHg)	No. of true	No. of false
Healthy volunteers	5.9 ± 2.3	9 (≤ 10 mmHg)	1 (≥ 10 mmHg)
Patients with liver cirrhosis	21.5 ± 8.8	4 (≥ 10 mmHg)	1 (≤ 10 mmHg)

As for the next step, the interaction between structures and the fluid will be considered, resulting in a more precise model than the current one of simple single-phase laminar physics. Also, in order to virtually model the invasive measurement as close as possible, the geometry ought to be improved as well.

Although the spatial resolution of CT is usually higher than MRI for liver imaging, the high-resolution 3D GRE sequence used in this study achieves resolution of $0.87 \times 0.87 \times 2.0\,\mathrm{mm}^3$. We believe that the resolution is comparable to conventional CT and adequate for CFD modelling. Interestingly, although the fine-scale capillaries are not modelled, the vHVPG result is consistent with histopathology diagnosis. It has been reported that the side branches do affect some specific wall variables of CFD simulation results, e.g. wall shear stress, although it is in the cardiovascular system [8,11,15]. However, it was suggested that including side blood vessels up to 1 mm in diameter is appropriate [15]. While the vessel diameter that can be observed in our study is no less than 0.87 mm corresponding to the spatial resolution, we believe such simplification has a limited influence on the calculation of pressure gradient. However, we still believe that the way forward lies in some modified CFD model including nearly all the vessels or at least their effects, constructed automatically with as little manual interference as possible.

4   Conclusion

In this article, we introduced a one-stop solution to acquire an accurate estimate of HVPG non-invasively, assisted by PC-MRI and CFD. For the first time, CFD model is established entirely based on MRI for the geometry and boundary conditions of velocity in the liver system, as we take advantage of the latest technology of PC-MRI. Since the limit of resolution forbids a complete map of entire blood vessels in the liver, trunks of hepatic vein and portal vein are preserved and connected automatically by our original programming method, which demonstrates complex networking very similar to the real vessels. FEM solver is applied in the CFD model accelerating the calculation speed. By now, it only takes less than twenty minutes to compute HVPG from the raw image, while it can even be faster in our further study. Testing in a data set of ten healthy people and five patients with liver cirrhosis, we verify the accuracy of our method in diagnosing people with CSPH. Our study facilitates the process of non-invasive and patient-specific treatment.

Acknowledgements. This study was funded by the National Natural Science Foundation of China (No. 81971583, No. 62001120), National Key R&D Program of China (No. 2018YFC1312900, No. 2019YFA0709502), Shanghai Natural Science Foundation (No. 20ZR1406400), Shanghai Municipal Science and Technology Major Project (No. 2017SHZDZX01, No. 2018SHZDZX01), Shanghai Municipal Science and Technology (No. 17411953600), Shanghai Sailing Program (No. 20YF1402400), ZJLab and Key Laboratory of Computational Neuroscience and BrainInspired Intelligence (Fudan University), Ministry of Education, China.

References

1. Castéra, L., et al.: Early detection in routine clinical practice of cirrhosis and oesophageal varices in chronic hepatitis c: comparison of transient elastography (fibroscan) with standard laboratory tests and non-invasive scores. J. Hepatol. **50**(1), 59–68 (2009)
2. Cito, S., Pallarés, J., Vernet, A.: Sensitivity analysis of the boundary conditions in simulations of the flow in an aortic coarctation under rest and stress conditions. In: Camara, O., Mansi, T., Pop, M., Rhode, K., Sermesant, M., Young, A. (eds.) STACOM 2013. LNCS, vol. 8330, pp. 74–82. Springer, Heidelberg (2014). https://doi.org/10.1007/978-3-642-54268-8_9
3. De, F.R., Faculty, B.V.: Expanding consensus in portal hypertension: report of the Baveno vi consensus workshop: stratifying risk and individualizing care for portal hypertension. J. Hepatol. **63**(3), 743–752 (2015)
4. Dong, J., Qi, X.: Liver imaging in precision medicine. Ebiomedicine **32**, 1–2 (2018). S2352396418301816
5. Falgout, R.D.: An introduction to algebraic multigrid computing. Comput. Sci. Eng. **8**(6), 24–33 (2006)
6. Fang, C., et al.: Consensus recommendations of three-dimensional visualization for diagnosis and management of liver diseases. Hepatol. Int. **14**(4), 437–453 (2020). https://doi.org/10.1007/s12072-020-10052-y
7. Franchis, R.D.: Portal Hypertension VI: Proceedings of the Sixth Baveno Consensus Workshop: Stratifying Risk and Individualizing Care. Springer, Cham (2016). https://doi.org/10.1007/978-3-319-23018-4
8. Giannopoulos, A.A., et al.: Quantifying the effect of side branches in endothelial shear stress estimates. Atherosclerosis **251**, 213–218 (2016)
9. Gorelick, P.B., Wong, K.S., Bae, H.J., Pandey, D.K.: Large artery intracranial occlusive disease: a large worldwide burden but a relatively neglected frontier. Stroke J. Cereb. Circ. **39**(8), 2396 (2008)
10. Jenkins, R.H., Mahal, R., Maceneaney, P.M.: Noninvasive imaging of carotid artery disease: critically appraised topic. Can. Assoc. Radiol. J. **54**(2), 121–123 (2003)
11. Li, Y., et al.: Impact of side branch modeling on computation of endothelial shear stress in coronary artery disease: Coronary tree reconstruction by fusion of 3D angiography and oct. J. Am. Coll. Cardiol. **66**(2), 125–135 (2015)
12. Liu, F., et al.: Development and validation of a radiomics signature for clinically significant portal hypertension in cirrhosis (chess1701): a prospective multicenter study. Ebiomedicine **36**, 151–158 (2018)
13. Schaller, J., Goubergrits, L., Yevtushenko, P., Kertzscher, U., Riesenkampff, E., Kuehne, T.: Hemodynamic in aortic coarctation using MRI-based inflow condition. In: Camara, O., Mansi, T., Pop, M., Rhode, K., Sermesant, M., Young, A. (eds.) STACOM 2013. LNCS, vol. 8330, pp. 65–73. Springer, Heidelberg (2014). https://doi.org/10.1007/978-3-642-54268-8_8
14. Tana, M.M., Muir, A.J.: Diagnosing liver fibrosis and cirrhosis: serum, imaging, or tissue? Clin. Gastroenterol. Hepatol. **16**(1), 16–18 (2018)
15. Vardhan, M., Gounley, J., Chen, S.J., Kahn, A.M., Leopold, J.A., Randles, A.: The importance of side branches in modeling 3D hemodynamics from angiograms for patients with coronary artery disease. Sci. Rep. **9**(1), 8854 (2019)
16. Wadhawan, M., Dubey, S., Sharma, B.C., Sarin, S.K., Sarin, S.K.: Hepatic venous pressure gradient in cirrhosis: correlation with the size of varices, bleeding, ascites, and child's status. Digest. Dis. Sci. **51**(12), 2264–2269 (2006)

17. Zhang, X., Zhao, X., Tan, T.: Robust dialog state tracker with contextual-feature augmentation. Appl. Intell. **51**(4), 2377–2392 (2020). https://doi.org/10.1007/s10489-020-01991-y
18. Zhao, Xi., et al.: Multiscale study on hemodynamics in patient-specific thoracic aortic coarctation. In: Camara, Oscar, Mansi, Tommaso, Pop, Mihaela, Rhode, Kawal, Sermesant, Maxime, Young, Alistair (eds.) STACOM 2013. LNCS, vol. 8330, pp. 57–64. Springer, Heidelberg (2014). https://doi.org/10.1007/978-3-642-54268-8_7

Deep-Cleansing: Deep-Learning Based Electronic Cleansing in Dual-Energy CT Colonography

Guibo Luo, Tianyu Liu, Bin Li, Michael Zalis, and Wenli Cai$^{(\boxtimes)}$

Massachusetts General Hospital and Harvard Medical School, Boston, USA
Cai.Wenli@mgh.harvard.edu

Abstract. Electronic cleansing (EC) is an image-processing technique for subtraction of tagged fecal regions in the colon for visualization of the entire colonic surface in CT colonography (CTC). This paper introduced a deep-learning based EC method in dual-energy CTC (DE-CTC), named "Deep-Cleansing". First, we calculated the "effective" atomic number (EAN) by fractions of atomic mass number using the low- and high-energy images in DE-CT. Second, multiple types of combinations of input channels (images), and multiple backbone networks were investigated in U-Net architecture for the optimal performance of EC. Finally, we trained and evaluated our Deep-Cleansing by a total of 139 DE-CTC cases (approximately 80K DE-CTC images), which were randomly divided into training, validation and testing set at an approximate ratio of 55%:20%:25%. Our Deep-Cleansing method achieved DICE coefficients of 0.953 in validation and 0.956 in testing datasets, respectively. Overall, the average cleansing ratio was 96.14%, and the soft-tissue preservation ratio was 97.74%, both were significantly higher than EC without EAN maps. Our results indicated that the use of Deep-Cleansing substantially improved the accuracy of EC in the subtraction of tagged fecal regions in CTC.

Keywords: Medical Physics · Dual-energy CT · CT colonography · Electronic cleaning

1 Introduction

Computed tomographic colonography (CTC), or virtual colonoscopy, has been confirmed as a viable and cost-effective alternative to optical colonoscopy (OC) for colon cancer screening by large-scale and multi-center clinical trials [1]. Fecal tagging is a means of marking the residual fecal materials by using the oral administration of a positive contrast agent (such as barium or iodine) [2]. Electronic cleansing (EC), or digital bowel cleansing, is an image-processing technique for subtraction of tagged fecal regions in CTC images after image acquisition [3]. The goal of EC is to "virtually cleanse" tagged fecal materials in the colon that can obscure the colonic surface, and by preservation of colonic soft-tissue structures such as folds and polyps [4].

© Springer Nature Switzerland AG 2021
M. de Bruijne et al. (Eds.): MICCAI 2021, LNCS 12907, pp. 43–53, 2021.
https://doi.org/10.1007/978-3-030-87234-2_5

Conventional EC solution is to apply an image classification or segmentation method for explicit removal of tagged fecal regions in CTC images. However, accurate estimation of individual components and determination of boundaries of tagged fecal regions is highly challenging due to the highly variable in terms of bowel contents, the locations and thickness of air-tagging mixture layer. Applying dual-energy CT (DE-CT) to CTC image acquisition (DE-CTC) provides a promising means to differentiate tagged fecal materials from soft-tissue structures in the colon. However, current EC methods for DE-CTC such as spectral analysis [5], material decomposition [6], and multi-spectral profile learning [7] are merely based on the difference in CT value at two photon energies, which are heuristic and incomplete to solve the EC artifacts caused by partial volume effect (PVE) and tagging inhomogeneity. The use of DE-CT physical quantities to characterize the molecule of a biological material is more reliable for classification of tagged fecal regions. In addition, the determination of colonic surface layer, which is a soft-tissue mucosa, is more reliable and consistent in CTC images than that of air-tagging mixture layer. Thus, instead of segmentation of multiple tagged fecal regions, it is more robust to directly segment the colon lumen (including air, tagged materials, and air-tagging mixtures) as a whole to subtract the tagged fecal regions to reduce the EC artifacts cause by air-tagging mixture layer.

In this study, we developed a deep-learning based EC method in DE-CTC by using effective atomic number (EAN) calculated in DE-CT and U-Net segmentation model, named "Deep-Cleansing", for an accurate segmentation of colon lumen for subtraction of tagged fecal regions in the colon in DE-CTC.

The main contributions of the study are: (1) the collection of a large DE-CTC data set with radiologists-confirmed contouring of colon lumen for training of deep-learning based EC method, (2) the calculation of EAN in DE-CT to improve the material classification in DE-CTC, and (3) the comparison of multiple types of input channel combinations in U-Net models for an optimal segmentation of colon lumen in DE-CTC.

2 Problems and Motivations

Since EC was firstly introduced in later 90's [8], various EC methods have been developed including voxel classification [9], gradient-based edge detection [10], material transition [11], structure analysis [12], mosaic decomposition [13], hybrid methods [14–16], and recent convolutional neural network (CNN) based method [7]. However, conventional EC methods tend to generate severe artifacts mainly caused by image pseudo-enhancement, partial volume effect, and fecal-tagging inhomogeneity [7, 17].

These problems in conventional EC methods stem from image classification of multiple-material mixtures in CTC such as air, soft tissue, and tagged fecal residuals. DE-CT provides an effective means to estimate material composition by analysis of the difference of two CT values acquired at two different X-ray tube voltages (such as 80 kVp and 140 kVp). Our recent finding, that the effective atomic numbers (EAN, Z_{eff}) by fractions of atomic mass number are substantially higher (nearly double) in air and iodine-tagging materials (major substitutes in colonic lumen) than those in water-like materials such as soft tissue or fat, as demonstrated in Fig. 1(c). EAN is a very effective physical parameter to classify colonic structures (mucosa surface, colonic fold, and polyps) from colon lumen (air, tagging residuals, and air-tagging mixtures) in DE-CTC images. This motivates us to investigate a deep-learning based DE-EC solution for an artifact-free classification of tagged fecal residuals and accurate segmentation of entire colon lumen by integrating high- and low-energy CT images and EAN maps to overcome the major EC artifacts caused by PVE and inhomogeneity.

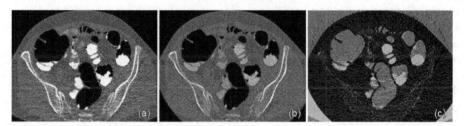

Fig. 1. DE-CTC images and Z_{eff} image. (a, b) A pair of transverse images at tube voltage of (a) 80 kVp and (b) 140 kVp. (c) The Z_{eff} image of (a) and (b).

Existing CNN-based EC methods can be categorized as CNN classifier or GAN model [18, 19]. CNN classifier [18] differentiates each voxel as one type of materials/mixtures for instance air, soft-tissue (ST), fat, tagged fecal material, air-tagging mixture, air-ST mixture, tagging-ST mixture, etc., and then replaces the tagging materials or tagging component in mixtures with "air", as conventional EC methods typically do. Multiple-material classification is intricate as some studies indicated that there are 15 possible material mixtures in CTC images [9], and accurate estimation of tagging components in mixtures is challenging. On the other hand, GAN models [19] train a 3D-vox2vox model for EC by using paired uncleansed and cleansed CTC images, which rely on phantom or pre-cleansed CTC images with and without tagging materials. Due to the limited GPU memory, these GAN models tend to down-sampling the voxel resolution, which will inevitably blur or eliminate some important details along the contours between soft-tissue and luminal air.

In contrast, we proposed the direct segmentation of colon lumen by using a U-Net model regardless of bowel contents (such as air, tagged fecal materials, air-tagging mixtures, etc.) because the colonic surface layer and its boundary are more reliable and consistent than tagged fecal materials/mixtures. Segmentation of colon lumen instead of multiple tagged fecal materials and classification of colon bowel contents using EAN are two major innovations of our proposed method.

3   Methods

Atomic number Z is the number of protons or electrons of an atom, and an "effective" atomic number of a compound (e.g. water) or a mixture of different materials (such as tissue and bone) is defined as the weighted sum of atomic number by their elemental compositions, in which the weighting fractions may be defined either by elemental electron number (for estimation of electron density) or by atomic mass number (the atomic mass number is equal to the number of protons plus the number of neutrons that it contains) [20]. EAN is commonly utilized in characterization of different materials in radiation physics [21].

3.1 Calculation of EAN

The numerical value of EAN is not unique, but hinges upon the mathematical formulation chosen. Conventionally, Mayneord's power-law formula is commonly used to define EAN for materials with known elemental compositions: $Z_{eff} = \sqrt[m]{\sum f_i Z_i^m}$, where the relative electron fraction of the i-th element Z_i is given by f_i such that $\sum f_i = 1$; a value of 2.94 for exponent m is used for low photon energies [22].

In this study, we chose to use a formalism recently proposed by Taylor [23], where the EAN of a compound or mixture is associated with its average atomic microscopic cross-section $\sigma_{a,m}$, defined in Eq. (1),

$$\sigma_{a,m} = \sum_i f_i \sigma_i \tag{1}$$

where f_i is the fraction of the i-th element by the number of atoms.

Under Taylor's formalism, the EAN of a material with known elemental composition can then be formally defined by Eq. (2),

$$Z_{eff} = \Psi^{-1}(\sigma_{a,m}) \tag{2}$$

where $\Psi(Z)$ is a continuous, monotonic function constructed by given discrete data points (Z, σ_a) for elements at a certain energy.

Taylor's formalism is distinct from the classic Mayneord's power-law formalism in that the EANs of air and iodine-tagged fecal materials have greater difference than those of soft-tissues or fat, as shown in Table 1, making it favorable to detection of colon lumen and air-tagging mixture layer in EC. However, Taylor's EANs are weakly energy dependent within the DE-CT energy range. For that reason, the average EANs: $Z_{eff,ave} = (Z_{eff,high} + Z_{eff,low})/2$ is defined as the EANs of DECT.

Table 1. Comparison of EANs of materials in CTC calculated by the classic Mayneord's power-law formalism and by Taylor's formalism at 80/140 kVp.

Material	Mayneord's EAN	Taylor's EAN
Adipose tissue (ICRU-44)	6.33	3.25
Water, Liquid	7.42	3.66
Tissue, Soft (ICRU-44)	7.46	3.77
Air, Dry (near sea level)	7.63	7.37
Bone, Cortical (ICRU-44)	13.23	8.51
Iodine-water (20 mg/ml)	9.86	6.04
Iodine-water (40 mg/ml)	11.49	8.59

In order to effectively apply Taylor's EANs to DE-CTC images, we devised a lookup table (HU_{high}, HU_{low}, Z_{eff}) corresponding to dual-energies of 140/80 kVp. The resulting EAN surface is a narrow, triangular region, which laid on the upper diagonal of the square forming by the two energy axes. Combining the collection of the 16 NIST reference materials (http://www.nist.gov/), iodine-water solution with different concentrations (such as 20 mg/ml, 40 mg/ml), and mixed with 10%, 20% ... 90% air by volume as "collocation points", EANs were generated using a radial basis function ϕ with y_k as the k-th collocation point: $Z_{eff}(x) = \sum_k \alpha_k \phi(\|x - y_k\|^2)$. The thin-plate biharmonic spline was chosen as the radial basis function such that $\phi(r) = r^2 \log(r)$ for $r \neq 0$ and $\phi(0) = 0$. As shown in Fig. 2, the EAN interpolator enables an EAN look up table for the direct mapping between a pair of (HU_{high}, HU_{low}) and Z_{eff} values.

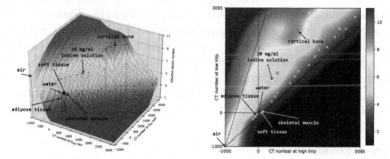

Fig. 2. Generation of Z_{eff} (E_{high}, E_{low}) table using biharmonic spline algorithm by fitting 16 tissues in NIST table and a set of iodine solutions with different concentrations, and mixed with air by volumes in $E_{high} = 89$ keV (140 kVp) and $E_{low} = 51.93$ keV (80 kVp).

3.2 DE-CT Segmentation of Colon Lumen

The segmentation of colon lumen including lumen air and tagged regions was carried out by using the U-Net architecture [24]. Compared with the original U-Net, we replaced the original two plain CNN layers of each resolution with more advanced backbones network, which were supposed to be more effective for image features extracting. Specifically, two types of backbone networks including VGG16 [25] and EfficientNet [26] were investigated in U-Net architecture.

As each DE-CTC volume contains more than 2×600 slices, we adopted 2D or 2.5D instead of 3D U-Net architecture. In the context of DE-CTC, we investigated multiple combinations of input channels (images), and multiple backbone networks in U-Net architecture for the optimal performance of segmentation.

Specifically, five types of input channel combinations were investigated including combinations of the high-energy (High), low-energy (Low), mixed-energy (Mix) CT images, and EAN maps, as shown in Fig. 3. The mixed-energy image is fused by the high-energy and low-energy CT images to simulate the single energy (120kVP) image and serve as a baseline input type (a). As each CTC volume has too many slices, 2.5D is a trade-off between 2D and 3D and maybe better for deep-learning training. Input type (b) is composed with 2.5D strategy as follows: each slice of mixed-energy images was

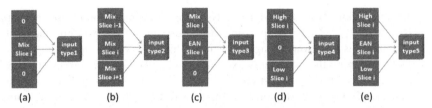

Fig. 3. Illustration of the five types of inputs channel combinations: (a) only the mixed-energy images, (b) concatenate the mixed-energy images with 2.5D strategy, (c) combination of mixed-energy image and the EAN map, (d) combination of dual-energy images, and (e) combination of dual-energy images and the EAN maps.

concatenated with its two nearest neighbor slices. Based on mixed-energy images, we added another channel with EAN map input type (c) to see the effectiveness of EAN. The combination of high-energy and low-energy images could provide more information to classification, which was directly merged as the input type (d). Upon the dual energy, EAN may provide further features to the segmentation, so we constructed input type (e) to merge both dual energy images and EAN map together. In order to utilize the pre-trained models of VGG16 and EfficientNet, all the inputs were maintained three channels and the empty channels were filled with zeros.

3.3 Deep-Cleaning Pipeline

Based on the DE-EC scheme of virtual colon tagging in [6], we have developed the Deep-Cleansing scheme by using the EAN and the U-Net architecture. The major steps in the Deep-Cleansing scheme are: (1) Calculation of Z_{eff} maps, (2) Segmentation of

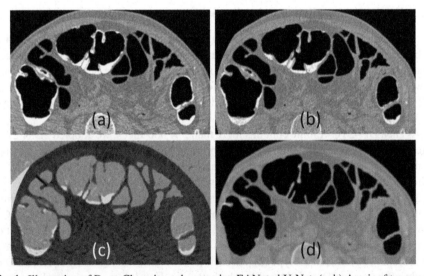

Fig. 4. Illustration of Deep-Cleansing scheme using EAN and U-Net. (a, b) A pair of transverse images at tube voltage of (a) 80 kVp and (b) 140 kVp. (c) The EAN image calculated from (a) and (b). (d) The cleansed image by using "Deep-Cleansed" scheme.

colon lumen and tagged regions using U-Net, and (3) Subtraction of entire colon lumen. The segmented entire colonic lumen including tagged fecal regions were subtracted and replaced with air, followed by a mucosal reconstruction filter to reconstruct the PVE between lumen and the original colonic soft-tissue surface. Figure 4 demonstrates one of the Deep-Cleansing examples.

4   Evaluation and Results

To evaluate the proposed Deep-Cleansing scheme, we retrospectively collected 139 DE-CTC cases with an isotropic resolution of $0.6 \times 0.6 \times 0.6$ mm. Each patient underwent a 24-h low-fiber, low-residue dietary bowel preparation with oral administration of 150 ml of iodinated contrast agent Gastrografin (Bayer Schering Pharma). These patients were imaged by our low-dose DE-CTC imaging protocol on a DE-CT scanner (SOMOTOM Definition Flash) at tube voltages of 80/140Sn kVp with the automatic dose exposure control module (CARE Dose 4D) in both supine and prone positions. The average radiation dose for each position was 0.70 ± 0.11 mSv, approximately 70% less than the recommended radiation dose for CTC screening (2.5–2.8 mSv). The ROIs of the colon lumen and tagged regions were interactively segmented on the mixed CT images on our in-house CTC platform by radiologists. Our new defined ROIs are the entire colon lumen (including air, tagged materials, and air-tagging mixtures), whose contours of colon lumen or surface layer are more reliable than ROIs in conventional methods. Moreover, we utilized intelligent scissors to optimize the contours. Therefore, our method significantly reduced the inter-annotator variations and improved the reliability of the ground truth. All DE-CTC scans (approximately 80K images) were randomly divided into 76 scans (55%) for training, 28 scans (20%) for validation, and 35 scans (25%) for testing.

4.1   Implementation Details and Evaluation Metrics

In the preprocessing step of the segmentation, 12-bits CT images were first mapped to 8-bit images by using standard CTC window-level (WW: 2000 HU, WL:0 HU). The original size of images (512×512) was used without down-sampling because resizing the images may reduce the segmentation accuracy.

The network was supervised by the loss function of dice coefficient. We used Adam optimizer to train the U-Net model. The learning rate was set to $1e-4$ and the batch size was 8. The network parameters of the model were initialized with weights pre-trained on ImageNet. The training process stopped when the validation loss does not improve in 10 epochs or the training epochs exceeded 100. We trained the models on a NVIDIA Tesla P40 GPU cluster with 24 GB memory.

During validation and testing, we compared the segmentation output with ground truth in following metrics: Dice coefficient (Dice), Jaccard Index (Jaccard), and two popular EC metrics: cleansing ratio (Rec) and soft-tissue preservation ratio (Rst) [13], which measures the less-segmentation and over-segmentation (or erosion) of the colon lumen, respectively.

4.2   Quantitative Evaluation and Comparison

We evaluated the effectiveness of EAN in the segmentation of colon lumen and tagged regions of five types of input channel combinations and two types of backbone networks (VGG16-UNet and EfficientNetB4-UNet), respectively. The results are displayed in Table 2 and Table 3, and the best results are highlighted in bold. The results of both validation and testing showed that the use of EAN map may significantly improve the segmentation accuracy, as well as the Rec and Rst. Considering the average size of the entire colon lumen and tagged regions is approximately 2000–3000 CC, 0.1% improvement indicates approximately 2–3 CC volume improvements. We may notice that VGG16-UNet tends to work better than EfficientNetB4-UNet, because the first CNN layer of VGG16 uses the same resolution (512×512) as input while that of EfficientNetB4 uses half of the resolution (256×256) of input. EfficientNet [26] utilizes a compound scaling method that uniformly scales all dimensions of depth/width/resolution. With deeper layers, the resolution of EfficientNet will sacrifice, which is not suitable for the precise segmentation required in our DE-EC scheme.

The segmentation results of different types of inputs are shown in Fig. 5 for visual assessment. The results indicate that large part of tagged regions will be segmented by using the EAN map, and the using of dual energy and EAN map could achieve the closest segmentation compared to the ground truth. Compared with the mosaic decomposition [13] that used SVM classifier for differentiation of tagged fecal residuals, which achieved

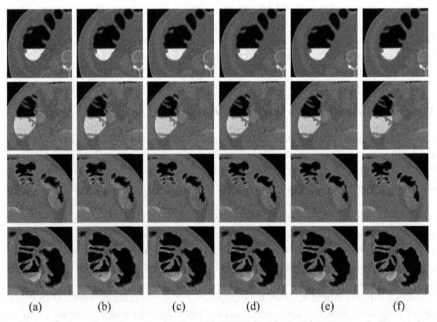

| | | | | | |
| (a) | (b) | (c) | (d) | (e) | (f) |

Fig. 5. Detail view of segmentation results with VGG16 as the backbone network. From column 1 to column 6 are inputs of: (a) only the mixed-energy images, (b) 2.5D mixed-energy images, (c) combination of mixed-energy image and the EAN map, (d) combination of dual-energy images, (e) combination of dual-energy images and EAN maps, and (f) the ground truth.

Rec of 90.6% and Rst of 97.8%, our Deep-Cleansing scheme demonstrated a significant improvement in the accurate segmentation of tagged fecal regions in DE-CTC images. We observed that EC artifacts such as pseudo colonic structures, incomplete cleansing, and erosion of soft tissue structures were significantly reduced after we subtracted the entire colon lumen as they tend to be caused by inhomogeneity or mixtures within the colon lumen.

Table 2. Effectiveness of EAN for segmentation of colon lumen on validation set

Input	VGG16-UNet				EfficientNetB4-UNet			
	Dice	Jaccard	Rec	Rst	Dice	Jaccard	Rec	Rst
M	0.940	0.891	92.50%	98.15%	0.941	0.893	92.36%	98.36%
M2.5D	0.943	0.896	92.97%	98.25%	0.943	0.895	93.24%	98.05%
M,E	0.950	0.900	94.09%	98.42%	0.947	0.902	**93.87%**	98.26%
H,L	0.948	0.904	93.65%	98.48%	0.945	0.898	92.41%	97.70%
H,L,E	**0.953**	**0.910**	**94.18%**	**98.55%**	**0.949**	**0.905**	93.68%	**98.54%**

Table 3. Effectiveness of EAN for segmentation of colon lumen on test set

Input	VGG16-UNet				EfficientNetB4-UNet			
	Dice	Jaccard	Rec	Rst	Dice	Jaccard	Rec	Rst
M	0.945	0.896	95.53%	96.86%	0.944	0.896	94.79%	97.22%
M2.5D	0.947	0.900	95.79%	96.96%	0.947	0.900	**95.94%**	96.88%
M,E	0.955	0.914	95.95%	97.72%	0.949	0.903	94.65%	97.84%
H,L	0.955	0.915	95.74%	**97.86%**	0.953	0.911	94.64%	97.00%
H,L,E	**0.956**	**0.916**	**96.14%**	97.74%	**0.955**	**0.914**	95.20%	**98.08%**

5 Conclusions

In this study, we investigated the EAN and U-Net model for segmentation of colon lumen to subtraction of tagged fecal regions and its air-tagging mixtures in DE-CTC. We calculated the EAN map by a look up table from DE-CTC values, and use EAN map as one of the inputs of U-Net segmentation. To evaluate the effectiveness of EAN map on colon lumen segmentation, multiple types of combinations of input channels (images), and multiple backbone networks were investigated in U-Net architecture. The experiments demonstrated that EAN provides an effective solution to identify tagged fecal materials and U-Net can significantly improve the segmentation of colon lumen for subtraction of tagged fecal regions in DE-CTC images.

Acknowledgement. This research was supported by National Cancer Institute (NCI) under award number R03CA223711.

References

1. Johnson, C.D., et al.: Accuracy of CT colonography for detection of large adenomas and cancers. N. Engl. J. Med. **359**(12), 1207–1217 (2008)
2. Lefere, P., et al.: CT colonography after fecal tagging with a reduced cathartic cleansing and a reduced volume of barium. Am. J. Roentgenol. **184**(6), 1836–1842 (2005)
3. Zalis, M.E., Perumpillichira, J., Hahn, P.F.: Digital subtraction bowel cleansing for CT colonography using morphological and linear filtration methods. IEEE Trans. Med. Imaging **23**(11), 1335–1343 (2004)
4. Zalis, M.E., et al.: Diagnostic accuracy of laxative-free computed tomographic colonography for detection of adenomatous polyps in asymptomatic adults: a prospective evaluation. Ann. Intern. Med. **156**(10), 692–702 (2012)
5. Eliahou, R., et al.: Dual-energy based spectral electronic cleansing in non-cathartic computed tomography colonography: an emerging novel technique. Semin. Ultrasound CT MRI **31**, 309–314 (2010)
6. Cai, W., et al.: Electronic cleansing in fecal-tagging dual-energy CT colonography based on material decomposition and virtual colon tagging. IEEE Trans. Biomed. Eng. **62**(2), 754–765 (2015)
7. Tachibana, R., et al.: Deep learning electronic cleansing for single- and dual-energy CT colonography. Radiographics **38**(7), 2034–2050 (2018)
8. Wax, M., et al.: Electronic colon cleansing for virtual colonscopy. In: The 1st Symposium on Virtual Colonscopy (1998)
9. Wang, S., et al.: An EM approach to MAP solution of segmenting tissue mixture percentages with application to CT-based virtual colonoscopy. Med. Phys. **35**(12), 5787–5798 (2008)
10. Lakare, S., et al.: Electronic colon cleansing using segmentation rays for virtual colonoscopy. SPIE Med. Imaging **4683**, 412–418 (2002)
11. Serlie, I., et al.: Electronic cleansing for computed tomography (CT) colonography using a scale-invariant three-material model. IEEE Trans. Biomed. Eng. **57**, 1306–1317 (2010)
12. Cai, W., et al.: Structure-analysis method for electronic cleansing in cathartic and non-cathartic CT colonography. Med. Phys. **35**(7), 3259–3277 (2008)
13. Cai, W., et al.: Mosaic decomposition: an electronic cleansing method for inhomogeneously tagged regions in noncathartic CT colonography. IEEE Trans. Med. Imaging **30**(3), 559–574 (2011)
14. Carston, M.J., Manduca, A., Johnson, D.: Electronic stool subtraction using quadratic regression, morphological operations, and distance transform. In: SPIE Medical Imaging 2007: Physiology, Function, and Structure from Medical Images (2007)
15. Franaszek, M., et al.: Hybrid segmentation of colon filled with air and opacified fluid for CT colonography. IEEE Trans. Med. Imaging **25**(3), 358–368 (2006)
16. George Linguraru, M., et al.: Automated image-based colon cleansing for laxative-free CT colonography computer-aided polyp detection. Med. Phys. **38**(12), 6633 (2011)
17. Cai, W., et al.: Informatics in radiology: electronic cleansing for noncathartic CT colonography: a structure-analysis scheme. Radiographics **30**(3), 585–602 (2010)
18. Tourassi, G.D., et al.: Performance evaluation of multi-material electronic cleansing for ultra-low-dose dual-energy CT colonography. In: SPIE Medical Imaging 2016: Computer-Aided Diagnosis (2016)

19. Tachibana, R., et al.: Electronic cleansing in CT colonography using a generative adversarial network. In: SPIE Medical Imaging 2019: Imaging Informatics for Healthcare, Research, and Applications (2019)

20. Van, A., Joanne, K., et al.: Feasibility and accuracy of tissue characterization with dual source computed tomography. Phys. Med. **28**(1), 25–32 (2012)

21. Kurudirek, M., Aksakal, O., Akkus, T.: Investigation of the effective atomic numbers of dosimetric materials for electrons, protons and alpha particles using a direct method in the energy region 10 keV-1 GeV: a comparative study. Radiat. Environ. Biophys. **54**(4), 481–492 (2015)

22. Khan, F.M.: The Physics of Radiation Therapy, 4th edn. Lippincott Williams & Wilkins, Philadelphia (2012)

23. Taylor, M.L., et al.: Robust calculation of effective atomic numbers: the auto-zeff software. Med Phys **39**(4), 1769–1778 (2012)

24. Ronneberger, O., Fischer, P., Brox, T.: U-net: convolutional networks for biomedical image segmentation. In: Navab, N., Hornegger, J., Wells, W.M., Frangi, A.F. (eds.) Medical Image Computing and Computer-Assisted Intervention – MICCAI 2015: 18th International Conference, Munich, Germany, October 5-9, 2015, Proceedings, Part III, pp. 234–241. Springer International Publishing, Cham (2015). https://doi.org/10.1007/978-3-319-24574-4_28

25. Simonyan, K., Zisserman, A.: Very deep convolutional networks for large-scale image recognition. arXiv:1409.1556v6 (2015)

26. Tan, M., Le, Q.: EfficientNet: rethinking model scaling for convolutional neural networks. International Conference on Machine Learning, pp. 6105–6114 (2019)

Clinical Applications - Breast

Clinical Applications – Breast

Interactive Smoothing Parameter Optimization in DBT Reconstruction Using Deep Learning

Pranjal Sahu[1]([envelope]), Hailiang Huang[2], Wei Zhao[2], and Hong Qin[1]

[1] Computer Science Department, Stony Brook University, New York, USA
{psahu,qin}@cs.stonybrook.edu
[2] Department of Radiology, Stony Brook School of Medicine, New York, USA
hailiang.huang@stonybrook.edu, wei.zhao@stonybrookmedicine.edu

Abstract. Medical image reconstruction algorithms such as Penalized Weighted Least Squares (PWLS) typically rely on a good choice of tuning parameters such as the number of iterations, the strength of regularizar, etc. However, obtaining a good estimate of such parameters is often done using trial and error methods. This process is very time consuming and laborious especially for high resolution images. To solve this problem we propose an interactive framework. We focus on the regularization parameter and train a CNN to imitate its impact on image for varying values. The trained CNN can be used by a human practitioner to tune the regularization strength on-the-fly as per the requirements. Taking the example of Digital Breast Tomosynthesis reconstruction, we demonstrate the feasibility of our approach and also discuss the future applications of this interactive reconstruction approach. We also test the proposed methodology on public Walnut and Lodopab CT reconstruction datasets to show it can be generalized to CT reconstruction as well.

Keywords: CNN · Breast Tomosynthesis reconstruction · Deep Learning

1 Introduction

Medical image reconstruction is typically formulated as an iterative optimization problem. Examples of such reconstruction algorithms include penalized weighted least squares (PWLS) etc. These algorithms employ a regularization term in the objective function to incorporate prior information in the algorithm. Example of such prior functions are Total Variation, Huber function, Quadratic function etc. A trade-off between the projection data consistency and regularization term is then decided to obtain the best looking image for a particular task with the help of smoothing parameter β (see Eq. 1). However, tuning this parameter is often a manual process which is done by trial and error. This is a time consuming and tedious process specially when the image to be reconstructed has large

M. de Bruijne et al. (Eds.): MICCAI 2021, LNCS 12907, pp. 57–67, 2021.
https://doi.org/10.1007/978-3-030-87234-2_6

dimension for example in Digital Breast Tomosynthesis (DBT) where the volume to be reconstructed has number of voxels of the order of millions. In DBT, limited angle acquisition is performed at a low dose to take 15–25 projections of the breast volume. Masses and calcifications are the two important features which radiologists look for in a DBT image. Masses are relatively large and low contrast lesions whose visibility can be enhanced via removal of overlapping tissue structure in tomosynthesis. Calcifications are small 150–300 micron lesions and their visibility is hampered by the quantum noise [1]. To enhance microcalcification visibility and to reduce the quantum noise, a smoothing prior is used in iterative reconstruction algorithms. However, a careful tuning is needed to obtain a good quality image. Overly smoothing will blur out the calcifications while under smoothing will result in quantum noise over-powering the calcification visibility. Similarly, overly smoothing will smooth out the spiculations and characteristic features in masses. To solve this problem, we propose to use a CNN model to imitate the reconstruction algorithm's output. The CNN model acts as a non-linear function that outputs the result of reconstruction algorithm for varying values of β on the fly.

When a human practitioner looks at a particular suspicious section of the entire reconstructed volume they might want to see the result with a different β parameter to obtain a right balance of noise vs resolution. In such situation, the trained CNN model can be used to obtain the result instantly which obviates the need for performing time-consuming reconstruction from scratch. Taking the example of DBT reconstruction we demonstrate the feasibility of our approach for the penalized weighted least squares reconstruction algorithm with Huber and Quadratic prior objectives. Later, we also test our method on CT reconstruction datasets such as Walnut CT [2] and Lodopab CT [3].

2   Related Work

Tuning of reconstruction parameters is not entirely a new topic and there have been previous works in this domain. For example, in [4], authors proposed a reinforcement learning based approach for intelligent tuning of parameters in iterative CT reconstruction algorithms. Their approach relied on the availability of higher dose (ground truth) data for training a CNN model which could automatically decide the change in the tuning parameter while performing reconstruction. However, their approach cannot be utilized in cases where the higher dose data is not available. Previously, methods have been proposed to optimize the reconstruction parameters using Model observers [5–10]. In [11], authors studied the reconstruction algorithms' performance for low contrast lesion detectability in CT using the Channelized Hotelling Observer (CHO) [12]. Recently supervised machine learning approaches have also been adopted to train the Model observers, for example, SVM [13] or CNN [14–16]. In [17], authors proposed a novel approach where they incorporate an additional objective in the reconstruction cost function to enhance calcification detectability. They formulated the new objective using the lesion's detectability index obtained using the CHO.

Image quality has been shown to be subjective and varies with each radiologists [18], therefore, a reconstruction method that can adapt with the radiologists preferences is preferable [19]. Till now methods have obtained the optimized smoothing parameter β for the entire dataset [5–10]. In this work, we propose a novel approach where the parameter β can be optimized for each case by giving the tuning capability into practitioner hands' made possible by harnessing the function approximation capability of neural networks.

3   Methodology

The objective function $\Phi(\boldsymbol{\mu})$ which is minimized for a typical PWLS reconstruction takes the following form:

$$\Phi(\boldsymbol{\mu}) = L(\boldsymbol{\mu}) - \beta R(\boldsymbol{\mu}), \quad \text{where,} \quad L(\boldsymbol{\mu}) = (\mathbf{g} - A\boldsymbol{\mu})^T \mathbf{W}(\mathbf{g} - A\boldsymbol{\mu}), \quad (1)$$

where, $\boldsymbol{\mu}$ is a vector of voxel attenuation values to be determined, $L(\boldsymbol{\mu})$ is the projection data consistency term, $R(\boldsymbol{\mu})$ is the penalty function that penalizes roughness, β is the penalty (regularization) strength, $\mathbf{g}$ is projection data and weighing matrix $\mathbf{W}$ in special case of penalized SART reconstruction is a diagonal matrix whose diagonal element is the reciprocal of the m-th row sum (a_{m+}) of the forward projection matrix A [20].

The penalty $R(\boldsymbol{\mu})$ is a function of neighbouring voxels as shown below, where, $\psi(t)$ is the potential function and

$$R(\boldsymbol{\mu}) = \frac{1}{2} \sum_{j=1}^{p} \sum_{k \in N_j} \omega_{jk} \psi \left(\mu_j - \mu_k \right) \tag{2}$$

$k \in N_j$ are the indices of neighbouring voxels for a voxel index j, p is the total number of voxels and w_{jk} are the weights (1/distance). Two popular $\psi(t)$ functions are Huber and Quadratic prior which are defined below:

$$\text{Huber, } \Psi(t) \triangleq \begin{cases} t^2/2, & |t| \leq \delta \\ \delta|t| - \delta^2/2, & |t| > \delta \end{cases}, \text{ and, Quadratic, } \Psi(t) = t^2/2, \quad (3)$$

where δ is a parameter that determines the signal preserving threshold. In case of huber prior, the penalty for voxel difference lesser than δ is quadratic while it is linear for values greater than δ, thereby, preserving signal such as calcifications. The update rule for the reconstruction then becomes,

$$\boldsymbol{\mu}_{i+1} = \boldsymbol{\mu}_i + \lambda \left(\frac{A^T \left(\frac{\mathbf{g} - A\mu_i}{AI} \right)}{A^T I} - \beta R'(\mu_i) \right), \tag{4}$$

where, λ is a fixed constant, $R'(\mu_i)$ is the derivative of $R(\mu_i)$, I is the identity matrix and i is the iteration number.

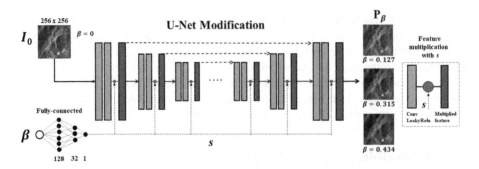

Fig. 1. Modified U-Net to perform filtering conditioned on the input regularization strength β. A single model is trained to predict the output image P_β based on the input image I_0 for a wide range of β values.

3.1 Modified U-Net for Imitating Reconstruction

Varying β can significantly alter the image appearance and lesion detectability, however, performing reconstruction on a new β from scratch is time consuming. Therefore, we approximate this behaviour using a CNN model that is able to generate the images with good accuracy for a range of β values. One way could be to train separate network for a discreet set of β values for example [0.1, 0.2, 0.5] etc. However, this approach is not practical. Therefore, we introduce a conditioning variable β as input to the CNN that instructs the model to adapt image for a given β. By this approach, we only need to train a single CNN that can work for a range of β values, see Fig. 1. U-Net based architectures have been very successful in image-to-image translation tasks [21] and hence we use it as the backbone in our work. The CNN takes as input two things, one the image patch I_0 from the reconstructed volume to be processed and second the regularization strength β which is to be applied. The scalar value β passes through a fully-connected layer to output a scalar value s. The feature maps obtained after each layer in the encoder and decoder are then multiplied element wise with the scalar variable s. In our experiments, we observed that the fully-connected layer is necessary and directly multiplying β value with feature maps does not perform well. It should be noted that there can be other ways to introduce this conditioning variable in the CNN but for our purpose this modification was sufficient. The Modified U-Net can be interpreted as a non-linear function $f(I_0, \beta)$ which predicts the image patch P_β corresponding to I_0 when reconstructed by an algorithm A with the regularization strength β. The input patch to CNN is always the image patch obtained from volume reconstructed with zero regularization ($\beta = 0$) strength i.e. I_0. In addition to the introduction of a conditioning variable, the activation function is changed everywhere to LeakyRelu with the parameter $\alpha = 0.2$. Also the number of filters in each layer is reduced to half of that of original U-Net i.e. 16, 32, 64, 128, 256. No softmax or sigmoid activation layer is used at the end. The output of U-Net is subtracted from the input to obtain the final output. Rest everything remains the same as original U-Net architecture. The implementation of the model in Pytorch is shared at https://github.com/PranjalSahu/InteractiveSmoothingDBT. The network is trained in a supervised manner by creating a dataset of reconstructed volumes for varying values of β as described next.

Table 1. Train, validation and test set distribution for all the experiments.

Dataset	Prior	β range	Train	Validation	Test
DBT	Huber	(0.1, 0.5]	6900 (26)	2750 (10)	3250 (12)
DBT	Quadratic	(0, 0.2]	10100 (26)	4000 (10)	4750 (12)
LodoPab	Huber	(0, 0.02]	4000	1000	1000
LodoPab	Quadratic	(0, 0.01]	4000	1000	1000
Walnut	Huber	(0, 0.05]	5560 (28)	800 (4)	1600 (8)
Walnut	Quadratic	(0, 0.025]	5440 (28)	800 (4)	1600 (8)

3.2 Datasets Preparation and Details

We train and test our method on three datasets 1.) Private Digital Breast Tomosynthesis dataset 2.) Public Walnut CT dataset 3.) Public Lodopab CT dataset. Here, we describe the data generation details for training the Modified U-Net for each dataset.

Digital Breast Tomosynthesis Dataset: The DBT volumes are reconstructed at full resolution with $0.085 \times 0.085 \times 1\,\mathrm{mm}$ voxel size following the acquisition geometry of Siemens Inspiration DBT system. The dataset comprises 48 cases acquired using the same DBT system which is randomly divided into train, validation and test splits comprising 26, 10 and 12 cases respectively. Corresponding to each case, 6 different values of β is uniformly sampled in $[0.1, 0.5]$ for Huber prior and 8 different values of β is uniformly sampled in $[0.0, 0.2]$ for Quadratic prior. We perform reconstruction using the Ordered Subset version which speeds up the convergence [22] and also limits the GPU memory requirement which is very important since the projections are high resolution in case of DBT. The number of subsets is equal to 5. Siemens Inspiration System takes 25 projections therefore the subset size becomes $25/5 = 5$. Subsets are formed by taking every alternate 5th projection. Reconstruction is performed for a total of 5 iterations and the value of $\lambda = 0.9$, see Eq. 4. Each volume takes around 1 min to reconstruct with GPU acceleration where forward and backward projection operators follow Siddon's method. Image patches of size 256×256 are then randomly extracted having at-least 90% breast region occupancy from the reconstructed volumes in train, validation and test splits to obtain the data pairs $\{(I_0^k, \beta), G_\beta^k\}$, where $k \in N$ is the patch index and G_β^k is the ground truth patch taken from the volume reconstructed with regularization strength β. The distribution of volumes and image samples extracted for each split is summarized in Table 1 as Image count (Volume count).

Walnut CT Dataset: In [2] authors introduced the Walnut CT dataset for research on machine learning based reconstruction algorithms. It comprises projections for 42 different walnuts scanned in a laboratory X-ray. The dataset is randomly split into train, validation and test splits. For each walnut, CB (conebeam) projections on three different source orbits is provided along with projection geometry. In our experiments, we only use the central orbit projections. Reconstruction is performed using the GPU accelerated Astra reconstruction toolkit [23]. Here, we take all the projections together in any update step since we could fit all of them in the GPU. Volume of size $251 \times 251 \times 251$ is reconstructed with a voxel size of $0.2 \times 0.2 \times 0.2$ mm. Total of 50 iterations is used and $\lambda = 0.9$ for both Huber and Quadratic prior. Value of $\delta = 0.0005$ when using Huber prior. Corresponding to each volume, 5 different values of β is uniformly sampled in $[0, 0.05]$ and $[0, 0.025]$ for Huber and Quadratic prior respectively. From each volume, slices are randomly selected. For this dataset, unlike DBT, the reconstructed volume is relatively small ($251 \times 251 \times 251$), therefore we take the entire slice and perform zero-padding to obtain 256×256 size image. The count of image slices (and volume) for each split is shown in Table 1.

Lodopab CT Dataset: To showcase that our method also works on real CT data, we use the publicly available Lodopab lung CT dataset [3]. It comprises 40000 scan slices from around 800 patients selected from the LIDC/IDRI database [24]. A total of 6000 scans are randomly selected from the entire dataset and divided into train, validation and test splits as shown in Table 1. Poisson noise corresponding to 4096 incident photons is applied to the projections before reconstruction. For this dataset, 362×362 image is reconstructed with a 2D parallel beam geometry using 1000 angles and 513 detector pixels. Value of $\delta = 0.0002$ when using Huber prior and $\lambda = 0.9$ for both priors. Reconstruction runs for 100 iterations for Quadratic prior and 200 iterations for the Huber prior. The value of β is uniformly sampled from $(0, 0.02]$ and $(0, 0.01]$ for Huber and Quadratic priors respectively. Images are zero padded to obtain 384×384 size before passing through the Modified U-Net.

The network architecture is same for the three datasets as shown in Fig. 1 and experiments are conducted separately for each dataset. Modified U-Net is trained with Mean Absolute Error loss calculated between the network's prediction P_β^k and the ground truth G_β^k. Adam Optimizer with a learning rate of 0.0001 is used to optimize the network's parameters using back-propagation algorithm. Network is implemented in PyTorch and training is performed on a system with a 2080 Ti GPU with a batch size of 8 for a maximum of 1000 epochs. Network weights with the best mean validation set performance in terms of SSIM metric is selected to perform inference on the test split. Next we discuss the performance of Modified U-Net's on test images.

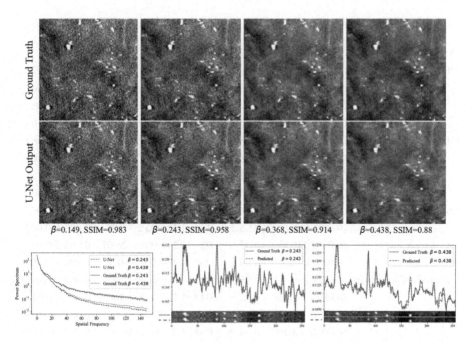

Fig. 2. Top: Qualitative comparison of U-Net output P_β and ground truth G_β for DBT test sample, Huber prior (window = 0.02, level = 0.1). Bottom: NPS comparison of P_β and G_β. Line profile comparison over calcifications for $\beta = 0.438, 0.243$.

4   Results and Discussion

The feasibility of the proposed interactive smoothing parameter determination approach hinges upon accurate imitation by CNN model of the reconstruction algorithm. Due to this we test the performance of Modified U-Net both quantitatively and qualitatively. The output P_β from Modified U-Net obtained using (I_0, β) is compared with the ground truth G_β obtained from PWLS reconstruction algorithm. All the results shown here are from the test split. We compare the similarity between P_β and G_β using SSIM (Structural Similarity Index), MAE (Mean Absolute Error) and HaarPSI (C = 100) [25]. The results are shown in Table 2 for different β value range. It can be observed that the network is able to imitate the output with very high accuracy for a wide range of β, however, we also observe that performance degrades as we increase the β value i.e. as the degree of smoothness is increased. We also observe that the network has higher accuracy for CT datasets in comparison to DBT which is possibly due to access to the entire slice in case of CT while only having access to a local image patch in case of DBT. Another thing to note is that the performance of CNN is higher for Huber prior in comparison to Quadratic prior across all datasets. Qualitative comparison of a test set DBT image patch is shown in Fig. 2 for Huber prior. It can be seen in Fig. 2, that the predicted image (P_β) is perceptually indistinguishable from the ground truth image (G_β). We also compare the

Table 2. Quantitative results on DBT, LodoPab and Walnut CT (mean±*std*)

Dataset	Prior	β range	SSIM	MAE	HaarPSI
DBT	Huber	(0.1, 0.2]	0.99 ± 0.006	1.8e-04 ± 4.6e-05	0.98 ± 0.02
		(0.2, 0.3]	0.98 ± 0.013	4.0e-04 ± 6e-05	0.92 ± 0.04
		(0.3, 0.4]	0.96 ± 0.026	4.4e−04 ± 9e-05	0.85 ± 0.08
		(0.4, 0.5]	0.89 ± 0.070	4.6e-04 ± 12e-05	0.81 ± 0.10
DBT	Quadratic	(0.0, 0.05]	0.96 ± 0.027	4.1e-04 ± 2.2e-04	0.84 ± 0.082
		(0.05, 0.10]	0.88 ± 0.056	4.6e-04 ± 2.5e-04	0.74 ± 0.108
		(0.10, 0.15]	0.82 ± 0.092	5.0e-04 ± 3.3e-04	0.71 ± 0.124
		(0.15, 0.20]	0.75 ± 0.135	5.5e-04 ± 3.9e-04	0.7 ± 0.130
LodoPab	Huber	(0, 0.005]	0.96 ± 0.037	1.3e-05 ± 1.6e-06	0.994 ± 0.003
		(0.005, 0.01]	0.98 ± 0.014	1.11e-05 ± 4e-07	0.995 ± 0.002
		(0.010, 0.015]	0.97 ± 0.02	1.11e-05 ± 3e-07	0.993 ± 0.004
		(0.015, 0.020]	0.96 ± 0.026	1.1e-05 ± 4e-07	0.990 ± 0.007
LodoPab	Quadratic	(0, 0.0025]	0.98 ± 0.02	1.8e-05 ± 1e-06	0.993 ± 0.011
		(0.0025, 0.005]	0.96 ± 0.043	1.7e-05 ± 1.1e-06	0.99 ± 0.014
		(0.0050, 0.0075]	0.95 ± 0.048	1.7e-05 ± 9e-07	0.983 ± 0.021
		(0.0075, 0.0100]	0.94 ± 0.052	1.7e-05 ± 7e-07	0.981 ± 0.021
Walnut	Huber	(0, 0.01]	0.99 ± 0.002	1e-04 ± 3e-05	0.996 ± 0.003
		(0.01, 0.02]	0.99 ± 0.008	2e-04 ± 7e-04	0.973 ± 0.011
		(0.02, 0.03]	0.97 ± 0.016	3.1e-04 ± 9e-04	0.939 ± 0.013
		(0.03, 0.04]	0.95 ± 0.044	3.4e-04 ± 1.1e-04	0.911 ± 0.014
		(0.04, 0.05]	0.95 ± 0.026	3.7e-04 ± 1.2e-04	0.893 ± 0.015
Walnut	Quadratic	(0, 0.005]	0.98 ± 0.003	3.3e-04 ± 9.6e-05	0.908 ± 0.020
		(0.005, 0.01]	0.96 ± 0.001	3.7e-04 ± 1.2e-04	0.883 ± 0.017
		(0.01, 0.015]	0.96 ± 0.016	3.8e-04 ± 1.2e-04	0.877 ± 0.018
		(0.015, 0.020]	0.96 ± 0.017	3.7e-04 ± 1.2e-04	0.879 ± 0.023
		(0.020, 0.025]	0.96 ± 0.012	3.6e-04 ± 1.2e-04	0.878 ± 0.025

Noise Power Spectrum (NPS) of the test image in bottom row of Fig. 2. It can be seen that the NPS is very well aligned for P_β and G_β. The difference becomes more apparent for higher frequency components. We also notice that the difference between P_β and G_β grows for high frequency details as we increase the β which is consistent with the quantitative results shown in Table 2. The image generation by CNN should be fast enough to provide an interactive capability to the practitioner. The filtering of entire DBT slice of size 3000×1200 using the Modified U-Net for a particular β takes around 0.22 ± 0.01 (mean±std) seconds (averaged over 5 different runs) on a 2080Ti GPU. If the entire DBT volume of size $3000 \times 1200 \times 40$ is filtered then it takes on an average 10.10 ± 0.10 s which is significantly lesser than performing reconstruction from scratch which on an average takes 63 ± 3.4 s on the same GPU. Reconstruction only needs to be performed for $\beta = 0$ and the rest other can be obtained satisfactorily for a wide range of β by the Modified U-Net. This can significantly improve the workflow of practitioners. In addition to this, the proposed method opens up new possibilities. The Modified U-Net along with a CNN model observer could be used to tune the reconstruction parameters automatically. Recently few works have introduced CNN model observers that can determine the detectability index for a lesion in an image when observed by a human observer with very high accuracy [16]. Such a CNN model observer and the Modified U-Net can work in tandem to obtain the best looking image for a human observer. Due to GPU memory constraints, currently we have only considered image patches in DBT, however, this can be extended for entire slice in future which might improve the performance. For further validation, we obtain the mean CNR (co-relates with detectability) of 10 known calcification (0.196 mm) in the CIRS phantom for few β values (β: 0.101, 0.243, 0.338, 0.406, U-Net: 6.43, 6.71, 6.13, 5.88, PWLS (Huber): 7.22, 7.86, 6.54, 5.67). U-Net based model produces a smoother image compared to PWLS (Fig. 2 NPS) and results in lower CNR, however, the trend with varying β is the same for both, which is useful for determining the peak beta without performing reconstruction again. Our work provides a proof-of-concept and further improvements could be seen with sophisticated architectures and loss functions. In future, we would also conduct a user study to quantify the advantages of proposed approach in clinical workflow.

Acknowledgments. This material is based upon work supported by the National Science Foundation under Grant No. IIS-1715985 and IIS-1812606.

References

1. Sahu, P., Huang, H., Zhao, W., Qin, H.: Using virtual digital breast tomosynthesis for de-noising of low-dose projection images. In: 2019 IEEE 16th International Symposium on Biomedical Imaging (ISBI 2019), pp. 1647–1651. IEEE (2019)
2. Der Sarkissian, H., Lucka, F., van Eijnatten, M., Colacicco, G., Coban, S.B., Batenburg, K.J.: A cone-beam x-ray computed tomography data collection designed for machine learning. Sci. Data **6**(1), 1–8 (2019)
3. Leuschner, J., Schmidt, M., Baguer, D.O., Maass, P.: LoDoPaB-CT, a benchmark dataset for low-dose computed tomography reconstruction. Sci. Data **8**(1), 1–12 (2021)
4. Shen, C., Gonzalez, Y., Chen, L., Jiang, S.B., Jia, X.: Intelligent parameter tuning in optimization-based iterative CT reconstruction via deep reinforcement learning. IEEE Trans. Med. Imaging **37**(6), 1430–1439 (2018)
5. Barrett, H.H., Yao, J., Rolland, J.P., Myers, K.J.: Model observers for assessment of image quality. Proc. Nat. Acad. Sci. **90**(21), 9758–9765 (1993)
6. Rose, S.D., Roth, J., Zimmerman, C., Reiser, I., Sidky, E.Y., Pan, X.: Parameter selection with the hotelling observer in linear iterative image reconstruction for breast tomosynthesis. In: Medical Imaging 2018: Image Perception, Observer Performance, and Technology Assessment, voL. 10577, p. 105770P. International Society for Optics and Photonics (2018)

7. Sidky, E.Y., Duchin, Y., Reiser, I., Ullberg, C., Pan, X.: Optimizing algorithm parameters based on a model observer detection task for image reconstruction in digital breast tomosynthesis. In: 2011 IEEE Nuclear Science Symposium Conference Record, pp. 4230–4232. IEEE (2011)
8. Michielsen, K., Nuyts, J., Cockmartin, L., Marshall, N., Bosmans, H.: Design of a model observer to evaluate calcification detectability in breast tomosynthesis and application to smoothing prior optimization. Med. Phys. **43**(12), 6577–6587 (2016)
9. Zeng, R., Park, S., Bakic, P., Myers, K.J.: Evaluating the sensitivity of the optimization of acquisition geometry to the choice of reconstruction algorithm in digital breast tomosynthesis through a simulation study. Phys. Med. Biol. **60**(3), 1259 (2015)
10. Makeev, A., Glick, S.J.: Investigation of statistical iterative reconstruction for dedicated breast CT. Med. Phys. **40**(8), 081904 (2013)
11. Racine, D., Ba, A.H., Ott, J.G., Bochud, F.O., Verdun, F.R.: Objective assessment of low contrast detectability in computed tomography with channelized hotelling observer. Physica Medica **32**(1), 76–83 (2016)
12. Platiša, L.: Channelized hotelling observers for the assessment of volumetric imaging data sets. JOSA A **28**(6), 1145–1163 (2011)
13. Brankov, J.G., Yang, Y., Wei, L., El Naqa, I., Wernick, M.N.: Learning a channelized observer for image quality assessment. IEEE Trans. Med. Imaging **28**(7), 991–999 (2009)
14. Kopp, F.K., Catalano, M., Pfeiffer, D., Fingerle, A.A., Rummeny, E.J., Noël, P.B.: CNN as model observer in a liver lesion detection task for x-ray computed tomography: a phantom study. Med. Phys. **45**(10), 4439–4447 (2018)
15. Massanes, F., Brankov, J.G.: Evaluation of CNN as anthropomorphic model observer. In Medical Imaging 2017: Image Perception, Observer Performance, and Technology Assessment, vol. 10136, p. 101360Q. International Society for Optics and Photonics (2017)
16. Kim, B., Han, M., Baek, J.: A convolutional neural network-based anthropomorphic model observer for signal detection in breast CT images without human-labeled data. IEEE Access **8**, 162122–162131 (2020)
17. Sghaier, M., Chouzenoux, E., Palma, G., Pesquet, J.-C., Muller, S.: A new approach for microcalcification enhancement in digital breast tomosynthesis reconstruction. In: 2019 IEEE 16th International Symposium on Biomedical Imaging (ISBI 2019), pp. 1450–1454. IEEE (2019)
18. Brankov, J.G., Pretorius, P.H.: Personalized numerical observer. In: Medical Imaging 2010: Image Perception, Observer Performance, and Technology Assessment, vol. 7627, p. 76270T. International Society for Optics and Photonics (2010)
19. Cheng, Y., et al.: Validation of algorithmic CT image quality metrics with preferences of radiologists. Med. Phys. **46**(11), 4837–4846 (2019)
20. Lee, H.C., et al.: Variable step size methods for solving simultaneous algebraic reconstruction technique (SART)-type CBCT reconstructions. Oncotarget **8**(20), 33827 (2017)
21. Isola, P., Zhu, J.-Y., Zhou, T., Efros, A.A.: Image-to-image translation with conditional adversarial networks. In: Proceedings of the IEEE Conference on Computer Vision and Pattern Recognition, pp. 1125–1134 (2017)
22. Wang, G., Jiang, M.: Ordered-subset simultaneous algebraic reconstruction techniques (OS-SART). J. X-ray Sci. Technol. **12**(3), 169–177 (2004)
23. Van Aarle, W.: Fast and flexible x-ray tomography using the Astra toolbox. Opt. Express **24**(22), 25129–25147 (2016)

24. Armato, S.G., III., et al.: The lung image database consortium (LIDC) and image database resource initiative (IDRI): a completed reference database of lung nodules on CT scans. Med. Phys. **38**(2), 915–931 (2011)
25. Reisenhofer, R., Bosse, S., Kutyniok, G., Wiegand, T.: A HAAR wavelet-based perceptual similarity index for image quality assessment. Signal Process. Image Commun. **61**, 33–43 (2018)

Synthesis of Contrast-Enhanced Spectral Mammograms from Low-Energy Mammograms Using cGAN-Based Synthesis Network

Yanyun Jiang[1], Yuanjie Zheng[1,2(✉)], Weikuan Jia[1], Sutao Song[1], and Yanhui Ding[1(✉)]

[1] School of Information Science and Engineering, Shandong Normal University, Jinan, China
yjzheng@sdnu.edu.cn

[2] Shandong Provincial Key Lab for Distributed Computer Software Novel Technology, Jinan, China

Abstract. Contrast-enhanced spectral mammography (CESM) is a valuable tool in the diagnosis and staging of primary breast cancer, for which it has an extremely high predictive value. However, the iodinated contrast media injected during CESM examination can cause adverse reactions, such as allergic reactions, and even cause contra-induced nephropathy. Therefore, iodinated contrast media cannot be used for some patients. To address this problem, we developed a cGAN-based Synthesis Network (cGSNT) that uses mammogram images to synthesize corresponding CESM images. Key points of this study are that cGSNT utilizes the cycle-consistent approach to reduce information loss when converting from high to low tissue contrast images, and the introduction of concatenation layers for dual-view information fusion. The experimental results on paired images of low-energy CESM (mammogram) and recombined CESM demonstrated that the synthetic image was very similar to the real CESM image, and the proposed cGSNT qualitatively and quantitatively outperforms typical non-learning and other popular deep learning methods.

Keywords: Contrast-enhanced spectral mammography (CESM) · Mammography · Image synthesis

1 Introduction

Contrast-enhanced spectral mammography (CESM) is a valuable tool in the diagnosis and staging of primary breast cancer, which provides contrast-enhanced images of abnormal vascular development around lesions as an alternative to MRI

Electronic supplementary material The online version of this chapter (https://doi.org/10.1007/978-3-030-87234-2_7) contains supplementary material, which is available to authorized users.

M. de Bruijne et al. (Eds.): MICCAI 2021, LNCS 12907, pp. 68–77, 2021.
https://doi.org/10.1007/978-3-030-87234-2_7

in determining breast cancer areas [5,6,15]. Compared with full field digital mammography images [2,4], using low-energy X-ray (usually around 30 kVp) exposures, the CESM acquisition equipment uses dual-energy breast exposure (about 26–33 kVp and 44–50 kVp) to generate two types of images: low-energy and high-energy, for which patients are pre-injected with nonionic low-osmolar iodinated contrast material. Then, the low-energy image and the high-energy image are operated to generate a recombined image which suppresses the background of breast tissue and highlights the areas of iodine uptake [18].

Figure 1 shows a set of CESM images, including low-energy and recombined images, each of which has two perspectives: craniocaudal (CC) view and mediolateral oblique (MLO) view. The appearance of the low-energy image is similar to the standard digital mammography image that represents tissue density [11]. The low- and high-energy images are combined through a specific dual-energy recombination algorithm to eliminate texture and generate a recombined CESM image. The recombined CESM images highlight areas of unusual blood flow, showing a noticeable intensity difference between cancerous areas and the background of breast tissue. However, in addition to the cost of contrast mediums, the intravenous (IV) administration of contrast material may also cause the occasional anaphylactic reaction. Of which, contrast-induced nephropathy (CIN) is a severe complication caused by the application of iodinated contrast material (CM) [3,18]. For patients with a history of contrast medium allergy and a high risk of CIN, it is not possible to use contrast medium injection angiography [1]. Hence, mapping from a single-energy image (low-energy CESM image, or mammogram) to a recombined CESM image is very meaningful for breast cancer screening and determining the range of abnormalities.

Image synthesis is a common problem in medical image analysis, and many methods have been proposed to solve this issue for translating the source image to the target image. Recently, conditional generative adversarial networks (cGANs) have become very popular for image translation [10,20]. Isola et al. [10] proposed using conditional adversarial networks for synthesizing photos from label maps. GAN-based methods have also shown great success in medical image synthesis [16,17,19]. Theoretically speaking, some of the existing image-to-image translation techniques can be used for the translation from the low-energy image (mammogram image) to the recombined CESM image. Recently, Gao et al. proposed SD-CNN [8], and residual inception encoder-decoder neural network (RIED-Net) [7], taught to predict virtual recombined images from low-energy images, which can then be used to classify the cases as benign vs. cancer.

In this work, our goal is to estimate the recombined CESM image from the low-energy CESM image (mammogram image) that does not require the injection of iodinated contrast media. Specifically, we proposed imposing dual-view information interaction in conditional generative adversarial networks (cGAN) [10] to simulate a doctor analyzing images. A description of cGAN can be found in the article [10], which can be used for image-to-image translation and can be widely adopted in multi-modality image synthesis in medical image analysis [13,19]. To reduce the loss of information, cGSNT makes use of a cycle consistent

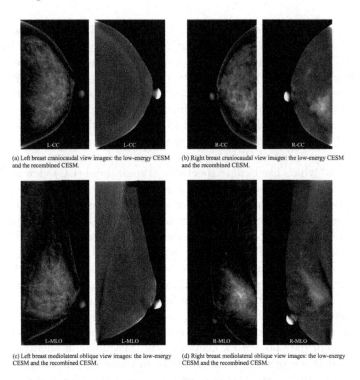

(a) Left breast craniocaudal view images: the low-energy CESM and the recombined CESM.

(b) Right breast craniocaudal view images: the low-energy CESM and the recombined CESM.

(c) Left breast mediolateral oblique view images: the low-energy CESM and the recombined CESM.

(d) Right breast mediolateral oblique view images: the low-energy CESM and the recombined CESM.

Fig. 1. CESM images. In (a), (b), (c), and (d), the low-energy CESM images are on the left, and the CESM images are on the right.

approach, which is inspired by CycleGAN [20], which uses a cycle consistency loss to constrain the difference between the input image and the cGSNT generated image.

In general, the innovations of our work can be summarized as follows: (1) To the best of our knowledge, this method is the first to realize the full image synthesis of recombined CESM images from low-energy images in the field of breast imaging. (2) We propose an improved version, cGSNT, which is based on the CycleGAN model, which utilizes the cycle-consistent approach to reduce the information loss during the image synthesis process, and fuses the dual-view attribute information, from the craniocaudal (CC) view and mediolateral oblique (MLO) view, to produce reasonable translation results.

2   Materials and Methods

2.1   System Overview

Standard CESM data are CC view and MLO view of the left and right breast, each view contains low-energy CESM and recombined CESM image. For each set of images, the combined CESM image is used as the regression target for the low-energy CESM image. Low-energy CESM and recombined CESM are aligned, no

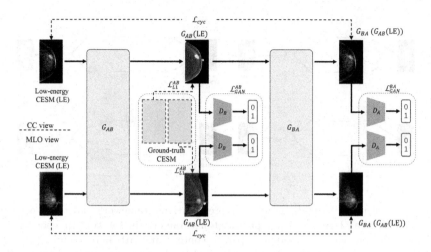

Fig. 2. The framework of the proposed cGAN-based synthesis network (cGSNT).

additional pre-alignment steps are required. Figure 2 shows the structure of the proposed cGSNT, which is inspired by CycleGAN [20]. The generator networks G_{AB} and G_{BA} are applied to learn the distribution of the real images from two different domains (low-energy CESM image as domain A and recombined CESM image as domain B) to estimate the possible mapping of the input images. The discriminators D_A and D_B are used to distinguish whether each input $N \times N$ overlapping patch is "real" or "fake". In the testing stage, we feed the low-energy CESM image pair (CC view and MLO view images) into the trained generator networks G_{AB} and G_{BA} for the recombined CESM image pair prediction.

2.2 Network Structure

Generator. We design the generator network based on U-Net [14], which is a multilayer convolutional encoder-decoder neural network that can be used for image-to-image translation [10]. In particular, we employ a variety of different image-to-image translation (DRIT) [12] methods to improve the generator, that disentangles content and attribute representations. Then, the dual information fusion layer stitches the attribute features from the two views, and the merged attribute features are fed to the hybrid decoder together with the original features.

To achieve image content and attribute representations disentanglement, we utilize the DRIT network (in Fig. 3 (a)) to train the content encoder $\{E^c_{CC}, E^c_{MLO}\}$ and the attribute encoder $\{E^a_{CC}, E^a_{MLO}\}$. Take CC view images as an example, content encoder E^c_{CC} captures the underlying spatial structure and the attribute encoder E^a_{CC} extracts the rendering information in the image corresponding to the tumor features. The image reconstruction decoder R_{CC}

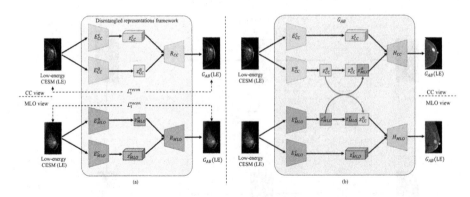

Fig. 3. The details of the disentangled representations framework and the generator network G_{AB}. (a) Disentangled representations framework: the content encoder E^c and attribute encoder E^a used to separate content and attribute vectors $\{z^c, z^a\}$, and the image reconstruction decoder $\{R_{CC}, R_{MLO}\}$ restore the original image. (b) Generator network G_{AB}: the trained content encoder E^c and the attribute encoder E^a are employed to extract the content and attribute vectors $\{z^c, z^a\}$, and the attribute vectors from two views are aggregated together with content features for generating transformed images. Moreover, a dual-view information fusion layer is also proposed to fuse attribute vectors $\{z^a_{CC}, z^a_{MLO}\}$ from the CC view and MLO view, for improving the translation performance.

restores images conditioned on both content and attribute vectors $\{z^c_{CC}, z^a_{CC}\}$. A self-reconstruction loss $\mathcal{L}^{recon}_1$ is applied to constrain the image to be restored. In addition, content discriminator D^c is used to help generate the content representation of the image [12].

Figure 3 (b) shows the generator networks G_{AB} architecture. We operate the pre-trained content encoder $\{E^c_{CC}, E^c_{MLO}\}$ and attribute encoder $\{E^a_{CC}, E^a_{MLO}\}$ to separate CC view and MLO view content and attribute representation $\{z^c_{CC}, z^a_{CC}\}$, $\{z^c_{MLO}, z^a_{MLO}\}$. The dual-view attribute fusion layer combines the attribute vectors z^a_{CC} and z^a_{MLO} generated by the attribute encoders of the two views into a composite vector $\{z^a_{CC}, z^a_{MLO}\}$. Unlike DRIT's reconstruction decoder receives a content vector and an attribute vector$\{z^c, z^a\}$, the proposed hybrid decoder $\{H_{CC}, H_{MLO}\}$ generates image based on a single content vector and composite attribute vectors $\{z^a_{CC}, z^a_{MLO}\}$ obtained from two views.

Discriminator. The discriminator $\{D_A, D_B\}$ in our proposed cGSNT based on PatchGAN, which are designed to perceive the difference between the synthesized image and the ground-truth domain, and aim to distinguish "real" or "fake" CESM images.

2.3   Losses and Objectives

To achieve the objective of learning generators for mapping between A and B domains, we adopt the following adversarial loss function that can be hold to mapping function G_{AB} $(A \rightarrow B)$:

$$
\begin{aligned}
\mathcal{L}_{\text{GAN}}^{AB}(G_{AB}, D_B, A, B) = & \mathbb{E}_{b \sim p_{data}(b)}[\log D_B(b)] \\
& + \mathbb{E}_{a \sim p_{data}(a)}[\log(1 - D_B(G_{AB}(a)))].
\end{aligned}
\tag{1}
$$

Similarly, the adversarial loss for mapping from domain B to A is $\mathcal{L}_{\text{GAN}}^{BA}(G_{BA}, D_A, B, A)$.

In addition, a typical loss $L1$ is operated to train generator G_{AB} and G_{BA}:

$$
\mathcal{L}_{L1}^{AB}(G_{AB}, A, B) = \mathbb{E}_{b \sim p_{data}(b)}[\|G_{AB}(a) - b\|_1],
\tag{2}
$$

$$
\mathcal{L}_{L1}^{BA}(G_{BA}, B, A) = \mathbb{E}_{a \sim p_{data}(a)}[\|G_{BA}(b) - a\|_1].
\tag{3}
$$

The above networks are trained simultaneously, the discriminator D aims to distinguish between real and synthesized data correctly, and generator G tries to produce a realistic image that confuses D. The generator $G_{AB}(\cdot)$ builds a non-linear mapping from domain A to domain B that can translate A to $G_{AB}(A)$. $G_{BA}(\cdot)$ also resolves a non-linear translation from the output of $G_{AB}(\cdot)$ to domain A. In addition, we use the cycle-consistent constraint that argues that the learned mapping functions should be able to bring an image back to its original image, to achieve self-supervision (e.g., $G_{BA}(G_{AB}(a)) \approx a$ and $G_{AB}(G_{BA}(b)) \approx b$). The objective function can be defined as:

$$
\begin{aligned}
\mathcal{L}_{\text{cyc}}(G_{AB}, G_{BA}, A, B) = & \mathbb{E}_{a \sim p_{data}(a)}[\|G_{BA}(G_{AB}(a)) - a\|_1] \\
& + \mathbb{E}_{b \sim p_{data}(b)}[\|G_{AB}(G_{BA}(b)) - b\|_1].
\end{aligned}
\tag{4}
$$

This constraint implements the hypothesis that if we convert the image to another domain and then back again, the image after the two conversions is identical to the original image.

Our final objective is

$$
F = \arg \min_{G} \max_{D} (\mathcal{L}_{\text{GAN}}^{AB} + \mathcal{L}_{\text{GAN}}^{BA}) + \lambda_1 \mathcal{L}_{\text{cyc}} + \lambda_2 (\mathcal{L}_{L1}^{AB} + \mathcal{L}_{L1}^{BA}).
\tag{5}
$$

Here, λ_1 and λ_2 are used to balance the weight for each term. The model was trained using $\lambda_1 = 0.1$ and $\lambda_2 = 0.01$ in our experiment.

3   Dataset Description

We collected 392 sets of low-energy and recombined CESM breast images acquired from 196 patients at Yantai Yuhuangding Hospital. All the patients are females who come to the hospital for treatment, including those who have symptoms and those who needed further confirmation after screening, the mean age was 47.8 years (range: 21–69 years). For these 196 patients, on the basis of imaging examinations, doctors recommended a biopsy for some patients for further confirmation. In this dataset, there are 117 positive samples from 117 patients. Every set of breast images, each containing one mammogram image and the corresponding recombined CESM image, was taken in CC and MLO views by a SenoBright™ HD device (GE Healthcare, Chicago, IL, USA), and the image size is 2394 × 3062. The low-energy CESM images with intensity range from 0–3164, and the recombined CESM images with intensity range from 0–3887. We split our dataset into three sets: training, validation, and testing; they each contained 240, 60, and 92 image sets, respectively. For the training dataset, validation dataset and test dataset, we adopt a random division method while ensuring that the ratio of positive and negative samples is roughly the same.

4   Experimental Setup

Based on the proposed method in Fig. 2, we trained and tested our algorithm to predict the recombined CESM images from the low-energy image. Unlike the previous work [8], our synthesis model performed image translation tasks on the whole image and did not need an experienced doctor to identify the cancer region bounding box: for each image, we performed image synthesis tasks to utilize the proposed model.

We implemented our neural network using PyTorch with four NVIDIA Tesla V100 GPUs with 32 GB of video memory to train and evaluate the networks. The learning rates are initialized as 2e−4, followed by decreasing the learning rates by cosine annealing schedule during the training.

5   Results

Figure 4 shows a set of low-energy CESM images and the corresponding synthesized recombined CESM images and ground-truth images. (a) is a negative sample, the BI-RADS score for this is 1, and no biopsy is done; (b) is a positive sample, the BI-RADS score for this is 6, and the biopsy result proves to be positive. The proposed model produces high-quality recombined CESM images where the areas of abnormal blood flow (cancer areas) are prominently highlighted. The prominence relation is consistent between the synthesized recombined CESM images and the actual recombined CESM images.

Table 1 lists the mean value (Mean), standard deviation (std) and median (Med.) of mean absolute error (MAE), peak signal-to-noise ratio (PSNR), structured similarity index (SSIM). Overall, the proposed method resulted in the highest performance with respect to MAE, PSNR, and SSIM. The proposed algorithm resulted in a mean PSNR of 26.7, a gain of 4.5 compared to the results obtained by Huynh et al. [9] using SRF with ACM. Compared with pix2pix GAN, the mean PSNR value is further improved by 0.9 when using the CycleGAN model. The mean MAE, PSNR, and SSIM values were improved by 0.3, 0.5 and 0.004, respectively, for cGSNT using the network with dual-view information fusion.

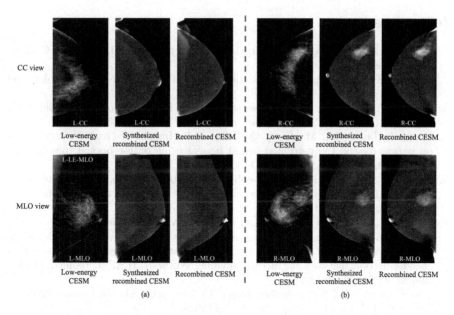

Fig. 4. Visualization of the recombined CESM images from the low-energy images for one subject. (a) BI-RADS: 1; No biopsy. (b)BI-RADS: 6; Positive biopsy.

Table 1. The quantitative results of the recombined CESM images from low-energy images by cGSNT and alternative methods.

Method	MAE			PSNR		SSIM	
	Mean (std)	Med.	MAX	Mean (std)	Med.	Mean (std)	Med.
SR	16.2 (2.8)	16.0	34.3	22.8 (1.8)	23.1	0.884 (0.028)	0.882
SRF+	16.1 (2.5)	15.5	26.7	22.2 (1.6)	22.9	0.882 (0.023)	0.887
RIED-Net	12.3 (2.5)	12.2	23.5	25.7 (1.6)	25.8	0.913 (0.023)	0.915
pix2pix	12.7 (2.1)	12.6	26.1	25.3 (1.5)	25.7	0.911 (0.021)	0.910
CycleGAN	11.9 (1.8)	11.8	21.2	26.2 (1.4)	26.3	0.920 (0.018)	0.921
Proposed	**11.6 (1.5)**	**11.4**	**18.3**	**26.7 (1.3)**	**26.7**	**0.924 (0.016)**	**0.926**

6   Conclusion

The proposed cGSNT using a conditional generative adversarial network architecture for the synthesis task of the mammogram image to the recombined CESM image is an effective method to improve the accuracy of synthesis. In addition, our model with cycle-consistent constraint outperformed cGAN trained using the same dataset. Further, the disentangled representations method could encode the content and attribute features separately, and then fuse the attribute vectors from two views to improve their results.

References

1. Bartorelli, A.L., Marenzi, G.: Contrast-induced nephropathy. J. Intervent. Cardiol. **21**(1), 74–85 (2008)
2. Bick, U.: Full-field digital mammography. Breast Cancer Res. **2**(2 Supplement) (2000)
3. Morcos, S., Thomsen, H.: Adverse reactions to iodinated contrast media. Eur. Radiol. **11**(7), 1267–1275 (2001). https://doi.org/10.1007/s003300000729
4. Deng, B., Lundqvist, M., Fang, Q., Carp, S.A.: Impact of errors in experimental parameters on reconstructed breast images using diffuse optical tomography. Biomed. Opt. Express **9**(3), 1130–1150 (2018)
5. Fallenberg, E.M., et al.: Contrast-enhanced spectral mammography versus MRI: initial results in the detection of breast cancer and assessment of tumour size. Eur. Radiol. **24**(1), 256–264 (2014)
6. Fanizzi, A., et al.: Fully automated support system for diagnosis of breast cancer in contrast-enhanced spectral mammography images. J. Clin. Med. **8**(6), 891 (2019)
7. Gao, F., Wu, T., Chu, X., Yoon, H., Xu, Y., Patel, B.: Deep residual inception encoder-decoder network for medical imaging synthesis. IEEE J. Biomed. Health Inform. **24**(1), 39–49 (2020)
8. Gao, F., et al.: SD-CNN: a shallow-deep CNN for improved breast cancer diagnosis. Comput. Med. Imaging Graph. **70**, 53–62 (2018)
9. Huynh, T., et al.: Estimating CT image from MRI data using structured random forest and auto-context model. IEEE Trans. Med. Imaging **35**(1), 174–183 (2015)
10. Isola, P., Zhu, J.Y., Zhou, T., Efros, A.A.: Image-to-image translation with conditional adversarial networks. In: Proceedings of the IEEE Conference on Computer Vision and Pattern Recognition, pp. 1125–1134 (2017)
11. Lalji, U., Jeukens, C., Houben, I., Nelemans, P., van Engen, R., van Wylick, E.: Evaluation of low-energy contrast-enhanced spectral mammography images by comparing them to full-field digital mammography using EUREF image quality criteria. Eur. Radiol. **25**(10), 2813–2820 (2015)
12. Lee, H.Y., Tseng, H.Y., Huang, J.B., Singh, M., Yang, M.H.: Diverse image-to-image translation via disentangled representations. In: Proceedings of the European conference on Computer Vision, pp. 35–51 (2018)
13. Nie, D., et al.: Medical image synthesis with deep convolutional adversarial networks. IEEE Trans. Biomed. Eng. **65**(12), 2720–2730 (2018)
14. Ronneberger, O., Fischer, P., Brox, T.: U-Net: convolutional networks for biomedical image segmentation. In: Navab, N., Hornegger, J., Wells, W.M., Frangi, A.F. (eds.) MICCAI 2015. LNCS, vol. 9351, pp. 234–241. Springer, Cham (2015). https://doi.org/10.1007/978-3-319-24574-4_28
15. Sorin, V., et al.: Contrast-enhanced spectral mammography in women with intermediate breast cancer risk and dense breasts. Am. J. Roentgenol. **211**(5), W267–W274 (2018)
16. Yang, Q., et al.: MRI cross-modality image-to-image translation. Sci. Rep. **10**(1), 1–18 (2020)
17. Yu, B., Wang, Y., Wang, L., Shen, D., Zhou, L.: Medical image synthesis via deep learning. In: Lee, G., Fujita, H. (eds.) Deep Learning in Medical Image Analysis. AEMB, vol. 1213, pp. 23–44. Springer, Cham (2020). https://doi.org/10.1007/978-3-030-33128-3_2
18. Zanardo, M., et al.: Technique, protocols and adverse reactions for contrast-enhanced spectral mammography (CESM): a systematic review. Insights Imaging **10**(1), 1–15 (2019). https://doi.org/10.1186/s13244-019-0756-0

19. Zhang, Z., Yang, L., Zheng, Y.: Translating and segmenting multimodal medical volumes with cycle-and shape-consistency generative adversarial network. In: Proceedings of the IEEE Conference on Computer Vision and Pattern Recognition, pp. 9242–9251 (2018)
20. Zhu, J.Y., Park, T., Isola, P., Efros, A.A.: Unpaired image-to-image translation using cycle-consistent adversarial networks. In: Proceedings of the IEEE International Conference on Computer Vision, pp. 2223–2232 (2017)

Self-adversarial Learning for Detection of Clustered Microcalcifications in Mammograms

Xi Ouyang[1,2], Jifei Che[1], Qitian Chen[1], Zheren Li[1], Yiqiang Zhan[1], Zhong Xue[1], Qian Wang[2(✉)], Jie-Zhi Cheng[1], and Dinggang Shen[1,3(✉)]

[1] Shanghai United Imaging Intelligence Co., Ltd., Shanghai, China
[2] School of Biomedical Engineering, Shanghai Jiao Tong University, Shanghai, China
wang.qian@sjtu.edu.cn
[3] School of Biomedical Engineering, ShanghaiTech University, Shanghai, China
dgshen@shanghaitech.edu.cn

Abstract. Microcalcification (MC) clusters in mammograms are one of the primary signs of breast cancer. In the literature, most MC detection methods follow a two-step paradigm: segmenting each MC and analyzing their spatial distributions to form MC clusters. However, segmentation of MCs cannot avoid low sensitivity or high false positive rate due to their variability in size (sometimes $<0.1\,\mathrm{mm}$), brightness, and shape (with diverse surroundings). In this paper, we propose a novel self-adversarial learning framework to differentiate and delineate the MC clusters in an end-to-end manner. The class activation mapping (CAM) mechanism is employed to directly generate the contours of MC clusters with the guidance of MC cluster classification and box annotations. We also propose the self-adversarial learning strategy to equip CAM with better detection capability of MC clusters by using the backbone network itself as a discriminator. Experimental results suggest that our method can achieve better performance for MC cluster detection with the contouring of MC clusters and classification of MC types.

Keywords: Clustered microcalcifications · Class activation mapping · Self-adversarial learning

1 Introduction

Breast cancer is the worldwide leading cause of death for women [3]. Early detection of breast cancer can reduce the mortality rate and increase the survival rate [18]. Mammography is the primary diagnostic tool to detect breast cancer at the early stage, and enjoys the advantages of low cost and high sensitivity [16]. In mammography screening, the presence of clustered microcalcifications (MCs) is a very crucial sign of breast cancer. However, calcifications in mammograms can also be benign. Calcifications associated with malignant lesions are proved to be smaller in size and more densely distributed (clustered in groups) [15]. In

---

X. Ouyang and J. Che—Contributed equally to this work.

© Springer Nature Switzerland AG 2021
M. de Bruijne et al. (Eds.): MICCAI 2021, LNCS 12907, pp. 78–87, 2021.
https://doi.org/10.1007/978-3-030-87234-2_8

mammographic images, breast calcifications in the early stages of breast cancer appear like scattered spots that range mostly from 0.1 to 1.0 mm in size, even sometimes <0.1 mm [11]. Such tiny spots can be also easily ignored by experienced radiologists. Therefore, there is always a strong demand of a computer-aided detection (CADe) system to assist the reading of mammography.

In the literature, most clustered MC detection methods consist of two steps: single MC segmentation and a post-processing step for clustering [2]. To achieve segmentation of MC, different methods including the traditional methods and deep networks are proposed. Traditional methods usually extract various types of features, (e.g., harr-like features [4], shape and texture features [9]) from images for the training of a binary classifier to demarcate the calcification pixels. With the rapid development of deep learning, deep networks have also made great progress in MC segmentation [1,5,20]. Zhang et al. [20] achieved 85.31% recall at one false per image with their U-Net [14] based reconstruction network. After segmentation of MCs, post-processing methods are applied to analyze the spatial distribution of MCs for clustering. Zhang et al. [21] proposed an adaptive Gaussian mixture model (GMM) method to classify clustered MCs. Basile et al. [2] explored a set of codified domain expert rules to automatically group the single segmented MCs into clusters.

However, there exist several issues in these existing methods. First, accurate MC segmentation is still a quite challenging task due to the tiny size (sometimes <0.1 mm) and variation in brightness, shape with diverse surroundings, leading to the miss detection. Imperfect MC segmentation may compromise the performance of the latter clustering step. Second, most MC clustering algorithms are designed by the manual rules or traditional models, which have limited performance and also may not delineate high-quality contours for the MC clusters.

To address the aforementioned issues, we propose an end-to-end network to segment MCs, identify the malignancy of MC and delineate the MC clusters at image level in one shot. With our network design, the segmentation task can assist the task of cluster malignancy classification and detection in a multi-task manner. Meanwhile, we employ the Class Activation Mapping (CAM) [22] of the classification task for the contouring of MC clusters, which makes the MC clustering task trainable in our network. We adopt the online CAM mechanism [13] here, which can be refined with the attention loss under the supervision of the bounding boxes that enclose malignant lesions. Moreover, we propose self-adversarial learning to take the network itself as a discriminator in a self-guidance fashion. It can assure that CAM better covers entire MC clusters by not only on focusing on most discriminative regions. The self-adversarial learning can also improve localization performance for difficult cases. Experimental results suggest that our approach can achieve great performance to identify and delineate MC clusters.

2   Method

The overall framework is shown in Fig. 1. Specifically, for each training sample, random cropping of 512 × 512 image patches from the original mammography

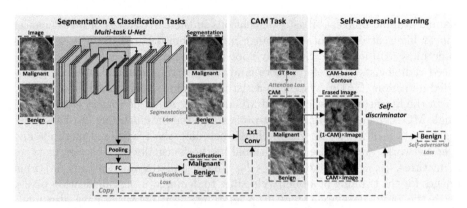

Fig. 1. The pipeline of the proposed algorithm for detection of clustered MCs. We propose a multi-task U-Net for the single MC segmentation, MC type classification, and MC clustering. Moreover, we propose the self-adversarial learning to use the network itself as a self-discriminator based on the images with different erasing schemes for malignant and benign images. For malignant images, the detected MC cluster contour generated from the CAM attention map is shown in red while the green box refers to the MC cluster regions annotated by an experienced radiologist (Color figure online).

is implemented to ensure the input has an appropriate resolution for the learning of MC features. We adopt the U-Net equipped with GC blocks [6] after each skip connection as the backbone. In this paper, the network is trained with three constraints (*i.e.*, segmentation loss, classification loss, and attention loss) to simultaneously learn the MC segmentation, image-level MC malignancy classification, and MC clustering. Here, we adopt the online CAM mechanism for contouring of the MC clusters. Moreover, to endow the CAM to better demarcate the entire MC clusters, we propose self-adversarial learning to employ the network itself as a discriminator in a self-guidance fashion.

2.1 Multi-task Mechanism

Different from the conventional two-step methods, we directly identify MC clusters with the attentive regions of CAM derived from the image-level MC classification task. The CAM highlights the cluster regions when the network classifies the image patch as the presence of malignant MC. The MC segmentation in our framework is an auxiliary task to assist the classification and CAM attention task.

The feature maps at the first up-sampling block of the U-Net are served as the input for the classification and attention task. These feature maps will pass through a *log-sum-exp* (LSE) pooling [17] layer and fully-connected (FC) layer with Sigmoid to produce the prediction for the binary classification task (Benign/Malignant). At the same time, following the online CAM strategy in [13], we share the weights of the FC layers to a 1×1 convolutional layer to aggregate the feature maps. It generates the CAM attention maps for the

MC cluster localization, which can be refined by the attention loss to regularize the CAM attention maps to match the bounding boxes of malignant regions.

2.2 Self-adversarial Learning

Although the online-CAM mechanism can generate accurate attention maps for the network, the CAM regions tend to focus on small and sparse discriminative regions from the object of interest [19]. Since each cluster is composed of many scattered points, it will miss many single MCs for the CAM attention maps and fails to cover the complete MC clusters. Meanwhile, there exist some heterogeneously dense or extremely dense compacted glands especially in dense breasts, which may contain some white spots. It can easily be mistaken for false positives.

To solve these issues, we propose self-adversarial learning to provide more self-guidance for the CAM attention maps. As shown in Fig. 1, we build up a self-discriminator with part of the multi-task U-Net model. It copies parameters from part of the multi-task U-Net model, $i.e.$, the encoder, the bottleneck layer, the first up-sampling block of the decoder, and the FC layer with Sigmoid. The proposed self-adversarial learning follows two schemes for malignant and benign images. For malignant images, we expect that the CAM attention map of the network should contain all the MC cluster regions. To this end, assuming that the input malignant image as $image$ and the CAM attention map as CAM (the value ranges from $[0-1]$), we firstly get an erased image as $(1 - CAM) \times image$, and input it into the self-discriminator. Since we expect that this operation could erase all the pixels associated with MC clusters, we force the self-discriminator to classify the erased image into benign type, which can be formulated as:

$$l_{sa}^{malignant} = -\log(1 - W_{SD}\{(1 - CAM) \times image\}), \tag{1}$$

where W_{SD} is the self-discriminator network and $W_{SD}\{(1 - CAM) \times image\}$ refers to network prediction of the input erased image. It is the cross-entropy constraint to ensure the prediction to be the benign type. This loss function helps the network to pay attention to the complete MC cluster regions but not only small discriminative regions.

For the benign images, the CAM attention regions localize in the dense compacted glands which have some similar patterns with MC clusters. Therefore, to better distinguish these region and MC clusters, we firstly erase regions that the network do not pay attention instead of the regions of interest, formulating as $CAM \times image$. Then, we ensure that the erased image pass through the self-discriminator to be classify into benign type. It can be defined as:

$$l_{sa}^{benign} = -\log(1 - W_{SD}\{CAM \times image\}), \tag{2}$$

where $W_{SD}\{CAM \times image\}$ refers to network prediction of the input $CAM \times image$. It can force the network to learn more fine-grained differences between benign regions with dense compacted glands and malignant MC clusters.

2.3   Total Loss Function

The total loss function consists of four items: segmentation loss, classification loss, attention loss and self-adversarial loss. It can be formulated as:

$$l_{total} = \lambda_1 l_{seg} + \lambda_2 l_{cls} + \lambda_3 l_{att} + \lambda_4 l_{sa}^{malignant} + \lambda_5 l_{sa}^{benign}, \qquad (3)$$

where λ_1, λ_2, λ_3, λ_4 and λ_5 are weighting factors. In our experiments, these hyper-parameters are determined with a search on validation set.

For the segmentation loss l_{seg}, since it is a binary segmentation problem, we use the Sigmoid layer for the segmentation predictions of the multi-task U-Net and take the binary cross-entropy (BCE) loss for each pixel. We formulate it as:

$$l_{seg} = \frac{\sum_{ij} BCE(seg, mask)}{\sum_{ij} seg + \sum_{ij} mask}, \qquad (4)$$

where i and j represent the $(i, j)^{th}$ pixel in the corresponding segmentation map. seg is the segmentation result of the U-Net and $mask$ is the ground-truth mask of MCs. Here, due to the tiny size of MCs, we use the sum of regions of segmentation prediction and ground-truth mask of MCs as an adaptive normalization factor instead of the image size for the final segmentation loss of each image.

Meanwhile, l_{cls} is BCE constraints for the malignant/benign classification. We also invite an experienced radiologist to annotate the MC clusters of malignant regions in mammography images with boxes. We use Dice coefficient [12] as the attention loss l_{att}, regularizing the CAM attention maps to conform to the ground-truth bounding boxes.

3   Experiments

3.1   Dataset

In this study, we collect 603 mammography images of 383 patients, from our collaborative hospitals with the IRB approvals. Specifically, 307 images belong to bilateral craniocaudal (CC) view, whereas 296 images belong to mediolateral oblique (MLO) view. Mammography scanners include UIH and Hologic machines. For images from UIH machines, the spacing is 0.0495 mm and resolution is 4604 × 5859. The spacing is 0.07 and resolution is 3328 × 4096 for images from Hologic machines. All the images are well-annotated with bounding boxes for MC clusters by an experienced radiologist, resulting in 773 boxes. We randomly split 353, 50 and 200 images into the training, validation and testing set, respectively.

3.2   Experimental Settings

Adam optimization [10] is adopted and the momentum is set as 0.9, and the learning rate is set to 0.0001 that is reduced by a factor of 0.5 after every 100

Table 1. Comparison of performance of detection of clustered MCs from different models.

Method of cluster detection	Model	$R@0.2$	$R@0.4$	$R@0.6$
Segmentation	U-Net	0.816	0.870	0.900
	Ours-Seg	0.816	0.887	0.921
CAM	Multi-task U-Net w/o seg	–	–	0.594
	Multi-task U-Net w/o sa	0.736	0.795	0.837
	Multi-task U-Net w/ sa-malignant	0.904	0.958	0.971
	Ours-CAM	0.925	**0.967**	0.979
Segmentation & CAM	Ours-Seg + Ours-CAM	**0.946**	**0.967**	**0.983**

epochs. The batch size is set to 4. The loss weight factors λ_1, λ_2, λ_3, λ_4 and λ_5 are set to 1, 0.5, 0.5, 0.2, 0.1, respectively. The best checkpoint model with the best evaluation performance within 500 epochs is used as the final model and then evaluated on the testing set. All following experimental results are from the testing set. In our experiments, we resample all images into the same spacing of 0.1 mm. For the input of our network, we randomly sample the 512×512 image patch from the entire mammography images. The image patch will be annotated as malignant type when it covers over more than 20% of the area of any bounding boxes of MC clusters. Otherwise, it will be annotated as benign type when not covering any bounding boxes.

3.3 Evaluation Metrics

We evaluate the performance of the following tasks: MC type classification and detection of MC clusters. We use four metrics to measure the classification results, i.e., area under the receiver operating characteristic curve (AUC), sensitivity, specificity, and F1-score. Sensitivity, specificity, and F1-score calculated at the threshold of 0.5. For the detection of MC clusters, we report the recalls at k false positive per image (simplified as $R@k$), where $k \in \{0.2, 0.4, 0.6\}$ for all models. A bounding box is considered as recalled if there is at least one prediction cluster contour with IoU >0.5.

3.4 Experimental Results

Comparison of Performance of Detection of Clustered MCs. For better comparison, we will show the performance of detection of clustered MCs based on the segmentation and CAM results in our proposed method.

Results are shown in Table 1, including different baselines with the segmentation and CAM-based cluster detection methods. In our experiments, we compare our proposed method with four baselines: "U-Net", "Multi-task U-Net w/o seg", "Multi-task U-Net w/o sa", "Multi-task U-Net w/ sa-malignant". "U-Net" refers to the U-Net model trained with only single MC segmentation (l_{seg}), which refers to the state-of-the-art single MC segmentation performance. To get the detection results of clustered MCs based on the segmentation results, as the work [2], we also use a set of codified domain expert rules to group the MC segmentation results into the cluster. Basically, we follow the definition of MC clusters in BI-RADs [7]: five calcifications are grouped within 1 cm (no more than 2 cm) of each other. Then, we get cluster contours by using *Graham's scan* method [8] to calculate the convex hull for each cluster. It can be observed that the cluster results based segmentation from "U-Net" can achieve 0.816 recalls at 0.2 false positive per image. Also, we use the same post-processing method to generate cluster contours by using the segmentation results of our proposed method ("Ours-Seg"). With the assistance of classification and attention tasks, we can achieve better performance of $R@0.4$ and $R@0.6$. It proves that the multi-task mechanism can help to improve the MC segmentation performance.

Moreover, we show the performance of the models using the CAM attention result for cluster detection in Table 1. Here, we directly use the edge of the CAM regions without any complex post-processing method as the cluster contours. "Multi-task U-Net w/o seg" refers to the model trained with only l_{cls}, and l_{att}, which achieve only 0.594 of $R@0.4$. We find that the CAM of this model appears a lot in the high-density gland areas but not focus on MCs, proving that MC segmentation is the important guidance. "Multi-task U-Net w/o sa" model is defined as the multi-task U-Net model trained with l_{seg}, l_{cls}, and l_{att} but without self-adversarial learning loss. This model shows the naive CAM performance of cluster detection, which is still worse than the segmentation-based cluster detection methods. Then, we add the part of self-adversarial learning loss for malignant images ($l_{sa}^{malignant}$) to define the "Multi-task U-Net w/ sa-malignant" model. We can see that it can improve the score of $R@0.4$ to 0.904, which obviously outperforms all segmentation-based cluster detection methods.

When using the CAM results from our proposed method ("Ours-CAM") for cluster detection, it can further achieve the 0.925 scores at $R@0.4$. It proves that the self-adversarial learning for both malignant and benign images can make the CAM better identify the MC clusters at the same false-positive rate. Moreover, we show the detection performance of the model ("Our-Seg + Our-CAM") which combines the results from the segmentation and CAM of our proposed method. We add the detected cluster contours with high confidence by using the segmentation results into the detection results of CAM attention, which can help to reduce the missed detection rate. It can further achieve 0.983 recalls at 0.6 false positive per image.

Comparison of Classification Performance. Here we show the classification performance of different models in Table 2. In this experiment, all models are tested at the patch level (total 305 patches with 512×512 size). It can be observed that the scores of all classification metrics are improved after adding the self-adversarial learning for malignant images ($l_{sa}^{malignant}$). Moreover, compared with the classification performance of the multi-task U-Net with the only sa-malignant part, we can see the sa-benign part of self-adversarial learning in our proposed method can further improve the recalls while keeping precision rate.

Table 2. Comparison of classification performance of different models. All the models use the classification prediction for the comparison.

Model	AUC	Precision	Recall	F1-score
Multi-task U-Net w/o sa	0.905	0.949	0.864	0.877
Multi-task U-Net w/ sa-malignant	0.943	0.961	0.925	0.910
Ours-Cls	**0.950**	**0.966**	**0.935**	**0.929**

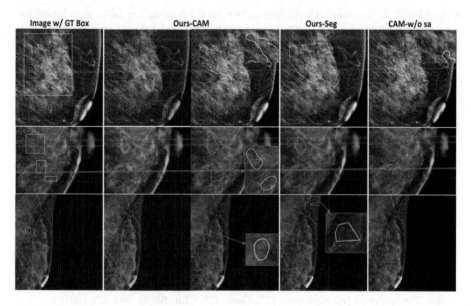

Fig. 2. Visualization results from different models. All the images are cropped to better show the lesion regions. The green boxes refer to the lesion regions labeled by the invited radiologist. The contours from the different models are shown in yellow (Color figure online).

Visualization Results

The visualization results are shown in Fig. 2, where we show the original CAM heatmap from our proposed method and the contours from different methods. We can see that the contours of MC clusters can be directly generated based on the edge of highlight regions in CAM heat maps. In the first row, the contour based on segmentation results is rough and contains too many non-related regions. The contour of CAM without the proposed self-adversarial learning only delineates partial MC clusters. It can be observed that the contour of our proposed method can fit well to the malignant regions. In the second row, there are three MC clusters in this case, and some MCs are quite tiny and densely grouped in gland regions. It is hard to segment all single MCs in such a hard case, leading to the miss detection of the segmentation-based cluster detection method. Our proposed method can still generate the accurate contour while the CAM model without self-adversarial learning also cannot identify all the MC clusters. The segmentation-based method also generates some false positives in the case of the third row, whereas the CAM-based methods can give extract detection results even for amorphous MC clusters.

4   Conclusion

A multi-task U-Net model with self-adversarial learning has been developed for detection of clustered MCs. Comparing with the traditional two-step method with single MC segmentation, it uses the CAM attention region to directly generate the cluster contour in an end-to-end manner. The extensive experimental results show that our proposed method can generate high-quality contours and also effective for detecting very difficult small calcification clusters.

Acknowledgement. This work was supported by STCSM grants (19QC1400600).

References

1. AlGhamdi, M., Abdel-Mottaleb, M., Collado-Mesa, F.: DU-Net: convolutional network for the detection of arterial calcifications in mammograms. IEEE Trans. Med. Imaging **39**(10), 3240–3249 (2020)
2. Basile, T., et al.: Microcalcification detection in full-field digital mammograms: a fully automated computer-aided system. Physica Medica **64**, 1–9 (2019)
3. Bray, F., Ferlay, J., Soerjomataram, I., Siegel, R.L., Torre, L.A., Jemal, A.: Global cancer statistics 2018: Globocan estimates of incidence and mortality worldwide for 36 cancers in 185 countries. CA: Cancer J. Clin. **68**(6), 394–424 (2018)
4. Bria, A., Karssemeijer, N., Tortorella, F.: Learning from unbalanced data: a cascade-based approach for detecting clustered microcalcifications. Med. Image Anal. **18**(2), 241–252 (2014)
5. Cai, G., Guo, Y., Zhang, Y., Qin, G., Zhou, Y., Lu, Y.: A fully automatic microcalcification detection approach based on deep convolution neural network. In: Medical Imaging 2018: Computer-Aided Diagnosis, vol. 10575, p. 105752Q. International Society for Optics and Photonics (2018)
6. Cao, Y., Xu, J., Lin, S., Wei, F., Hu, H.: GCNet: non-local networks meet squeeze-excitation networks and beyond. In: Proceedings of the IEEE/CVF International Conference on Computer Vision Workshops (2019)
7. D'Orsi, C.J., et al.: ACR BI-RADS Atlas: Breast Imaging Reporting and Data System; Mammography, Ultrasound, Magnetic Resonance Imaging, Follow-up and Outcome Monitoring. Data Dictionary. ACR, American College of Radiology (2013)
8. Graham, R.L.: An efficient algorithm for determining the convex hull of a finite planar set. Info. Pro. Lett. **1**, 132–133 (1972)
9. Khalaf, A.F., Yassine, I.A.: Novel features for microcalcification detection in digital mammogram images based on wavelet and statistical analysis. In: 2015 IEEE International Conference on Image Processing (ICIP), pp. 1825–1829. IEEE (2015)
10. Kingma, D.P., Ba, J.: Adam: a method for stochastic optimization. arXiv preprint arXiv:1412.6980 (2014)

11. Ma, Y., Tay, P.C., Adams, R.D., Zhang, J.Z.: A novel shape feature to classify microcalcifications. In: 2010 IEEE International Conference on Image Processing, pp. 2265–2268. IEEE (2010)
12. Milletari, F., Navab, N., Ahmadi, S.A.: V-Net: fully convolutional neural networks for volumetric medical image segmentation. In: 2016 Fourth International Conference on 3D Vision (3DV), pp. 565–571. IEEE (2016)
13. Ouyang, X., et al.: Learning hierarchical attention for weakly-supervised chest x-ray abnormality localization and diagnosis. IEEE Trans. Med. Imaging (2020)
14. Ronneberger, O., Fischer, P., Brox, T.: U-Net: convolutional networks for biomedical image segmentation. In: Navab, N., Hornegger, J., Wells, W.M., Frangi, A.F. (eds.) MICCAI 2015. LNCS, vol. 9351, pp. 234–241. Springer, Cham (2015). https://doi.org/10.1007/978-3-319-24574-4_28
15. Sankar, D., Thomas, T.: A new fast fractal modeling approach for the detection of microcalcifications in mammograms. J. Digit. Imaging 23(5), 538–546 (2010)
16. American Cancer Society: Cancer facts & figures. The Society (2008)
17. Sun, C., Paluri, M., Collobert, R., Nevatia, R., Bourdev, L.: ProNet: learning to propose object-specific boxes for cascaded neural networks. In: CVPR (2016)
18. Tabar, L., Yen, M.F., Vitak, B., Chen, H.H.T., Smith, R.A., Duffy, S.W.: Mammography service screening and mortality in breast cancer patients: 20-year follow-up before and after introduction of screening. Lancet 361(9367), 1405–1410 (2003)
19. Wei, Y., Feng, J., Liang, X., Cheng, M.M., Zhao, Y., Yan, S.: Object region mining with adversarial erasing: a simple classification to semantic segmentation approach. In: Proceedings of the IEEE Conference on Computer Vision and Pattern Recognition, pp. 1568–1576 (2017)
20. Zhang, F., et al.: Cascaded generative and discriminative learning for microcalcification detection in breast mammograms. In: Proceedings of the IEEE/CVF Conference on Computer Vision and Pattern Recognition, pp. 12578–12586 (2019)
21. Zhang, Z., et al.: Adaptive gaussian mixture model-based statistical feature extraction for computer-aided diagnosis of micro-calcification clusters in mammograms. SICE J. Control Meas. Syst. Integr. 13(4), 183–190 (2020)
22. Zhou, B., Khosla, A., Lapedriza, A., Oliva, A., Torralba, A.: Learning deep features for discriminative localization. In: Proceedings of the IEEE Conference on Computer Vision and Pattern Recognition, pp. 2921–2929 (2016)

Graph Transformers for Characterization and Interpretation of Surgical Margins

Amoon Jamzad[1][✉], Alice Santilli[1], Faranak Akbarifar[1], Martin Kaufmann[2], Kathryn Logan[3], Julie Wallis[3], Kevin Ren[3], Shaila Merchant[2], Jay Engel[2], Sonal Varma[3], Gabor Fichtinger[1], John Rudan[2], and Parvin Mousavi[1]

[1] School of Computing, Queen's University, Kingston, Canada
a.jamzad@queensu.ca
[2] Department of Surgery, Queen's University, Kingston, Canada
[3] Department of Pathology and Molecular Medicine, Queen's University, Kingston, Canada

Abstract. PURPOSE: Deployment of deep models for clinical decision making should not only provide predicted outcomes, but also insights on how decisions are made. Considering the interpretability of Transformer models, and the power of graph networks in analyzing the inherent hierarchy of biological signals, a combined approach would be the next generation solution in computer aided interventions. In this study, we propose a framework for classification and visualization of surgical mass spectrometry data using Graph Transformer model to empower the interpretability of breast surgical margin assessment. METHODS: Using the iKnife, 144 burns (103 normal, 41 cancer) were collected and converted to multi-level graph structures. A Graph Transformer model was modified to output the intermediate attention parameters of the network. Beside ablation and prospective study, we propose multiple attention visualization approaches to facilitate the interpretability. RESULTS: In a 4-fold cross validation experiment, an average classification AUC of 95.6% was achieved, outperforming baseline models. We could also distinguish and visualize clear pattern of attention difference between burns. For instance, cancerous and normal burns gather more attention in the lower and higher subbands of the spectra respectively. Looking at cancer subtype prospectively, a pattern of cancer progression was also observed in the attention features. CONCLUSION: Graph Transformers are powerful in providing high network interpretability. When paired with proper visualization, they can be deployed for computer assisted interventions.

Keywords: Graph networks · Graph transformers · Breast cancer · Attention · Interpretation

1 Introduction

A significant challenge in deploying deep learning models in computer assisted interventions is the need for methodology that supports interpretability of the

© Springer Nature Switzerland AG 2021
M. de Bruijne et al. (Eds.): MICCAI 2021, LNCS 12907, pp. 88–97, 2021.
https://doi.org/10.1007/978-3-030-87234-2_9

models. For clinical decision making in particular, it is key to not only provide predicted outcomes from models, but also insights as to how decisions are made. Concept activation vectors and class activation maps are among recently developed approaches to address this [7]. Another solution is Transformers, a powerful class of neural networks that yield state-of-the-art performance, while providing interpretability for the predictions [12]. Self-attention is the core component of Transformers where weights are calculated for different segments of input sequences and how they influence the final prediction. Although originally developed for text analysis [12], Transformers have recently gained attention in biological applications such as protein sequencing [8].

Biological systems have inherent hierarchy and inter-connectivity and lend themselves well to graph representation and analysis. With the rapid evolution of deep learning, Graph Neural Networks (GNN) have been applied for examination of biological images and signals. For certain applications such as analysis of brain connectivity [14], the translation of data into graphs is straightforward and intuitive while for medical imaging [13] and tissue characterization [1] a creative conversion step is required to represent the problem. Despite the power of GNNs, their predictions are not directly explainable due to the complex structure of data. Several methods like GNNExplainer [15] have been proposed recently to address this issue, but the field still lacks interpretable visualization of the results.

Considering the power of GNNs and interpretability of Transformers, a combined approach would be the next generation solution for biomedical problems. Very recently, graph transformer networks (GTN) have shown competitive performance against other attention-based GNNs on real-world and synthetic graph datasets [4]. Given the hierarchical and interconnected architecture of graphs and the interpretation abilities of GTNs, their application to intraoperative tissue characterization using chemical profiles of tissue (sequential and hierarchical in nature) could help detect residual cancer left on surgical margins. Chemical signatures of tissue can be produced intraoperatively using iKnife, a real-time mass spectrometry device that records metabolite information through contact of a cautery tip to tissue [2]. In general, mass spectra of tissue contain several classes of molecules within mass to charge ratio (m/z) subbands: fatty acids & small molecules m/z 50–500, phospholipids m/z 500–850, and triglycerides m/z 850–1000. Molecules that are specific to tissues or tumors have known signatures of mass spectra.

In this paper, we propose to extend the concept of Transformers and graph neural networks for surgical margins detection, and to better interpret the classification predictions. By representing spectrum subbands as nodes in a multi-level hierarchical structure, we generated graphs with rich node features corresponding to specific classes of molecules, and preserve critical positional information of subbands. The Graph Transformer network is modified to handle the rich node features and output the intermediate attention parameters in addition to conventional output for later visualization. We compare our results with three different graph and non-graph baseline deep models and utilize prospective data to test the model robustness. Beside novel graph conversion and state-of-the-art performance, we also propose multiple attention visualization approaches to

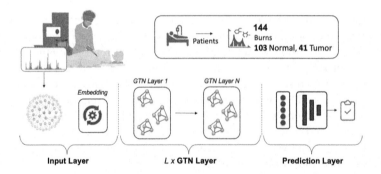

Fig. 1. Overview of our proposed approach. Data is collected from breast cancer patients with iKnife. Each mass spectra is converted into a graph. The data are used to train a network consisting of N graph transformer layers, and dense layers for graph classification.

facilitate the interpretability of the classification models for clinicians. By uncovering the attention of the network, we display their associated subbands in the spectra corresponding to biological molecules.

2   Materials and Methods

Figure 1 depicts the workflow of the proposed method. Each spectrum is collected through an iKnife burn from an excised specimen of a patient with breast cancer. The burns are each converted to graphs using a hierarchical approach [1]. These graphs are then passed through a deep graph network consisting of graph transformer layers (GTL), and dense layers for graph classification.

2.1   Data Collection and Curation

We collect iKnife data from both normal and cancer breast tissue. Fresh specimens are obtained from lumpectomy patients at our tertiary clinical center. To avoid compromising routine pathology due to destructiveness of iKnife, we are given sampling opportunities from excised tissue at homogeneous locations at the discretion of a pathologist, which is an ideal surrogate for true mixed-margin. Burns from non-cancerous tissues are made on the posterior and anterior distant margins of the specimen. The tissue is then breadloafed by the pathologist who indicates where tumor burns can be acquired. All burns are collected using a standard cautery tip from homogeneous cancer or benign tissue locations.

A total of 144 burns (41 tumor, 103 normal), accumulation of 2 years clinical acquisition, are stratified into 4 folds with similar cancer to normal ratio in each fold, and ensuring that all data from a patient stayed within the same fold. Each spectra (burn) is preprocessed through de-noising, lock mass correction, binning, and normalization; the subband of 100 to 1000 m/z in each burn is selected for further analysis. Following preprocessing, each data sample representing a burn has 900 intensity values (features) in the range of m/z 100–1000.

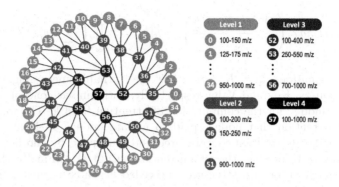

Fig. 2. Visualization of a 4-level hierarchical graph created from a sample spectra. At the lowest level (orange), the spectrum is divided into 50 m/z subbands with 50% overlap. In the next level, subbands of 100 m/z are max-pooled to preserve the number of features per node while increasing the width of the subband. Level 3 and 4 are created from subbands of 300 m/z and 900 m/z respectively (Color figure online).

2.2 Graph Conversion

In this study, a 4-level hierarchy with 50 features per node, and 50% intra-level subband overlap is considered as the structure for graph conversion. As demonstrated in Fig. 2, a sample burn is divided into 50 m/z subbands to generate nodes at the lowest level of a graph. In the next level, subbands of 100 m/z were max-pooled (size and stride of 2) to extract 50 features for each node at this level. To cover the whole spectrum, max-pooling with size and stride of 3 was used for 3rd and 4th levels which yields to nodes with subbands of 300 and 900 respectively. Intra-level edges connects neighbouring nodes with overlap within a level, while inter-level edges exist in consecutive levels if one node is the subset of the other [1]. At the end of this step, each spectrum is converted to a graph with 58 nodes and 135 edges.

2.3 Network Architecture

The network consists of initial embedding layer, GTLs, a node aggregation, dense layers, and the final predictor [4]. Let G be a graph with N nodes and the feature vector $a_i \in R^{d_n \times 1}$ for each node i. the initial embedding is a linear layer that projects the node features to a $d \times 1$ dimensional hidden space represented by h_i^0. Afterwards, in each GTL l, the features of node i are updated through attention mechanism based on the features of all of neighboring nodes, h_j, connected directly via edges e_{ij}. The overall structure of the graph is preserved and the output is still a graph with the same number of nodes and edges. This layer uses multiple attention "heads" (parallel attention mechanisms) to improve the stability. The update equation for node features at layer l is as follow:

$$w_{ij}^{k,l} = softmax_j \left(\frac{Q^{k,l} h_i^l \cdot K^{k,l} h_j^l}{\sqrt{d}} \right), \quad \hat{h}_i^{l+1} = O^l \left\|_{k=1}^{H} \left(\sum_{j \in N_i} w_{ij}^{k,l} V^{k,l} h_j^l \right) \right. \quad (1)$$

In these equations, $Q^{k,l}$, $K^{k,l}$, and $V^{k,l}$ are learnable linear weights at layer l denoting Query, Key and Value respectively, with similar concept as normal transformer model, for the computation of attention head k. The weights $w_{ij}^{k,l}$ denote the attention that node j paid to update node i. In Eq. 1, all of the H attentions heads are then concatenated and multiplied by another learnable linear weight O^l to produce the final attention output $\hat{h}_i^{l+1}$. Finally, this passed to a two-layer feed forward structure, succeeded by batch normalization layers and residual connections to generate the layer output h_i^{l+1}.

After the last GTL, feature vectors from all N nodes are aggregated through summation to generate a flat readout representation of the whole graph. This graph embedding can then be passed through feed forward structure with multiple dense layers to generate the prediction output and the loss to train the model parameters. We used weighted cross-entropy loss function since we have defined the tissue characterization problem as a graph classification task.

Based on the proposed conversion approach, our graphs are rich in node features and it is crucial to avoid training that is invariant to node position. Although the graph is symmetric, each node represents a specific subband that has different biological meaning. To help the model to learn both structural and positional information, the Laplacian eigenvectors from adjacency and degree matrices of the graph was also used at the input for positional encoding [4].

2.4   Experiments

Ablation Study: In the first step, we performed an ablation study of the structure parameters to establish the best architecture for following attention investigation. Using two folds for training and one for validation, we trained models with different sets of parameters and measure the test set performance. We investigated the number of graph transformer layers L in range of 1 to 4, number of attention heads H between 2 and 16 in steps of 2, and number of hidden features per attention heads d between 5 and 12. The number of dense layers in feed forward structure was fixed to 3 in the ablation study. The network performance of the winner model was then compared with 3 baseline models in a 4-fold cross-validation scheme: a 3-layer dense network, a non-graph convolutional neural network (CNN), and a non-transformer graph convolutional network (GCN) [5]. Adam optimizer with dynamic learning rate with initial value of 10^{-4}, batch size of 16, and early stopping on the validation loss, was used for all experiments.

Attention Interpretation: To explore the interpretability of the network, the weights $w_{ij}^{k,l}$ within the attention mechanism of the last GTL of a high performing model were extracted and visualized for the whole dataset. Although usually represented as edges, we also aggregated the attention weights at source

Table 1. The average (standard deviation) performance metrics of proposed graph transformer model in comparison with dense, CNN, and GCN baseline models in 4-fold cross validation scheme.

	Accuracy %	Sensitivity %	Specificity %	AUC%
Graph Transformer model	90.4 (6.3)	92.6 (6.4)	84.8 (15.1)	**95.6 (4.6)**
Dense model (baseline)	90.5 (4.2)	72.6 (18.2)	97.4 (4.0)	85.0 (9.0)
CNN model (baseline)	91.5 (4.3)	79.0 (16.1)	96.2 (4.4)	87.6 (7.5)
GCN model (baseline)	91.6 (4.0)	96.3 (4.0)	80.0 (11.7)	91.8 (7.4)

nodes to expand the visualization to nodes as well. Finally, in addition to attention comparison between cancer and normal labels (that were used for model training), we also visualized the network attention for cancer subtypes.

Prospective Validation: To show the robustness of the model, the performance of an ensemble model trained based on the mentioned data was calculated on a prospective test set of 4 patients with 46 normal and 10 cancer burns. We evaluate the significance of each level of the hierarchical graph for classification through ablation of low- and high-level nodes during test.

3   Results and Discussion

3.1   Ablation Study

We first perform an ablation study on three structural network parameters as described in the previous section. We observe that a higher number of attention heads and hidden features improves the performance. Furthermore, limiting the number of attention heads to two, results in low performance regardless of the number of hidden features. This was expected as in standard transformers, the use of more attention heads are reported to increase the training stability and robustness to random initialization [12]. We also note that as the number of GTLs increases, the overall AUC performance increases but plateaus after 3 GTLs. All models performed well with AUC 80%–98%, and the final selected structure consisted of 14 attention heads, 10 hidden features, and 3 GTLs, with 490K parameters. This configuration was used for further cross-validation analysis.

Table 1 summarizes the average and standard deviation of cross validated performance metrics of the proposed model in comparison with different baselines. We kept the number of layers in baselines equivalent to the selected GTN i.e. dense baseline with 3 layers, CNN baseline with 3-layer convolution followed by 3-layer dense, and GCN baseline with 3 graph convolutional layers followed by a 3-layer dense structure. Although we ran a separate ablation study for the baseline models, the results were almost in the same range. The proposed GTN (AUC = 95.6%) outperforms all 3 baselines statistically significantly (*p*-values of 0.005, 0.008, and 0.02 in one-tailed Wilcoxon Signed-Rank test, respectively).

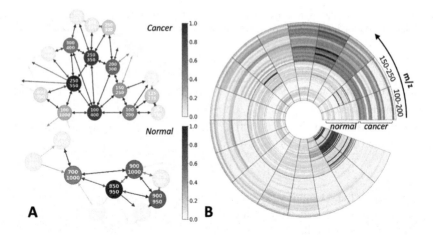

Fig. 3. Visualization of attention distribution in nodes and edges, extracted from the inner mechanism of the last GTL. **A:** Normalized attention for two sample burns from cancerous (top) and normal tissue (bottom) **B:** This cyclic visualization represents the node attention distribution in cancerous and healthy burns in 2nd hierarchical level of the graph structure. Each sector represents 100 m/z subband and each ring represents one spectrum. The cancerous burns are in red, and the normal ones in blue (Color figure online).

3.2   Attention Interpretation

As mentioned, the motivation for using graph transformer was to increase the interpretability of the results. It can give insight into the mass charge ratio subbands that are most important in differentiating between cancer and non cancer burns. In addition to contributing to our goal of real-time classification, understanding the importance of chemical bands can provide a degree of certainty for the user, and also help to link other properties, like cancer subtypes, back to ion groups that are related to certain metabolomic pathways [3]. A graph transformer network with mentioned parameters from the ablation study was trained and the attention weights for each pair of connected nodes were extracted for each graph in the dataset from the last GTL. The results are visualized in Fig. 3. Part (A) of the figure shows the normalized nodes and edges attention for two sample burns with cancer (top red) and normal (bottom blue) labels. Nodes/edges with higher attention appear darker in color and ones with near zero attention appear white. As can be seen in the figure, the network pays more attention to the subbands higher than 700 m/z at the end of the range for the normal burns, while for detecting cancer, the attentions are more distributed on the lower bands. We can also see that the nodes with the highest attention seem to be in level 2 and 3, as shown in the multi-level structure of Fig. 2.

A similar cancer attention pattern can be observed in part (B) of Fig. 3. To observe the normalized attention distribution at level 2, all burns were visualized in Fig. 3 part (B). In this figure, each sector represents one node (subband) and

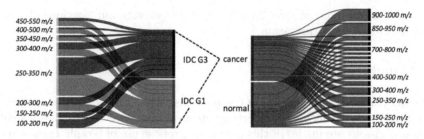

Fig. 4. Comparison of graph transformer attention distribution on the 2nd graph layer for: *Left:* low grade (G1) vs. high grade (G3) cancer and *Right:* cancer and normal burns of the whole dataset.

each ring represent one spectrum. The attention value is coded in the intensity of the label color (the darker the color, the higher the attention). We can see a general pattern emerge in this figure where the cancerous and normal burns gather more attention in the lower and higher subbands respectively. As also reported by other groups, normal breast tissue (adipose) is very fatty and known to have a high concentration of triglycerides, m/z 850–1000 [9,11], which is consistent with the attention findings here. Also, a study on breast data using DESI found that the lower subbands were important for differentiation between normal and cancerous tissue [10].

To further examine the patterns that exist within breast tumor burns, we regrouped all cancer burns by grade. In breast cancer, Grade 1 is associated with cells that tend to be slow growing and less likely to spread while Grade 3 cells are abnormal, grow quickly and more commonly spread. The left Sankey diagram in Fig. 4 displays the attention distribution of these two groups. We used the same trained model, extracted and grouped the attentions based on the the cancer subtype. The attention values were neglectable in higher subbands for both subtypes, as only the cancer burns were visualized. As can be seen, the attention paid to middle subbands are similar for both subtypes. However, low grade cancers received more attention at the higher subbands versus high grade received more lower band attentions. It can be hypothesised that the attention mechanism detecting the transition in cancer progression, without having a priori knowledge of the subtype labels during training.

3.3 Prospective Study

Finally, to confirm the robustness of the proposed workflow, we created an ensemble model from all variations of trained models in retrospective 4-fold analysis. The ensemble decision is made based on the average of all models' predictions. The AUC of 91.09% was achieved which proves the models robustness to prospective data. To further examine the importance of nodes in different levels in the performance, we also tested the ensemble model with input graphs with ablated nodes in lowest and highest levels. The achieved AUC of 83.26% demonstrated the higher importance of level 2 an 3, which agrees with our previous observations in attention interpretation.

4   Conclusion

For deployment of deep models in computer assisted interventions, it is crucial to not only provide accurate predicted outcomes from models, but also insights on how decisions are made. In this paper, we propose use of Graph Transformer networks to empower the interpretability of breast surgical margins assessment using iKnife modality. We capitalize on the attention capabilities of transformers and, by dividing our rich chemical information into nodes, are able to decipher the importance of certain subbands. The proposed visualization of mass spectrometry data provides higher level of interpretability which makes the research exploration more convenient. In breast cancer, where follow-up treatments are numerous, time consuming and often invasive, the correct treatment plan decision is not always trivial. It has been shown that chemical information about the tumor and tissue type may help provide insight to those decisions, helping patients get effective treatment faster [6]. In the future, it would be important to investigate other patient and tumor characteristics to uncover patterns in their chemical ranges. With this information, we would be able to gain further understanding of the process of carcinogenesis and its association with tissue chemical signatures.

References

1. Akbarifar, F., et al.: Graph-based analysis of mass spectrometry data for tissue characterization with application in basal cell carcinoma surgery. In: SPIE Medical Imaging: Image-Guided Procedures, Robotic Interventions, and Modeling, p. 11598 (2021)
2. Balog, J., et al.: Intraoperative tissue identification using rapid evaporative ionization mass spectrometry. Sci. Transl. Med. **5**(2), 194 (2013)
3. DeBerardinis, R., Lum, J., Hatzivassiliou, G., Thompson, C.: The biology of cancer: metabolic reprogramming fuels cell growth and proliferation. Cell Metab. **7**, 11–20 (2008)
4. Dwivedi, V., Bresson, X.: A generalization of transformer networks to graphs. In: AAAI 2021 Workshop on Deep Learning on Graphs. arXiv:2012.09699 (2021)
5. Kipf, T.N., Welling, M.: Semi-supervised classification with graph convolutional networks. In: 5th International Conference on Learning Representations, ICLR (2017)
6. Koundouros, N., et al.: Metabolic fingerprinting links oncogenic PIK3CA with enhanced arachidonic acid-derived eicosanoids. Cell **181**, 1596–1611 (2020)
7. Linardatos, P., Papastefanopoulos, V., Kotsiantis, S.: Explainable AI: a review of machine learning interpretability methods. Entropy (Basel, Switzerland) **23**(1), 18 (2020)
8. Nambiar, A., Liu, S., Hopkins, M., Heflin, M., Maslov, S., Ritz, A.: Transforming the language of life: transformer neural networks for protein prediction tasks. bioRxiv (2020). https://doi.org/10.1101/2020.06.15.153643

9. Santilli, A.M.L., et al.: Domain adaptation and self-supervised learning for surgical margin detection. Int. J. Comput. Assist. Radiol. Surg. **16**(5), 861–869 (2021). https://doi.org/10.1007/s11548-021-02381-6

10. Santoro, A., et al.: In situ DESI-MSI lipidomic profiles of breast cancer molecular subtypes and precursor lesions. Cancer Res. **80**, 1246–1257 (2020)

11. St-John, E., et al.: Diagnostic accuracy of intraoperative techniques for margin assessment in breast cancer surgery. Anal. Surg. **265**(2), 300–310 (2017)

12. Vaswani, A., et al.: Attention is all you need. In: NIPS. arXiv:1706.03762 (2017)

13. Wu, J., Zhong, J., Chen, E., Zhang, J., Ye, J., Yu, L.: Weakly- and semi-supervised graph CNN for identifying basal cell carcinoma on pathological images. Graph Learn. Med. Imaging **11849**, 112–119 (2019)

14. Yao, D., et al.: Triplet graph convolutional network for multi-scale analysis of functional connectivity using functional MRI. In: Zhang, D., Zhou, L., Jie, B., Liu, M. (eds.) GLMI 2019. LNCS, vol. 11849, pp. 70–78. Springer, Cham (2019). https://doi.org/10.1007/978-3-030-35817-4_9

15. Ying, Z., Bourgeois, D., You, J., Zitnik, M., Leskovec, J.: GNNExplainer: generating explanations for graph neural networks. In: Advances in Neural Information Processing Systems NeurIPS 2019, vol. 32, pp. 9240–9251. Curran Associates, Inc. (2019)

Domain Generalization for Mammography Detection via Multi-style and Multi-view Contrastive Learning

Zheren Li[1,2]🅾, Zhiming Cui[1,3], Sheng Wang[1,2], Yuji Qi[4], Xi Ouyang[1,2],
Qitian Chen[1], Yuezhi Yang[3], Zhong Xue[1], Dinggang Shen[1,5],
and Jie-Zhi Cheng[1(✉)]

[1] Shanghai United Imaging Intelligence Co., Ltd., Shanghai, China
`jzcheng@ntu.edu.tw`
[2] Institute for Medical Imaging Technology, School of Biomedical Engineering,
Shanghai Jiao Tong University, Shanghai, China
[3] Department of Computer Science, The University of Hong Kong,
Pok Fu Lam, Hong Kong
[4] Department of Biomedical Engineering, Yale University, New Haven, USA
[5] School of BME, ShanghaiTech University, Shanghai, China

Abstract. Lesion detection is a fundamental problem in the computer-aided diagnosis scheme for mammography. The advance of deep learning techniques have made a remarkable progress for this task, provided that the training data are large and sufficiently diverse in terms of image style and quality. In particular, the diversity of image style may be majorly attributed to the vendor factor. However, the collection of mammograms from vendors as many as possible is very expensive and sometimes impractical for laboratory-scale studies. Accordingly, to further augment the generalization capability of deep learning model to various vendors with limited resources, a new contrastive learning scheme is developed. Specifically, the backbone network is firstly trained with a multi-style and multi-view unsupervised self-learning scheme for the embedding of invariant features to various vendor-styles. Afterward, the backbone network is then recalibrated to the downstream task of lesion detection with the specific supervised learning. The proposed method is evaluated with mammograms from four vendors and one unseen public dataset. The experimental results suggest that our approach can effectively improve detection performance on both seen and unseen domains, and outperforms many state-of-the-art (SOTA) generalization methods.

Keywords: Domain generalization · Breast lesion detection · Contrastive learning

---

Z. Li and Z. Cui—Equal contribution.

Electronic supplementary material The online version of this chapter (https://doi.org/10.1007/978-3-030-87234-2_10) contains supplementary material, which is available to authorized users.

M. de Bruijne et al. (Eds.): MICCAI 2021, LNCS 12907, pp. 98–108, 2021.
https://doi.org/10.1007/978-3-030-87234-2_10

1 Introduction

The advance of deep learning (DL) techniques have remarkably improved the computer-aided detection (CADe) of breast lesions in mammography. It has been shown in many studies [10,12,15] that the incorporation of DL-based CADe software in the reading workflow of mammography can effectively improve the detection accuracy. Promising DL-based CADe performance requires large and diverse training data. In particular, the inclusion of wide variety of vendors is very important to equip the DL-based CADe system with prominent generalization capability. As shown in Fig. 1, the styles of images from various vendors vary significantly. Accordingly, the generalization of DL-based CADe system may be limited if the data of each vendor are not sufficiently included in the training stage. However, the collection of large and diverse data with various vendors is very expensive and is impractical. Meanwhile, it is also well-known that there exists domain gap between datasets from various institutes. Therefore, a domain generalization method is needed to alleviate the demands of large and diverse data from various vendors for the DL-based CADe scheme.

In the literature, the domain generalization for DL technique can be classified into three categories: 1) conventional data argumentation methods, e.g. rotation, flip, deformation and jittering [14,18,23], 2) learning based methods with the generative deep neural networks [6,19,21,22], 3) learning based methods for the exploration of task-specific and domain-invariant features [2,4,5,7–9,20]. However, all these learning based methods relied on annotated data, which are costly obtain and availability is relatively limited. A new method to automatically mine domain-invariant features from large unannotated data is very needed.

In this study, to address the above mentioned issues, we explore the contrastive learning technique to augment the generalization capability of DL-based CADe in self-supervised fashion. The contrastive learning has also been shown to generate better pretrained models for several medical image problems, e.g., diagnosis of chest radiography [1,16,25], dermatology images [1], etc., but less exploited for extracting domain invariant features. To specifically address the issue of vendor domain gap, we propose a multi-style and multi-view contrastive learning method to boost the lesion detection performance in mammography. Specifically, to attain the goal of generalization robustness to multiple vendor styles, the CycleGAN [26] technique is employed to generate multiple vendor-style images from a single vendor style image. The generated multiple style images from the same source are randomly paired as positive samples for the multi-style contrastive learning. Meanwhile, for the multi-view contrastive learning, the CC and MLO views of the same breast are also paired as positive samples. After self-supervised training, the backbone of the contrastive learning model is employed for the downstream detection task.

To our best knowledge, this is the first study that explores the contrastive learning to explicitly address the issue of vendor-style domain gap. We specifically train our method on three vendor domains, and then test on images from both seen and unseen domains. The results demonstrated that our method can effectively boost the lesion detection performance on either seen or unseen

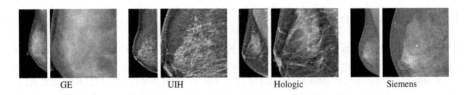

GE   UIH   Hologic   Siemens

Fig. 1. Style differences among four mammography vendors, i.e., GE, United Imaging Healthcare(UIH), Hologic, and Siemens. All images of the four vendors are MLO views. To illustrate the detailed image variation, a zoom-in image is provided on the right of each MLO view.

domain. We also compare our multi-style and multi-view contrastive learning to several state-of-the-art domain generalization methods and suggest our pretrained model can yield better generalization effect on both seen and unseen data.

2   Method

Figure 2 illustrates our domain generalization framework for lesion detection in mammography. We first conduct multi-style and multi-view contrastive learning scheme to extract domain invariant features on the data from different vendors. Afterward, the pretrained network backbone is transferred to the downstream task of lesion detection, especially on the data from unseen vendors.

2.1   Contrastive Learning Scheme

Contrastive learning is a self-supervised learning method, which trains an encoder to quantify the image representation into a proper vector space without the supervision of annotations. The derived model from contrastive learning can be served as a pretrained model for various downstream tasks like segmentation, detection, etc. The major advantage may lie in the flexibility to the downstream tasks.

The basic idea of contrastive learning is to bundle diversified images of the same class/object/subject as positive pairs for the exploration of reliable representation in feature space by the self-supervised learning manner. Specifically, given a mini-batch of N images, each example is randomly augmented twice with the diversifying operations, e.g., cropping, rotation, style-transferring, etc., to generate an augmented mini-batch with $2N$ samples. In the augmented mini-batch, two samples from the same image source are treated as a positive pair (i, j), whereas the other $2(N-1)$ samples within the mini-batch are regarded as negative pairs. With the positive and negative pairs, the contrastive learning is driven with the contrastive loss to maximize the agreement for the positive pairs. The contrastive loss is defined as:

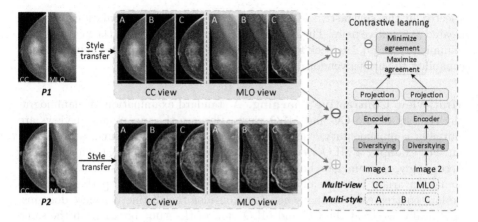

Fig. 2. Illustration of multi-style and multi-view contrastive learning.

$$\ell_{i,j} = -\log \frac{\exp(\mathrm{sim}(z_i, z_j)/\tau)}{\sum_{k=1}^{2N} \mathbb{1}_{[k \neq i]} \exp(\mathrm{sim}(z_i, z_k)/\tau)}, \tag{1}$$

where $\mathrm{sim}(\cdot)$ is the dot product and z refers to the extracted features. $\mathbb{1}_{[k \neq i]} \in \{0, 1\}$ is an indicator function equaling 1 when $k \neq i$ and τ is a temperature parameter.

Via maximizing the agreement for the positive pairs, the learnt features are supposed to be more general and robust to the diversifying operations. We further leverage the concept of contrastive learning to explore the generalization for various vendor styles and the domains of CC and MLO views. The details of multi-style and multi-view contrastive learning are elaborated in the following section.

2.2 Multi-style and Multi-view Contrastive Learning

Multi-style Contrastive Learning. Referring to Fig. 2, the image styles vary to different vendors and are regarded as distinctive domains. To endow the backbones of the lesion detectors better vendor-style generalization capability, we exploit the CycleGAN [26] technique as the diversifying operation for the contrastive learning. Specifically, given M seen vendor-style domain, we train $\binom{M}{2}$ generators, which map the data distribution of source domain Ω_i to target Ω_j domain, $\forall i, j \in M$. Comparing to the method used in [19], the CycleGAN realizes the style transfer with bidirectional learning process. The work [19] unidirectionally takes a few references images may attain limited transferring effect.

With the $\binom{M}{2}$ generators, each of the original N images in the training set can be diversified into $M - 1$ transferred images. Afterward, a mini-batch with $M \times N$ images can be formed. The positive pairs for the contrastive learning are constituted with any two images of the same image source in the original N image set. Therefore, there are possible $N \times \binom{M}{2}$ positive pairs that be randomly selected in the contrastive learning, whereas the other combination pairs are treated as negative pairs. The multi-style contrastive learning is then carried with minimizing the Eq. 1 to seek feature embedding space with better generalization capability to various vendor domains.

Multi-view Contrastive Learning. A standard examination of mammography consists of two CC and MLO views. Because the two standard views are mutually complementary, the appearance of CC and MLO images are different. For example, a MLO view includes axilla region, while a CC view doesn't. Accordingly, domain gap between these two views may exist. To seek view domain-invariant feature embedding for better lesion detection performance, we explore contrastive learning scheme to consider the distinctive view domains. Specifically, we treat the CC and MLO view of the same breast from the same patient as positive pair, whereas the other combination of CC and MLO is a negative pair. To further enrich the sample diversity, we further implement diversifying operations including random cropping, random rotation, horizontal flipping, and adjustment of brightness, contrast, and saturation for the CC and MLO in each positive or negative pair. With the prepared sample pairs, the multi-style contrastive learning can be carried out for the embedding of view-invariant features.

The synergy of the multi-style and multi-view contrastive learning is realized as illustrated in Fig. 2. Specifically, we firstly employ the style transferring CycleGAN to derive diversified vendor styles and perform the random diversifying operations on the CC and MLO views as the sample pairs for contrastive learning. We will show that our multi-style and multi-view contrastive learning can boost the lesion detection tasks effectively on either seen and unseen domains.

2.3 Lesion Detection Detection with Contrastive Learning

In this study, we employ the classic detection network of FCOS [17] to identify mass and clustered calcifications in mammography. The derived pretrained model from the contrastive learning is adopted as the backbone of the FCOS architecture for the realization of the downstream lesion detection tasks.

3   Experiments

Implementation Details. Our method was developed based on Pytorch package and trained with NIVIDA Titan RTX GPUs. We train CycleGAN [24,26] models with the settings of 50, 100, 200 epochs, and choose the setting, i.e., 100 epochs, that can achieve best lesion detection performance on the validation set. The backbone of the generator is ResNet with 9 blocks with 20 conv layers, while the discriminator is PatchGAN with 6 conv layers for binary classifications. In training, the sizes of inputs and outputs are 512 * 512 by random cropping the images. MSE and L1 losses are used for classification and reconstruction. For a style transfer, e.g., A to B, the involved training data from A and B are both 1000 for balanced training. Code and models of style transfer are available at: https://github.com/lizheren/MSVCL_MICCAI2021.

We adopt ResNet-50 as the backbone model for contrastive learning and FCOS detector. For fair comparison, the learning rate and batch size for all contrastive learning schemes are set the same as 0.3 and 256, respectively. Meanwhile, all contrastive learning schemes in all experiments use the same diversifying operations, including random cropping, random rotation in $\pm10°$, horizontal flipping and random color jittering (strength = 0.2). For the training of FCOS models, the SGD method is adopted with the parameters of learning rate, weight decay and momentum set as 0.005, 10^4, 0.9, respectively. The epoch and batch size are set to 50 and 8, respectively, throughout all experiments. Several augmentation methods, e.g. random flipping, scaling, etc., are also implemented in the training of FCOS.

3.1 Datasets

In this paper, mammograms of four vendors, i.e., GE, United Imaging Healthcare(UIH), Hologic, and Siemens, denoted as A, B, C, and D, respectively, are involved. All the data of the four vendors were collected from Asian women. The data of vendors A, B and C are set as seen domains whereas the vendor D is treated as unseen domain. To evaluate the generalization capability of our method, we also involve the public dataset, INbreast [13], denoted as E as unseen dataset. In total, this study involves 28,700 mammograms, where 27,000 unannotated images are used for style transfer and contrastive learning. The remaining 1,700 annotated images are adopted for the training/validation/testing of the detection tasks. We conduct a preprocessing step to align various mammograms from A, B, C, D, and E into the same pixel spacing of 0.1 mm to facilitate the training of deep learning methods. For the assessment of detector, we adopt the mean average precision(mAP) metrics for the quantitative comparison (Table 1).

Table 1. Details of usage for the data involved in this study. Specifically, the columns "Style Transfer" and "Self-Supervision" suggest the number of images involved in the training of CycleGAN and contrastive learning, respectively.

Domain	Dataset	Style transfer	Self-supervision	Detection (train/val/test)	Vendor
Seen	Style A	1000	8000	360/40/100	GE
	Style B	1000	8000	360/40/100	UIH
	Style C	1000	8000	360/40/100	HOLOGIC
Unseen	Style D	0	0	0/0/100	SIEMENS
	Style E	0	0	0/0/100	INbreast

3.2 Ablation Study

The ablation experiments are composed of two parts. First, we evaluate efficacy of original simple contrastive learning (SimCLR) [3] on the lesion detection tasks. Specifically, we compare lesion detection performance with 1) no pretraining, 2) ImageNet, 3) SimCLR on MammoPre, and 4) SimCLR on ImageNet → MammoPre pretrained models and report the corresponding performance in the first to fourth rows of Table 2. The third row suggests that SimCLR is trained from scratch with the unlabeled images of vendors A, B and C, where these unlabeled set is denoted as MammoPre. The fourth row ImageNet → MammoPre indicates that SimCLR is initialized with ImageNet parameters and then trained with the MammoPre set. In this part, the pretrained model of ImageNet → MammoPre can yield better detection performance.

The second part is to assess the effectiveness of each component of our framework. Specifically, we compare the pretrained models derived from multi-style contrastive learning (MSCL), multi-view contrastive learning (MVCL) and the combination of multi-style and multi-view contrastive learning (MSVCL) w.r.t. the lesion detection performances, which are listed in the fifth to seventh rows in Table 2, respectively. As can be found in Table 2, the effectiveness of our contrastive learning for the lesion detection tasks on both seen and unseen domains is corroborated. The detailed ablation analysis w.r.t. the mass and clustered calcifications detection tasks can be found in the Tables 1 and 2 of the supplement.

Table 2. Ablation analysis w.r.t. the pre-training strategy and our methods. The results on mAP of the mass and clustered calcifications detection tasks are reported. The performances w.r.t. the mass and clustered calcifications detection tasks are shown in the Table 1 and Table 2 of the Supplement, respectively.

Method	Pretrain strategy	Seen domain				Unseen domain		
		Style A	Style B	Style C	Avg.	Style D	Style E	Avg.
Random	None	0.427	0.563	0.487	0.492	0.345	0.525	0.435
Supervised	ImageNet	0.731	0.7805	0.728	0.746	0.673	0.811	0.742
SimCLR	MammoPre	0.740	0.772	0.742	0.751	0.665	0.794	0.729
SimCLR	ImageNet → MammoPre	**0.749**	**0.774**	**0.759**	**0.761**	**0.687**	**0.819**	**0.753**
MSCL	ImageNet → MammoPre	0.768	0.780	0.775	0.774	0.703	0.845	0.774
MVCL	ImageNet → MammoPre	0.756	0.785	0.771	0.771	0.706	0.830	0.768
MSVCL	ImageNet → MammoPre	**0.779**	**0.812**	**0.784**	**0.792**	**0.717**	**0.862**	**0.789**

3.3 Comparison with State-of-the-Art (SOTA) Methods

To further compare with other domain generalization methods, three SOTA methods, i.e., BigAug [23], Domain Diversification(DD) [6], and EISNet [20] are implemented. The BigAug [23] is a conventional data augmentation method, whereas the DD [6] proposed a generative learning method for domain generalization. The EISNet [20] is a learning-based method that explores task-specific and domain-invariant features. Our method on the other hand can decouple the downstream tasks intrinsically and provide task- and domain-invariant features. We trained all methods with the multi-view images from various vendors and explored the domain generalization correspondingly.

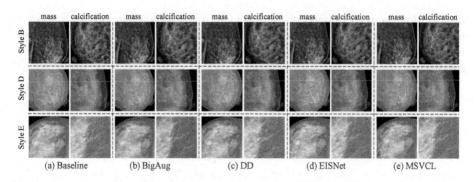

Fig. 3. Qualitative results for mass and clustered calcifications detection from different methods on one seen domain (style B) and two unseen domains (style D, E). The green boxes refer to the lesion regions labeled by the radiologist. The red boxes refer to the lesion regions detected by different methods. (Color figure online)

The comparison results are shown in Table 3, where the Baseline row suggests the result of SimCLR with ImageNet → MammoPre training strategy. As it can be found, our MSVCL can promise the best lesion detection performance on either seen or unseen domains. In particular, our method outperformed the EIS-Net by 1.8% on seen domains and 2.2% on unseen domains and data. Therefore, our MSVCL can be a new referential method for the pretraining of useful backbone for the downstream mammographic image analysis problems. The details performance comparison w.r.t. the mass and clustered calcifications detection tasks can be found in the Tables 3 and 4 of the supplement.

Table 3. Performance comparison between our MSVCL and other SOTA domain generalization methods. The results on mAP of the mass and clustered calcifications detection tasks are reported. The performances w.r.t. the mass and clustered calcifications detection tasks are shown in the Table 3 and Table 4 of the Supplement, respectively.

Method	Seen domain				Unseen domain		
	Style A	Style B	Style C	Avg.	Style D	Style E	Avg.
Baseline	0.749	0.774	0.759	0.761	0.687	0.819	0.753
BigAug [23]	0.756	0.792	0.743	0.763	0.692	0.829	0.760
DD [6]	0.754	0.792	0.770	0.772	0.695	0.821	0.758
EISNet [20]	0.765	0.781	0.775	0.774	0.690	0.845	0.767
MSVCL (ours)	**0.779**	**0.812**	**0.784**	**0.792**	**0.717**	**0.862**	**0.789**

To further visually illustrate the efficacy of our MSVCL pretrained model, the t-SNE [11] is employed to visualized data distribution of various vendor domains in the embedded feature space. Figure 4 compares the pretrained models of Baseline and MSVCL. As can be found, our MSVCL can better break the distribution boundaries between various vendor domains. Therefore, the encoded features shall be more style-invariant. We also demonstrate the detection results visually in the Fig. 3 and the style transfer results of CycleGAN in Fig. 1 of the supplement for more visual assessment.

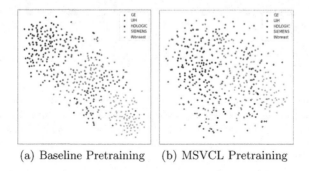

(a) Baseline Pretraining   (b) MSVCL Pretraining

Fig. 4. t-SNE visualization for the pre-trained models from Baseline and MSVCL.

4   Conclusion

A new domain generalization method is proposed to assist the lesion detection schemes in mammography. Specifically, we conduct multi-style and multi-view contrastive learning scheme to embed domain-invariant features to various vendors. The experimental results suggest that our domain generalization method can help to significantly improve the lesion detection tasks on both seen and unseen domains. In particular, for unseen domain of INbreast dataset, we also can achieve the best performance, compared to the three implemented SOTA domain generalization methods. Therefore, the efficacy of our method is corroborated.

References

1. Azizi, S., et al.: Big self-supervised models advance medical image classification. arXiv preprint arXiv:2101.05224 (2021)
2. Chen, N., et al.: Unsupervised learning of intrinsic structural representation points. In: Proceedings of the IEEE/CVF Conference on Computer Vision and Pattern Recognition, pp. 9121–9130 (2020)
3. Chen, T., Kornblith, S., Norouzi, M., Hinton, G.: A simple framework for contrastive learning of visual representations. In: International Conference on Machine Learning, pp. 1597–1607. PMLR (2020)
4. Cui, Z., et al.: Structure-driven unsupervised domain adaptation for cross-modality cardiac segmentation. IEEE Trans. Med. Imaging (2021)
5. Dou, Q., Castro, D.C., Kamnitsas, K., Glocker, B.: Domain generalization via model-agnostic learning of semantic features. arXiv preprint arXiv:1910.13580 (2019)

6. Kim, T., Jeong, M., Kim, S., Choi, S., Kim, C.: Diversify and match: a domain adaptive representation learning paradigm for object detection. In: Proceedings of the IEEE/CVF Conference on Computer Vision and Pattern Recognition, pp. 12456–12465 (2019)
7. Li, H., Pan, S.J., Wang, S., Kot, A.C.: Domain generalization with adversarial feature learning. In: Proceedings of the IEEE Conference on Computer Vision and Pattern Recognition, pp. 5400–5409 (2018)
8. Li, Y., et al.: Deep domain generalization via conditional invariant adversarial networks. In: Proceedings of the European Conference on Computer Vision (ECCV), pp. 624–639 (2018)
9. Liu, Q., Dou, Q., Heng, P.-A.: Shape-aware meta-learning for generalizing prostate MRI segmentation to unseen domains. In: Martel, A.L., et al. (eds.) MICCAI 2020. LNCS, vol. 12262, pp. 475–485. Springer, Cham (2020). https://doi.org/10.1007/978-3-030-59713-9_46
10. Lotter, W., et al.: Robust breast cancer detection in mammography and digital breast tomosynthesis using an annotation-efficient deep learning approach. Nat. Med. **27**(2), 244–249 (2021)
11. Van der Maaten, L., Hinton, G.: Visualizing data using t-SNE. J. Mach. Learn. Res. **9**(11) (2008)
12. McKinney, S.M., et al.: International evaluation of an AI system for breast cancer screening. Nature **577**(7788), 89–94 (2020)
13. Moreira, I.C., Amaral, I., Domingues, I., Cardoso, A., Cardoso, M.J., Cardoso, J.S.: INbreast: toward a full-field digital mammographic database. Acad. Radiol. **19**(2), 236–248 (2012)
14. Romera, E., Bergasa, L.M., Alvarez, J.M., Trivedi, M.: Train here, deploy there: robust segmentation in unseen domains. In: 2018 IEEE Intelligent Vehicles Symposium (IV), pp. 1828–1833. IEEE (2018)
15. Salim, M., et al.: External evaluation of 3 commercial artificial intelligence algorithms for independent assessment of screening mammograms. JAMA Oncol. **6**(10), 1581–1588 (2020)
16. Sowrirajan, H., Yang, J., Ng, A.Y., Rajpurkar, P.: MoCo pretraining improves representation and transferability of chest x-ray models. arXiv preprint arXiv:2010.05352 (2020)
17. Tian, Z., Shen, C., Chen, H., He, T.: FCOS: fully convolutional one-stage object detection. In: Proceedings of the IEEE/CVF International Conference on Computer Vision, pp. 9627–9636 (2019)
18. Volpi, R., Namkoong, H., Sener, O., Duchi, J.C., Murino, V., Savarese, S.: Generalizing to unseen domains via adversarial data augmentation. In: NeurIPS (2018)
19. Wang, S., et al.: mr^{2}NST: multi-resolution and multi-reference neural style transfer for mammography. In: Rekik, I., Adeli, E., Park, S.H., Valdés Hernández, M.C. (eds.) PRIME 2020. LNCS, vol. 12329, pp. 169–177. Springer, Cham (2020). https://doi.org/10.1007/978-3-030-59354-4_16
20. Wang, S., Yu, L., Li, C., Fu, C.-W., Heng, P.-A.: Learning from extrinsic and intrinsic supervisions for domain generalization. In: Vedaldi, A., Bischof, H., Brox, T., Frahm, J.-M. (eds.) ECCV 2020. LNCS, vol. 12354, pp. 159–176. Springer, Cham (2020). https://doi.org/10.1007/978-3-030-58545-7_10
21. Yue, X., Zhang, Y., Zhao, S., Sangiovanni-Vincentelli, A., Keutzer, K., Gong, B.: Domain randomization and pyramid consistency: simulation-to-real generalization without accessing target domain data. In: Proceedings of the IEEE/CVF International Conference on Computer Vision, pp. 2100–2110 (2019)

22. Zakharov, S., Kehl, W., Ilic, S.: DeceptionNet: network-driven domain randomization. In: Proceedings of the IEEE/CVF International Conference on Computer Vision, pp. 532–541 (2019)

23. Zhang, L., et al.: Generalizing deep learning for medical image segmentation to unseen domains via deep stacked transformation. IEEE Trans. Med. Imaging **39**(7), 2531–2540 (2020)

24. Zhang, Y., Miao, S., Mansi, T., Liao, R.: Task driven generative modeling for unsupervised domain adaptation: application to X-ray image segmentation. In: Frangi, A.F., Schnabel, J.A., Davatzikos, C., Alberola-López, C., Fichtinger, G. (eds.) MICCAI 2018. LNCS, vol. 11071, pp. 599–607. Springer, Cham (2018). https://doi.org/10.1007/978-3-030-00934-2_67

25. Zhou, H.-Y., Yu, S., Bian, C., Hu, Y., Ma, K., Zheng, Y.: Comparing to learn: surpassing ImageNet pretraining on radiographs by comparing image representations. In: Martel, A.L., et al. (eds.) MICCAI 2020. LNCS, vol. 12261, pp. 398–407. Springer, Cham (2020). https://doi.org/10.1007/978-3-030-59710-8_39

26. Zhu, J.Y., Park, T., Isola, P., Efros, A.A.: Unpaired image-to-image translation using cycle-consistent adversarial networks. In: Proceedings of the IEEE International Conference on Computer Vision, pp. 2223–2232 (2017)

Learned Super Resolution Ultrasound for Improved Breast Lesion Characterization

Or Bar-Shira[1]([✉]), Ahuva Grubstein[2,3], Yael Rapson[2,3], Dror Suhami[2,3], Eli Atar[2,3], Keren Peri-Hanania[1], Ronnie Rosen[1], and Yonina C. Eldar[1]

[1] Department of Computer Science and Applied Mathematics, Weizmann Institute of Science, Rehovot, Israel
or.barshira@weizmann.ac.il
[2] Radiology Department, Beilinson Campus, Rabin Medical Center, Petah Tikva, Israel
[3] Sackler Faculty of Medicine, Tel Aviv University, Tel Aviv, Israel

Abstract. Breast cancer is the most common malignancy in women. Mammographic findings such as microcalcifications and masses, as well as morphologic features of masses in sonographic scans, are the main diagnostic targets for tumor detection. However, improved specificity of these imaging modalities is required. A leading alternative target is neoangiogenesis. When pathological, it contributes to the development of numerous types of tumors, and the formation of metastases. Hence, demonstrating neoangiogenesis by visualization of the microvasculature may be of great importance. Super resolution ultrasound localization microscopy enables imaging of the microvasculature at the capillary level. Yet, challenges such as long reconstruction time, dependency on prior knowledge of the system Point Spread Function (PSF), and separability of the Ultrasound Contrast Agents (UCAs), need to be addressed for translation of super-resolution US into the clinic. In this work we use a deep neural network architecture that makes effective use of signal structure to address these challenges. We present in vivo human results of three different breast lesions acquired with a clinical US scanner. By leveraging our trained network, the microvasculature structure is recovered in a short time, without prior PSF knowledge, and without requiring separability of the UCAs. Each of the recoveries exhibits a different structure that corresponds with the known histological structure. This study demonstrates the feasibility of in vivo human super resolution, based on a clinical scanner, to increase US specificity for different breast lesions and promotes the use of US in the diagnosis of breast pathologies.

Keywords: Breast cancer · Super resolution ultrasound · Deep learning

1 Introduction

Breast cancer is the most commonly occurring cancer in women and the second most common cancer overall. Breast cancer diagnosis in its early stages plays a

© Springer Nature Switzerland AG 2021
M. de Bruijne et al. (Eds.): MICCAI 2021, LNCS 12907, pp. 109–118, 2021.
https://doi.org/10.1007/978-3-030-87234-2_11

critical role in patient survival. Mammographic findings such as microcalcifications and masses, as well as morphologic features of masses in sonographic scans, are the main diagnostic targets for tumor detection. However, there is a continued need to improve the sensitivity and specificity of these imaging modalities. A leading alternative target is neoangiogenesis. Neoangiogenesis is a process of development and growth of new capillary blood vessels from pre-existing vessels. When pathological, it contributes to the development of numerous types of tumors, and the formation of metastases [9,11,23]. Robust, precise, fast, and cost-effective in-vivo microvascular imaging can demonstrate the impaired or remodeled microvasculature, thus, it may be of great importance for early detection and clinical management of breast pathologies [20].

Diagnostic imaging plays a critical role in healthcare, serving as a fundamental asset for timely diagnosis, disease staging, and management as well as for treatment strategy and follow-up. Among the diagnostic imaging options, US imaging [4] is uniquely positioned, being a highly cost-effective modality that offers the clinician and the radiologist a high level of interaction, enabled by its real-time nature and portability [19]. The conventional US is limited in resolution by diffraction, hence, it does not resolve the microvascular architecture. Using encapsulated gas microbubbles with size similar to red blood cells as UCAs, extends the imaging capabilities of US, allowing imaging of fine vessels with low flow velocities. Specifically, Contrast-Enhanced US (CEUS) enables real-time hemodynamic and noninvasive perfusion measurements with high-penetration depth. However, as the spatial resolution of conventional CEUS imaging is bounded by diffraction, US measurements are still limited in their capability to resolve the microvasculature [1].

The diffraction limited spatial resolution was recently surpassed with the introduction of super resolution Ultrasound Localization Microscopy (ULM). This technology facilitated fine visualization and detailed assessment of capillary blood vessels. ULM relies on concepts borrowed from super-resolution fluorescence microscopy techniques such as Photo-Activated Localization Microscopy (PALM) and Stochastic Optical Reconstruction Microscopy (STORM) [2,18], which localize individual fluorescing molecules with subpixel precision over many frames and sum all localizations to produce a super-resolved image. In the ultrasonic version, CEUS is used [5,6], where the fluorescent beacons are replaced with UCAs which are scanned with an ultrasonic scanner. When the concentration of the UCAs is sparse, individual UCAs in each diffraction-limited ultrasound frame are resolvable. Thus, when the system PSF is known, localization with micrometric precision can be obtained for each UCA. As these contrast agents flow solely inside the blood vessels, the accumulation of these subwavelength localizations facilitates the recovery of a super-resolved map of the microvasculature [1,3].

Various works based on the above idea were illustrated in vitro, and in vivo on different animal models [3]. However, most of the super resolution US implementations to date are still limited to research ultrasound scanners with high framerate (HFR) imaging capability [13]. First in vivo human demonstrations using clinical scanners with low imaging frame-rates (<15 Hz) were shown for breast cancer [8,17], lower limb [12], and prostate cancer [14]. However, all methods rely on prior parameters calibration that characterize the system PSF to facilitate accurate identification of UCA signals. Further in human demonstrations were recently achieved for different internal organs and tumors [13]. Nevertheless, the processing was performed on in-phase/quadrature (IQ) data that was acquired with a high frame rate US scanner; both are not commonly available in clinical practice.

While super resolution ULM avoids the trade-off between resolution and penetration depth, it gives rise to a new trade-off that balances localization precision, UCA concentration, acquisition time, reconstruction time, and dependency on prior knowledge such as the PSF of the system. These challenges need to be addressed for translation of super-resolution US into the clinic where high UCAs concentrations, limited time, significant organ motion and lower frame-rate imaging are common [19].

In this work, we suggest a new approach to enable increased specificity in characterization of breast lesions by relying on a model-based convolutional neural network called deep unrolled ULM, suggested by van Sloun et al. [19]. Although the method was used before for an in vivo animal model with a high frame rate scanner, here it is used for the first time for in vivo human scans with a clinical US scanner operating at low frame rates. The network makes effective use of structural signal priors to perform localization microscopy in dense scenarios. Furthermore, no prior knowledge about the system PSF is required at inference. A learned PSF alleviates the dependency on the user experience thus making the process of super-resolution more accessible. We present preliminary in vivo human results on a standard clinical US scanner for three lesions in breasts of three patients. The results demonstrate a 31.25 μm spatial resolution. The three recoveries exhibit three different vasculature patterns, one for each lesion, that correspond with the histological known structure. To the best of our knowledge, this is the first in vivo human super resolution imaging leveraging deep learning using a standard clinical scanner with low frame rates to help differentiate between breast lesions. This study demonstrates the feasibility of a learning-based approach for in human super resolution, based on a clinical scanner, to increase the specificity of US for characterization of different lesion types and promotes the use of US in the diagnosis of breast pathologies.

2   Materials and Methods

2.1   Clinical Measurements

The clinical CEUS data were acquired at the Department of Radiology, Rabin Medical Center, Petah Tikva, Israel. The study was approved by the Helsinki committee of Rabin Medical Center, under number 0085-19-RMC. Written informed consent was obtained from all participants for CEUS imaging and the use of data for the study of improving US breast imaging. Twenty-one women aged 35–64 years with breast lesions were enrolled into this study. Measurement data of three different patients having three types of breast lesions were retrospectively evaluated. A fibroadenoma- a benign solid mass of the breast, A cyst- a benign fluid filled lesion, and an invasive ductal carcinoma- a malignant mass.

For the measurements, the patient was lying supine in a stable position. Each patient was intravenously administered 5 mL of contrast material containing 40 µL of sulphur hexafluoride microbubbles (SonoVue, Bracco, Milan, Italy), followed by a 5 mL saline flush. Real-time B-mode was used to guide the image plane and real time CEUS was used to monitor the UCAs signal right after the injection. The B-mode and CEUS images were saved for post-processing offline. During data acquisition, the patients and the dedicated breast radiologist were given oral and written instructions to be exceedingly stable to reduce out-of-plane motion.

2.2   Ultrasound Imaging Settings

The CEUS measurements were performed in a contrast specific mode to enhance sensitivity to UCAs while suppressing backscattering from tissue. A SL10-2 linear transducer (bandwidth 2–10 MHz) connected to a Hologic SuperSonic Mach 30 (SuperSonic Imagine, Aix-en-Provence, France) was used. The mechanical index during the examinations was 0.07. Both the B-mode images and contrast mode images were recorded with frame rate 25 Hz; 6286 frames were recorded for each measurement (about 4 min).

2.3   Image Preprocessing

All the following procedures were implemented in MATLAB 2019a (Mathworks, Natick, MA, USA).

The total number of detected vessels within the acquisition time is influenced by the flow-rate of UCAs in the vessels which depends on the blood flow in the vessels and on the UCA concentration in the blood [8]. To increase the number of blood vessels detected we looked at the time intensity curve (TIC) calculated as the mean intensity at each frame from the CEUS sequence and viewed frames after the maximum intensity was reached. This was used as an indication that the UCAs are in wash out phase.

To reduce out-of-plane motion, which can severely hinder recovery of the vasculature [8], we divided the measurements into subsequences of similar frames. This was achieved by computing the cross-correlation of the B-mode images across a manually selected region of interest (ROI) with sufficient contrast. Consecutive frames were assigned to the same subsequence if their cross-correlation was above 90%. We considered subsequences containing above 1000 frames. Small motions were corrected for the selected subsequences using image registration that accounts for translation. The transformation matrix was computed using the B-mode images, where the first frame of a sequence was the reference frame. Because the UCAs were not visible in the B-mode sequences, the motion estimation was not disturbed. A spatiotemporal singular-value-decomposition (SVD) based filter was applied to the CEUS frames to extract moving UCAs signals, which were then used for the super resolution recovery.

2.4 Localization of UCAs

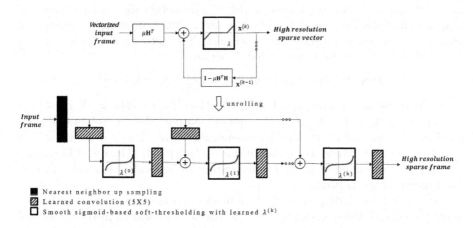

Fig. 1. Illustration of the deep unrolled ULM architecture. Top: block diagram of the ISTA algorithm; The block with the blue graph is the soft thresholding operator with parameter λ; The other blocks denote matrix multiplication (from the left side), where μ is a constant parameter that controls the step size of each iteration and H is a dictionary matrix based on the PSF. Bottom: deep unrolled ULM. Each iteration of the algorithm step is represented as one block of the network. Concatenating these blocks forms a deep neural network. In the unrolled network the parameters of the iterative algorithm are substituted with trainable parameters and convolutional filters. The network is trained through back propagation (Color figure online).

The following procedure was implemented in Python using the Tensorflow framework on a desktop with a 3.59 GHz AMD Ryzen 7 3700X 8-core processor and a NVIDIA GeForce GTX 1650 graphics card.

Detection was performed by applying deep unfolded ULM that makes effective use of sparsity to promote localization with dense concentration of UCAs. UCAs distribution within a frame is highly sparse on a high-resolution grid [21]. This implies that the support of the resolved image contains only a few non-zero values, allowing to pose the localization task as a sparse image recovery problem [1,22]. The architecture of the network, illustrated in Fig. 1, is devised by unrolling the Iterative Soft Threshold Algorithm (ISTA) [7], a popular method for sparse recovery, into a deep neural network where the parameters of the algorithm are replaced with learnable parameters and convolutional filters. Since each layer of the network assimilates an action from the algorithm, the unrolled network naturally inherits prior structures and domain knowledge, rather than learn them from intensive training data, prompting its generalization ability.

Training is performed in a supervised manner. To overcome the lack of sufficient amount of available data, the network is trained using on-line synthesized data of corresponding low-resolution inputs and high resolution targets. Specifically, the targets contain the most basic primitives of the CEUS image, point sources, on a high resolution grid, while the diffraction-limited input consists of point sources convolved with a PSF on a low resolution grid. The density, locations, intensities, background noise, and PSF parameters are randomly sampled from distributions that are defined by the user. Thus, robust inference under a wide variety of imaging conditions is achieved. The network is trained through loss minimization via the following loss function:

$$L(\mathbf{X}, \mathbf{Y} \mid \theta) = \|f(\mathbf{X} \mid \theta) - \mathbf{G} * \mathbf{Y}\|_2^2 + \lambda \|f(\mathbf{X} \mid \theta)\|_1^1 \qquad (1)$$

where $\mathbf{Y}$ is the target image containing the true UCA locations, $\mathbf{X}$ is the low resolution input, and $f(\mathbf{X} \mid \theta)$ is the network output. Here λ (set to 0.01) is a regularization parameter that promotes sparsity of the recovered image, and $\mathbf{G}$ is a Gaussian filter (standard deviation was set to 1 pixel). The use of $\mathbf{G}$ enables the network to be more forgiving to small errors in the localization and promotes convergence of the network.

We used a 9 block deep network, each block consists of the operations applied in one iteration of ISTA as shown in Fig. 1. The network was trained with ADAM optimizer ($\beta_1 = 0.9$, $\beta_2 = 0.999$, and an initial learning rate of 5e−4) [15]. The batch size and number of epochs were set to 64 and 1000 respectively.

3   Results

The cross correlation revealed different out-of-plane motions across the scans. Consequently, subsequences of different sizes were formed. For consistency, from each scan, we chose a subsequence of 1000 frames for further evaluation. Each of the chosen sequences were aligned and filtered to extract moving UCAs signals. Next, the filtered data was used as input to the network in order to recover the super resolved image.

Figure 2 presents the super resolution recoveries. The top row displays a fibroadenoma, the middle row displays a cyst, and the bottom row displays an invasive ductal carcinoma. The super resolution recoveries are shown together with the corresponding B-mode (standard) scans. The fibroadenoma recovery depicts an oval, well circumscribed mass with homogeneous high vascularization, the cyst displays a round structure with high concentration of blood vessels at the periphery of the lesion, and the malignant mass recovery displays irregular mass with ill-defined margins, high concentration of blood vessels at the periphery of the mass, and a hypoechoic region at the center of the mass. The correspondence between the structural characteristics of the recovered vasculature of the different lesions and their histological known structure was authenticated by dedicated breast radiologists.

Figure 3 compares between two super resolved images of the fibroadenoma. The left image was recovered with our method, while the right image was recovered with the classical ULM technique in which local maxima are computed for each frame. Since the computation is applied on the preprocessed data (as described in the previous section), the detected maxima are assumed to correspond to UCAs. Then, the intensity-weighted center of mass of each UCA signal is calculated to obtain coordinates of its localized position. Finally, all localizations are summed to obtain the final image. While both methods reveal similar patterns, a higher density of UCA localizations is observed in the left image via deep unrolled ULM.

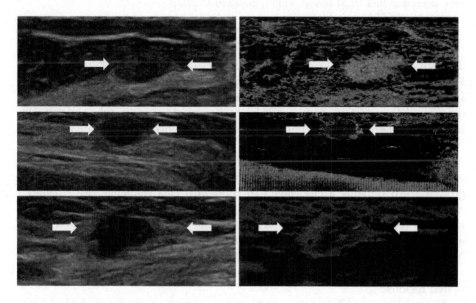

Fig. 2. Super resolution demonstrations in human scans of three lesions in breasts of three patients. Left: B-mode images. Right: super resolution recoveries. The white arrows point at the lesions; Top: fibroadenoma (benign). The super resolution recovery shows an oval, well circumscribed mass with homogeneous high vascularization. Middle: cyst (benign). The super resolution recovery shows a round structure with high concentration of blood vessels at the periphery of the lesion. Bottom: invasive ductal carcinoma (malignant). The super resolution recovery shows an irregular mass with ill-defined margins, high concentration of blood vessels at the periphery of the mass, and a low concentration of blood vessels at the center of the mass.

4 Discussion

This work attempts to enhance the diagnosis and monitoring of breast cancer using US imaging, in order to enable faster, safer, and more accessible treatment using a non-ionizing US device. We demonstrated the feasibility of implementing super resolution US in vivo in humans using a clinical US scanner and standard clinical procedure of contrast administration, promoting simple assimilation of

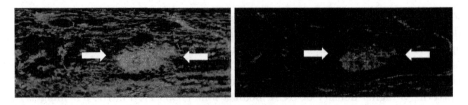

Fig. 3. Super resolution reconstructions of (the same) in human scan of fibroadenoma. Left: deep unrolled ULM (ours). Right: classical ULM.

the methods into clinical practice. We showed super resolution recoveries of three different breast lesions. The results successfully demonstrated different morphological features of the various lesions, assisting the differentiation between them. In clinical practice, the physicians distinguish between the lesions according to different characteristics, such as the lesion boundaries and its echogenicity [10]. Identifying these differences might be hard for the untrained eye as can be seen by viewing the B-mode images (Fig. 2, left column). Exploring the super resolution recoveries (Fig. 2, right column) reveals a different vascular profile for each lesion and improves lesion characterization using US. Still, the lack of ground truth remains a challenge. To cope with this challenge, we consulted with experts to confirm the histological reasoning of the microvascular profiles that were recovered. Furthermore, a non-learning-based method was used as reference. Both methods showed similar profiles further supporting the findings. Nonetheless, exploring Fig. 3 reveals that denser UCA localizations are achieved via the learning-based method (ours) which promotes recoveries in highly populated regions.

Super-resolution US imaging that leverages deep learning for clinical applications is an exciting opportunity, enabling to harness state of the art technology for noninvasive, robust, precise, and fast in-vivo microvascular imaging. Yet, in clinical application, the black box nature of deep neural networks can hinder the trust of experts in the recoveries. The technique of unrolling [16], used to devise the network architecture, facilitates a highly interpretable network whose reasoning can be easily reflected to the physician. Hence, allowing the physician a better control of the diagnostic process and a means to understand the origin of the artifacts if introduced by the algorithm and identify potential failure cases. While the use of deep learning enables to devise a tool that is fast and has a high generalization ability, the lack of a sufficient amount of in vivo data imposes a great challenge when training. To overcome this difficulty we synthesized data of the CEUS image basic building blocks. The synthesized data enables access to an abundant amount of data required for training, without dependency on prior in vivo data, while training on the image basic building blocks helps the network to avoid overfitting to specific vascular structures.

When tuning the pixel size, the clinical requirements must be considered. In this work, the recovery was performed using a high resolution grid with a 31.25 μm axial/lateral pixel resolution. This enabled the visualization of blood vessels with a resolution under 100 μm, corresponding to venules and arterioles. Recovery at the chosen resolution enabled to address the clinical need of increasing specificity of the lesions.

We hope that this research will contribute to the ongoing efforts for breast cancer early detection and treatment. The impact of this work can go beyond the field of breast imaging by enhancing the treatment capabilities of vasculature effected pathologies (e.g., enhancing the differentiation between inflammatory and fibrotic phases of disease, and thus effecting treatment choices) by using the proposed technology that makes effective use of clinical US scanners and an interpretable artificial intelligence framework.

Acknowledgements. The authors acknowledge the contribution of prof. Ruud van Sloun to the development of the algorithm used in this work.

References

1. Bar-Zion, A., Solomon, O., Tremblay-Darveau, C., Adam, D., Eldar, Y.C.: SUSHI: sparsity-based ultrasound super-resolution hemodynamic imaging. IEEE Trans. Ultrason. Ferroelectr. Freq. Control **65**(12), 2365–2380 (2018)
2. Betzig, E., et al.: Imaging intracellular fluorescent proteins at nanometer resolution. Science **313**(5793), 1642–1645 (2006)
3. Christensen-Jeffries, K., et al.: Super-resolution ultrasound imaging. Ultrasound Med. Biol. **46**(4), 865–891 (2020)
4. Cosgrove, D., Lassau, N.: Imaging of perfusion using ultrasound. Eur. J. Nucl. Med. Mol. Imaging **37**(1), 65–85 (2010)
5. Couture, O., Bannouf, S., Montaldo, G., Aubry, J.F., Fink, M., Tanter, M.: Ultrafast imaging of ultrasound contrast agents. Ultrasound Med. Biol. **35**(11), 1908–1916 (2009)
6. Couture, O., Fink, M., Tanter, M.: Ultrasound contrast plane wave imaging. IEEE Trans. Ultrason. Ferroelectr. Freq. Control **59**(12), 2676–2683 (2012)
7. Daubechies, I., Defrise, M., De Mol, C.: An iterative thresholding algorithm for linear inverse problems with a sparsity constraint. Commun. Pure Appl. Math.: J. Courant Inst. Math. Sci. **57**(11), 1413–1457 (2004)
8. Dencks, S., et al.: Clinical pilot application of super-resolution us imaging in breast cancer. IEEE Trans. Ultrason. Ferroelectr. Freq. Control **66**(3), 517–526 (2018)
9. Fox, S.B., Generali, D.G., Harris, A.L.: Breast tumour angiogenesis. Breast Cancer Res. **9**(6), 1–11 (2007)
10. Gokhale, S.: Ultrasound characterization of breast masses. Indian J. Radiol. Imaging **19**(3), 242 (2009)
11. Goussia, A., et al.: Associations of angiogenesis-related proteins with specific prognostic factors, breast cancer subtypes and survival outcome in early-stage breast cancer patients. A Hellenic Cooperative Oncology Group (HeCOG) trial. PLoS ONE **13**(7), e0200302 (2018)
12. Harput, S.: Two-stage motion correction for super-resolution ultrasound imaging in human lower limb. IEEE Trans. Ultrason. Ferroelectr. Freq. Control **65**(5), 803–814 (2018)
13. Huang, C., et al.: Super-resolution ultrasound localization microscopy based on a high frame-rate clinical ultrasound scanner: an in-human feasibility study. arXiv preprint arXiv:2009.13477 (2020)

14. Kanoulas, E., et al.: Super-resolution contrast-enhanced ultrasound methodology for the identification of in vivo vascular dynamics in 2D. Invest. Radiol. **54**(8), 500 (2019)
15. Kingma, D.P., Ba, J.: Adam: a method for stochastic optimization. arXiv preprint arXiv:1412.6980 (2014)
16. Monga, V., Li, Y., Eldar, Y.C.: Algorithm unrolling: interpretable, efficient deep learning for signal and image processing. IEEE Sig. Process. Mag. **38**(2), 18–44 (2021)
17. Opacic, T., et al.: Motion model ultrasound localization microscopy for preclinical and clinical multiparametric tumor characterization. Nat. Commun. **9**(1), 1–13 (2018)
18. Rust, M.J., Bates, M., Zhuang, X.: Sub-diffraction-limit imaging by stochastic optical reconstruction microscopy (STORM). Nat. Meth. **3**(10), 793–796 (2006)
19. van Sloun, R.J., Cohen, R., Eldar, Y.C.: Deep learning in ultrasound imaging. Proc. IEEE **108**(1), 11–29 (2019)
20. van Sloun, R.J., et al.: Super-resolution ultrasound localization microscopy through deep learning. IEEE Trans. Med. Imaging **40**, 829–839 (2020)
21. van Sloun, R.J., Solomon, O., Eldar, Y.C., Wijkstra, H., Mischi, M.: Sparsity-driven super-resolution in clinical contrast-enhanced ultrasound. In: 2017 IEEE International Ultrasonics Symposium (IUS), pp. 1–4. IEEE (2017)
22. Solomon, O., van Sloun, R.J., Wijkstra, H., Mischi, M., Eldar, Y.C.: Exploiting flow dynamics for superresolution in contrast-enhanced ultrasound. IEEE Trans. Ultrason. Ferroelectr. Freq. Control **66**(10), 1573–1586 (2019)
23. Toi, M., Inada, K., Suzuki, H., Tominaga, T.: Tumor angiogenesis in breast cancer: its importance as a prognostic indicator and the association with vascular endothelial growth factor expression. Breast Cancer Res. Treat. **36**(2), 193–204 (1995)

BI-RADS Classification of Calcification on Mammograms

Yanbo Zhang[1], Yuxing Tang[1], Zhenjie Cao[1], Mei Han[1], Jing Xiao[2], Jie Ma[3], and Peng Chang[1(✉)]

[1] PAII Inc., Palo Alto, USA
[2] Ping An Insurance (Group) Company of China, Shenzhen, China
[3] Department of Radiology, Shenzhen People's Hospital, Shenzhen, China

Abstract. Calcification is one of the most common and important lesions in mammograms, and a higher BI-RADS category indicates a higher cancer risk. In this paper, we present the first deep learning-based six-class BI-RADS classification for each individual calcification in mammograms. We propose an attention ROI generation strategy to highlight calcification features. Moreover, by incorporating malignancy information, the designed new loss function effectively boosts the performance of the model. We also design a novel evaluation metric for BI-RADS classification, which considers the severity of malignancy. Experimental results have demonstrated the superior classification performance of the proposed approach to the competing methods.

Keywords: Mammogram · Deep learning · BI-RADS · Calcification · Quadratic weighted kappa

1 Introduction

According to the World Health Organization, breast cancer became the most commonly diagnosed type of cancer worldwide, with 2.3 million cases in 2020, surpassing the number of new cases of lung cancer for the first time. Breast cancer now accounts for 11.7% of all new annual cancer cases globally [20]. Mammography is the best and widely used approach for early detection of breast cancer [4], with about 39 million mammograms performed each year in the United States.

Breast calcifications are common findings on mammograms. While most breast calcifications are benign (noncancerous), some calcification patterns, such as tight clusters with irregular shapes and fine appearance, may indicate breast cancer or precancerous changes in breast tissue. To assess the cancer risk and assure the imaging quality, the American College of Radiology has established the Breast Imaging-Reporting and Data System (BI-RADS) [1]. BI-RADS category or level reporting enables radiologists to standardize mammogram interpretation and demonstrates a close correlation with the risk of breast malignancy: 1-healthy, 2-benign, 3-probably benign, 4-suspicious abnormality, 5-highly suspicious of malignancy, 6-biopsy proven malignant.

© Springer Nature Switzerland AG 2021
M. de Bruijne et al. (Eds.): MICCAI 2021, LNCS 12907, pp. 119–128, 2021.
https://doi.org/10.1007/978-3-030-87234-2_12

Despite the recent advances in deep neural networks (DNN)-based approaches for computer-aided diagnosis/detection (CADx/CADe) in mammography [6, 11–14, 21–23], automated BI-RADS categorization of breast calcifications with robust performance remains a challenging issue, due to the fuzzy nature of the calcifications. Previous works either classified calcifications simply as benign or malignant [17], or classified them into incomplete BI-RADS categories. For example, Avalos-Rivera et al. [2] developed an artificial neural network to classify calcification ROIs (Region of Interest) into BI-RADS categories of 2, 3, and 4 (ignored 5), which does not meet the current BI-RADS standard. Some existing works have studied the BI-RADS categorization of whole mammographic images [7, 10, 18] instead of individual lesions or a certain lesion type (e.g., mass or calcification), which posed significant difficulties on the interpretability of the CADx/CADe system. Generally, radiologists evaluate each lesion in the breast and report the largest BI-RADS score for the breast. Hence, BI-RADS category prediction at the lesion-level is more advantageous to assist radiologists in clinical practice. Recently, there have been growing research interests on mass classification [5]. Yet hardly any work has been done for the classification of breast calcifications using the latest BI-RADS standard.

BI-RADS categorization is different from the traditional multi-class classification problem (e.g., natural object or lesion classification), in the sense that the BI-RADS levels are essentially a series of ordinal and discrete labels, which are inherently ordered according to the likelihood of malignancy. Directly applying multi-class classification models with cross-entropy loss that does not consider ordinal information is thus of inferior performance. In addition to the subjective BI-RADS labels assessed by the radiologists, binary biopsy results are considered to be the gold standard for lesion malignancy. Generally, these two sources of labels are highly correlated: biopsy-proven benign/malignant calcifications are expected to have smaller/larger BI-RADS scores.

In this work, we consider this label consistency and explicitly model it into our ordinal classification loss function, to penalize BI-RADS predictions that are inconsistent with biopsy results. More specifically, we design a malignancy adjusted, weighted BI-RADS classification loss that penalizes heavily for inconsistent predictions (e.g., predicting BI-RADS 5 for a biopsy-proven benign lesion, or BI-RADS 2 for a malignant lesion) and lightly for consistent predictions (e.g., predicting BI-RADS 2 for a benign lesion and 5 for a malignant lesion). Besides, since there might be multiple calcifications in an image patch, we design an attention mechanism to pre-process the input patch such that the network could focus on the specific calcification region to extract more discriminative features. To evaluate the performance of the proposed approach, we also propose a new metric that considers the malignancy of the calcification for a fair evaluation.

The main contribution of this work is three-fold: 1) we develop the first six-class BI-RADS classification algorithm for each calcification in mammograms, and 2) we design a specific attention-based ROIs and the malignancy adjusted loss that effectively boosts the feature learning and model optimization; 3) we introduce a novel metric, malignancy adjusted quadratic weighted

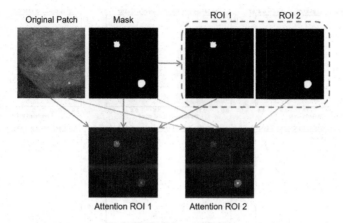

Fig. 1. The generation of calcification attention ROIs.

Kappa (MAQWK), which is a general evaluation metric for rating tasks that can be used in the medical imaging domain.

2 Methods

2.1 Method Overview

We propose a six-class (BI-RADS 2, 3, 4A, 4B, 4C, and 5) classification method for each individual calcification in mammograms. We first obtain calcification masks and BI-RADS categories annotated by radiologists, together with biopsy results. Image patches containing calcifications are then extracted, and we use masks to highlight the calcifications. In this way, the calcification patches with attention ROIs are generated as input data as described in Sect. 2.2. With malignancy information, we introduce a weighting coefficient to the loss function as derived in Sect. 2.3. The malignancy adjusted loss makes predicted BI-RADS more acceptable in practical applications. Finally, we design a novel evaluation metric for medical imaging rating tasks in Sect. 2.4.

2.2 Attention-Based Pre-processing

Figure 1 shows the generation of calcification attention ROI patches. For an image patch extracted from the original image and the corresponding binary mask, we aim to predict the BI-RADS category for each individual calcification. As shown in Fig. 1, an ROI mask for each calcification is split from the mask of each patch. Each generated attention ROI consists of three channels: the first channel is the original image patch, providing all breast tissue information within this patch; the second channel is the element-wise product of the original patch and the mask, highlighting all the calcifications and their spatial relations; and the third channel is the element-wise product of the original patch and one ROI

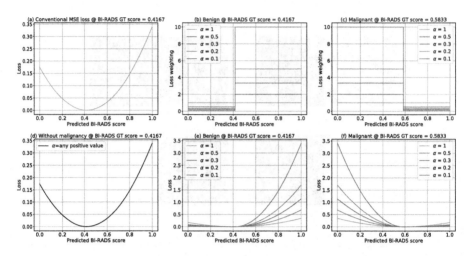

Fig. 2. Malignancy adjusted BI-RADS loss. (a) is the conventional MSE loss, (b) and (c) present the loss weighting of benign and malignant cases, respectively, and (e) and (f) are the corresponding loss curves. For the case without malignancy, the loss curve in (d) is identical to (a).

mask, focusing on the calcification to be classified. The attention-based ROIs with such a design contain richer information from various perspectives, therefore, the features extracted from these patches are more meaningful spatially to enhance the classification performance.

We treat the classification for ratings as a regression task. The inputs to the classification model are the generated attention ROI patches, and the output is a scalar between $[0, 1]$, representing the approximate risk probability. Consequently, this scalar is linearly mapped to predict a BI-RADS category.

2.3 Malignancy Adjusted Loss

According to the BI-RADS standard, score 1 means no lesions, 0 indicates incomplete information and requires follow-up, and 6 represents biopsy-proven breast cancer. We exclude the above three clearly defined BI-RADS categories and only consider the remaining six BI-RADS categories: 2, 3, 4A, 4B, 4C, and 5, for calcification classification as a rating problem. Let the number of calcification classes be N^1, and the cancer risk consistently increases from the first class to the Nth class. Thus, we convert the classification task into a linear regression task. We uniformly divide the risk range $[0, 1]$ into N segments, thus the value range and center for the nth class range are $[(n-1)/N, n/N]$ and $(2n-1)/2N$.

---

[1] Previous BI-RADS standard utilizes a single category 4 instead of subcategories 4A, 4B, 4C, hence the number of calcification classes reduces to $N = 4$.

The regression loss using the conventional Mean Squared Error (MSE) is defined as:

$$Loss(b^{gt}, b^{pred}) = \left(b^{gt} - b^{pred}\right)^2, \tag{1}$$

where b^{gt} and b^{pred} denote the BI-RADS risk score of ground truth and prediction, respectively, and b^{gt} is the center of the ground truth segment. We use the following formula to measure if the bias direction of the predicted BI-RADS score is consistent with the biopsy malignancy:

$$C(b^{gt}, b^{pred}, m^{gt}) = -\left(b^{gt} - b^{pred}\right)\left(m^{gt} - m^{mid}\right), \tag{2}$$

where m^{gt} is the binary biopsy malignancy, with 0 and 1 indicate benign and malignant, respectively, m^{mid} is an auxiliary constant, which is set to be 0.5. If a mammogram does not have a biopsy result, m^{gt} is assigned with the same value as m^{mid}. A positive $C(\cdot)$ indicates consistency and a negative $C(\cdot)$ represents inconsistency between the BI-RADS prediction and the biopsy label, while $C(\cdot) = 0$ indicates unavailable biopsy malignancy or an accurate BI-RADS prediction. We give a higher penalty to the inconsistent case and a smaller one to the consistent case. Therefore, we propose a malignancy-adjusted loss weight:

$$W(b^{gt}, b^{pred}, m^{gt}) = \alpha^{\frac{C(b^{gt}, b^{pred}, m^{gt})}{|C(b^{gt}, b^{pred}, m^{gt})| + \epsilon}}, \tag{3}$$

where α is a weighting coefficient with the value range $(0, 1]$. ϵ is a very small positive constant to ensure that $W(\cdot)$ in Eq. 3 is continuous and differentiable.

The final weighted loss is denoted as:

$$Loss(b^{gt}, b^{pred}, m^{gt}) = \alpha^{\frac{C(b^{gt}, b^{pred}, m^{gt})}{|C(b^{gt}, b^{pred}, m^{gt})| + \epsilon}} \left(b^{gt} - b^{pred}\right)^2. \tag{4}$$

Figure 2 shows some examples of the proposed malignancy-adjusted MSE loss for BI-RADS classification. We set loss weighting parameter α to 0.2 as default.

2.4   Evaluation Metric

Quadratic weighted kappa (QWK) measures the agreement between two ratings [19]. This metric typically varies from 0 (random agreement between raters) to 1 (complete agreement between raters). A negative value means the classifier performs worse than random choice. The quadratic weighted kappa is calculated between the scores assigned by the human rater and the predicted scores. QWK is defined as:

$$k = 1 - \frac{\sum_{i,j} \mathbf{W}_{i,j} \mathbf{O}_{i,j}}{\sum_{i,j} \mathbf{W}_{i,j} \mathbf{E}_{i,j}}, \tag{5}$$

where $\mathbf{W}$ is a $N \times N$ matrix and its element $\mathbf{W}_{i,j} = (i - j)^2$ is the weighted cost associated with misclassifying label i as label j, and matrices $\mathbf{O}$ and $\mathbf{E}$ are the confusion matrix and the expected rating matrix [10], respectively.

Although QWK has been commonly used in medical image rating tasks, it is still not a proper metric for malignancy rating. As the examples presented

Table 1. A comparison of evaluation metrics on two classifiers. Consider an example list of six samples with various BI-RADS categories, all samples are correctly predicted except for the one with BI-RADS 4A. Both calcification example cases 1 and 2 are with the BI-RADS category 4A, and final confirmed as malignant and benign, respectively.

Cases	Classifiers	Predicted BI-RADS	QWK	MAQWK	
				MAQWK-M	MAQWK-B
Case 1 (4A, Malignant)	Classifier 1	2	0.9024	0.8356	–
	Classifier 2	4C	0.8919	**1.0**	–
Case 2 (4A, Benign)	Classifier 1	4C	0.8919	–	0.8153
	Classifier 2	2	0.9024	–	**1.0**

in Table 1, the classifiers 1 and 2 predict case 1 (malignant, BI-RADS 4A) as BI-RADS 2 and 4C. Because these two predictions have the same distance to the ground truth, they have very close QWK values. However, in clinical settings, BI-RADS 2 is considered as benign and patients with BI-RADS 4C are required to conduct further actions to confirm its malignancy. Thus, classifier 1 may result in missed diagnosis for case 1. Likewise, classifier 1 may also lead to a false alarm for case 2. Although classifier 2 performs much better than classifier 1, they have extremely close QWK values.

For a more appropriate evaluation, we propose a malignancy-adjusted quadratic weighted kappa (MAQWK) and define the malignancy adjusted weighting matrix as:

$$\mathbf{W}^c = \mathbf{W} * \mathbf{I}^{u/l}, \tag{6}$$

where "c" is an indicator of either "m" or "b", indicating malignant/benign case, respectively. $\mathbf{W}^c$ represents the designed weighting matrix, $\mathbf{I}^{u/l}$ is an unit upper triangular matrix $\mathbf{I}^u$ for malignant or an unit lower triangular matrix $\mathbf{I}^l$ for benign. The matrices $\mathbf{I}^u$ and $\mathbf{I}^l$ find out the elements of higher/lower predicted BI-RADS cases in the confusion matrix, respectively. The operator $*$ means the element-wise product, therefore, $\mathbf{W}^c$ is either an upper or lower triangular part of the original quadratic weight matrix $\mathbf{W}$. In this way, $\mathbf{W}^m$ considers only the cases where the predicted BI-RADS is lower than the ground truth for malignant cases, and $\mathbf{W}^b$ considers only the cases where the predicted BI-RADS is higher than the ground truth for benign cases. MAQWK is defined as:

$$k^c = 1 - \frac{\sum_{i,j} \mathbf{W}^c_{i,j} \mathbf{O}^c_{i,j}}{\sum_{i,j} \mathbf{W}^c_{i,j} \mathbf{E}^c_{i,j}}, \tag{7}$$

where $\mathbf{O}^c$ represents either $\mathbf{O}^m$, which is the confusion matrix computed with only malignant cases, or $\mathbf{O}^b$, which is the counterpart of benign cases. Likewise, $\mathbf{E}^m$ and $\mathbf{E}^b$ are corresponding expected matrices for malignant and benign cases, respectively. We denote k^m and k^b as the malignancy adjusted quadratic

Table 2. Number of calcification in each BI-RADS categories.

BI-RADS	2	3	4A	4B	4C	5	**Total**
Number	3180	1751	323	241	778	902	**7175**

weighted kappa for malignant (MAQWK-M) and benign (MAQWK-B), respectively. Table 1 shows that classifier 2 achieves higher values than classifier 1 in terms of both MAQWK-M and MAQWK-B, implying a more reasonable metric for this task.

3 Experimental Setting

3.1 Datasets

We collaborated with a hospital to build a dataset for this task[2]. Mammograms were collected with two vendors' digital mammography machines, the SIEMENS Mammomat Inspiration (Germany) and the GIOTTO Image MD (Italy). Two radiologists delineated calcification regions and labeled the corresponding BI-RADS categories, before being finally checked by an experienced radiologist. The biopsy results indicating benign or malignant were confirmed by histopathology. The dataset consists of 708 patients with 1776 mammograms containing calcifications. There were 2731 malignant and 5426 benign calcifications, and the numbers of BI-RADS categories are listed in Table 2, where BI-RADS 0 and 6 are excluded. The patients were randomly split by 3:1:1 as the training, validation and test sets. A sliding window moved in mammograms with a step of 100 pixels to extract the image and mask patches, each patch was 400×400 pixels.

3.2 Implementation Details

In this work, we use the ResNet-18 [9] as our DNN backbone (pre-trained on ImageNet). First, an attention ROI is created from original image and mask as the input of backbone, and the output of backbone is a cancer risk score. In the training stage, the obtained risk score is directly used for computing the loss. In the inference stage, if this score falls in the range $[(n-1)/6, n/6]$, then the predicted BI-RADS category is the nth category of the list [2, 3, 4A, 4B, 4C, 5].

We applied SGD with momentum as the optimizer for training. The initial learning rate was 0.001 and decreases by 0.3 every 10 epochs, with momentum as 0.99 and batch size as 32. A sampling strategy was used to balance the number of various BI-RADS classes in each mini-batch. The training stopped after 50 epochs. The method was implemented on a Linux workstation with two NVIDIA V100 GPUs (16G memory each).

---

[2] This retrospective case-control study was approved by the ethics review and institutional review board, which waived the requirement for individual informed consent.

Table 3. Performance comparison of various methods: standard multi-class (MC) classification, the regression-based method with MSE loss and the proposed MAMSE loss, and the impact of attention ROIs.

Method	QWK	MAQWK	
		MAQWK-M	MAQWK-B
MC	0.6583	0.1121	−0.0027
MSE	0.6877	0.1544	0.0302
MAMSE	0.7657	0.1468	0.1054
MC + attention ROI	0.8216	0.2517	0.2493
MSE + attention ROI	0.8650	0.2858	0.3242
MAMSE + attention ROI (Proposed)	**0.8870**	**0.3489**	**0.3786**

4   Experimental Results

Table 3 presents the performance comparison of standard multi-class (MC) classification, the regression-based method with standard MSE loss and with the proposed MAMSE loss. To show the impact of the proposed attention ROIs, we also compare them with the original gray image patches as the input for all models. For a fair comparison, all methods used the same backbone and training settings. It is clear that the introduced attention strategy remarkably promotes classification performance in terms of QWK and MAQWK for all methods. Moreover, the designed specific loss can further improve the performance, achieving the best results.

Figure 3 shows the performance of the proposed method with different choices of loss weighting parameter α. The proposed approach is able to obtain relatively stable high performance when α is between 0.2 and 0.5. QWK and MAQWK fluctuate when α is approaching 0, and both decline steadily with the increase of α toward 1. The proposed method degrades to conventional MSE-based regression when α is 1.

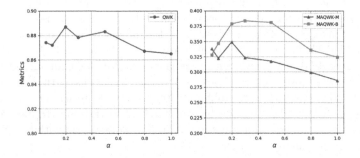

Fig. 3. Plots of metrics of the proposed method with different choices of α.

5   Discussion and Conclusion

The core contributions of the proposed model are the designed loss and attention mechanism, which are independent of architecture. Hence, without loss of generality, we only used ResNet as the backbone to evaluate the efficacy of our method in this work. Other advanced architectures can be directly employed as a backbone, and similar advantages of the proposed modules can be expected.

The designed malignancy adjusted loss and the corresponding evaluation metrics consider the practical needs in clinical settings. They can be easily applied to other medical image rating tasks, such as for liver (LI-RADS) [15], gynecology (GI-RADS) [3], colonography (C-RADS) [16], diabetic retinopathy diagnosis [8], etc. The incorporation of the malignancy-based adjustment is an intuitive and interpretable way to transfer the domain knowledge and special request from the professionals to the DL framework. The loss is also flexible, accommodating the samples without malignancy information by assigning the pseudo label m^{mid}.

In this paper, we proposed a specific BI-RADS classification method and an evaluation metric for mammographic calcifications. The developed attention strategy and malignancy adjusted MSE loss effectively improve the classification performance. It shows great potential to be expanded to other tasks and domains. We have invited the radiologists from our collaborating hospitals to further evaluate its practical efficacy and other use cases.

References

1. American College of Radiology: ACR BI-RADS Atlas: Breast Imaging Reporting and Data System; Mammography, Ultrasound, Magnetic Resonance Imaging, Follow-up and Outcome Monitoring. Data Dictionary. ACR, American College of Radiology (2013)
2. Avalos-Rivera, E.D., Pastrana-Palma, A.: Classifying region of interests from mammograms with breast cancer into BIRADS using artificial neural networks. Adv. Sci. Technol. Eng. Syst. J. **2**(3), 233–240 (2017)
3. Basha, M.A.A., et al.: Gynecology imaging reporting and data system (GI-RADS): diagnostic performance and inter-reviewer agreement. Eur. Radiol. **29**(11), 5981–5990 (2019)
4. Bleyer, A., Welch, H.G.: Effect of three decades of screening mammography on breast-cancer incidence. N. Engl. J. Med. **367**(21), 1998–2005 (2012)
5. Boumaraf, S., Liu, X., Ferkous, C., Ma, X.: A new computer-aided diagnosis system with modified genetic feature selection for BI-RADS classification of breast masses in mammograms. BioMed Res. Int. **2020**, 17 (2020)
6. Sainz de Cea, M.V., Diedrich, K., Bakalo, R., Ness, L., Richmond, D.: Multi-task learning for detection and classification of cancer in screening mammography. In: Martel, A.L., et al. (eds.) MICCAI 2020. LNCS, vol. 12266, pp. 241–250. Springer, Cham (2020). https://doi.org/10.1007/978-3-030-59725-2_24
7. Geras, K.J., et al.: High-resolution breast cancer screening with multi-view deep convolutional neural networks. arXiv preprint arXiv:1703.07047 (2017)

8. Gulshan, V., et al.: Development and validation of a deep learning algorithm for detection of diabetic retinopathy in retinal fundus photographs. JAMA **316**(22), 2402–2410 (2016)
9. He, K., Zhang, X., Ren, S., Sun, J.: Deep residual learning for image recognition. In: Proceedings of the IEEE Conference on Computer Vision and Pattern Recognition, pp. 770–778 (2016)
10. Liu, X., et al.: Unimodal regularized neuron stick-breaking for ordinal classification. Neurocomputing **388**, 34–44 (2020)
11. Liu, Y., Zhang, F., Zhang, Q., Wang, S., Wang, Y., Yu, Y.: Cross-view correspondence reasoning based on bipartite graph convolutional network for mammogram mass detection. In: Proceedings of the IEEE/CVF Conference on Computer Vision and Pattern Recognition, pp. 3812–3822 (2020)
12. Liu, Y., et al.: From unilateral to bilateral learning: detecting mammogram masses with contrasted bilateral network. In: Shen, D., et al. (eds.) MICCAI 2019. LNCS, vol. 11769, pp. 477–485. Springer, Cham (2019). https://doi.org/10.1007/978-3-030-32226-7_53
13. Lotter, W., et al.: Robust breast cancer detection in mammography and digital breast tomosynthesis using an annotation-efficient deep learning approach. Nature Med. **27**, 1–6 (2021)
14. McKinney, S., et al.: International evaluation of an AI system for breast cancer screening. Nature **577**, 89–94 (2020)
15. Mitchell, D.G., Bruix, J., Sherman, M., Sirlin, C.B.: LI-RADS (liver imaging reporting and data system): summary, discussion, and consensus of the LI-RADS management working group and future directions. Hepatology **61**(3), 1056–1065 (2015)
16. Pooler, B.D., Kim, D.H., Lam, V.P., Burnside, E.S., Pickhardt, P.J.: Ct colonography reporting and data system (C-RADS): benchmark values from a clinical screening program. Am. J. Roentgenol. **202**(6), 1232–1237 (2014)
17. Shen, L., Rangayyan, R.M., Desautels, J.L.: Detection and classification of mammographic calcifications. Int. J. Pattern Recognit Artif Intell. **7**(06), 1403–1416 (1993)
18. Vanderheyden, R., Xie, Y.: Mammography image BI-RADS classification using ohplall. In: 2020 IEEE Sixth International Conference on Big Data Computing Service and Applications (BigDataService), pp. 120–127. IEEE (2020)
19. Warrens, M.J.: Some paradoxical results for the quadratically weighted kappa. Psychometrika **77**(2), 315–323 (2012)
20. World health Organization: Breast cancer now most common form of cancer: WHO taking action. https://www.who.int/news/item/03-02-2021-breast-cancer-now-most-common-form-of-cancer-who-taking-action
21. Wu, N., et al.: Deep neural networks improve radiologists' performance in breast cancer screening. IEEE Trans. Med. Imaging **39**(4), 1184–1194 (2019)
22. Yang, Z., et al.: MommiNet: mammographic multi-view mass identification networks. In: Martel, A.L., et al. (eds.) MICCAI 2020. LNCS, vol. 12266, pp. 200–210. Springer, Cham (2020). https://doi.org/10.1007/978-3-030-59725-2_20
23. Zhang, F., et al.: Cascaded generative and discriminative learning for microcalcification detection in breast mammograms. In: Proceedings of the IEEE Conference on Computer Vision and Pattern Recognition, pp. 12578–12586 (2019)

Supervised Contrastive Pre-training for Mammographic Triage Screening Models

Zhenjie Cao[1], Zhicheng Yang[1], Yuxing Tang[1], Yanbo Zhang[1], Mei Han[1], Jing Xiao[2], Jie Ma[3], and Peng Chang[1(✉)]

[1] PingAn Tech, US Research Lab, Palo Alto, USA
[2] Ping An Technology, Shenzhen, China
[3] Shenzhen People's Hospital, Shenzhen, China

Abstract. Inspired by the recent success of self-supervised contrastive pre-training on ImageNet, this paper presents a novel framework of Supervised Contrastive Pre-training (SCP) followed by Supervised Fine-tuning (SF) to improve mammographic triage screening models. Our experiments on a large-scale dataset show that the SCP step can effectively learn a better embedding and subsequently improve the final model performance in comparison with the direct supervised training approach. Superior results of AUC and specificity/sensitivity have been achieved for our mammographic screening task compared to previously reported SOTA approaches.

Keywords: Mammogram · Contrastive learning · Screening · Dual-view

1 Introduction

Mammographic screening is a cost-effective method for early detection of breast cancer, with approximately 39 million mammograms performed annually in the United States [1]. It has been reported that the U.S. radiologists ranged from 66.7% to 98.6% for sensitivity and from 71.2% to 96.9% for specificity in mammogram-based breast cancer diagnosis [19]. While many previous works proposed deep learning (DL) models to identify cancer patients and help improve the radiologists' performance [21], in this paper, we focus on training DL models to triage a portion of mammograms as cancer-free to reduce radiologists' workload, and therefore improve their efficiency and specificity, without sacrificing sensitivity.

Unlike the previous deep neural nets (DNN)-based mammographic screening systems, which are trained directly by supervised learning [4,16,20,21,25], we propose a Supervised Contrastive Pre-training + Supervised Fine-tuning (SCP+SF) framework. It first performs the SCP pre-training through a carefully designed Siamese contrastive learning module, searching for an ideal embedding space, then transfers the pre-trained encoder to the SF module for the supervised fine-tuning phase.

© Springer Nature Switzerland AG 2021
M. de Bruijne et al. (Eds.): MICCAI 2021, LNCS 12907, pp. 129–139, 2021.
https://doi.org/10.1007/978-3-030-87234-2_13

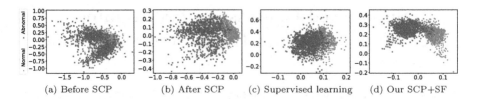

Fig. 1. Visualization of the normal and abnormal sample projections. (a)(b) are from the Siamese contrastive learning module, and (c)(d) are from the final dual-view model.

Contrastive learning has been applied to self-supervised visual representation learning [6,7,10,11,29], exemplified by the recent success of *SimCLR/SimCLR-v2* [6,7], which shows that self-supervised pre-training on ImageNet with a simple contrastive learning framework can generate competitive results on downstream image classification tasks comparing with fully supervised learning. The follow-up work [14] shows that contrastive pre-training can also be applied to supervised settings and further improve the SOTA performance on ImageNet.

Contrastive pre-training is fundamentally a guided clustering process with the objective of learning an embedding space to better separate the samples from different classes, and in turn, the following supervised fine-tuning can be carried out more effectively.

In this paper, we demonstrate that the proposed SCP+SF framework can be effectively applied to medical imaging and boost the performance of the triage screening task. Figure 1a and 1b (best viewed in color) visualize the sample projections from the proposed contrastive learning module before and after the SCP phase, clearly illustrating the improvement in the separability of the two clusters representing the healthy and at-risk populations. Figure 1c and 1d are the sample projections from our proposed dual-view model, with the direct supervised learning, and with the proposed SCP+SF training framework, further demonstrating that the SCP+SF results in better clustering quality.

Our experiments show that when trained on our in-house dataset of 134,488 images from 30,487 patients and tested on 2,538 images from 640 patients with biopsy-proven ground truth, our screening models trained with SCP+SF surpass the previously reported SOTA approaches [21,28,30] by a large margin, in terms of AUC and specificity/sensitivity.

The main contributions of this paper include: 1) we present a novel framework of SCP+SF, with a carefully designed Siamese contrastive learning module, including details of the network architecture and loss design, and 2) we show that for our mammographic triage screening task, models trained with SCP+SF consistently outperform their directly supervised counterparts and achieve superior performance over previously reported SOTA approaches.

2   Related Work

Contrastive Pre-training. Most contrastive pre-training works have been conducted within the realm of self-supervised learning on ImageNet data [2,6,7,10, 11,23,29,31], involving different forms of contrastive loss [9]. The recent work of *SimCLR-v2* [7] shows that self-supervised contrastive pre-training can compete with its fully supervised counterpart after fine-tuning on downstream tasks. The work of *SupCon* [14] generalizes the contrastive loss to the supervised setting. In medical imaging, the work of [12] proposes to carry out self-supervised contrastive pre-training at both global and local levels on the Magnetic Resonance Imaging (MRI) dataset before fine-tuning for MRI image segmentation.

Mammographic Screening. Previous works on deep learning based mammographic screening include two types of triage tasks: 1) identifying the healthy patients to reduce workload [16,17,25,30], and 2) identifying the mammograms with malignant findings [4,13,21,26–28]. We focus on the first task in this paper and treat the BI-RADS 1^1 category mammograms as healthy/normal.

Most of the above screening methods take the direct supervised learning approach, except for [28], which pre-trained the screening model on a large amount of data with BI-RADS labels before fine-tuning it with biopsy ground truth. However, this pre-training phase is *not* based on the contrastive learning principle, therefore different from our approach.

In particular, the approach in [30] is a classic single-view based method, and the approaches in [21,28] represent the latest SOTA multi-view mammographic screening methods. All three approaches have been tested on large-scale datasets. Therefore they are selected for comparison with our proposed method in Sect. 5.

3   Method

SCP+SF Framework. The overall architecture of the SCP+SF framework is illustrated in Fig. 2. The *Siamese contrastive learning module* is designed to carry out the SCP phase, and the resulting Siamese encoders are then transferred to the *single-view learning module* and the *dual-view learning module* to continue the SF phase, respectively, as shown by the magenta arrows in Fig. 2. We further elaborate on both phases in the following subsections.

SCP Phase. The SCP phase is carried out by the Siamese contrastive learning module, which consists of a Siamese encoding block and a Siamese projection block. In the encoding block, one pair of the input mammographic images are simultaneously fed into the shared-weight encoders. The encoded features are then projected into a lower dimensional space by max-pooling and 1×1 conv operations before flattened into two 1-dimensional vectors. The 1-D vectors are further reduced to 2×1 output vectors through fully connected layers and sigmoid operation, representing the likelihood for each class.

---

[1] Details regarding the BI-RADS standard can be found in [8].

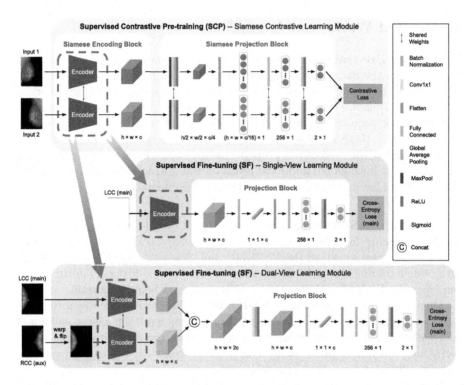

Fig. 2. The architecture of SCP+SF framework, consisting of the Siamese contrastive learning module and SF modules for both the single-view and the dual-view model.

The contrastive loss [9] is designed to draw the samples from the same class closer and separate the samples from different classes farther apart in the projected space. Given a pair of input images (I, I'), we use the regular L2 distance in the loss function and set *margin* as 1:

$$L(I, I') = \begin{cases} D^2 & \text{if } l_I = l_{I'} \\ \max\left(0, margin - D\right)^2 & \text{if } l_I \neq l_{I'}, \end{cases} \tag{1}$$

where

$$D = \left\| P_{\text{sia}}(E_{\text{sia}}(I)) - P_{\text{sia}}(E_{\text{sia}}(I')) \right\|_{L2}, \tag{2}$$

and $E_{\text{sia}}(\cdot)$ and $P_{\text{sia}}(\cdot)$ denote the Siamese encoder block and nonlinear projection block, respectively; l_I and $l_{I'}$ indicate the corresponding BI-RADS labels. The loss for a batch of N image pairs can be simply defined as $\mathcal{L}_{\text{batch}} = \sum_{i=1}^{N} L(I_i, I'_i)$.

Other types of loss function (e.g., inner product based) as in [6,14] are also experimented and compared in Sect. 5. The SCP phase is completed once the training of this module ends.

SF Phase. Each mammogram typically includes four views, the left and right craniocaudal (LCC/RCC), and mediolateral-oblique (LMLO/RMLO), and the triage screening model can take one or multiple views as input.

Single-View Model. The *Single-view learning module* in Fig. 2 illustrates the network architecture of the single-view model. During the SF phase, its encoder block is directly transferred from the Siamese encoding block trained in SCP phase and kept intact, while the projection block after the encoder is fine-tuned based on the regular cross-entropy loss.

Dual-View Model. In practice, radiologists routinely identify the abnormalities through bilateral analysis of mammography image pairs (i.e., LCC/RCC, or LMLO/RMLO). Therefore we also experimented with the bilateral views as the input for a dual-view model, in addition to the single-view model.

The *Dual-view learning module* in the SF phase comprises a dual-view based input structure, a Siamese encoder, and a projection block, as shown in Fig. 2. Since our screening model output is for each image, we designate one image of the bilateral pair as the main input, and the other image serves as the auxiliary input. For the example shown in Fig. 2, the LCC view is the main input, and the RCC view from the same patient serves as the auxiliary input. The RCC input will first be registered and warped according to the LCC view before being fed into the shared-weight pre-trained encoder, in tandem with the LCC view. The output encoded features are then concatenated before being projected into a lower dimension and further reduced to a 2×1 vector. Similar to the single-view model, the encoder block of the dual-view model is directly transferred from the SCP phase and fixed during the SF phase.

Sample Selection Strategy. During supervised contrastive learning, a batch of images is first randomly selected from the training set, and then the positive and negative pairs are identified according to the sample labels within this selected batch [14]. Limited by the affordable batch size, we experimented with two slightly different sampling strategies. One method is random sampling, where the training batch includes N pairs of positive and negative image pairs directly sampled from *the entire training set* with the corresponding labels. Since our dual-view model takes input from a pair of images from the same patient, we also experimented with a patient-constrained sampling method, where each randomly sampled positive or negative pair must come from *the same patient*.

4   Experiment Design

Our triage screening task aims to identify normal mammograms (BI-RADS 1) with near-perfect accuracy in physical screening scenario. Patients with any suspicious regions in the breasts should not be screened out as normal patients. This task can be further defined as a binary classification problem of BI-RADS 1 (normal/healthy) vs. other BI-RADS categories (abnormal). The majority of screening mammograms belongs to BI-RADS 1, and thus screen them off can assist radiologists to reduce their workload.

Table 1. Number of mammography images in each BI-RADS category.

BI-RADS	Abnormal						Subtotal	Normal	Total
	2	3	0	4	5	6		1	
Train & Val	35290	14508	1570	3288	732	174	55562	78926	134488
Test	728	226	136	88	6	8	1192	1346	2538

Datasets. Our data for training and validation is collected from three collaborative hospitals at distinct geographical locations using Siemens and Giotto equipment in accordance with the ACR standard (American College of Radiology) and dated from 2011 to 2018.[2] This dataset contains both image data and diagnosis reports that are all from screening exams. It includes 30,487 patients, among which 13,931 patients have at least one breast diagnosed as abnormal (other than BI-RADS 1), and 16,556 patients have both breasts diagnosed as normal (BI-RADS 1). Our test set includes 640 patients collected within 31 consecutive days (March 2019) from one of those three hospitals, among which 405 patients have at least one breast diagnosed as abnormal, and 235 patients have both breasts diagnosed as normal. Mammograms in test set come with biopsy proven malignancy results. Table 1 shows the number of images in each BI-RADS category for these two datasets. For abnormal cases, the BI-RADS categories are listed by increasing risk level, where BI-RADS 0 is often regarded as between BI-RADS 3 and 4 by radiologists [3,22]. In the literature [21,28,30], mammographic screening models have been developed and tested only with large-scale private datasets conforming to the realistic patient distribution. Public datasets like DDSM [18] have BI-RADS 1 patients' data marked as 'normal'. But they account for less than 30% of the complete dataset, of which the data distribution is not consistent with screening mammography scenarios. Other public mammography datasets either contain no BI-RADS 1 patients' data or only have diagnostic mammograms (when a screening mammogram does show an abnormality, a diagnostic mammogram may be needed). Since no similar public dataset is available, we follow the convention and perform experiments on our own large-scale datasets.

Implementation Details.[3] The dataset is split into the training and validation sets by 8:1 ratio. All input images are resized to 1008 × 800 and retain the original aspect ratio. The SCP and SF phases share the following training

---

[2] This retrospective case-control study was approved by the ethics review and institutional review board, which waived the requirement for individual informed consent.

[3] All implementation is with python3.7 and pytorch 1.1.0.

Table 2. Performance comparison of different approaches on the triage screening task. *SCP+SF(R)* refers to SCP with random sample selection and *SCP+SF(P)* indicates SCP with patient-constrained sample selection as in Sect. 3. *SV, DV* and *4V* stand for single-view, dual-view and 4-view method, respectively. The last column shows the number of incorrectly screened abnormal images with a breakdown according to their BI-RADS levels (in the order of 2, 3, 0, 4, 5, 6). The 95% confidence intervals (CI) are shown in the square brackets.

Method	AUC	Sensitivity = 20%	
		Specificity	# of *incorrectly* screened out images out of 1192 abnormal images
[30] SV (2019)	0.8438 [.8418, .8462]	0.9723 [.9687, .9753]	33 (10, 8, 12, 2, 0, 1)
[28] 4V (2019)	0.8702 [.8676, .8734]	0.9773 [.9749, .9797]	27 (9, 8, 9, 1, 0, 0)
[21] 4V (2020)	0.8617 [.8544, .8696]	0.9765 [.9737, .9793]	28 (8, 12, 7, 1, 0, 0)
SV	0.8349 [.8329, .8373]	0.9715 [.9691, .9741]	34 (7, 9, 14, 2, 1, 1)
SV SCP+SF(R)	0.8518 [.8492, .8550]	0.9757 [.9743, .9771]	29 (7, 9, 11, 2, 0, 0)
SV SCP+SF(P)	0.8429 [.8368, .8490]	0.9748 [.9712, .9784]	30 (10, 6, 11, 2, 0, 1)
DV	0.8554 [.8528, .8580]	0.9765 [.9747, .9783]	28 (7, 8, 12, 1, 0, 0)
DV SCP+SF(R)	0.8805 [.8746, .8864]	0.9799 [.9780, .9818]	24 (9, 5, 10, 0, 0, 0)
DV SCP+SF(P)	**0.9040 [.9001, .9079]**	**0.9832 [.9816, .9850]**	**20 (7, 5, 8, 0, 0, 0)**

parameter settings. The initial learning rate is 1×10^{-5} with 4 warming-up steps and reduced to 1×10^{-6} after 100 epochs. Adam is used [15], with a weight decay of 5×10^{-4}. Two NVIDIA V100 GPUs (16G memory each) are used, and the batch size for contrastive learning is set to 6 due to the computation limit. The model training normally completes within 300 epochs. Our final model's runtime is less than 3 s on our machine. One limitation is the model does require GPU to run, and it has not been fully tested on data outside our collaborators. For comparison purposes, we re-implemented methods as in [21,28,30] and applied them to our datasets.

Evaluation Metric. Since the goal is to screen out a portion of normal mammograms with near-perfect accuracy, we set the sensitivity (recall rate of normal images) at 20%, which is commonly used in clinical studies for mammogram triage screening [30], and compare the specificity rate (percentage of correctly classified abnormal images) of different approaches. In addition, AUC is used to compare the overall performance of the classification models.

5   Experimental Results

Table 2 shows the performance comparison of the triage screening task with different approaches, where Yala [30], Wu [28] and McKinney [21] are all previous SOTA models, alongside our proposed single-view and dual-view models with different SCP+SF training strategies. We also scrutinize the number of abnormal

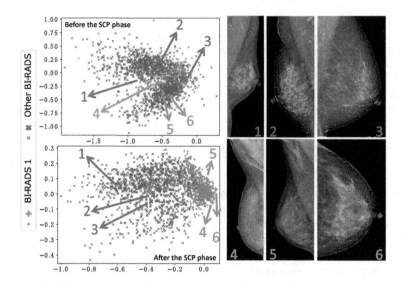

Fig. 3. Visualization of the normal (BI-RADS 1) and abnormal (other BI-RADS) sample projections from the test set data before (top-left) and after (bottom-left) the SCP phase, along with 6 samples and their corresponding images.

images that are incorrectly screened out from the 1,192 abnormal images in the test set, and further break it down according to the BI-RADS level.

Effectiveness of SCP+SF. As shown in Table 2, for both single-view and dual-view models, the SCP+SF framework effectively improves the overall performance of the models, including the AUC and the specificity at given sensitivity (20%). In turn, the number of total incorrectly screened abnormal images is reduced. In addition, for the dual-view model, the SCP+SF framework can completely remove the error made for BI-RADS 4,5,6 images, which is critical in practice since those images often correspond to higher cancer risk. In general, most misclassified abnormal mammograms are due to small scattered benign calcifications, which are occasionally misclassified into BI-RADS 1 category. For the single-view model, the SCP+SF framework can also reduce the error for BI-RADS 5,6 images to near zero. We further confirm from the separate biopsy reports that the incorrectly screened images from the dual-view methods with SCP+SF do not include any malignant findings.

Compared with the previous SOTA approaches, our single-view model with SCP+SF and random sample selection generates the best single-view performance. Our dual-view model trained with SCP+SF and patient-constrained sample selection generates the best result overall, outperforming Wu [28] and McKinney [21], which are both 4-view based models. At higher sensitivity, such as 0.8, our SCP+SF with patient-constrained sample selection method improves the specificity from 0.805 to 0.858 comparing to our vanilla dual view model.

Figure 3 shows that the normal and abnormal sample projections from the Siamese contrastive learning module on the test data are much better clustered after the SCP phase, further illustrating its efficacy. Three sample images from each class are also given.

Regarding sample selection strategy, the patient-constrained sampling further improves the dual-view model over the random sampling method, while random sampling is slightly better than the patient-constrained sampling for the single-view model, both are consistent with our expectation.

Table 3. Ablation study on different encoders and contrastive loss.

Loss	Inner Product	L2 Loss		
Encoder	ResNet-22	ResNet-34	ResNet-50	ResNet-22
AUC	0.9031	0.9013	0.8993	**0.9040**
Specificity (Sen. = 20%)	98.23%	98.15%	98.15%	**98.32%**

Ablation Study on SCP. Table 3 shows the ablation study results. For the backbone, there is no significant difference in terms of the size of ResNet on our task, and ResNet-22 (as in [28]) is selected to serve our encoder. For the loss function, the inner product distance gives a comparable result (slightly worse) as the L2 distance we use in the contrastive loss.

6   Conclusions

We present a novel framework of Supervised Contrastive Pre-training followed by Supervised Fine-tuning (SCP+SF) for mammographic triage screening task. Our experiments with a total of 137,026 images have demonstrated that the SCP+SF framework substantially improved the final model performance, comparing with the direct supervised training. Superior results have also been achieved in comparison with previously reported SOTA approaches. One limitation of this study is that we only applied regular L2 distance for contrastive loss calculation. We plan to experiment with other loss functions, such as triplet loss [5] and magnet loss [24] formulations. We also plan to apply the SCP+SF approach to other medical imaging classification tasks in the future.

References

1. MQSA national statistics. https://www.fda.gov/radiation-emitting-products/mqsa-insights/mqsa-national-statistics
2. Bachman, P., Hjelm, R.D., Buchwalter, W.: Learning representations by maximizing mutual information across views. In: Advances in Neural Information Processing Systems, vol. 32, pp. 15535–15545. Curran Associates, Inc. (2019)

3. Castells, X., et al.: Risk of breast cancer in women with false-positive results according to mammographic features. Radiology **280**(2), 379–386 (2016)
4. Sainz de Cea, M.V., Diedrich, K., Bakalo, R., Ness, L., Richmond, D.: Multi-task learning for detection and classification of cancer in screening mammography. In: Martel, A.L., et al. (eds.) MICCAI 2020. LNCS, vol. 12266, pp. 241–250. Springer, Cham (2020). https://doi.org/10.1007/978-3-030-59725-2_24
5. Chechik, G., Sharma, V., Shalit, U., Bengio, S.: Large scale online learning of image similarity through ranking. J. Mach. Learn. Res. **11**(3) (2010)
6. Chen, T., Kornblith, S., Norouzi, M., Hinton, G.: A simple framework for contrastive learning of visual representations. In: Proceedings of the 37th International Conference on Machine Learning, vol. 119, pp. 1597–1607. PRML (2020)
7. Chen, T., Kornblith, S., Swersky, K., Norouzi, M., Hinton, G.E.: Big self-supervised models are strong semi-supervised learners. In: Advances in Neural Information Processing Systems, vol. 33, pp. 22243–22255. Curran Associates, Inc. (2020)
8. D'Orsi, C.: 2013 ACR BI-RADS Atlas: Breast Imaging Reporting and Data System. American College of Radiology (2014)
9. Hadsell, R., Chopra, S., LeCun, Y.: Dimensionality reduction by learning an invariant mapping. Proceedings of the IEEE Conference on Computer Vision and Pattern Recognition, vol. 2, pp. 1735–1742 (2006)
10. He, K., Fan, H., Wu, Y., Xie, S., Girshick, R.: Momentum contrast for unsupervised visual representation learning. In: Proceedings of the IEEE Conference on Computer Vision and Pattern Recognition, pp. 9729–9738 (2020)
11. Hjelm, R.D., Fedorov, A., Lavoie-Marchildon, S., Grewal, K., Trischler, A., Bengio, Y.: Learning deep representations by mutual information estimation and maximization. In: International Conference on Learning Representations (2019)
12. Karani, K.C.E.E.N., Konukoglu, E.: Contrastive learning of global and local features for medical image segmentation with limited annotations. In: Advances in Neural Information Processing Systems, vol. 33. Curran Associates, Inc. (2020)
13. Khan, H.N., Shahid, A.R., Raza, B., Dar, A.H., Alquhayz, H.: Multi-view feature fusion based four views model for mammogram classification using convolutional neural network. IEEE Access **7**, 165724–165733 (2019)
14. Khosla, P., et al.: Supervised contrastive learning. In: Advances in Neural Information Processing Systems, vol. 33, pp. 18661–18673. Curran Associates, Inc. (2020)
15. Kingma, D.P., Ba, J.: Adam: a method for stochastic optimization. In: International Conference on Learning Representations (2015)
16. Kontos, D., Conant, E.F.: Can AI help make screening mammography "lean"? Radiology **293**(1), 47–48 (2019)
17. Lång, K., Dustler, M., Dahlblom, V., Åkesson, A., Andersson, I., Zackrisson, S.: Identifying normal mammograms in a large screening population using artificial intelligence. Eur. Radiol. **31**(3), 1687–1692 (2021)
18. Lee, R.S., Gimenez, F., Hoogi, A., Miyake, K.K., Gorovoy, M., Rubin, D.L.: A curated mammography data set for use in computer-aided detection and diagnosis research. Sci. Data **4**, 170177 (2017)
19. Lehman, C.D., et al.: National performance benchmarks for modern screening digital mammography: update from the breast cancer surveillance consortium. Radiology **283**(1), 49–58 (2017)
20. Lehman, C.D.: Artificial intelligence to support independent assessment of screening mammograms-the time has come. JAMA Oncol. **6**(10), 1588–1589 (2020)
21. McKinney, S.M., et al.: International evaluation of an AI system for breast cancer screening. Nature **577**(7788), 89–94 (2020)

22. Nelson, H.D., O'Meara, E.S., Kerlikowske, K., Balch, S., Miglioretti, D.: Factors associated with rates of false-positive and false-negative results from digital mammography screening: an analysis of registry data. Ann. Intern. Med. **164**(4), 226–235 (2016)

23. Oord, A.V.D., Li, Y., Vinyals, O.: Representation learning with contrastive predictive coding. arXiv preprint arXiv:1807.03748 (2018)

24. Rippel, O., Paluri, M., Dollar, P., Bourdev, L.: Metric learning with adaptive density discrimination. arXiv preprint arXiv:1511.05939 (2015)

25. Rodriguez-Ruiz, A., et al.: Can we reduce the workload of mammographic screening by automatic identification of normal exams with artificial intelligence? A feasibility study. Eur. Radiol. **29**(9), 4825–4832 (2019)

26. Shen, L., Margolies, L.R., Rothstein, J.H., Fluder, E., McBride, R., Sieh, W.: Deep learning to improve breast cancer detection on screening mammography. Sci. Rep. **9**(1), 1–12 (2019)

27. Sun, L., Wang, J., Hu, Z., Xu, Y., Cui, Z.: Multi-view convolutional neural networks for mammographic image classification. IEEE Access **7**, 126273–126282 (2019)

28. Wu, N., et al.: Deep neural networks improve radiologists' performance in breast cancer screening. IEEE Trans. Med. Imaging **39**(4), 1184–1194 (2019)

29. Wu, Z., Xiong, Y., Yu, S.X., Lin, D.: Unsupervised feature learning via non-parametric instance discrimination. In: Proceedings of the IEEE Conference on Computer Vision and Pattern Recognition, pp. 3733–3742 (2018)

30. Yala, A., Schuster, T., Miles, R., Barzilay, R., Lehman, C.: A deep learning model to triage screening mammograms: a simulation study. Radiology **293**(1), 38–46 (2019)

31. Zhuang, C., Zhai, A.L., Yamins, D.: Local aggregation for unsupervised learning of visual embeddings. In: Proceedings of the IEEE International Conference on Computer Vision, pp. 6002–6012 (2019)

Trainable Summarization to Improve Breast Tomosynthesis Classification

Mickael Tardy[1,2]([✉]) [ID] and Diana Mateus[1] [ID]

[1] Ecole Centrale de Nantes, LS2N, UMR CNRS 6004, Nantes, France
[2] Hera-MI, SAS, Nantes, France

Abstract. Digital Breast Tomosynthesis (DBT) is an emerging imaging technique for breast cancer screening aiming to overcome certain limitations of traditional mammography, such as the superimposition of tissues. On the downside, DBT increases the radiologists' workload as it generates stacks of high-resolution images, which are time-consuming to review and annotate. In this work, we propose a deep- multiple-instance-based method for DBT volume classification that relies on the local summarization of DBT slices (referred to as slabbing) and only requires volume-wise labels for training. Slabbing offers several advantages: i) it reduces the classifier's computational complexity across the depth, letting it focus on the higher transversal resolution. Thanks to this strategy, we are the first to train a method at almost full-resolution (as high as 120 × 2500 × 2000); ii) it produces slabs that are closer to standard mammography, favoring an efficient transfer from classifiers trained on larger mammography databases; and iii) the slabs combined with a Multiple-Instance Learning (MIL) classifier result in localized information favoring interpretability. The proposed slabbing MIL approach is also novel for the automatic classification of DBTs. Moreover, we propose a trainable alternative to the handcrafted slabbing algorithms based on slice-wise attention that improves performance. We perform an experimental validation on a subset of the public BCS-DBT dataset and achieve an AUC of 0.73 with five-fold cross-validation. On a private multi-vendor dataset we obtain a similar AUC of 0.73, demonstrating an excellent performance consistency.

Keywords: Digital breast tomosynthesis · Breast cancer · Classification · Transfer learning · Summarization

1 Introduction and Related Work

Breast cancer is one of the most common in the women population, and one of the leading causes of cancer death [18]. Recent clinical advances have allowed to decrease the death rate and increase the chances of recovery [8]. Such advances

Supported by Hera-MI SAS, Nantes, France, European Regional Development Fund, Pays de la Loire and Nantes Métropole (Connect Talent MILCOM).

© Springer Nature Switzerland AG 2021
M. de Bruijne et al. (Eds.): MICCAI 2021, LNCS 12907, pp. 140–149, 2021.
https://doi.org/10.1007/978-3-030-87234-2_14

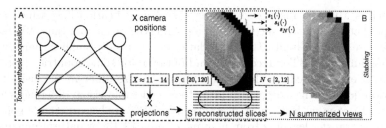

Fig. 1. Illustration of DBT pipeline: (A) acquisition process, generating S slices from X projections, and (B) generation of N summarized views from S reconstructed slices.

rely on timely and early detection, usually done through regular clinical surveillance as well as complementary imaging exams. While mammography remains the most used modality, Digital Breast Tomosynthesis (DBT) is an emerging imaging technique [20] based on a limited-angle tomographic reconstruction (see Fig. 1-A), which reduces the depth ambiguities caused by tissue superimposition in mammography and has, therefore, the potential to reduce false positives and false negatives detections [10,11,16]. However, DBT produces a stack of high-resolution images. Each slice has usually a resolution of $\approx 2500 \times 2000$ pixels and the whole stack contains between 30 and 120 slices or more. Such large volumes increase the imagers' workload, as each volume needs to be scrolled through for evaluation. Current mammography systems' vendors propose synthesized 2D images (along with DBT stacks), whose quality is comparable to the traditional mammography [14]. However, recent studies show that to achieve higher performances, the whole stack shall still be reviewed [15].

Recently, Computer Aided Diagnosis (CAD) solutions have been proposed to reduce the reading time and facilitate the review of DBTs [4,9]. To that end, several deep-learning-based methods have been proposed for the binary classification (i.e., benign/malignant) [7,12,13] of DBT volumes.

Several challenges arise when designing deep-learning CAD methods for DBT analysis. First, processing high-resolution images is resource-consuming[1]. Thus, current methods require the rescaling of each slice to 1024×1024 [23], or even as low as 256×256 [7]. However, recent works in mammography [17,19,22] have shown that keeping a high resolution is advantageous, as it allows to capture the smallest findings (e.g., microcalcifications <1 mm) [2]. Second, the ground truth is usually scarce. In case of a whole volume being classified as malignant, more than a half of slices may not be related to the pathology, while the explicit ground truth for each slice may be unavailable. Such imbalance is increased even more with malignant cases usually being only $\approx 10\%$ of the whole dataset [3,23]. Recent works exploit more precise annotations, such as the most representative slice [7], or a bounding box around the Region of Interest (ROI) [13], which both require further involvement from the experts. A more annotation-efficient

---

[1] Volumes can exceed $\approx 120 \times 2500 \times 2000$ pixels, i.e., $\approx 600M$ pixels.

approach is Multiple Instance Learning (MIL) classification [23], which aggregates predictions for each slice through a pooling operation and thus only require volume-wise labels. Finally, the DBT datasets are usually smaller than those of mammograms, as DBT is recent [20] and optional for some countries [16]. Data scarcity is exacerbated by the DBT systems having different acquisition and reconstruction settings [20], which leads to considerable visual differences across vendors. To cope with it, most deep-learning DBT analysis methods generally build upon transfer learning from mammography [12,13] or natural images [23].

In this work, we focus on the classification of DBT volumes in the context of breast cancer screening. To deal with the challenges above, we propose creating an interpretable intermediate representation that condenses high-resolution information to ease the volume processing. To this end, we devise a method that summarizes the volume into a small number of views (see Fig. 1, B), generated from a group of contiguous slices (i.e., slabbing) [6]. In this way, each volume is resumed into 5–10 high-resolution slab images. Such summarizing offers several benefits. First, it does not interfere with the original image resolution. Second, it facilitates a MIL training, as it reduces the number of samples in each bag by around 90%, which in turn improves the classification performance, as we show in the experimental results. Finally, it enhances the transferability from mammography classifiers.

The most common summarization strategy consists of a Maximum Intensity Projection (MIP), keeping the most intense pixel across the depth. MIP has been successfully applied in the context of deep-learning-based methods to CT [24] and MRI imaging [1]. Summarizing DBT volumes is more difficult due to noise and contrast issues, for which Diekmann et al.[6] propose several strategies: i) Maximum Intensity Projection (MIP), ii) simple averaging (i.e., retaining the average of intensity values); and iii) SoftMIP, a custom weighted average. MIP results in the highest noise and contrast values. While averaging reduces the amount of noise, it also decreases contrast (problematic when it comes to visualizing microcalcifications). SoftMIP achieves a compromise of the two. In our work, instead of a handcrafted summarizing algorithm [6], we propose a trainable model for slabs generation. Our method uses spatial attention in calculating the slabs, which leads to a performance increase compared to handcrafted methods.

Our contributions are as follows:

– Proposing the use of slabbing for DBT classification;
– A novel trainable attention-based model for slab generation;
– An end-to-end method capable of processing full-resolution DBT volumes.

As a result, our method improves the performance over plane MIL and simple slabbing strategies, efficiently reuses classifiers trained on mammography multi-vendor data, and achieves consistent performances over multi-vendor and multi-center DBT datasets.

2 Methods

In this work, we propose a method for the classification of DBT volumes complying with the following requirements: i) enabling the processing of full-resolution volumes; ii) learning from volume-wise ground truth only; and iii) allowing transfer learning from mammography. To fulfill these requirements, we propose to summarize the stack of slices of the DBT volumes to a smaller number of interpretable slabs and process them with a classifier. An overview of the method is shown in Fig. 2.

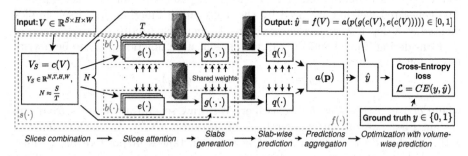

Fig. 2. Overview of the proposed method: the function $f(\cdot)$ takes the volume V as the input, generates the output prediction $\hat{y}$, and is optimized with the cross-entropy loss against the volume-wise ground truth y.

Let $\mathcal{D} = \{V_i, y_i\}_{i=1}^{M}$ be a dataset composed of DBT volumes $V_i \in \mathbb{R}^{S_i \times H_i \times W_i}$ and volume-wise labels, $y_i \in \{\mathcal{C}_k\}_{k=1}^{K}$ for a K-class classification. We design a classification function $f(\cdot)$, yielding a class probability prediction $\hat{y}_i$ for a given volume V_i, that is, $\hat{y}_i = f(V_i)$.

Having only volume-wise ground-truth available, we rely on a MIL approach, which allows us to build upon more resource-efficient 2D classifiers, and therefore, use transfer learning from mammography. For a given volume V having S instances (i.e., slices)[2], the volume-wise prediction $\hat{y}$ is obtained by aggregating with function $a(\cdot)$, e.g., a $\max(\cdot)$ operation, the individual predictions $\hat{p}_j$ of its instances:

$$\hat{y} = a(\mathbf{p}), \quad \text{where} \quad \mathbf{p} = [\hat{p}_1, \dots, \hat{p}_j, \dots, \hat{p}_S], \tag{1}$$

and $\hat{p}_j$ is the prediction of j-th instance in the bag. Instead of composing the MIL bag with the whole set of DBT slices such as in [23], which leads to a large number of instances $S \gg 1$, we propose to use a smaller number of summarized slabs [6]. To this end, the whole stack of slices $V \in \mathbb{R}^{S \times H \times W}$ is, first, partitioned into N groups $V_j \in \mathbb{R}^{T \times H \times W}$ s.t. $V = \cup_{j=1}^{N} V_j$, where $N = \lceil \frac{S}{T} \rceil$ is an integer representing the number of groups, and T is the hyper-parameter representing the slab thickness. The j-th slab $s_j \in \mathbb{R}^{H \times W}$ is defined as follows:

$$s_j(V) = b(V_j), \tag{2}$$

---

[2] Hereafter we omit the index i from V_i and y_i to simplify the notation.

where $b(\cdot)$ is the slab-generating function for the group of slices V_j. For example, in case of a MIP slabbing algorithm, the value of the j-th slab at pixel (x, y) is computed as $s_j(x, y) = \max_{z \in \{1,\dots,T\}} V_j(x, y, z)$. In our work, as an alternative to handcrafted algorithms, we propose a trainable implementation of $b(\cdot)$ referred to as AttIP (for Attentive Intensity Projection), as follows:

$$s_j = b(V_j) = g(V_j, e(V_j)) \quad \text{with} \quad s_j(x, y) = \frac{1}{T} \sum_{z=1}^{T} V_j(x, y, z) \odot e(V_j(x, y, z)) \quad (3)$$

where $e(\cdot)$ is a trainable function generating a pixel-wise ponderation support. The generated weights are used by the aggregation function $g(\cdot)$, itself implemented as a depth-wise average over the element-wise product $\odot$.

Considering Eq. (1) and Eq. (3), the classification function $f(\cdot)$ can be rewritten as follows:

$$\hat{y} = f(V) = \max_{j \in \{1,\dots,N\}} q(s_j) \quad (4)$$

where $q(\cdot)$ is a trainable slab classifier. We draw several advantages from our design. First, we obtain smaller bags of instances $|N| < |S|$ facilitating the MIL, which reduces the number of negative instances in positive bags, thus, the imbalance of the training data. Second, the trainable $e(\cdot)$ function allows for feature-based slabs-generation, unlike the handcrafted algorithms that are intensity-based. Third, the $e(\cdot)$ function can benefit from transfer learning, as it is used to extract meaningful features from 2D DBT slices. In summary, the DBT classifier $f(\cdot)$ is a deep learning model composed of trainable attention $e(\cdot)$ modulating a slab generator $g(\cdot)$, and a slab classifier $q(\cdot)$, which can be optimized end-to-end with a loss $\mathcal{L}(y, \hat{y})$ exploiting volume-wise ground truth.

3   Experimental Validation

Dataset. We evaluate the proposed method on a subset of the BCS-DBT dataset [3]. At the time of writing, only a part of the whole dataset has been released with ground truth. For our experiments in the binary classification task, we extracted a subset of volumes composed of 100 normal cases and 75 biopsy-proven cancer cases (i.e., all but one malignant case from the BCS-DBT training dataset, to keep equally balanced five folds for cross-validation). Moreover, to evaluate the performance consistency in the binary classification task over different datasets [21], we also used a private multivendor dataset (denoted PMV-DBT) containing 58 normal and 58 proven malignant DBT volumes coming from two different vendors and three different imaging centers (board approvals were obtained from each center). Finally, we used a private multi-vendor mammography dataset (denoted PMV-MG) for network pretraining and in the performance consistency experiments. The training set of PMV-MG contains 1000 benign and 1000 proven malignant images, the test set contains 250 benign and 250 malignant images. These datasets can be shared upon justified requests and right-holder approvals.

Data Preparation. We keep the original resolution of the images to prevent the loss of information. To that end, we do not resize the input images but crop them to a bounding box around the breast, i.e., excluding the surrounding background pixels (see Fig. 1 for illustration). We also rescale the intensity values to the range of $[0, 1]$. We used randomized augmentation techniques including vertical flipping, as well as horizontal and vertical shifting.

Implementation Details. To process images of arbitrary size, we rely on a Fully Convolutional Network (FCN). For the class prediction function $q(\cdot)$ we use the Resnet22 [19]. For the slices attention function $e(\cdot)$ we use a shallower Resnet10 network, applying spatial pooling, followed by a sigmoid activation over the last convolutional layer output. Both networks use separable convolutions instead of regular ones and the weights are shared over the N branches (see Fig. 2), which result in the entire network having $\approx 500K$ parameters. $f(\cdot)$ is trained end-to-end with a cross-entropy loss, an Adam optimizer, and a learning rate of $2 \cdot 10^{-5}$. We train the networks for 20 epochs with early stopping, whenever a performance decrease is observed (e.g., overfitting).

Hyper-Parameters. Our slabs generation method relies on the partitioning of a volume V on N slabs and uses parameter T to determine the number of slices per slab. Hereafter, we set $T = 10$, which is similar to the 1cm slab thickness in [6] as most commonly slice spacing is equal to $1mm$ for the majority of vendors.

Transfer Learning. The Deep Neural Network (DNN) used in our experiments are pre-trained on a multivendor mammography dataset PMV-MG with a binary classification task.

Comparison. To illustrate the contribution of our approach, we compare it to the baseline MIL slice-wise classifier using maximum pooling across predictions similar to [23]. We also compare our proposed attention-based slabbing to the handcrafted MIP and SoftMIP techniques [6]. We evaluate all the methods with transfer learning from PMV-MG without and with fine-tuning on BCS-DBT.

Metrics. We report the Area Under Curve (AUC) to evaluate the performance of the binary classification task (i.e., normal vs. malignant). On BCS-DBT subset, we use 5-fold cross-validation. We use DeLong test [5] for the statistical significance of the AUC metrics.

3.1   Experiments

Transferring Knowledge from Mammography to DBT. We have two trainable components in our method: the slabbing $e(\cdot)$ and the classifier $q(\cdot)$, Eq. (3) and Eq. (4), which are implemented as DNNs. Here, we investigate the effectiveness of our method in facilitating the knowledge transfer from mammography to DBT. We also evaluate the effectiveness of additional training of $e(\cdot)$ and $q(\cdot)$ on DBT

Table 1. AUC results of the study of Transferring knowledge from mammography to DBT. FT: Fully trainable, PF: Partially Frozen, FF: Fully Frozen

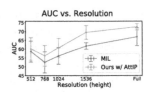

AUC vs. Resolution

Method	Initial	FT	PF	FF
MIL	63.80 ± 9.74	67.03 ± 6.72	64.30 ± 8.62	NA
Zhang [23]	NA	NA	62.27 ± 10.62	NA
Doganay [7]	NA	62.84 ± 11.62	NA	NA
Ours w/ MIP	66.83 ± 4.44	67.21 ± 3.97	66.91 ±4.72	NA
Ours w/ SoftMIP	68.43 ± 4.16	67.52 ± 4.41	68.63 ±4.94	NA
Ours w/ AttIP	**71.13 ± 4.76**	**69.91 ± 3.98**	**70.97 ± 4.91**	**72.66 ± 3.59**

Fig. 3. Evaluation of the different values of image heights from 512 to full resolution on BCS-DBT data.

data to refine that knowledge. In this experiment, we use the subset of the BCS-DBT dataset with 5-fold cross-validation. We evaluate the following settings: i) "Fully Trainable" (FT), where all the weights trainable; ii) "Partially Frozen" (PT), where slabs generator and the dense layer of the classifier are trainable; and iii) "Fully Frozen", (FF), where only slabs generator is trainable. Initial performances without fine-tuning are also reported. See Table 1 for results. The first column confirms that any type of slabbing ("ours w/") increases the chances of a MIL classifier to succeed. We hypothesize this is due to the higher resemblance of the slabs to the mammography data we transfer knowledge from. The gain is the most important when using an attentive-DNN to weigh the individual slices during the slab generation, instead of the simpler MIP and SoftMIP operations. We also note the high variability of the baseline MIL method, $\sigma = 9.74$ over the 5 folds, while all slab-based methods have $\sigma < 5.0$. Regarding the fine-tuning with DBT data, we remark that this additional domain knowledge only improves the performance in the "frozen setting", i.e., where only the slabbing network is trainable, although without statistical significance ($p > 0.1$).

Image Resolution. To illustrate the advantages of using high-resolution imaging, we report the performance with images from BCS-DBT dataset of several resolutions: full original resolution, and images resized to 1536, 1024, 768, 512 height. See results on Fig. 3. We note that the highest performances are achieved with the full resolution. We observe that as the resolution decreases, the performances drop and the differences between the two methods become less noticeable.

Table 2. Results of the study of performance consistency across datasets before and after fine-tuning on BCS-DBT data. AUC are reported.

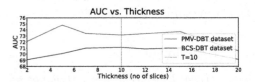

AUC vs. Thickness

Method	PMV-DBT		PMV-MG	
	Before	After	Before	After
MIL	65.17	66.87	*85.14*	67.28
Ours w/ AttIP	**73.15**	**73.94**	*85.14*	**85.14**

Fig. 4. Evaluation of the different values of slab thickness T on two datasets: BCS-DBT and PMV-DBT.

Performance Consistency Across Multi-modal and Multi-vendor Datasets. In this experiment, we use the PMV-MG and PMV-DBT datasets to investigate if there is forgetting on the mammography database, as well as to evaluate the generalization of our approach to unseen multi-vendor data. First, using the pre-trained mammography classifiers, we perform the fine-tuning on DBT data. We then explore if such fine-tuning induces a performance decrease on the mammography dataset PMV-MG. We also study the performance on the PMV-DBT dataset before and after fine-tuning with BCS-DBT data (denoted "before" and "after" respectively). The results are shown in Table 2. In the case of fine-tuning, both, the baseline and our method improve, however, the improvements are not statistically significant ($p > 0.1$). More importantly, our method allows keeping the performances on the mammography images while improving on DBT data. The absence of forgetting suggests that our method is effective in fusing multi-modal DBT and mammographic data to build a richer common binary classifier.

Slab Thickness Study. Our method involves choosing a fixed slab thickness T. We explore different values of T and report the classification performances on the BCS-DBT subset and PVR-DBT set. The results are shown in Fig. 4. One can see that $T = 10$ is an optimal choice for the two studied datasets.

4   Discussion and Conclusion

In this work, we proposed and evaluated a novel trainable slab-based classification method for DBT volumes. Our experiments using weak annotations (i.e., volume-wise) have shown both, the advantages over the baseline MIL approach and handcrafted slabbing techniques. We note, that the slabs classifier does not significantly benefit from fine-tuning. That is probably due to the DBT dataset having a low number of contributive malignant cases compared to the mammography dataset used for pre-training (i.e., 60 vs. 1000 malignant samples).

Our transfer learning experiments have shown the ability of our method to maximize the performance on the DBT data without losing the knowledge from mammography. Such behavior allows training the classifier concurrently from both, mammograms, and DBT slabs without the need to fine-tune the models for one modality. Moreover, our method allows for the independent training of the slabs generator, letting free the choice of the mammography classifier.

When evaluating on two different DBT datasets (i.e., PMV-DBT and BCS-DBT), we obtain comparable classification results ($p > 0.1$), which further confirms the performance consistency of the proposed method.

We note, that our results $AUC \approx 73.00$ are lower than those in some other works (e.g., $AUC = 85.40$ for Zhang *et al.* [23]). However, when training the method from [23] on our dataset we obtain lower $AUC = 62.27$, which is also valid for [7] that in our case yields $AUC = 62.84$. We attribute the loss in performances of the state-of-the-art method to a different size training dataset. That is, both methods require training of some weights from scratch, while our method allows the transfer of learning from mammography.

Our method has the advantage of handling high-resolution imaging. As a drawback, its training requires performant hardware, e.g., a GPU with 32 Gb of memory and sufficient amount of RAM or/and swap (i.e., at least 60 Gb).

Our method can apply to other types of 3D imaging, e.g., CT, MRI, or US. Moreover, the prediction and aggregation parts of our classification network can be replaced by different objectives (e.g., segmentation), further extending the application fields of the method.

Overall, the experiments have shown our method to be both, robust and generalizable, which could be appealing for clinical application as it promises less variability across different clinical settings and vendors.

References

1. Antropova, N., Abe, H., Giger, M.L.: Use of clinical MRI maximum intensity projections for improved breast lesion classification with deep convolutional neural networks. J. Med. Imaging **5**(01), 1 (2018). https://doi.org/10.1117/1.jmi.5.1.014503
2. Balleyguier, C., Ayadi, S., Nguyen, K.V., Vanel, D., Dromain, C., Sigal, R.: BIRADS^{TM} classification in mammography. Eur. J. Radiol. **61**(2), 192–194 (2007). https://doi.org/10.1016/j.ejrad.2006.08.033
3. Buda, M., et al.: Detection of masses and architectural distortions in digital breast tomosynthesis: a publicly available dataset of 5,060 patients and a deep learning model. arXiv:eess.IV/2011.07995 (2021)
4. Conant, E.F., et al.: Improving accuracy and efficiency with concurrent use of artificial intelligence for digital breast tomosynthesis. Radiol. Artif. Intell. **1**(4), e180096 (2019). https://doi.org/10.1148/ryai.2019180096
5. DeLong, E.R., DeLong, D.M., Clarke-Pearson, D.L.: Comparing the areas under two or more correlated receiver operating characteristic curves: a nonparametric approach. Biometrics **44**(3), 837–845 (1988)
6. Diekmann, F., et al.: Thick slices from tomosynthesis data sets: phantom study for the evaluation of different algorithms. J. Digital Imaging **22**(5), 519–526 (2009). https://doi.org/10.1007/s10278-007-9075-y
7. Doganay, E., Li, P., Luo, Y., Chai, R., Guo, Y., Wu, S.: Breast cancer classification from digital breast tomosynthesis using 3D multi-subvolume approach. In: Deserno, T.M., Chen, P.H. (eds.) Medical Imaging 2020: Imaging Informatics for Healthcare, Research, and Applications, vol. 11318, p. 12. SPIE (2020). https://doi.org/10.1117/12.2551376
8. Fisher, B., et al.: Twenty-year follow-up of a randomized trial comparing total mastectomy, lumpectomy, and lumpectomy plus irradiation for the treatment of invasive breast cancer. N. Engl. J. Med. **347**(16), 1233–1241 (2002). https://doi.org/10.1056/NEJMoa022152
9. Geras, K.J., Mann, R.M., Moy, L.: Artificial intelligence for mammography and digital breast tomosynthesis: current concepts and future perspectives. Radiology **293**(2), 246–259 (2019). https://doi.org/10.1148/radiol.2019182627
10. Hans Kleinknecht, J., Ileana Ciurea, A., Augusta Ciortea, C.: Pros and cons for breast cancer screening with tomosynthesis - a review of the literature. Med. Pharm. Rep. **93**(4), 335–341 (2020). https://doi.org/10.15386/mpr-1698
11. Houssami, N., Skaane, P.: Overview of the evidence on digital breast tomosynthesis in breast cancer detection. Breast (Edinburgh, Scotland) **22**(2), 101–108 (2013). https://doi.org/10.1016/j.breast.2013.01.017

12. Lotter, W., et al.: Robust breast cancer detection in mammography and digital breast tomosynthesis using an annotation-efficient deep learning approach. Nature Med. 1–6 (2021). https://doi.org/10.1038/s41591-020-01174-9

13. Mendel, K., Li, H., Sheth, D., Giger, M.: Transfer learning from convolutional neural networks for computer-aided diagnosis: a comparison of digital breast tomosynthesis and full-field digital mammography. Acad. Radiol. **26**(6), 735–743 (2019). https://doi.org/10.1016/j.acra.2018.06.019

14. Murakami, R., Uchiyama, N., Tani, H., Yoshida, T., Kumita, S.: Comparative analysis between synthetic mammography reconstructed from digital breast tomosynthesis and full-field digital mammography for breast cancer detection and visibility. Eur. J. Radiol. Open **7**, 100207 (2020). https://doi.org/10.1016/j.ejro.2019.12.001

15. Murphy, M.C., Coffey, L., O'Neill, A.C., Quinn, C., Prichard, R., McNally, S.: Can the synthetic C view images be used in isolation for diagnosing breast malignancy without reviewing the entire digital breast tomosynthesis data set? Irish J. Med. Sci. **187**(4), 1077–1081 (2018). https://doi.org/10.1007/s11845-018-1748-7

16. Nguyen, T., et al.: Overview of digital breast tomosynthesis: clinical cases, benefits and disadvantages. Diagn. Intervent. Imaging **96**(9), 843–859 (2015). https://doi.org/10.1016/j.diii.2015.03.003

17. Ribli, D., Horváth, A., Unger, Z., Pollner, P., Csabai, I.: Detecting and classifying lesions in mammograms with deep learning. Sci. Rep. **8**(1), 1–7 (2018). https://doi.org/10.1038/s41598-018-22437-z

18. Siegel, R.L., Miller, K.D., Fuchs, H.E., Jemal, A.: Cancer statistics, 2021. CA Cancer J. Clin. **71**(1), 7–33 (2021). https://doi.org/10.3322/caac.21654

19. Tardy, M., Mateus, D.: Looking for abnormalities in mammograms with self-and weakly supervised reconstruction. IEEE Trans. Med. Imaging 1 (2021). https://doi.org/10.1109/TMI.2021.3050040

20. Vedantham, S., Karellas, A., Vijayaraghavan, G.R., Kopans, D.B.: Digital breast tomosynthesis: state of the art. Radiology **277**(3), 663–684 (2015). https://doi.org/10.1148/radiol.2015141303

21. Wang, X., Liang, G., Zhang, Y., Blanton, H., Bessinger, Z., Jacobs, N.: Inconsistent performance of deep learning models on mammogram classification. J. Am. Coll. Radiol. **17**(6), 796–803 (2020). https://doi.org/10.1016/j.jacr.2020.01.006

22. Wu, N., et al.: Deep neural networks improve radiologists' performance in breast cancer screening. IEEE Trans. Med. Imaging **39**(4), 1184–1194 (2020). https://doi.org/10.1109/TMI.2019.2945514

23. Zhang, Y., Wang, X., Blanton, H., Liang, G., Xing, X., Jacobs, N.: 2D convolutional neural networks for 3D digital breast tomosynthesis classification. In: Proceedings - 2019 IEEE International Conference on Bioinformatics and Biomedicine, BIBM 2019, pp. 1013–1017. Institute of Electrical and Electronics Engineers Inc., November 2019. https://doi.org/10.1109/BIBM47256.2019.8983097

24. Zheng, S., Guo, J., Cui, X., Veldhuis, R.N., Oudkerk, M., Van Ooijen, P.M.: Automatic pulmonary nodule detection in CT scans using convolutional neural networks based on maximum intensity projection. IEEE Trans. Med. Imaging **39**(3), 797–805 (2020). https://doi.org/10.1109/TMI.2019.2935553

Clinical Applications - Dermatology

Multi-level Relationship Capture Network for Automated Skin Lesion Recognition

Zihao Liu[1,2], Ruiqin Xiong[1], and Tingting Jiang[1(✉)]

[1] NELVT, Department of Computer Science, Peking University, Beijing, China
{lzh19961031,ttjiang}@pku.edu.cn
[2] Advanced Institute of Information Technology, Peking University, Hangzhou, China

Abstract. Automated skin lesion recognition of dermoscopy images is effective for improving diagnostic performance. Current popular solutions either leverage a single image to learn better feature representations or take advantage of pairwise images for more discriminative recognition. However, they ignore modeling the relationship between important regions within the central lesion area, or mining the deeper semantic correlation between different images. In this paper, we propose a novel Multi-level Relationship Capture Network (MRCN), which focuses on relationship mining at two different levels, the region level and the image level. Specifically, a region-correlation learning module is proposed to model the relationship between different important regions in the central lesion area. Meanwhile, a cross-image learning module is designed to model the deep semantic correlation between multiple images. Besides, a lesion discerning module and a consistency regularization module are adopted to extract the feature of the lesion area and to serve as an extra consistency constraint, respectively. Comprehensive experiments are conducted on three challenging datasets, and the experimental results show that our MRCN can achieve the state-of-the-art performance compared to previous work, which demonstrates its advantages and superiority.

1 Introduction

Skin disease is one of the most common diseases in the world, which aroused public attention [13,15]. A large number of methods have been proposed for the automated recognition of dermoscopy images since the manual inspection is subjective.

Most methods utilize a single image for the final recognition. Early approaches apply hand-crafted features to solve this problem [2,10,17]. Recently, many CNN-based methods are also proposed. One stream of them is mainly designed for learning better feature representations [8,27]. Nevertheless, it is not enough to work at the feature level. For dermoscopy images, only the lesion area located in the center of the image is valuable for the diagnosis, which is called the "central lesion area", as shown in Fig. 1. Regarding this, another stream aims to take advantage of this characteristic. Some crop out the lesion area before the classification [14,26], the others utilize attention mechanisms to focus on the lesion area [30]. However, all the above methods ignore to mine the hidden information within the central lesion area. During the diagnosis process of dermatologists, different regions within

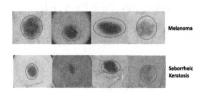

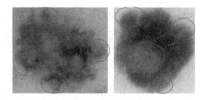

Fig. 1. Some examples of melanoma and seborrheic keratosis. The central area circled by the red circle is called the "central lesion area". (Color figure online)

Fig. 2. The illustration of important regions. The red circle indicates the important sub-regions within the central lesion area. They are usually located in the center or on the edges and their relationship is evaluated by doctors. (Color figure online)

the central lesion area are examined by doctors. These regions are viewed by different importance, and their relationship is evaluated for a more in-depth analysis, as illustrated in Fig. 2. Note that here "region" denotes local attended regions within the central lesion area. Thus efficiently modeling the relationship between these meaningful and important regions is important for the classification, which is called **"Region level relationship challenge"**.

Besides the region-level relationship challenge with a single image, there is another challenge at the image level. For the dermoscopy image, the visual difference within the same class could be even more notable than that between different classes, as shown in Fig. 1. How to effectively explore the semantic similarities and discriminations between different images, no matter whether they are of the same category or not, is a big challenge of this task, which is called **"Image level relationship challenge"**. To tackle this, a few recent approaches propose to utilize image pairs instead of a single image [20, 22, 28], and discriminate whether they are from the same class. However, they just simply concatenate the two features, ignoring to model the deeper semantic correlation between the two images for more abundant messages, which could facilitate each other. For doctors, it is commonly adopted to mine complementary information and summarize contrastive visual appearances, *e.g.*, semantic similarity and the discrimination positions with different scales and locations, as for a more effective joint judgment. Thus, there is still much room to improve the solution for the "Image level relationship challenge".

To address the above two challenges, we propose a novel Multi-level Relationship Capture Network (MRCN), which **focuses on relationship mining at two different levels, the region level and the image level**. At the region level, inspired by the attention mechanism [21] and guided by doctors' expertise, a region-correlation learning module is proposed to model the relationship between different important regions within the central lesion area. At the image level, inspired by the doctors' practice, a cross-image learning module is introduced to learn the deep semantic correlation between multiple images for complementary information. Besides, a lesion discerning module and a consistency regularization module are proposed to extract the feature of the central lesion area and serve as an extra regularization, respectively.

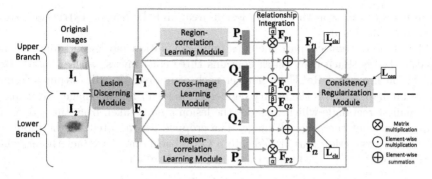

Fig. 3. An overview of the MRCN. There are two branches, the upper and the lower branch.

Experiments are conducted on three public datasets to demonstrate the effectiveness of our MRCN. We achieve state-of-the-art performance on all of them. To sum up, the main contributions are:

(1) To our best knowledge, this is the first paper to mine the relationship both at the image level and the region level for this task. The correlation between different important regions of the lesion area is modeled, and the deep semantic correlation between multiple images is learned to facilitate each other.
(2) A new architecture of MRCN is proposed, including two newly designed modules: region-correlation learning module and cross-image learning module, which are deeply in line with the intuition of doctors and integrates their expertise.
(3) Our MRCN achieves state-of-the-art performance on three public datasets.

2   Methodology

In this section, we elaborate on the whole architecture of MRCN, which is illustrated in Fig. 3. Given an image pair I_1 and I_2, they are first processed by the lesion discerning module, generating the features corresponding to the central lesion areas for each branch, denoted as F_1 and F_2. The following region-correlation learning module is equipped by each branch. It takes F of each image as input and outputs an attention feature P for each image, denoted as P_1 and P_2. In parallel, a cross-image learning module is proposed, which synergically utilizes F_1 and F_2 as input. An attention feature Q is generated for each image, denoted as Q_1 and Q_2. After that, for each branch, P and Q are aggregated with F for relationship integration, obtaining the final feature F_f, which is used for the final recognition for each image. Finally, serving as an extra regularization, the consistency regularization module takes F_1, F_2, F_{f1}, and F_{f2} as input, and evaluates the consistency, *i.e.*, whether they belong to the same image.

Lesion Discerning Module. This module is introduced to extract the feature of the "central lesion area", as shown in Fig. 4. It contains two parts: the lesion attention part, to crop the central lesion area; and the feature extraction part, to extract the feature.

The lesion attention part includes four convolution blocks and four deconvolution layers. Each conv block contains three conv layers, with a batch normalization layer and ReLU layer after each conv layer. It will output a lesion attention map, which is the binary segmentation result for the image. After that, the smallest rectangle which includes the lesion area is taken from the attention map to crop out the original image. The cropped lesion patch, as the input of the feature extraction part, is fed into a CNN backbone to extract the original feature $F \in \mathbb{R}^{H \times W \times C}$, which is the output of this module.

Region-Correlation Learning Module. This module is designed to mine the region level relationship, as shown in Fig. 5. Previous works have demonstrated that in CNNs, different feature channels correspond to different locations and regions of an image [16,25]. Thus, we design two channel-wise based attention blocks. The intra-channel attention learning block evaluates the importance of each channel itself, and the inter-channel correlation block models the correlation between channels. After that, the information of these two blocks is integrated by a region-correlation modeling layer.

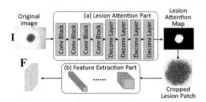 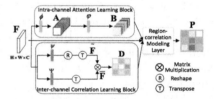

Fig. 4. Lesion discerning module.   **Fig. 5.** Region-correlation learning module.

For intra-channel attention learning block, firstly, an average pooling function ϕ is applied on F, which generates $A \in \mathbb{R}^{C \times 1}$. Then a learning function φ will be applied to A to further study the importance of each channel and generate $B \in \mathbb{R}^{C \times 1}$. In B, each element b_k represents the importance of kth channel.

$$A = \phi(F) = \frac{1}{H \times W} \sum_{i=1}^{H} \sum_{j=1}^{W} F_{ij}, \quad B = \varphi(A) \tag{1}$$

On the other hand, for inter-channel correlation block, F will go through two learning functions ψ and Ψ respectively, then a matrix multiplication between the two results is conducted to generate an attention map $D \in \mathbb{R}^{C \times C}$:

$$D = \psi(F) * \Psi(F) = \widetilde{F} * \overline{F} \tag{2}$$

where "*" means matrix multiplication. In D, each element d_{ij} measures the jth channel's impact on the ith channel.

Next, the region-correlation modeling layer is introduced to merge the messages of B and D and outputs $P \in \mathbb{R}^{C \times C}$. Each element p_{ij} of P is calculated by Eq. (3):

$$p_{ij} = d_{ij}(b_i b_j) \tag{3}$$

In this way, the importance of each channel itself will be integrated into the relationship between channels. It is worth noting that [21] simply models the spatial and channel relationships by pooling. But this module focuses on the correlation between important regions by integrating the importance of each channel and the relationship between them. The motivation, perspective and architecture are different.

Cross-Image Learning Module. This module is designed to model the image level relationship, as illustrated in Fig. 6. Firstly, F_1 and F_2 is processed by two learning functions T_1 and T_2 respectively and obtain $U \in \mathbb{R}^{N \times C}$ and $V \in \mathbb{R}^{N \times C}$, where $N(N = HW)$ is the number of spatial positions. Then, the spatial correlation modeling layer further measures the contextual semantic relevance. Lastly, the complementary learner encodes the learned correlation and generates the final attention maps.

The spatial modeling layer takes U and V as input, measures the correlation by cosine distance. It obtains two spatial correlation maps $S^{2 \to 1}$, $S^{1 \to 2} \in \mathbb{R}^{N \times N}$:

$$S_{ij}^{2 \to 1} = (\frac{u_i}{||u_i||_2})(\frac{v_j}{||v_j||_2})^T, \quad S_{ij}^{1 \to 2} = (\frac{v_i}{||v_i||_2})(\frac{u_j}{||u_j||_2})^T, \quad i, j = 1, ..., N \tag{4}$$

in which u_i denotes the ith row in U, similarly with v_i. Each $S_{ij}^{2 \to 1}$ is an affinity score reflects the message from j^{th} element in V to the ith element in U. Similarly with $S^{1 \to 2}$. Therefore, for one image, the semantic relevant elements in the other image are highlighted, results in higher values for the corresponding elements in $S^{2 \to 1}$ and $S^{1 \to 2}$.

The following complementary learner consists of two parts. The first part takes $S^{2 \to 1}$ and $S^{1 \to 2}$ as input, further learns the relevance between different elements by a learning function ϑ, encodes the message and generates attention map $Q_1 \in \mathbb{R}^{N \times 1}$ and $Q_2 \in \mathbb{R}^{N \times 1}$, which is the output of this module.

$$Q_1(i) = \sum_j^N \vartheta(S_{ij}^{2 \to 1}), \qquad Q_2(i) = \sum_j^N \vartheta(S_{ij}^{1 \to 2}) \tag{5}$$

For Q_1, the ith element represents the integration of the semantic messages from all the elements in F_2 to the ith element in F_1. Similarly with Q_2. The learning functions, which could be seen as the combination of convolution functions and activation functions, are illustrated specifically in Sect. 3.2. Therefore, those elements with a higher response, indicating more correlation with the other image, will correspond to a higher final score, emphasizing the complementary information.

Relationship Intergration. For each branch, P and Q will be integrated to F, as illustrated in Fig. 3. A matrix multiplication is applied between P and

F, meanwhile Q will be fused to original feature F by applying an element-wise multiplication. Then the results will be multiplied by a learnable weighting factor α and β respectively, then conduct a matrix summation to F, to generate the final feature F_f for each branch:

$$F_f = F + \alpha \cdot (P * F) + \beta \cdot (Q \cdot F) \tag{6}$$

where "$\cdot$" is element-wise multiplication and "$*$" denotes matrix multiplication.

F_f will pass through two fully connected layers to obtain the classification probabilistic prediction, which is supervised by normal cross-entropy loss $\mathcal{L}_{cls}$.

Consistency Regularization Module. An extra regularization is needed to impose extra constraints in addition to the effect of the Cross-image learning module and Region-correlation learning module for better feature learning. Specifically, the constraints are to ensure that after integrating messages from other images, the network can still correctly discriminate which pair of F and F_f is from the same image and which is from different images, illustrated as Fig. 7.

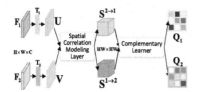
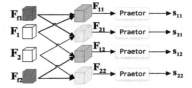

Fig. 6. Cross-image learning module.   **Fig. 7.** Consistency regularization module.

Taking F_1, F_2, F_{f1}, F_{f2} as input, the concatenation of F_1 and F_{f1} (F_{11}); F_2 and F_{f1} (F_{21}); F_1 and F_{f2} (F_{12}); F_2 and F_{f2} (F_{22}), are fed into an adaptive "praetor" respectively, and output four consistency scores. The praetor consists of two convolutional layers and two fully connected layers. The consistency score s_{11}, which corresponds to F_{11} is optimized to as close as possible to 1 since F_1 and F_{f1} are from the same image. The same with s_{22}. On the other hand, the consistency scores s_{12} and s_{21} are expected to be close to 0. These four scores will be supervised by binary cross-entropy loss:

$$\mathcal{L}_{con} = -\frac{1}{N_p}\sum_{z=1}^{N_p}\sum_{i=1}^{2}\sum_{j=1}^{2}(y_{ij}^z\log(s_{ij}^z) + (1 - y_{ij}^z)\log(1 - s_{ij}^z)) \tag{7}$$

where y_{ij}^z is the consistency label of zth image pair, N_p is the total number of image pair. For each image pair, y_{11}^z and y_{22}^z are 1, and y_{12}^z and y_{21}^z are 0.

The final loss is computed as $\mathcal{L} = \mathcal{L}_{cls} + \gamma\mathcal{L}_{con}$, where γ is a hyper-parameter.

3   Experiments

3.1   Datasets

We employ three benchmark datasets for experiments: the ISIC 2016 challenge dataset [7] consisting of 1279 images from 2 categories, the ISIC 2017 challenge dataset [3] including 2750 images from 3 categories and the ISIC 2019 challenge dataset [4,18] including 33569 images from 9 classes. We use the official training set, validation set and test set for evaluation. Note that ISIC 2016 and ISIC 2017 are two ended challenges, ISIC 2019 is an ongoing challenge, the results are obtained by submitting the predictions to the platform [19], which will be published on the leaderboard.

3.2   Implementation Details

ResNet50 is chosen as the backbone for the feature extraction part. For each image, its pair image is randomly chosen, with each resized to 448×448. ψ and Ψ are 1×1 convolution layers, T_1 and T_2 are conducted by 3×3 convolution layers. φ is the combination of a convolution layer and a ReLU layer, and ϑ is the combination of two convolution layers, with a ReLU layer between them. The learning rate is initialized to 0.001 and annealed by 0.5 every 10 epochs. The batch size is set to 40 on four NVIDIA GTX 2080Ti GPUs. γ is set to 0.05. As for evaluation metrics, we utilize Area Under Receiver operation Curve (AUC), Average Precision (AP), Accuracy (ACC), Sensitivity (SE) and Specificity (SP).

Training Phase: We follow the tradition in the other methods [6] and use the data with segmentation maps to train the lesion attention part of the lesion discerning module separately first at the training phase. After that, taking an image pair as inputs, the lesion discerning module outputs two corresponding features. These two features are the input of later architecture.

Testing Phase: The cross-image learning module and consistency regularization module will be removed during inference. Taking a single test image for the input, only the upper branch is used to obtain the final result.

3.3   Ablation Study

To investigate the impact of all the components in the network, we apply the ablation study on ISIC 2016 dataset. The performance is shown in Table 1. We denote "LD" as lesion discerning module, "RC" as region-correlation learning module, "CL" as cross-image learning module and "CR" as consistency regularization module. The pre-trained ResNet50 is used as the baseline model, denoted as "Baseline", which obtains an AP of 0.698, ACC of 0.843 and AUC of 0.814. The third, fourth and fifth rows are respectively the results of adding "LD", "RC" and "CL" to the baseline. The consistency regularization module will only work when there is at least one of the region-correlation learning module and the cross-image learning module. The AP value is improved by 3.2%, 3.5% and

3.3% respectively. The experimental result proves that individually adopting the three modules can benefit the model. When two modules are combined with baseline, illustrated as "LD+RC", "LD+CL", "RC+CR" and "CL+CR", the results are better. This demonstrates that combining two modules performs better than only combining one module with baseline. On this basis, when three modules are adopted, illustrated as "LD+RC+CL", "LD+RC+CR", "LD+CL+CR", and "RC+CL+CR", the performance further improves, proving that compared to combining two modules, adopting three modules gains a further improved performance. Finally, when the full model is adopted, represented as "MRCN (full)", the performance is the best, improving over the baseline by 10.9%, 5.5%, 7.6% in AP, ACC and AUC. Besides, the improvement of CR is smaller compared to CL and RC. For example, "LD+RC+CR" is worse than "LD+RC+CL"; "LD+CL+CR" is worse than "LD+RC+CL". This demonstrates the effectiveness of CL, RC themselves, and that CR only serves as an extra regularization.

3.4   Comparison with Other Methods

ISIC2016. To follow the tradition of other methods, we compare the performance of AP, ACC and AUC with five recent methods and top-five ranking methods on the challenge leaderboard. The results are shown in Table 2. This challenge is ranked based only on AP. Also, we do not use any extra data. Our method achieves the best performance in all three metrics, in which the AP is 0.807, significantly surpass the second place by 6.7%. Our ACC and AUC also exceed DCNN-FV by 1.1% and 2.7%.

Table 1. Ablation study on ISIC 2016 dataset.

Method	AP	ACC	AUC
Baseline	0.698	0.843	0.824
LD	0.730	0.848	0.839
RC	0.733	0.852	0.834
CL	0.731	0.861	0.846
LD+RC	0.757	0.871	0.862
LD+CL	0.760	0.872	0.863
RC+CR	0.749	0.873	0.868
CL+CR	0.752	0.873	0.868
LD+RC+CL	0.784	0.880	0.891
LD+RC+CR	0.776	0.882	0.879
LD+CL+CR	0.781	0.881	0.883
RC+CL+CR	0.779	0.883	0.886
MRCN (full)	**0.807**	**0.898**	**0.900**

Table 2. Results of our method, five recent methods and top five ranking methods on ISIC 2016. "AP" is the only ranking metric.

Method	AP^*	ACC	AUC
Our MRCN	**0.807**	**0.898**	**0.900**
CIN [9]	0.740	0.887	0.873
L-CNN [20]	0.724	0.876	0.854
AttnMel-CNN [24]	0.693	–	0.852
DCNN-FV [27]	0.685	0.868	0.852
SDL [28]	0.664	0.858	0.818
CUMED [26]	0.637	0.855	0.804
GTDL [7]	0.619	0.813	0.802
Result2 [7]	0.615	0.844	0.808
USYD [7]	0.580	0.686	0.793
Mufic-IT [7]	0.534	0.760	0.685

ISIC2017. We compare our method with six recent methods and top-five ranking methods on the challenge leaderboard. The results are shown in Table 3. Note that the Average AUC of the two sub-tasks is the ranking metric of this challenge. The AUC are 0.947 and 0.988 respectively for the two sub-tasks, which improve the second place by 2.7% and 0.7%, and the average AUC improves by 1.8%. In addition, the results of AP, ACC in two sub-tasks, and SE in sub-task1, are also best, comparing to other approaches. To sum up, without using any extra data, our method achieves the best performance on the ranking metric and most of the other metrics.

Table 3. Results of our method, six recent methods and top five ranking methods on ISIC 2017 dataset. Note that "Average AUC" is the only ranking metric, which is highlighted by "*".

Methods	External data	Melanoma classification					Seborrheic Keratosis					Average AUC*
		AUC*	AP	ACC	SE	SP	AUC*	AP	ACC	SE	SP	
Our MRCN	0	**0.947**	**0.864**	**0.906**	**0.796**	0.921	**0.988**	**0.917**	**0.949**	0.918	0.946	**0.968**
CIN [9]	0	0.920	0.814	0.894	0.645	0.948	0.981	0.902	0.943	0.829	0.965	0.951
MBDCNN [23]	1320	0.903	–	0.878	0.727	0.915	0.973	–	0.93	0.844	0.945	0.938
ARL-CNN [30]	1320	0.875	–	0.850	0.658	0.896	0.958	–	0.868	0.878	0.867	0.917
SSAC [22]	1320	0.873	–	0.835	0.556	0.903	0.959	–	0.912	0.889	0.916	0.916
SDL [29]	1320	0.868	0.689	0.872	–	–	0.955	0.818	0.917	–	–	0.912
RENI [11]	1444	0.868	0.710	0.828	0.735	0.851	0.953	0.786	0.803	**0.978**	0.773	0.911
gpm-LSSSD [5]	900	0.856	0.747	0.823	0.103	**0.998**	0.963	0.839	0.875	0.178	**0.998**	0.910
Alea-Jacta-Est [12]	7544	0.874	0.715	0.872	0.547	0.950	0.943	0.790	0.895	0.356	0.990	0.908
EResNet [1]	1600	0.870	0.732	0.858	0.427	0.963	0.921	0.770	0.918	0.589	0.976	0.896

ISIC2019. This challenge dataset is ranked based only on the balanced multi-class accuracy (BMCA), which is the average recall score. We compare with the methods on the challenge leaderboard. The results are shown on the platform [19]. Our method obtains the highest BMCA with 0.635, noticeably improved the second place by 1.2%. Besides, we yield the best result on SE, NPV and the second in AUC.

3.5   Visualization Results

To better understand how region-correlation learning module and cross-image learning module work, we visualize the attention maps of P and Q, and show three examples in Fig. 8. As shown, the attention maps P successfully pay attention to the regions on the edges, as well as those located in the center, which are both crucial for the diagnosis. Besides, comparing the attention maps of P and Q for the same image, the attention map of Q can identify some central and edge regions which have not been highlighted by P. This result suggests that the cross-image learning module can learn useful supplementary messages between image pairs.

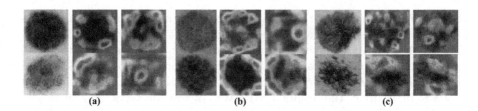

Fig. 8. Visualization results of attention map P and Q. There are three examples, each example contains three columns: the original image pair, the corresponding activation of P and the corresponding activation of Q.

4   Conclusion and Future Work

In this paper, we propose a novel Multi-level Relationship Capture Network (MRCN), which focuses on relationship mining from two levels, the region and the image level. Specifically, it contains four modules, a lesion discerning module, a region-relation learning module, a cross-image learning module and a consistency regularization module. The proposed method achieves state-of-the-art performance on three benchmark datasets. In future works, we will give more qualitative and quantitative results, including discussions of each module in our method, and of the top-ranking methods.

Acknowledgement. This work was partially supported by the Natural Science Foundation of China under contracts 62088102 and 62072009. We also acknowledge the Clinical Medicine Plus X-Young Scholars Project, and High-Performance Computing Platform of Peking University for providing computational resources.

References

1. Bi, L., Kim, J., Ahn, E., Feng, D.: Automated skin lesion analysis using large-scale dermoscopy images and deep residual networks. arXiv preprint arXiv:1703.04197 (2017)
2. Catarina, B., Emre, C.M., Jorge, S.M.: Improving dermoscopy image classification using color constancy. IEEE J. Biomed. Health Inform. **19**(3), 1146–1152 (2014)
3. Codella, N.C., et al.: Skin lesion analysis toward melanoma detection: a challenge at the 2017 international symposium on biomedical imaging (ISBI), hosted by the international skin imaging collaboration (ISIC). In: International Symposium on Biomedical Imaging, pp. 168–172. IEEE (2018)
4. Combalia, M., Codella, N.C., Rotemberg, V., et al.: BCN20000: dermoscopic lesions in the wild. arXiv preprint arXiv:1908.02288 (2019)
5. Díaz, I.G.: Incorporating the knowledge of dermatologists to convolutional neural networks for the diagnosis of skin lesions. arXiv preprint arXiv:1703.01976 (2017)
6. Gutman, D., et al.: https://challenge.isic-archive.com/landing/2016/41 (2016)
7. Gutman, D., et al.: Skin lesion analysis toward melanoma detection: a challenge at the international symposium on biomedical imaging (ISBI) 2016, hosted by the international skin imaging collaboration (ISIC). arXiv preprint arXiv:1605.01397 (2016)

8. Liang, R., Wu, Q., Yang, X.: Multi-pooling attention learning for melanoma recognition. In: 2019 Digital Image Computing: Techniques and Applications (DICTA), pp. 1–6. IEEE (2019)

9. Liu, Z., Xiong, R., Jiang, T.: Clinical-inspired network for skin lesion recognition. In: Martel, A.L., et al. (eds.) MICCAI 2020. LNCS, vol. 12266, pp. 340–350. Springer, Cham (2020). https://doi.org/10.1007/978-3-030-59725-2_33

10. Margarida, R., Catarina, B., Jorge S, M., Jorge, R.: A system for the detection of melanomas in dermoscopy images using shape and symmetry features. Comput. Methods Biomech. Biomed. Engineering: Imaging Visual. 5(2), 127–137 (2017)

11. Matsunaga, K., Hamada, A., Minagawa, A., Koga, H.: Image classification of melanoma, nevus and seborrheic keratosis by deep neural network ensemble. arXiv preprint arXiv:1703.03108 (2017)

12. Menegola, A., Tavares, J., Fornaciali, M., Li, L.T., Avila, S., Valle, E.: RECOD titans at ISIC challenge 2017. arXiv preprint arXiv:1703.04819 (2017)

13. Rebecca, L.S., Kimberly, D.M., Ahmedin, J.: Cancer statistics 2016. JAMA Dermatol. 66(1), 7–30 (2016)

14. ur Rehman, M., Khan, S.H., Rizvi, S.D., Abbas, Z., Zafar, A.: Classification of skin lesion by interference of segmentation and convolotion neural network. In: 2018 2nd International Conference on Engineering Innovation (ICEI), pp. 81–85. IEEE (2018)

15. Siegel, R.L., Miller, K.D., Jemal, A.: Cancer statistics, 2015. CA Cancer J. Clin. 65(1), 5–29 (2015)

16. Simon, M., Rodner, E.: Neural activation constellations: unsupervised part model discovery with convolutional networks. In: Proceedings of the IEEE International Conference on Computer Vision, pp. 1143–1151 (2015)

17. Tommasi, T., La Torre, E., Caputo, B.: Melanoma recognition using representative and discriminative kernel classifiers. In: Beichel, R.R., Sonka, M. (eds.) CVAMIA 2006. LNCS, vol. 4241, pp. 1–12. Springer, Heidelberg (2006). https://doi.org/10.1007/11889762_1

18. Tschandl, P., Rosendahl, C., Kittler, H.H: The HAM10000 dataset, a large collection of multi-source dermatoscopic images of common pigmented skin lesions. Sci. Data 5, 180161 (2018)

19. Tschandl, P., Rosendahl, C., Kittler, H.H: ISIC 2019 live leaderboard (2018). https://challenge.isic-archive.com/leaderboards/live

20. Wei, L., Ding, K., Hu, H.: Automatic skin cancer detection in dermoscopy images based on ensemble lightweight deep learning network. IEEE Access 8, 99633–99647 (2020)

21. Woo, S., Park, J., Lee, J.-Y., Kweon, I.S.: CBAM: convolutional block attention module. In: Ferrari, V., Hebert, M., Sminchisescu, C., Weiss, Y. (eds.) ECCV 2018. LNCS, vol. 11211, pp. 3–19. Springer, Cham (2018). https://doi.org/10.1007/978-3-030-01234-2_1

22. Xie, Y., Zhang, J., Xia, Y.: Semi-supervised adversarial model for benign-malignant lung nodule classification on chest CT. Med. Image Anal. 57, 237–248 (2019)

23. Xie, Y., Zhang, J., Xia, Y., Shen, C.: A mutual bootstrapping model for automated skin lesion segmentation and classification. IEEE Trans. Med. Imaging 39(7), 2482–2493 (2020)

24. Yan, Y., Kawahara, J., Hamarneh, G.: Melanoma recognition via visual attention. In: Chung, A.C.S., Gee, J.C., Yushkevich, P.A., Bao, S. (eds.) IPMI 2019. LNCS, vol. 11492, pp. 793–804. Springer, Cham (2019). https://doi.org/10.1007/978-3-030-20351-1_62

25. Yosinski, J., Clune, J., Nguyen, A., Fuchs, T., Lipson, H.: Understanding neural networks through deep visualization. arXiv preprint arXiv:1506.06579 (2015)
26. Yu, L., Chen, H., Dou, Q., Qin, J., Heng, P.A.: Automated melanoma recognition in dermoscopy images via very deep residual networks. IEEE Trans. Med. Imaging **36**(4), 994–1004 (2017)
27. Yu, Z., et al.: Melanoma recognition in dermoscopy images via aggregated deep convolutional features. IEEE Trans. Biomed. Eng. **66**(4), 1006–1016 (2019)
28. Zhang, J., Xie, Y., Wu, Q., Xia, Y.: Skin lesion classification in dermoscopy images using synergic deep learning. In: Frangi, A.F., Schnabel, J.A., Davatzikos, C., Alberola-López, C., Fichtinger, G. (eds.) MICCAI 2018. LNCS, vol. 11071, pp. 12–20. Springer, Cham (2018). https://doi.org/10.1007/978-3-030-00934-2_2
29. Zhang, J., Xie, Y., Wu, Q., Xia, Y.: Medical image classification using synergic deep learning. Med. Image Anal. **54**, 10–19 (2019)
30. Zhang, J., Xie, Y., Xia, Y., Shen, C.: Attention residual learning for skin lesion classification. IEEE Trans. Med. Imaging **38**(9), 2092–2103 (2019)

Culprit-Prune-Net: Efficient Continual Sequential Multi-domain Learning with Application to Skin Lesion Classification

Nourhan Bayasi[1]([envelope]) [ID], Ghassan Hamarneh[2] [ID], and Rafeef Garbi[1] [ID]

[1] BiSICL, University of British Columbia, Vancouver, BC, Canada
{nourhanb,rafeef}@ece.ubc.ca
[2] Medical Image Analysis Lab, Simon Fraser University, Burnaby, BC, Canada
hamarneh@sfu.ca

Abstract. Despite recent advances in deep learning based medical image computing, clinical implementations in patient-care settings have been limited with lack of sufficiently diverse data during training remaining a pivotal impediment to robust real-life model performance. Continual learning (CL) offers a desirable property of deep neural network models (DNNs), namely the ability to continually learn from new data to accumulate knowledge whilst retaining what has been previously learned. In this work we present a simple and effective CL approach for sequential multi-domain learning (MDL) and showcase its utility in the skin lesion image classification task. Specifically, we propose a new pruning criterion that allows for a fixed network to learn new data domains sequentially over time. Our MDL approach incrementally builds on knowledge gained from previously learned domains, without requiring access to their training data, while simultaneously avoiding catastrophic forgetting and maintaining accurate performance on all domain data learned. Our new pruning criterion detects *culprit units* associated with wrong classification in each domain and releases these units so they are dedicated for subsequent learning on new domains. To reduce the computational cost associated with retraining the network post pruning, we implement MergePrune, which efficiently merges the pruning and training stages into one step. Furthermore, at inference time, instead of using a test-time oracle, we design a smart gate using Siamese networks to assign a test image to the most appropriate domain and its corresponding learned model. We present extensive experiments on 6 skin lesion image databases, representing different domains with varying levels of data bias and class imbalance, including quantitative comparisons against multiple baselines and state-of-the-art methods, which demonstrate superior performance and efficient computations of our proposed method.

Keywords: Deep learning · Sequential learning · Multi-domain learning · Unit pruning · Dermatology

© Springer Nature Switzerland AG 2021
M. de Bruijne et al. (Eds.): MICCAI 2021, LNCS 12907, pp. 165–175, 2021.
https://doi.org/10.1007/978-3-030-87234-2_16

1   Introduction

Recent years witnessed exploding interest in deep learning approaches for performing medical image computing tasks. Nonetheless, real-life clinical deployment remains in its infancy due to practical limitations, most important of which is poor generalizability of deep learned models to new data/domains. Applicability of data-driven models is particularly challenging in medical imaging as availability and/or accessibility to sufficiently large and diverse training datasets is relatively more limited. Given that medical imaging data typically exhibit significant variability across sources (e.g., due to differences in scanner manufacturer, imaging protocol, patient population, or disease class), associated data bias and mismatch between training and test data frequently result in pronounced degradation in model performance at test time [21].

To help address poor model generalization, domain adaptation methods [8], which aim to transfer knowledge learned on one domain to other domains (i.e., data drawn from different distributions) became popular. A common issue in domain adaptation approaches is that they assume training data remains available post training. Another problem is that domain adaptation methods usually do not allow for accumulation of knowledge. This typically creates a static model that cannot adjust or adapt to new domains unless the training process is completely redone. In practice, this can cause challenges in the medical imaging field where data may not be accessible after the original training of a model due to proprietary or privacy constraints. A much desirable property is the ability of models to continually build knowledge over time by learning from heterogeneous multi-domain data, as well as retaining what is learned without requiring the data to be perpetually available.

Multi-domain learning (MDL), a subfield of continual learning (CL), has thus received growing interest [5]. A key consideration for effective MDL is avoiding catastrophic forgetting, where networks trained on new domains forget knowledge gained from previously learned domains [6]. In the literature, a common approach to tackle catastrophic forgetting is to form a (core) domain-agnostic model whose parameters are shared across all domains, and then add a set of extra parameters that are individually learned for each new domain, ensuring no or minimal performance drop across all domains [19,20,25]. However, such an approach introduces a scalability issue since adding per-domain model parameters would linearly increase with each additional domain. Another limiting problem is that the core model still needs to be trained on extensive data containing diverse classes and samples. An alternative approach to MDL is to use dynamic memory to maintain a subset of exemplars from previous domains and combine them with new data to form a more diverse training dataset for the model update [10]. This approach is impractical as the network has to be completely retrained every time a new domain is added, and data has to remain perpetually available for reuse.

Another difficulty in MDL approaches is addressing domain mapping at inference time. Domain mapping provides domain identification for each test image, without which the MDL model would not perform as anticipated. Most works deploy a task/domain oracle that assigns a test image to the correct learned model/sub-model [11,15]. However, a well-designed CL/MDL approach should not rely on a task/domain oracle to perform prediction, as argued by AlJundi et al. [2], where they proposed a gate that automatically forwards the test image to the appropriate learned model using a set of autoencoders.

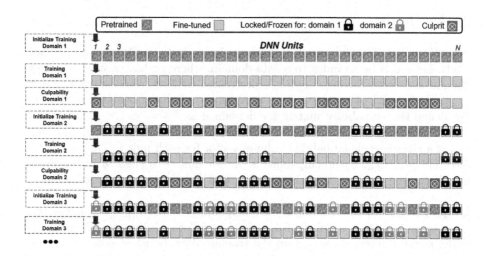

Fig. 1. Schematic visualization of the proposed Culprit-Prune-Net, where a single network learns multiple domains sequentially based on culprit unit pruning. From top to bottom: The top DNN units are all fine-tuned on the first domain, then culprit units are released to learn domain 2 while non-culprit units are reserved for the first domain. The process is repeated for subsequent domains.

To overcome the aforementioned limitations, in this paper we propose an effective approach for sequential MDL that is also computationally efficient and scalable, as it is based on a fixed-sized architecture. Since different units in a network make different contributions towards a classification decision [26], we assign a subset of the network units to learn each domain separately [15]. This is achieved using a novel unit pruning scheme that identifies *culprit units* associated with wrong decisions when learning a certain domain data. Identified culprit network units, associated with high 'culpability' scores, are pruned for one domain but subsequently reset and offered for learning new domains. Non-culprit units are however preserved (locked) for each domain, as illustrated in Fig. 1. While other works have put efforts in devising methods for scoring units during pruning, our method does not require any hyperparameters [7,12], and it does not add additional layers to the network [24], as detailed in the following section. Furthermore, we implement MergePrune, a 'gradual pruning-while-training' approach that allows for the pruning and training stages to work in parallel to save computations [1], which further improves the efficiency of our method. Our MergePrune prunes away the worst culprit units gradually at fixed epochs during the training stage until reaching a pre-defined pruning ratio. Finally, to avoid using domain-oracles at inference time for domain mapping, we deploy a smart gate based on Siamese networks that maps a test image to the domain with the most similar data. We evaluate and demonstrate the effectiveness of our approach in the context of skin lesion image classification.

2   Methods

Notation: Let $\mathcal{X} \in \mathbb{R}^D$ and $\mathcal{Y} \in \mathbb{N}$ be the input and class label space, respectively. Let $d \in \mathbb{N}$ be an index to a data domain, where we have access to a single domain at a time. During training, a pretrained CNN, θ, receives training data $\{x, y, d\}$ where $\{x, y\}$ are the training samples and d is the corresponding domain. We define the culpability matrix for domain d as $C_d = [c_{d,1}, c_{d,2}, \ldots, c_{d,N}] \in \mathbb{R}^{M \times N}$, where M is the number of classes, N is the number of units in θ, and $c_{d,n}$ is a class-specific culpability column vector for each unit n, i.e. $[c_{d,n}]_m$, the m-th entry of $c_{d,n}$, is the culpability score for unit n to class m in domain d, identifying the units most incriminated in that class erroneous classification.

Culpability Score: We compute the culpability score $[C_d]_{m,n}$ of unit n to a given class m during training of domain d as follows: After every λ epochs (λ is the pruning interval), we group the training data $\{x, y\}$ of domain d into 2 groups per ground truth class m: <u>R</u>ightly (correctly) classified: $\mathcal{R}_m^d$ is the set of (x, y) pairs in domain d with ground truth class m and are predicted as class m; and <u>W</u>rongly classified: $\mathcal{W}_m^d$ is the set of (x, y) pairs in domain d with ground truth class m but with a predicted class other than m. We analyze the rectified activations a_i^n of the input data sample i for each unit n in a layer in order to detect the units most culpable with misclassification and prune them. Inspired by neuroscience studies [18], those units with stronger activations associated with wrong classifications (c.f. $\mathcal{W}_m^d$) than with correct classifications (c.f. $\mathcal{R}_m^d$) are assigned higher culpability scores in $\hat{C}_d \in \mathbb{R}^{M \times N}$, i.e.,

$$\left[\hat{C}_d\right]_{m,n} = \frac{1}{|\mathcal{W}_m^d|} \sum_{i \in \mathcal{W}_m^d} \sum_{w,h} a_i^n(w, h) \quad - \quad \frac{1}{|\mathcal{R}_m^d|} \sum_{i \in \mathcal{R}_m^d} \sum_{w,h} a_i^n(w, h). \quad (1)$$

Since, in convolutional layers, the whole activation map is computed from a single unit, we aggregate the activation map a_i^n across the whole spatial dimensions, hence the summations over w, h in (1). The final $[C_d]_{m,n}$ is $[\hat{C}_d]_{m,n}$ normalized by the sum of all activation values, $[C_d]_{m,n} = [\hat{C}_d]_{m,n} / \sum_{m,n} [\hat{C}_d]_{m,n}$.

MergePrune: Instead of performing pruning and training separately, we merge the two stages into 'pruning-while-training'. We define N_o as the total number of training epochs, $p \in [0, 1]$ as the pruning percentage, l as the maximum pruning epoch defined as $l < N_o$, z as an integer defined as $z = \lfloor \frac{l}{\lambda} \rfloor$, and w as the percentage of pruned units per interval defined as $w = \lfloor \frac{p}{z} \rfloor$. Given θ and the first domain $d = 1$, we aim to compress the network by pruning away the culprit units associated with the highest culpability scores while training the network until the target pruning ratio p is achieved. MergePrune works as follows: We prune w of culprit units (and corresponding weights) after each pruning interval of epochs: $\lambda, 2\lambda, \ldots, z\lambda$, which enables successive epochs to compensate for possible accuracy drops associated with the pruning. Such an approach leads to a comparable performance to that of regular pruning followed by retraining techniques but with improved computational efficiency. To learn a new domain (e.g., $d = 2$), we release the identified culprit units and reuse the locked/frozen units from domain $d = 1$ for effective transfer learning and knowledge accumulation throughout the network (the units of domain $d = 1$ are preserved and not updated). We repeat the 'pruning-while-training' approach to free units for domain $d = 2$ and prune away culprit units associated with it. The whole process is repeated for all domains.

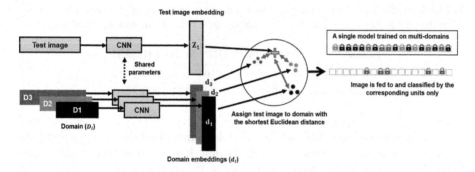

Fig. 2. Illustration of our Culprit-Prune-Net at inference time assuming test image belongs to the 3rd domain period.

Inference: We design a smart gate using Siamese networks [4] to tackle the problem of inference-time domain mapping. We use Siamese networks to encode each data domain into a unique output vector (by averaging all per-domain image vectors). At test time, we feed a test image into the same network, without performing any updates on weights or biases, calculate the test image's output vector and it's associated Euclidean distance to each domain's stored vector. The domain associated with the smallest distance is then assigned as the test image domain. Culprit-Prune-Net subsequently uses the corresponding units of the assigned domain to predict the class of the test image, as shown in Fig. 2. The main advantage of our Siamese networks is that it can be trained using a minimal number of epochs to get the output vectors, which saves computations. This is because we are using it to measure distances only and not to evaluate a function.

3   Experiments and Results

Datasets and Implementation Details: We use 6 public skin lesion datasets each representing a unique data domain: HAM10000 (HAM) [22], Dermofit (DMF) [3], Derm7pt (D7P) [13], MSK [9], PH2 [16], and UDA [9], each comprising skin lesion images collected using different equipment at different clinical sites. We follow the same protocol used in [23] to augment and partition each dataset (Table 1). In all our experiments, we use ResNet-50 pretrained on ImageNet as our base backbone model. We train the model using cross-entropy loss for $N_o = 100$ epochs with a constant learning rate of $1e-5$ and a batch size of 30. As for the Siamese networks, the backbone is a regular convnet with architecture conv-pool-conv-pool-conv-pool-conv-pool-fc-fc-softmax. We train it using cross-entropy loss for 20 epochs with a constant learning rate of $1e-5$ and a batch size of 32.

Table 1. Statistics of the 6 public skin lesion datasets used to validate our method. Each dataset is partitioned into 50% training, 20% validation, and 30% test sets.

Datasets	HAM	DMF	D7P	MSK	PH2	UDA
No. of classes	7	7	6	4	2	2
No. of images	10015	1212	1926	3551	200	601

In our Culprit-Prune-Net, we start with learning one of the domains and then sequentially add the other domains. For the first domain, we set the pruning ratio to $p = 70\%$ and train the model for 10 epochs. We start to prune $w = 17\%$ of the worst culprit units after every $\lambda = 15$ epochs while training the model until epoch $l = 70$. The rest of the epochs are used to ensure the convergence of the network (we note that p, λ and l are user-chosen whereas w is calculated as described in the Methods section). For every new domain, we fine-tune the network for 10 more epochs, set a new pruning ratio (z and w change accordingly), release culprit units from prior domains and repeat the same MergePrune protocol to assign specific set of units for each domain. We note that our reported accuracy results in experiments labelled with ($\star$) are averaged over all possible runs ($5! = 120$ permutations of domain ordering). This is done to eliminate any potential effect of domain training order sensitivity, i.e., discrepancy in performance with respect to the sequence of domain processing used.

Comparing Against Baselines: In experiments $\mathcal{A}$-$\mathcal{D}$, we compare the results of our Culprit-Prune-Net against 3 baselines on 5 skin lesion datasets. In Exp $\mathcal{A}$: joint training model, we train a single model on all domains data, depicting an ideal scenario where all data is available and can be pooled together for training. In Exp $\mathcal{B}$: multiple fine-tuned models, we train 5 models by fine-tuning our base model to one of the domain data, separately. In Exp $\mathcal{C}$: single fine-tuned model, we train a single model by *sequentially* fine-tuning it on each domain data (Exp $\mathcal{C}$-Row.1: Model fine-tuned on HAM, tested on in-distribution HAM and unseen domains DMF, D7P, MSK, etc. Exp $\mathcal{C}$-Row.2: Same model is fine-tuned on DMF, tested on in-distribution DMF, previously learned domain HAM, and unseen domains D7P, MSK, etc.). Finally, in Exp $\mathcal{D}$, we train our Culprit-Prune-Net as discussed in the previous paragraph. Table 2 shows the classification accuracy achieved on the test sets of each dataset. We observe that the highest accuracy was achieved in Exp $\mathcal{B}$, not surprisingly since training/learning of the 5 models in that experiment were optimized exclusively to learn each domain separately. The results in Exp $\mathcal{A}$ were the second best, also expected but that scenario requires access to all domain data during training, which is rarely possible in real-life applications. Our Culprit-Prune-Net in Exp $\mathcal{D}$ achieves comparable results to baselines $\mathcal{A}$ and $\mathcal{B}$ despite our high pruning ratios, which indicates effective pruning and multi-domain learning. Our network removes culprit units per class in each domain and transfers this knowledge to new domains, which seems to significantly improve the learning process especially when different domains have common classes. In Exp $\mathcal{C}$, as anticipated, we observe an increasing drop in performance on older datasets, i.e., sequentially fine-tuning a single model on new domains shows catastrophic forgetting. We also detect poor performance across other datasets which were not part of the training process. This highlights the importance of having a model capable of adapting to different domain datasets.

Comparing Against Other Pruning Methods: In experiments $\mathcal{E}$-$\mathcal{I}$, we investigate the role of sparsity in the context of our MDL. In Exp $\mathcal{E}$, we prune weights with the smallest absolute magnitude as in [15]. In Exp $\mathcal{F}$, we prune filters whose weights have the smallest L1-norm in a given layer as in [14]. The idea behind these two pruning methods is that weights or filters with small magnitudes are unlikely to be critical to the final classification. In Exp $\mathcal{G}$, we adopt the method in [17] to prune filters based on their first order Taylor expansion (1st-TE), which is used to approximate the contribution of a filter to the loss function). Finally, in Exp $\mathcal{H}$ and $\mathcal{I}$, we prune random weights and random units respectively. We compare all the above mentioned methods of the literature against our proposed culprit-based pruning technique, and show the classification results in Table 2. We observe that our culprit-based and the 1st-TE pruning methods outperform all other techniques. Also, we note that employing unit pruning appears to be more beneficial for MDL than employing weight pruning as it helps in reducing learning interference with previous tasks and thus overcoming the forgetting problem (Exp $\mathcal{D}, \mathcal{F}, \mathcal{G}, \mathcal{I}$ vs Exp $\mathcal{E}, \mathcal{H}$).

To Retrain or Not to Retrain: In experiments $\mathcal{J}$-$\mathcal{M}$, we further compare the performance of our culprit-based pruning against 1st-TE pruning by replacing the MergePrune stage (pruning-while-training). First we prune with retraining (Exp $\mathcal{J}$ and $\mathcal{K}$), and then we prune without retraining (Exp $\mathcal{L}$ and $\mathcal{M}$). Pruning with retraining prunes the network in one-shot then retrains the pruned network over a number of epochs (we used 40 epochs). Pruning without retraining does not retrain the pruned network at all. The results are displayed in Table 2. As expected, both our culprit-based and 1st-TE pruning performed better when retraining the one-shot pruned network as indicated in Exp $\mathcal{J}$ and $\mathcal{K}$ vs Exp $\mathcal{D}$

Table 2. Experiments $\mathcal{A}$ - $\mathcal{M}$ classification results. The pruning ratios in Exp $\mathcal{D}$ - $\mathcal{M}$ are: 70% in 1st task, 80% in 2nd and 3rd tasks, 85% in 4th task.

Exp	Method	Classification accuracy in test datasets (%)					Avg ± std%
		HAM	DMF	D7P	MSK	PH2	
Baselines							
$\mathcal{A}$	Joint learning	79.32	70.78	79.37	74.12	89.16	78.55 ± 6.96
$\mathcal{B}$	Separate learning	86.85	72.82	66.05	77.91	90.01	78.73±9.87
$\mathcal{C}$	Fine-tuning	86.85	30.86	39.68	54.99	85.00	59.47 ± 25.65
		41.43	71.23	12.79	14.34	17.5	31.46 ± 25.09
		39.62	35.39	70.31	58.93	80.0	56.85 ± 19.23
		31.34	32.92	49.69	75.71	50.12	47.95 ± 17.89
		29.59	23.86	45.61	57.94	90.64	49.53 ± 26.61
Our method							
$\mathcal{D}^*$	Culprit-Prune-Net	85.14	71.52	68.39	74.73	88.21	77.60 ± 8.64
Other pruning methods							
$\mathcal{E}^*$	Weights magnitude	81.51	69.71	67.31	71.68	86.96	75.43 ± 8.41
$\mathcal{F}^*$	L1—norm	82.46	70.52	68.54	71.91	86.76	76.03 ± 8.06
$\mathcal{G}^*$	1st-TE	84.54	72.03	67.18	73.84	87.93	77.10 ± 8.77
$\mathcal{H}^*$	Random weight pruning	82.41	69.93	66.62	71.57	84.38	74.98 ± 7.91
$\mathcal{I}^*$	Random unit pruning	84.73	70.53	71.23	73.26	84.97	76.94±7.29
Retrain or not to retrain							
$\mathcal{J}^*$	Pruning with retraining (Culprit)	86.76	73.95	70.67	79.21	91.49	81.42 ± 8.67
$\mathcal{K}^*$	Pruning with retraining (1st-TE)	87.27	74.68	69.64	78.93	91.96	80.49 ± 9.10
$\mathcal{L}^*$	Pruning without retraining (Culprit)	81.01	65.37	59.92	61.42	73.27	68.19 ± 8.84
$\mathcal{M}^*$	Pruning without retraining (1st-TE)	65.81	54.49	52.34	49.64	59.24	56.30 ± 7.37

and $\mathcal{G}$. However, when the network is not retrained after being pruned, the accuracy drops and gets even worse in the 1st-TE-based pruned network compared to our culprit-based technique suggesting our technique is more robust and practical on real world.

Generalizability to Unseen Domains: In experiments $\mathcal{N}$ and $\mathcal{O}$, we test our trained Culprit-Prune-Net from Exp $\mathcal{D}$ and the jointly trained model from Exp $\mathcal{A}$ on totally unseen data. The unseen held-out data belongs to the 6th dataset, UDA, which was not used until this point in any of our experiments. The classification accuracy is listed in Table 3. We observe that our Culprit-Prune-Net achieves higher accuracy on unseen data compared to the jointly trained network. This is likely because the jointly trained model was trained on multiple imbalanced and out-of-distribution datasets to UDA. In contrast, our Culprit-Prune-Net uses a smart gate that assigns the unseen test images to a specific domain based on a distance measurement, and classifies the test images using the units corresponding to the closest seen domain (in this specific experiment, the UDA test images were assigned to MSK domain/learned model).

Table 3. Generalizability to unseen domain results. The pruning ratios in Exp $\mathcal{N}$ are the same as in previous experiments.

Exp	Method	avg $\pm$ std (%)
$\mathcal{N}^*$	Our Culprit-Prune-Net (Exp $\mathcal{D}$)	71.90$\pm$7.84
$\mathcal{O}$	Joint Learning (Exp $\mathcal{A}$)	70.25$\pm$6.42

Performance of the Siamese Networks: We evaluate the performance of the Siamese networks by calculating the percentage of assigning a test image to a wrong domain (the lower, the better). We found the performance to be satisfactory, with only 2.18% of the time the Siamese networks were not able to send the test image to the corresponding domain. Although using a domain-oracle gate would have given us a 100% correct mapping always, it is not practical as it requires prior information that is not available at deployment.

4   Conclusions

We presented Culprit-Prune-Net, a novel approach for continual, sequential multi-domain learning that is based on per domain and class culpability scoring for network pruning. Pruned units constituting a subset of a fixed network are reset and sequentially retrained for each new domain, which ensures continual learning but also preserves previously attained knowledge resulting in better performance for each learned domain. The selection of units is based on identifying culprit units most responsible for wrong predictions in each class, per domain. We demonstrated that, even with high pruning ratios, the network is capable of performing accurately on real-life clinical image data from a half dozen different domains. In addition to outperforming competing methods in classification accuracy, our method tackles the problem of selecting the proper domain at test time and forwarding it to the corresponding learned units. Future work will focus on expanding the fixed network architecture to handle larger numbers of domains.

Acknowledgement. This work was funded by the NSERC Discovery program and Compute Canada.

References

1. Aketi, S.A., Roy, S., Raghunathan, A., Roy, K.: Gradual channel pruning while training using feature relevance scores for convolutional neural networks. IEEE Access **8**, 171924–171932 (2020)
2. Aljundi, R., Chakravarty, P., Tuytelaars, T.: Expert gate: lifelong learning with a network of experts. In: The IEEE Conference on Computer Vision and Pattern Recognition (CVPR), pp. 3366–3375 (2017)
3. Ballerini, L., Fisher, R.B., Aldridge, B., Rees, J.: A color and texture based hierarchical K-NN approach to the classification of non-melanoma skin lesions. In: Celebi, M., Schaefer, G. (eds.) Color Medical Image Analysis, pp. 63–86. Springer, Dordrecht (2013)

4. Bromley, J., Guyon, I., LeCun, Y., Säckinger, E., Shah, R.: Signature verification using a 'siamese' time delay neural network. In: Advances in Neural Information Processing Systems, p. 737 (1994)
5. Chen, Z., Liu, B.: Lifelong machine learning. In: Synthesis Lectures on Artificial Intelligence and Machine Learning, vol. 12, no. 3, pp. 1–207 (2018)
6. French, R.M.: Catastrophic forgetting in connectionist networks. Trends Cogn. Sci. **3**(4), 128–135 (1999)
7. Golkar, S., Kagan, M., Cho, K.: Continual learning via neural pruning. arXiv preprint arXiv:1903.04476 (2019)
8. Guan, H., Liu, M.: Domain adaptation for medical image analysis: a survey. arXiv preprint arXiv:2102.09508 (2021)
9. Gutman, D., et al.: Skin lesion analysis toward melanoma detection: a challenge at the international symposium on biomedical imaging (ISBI) 2016, hosted by the international skin imaging collaboration (ISIC). arXiv abs/1605.01397 (2016)
10. Hofmanninger, J., Perkonigg, M., Brink, J.A., Pianykh, O., Herold, C., Langs, G.: Dynamic memory to alleviate catastrophic forgetting in continuous learning settings. In: Martel, A.L., et al. (eds.) MICCAI 2020. LNCS, vol. 12262, pp. 359–368. Springer, Cham (2020). https://doi.org/10.1007/978-3-030-59713-9_35
11. Hung, S.C., Tu, C.H., Wu, C.E., Chen, C.H., Chan, Y.M., Chen, C.S.: Compacting, picking and growing for unforgetting continual learning. In: The Neural Information Processing Systems (NeurIPS) (2019)
12. Jung, S., Ahn, H., Cha, S., Moon, T.: Continual learning with node-importance based adaptive group sparse regularization. In: The Neural Information Processing Systems (NeurIPS) (2020)
13. Kawahara, J., Daneshvar, S., Argenziano, G., Hamarneh, G.: Seven-point checklist and skin lesion classification using multitask multimodal neural nets. IEEE J. Biomed. Health Inform. **23**(2), 538–546 (2018)
14. Li, H., Kadav, A., Durdanovic, I., Samet, H., Graf, H.P.: Pruning filters for efficient convnets. In: The International Conference on Learning Representations (ICLR) (2017)
15. Mallya, A., Lazebnik, S.: PackNet: adding multiple tasks to a single network by iterative pruning. In: The IEEE Conference on Computer Vision and Pattern Recognition (CVPR), pp. 7765–7773 (2018)
16. Mendonca, T., Ferreira, P.M., Marques, J.S., Marcal, A.R., Rozeira, J.: PH2 - a dermoscopic image database for research and benchmarking. In: The IEEE Engineering in Medicine and Biology Society (EMBS), pp. 5437–5440 (2013)
17. Molchanov, P., Mallya, A., Tyree, S., Frosio, I., Kautz, J.: Importance estimation for neural network pruning. In: The IEEE Conference on Computer Vision and Pattern Recognition (CVPR), pp. 11264–11272 (2019)
18. Morcos, A.S., Barrett, D.G., Rabinowitz, N.C., Botvinick, M.: On the importance of single directions for generalization. In: The International Conference on Learning Representations (ICLR) (2018)
19. Rebuffi, S.A., Bilen, H., Vedaldi, A.: Efficient parametrization of multi-domain deep neural networks. In: The IEEE Conference on Computer Vision and Pattern Recognition (CVPR), pp. 8119–8127 (2018)
20. Senhaji, A., Raitoharju, J., Gabbouj, M., Iosifidis, A.: Not all domains are equally complex: adaptive multi-domain learning. arXiv preprint arXiv:2003.11504 (2020)
21. Torralba, A., Efros, A.A.: Unbiased look at dataset bias. In: The IEEE Conference on Computer Vision and Pattern Recognition (CVPR), pp. 1521–1528 (2011)

22. Tschandl, P., Rosendahl, C., Kittler, H.: The HAM10000 dataset, a large collection of multi-source dermatoscopic images of common pigmented skin lesions. Sci. Data **5**(1), 1–9 (2018)
23. Yoon, C., Hamarneh, G., Garbi, R.: Generalizable feature learning in the presence of data bias and domain class imbalance with application to skin lesion classification. In: Shen, D., et al. (eds.) MICCAI 2019. LNCS, vol. 11767, pp. 365–373. Springer, Cham (2019). https://doi.org/10.1007/978-3-030-32251-9_40
24. Yu, R., et al.: NISP: pruning networks using neuron importance score propagation. In: The IEEE Conference on Computer Vision and Pattern Recognition (CVPR), pp. 9194–9203 (2018)
25. Zhao, H., et al.: What and where: learn to plug adapters via NAS for multi-domain learning. arXiv preprint arXiv:2007.12415 (2020)
26. Zhou, B., Sun, Y., Bau, D., Torralba, A.: Revisiting the importance of individual units in CNNs via ablation. CoRR abs/1806.02891 (2018)

End-to-End Ugly Duckling Sign Detection for Melanoma Identification with Transformers

Zhen Yu[1,6], Victoria Mar[2], Anders Eriksson[3], Shakes Chandra[3],
Paul Bonnington[4], Lei Zhang[1], and Zongyuan Ge[4,5,6(✉)]

[1] Central Clinical School, Faculty of Medicine, Nursing and Health Sciences,
Monash University, Melbourne, Australia
[2] Victorian Melanoma Service, Alfred Health, Melbourne, Australia
[3] School of Information Technology and Electrical Engineering,
The University of Queensland, Brisbane, Qld, Australia
[4] eResearch Centre, Monash University, Melbourne, Australia
zongyuan.ge@monash.edu
[5] Monash-Airdoc Research Centre, Monash University, Melbourne, Australia
[6] Monash Medical AI, Monash University, Melbourne, Australia
https://mmai.group

Abstract. The concept of ugly ducklings was introduced in dermatology
to improve the likelihood of detecting melanoma by comparing a suspicious lesion against its surrounding lesions. The ugly duckling sign suggests nevi in the same individual tend to resemble one another while malignant melanoma often deviates from this nevus pattern. Differentiating the
ugly duckling sign was more discriminatory between malignant melanoma
and other nevi than quantitatively assessing dermoscopic patterns. In this
study, we propose a framework for modeling ugly duckling context in
melanoma identification (called UDTR hereafter). To this end, we construct our model in three parts: Firstly, we extract multi-scale features
using a deep neural network from lesions in the same individuals; Then, we
learn lesion context by modeling the dependency among features of lesions
using a transformer encoder; Finally, we design a two branch architecture
for performing both patient-level prediction and lesion-level prediction
concurrently. Also, we propose a group contrastive learning strategy to
enforce a large margin between benign and malignant lesions in feature
space for better contextual feature learning. We evaluate our method on
ISIC 2020 dataset which consists of ∼30,000 images from ∼2,000 patients.
Extensive experiments evidence the effectiveness of our approach and
highlight the importance of detecting lesions with clues from surrounding lesions than that of only evaluating lesion in question.

Keywords: Melanoma diagnosis · Ugly duckling sign · Deep learning

1 Introduction

The incidence of malignant melanoma (MM) has been rising for several decades
[8]. Detecting melanoma at non-invasive stages remains key for preventing MM

© Springer Nature Switzerland AG 2021
M. de Bruijne et al. (Eds.): MICCAI 2021, LNCS 12907, pp. 176–184, 2021.
https://doi.org/10.1007/978-3-030-87234-2_17

related deaths. Visual dermoscopic examination with quantitative criteria such as "ABCD rule" and "7-point checklist" is a standard way for diagnosing MM. However, evaluating a lesion with dermosopic criteria is not always sufficient to rule out a melanoma as there is overlap in clinical features between benign nevi and melanoma. Grob et al. introduced the "ugly duckling sign" [6] to improve the accuracy of melanoma detection by using clues from surrounding lesions. The ugly duckling concept suggests nevi in the same individual tend to resemble one another while MM often deviates from this typical nevus pattern. This clinical realization highlights the importance of assessing the morphology of the lesion in question and comparing it to that of surrounding lesions, looking for an outlier in the context of similar-appearing nevi (see Fig. 1). Differentiating the ugly duckling sign was more discriminatory between MM and nevi than quantitatively evaluating dermoscopic patterns[4].

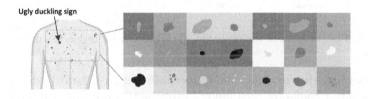

Fig. 1. The "ugly duckling" lesion tends to stick out and look different compared with the surrounding nevi. A suspicion of melanoma should be raised for this lesion.

Artificial intelligence (AI) algorithms have recently demonstrated remarkable performance in dermatology. Deep learning-based techniques are the most promising and have achieved at least equivalent performance when compared to experienced clinicians in image-based diagnosis under experimental conditions [2,3,5,10,11]. Although these studies show great potential to improve melanoma diagnosis, they all focus on evaluating lesion images in question. The presentation of an individual lesion can be problematic in the context of diagnosing small lesions or ambiguous melanoma as they lack dermoscopic signs for malignancy, e.g. nodular melanoma, for algorithms and clinicians alike. Identifying those atypical lesions requires more clues than dermoscopic patterns in an isolatedly presented lesion. Hence, we believe it is essential to model lesion contextual information from surrounding lesions to improve diagnostic algorithms and tools for melanoma detection.

In this study, we propose a framework to model ugly duckling context for melanoma identification. Our goal is to incorporate lesion contextual information by presenting a group of lesions for detecting melanoma and improving the overall diagnostic accuracy. To this end, we construct our method into three parts (see Fig. 2): Firstly, we extract multi-scale features for all lesions within the same individual using a deep neural network; Then, we learn lesion contextual embedding by modelling the dependencies among all lesions using a Transformer encoder; Finally, we perform both patient-level and lesion-level prediction concurrently

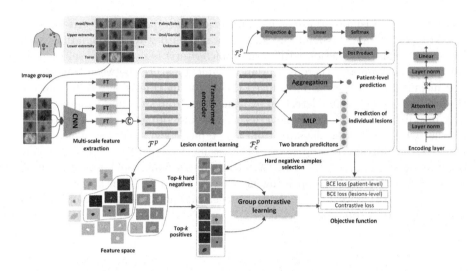

Fig. 2. The overview of the proposed UDTR model for melanoma detection.

with a two branches module. The former differentiates the ugly duckling sign while the latter computes each lesion's probability score. Apart from those, we propose a new group contrastive learning strategy to enforce a large margin between benign and malignant lesions in feature space for better contextual feature learning. Extensive experiments on the ISIC 2020 dataset demonstrate the effectiveness of our model. As far as we know, we are the first to employ the ugly duckling context for automated melanoma diagnosis.

2   Method

2.1   Lesion Context Learning with Transformer

Modelling the ugly duckling context enables distinguishing ambiguous MM from benign nevi and hence improving the diagnostic performance. We achieve this by applying a Transformer encoder to learning the relationship of various CNN features extracted from a group of lesion images belonging to the same patient.

Multi-scale CNN Feature Extraction. Each patient under dermoscopic examination contains a collection of lesion images. We define image group from p-th patient with $\mathbf{N}$ lesions as $\mathbf{X}^p = \{\mathbf{x}_1^p, \mathbf{x}_2^p, ..., \mathbf{x}_N^p\}$. We first extract deep features for each image via a ImageNet pre-trained CNN. As features from different layers of a CNN capture different scale information, we naturally adopt multi-scale feature for lesion context learning by concatenating multi-layer CNN outputs. In this study, we use off-the-shelf RegNet [7] and combine the intermediate feature maps after passing through a feature transformation (FT) module which consists of a global average pooling layer, a linear layer and a normalization layer. For p-th patient we denote deep features of $\mathbf{X}^p$ as $\mathbf{F}^p = \{\boldsymbol{f}_1^p, \boldsymbol{f}_2^p, ..., \boldsymbol{f}_N^p\} \in \mathbb{R}^{N \times D}$.

Transformer for Contextual Feature Encoding. Transformers have demonstrated to be a powerful full-attention model for capturing correlation between word embeddings for various nature language processing tasks [9], and it is desired for learning ugly duckling context from lesion images' features. The standard Transformer [9] comprises of an encoder-decoder structure and takes a sequence of features along with their position embedding as input. We use only the Transformer encoder because our task is to model the dependency among lesion image features for classification, which does not involve a generative process. Besides, we remove position encoding as we expect our model is invariant to lesion input ordering.

As shown in Fig. 2, the Transformer encoder consists of multiple encoding blocks constructed with alternating multi-head attention layer (MA), layer normalization layer (LN) and MLP layer. The multi-head attention layer comprises of multiple single attention layers which is based on self-attention mechanism. Supposing l-th single attention layer takes $\mathbf{F}_{l-1}^p$ as input, it first maps normalized features into query vectors of $\mathbf{F}_q^p \in \mathbb{R}^{N \times D_q}$, key vectors of $\mathbf{F}_k^p \in \mathbb{R}^{N \times D_q}$, and value vectors of $\mathbf{F}_v^p \in \mathbb{R}^{N \times D_v}$ separately with linear projections and then generates new feature vectors $\hat{\mathbf{F}}_l^p \in \mathbb{R}^{N \times D_v}$ using an attention function:

$$[\mathbf{F}_q^p, \mathbf{F}_k^p, \mathbf{F}_v^p] = \text{QKV}\left(\text{LN}\left(\mathbf{F}_{l-1}^p\right)\right) \tag{1}$$

$$\hat{\mathbf{F}}_l^p = \text{A}\left(\mathbf{F}_q^p, \mathbf{F}_k^p, \mathbf{F}_v^p\right) = \text{Softmax}\left(\frac{\mathbf{F}_q^p \cdot \mathbf{F}_k^{p\,T}}{\sqrt{D_q}}\right) \cdot \mathbf{F}_v^p \tag{2}$$

where QKV $(\cdot)$ denotes linear function realized by a fully-connected layer, LN $(\cdot)$ and A $(\cdot)$ represent a layer normalization function and self-attention function respectively. The pairwise dot product $\mathbf{F}_q^p \cdot \mathbf{F}_k^{p\,T} \in \mathbb{R}^{N \times N}$ measures how similar each pair of query vector and key vector is and the softmax function Softmax $(\cdot)$ computes a group of attention weights for every query vectors. The output of attention function A $\left(\mathbf{F}_q^p, \mathbf{F}_k^p, \mathbf{F}_v^p\right)$ is a weighted sum of $\mathbf{F}_v^p$ where a value vector gets a larger weight if its corresponding key vector has larger dot product with the query vectors. We can treat this process as comparing lesions in feature space and the attention weights to reflect on how closely a lesion is linking to another. The Multi-head attention layer performs Eq. (1) and (2) multiple times then concatenates the normalized attention results. We calculate the output of the transformer encoder with L stacked encoding blocks as:

$$\mathbf{F}_l^p = \text{MLP}\left(\text{LN}\left(\text{MA}\left(\mathbf{F}_{l-1}^p\right)\right)\right) + \text{MA}\left(\mathbf{F}_{l-1}^p\right), l = 1, 2, ..., L \tag{3}$$

$$\text{MA}\left(\mathbf{F}_{l-1}^p\right) = \text{LP}\left(\text{CAT}\left(\text{A}_1\left(\text{QKV}\left(\text{LN}\left(\mathbf{F}_{l-1}^p\right)\right)\right), \text{A}_2\left(\text{QKV}\left(\text{LN}\left(\mathbf{F}_{l-1}^p\right)\right)\right), \right.$$
$$\left. ..., \text{A}_h\left(\text{QKV}\left(\text{LN}\left(\mathbf{F}_{l-1}^p\right)\right)\right)\right)\right) + \mathbf{F}_{l-1}^p \tag{4}$$

where LP $(\cdot)$ is linear projection, CAT $(\cdot)$ denotes concatenation, h means number of attention operation. In this study, we set $D_q = D_k = D_v = 64$, $L = 4$ and $h = 6$. Since transformer encoder is capable of re-representing every individual feature of $\mathbf{F}^p$ with the context aggregated from entire feature sequence, we denote the final output features as $\mathbf{F}_c^p$ hereafter.

Two Branch Predictions. We design a two branch architecture to perform patient-level prediction and lesion-level prediction concurrently. As illustrated in Fig. 2, the global branch aggregates contextual feature $\mathbf{F}_c^p$ into an integral representation and then identifies ugly duckling sign from the input lesion group as a binary classification. The local branch directly computes the probability score for each lesion. We summarize the specific calculations as:

$$\mathbf{Pred}_{patient} = \mathrm{CLS}_{global}\left(\mathrm{Softmax}\left(\mathrm{LP}\left(\Psi\left(\mathbf{F}_c^p\right)\right)\right) * \mathbf{F}_c^p\right) \tag{5}$$

$$\mathbf{Pred}_{lesions} = \mathrm{CLS}_{local}\left(\mathrm{LP}\left(\mathrm{LN}\left(\mathbf{F}_c^p\right)\right)\right) \tag{6}$$

where $\Psi\left(\cdot\right)$ is a linear layer which projects input of $\mathbb{R}^{N \times D}$ into $\mathbb{R}^{1 \times D}$. Both CLS_{global} and CLS_{local} comprises of a normalization layer and a linear layer.

2.2   Group Contrastive Learning and Optimization

As our aim is to detect abnormal lesions or moles from normal ones by affinity, an intuitive thought is to use contrastive learning to form a representation space by maximizing distance among two categories of lesions. However, the extremely imbalanced setting for melignant melanoma versus benign nevi and the nuance in appearance of skin lesions complicate the feature contrastive learning. We therefore propose a group contrastive learning strategy by incorporating hard negative sample selection and a robust distance metric based on signal-to-noise ratio (SNR) [12]. Concretely, after obtaining predictions of lesion group from p-th patient with the UDTR model in a forward pass, we rank lesion features $\mathbf{F}^p$ according to corresponding probability scores from the local branch. Then, we calculate contrastive loss among top-k benign lesions' feature $\mathbf{F}_{ben}^p$ and malignant lesions' feature $\mathbf{F}_{mel}^p$ using SNR distance as follows:

$$\mathcal{L}_{gc}\left(\mathbf{F}_{ben}^p, \mathbf{F}_{mel}^p\right) = \max\left(0, m - \frac{1}{K_{mel}K_{ben}}\sum_i^{K_{mel}}\sum_j^{K_{ben}}\left(\frac{var\left(\boldsymbol{f}_i^p - \boldsymbol{f}_j^p\right)}{var\left(\boldsymbol{f}_i^p\right)}\right)\right)^2 \tag{7}$$

$$var\left(\boldsymbol{f}_i^p\right) = \frac{\sum_{d=1}^D\left(f_{i,d}^p - \mu_i\right)^2}{D} \tag{8}$$

where μ_i is the mean value of $\boldsymbol{f}_i^p$, m denotes margin which fixed as 1 and $K_{mel} = K_{ben}$ depends on minimum feature number of $\mathbf{F}_{ben}^p$ and $\mathbf{F}_{mel}^p$. The final objective function of for the model optimization is:

$$\mathcal{L} = \mathcal{L}_{local} + \alpha\mathcal{L}_{global} + \beta\mathcal{L}_{gc} \tag{9}$$

where $\mathcal{L}_{local}$ and $\mathcal{L}_{global}$ are binary cross entropy losses computed on predictions from the local branch and the global branch respectively. Here $\alpha = 0.2$ and $\beta = 0.02$.

3   Experiment and Results

3.1   Dataset and Implementation

In this study, we use the dataset from the SIIM-ISIC melanoma classification challenge. The challenge provides a skin lesion dataset containing ~30k images from ~2,000 patients. The images are organized with patient ID and the location information are also provided. The image number from same patients varies from 2 to 115. Within 2,056 patients, 428 patients are diagnosed with at least one melanoma. The number of melanoma detected in each patient varies between 1 to 8 with an average of 1.36.

We perform five-fold cross-validation to evaluate our method. Concretely, we first randomly partition the entire dataset into five folds. In each round of cross-validation, we select one fold as the testing set and further split the remaining part of the data into the training set and validation set (90% for training and 10% for validation). The hyper-parameters and models are only learned on the training set, and validation set, i.e. the testing set was never utilized to select a model. The standard data augmentation techniques such as random resized cropping, colour transformation, and flipping are used in all experiments. Each dermoscopic image is resized to a fixed size of 320 × 320. In the training stage, we randomly select a fixed number of lesion images from each patient as input[1]. During the test phase, we utilize ten crops augmentation and then average the final predictions. We do not use any other tricks for performance improvement.

3.2   Quantitative Results

Ablation Study: We first perform ablation study on our method to verify how different setting of each component affects the result. In Table 1, we give results of our UDTR model when trained with different sizes of input image groups. By increasing the input image number from 5 to 10, the AUC of lesion-level prediction increased with an improvement of 3.8%. The result remains relatively stable when we further enlarge the input image group size. Similar results were also observed on patient-level prediction with AUC improved from 75.97% to 79.34%. This is because the required input image number is greater than averaged lesion image number of patients and hence more images are repeatedly included without bringing new lesion context. Then, we use fixed size of 30 images as our model's input for following experiments.

In Table 2, we report result of our model on different output setting and training strategy. UDTR-L denotes only using local branch for lesion-level prediction, UDTR-LP is the model with two branches for both lesion-level and patient-level prediction. We can see UDTR-LP achieving slightly better result than UDTR-L with an improvement of ~0.5%, ~1.3% and ~1% in AUC, sensitivity and specificity respectively. Adding group contrastive learning (GCL) gains improvement of ~1.2% in AUC.

---

[1] We repeatedly sample images to the required input number of our model for those patients with insufficient lesion images.

Table 1. Ablation result of proposed model by varying input image numbers.

Image num	Lesion level prediction			Patient level prediction		
	AUC	Sensitivity	Specificity	AUC	Sensitivity	Specificity
5	82.89 ± 2.76	73.85 ± 3.25	74.01 ± 3.44	75.97 ± 2.51	69.36 ± 2.96	70.03 ± 2.69
10	85.05 ± 2.69	76.92 ± 3.59	77.06 ± 3.92	77.54 ± 2.45	69.82 ± 1.97	70.21 ± 2.56
15	85.09 ± 2.43	76.04 ± 3.14	76.85 ± 3.12	78.55 ± 2.34	70.76 ± 3.23	71.68 ± 3.23
20	85.22 ± 2.42	75.86 ± 2.98	76.19 ± 3.22	78.11 ± 2.48	70.54 ± 2.36	71.20 ± 1.74
30	**85.44 ± 2.75**	**76.55 ± 2.76**	**76.99 ± 2.75**	**79.34 ± 2.98**	**71.44 ± 3.52**	**71.38 ± 3.21**
40	85.06 ± 2.50	75.70 ± 2.87	76.32 ± 3.32	78.46 ± 3.46	71.00 ± 3.34	71.44 ± 3.37

Table 2. Ablation result on different output settings and group contrastive learning.

Models	AUC	Sensitivity	Specificity
UDTR-L	84.93 ± 2.89	75.22 ± 2.81	76.05 ± 3.01
UDTR-LP	85.44 ± 2.75	76.55 ± 2.76	76.99 ± 2.75
UDTR-LP + GCL	86.21 ± 1.44	76.96 ± 2.28	77.63 ± 2.01
UDTR-LP + GCL + ISIC pre-trained	**89.64 ± 0.95**	**81.64 ± 1.50**	**82.02 ± 1.36**

Table 3. Comparative result of the proposed model with other methods.

Methods		AUC
Without lesion context	Baseline CNN [7]	83.49 ± 2.91
With lesion context	Score post-processing	80.11 ± 2.19
	Feature post-processing	75.30 ± 3.12
	UDTR-L	84.93 ± 2.89
	UDTR-LP	86.21 ± 1.44
	UDTR-LP (ISIC pre-trained)	**89.64 ± 0.95**

Comparative Study: We then compare our model with that of models trained with and without lesion context in Table 3. The baseline CNN is built on ImageNet pre-trained RegNet by adding a new classification layer. Score post-processing denotes multiplying a coefficient ($c = 1.25$) to the highest prediction score for a lesion within a patient if the score exceeds the second highest score for a lesion from same patient by a certain margin ($m = 0.3$). Feature post-processing computes mean distances of features from baseline CNN for each lesion image to closest three neighbors within the same patient. Similar to the score post-processing, we increase the predication score of lesion image with highest mean distance, as outlier images are likely to be melanoma. The baseline CNN achieves an AUC of 83.49% but both two methods incorporating ugly duckling context by performing post-processing either on CNN's scores or features decrease the diagnostic performance. Adjusting CNN scores lead to an AUC decrease of ∼3.3% and feature post-processing reduces AUC to 75.3%. By contrast, our UDTR-L and UDTR-LP outperform the baseline CNN ∼1.5% and ∼3.2% in AUC respectively. UDTR with ISIC 2020 pre-trained CNN for feature extraction obtains an AUC of 89.64%.

These results evidence the superiority of our model in learning ugly duckling context for melanoma detection.

3.3   Visualization Results

In Fig. 3, we illustrate how our model gives predictions compared to that of the baseline CNN model learnt without using lesion context. In the top two groups of lesions, the proposed model shows more highlight on the melanoma class than baseline CNN. In the bottom two groups of benign lesion, both models output low probability scores for malignancy but our model has a more stable prediction score. For better understanding of our model in capturing the relationship among lesions' feature, we visualize weights from self-attention layers within the transformer encoder following [1]. As shown in Fig. 4, each node from top to bottom corresponds to a lesion's feature. At each layer, a node is connected to all feature from nodes in the previous layer and the line weight denotes how much an input feature contribute to the current node's output. In the first group of lesions (left figure), most of lesions have a strong correlation to the second lesion and the last lesion. In the second group, the fifth lesion links to more distinctly to other lesions.

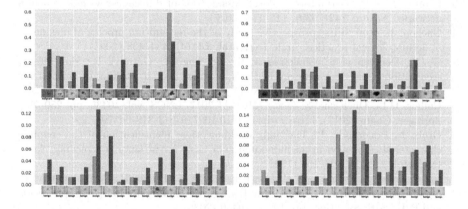

Fig. 3. Prediction scores of lesions from same patient. Yellow bars represent our model and green bars denote baseline CNN (best view in zoom mode). (Color figure online)

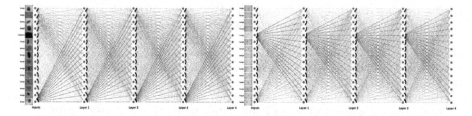

Fig. 4. Visualisation of attention weights from self-attention layers within our model. Line weights reflect the strength of correlation regarding a node to its input nodes.

4   Conclusion

In this study, we present a method based on CNN-Transformer and group contrastive learning to model ugly duckling context for melanoma detection. Our model is capable of concurrently performing patient-level prediction and lesions-level prediction from a group of lesions. Also, the proposed contrastive learning strategy enables generating more distinguishable features for learning the correlation among lesions. Experiments on the ISIC 2020 dataset demonstrate that our model can obtain better diagnostic result than baseline CNN trained without using lesion context and other methods incorporating ugly duckling context by post-processing.

References

1. Abnar, S., Zuidema, W.: Quantifying attention flow in transformers. arXiv preprint arXiv:2005.00928 (2020)
2. Brinker, T.J., et al.: Deep learning outperformed 136 of 157 dermatologists in a head-to-head dermoscopic melanoma image classification task. Eur. J. Cancer **113**, 47–54 (2019)
3. Esteva, A., et al.: Dermatologist-level classification of skin cancer with deep neural networks. Nature **542**(7639), 115–118 (2017)
4. Gachon, J., et al.: First prospective study of the recognition process of melanoma in dermatological practice. Arch. Dermatol. **141**(4), 434–438 (2005)
5. Ge, Z., Demyanov, S., Chakravorty, R., Bowling, A., Garnavi, R.: Skin disease recognition using deep saliency features and multimodal learning of dermoscopy and clinical images. In: Descoteaux, M., Maier-Hein, L., Franz, A., Jannin, P., Collins, D.L., Duchesne, S. (eds.) MICCAI 2017. LNCS, vol. 10435, pp. 250–258. Springer, Cham (2017). https://doi.org/10.1007/978-3-319-66179-7_29
6. Grob, J., Bonerandi, J.: The 'ugly duckling' sign: identification of the common characteristics of nevi in an individual as a basis for melanoma screening. Arch. Dermatol. **134**(1), 103–104 (1998)
7. Radosavovic, I., Kosaraju, R.P., Girshick, R., He, K., Dollár, P.: Designing network design spaces. In: Proceedings of the IEEE/CVF Conference on Computer Vision and Pattern Recognition, pp. 10428–10436 (2020)
8. Schadendorf, D., et al.: Melanoma. The Lancet **392**(10151), 971–984 (2018)
9. Vaswani, A., et al.: Attention is all you need. In: Advances in Neural Information Processing Systems, vol. 30, pp. 6000–6010 (2017)
10. Yu, L., Chen, H., Dou, Q., Qin, J., Heng, P.A.: Automated melanoma recognition in dermoscopy images via very deep residual networks. IEEE Trans. Med. Imaging **36**(4), 994–1004 (2016)
11. Yu, Z., et al.: Melanoma recognition in dermoscopy images via aggregated deep convolutional features. IEEE Trans. Biomed. Eng. **66**(4), 1006–1016 (2018)
12. Yuan, T., Deng, W., Tang, J., Tang, Y., Chen, B.: Signal-to-noise ratio: a robust distance metric for deep metric learning. In: Proceedings of the IEEE/CVF Conference on Computer Vision and Pattern Recognition, pp. 4815–4824 (2019)

Automatic Severity Rating for Improved Psoriasis Treatment

Xian Wu[1], Yangtian Yan[1], Shuang Zhao[2], Yehong Kuang[2], Shen Ge[1],
Kai Wang[1], and Xiang Chen[2(✉)]

[1] Tencent, Shenzhen, China
{kevinxwu,yangtianyan,shenge,ironswang}@tencent.com
[2] Xiangya Hospital of Central South University, Changsha, China
shuangxy@csu.edu.cn

Abstract. Psoriasis is a chronic skin disease which occurs to 2%–3% of the world's entire population. If treated properly, patients can still maintain a relatively high quality of life. Otherwise, Psoriasis could cause severe complications or even threat to life. Therefore, continuous tracking of severity degree is critical in Psoriasis treatment. However, due to the shortage of dermatologists, it's hard for patients to receive regular severity evaluation. Furthermore, evaluating the severity degree of Psoriasis is both time-consuming and error-prone which poses a heavy burden for dermatologists. To address this problem, we propose an automatic rating model which measures the severity degree quantitatively based on skin lesion pictures. The proposed rating model applies coarse to fine grained neural networks to evaluate skin lesions from multiple perspectives. According to experimental results, the proposed model outperforms experienced dermatologists.

Keywords: Severity rating · Psoriasis · PASI

1 Introduction

Psoriasis is a chronic skin disease which affects about 2%–3% of the world's entire population[1]. It is an immune system disease and cannot be fully cured. As a result, many patients have to suffer from Psoriasis throughout their entire lives. However if treated properly and timely, patients can still maintain a relatively high life quality. Otherwise, Psoriasis could cause severe complications such as diabetes and heart failure [3,19] or even threat to life.

Therefore, measuring the severity degree is a critical task in Psoriasis treatment. In current clinical practice, Psoriasis Area and Severity Index (PASI) [3] is the most frequently adopted indicator which uses a 0–72 score to measure the severity degree quantitatively. However, calculating PASI requires manually estimation of 16 sub-scores. This usually takes more than 30 min per patient

---

[1] http://www.worldpsoriasisday.com/.

X. Wu, Y. Yan, S. Zhao and Y. Kuang—Equal contribution.

M. de Bruijne et al. (Eds.): MICCAI 2021, LNCS 12907, pp. 185–194, 2021.
https://doi.org/10.1007/978-3-030-87234-2_18

[3] which poses a heavy burden on dermatologists. In addition, the PASI score highly relates to the subjective judgements of dermatologists. Different dermatologists or even the same dermatologist may come up with different PASI scores for the same case. The inconsistent PASI scores may mislead the judgement of severity progresses and result in improper therapies.

To address above problem, we propose a severity rating model which is designed to simulate the process of PASI score calculating. The proposed severity rating model adopts a siamese framework and includes two identical backbone neural networks. In this manner, we accept pairs of skin lesion photos as input and model their differences in severity degree. As a result, given N labelled skin lesion photos, we can generate $N \times (N-1)/2$ training instances which could reduce the dependency on labeled data. At different depths of the backbone neural network, we introduce a Lesion Attention Module (LAM) which can capture visual features of skin lesions from coarse to fine-grained granularity. We divide the 16 sub-scores to be estimated into two groups: 1) For the erythema, induration and desquamation that are only related to the skin lesion area, we propose a multiple task model and apply the same neural net for these sub-scores; 2) As for the area ratio, we set up a separate procedure which extracts visual features from the entire photo, since calculating the area ratio needs to consider both the lesion area as well as the normal skin.

To evaluate the proposed models in real clinical setting, we compare them with experienced dermatologists. The severity rating model achieves a 60% accuracy in erythema (5-class categorization), a 50% accuracy in induration (5-class categorization), a 54% accuracy in desquamation (5-class categorization), a 55% accuracy in area ratio (7-class categorization) and a 3.12 MAE for the overall PASI score. While experienced dermatologists achieve a 37% accuracy in erythema, a 43% accuracy in induration, a 43% accuracy in desquamation, a 43% accuracy in area ratio and a 4.67 MAE for the overall PASI score. Besides accuracy advantages, the proposed models can generate results in mill-seconds, while the dermatologist needs 30 min to make the decision.

2   PASI Score

Currently in clinical practice, Psoriasis Area and Severity Index (PASI) is the most frequently used indicator which can be calculated as follows:

1. Dermatologists examine skin lesions from four different body parts of a patient, including head, trunk, upper limbs, and lower limbs.
2. For each body part, dermatologists estimate erythema, induration and desquamation which are all represented by an integer score ranging from 0 to 4.
3. Then for each body part, dermatologists manually estimate the proportion of skin lesion area to normal skin area, resulting in a corresponding area ratio metric (range from 0 to 6 for 0% to 100%).
4. After acquiring four metrics for four body parts, the overall PASI score can be calculted via Eq. (1) and Eq. (2)

Table 1. How dermatologists from different levels of hospitals determine the severity degree of Psoriasis

	PASI score	Roughly estimate	Patient description	Do not know
Community-level	15.38%	76.92%	7.69%	0.00%
County-level	28.33%	65.00%	3.33%	3.33%
City-level	25.00%	72.62%	2.38%	0.00%
Province-level	48.48%	43.94%	4.55%	3.03%

$$PASI_{part_i} = (S_{ery_{part_i}} + S_{ind_{part_i}} + S_{des_{part_i}}) \times A_{part_i} \qquad (1)$$

$$PASI_{patient} = \sum_{i}^{parts} W_{part_i} \times PASI_{part_i} \qquad (2)$$

where $part_i \in \{\text{head}, \text{trunk}, \text{upper limbs}, \text{lower limbs}\}$ denotes the ith body part; $W_{part_i} \in \{0.1, 0.3, 0.2, 0.4\}$ denotes the corresponding weight; $S_{ery_{part_i}}$, $S_{ind_{part_i}}$, and $S_{des_{part_i}}$ denote the severity scores for the redness of erythema, the thickness of induration, and the scaling of desquamation, respectively, and A_{part_i} denotes the proportion score. $PASI_{patient}$ ranges from 0 to 72 [9].

Such a scoring process has two drawbacks. 1) Dermatologists have to estimate 4 scores $S_{ery_{part_i}}$, $S_{ind_{part_i}}$, $S_{des_{part_i}}$ and A_{part_i} for 4 body parts, that is 16 scores in total. The estimation heavily depends on the experience and expertise of dermatologists. Different dermatologists or even the same dermatologist could come up with varied scores for the same case. Therefore, manual PASI estimation may generate inconsistent scores. For patients whose severity needs to be tracked overtime, inconsistent measurement could mislead the judgment in Psoriasis severity progresses [8] and result in the adoption of an inappropriate therapy. 2) Since there are 16 scores to be estimated, the calculation process of PASI is quite time-consuming. An experienced dermatologist may need 30 min to calculate PASI for a single patient [3]. Considering the large number of Psoriasis patients, calculating PASI is a heavy burden for dermatologists.

We conduct a questionnaire on the usage of PASI score in clinical practice. We invite 223 dermatologists, among which, 13 dermatologists are from community level hospitals, 60 from county level, 84 from city level and 66 from province level. As shown in Table 1, even for dermatologists from province-level hospitals, less than half use PASI to measure Psoriasis severity, while others rely on rough estimation or let patients themselves to evaluate, which is quite error-prone and inconsistent. PASI adoption rate is even lower in the community-level, county-level and city-level hospitals. We also find that with the increase of the hospital level, the adoption rate of PASI is increasing, which shows from one perspective that the PASI is helpful in clinics.

To investigate the reason of the low PASI adoption rate, we ask the 223 dermatologists to choose the disadvantages of PASI (multiple answers). Only 13% are satisfied with PASI, and the major paint points are complex calculation, time-consuming and subjective. While the proposed severity rating model in this paper can alleviate all these pain points.

3   Approach

The severity rating model simulates the process of PASI calculation. For each
body part, the severity rating model predicts erythema, induration, desquamation and area ratio which are all discrete integers. In this manner, we could formulate severity rating as a classification task. However, considering the ordering
of different values, we formulate severity rating as a regression problem.

As shown in Fig. 1, the severity rating model consists of two identical sub-
networks. Each sub-network includes a backbone network and four Lesion Attention Modules (LAMs). The backbone network includes 33 convolutional layers
and 3 upsampling layers, residual shortcut connections are added to strengthen
skip-layer connections. The backbone network serves as a general visual encoder
and extracts features from the entire image. The four LAMs are equipped at different depth of the backbone network and combine features from intermediate
layers with skin lesion segmentation results. In this manner, the four LAMs can
attend to skin lesion regions and extract features at different granularity levels.
Among all sub-scores of PASI, the erythema, induration and desquamation are
mainly related to the skin lesion regions. Therefore, we employ features from
both LAMs and the backbone network to conduct the prediction. As to the area
ratio, we bypass the segmentation step and LAM module, and directly use the
features extracted by the backbone. This is because area ratio estimation needs
complete information from the entire image, including both the lesion regions
and the normal skin areas.

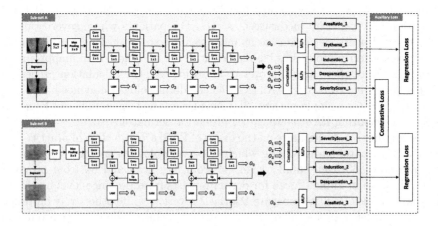

Fig. 1. The structure of severity rating network.

Lesion Attention Module. As shown in Fig. 2, LAM accepts X_s from skin
lesion segmentation models and X_b from the intermediate layers of the backbone
network as input. Both X_s and X_b are in the shape of $C \times W \times H$ where W and

H denote the width and height of the input image, C denotes the channel. For each pixel in X_s, if it belongs to a skin lesion region, we mark it as 1 otherwise we mark it as 0. We use Eq. (3) to transform X_b to attend to the skin lesion regions.

$$X = X_c \odot X_b \qquad (3)$$

Then the LAM passes X through two attention steps: pixel-wise attention and channel-wise attention. In pixel-wise attention step, inspired by the non-local operation [18], we first measure the relation between any two pixels with Eq. (4)

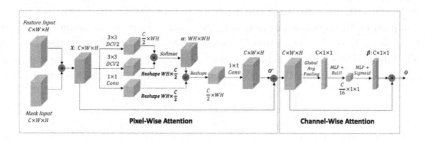

Fig. 2. The structure of LAM (Lesion Attention Module)

$$\alpha = Softmax(DCV2(X)^T \cdot DCV2(X)) \qquad (4)$$

where DCV2 denotes the deformable convolution V2 [20]. The DCV2 adds an offsets layer in both vertical and horizon directions, which enables the model to adapt to geometric variations. α is in the shape of $\mathbb{R}^{WH \times WH}$ and $\sum \alpha_{i,j} = 1$.

$$Z = Conv2((\alpha \cdot Conv1(X)^T)^T) \qquad (5)$$

where Conv1 denotes a 1×1 convolution layer which transforms X from $C \times W \times H$ to $C/2 \times W \times H$, Conv2 also denotes a 1×1 convolution layer which transforms $(\alpha \cdot Conv1(X)^T)^T$ from $C/2 \times W \times H$ back to $C \times W \times H$.

The output of the pixel-wise attention O' is obtained by adding Z with X as a residual connection to avoid over-fitting.

$$O' = X + Z \qquad (6)$$

where O' is in the shape of $C \times W \times H$.

Inspired by [10], we further process O' with the channel-wise attention. We first use the global average pooling to squeeze O', resulting in a feature map U with global receptive fields in each channel:

$$U = \frac{1}{W \times H} \sum_{i=1}^{W} \sum_{j=1}^{H} O'_{C,i,j} \qquad (7)$$

where U is in the shape of $C \times 1 \times 1$. We then apply MLPs on U in Eq. (8)

$$\beta = \sigma(W_2 Relu(W_1 U)))\tag{8}$$

where W_1 and W_2 denotes $C \times C$ matrix and σ denotes the sigmoid function. Then we acquire $\beta \in \mathbb{R}^{C \times 1 \times 1}$ and calculate the channel-wise attention with Eq. (9).

$$O = \beta \odot O'\tag{9}$$

where O is in the shape of $\mathbb{R}^{C \times W \times H}$:

Loss Function. In the proposed severity rating model, each sub-network is able to predict the score of erythema, induration, desquamation and area ratio. Since we formulate the score prediction as a regression model, we use Eq. (11) to calculate the loss. For an input image, let $\{y_e, y_i, y_d, y_a\}$ denote the true label and $\{s_e, s_i, s_d, s_a\}$ denote the predicted label, the regression loss is defined as:

$$l_t = \begin{cases} 0.5 \times (y_t - s_t)^2 & |y_t - s_t| < 1 \\ |y_t - s_t| - 0.5 & \text{otherwise} \end{cases}\tag{10}$$

$$L_1 = \sum_t^{e,i,d,a} l_t\tag{11}$$

where e, i, d, a denote the erythema, induration, desquamation and area ratio, respectively.

Since we have two sub-networks in the severity rating model, in addition to the loss of a single sub-network, we can use a pair of images as input and add their differences in the loss function. Let $t = \{(x_A, y_A, s_A), (x_B, y_B, s_B)\}$ denote a pair of images, where x denote an image, y denotes the actual PASI score and s denotes the predicted PASI score. Then we use Eq. (12) to calculate the loss between x_A and x_B,

$$L_2 = \begin{cases} (\Delta_p - \Delta_a)^2 & \Delta_p \times \Delta_a < 0 \\ 0.5 \times \max(|\Delta_p - \Delta_a| - m, 0)^2 & \text{otherwise} \end{cases}\tag{12}$$

where $\Delta_p = (s_A - s_B)$ and $\Delta_a = (y_A - y_B)$. In case of $\Delta_p \times \Delta_a < 0$, it means that the difference between two predictions is in the opposite direction to the difference between actual labels. In clinical practice, an incorrect ordering of two images in severity could lead to misjudgement in the severity tracking progresses of Psoriasis, which further causes an inappropriate therapy. Therefore, we exemplify the penalty on the incorrect direction predictions.

Combine the losses of a single sub-network and between two sub-networks, we define the overall loss as below:

$$L = \lambda \times L_1 + (1 - \lambda) \times L_2\tag{13}$$

where λ is introduced to balance between L_1 and L_2.

Table 2. The MAE of the severity rating model on sub-scores w/ and w/o LAM w/ and w/o siamse framework, lower MAE indicates better performance.

Model	W/LAM	W/Sia	Erythema	Induration	Desquamation	Area ratio
Baseline model [5]			0.495	0.545	0.555	1.11
The proposed model	No	No	0.563	0.612	0.588	0.665
	No	Yes	0.527	0.533	0.544	0.640
	Yes	No	0.503	0.539	0.523	0.655
	Yes	Yes	**0.463**	**0.481**	**0.487**	**0.634**

4 Experiment

4.1 Data Sets

We collect a data set of 1,915 Psoriasis patients with 5,205 images. For each body part of each patient, four metrics erythema, induration, desquamation and area ratio are all labeled. Therefore we can calculate the overall PASI score via Eq. (1). Each photo is also annotated which body part it belongs to. We build a separate data set of 200 Psoriasis patients for testing. Each labelling is a reached consensus among at least three experienced dermatologists.

4.2 The Performance of the Severity Rating Model

To the best of our knowledge, this is the first piece of work that automatically predicts the PASI score as well as all the 16 sub-scores. There are few automatic baselines for this particular task. Therefore we mainly conduct ablation studies and compare with experienced dermatologists.

We compare the accuracy of the severity models w/ and w/o the LAM (Fig. 2), w/ and w/o the siamese framework. For the model w/o LAM, we keep the input of LAM unchanged and send it to the next layer. For the model w/o the siamese framework, we only use one sub-network and optimize the loss in Eq. (11). As shown in Table 2, the LAM can improve the performance on Erythema, Induration and Desquamation significantly, as these sub-scores are mainly related to the skin lesion regions. As to the area ratio, LAM slightly improve the performance, as area ratio needs to consider both the lesion regions and normal skin. The siamese framework also reduce the MAE as it accepts image pairs as input, which better utilizes the training data set.

We also compare our approach with a state-of-the-art model [5] for pulmonary edema severity assessment of chest radiography images. Since this model is not originally designed for Psoriasis severity rating, we modify it by removing the textual part, as we do not have textual input. As shown in Table 2, the proposed model outperforms this baseline model.

To evaluate the severity rating model in clinical practice, we invite the Psoriasis patients who visited our collaborated hospitals in the second week of Jan 2020. 13 patients accepts the invitation. For each patient, we take pictures of their skin

Table 3. Comparison with dermatologists in accuracy, the predictions of the severity rating model are rounded to integers.

Annotator	Erythema	Induration	Desquamation	Area ratio	PASI score (MAE)
Dermatologists (Assistant & Assoicate)	36%	44%	44%	42%	4.78
Dermatologists (Professor)	39%	43%	42%	45%	4.34
Dermatologists (Avg)	37%	43%	43%	43%	4.67
Severity rating model	**60%**	**50%**	**54%**	**55%**	**3.12**

lesions and assign three dermatologists to label the erythema, desquamation and induration and area ratio for each body part. The labelling is conducted in a face-to-face manner and is a reached consensus among three dermatologists. Therefore, the labeling is trustworthy.

To collect the labelling from participant dermatologists, we build a labelling website and invite dermatologists to annotate these 13 patients online. The dermatologists are required to finish the scoring on their own. 78 dermatologists from 23 hospitals participated and 43 finished all the labelling for 13 patients.

As shown in Table 3, the proposed severity rating model outperforms dermatologists on all four metrics as well as the overall PASI score. Since the majority participant dermatologists are experienced in PASI scoring, the comparison proves the effectiveness of the proposed severity rating model. The severity rating model ranks the 8th among all 43 dermatologists.

5   Related Works

5.1   Skin Disease Classification

Many attempts have been made in the field of automatic diagnosis of skin diseases. For example, [2] develops a system which adjusts the decision-making process based on patient information; [4] conducts digital image analysis to diagnose the melanocyte lesion; [16] uses multiple databases to perform dermoscopic and digital dermoscopic examinations of cutaneous melanoma; [7] uses a deep convolution neural network framework Inception-V3 [17] to predict skin cancer types for lesion images, and achieve dermatologist-level performance; [12] proposed a deep learning model which predicts skin diseases of two layers: 27 classes in the first layer and 419 classes in the second layer.

5.2   Psoriasis Severity Rating

Due to the importance of automatic Psoriasis severity rating, some research efforts have been devoted to this area. [1] first maps skin lesions into a special-designed color space and then classified skin lesions into three different types; [13] scores erythema with K-nearest neighbor algorithm; [6] applies Gaussian mixture model to segment skin lesions into different color channels and then scored erythema by means of the trichromatic bands.

For deep learning based approaches, [14] estimates the severity levels of three indicators (erythema, induration and desquamation) by adding three output heads at one single deep neural network; [15] builds three different networks to evaluate these three indicators respectively. In the above methods, each indicator is classified into five discrete severity categories: from 0 to 4. A recent work PSENet [11] also uses convolutional networks to evaluate PSI scores. However, compared to the proposed approach, PSENet 1) only outputs the total score without the estimations of erythema, induration, desquamation and area ratio scores; 2) only calculates local PASI instead of the standard PASI considering all four body parts.

6 Conclusion

To improve the treatment of Psoriasis, we propose to automatically evaluate the severity scores. According to experimental results, the proposed model outperform experienced dermatologists. For patients, they are enabled to upload skin lesion photos and retrieve the results instantly. This is especially beneficial for patients in places where dermatologists are hard to reach; For dermatologists, the proposed model can relieve them from the dull PASI scoring.

Acknowledgments. We sincerely thank all the anonymous reviewers and chairs for their constructive comments and suggestions that substantially improved this paper. This work was supported by National Key R&D Program of China, No. 2018YFC0117000.

References

1. Ahmand, M., Ihtatho, D.: Objective assessment of psoriasis erythema for PASI scoring. J. Med. Eng. Technol. **33**, 514–516 (2009)
2. Alcón, J.F., et al.: Automatic imaging system with decision support for inspection of pigmented skin lesions and melanoma diagnosis. IEEE J. Sel. Top. Signal Process. **3**(1), 14–25 (2009)
3. Berth-Jones, J., et al.: A study examining inter-and intrarater reliability of three scales for measuring severity of psoriasis: psoriasis area and severity index, physician's global assessment and lattice system physician's global assessment. Br. J. Dermatol. **155**(4), 707–713 (2006)
4. Blum, A., Luedtke, H., Ellwanger, U., Schwabe, R., Rassner, G., Garbe, C.: Digital image analysis for diagnosis of cutaneous melanoma. development of a highly effective computer algorithm based on analysis of 837 melanocytic lesions. Br. J. Dermatol. **151**(5), 1029–1038 (2004)
5. Chauhan, G., et al.: Joint modeling of chest radiographs and radiology reports for pulmonary edema assessment. In: Martel, A.L., et al. (eds.) MICCAI 2020. LNCS, vol. 12262, pp. 529–539. Springer, Cham (2020). https://doi.org/10.1007/978-3-030-59713-9_51
6. Denmark, D.D.: An image based system to automatically and objectively score the degree of redness and scaling in psoriasis lesions. In: Proceedings FRA den 13. Danske Konference i, p. 130 (2004)

7. Berth-Jones, A., et al.: Dermatologist-level classification of skin cancer with deep neural networks. Nature **542**(7639), 115–118 (2017)
8. Fink, C., Fuchs, T., Enk, A., Haenssle, H.A.: Design of an algorithm for automated, computer-guided PASI measurements by digital image analysis. J. Med. Syst. **42**, 248 (2018)
9. George, Y., Aldeen, M., Garnavi, R.: Automatic psoriasis lesion segmentation in two-dimensional skin images using multiscale superpixel clustering. J. Med. Imaging **4**, 044004 (2017)
10. Hu, J., Shen, L., Sun, G.: Squeeze-and-excitation networks. In: Proceedings of the IEEE Conference on Computer Vision and Pattern Recognition, pp. 7132–7141 (2018)
11. Li, Y., et al.: PseNet: psoriasis severity evaluation network. In: Proceedings of the AAAI Conference on Artificial Intelligence, vol. 34, pp. 800–807 (2020)
12. Liu, Y., et al.: A deep learning system for differential diagnosis of skin diseases. Nature Med. **26**, 900–908 (2020)
13. Lu, J., Manton, J.H., Kazmierczak, E., Sinclair, R.: Erythema detection in digital skin images. In: 2010 IEEE International Conference on Image Processing, pp. 2545–2548. IEEE (2010)
14. Pal, A., Chaturvedi, A., Garain, U., Chandra, A., Chatterjee, R.: Severity grading of psoriatic plaques using deep CNN based multi-task learning. In: 2016 23rd International Conference on Pattern Recognition (ICPR), pp. 1478–1483. IEEE (2016)
15. Pal, A., Chaturvedi, A., Garain, U., Chandra, A., Chatterjee, R., Senapati, S.: Severity assessment of psoriatic plaques using deep CNN based ordinal classification. In: Stoyanov, D., et al. (eds.) CARE/CLIP/OR 2.0/ISIC -2018. LNCS, vol. 11041, pp. 252–259. Springer, Cham (2018). https://doi.org/10.1007/978-3-030-01201-4_27
16. Rajpara, S., Botello, A., Townend, J., Ormerod, A.: Systematic review of dermoscopy and digital dermoscopy/artificial intelligence for the diagnosis of melanoma. Br. J. Dermatol. **161**(3), 591–604 (2009)
17. Szegedy, C., Vanhoucke, V., Ioffe, S., Shlens, J., Wojna, Z.: Rethinking the inception architecture for computer vision. In: Proceedings of the IEEE Conference on Computer Vision and Pattern Recognition, pp. 2818–2826 (2016)
18. Wang, X., Girshick, R., Gupta, A., He, K.: Non-local neural networks. In: Proceedings of the IEEE Conference on Computer Vision and Pattern Recognition, pp. 7794–7803 (2018)
19. Zhou, Y., Sheng, Y., Gao, J., Zhang, X.: Dermatology in China. In: Journal of Investigative Dermatology Symposium Proceedings, vol. 17, pp. 12–14. Elsevier (2015)
20. Zhu, X., Hu, H., Lin, S., Dai, J.: Deformable convnets V2: more deformable, better results. In: Proceedings of the IEEE Conference on Computer Vision and Pattern Recognition, pp. 9308–9316 (2019)

Clinical Applications - Fetal Imaging

STRESS: Super-Resolution for Dynamic Fetal MRI Using Self-supervised Learning

Junshen Xu[1(✉)], Esra Abaci Turk[2], P. Ellen Grant[2,3], Polina Golland[1,4], and Elfar Adalsteinsson[1,5]

[1] Department of Electrical Engineering and Computer Science, MIT, Cambridge, MA, USA
junshen@mit.edu
[2] Fetal-Neonatal Neuroimaging and Developmental Science Center, Boston Children's Hospital, Boston, MA, USA
[3] Harvard Medical School, Boston, MA, USA
[4] Computer Science and Artificial Intelligence Laboratory, MIT, Cambridge, MA, USA
[5] Institute for Medical Engineering and Science, MIT, Cambridge, MA, USA

Abstract. Fetal motion is unpredictable and rapid on the scale of conventional MR scan times. Therefore, dynamic fetal MRI, which aims at capturing fetal motion and dynamics of fetal function, is limited to fast imaging techniques with compromises in image quality and resolution. Super-resolution for dynamic fetal MRI is still a challenge, especially when multi-oriented stacks of image slices for oversampling are not available and high temporal resolution for recording the dynamics of the fetus or placenta is desired. Further, fetal motion makes it difficult to acquire high-resolution images for supervised learning methods. To address this problem, in this work, we propose STRESS (**S**patio-**T**emporal **R**esolution **E**nhancement with **S**imulated **S**cans), a self-supervised super-resolution framework for dynamic fetal MRI with interleaved slice acquisitions. Our proposed method simulates an interleaved slice acquisition along the high-resolution axis on the originally acquired data to generate pairs of low- and high-resolution images. Then, it trains a super-resolution network by exploiting both spatial and temporal correlations in the MR time series, which is used to enhance the resolution of the original data. Evaluations on both simulated and *in utero* data show that our proposed method outperforms other self-supervised super-resolution methods and improves image quality, which is beneficial to other downstream tasks and evaluations.

Keywords: Fetal MRI · Image super-resolution · Self-supervised learning · Deep learning

Electronic supplementary material The online version of this chapter (https://doi.org/10.1007/978-3-030-87234-2_19) contains supplementary material, which is available to authorized users.

M. de Bruijne et al. (Eds.): MICCAI 2021, LNCS 12907, pp. 197–206, 2021.
https://doi.org/10.1007/978-3-030-87234-2_19

1   Introduction

Fetal magnetic resonance imaging (MRI) is an important approach for studying the development of fetal brain *in utero* [18] and monitoring fetal function [15]. Due to unpredictable and rapid fetal motion, dynamic fetal MRI, which aims at capturing fetal motion and dynamics of fetal function, is limited to fast imaging techniques, such as single-shot Echo-planar imaging (EPI) [2], with severe compromises in signal-to-noise ratio (SNR) and image resolution.

Super-resolution (SR) methods is frequently applied to fetal MRI to improve image quality. One well-established category of super-resolution methods for fetal MRI is based on slice-to-volume registration (SVR) [5,12,21]. In these methods, multiple stacks of slices at different orientations are acquired, which are then registered to reconstruct a static and motion-free volume of the chosen region of interest (ROI). However, multi-oriented stacks for oversampling the ROI may not available. Besides, in some applications, instead of a static ROI, a time series of MR volumes capturing the dynamics of fetal brain, body or placenta is of interest [11,15,20,23]. For example, in [23] and [15], interleaved multi-slice EPI time series are used for fetal body pose tracking and placental function analysis respectively. Thus, it is a still a challenge to enhance the resolution in dynamic fetal MRI.

Although supervised super-resolution methods achieved state-of-the-art results in natural images [14,24], the acquisition of HR MRI data with adequate SNR is time consuming and prone to motion artifacts, especially in fetal MRI. To avoid the need for HR data in supervised leanring, self-supervised super-resolution (SSR) methods have been developed, which utilize internal information from LR images for super-resolution. For instance, the ZSSR [19] method downsample the LR images to generate lower resolution (LR_2) images and train a network to learn a mapping from LR_2 to LR, which is then applied to the original LR images to estimate the HR images. Similar ideas are also explored in the field of MRI [9,25]. Zhao *et al.* extended [9] and proposed SMORE [25] for SSR of MR volume with anisotropic resolution where the information along the LR axis are learned from the other two HR axes. They blur the volume along the one of the HR axes, extract pairs of training samples to train a network and use it to enhance resolution along the LR axis. However, these methods only applied to a single slice or a stack of images and cannot utilize the temporal information in dynamic imaging.

In this work, we propose a SSR framework for dynamic fetal MRI with interleaved acquisition, named STRESS (**S**patio-**T**emporal **R**esolution **E**nhancement with **S**imulated **S**cans). Using the characteristic of interleaved slice acquisition, we perform simulated acquisitions on the originally acquired data to generate pairs of low- and high-resolution images. We then train a SR network on the extracted data, which exploits both internal spatial information within each frame and temporal correlation between adjacent frames. A optional self-denoising network is also introduced to this framework, when input images are of low SNR. We evaluate the STRESS framework on both simulated and *in utero* data to demonstrate that it can not only enhance resolution of dynamic fetal imaging but also improve performance of downstream tasks.

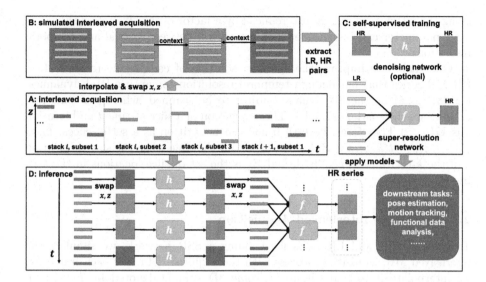

Fig. 1. The proposed STRESS workflow. A: Interleaved MRI acquisitions, e.g., $N_I = 3$. B: Acquired MR data are binned into different time frames. These frames are interpolated and transposed to produce a simulated object with motion. Then, we simulate a interleaved MR scan on this object and extract low- and high-resolution pairs from them C: We train the denoising network (optional) and super-resolution network in self-supervised manners. D: We apply the trained models to the originally or newly acquired data to generate a high-resolution MR volume series, which can be further used for other downstream tasks.

2   Methods

Figure 1 shows the workflow of the proposed STRESS method, which can be divided into four parts: 1) interleaved slice acquisition, 2) simulated acquisition, 3) self-supervised training, and 4) inference. The details of each part are described in the following sections.

2.1   Interleaved Acquisition

Interleaved slice acquisition is a widely used technique to avoid cross-excitation artifacts [3]. The number of slices skipped between two consecutive slice acquisitions is often referred to as the interleave parameter [16], N_I. For example, when $N_I = 2$, even slices are acquired after odd slices. Each image stack in interleaved acquisition are divided into N_I interleaved subsets. In dynamic imaging, multiple stacks are acquired. For simplicity, we refer to the i-th subset in the j-th stack as time frame F_k, where the index $k = N_I \times (i - 1) + j$. The acquisition time of each frame is only $1/N_I$ of the whole stack, making inter-slice motion artifacts within each frame milder. However, the spatial resolution of each frame along the interleaved axis is also reduced by a factor of N_I. Therefore the interleave parameter can be considered as a trade-off between spatial and temporal resolutions.

Our goal is to improve the spatial resolution of each frame to generate a HR MR series that has enough temporal resolution to capture fetal dynamics. Let $V_t(x, y, z)$ be the 3D dynamic object to be scanned, where t is time and (x, y, z) are the spatial variables. The acquisition of a slice at time t and location z is $V_t(\cdot, \cdot, z)$ Therefore, the k-th frame can be written as a set of slices, $F_k = \{V_t(\cdot, \cdot, z)|t = t(k, z), z \in \mathcal{Z}_k\}$, where $t(k, z)$ is the time when the slice at location z of the k-th frame is acquired, and $\mathcal{Z}_k$ is the set of slice locations in the k-th frame.

2.2 Simulated Interleaved Acquisition

To generate HR and LR pairs for training a SSR network, we simulate the interleaved MR acquisition process with the acquired data. For each frame F_k, we interpolate it to make it an isotropic 3D volume denoted by $\widetilde{F}_k(x, y, z)$. Then we swap the x- and z- axis[1] and result in a new 3D function $\widetilde{F}_k^T$, i.e., $\widetilde{F}_k^T(x, y, z) = \widetilde{F}_k(z, y, x)$. $\widetilde{F}_k^T(x, y, z)$ is an object of high resolution along the z-axis and having motion similar to V_t. Therefore, we can simulate interleaved acquisition along the z-axis to produce training pairs. The acquired frame in the simulated scan can be written as $S_k = \{\widetilde{F}_k^T(\cdot, \cdot, z)|z \in \mathcal{Z}_k\}$. Let $\widetilde{S}_k$ be the volume generated by interpolating S_k along the z-axis. We can see that the y-z planes of $\widetilde{S}_k$ and $\widetilde{F}_k$, i.e., $\widetilde{S}_{k+l}(x, \cdot, \cdot)$ and $\widetilde{F}_k^T(x, \cdot, \cdot)$ are pairs of LR and HR images. Besides, it is worth noting that the adjacent time frames provide contexts for estimating the missing slices in the target frame (Fig. 1 B). Therefore, it would be easier to learn a mapping from $\{\widetilde{S}_{k+l}(x, \cdot, \cdot)\}_{l=-L}^L$ to $\widetilde{F}_k^T(x, \cdot, \cdot)$, where L is the number of time frames used from each side.

2.3 Self-Supervised Training

Super-Resolution: We extract image patches with size of $P \times P$ from the series of images, $\{\widetilde{S}_{k+l}(x, \cdot, \cdot)\}_{l=-L}^L$, and concatenate them along the channel dimension to form input tensors $I_{LR} \in \mathbb{R}^{P \times P \times (2L+1)}$. Patches at the same spatial locations are also extracted from $\widetilde{F}_k^T(x, \cdot, \cdot)$ as targets and denoted as $I_{HR} \in \mathbb{R}^{P \times P}$. A network f is trained to learn the mapping between I_{LR} and I_{HR}. L1 loss is used to improve the output sharpness, i.e., $\mathcal{L} = ||f(I_{LR}) - I_{HR}||_1$. We adopt the EDSR [14] architecture for the SSR network f, with 16 residual blocks [8] and 64 feature channels.

Blind-Spot Denoising: Many fast imaging techniques for capturing fetal dynamics, e.g., EPI, suffer from low SNR [4]. Applying super-resolution algorithms to noisy images tends to emphasize image noise and results in images of

---

[1] We use x-axis here to keep the notation simple. In fact any axis within the x-y plane can be used.

low quality. To address this problem, we introduce an optional denoising network h to our framework, which can be apply when the original acquired images are of low SNR. The network h is a blind-spot denoising network (BDN) [13], i.e., the receptive field of h doesn't contain the central pixel. Therefore, when we train the network h to recover the input image I by minimizing the mean squared error, $||h(I) - I||_2^2$, the network will not become the identity function. Instead, $h(I)$ will approximate the mean of I, so that $h(I)$ can be considered as the denoised image. If BDN is enabled, we first train the denoising network h with images $I = \widetilde{F}_k^T(x, \cdot, \cdot)$. Then, when training the SSR network f, we replace the target I_{HR} with $h(I_{HR})$ and the loss becomes $\mathcal{L} = ||f(I_{LR}) - h(I_{HR})||_1$.

Training Details: We set $L = N_I/2$ and $P = 64$, if not specifically indicated. All neural networks are trained on a Nvidia Tesla V100 GPU using an Adam optimizer [10] with a learning rate of 1×10^{-4} for 30000 iterations. We use batch sizes of 64 and 16 for network f and h respectively, which depend on GPU memory. Training images are randomly flipped along the two axes for data augmentation. Our models are implemented with PyTorch 1.5 [17].

2.4 Inference

After training the models, we can apply them to the original or newly acquired data. If BDN is enabled, we first perform image denoising on each frame by applying h to each slice, such that F_k becomes $\{h(V_t(\cdot, \cdot, z))|t = t(k, z), z \in \mathcal{Z}_k\}$. Then, we interpolate it to generate a volume, $\widetilde{F}_k(x, y, z)$. Finally, the trained super-resolution network f is applied to the y-z plane of $\widetilde{F}_k(x, y, z)$ and its neighboring frames, which yields a super-resolved estimate $\hat{V}_k$, i.e., $\hat{V}_k(x, \cdot, \cdot) = f(\{\widetilde{F}_{k+l}(x, \cdot, \cdot)\}_{l=-L}^L)$. This process is repeated for all k until we get a HR estimation of the whole series, which can be used for other downstream tasks.

3 Experiments and Results

In the experiments, we apply the following methods to fetal MR volume series: 1) cubic B-spline interpolation along the interleaved axis; 2) interpolation along the temporal direction (TI); 3) spatio-temporal interpolation (STI); 4) SMORE [25] and 5) STRESS. In SMORE, we adopt the same super-resolution network architecture and the same training hyperparameters as STRESS for fair comparison. The reference PyTorch implementation for STRESS is available on GitHub[2]

3.1 CRL Fetal Dataset

The CRL fetal atlas [6] consist of T2-weighted fetal brain MRI with gestational age (GA) ranging from 21 to 38 weeks. The images are reconstructed to volume with size of $135 \times 189 \times 155$ and isotropic resolution of 1 mm. To simulate fetal

---

[2] https://github.com/daviddmc/STRESS.

motion, we use the fetal landmark time series in [23]. Specifically, we use two eyes and the midpoint of two shoulder to define the fetal pose and apply affine transformation to the MR volume to generate motion trajectories. There are 77 time series with length from 20 to 30 minutes in the landmark dataset. We randomly sample 10 1-min intervals from each series then apply the motion to the volumes, resulting in $18 \times 77 \times 10 = 13860$ data. We use 70% data for training and validation, 30% for test, data in the test set have different GAs from training and validation sets. We simulate MR scans with $N_I = 2, 4$ and 6, in-plane resolution of 1mm $\times$ 1mm and slice thickness of 1mm. SR methods are applied to the noise-free data and also noisy data corrupted by Rician noise [7] with standard deviation $\sigma = 3\%$ of the maximum intensity. BDN is enabled when there is noise.

Table 1 shows the peak signal-to-noise ratio (PSNR) and structural similarity index (SSIM) [22] comparing to the ground truth. PSNR and SSIM are computed within a mask of non-background voxels. The proposed STRESS method outperforms the competing methods at different interleave parameters, with and without noise. Figure 2 shows example slices of super-resolution results with $N_I = 4$ and Rician noise. Visual results also indicates that the outputs of STRESS have better image quality.

Table 1. PSNR and SSIM of the super-resolution results on the CRL dataset, where 'w/ noise' means adding Rician noise with $\sigma = 3\%$ of the maximum intensity. The best results are underlined.

Models	$N_I = 2$				$N_I = 4$				$N_I = 6$			
	w/o noise		w/ noise		w/o noise		w/ noise		w/o noise		w/ noise	
	PSNR	SSIM	PSNR	SSIM	PSNR	SSIM	PSNR	SSIM	PSNR	SSIM	PSNR	SSIM
SI	32.69	.9883	28.42	.8849	23.90	.9049	22.98	.8114	19.71	.7422	19.39	.6686
TI	29.01	.9111	25.31	.8258	29.21	.9076	25.48	.8273	28.60	.9084	25.52	.8288
STI	31.29	.9682	27.94	.8846	26.87	.9390	25.75	.8711	23.89	.8769	23.37	.8182
SMORE	36.19	.9895	30.38	.9006	31.36	.9687	28.57	.8916	25.29	.8703	24.27	.8093
STRESS	<u>36.77</u>	<u>.9921</u>	<u>33.51</u>	<u>.9702</u>	<u>34.56</u>	<u>.9873</u>	<u>32.81</u>	<u>.9655</u>	<u>28.98</u>	<u>.9480</u>	<u>28.24</u>	<u>.9213</u>

In addition, we also evaluate the performance of the STRESS method with and without BDN under different noise levels ($\sigma = 1\%, 3\%$, and 5% of the maximum intensity). The results are shown in Table 2. We can observe that the BDN makes a larger contribution to the performance of STRESS as the noise level increases.

3.2   Fetal EPI Dataset

We also evaluate our method with an *in utero* fetal EPI dataset in [15], which consist of 111 volumetric MRI time series at a gestational age ranging from 25 to 35 weeks. MRIs were acquired on a 3T Skyra scanner (Siemens Healthcare, Erlangen, Germany). Interleaved, multislice, single-shot, gradient echo EPI

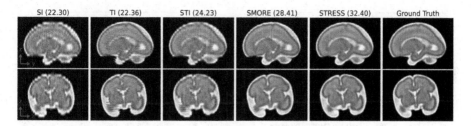

Fig. 2. Visual results from CRL fetal dataset ($N_I = 4$, Rician noise $\sigma = 3\%$ of the maximum intensity), numbers in the parentheses are PNSR with ground truth data as reference.

Table 2. Evaluations of STRESS with and without BDN under different noise levels ($N_I = 4$).

Models	$\sigma = 1\%$		$\sigma = 3\%$		$\sigma = 5\%$	
	PSNR	SSIM	PSNR	SSIM	PSNR	SSIM
STRESS w/o BDN	33.96	.9764	30.69	.9219	28.29	.8559
STRESS w/ BDN	33.99	.9826	32.81	.9655	31.09	.9425

sequence was used for acquisitions with in-plane resolution of 3mm × 3mm, slice thickness of 3 mm, average matrix size of $120 \times 120 \times 80$; TR = $5 - 8$ s, TE = $32-38$ ms, FA = $90°$, $N_I = 2$. Each subject was scanned for 10 to 30 min. We remove half of the slices at each frame to generate data with $N_I = 4$. We use 92 EPI series for training and 19 for testing. Due to the large voxel size in acquisition and the relatively high SNR, we disable BDN on this dataset. Besides, some volumes have matrix size less than 64, so we use $P = 32$ in this experiment.

Since ground truth is not available for the *in utero* dataset, we use the removed slices as reference to compute PSNR and SSIM. To further evaluate the quality of output images, we use fetal keypoint detection as a downstream task, where 15 fetal keypoints (ankles, knees, hips, bladder, shoulders, elbows, wrists and eyes) are detected from each time frame. Ground truth labels are manually annotated on the original data with $N_I = 2$. We apply a pretrained keypoint detection model [23] to the output volumes of each SR method. The percentage of correct keypoint (PCK) [1] are computed. PCK$(s) = N(s)/N \times 100\%$, where N is the total number of keypoints and $N(s)$ is the number of predicted keypoints with error less than threshold s.

Figure 3 shows the evaluation of super-resolution results on the fetal EPI dataset. The proposed STRESS method achieves the highest PSNR and SSIM among all competing methods, which is also shown by the t-test. Besides, when using the super-resolution results for fetal keypoint detection, the results of STRESS also have the best performance in terms of PCK, indicating that the STRESS method is able to generate MR time series with high image quality which is beneficial to downstream tasks.

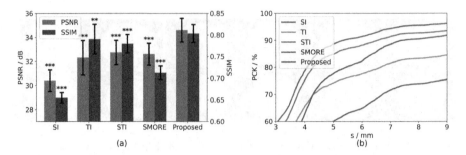

Fig. 3. Evaluation of super-resolution results from fetal EPI data with $N_I = 4$. Left: PSNR and SSIM comparing to the reference in the $N_I = 2$ data. Error bars show the corresponding standard deviations. $**$: p-value $< 10^{-2}$, $* * *$: p-value $< 10^{-3}$. Right: PCK curves for fetal landmark detection using a pretrained model.

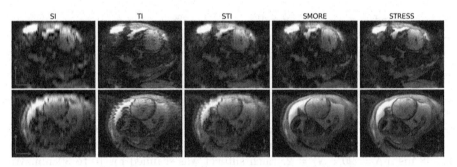

Fig. 4. Visual results from *in utero* fetal EPI dataset.

Figure 4 shows example slices of super-resolution results in one frame of the fetal MR series. We can see that the results of the proposed STRESS method have the best perceptual quality. The output of SI is very blurred, since it only interpolates along the z-axis. The TI and STI methods utilize temporal information with simple interpolation and therefore introduce severe inter-slice misalignment to the images. Although SMORE achieves better image quality than interpolation methods, the boundary of fetal brain is unclear in the outputs of SMORE. The reason is that SMORE only take a single frame as input without the temporal context, so that it cannot restore the details in the body parts that are corrupted by fetal motion, such as the fetal brain. STRESS, however, utilizes both spatial and temporal information of the scan data during the self-supervised training process, and therefore recovers more image details.

4   Conclusions

This paper presents STRESS, a self-supervised super-resolution framework for dynamic fetal imaging with interleaved slice acquisition. STRESS trains a SR network in a self-supervised manner, where low- and high-resolution training samples are extracted from simulated interleaved acquisitions. The SR network utilizes both internal spatial information within each frame and temporal correlation between adjacent frames to improve image quality and restore details corrupted by fetal motion. Evaluations on both simulated and *in utero* data shows that STRESS outperforms other competing methods. The experiments also demonstrate that STRESS is beneficial when serving as a data preprocessing step for further downstream analysis.

Acknowledgements. This research was supported by NIH U01HD087211, NIH R01EB01733 and NIH NIBIB NAC P41EB015902.

References

1. Andriluka, M., Pishchulin, L., Gehler, P., Schiele, B.: 2D human pose estimation: new benchmark and state of the art analysis. In: Proceedings of the IEEE Conference on Computer Vision and Pattern Recognition, pp. 3686–3693 (2014)
2. Diogi, M.C., et al.: Echo-planar flair sequence improves subplate visualization in fetal MRI of the brain. Radiology **292**(1), 159–169 (2019)
3. Dowling, J., et al.: Nonrigid correction of interleaving artefacts in pelvic MRI. In: Medical Imaging 2009: Image Processing, vol. 7259, p. 72592P. International Society for Optics and Photonics (2009)
4. Gholipour, A., et al.: Fetal MRI: a technical update with educational aspirations. Concepts Mag. Reson. Part A **43**(6), 237–266 (2014)
5. Gholipour, A., Estroff, J.A., Warfield, S.K.: Robust super-resolution volume reconstruction from slice acquisitions: application to fetal brain MRI. IEEE Trans. Med. Imaging **29**(10), 1739–1758 (2010)
6. Gholipour, A., et al.: A normative spatiotemporal MRI atlas of the fetal brain for automatic segmentation and analysis of early brain growth. Sci. Rep. **7**(1), 1–13 (2017)
7. Gudbjartsson, H., Patz, S.: The Rician distribution of noisy MRI data. Magn. Reson. Med. **34**(6), 910–914 (1995)
8. He, K., Zhang, X., Ren, S., Sun, J.: Deep residual learning for image recognition. In: Proceedings of the IEEE Conference on Computer Vision and Pattern Recognition, pp. 770–778 (2016)
9. Jog, A., Carass, A., Prince, J.L.: Self super-resolution for magnetic resonance images. In: Ourselin, S., Joskowicz, L., Sabuncu, M.R., Unal, G., Wells, W. (eds.) MICCAI 2016. LNCS, vol. 9902, pp. 553–560. Springer, Cham (2016). https://doi.org/10.1007/978-3-319-46726-9_64
10. Kingma, D.P., Ba, J.: Adam: a method for stochastic optimization (2017)
11. Kochunov, P., et al.: Fetal brain during a binge drinking episode: a dynamic susceptibility contrast MRI fetal brain perfusion study. Neuroreport **21**(10), 716 (2010)
12. Kuklisova-Murgasova, M., et al.: Distortion correction in fetal EPI using non-rigid registration with a Laplacian constraint. IEEE Trans. Med. Imaging **37**(1), 12–19 (2017)
13. Laine, S., Karras, T., Lehtinen, J., Aila, T.: High-quality self-supervised deep image denoising. arXiv preprint arXiv:1901.10277 (2019)
14. Lim, B., Son, S., Kim, H., Nah, S., Mu Lee, K.: Enhanced deep residual networks for single image super-resolution. In: Proceedings of the IEEE Conference on Computer Vision and Pattern Recognition Workshops, pp. 136–144 (2017)

15. Luo, J., et al.: In vivo quantification of placental insufficiency by bold MRI: a human study. Sci. Rep. **7**(1), 1–10 (2017)
16. Parker, D., Rotival, G., Laine, A., Razlighi, Q.R.: Retrospective detection of interleaved slice acquisition parameters from fMRI data. In: 2014 IEEE 11th International Symposium on Biomedical Imaging (ISBI), pp. 37–40. IEEE (2014)
17. Paszke, A., et al.: Automatic differentiation in pytorch (2017)
18. Saleem, N.S.: Fetal MRI: an approach to practice: a review. J. Adv. Res. **5**(5), 507–523 (2014)
19. Shocher, A., Cohen, N., Irani, M.: "zero-shot" super-resolution using deep internal learning. In: Proceedings of the IEEE Conference on Computer Vision and Pattern Recognition, pp. 3118–3126 (2018)
20. Turk, E.A., et al.: Placental MRI: effect of maternal position and uterine contractions on placental bold MRI measurements. Placenta **95**, 69–77 (2020)
21. Uus, A.: Deformable slice-to-volume registration for motion correction of fetal body and placenta MRI. IEEE Trans. Med. Imaging **39**(9), 2750–2759 (2020)
22. Wang, Z., Bovik, A.C., Sheikh, H.R., Simoncelli, E.P.: Image quality assessment: from error visibility to structural similarity. IEEE Trans. Image Process. **13**(4), 600–612 (2004)
23. Xu, J.: Fetal pose estimation in volumetric MRI using a 3D convolution neural network. In: Shen, D., et al. (eds.) MICCAI 2019. LNCS, vol. 11767, pp. 403–410. Springer, Cham (2019). https://doi.org/10.1007/978-3-030-32251-9_44
24. Zhang, Y., Tian, Y., Kong, Y., Zhong, B., Fu, Y.: Residual dense network for image super-resolution. In: Proceedings of the IEEE Conference on Computer Vision and Pattern Recognition, pp. 2472–2481 (2018)
25. Zhao, C., Dewey, B.E., Pham, D.L., Calabresi, P.A., Reich, D.S., Prince, J.L.: SMORE: a self-supervised anti-aliasing and super-resolution algorithm for MRI using deep learning. IEEE Trans. Med. Imaging **40**, 805–817 (2020)

Detecting Hypo-plastic Left Heart Syndrome in Fetal Ultrasound via Disease-Specific Atlas Maps

Samuel Budd[1]([⊠])(ID), Matthew Sinclair[1], Thomas Day[2,3], Athanasios Vlontzos[1],
Jeremy Tan[1], Tianrui Liu[1], Jacqueline Matthew[2,3], Emily Skelton[2,3,4],
John Simpson[2,3], Reza Razavi[2,3], Ben Glocker[1], Daniel Rueckert[1,5],
Emma C. Robinson[2], and Bernhard Kainz[1,6](ID)

[1] Imperial College London, Department of Computing, BioMedIA, London, UK
samuel.budd13@imperial.ac.uk
[2] King's College London, London, UK
[3] Guy's and St Thomas' NHS Foundation Trust, London, UK
[4] School of Health Sciences, City, University of London, London, UK
[5] Klinikum Rechts der Isar, Technical University of Munich, Munich, DE, Germany
[6] Friedrich–Alexander University Erlangen–Nürnberg, Baroque, DE, Germany

Abstract. Fetal ultrasound screening during pregnancy plays a vital role in the early detection of fetal malformations which have potential long-term health impacts. The level of skill required to diagnose such malformations from live ultrasound during examination is high and resources for screening are often limited. We present an interpretable, atlas-learning segmentation method for automatic diagnosis of Hypo-plastic Left Heart Syndrome (HLHS) from a single '4 Chamber Heart' view image. We propose to extend the recently introduced Image-and-Spatial Transformer Networks (Atlas-ISTN) into a framework that enables sensitising atlas generation to disease. In this framework we can jointly learn image segmentation, registration, atlas construction and disease prediction while providing a maximum level of clinical interpretability compared to direct image classification methods. As a result our segmentation allows diagnoses competitive with expert-derived manual diagnosis and yields an AUC-ROC of 0.978 (1043 cases for training, 260 for validation and 325 for testing).

Keywords: Segmentation · Atlas · Ultrasound

1 Introduction

Fetal Ultrasound (US) screening is a key part of ensuring the ongoing health of fetuses during pregnancy. Assessment of fetal development and accurate anomaly detection from US scans are integral in diagnosing potential fetal development issues at the earliest time possible to ensure the best care may be given. For these reasons a mid-trimester US scan is carried out between 18–22 weeks gestation in many countries as part of standard prenatal care procedures. During screening

© Springer Nature Switzerland AG 2021
M. de Bruijne et al. (Eds.): MICCAI 2021, LNCS 12907, pp. 207–217, 2021.
https://doi.org/10.1007/978-3-030-87234-2_20

'standard plane' views are used to acquire images in which key anatomical features may be examined, biometrics extracted and diagnosis of developmental issues may be made [15]. Several of these standard views and surrounding frames are used to make the diagnosis of Hypo-plastic Left Heart Syndrome (HLHS). Antenatal diagnosis of congenital heart disease such as HLHS has been shown to result in reduced mortality and morbidity of affected infants [7,10]. Unfortunately, antenatal detection of HLHS is not universal, due to the high level of skill required to make the diagnosis accurately from often noisy and inconsistent US views, which vary with gestational age, among other factors.

Recently, automatic ultrasound US scanning methods have been developed using deep learning, mitigating the difficulties of manual US screening through automatic detection of diagnostically relevant anatomical planes [5]. These systems have enabled the development of robust automated methods for estimation of anatomical biometrics and diagnosis of fetal structural malformations such as HLHS, under diverse acquisition scenarios with various imaging artefacts. Critically, these methods still provide limited interpretability of predictions, and as such reasoning about diagnosis and appropriate interventions remains a challenge even in the presence of accurate predictions of anatomical features and diagnosis of development issues [4,13,23,25].

Related Work: Automated segmentation of anatomical structures in US images has been the topic of significant research, with CNN based methods [6,20,24] often outperforming non-deep learning approaches [8,12,19]. Many of these methods perform well despite having no prior knowledge of the anatomical structure under consideration. However, in cases where performance drops, the resulting segmentations often bear no resemblance to the expected anatomical structure, resulting in segmentations that are not suitable for downstream analysis. As such, recent work to mitigate this fact has been introduced.

Methods such as Stochastic Segmentation Networks (SSNs) [14] aim to enforce continuity between anatomical structure segmentations (an assumption that holds in our case) to force a prediction to segment structures such that they remain connected and allow for sampling multiple plausible solutions to any given image segmentation. Similarly, [9,11] introduces topological priors to enforce continuity between segmented regions. Another recent approach aims to automatically learn an atlas of the anatomical structure under consideration during training of a segmentation model. Predicting both an image segmentation and a transformation between the automatically constructed atlas and the predicted segmentation, forces the resulting segmentation to retain the expected anatomical structure. In the presence of imaging artefacts or other image features the above behavior may result in a worse segmentation performance [21]. The aforementioned methods provide accurate segmentations of anatomical structures familiar to sonographers, however at present, these are not used to perform diagnosis of CHD or provide any means for disease-specific conditioning.

Deep Ultrasound Classification is currently the only option that has been explored in literature to perform automated diagnosis of CHD directly from US images. Deep classification methods achieve high accuracy [3,22,23], but rely on large curated datasets [3] or additional views of the heart to support multitask learning [23]. Unfortunately, conventional image classification is difficult to apply to fetal ultrasound because only very specific "standard planes" contain sufficient diagnostic information [15]. Classifying non-diagnostic frames (that should not be considered healthy or diseased) could lead to erroneous diagnosis or obscure the signal from the true diagnostic frames. As such, direct classification models have very little utility if clinicians cannot clearly interpret and assess the validity of the classification or find a view in pathological cases that would correspond to the defined anatomical standard [15].

Contribution: In this paper we introduce a novel method for the diagnosis of HLHS from US images using pathology-robust segmentation. To the best of our knowledge, we present for the first time a segmentation network that is able to jointly segment, register and build a labeled atlas that focuses on relevant features to robustly diagnose HLHS for fetal 4-chamber views in ultrasound imaging. By extending the recently proposed Atlas-ISTN framework [21] with an additional classification module and corresponding component in the loss function, our method provides an interpretable and accurate option for HLHS diagnosis compared to direct image classification approaches through segmentation of anatomical structures known to sonographers.

We assess the quality of our segmentations in the downstream task of inferring HLHS status from ventricular areas. From ground truth expert annotations we evaluate the possible correlation of this approximation to true disease status, which is confidently known from post-natal outcome records. We compare this correlation to using naive segmentation for area parameter extraction and our proposed method.

2   Method

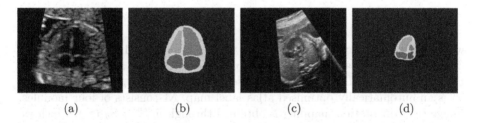

(a)   (b)   (c)   (d)

Fig. 1. Example ultrasound images and manual segmentations of anatomical areas. (a) Healthy patient 4CH ultrasound view; (b) manual segmentation of anatomical areas of healthy heart in (a); (c) HLHS patient's approximation of a 4CH ultrasound view; (d) manual segmentation of anatomical areas in (c).

We propose a new method for the automatic diagnosis of HLHS from a single US image of the '4-Chamber Heart View' (4CH view). Our system is inspired by current clinical practice and can be broken down into three major modules. First, for a given 4CH image, we seek a model that can provide accurate segmentations for 5 anatomical areas: *'Whole Heart', 'Left Ventricle', 'Right Ventricle', 'Left Atrium'* and *'Right Atrium'*. Figure 1 shows an example for the differences in these areas between a healthy fetus and a baby with HLHS.

Secondly, the resulting segmentation is used to extract simple image features informative of HLHS diagnosis. For each class, we calculate the ratio of that region's area to the area of every other segmented region to obtain a set of scalar features representative of that image. Finally, we use the quantitative features to construct a classifier for HLHS using: 1) class-weighted Logistic Regression Classifier as baseline; 2) Gaussian Process classifier with a radial basis function kernel. Figure 2 shows an overview over the diagnostic approach.

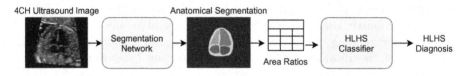

Fig. 2. Classification via deep segmentation and quantitative area ratio feature extraction

For the task of **Robust segmentation**, we adopt a recently proposed method, Atlas-ISTN [21], that generates a segmentation as well as learns a label atlas that both ensures robustness and can be used to inspect the inner beliefs of the network. Conditioning is used to sensitise the atlas generation to regions that are most relevant for the downstream task of disease classification.

Our detailed model is outlined in Fig. 3. As input we use uncropped 4-chamber ultrasound images, ground truth segmentation maps and a binary disease label in $X = \{x_i, x_i^{seg}, x_i^{HLHS}\}$. The model aims to learn:

$$\{\hat{y}_i^{seg}, \hat{y}_i^{HLHS}, y^a\} = \mathbf{M}(x_i),$$

where $\mathbf{M}$ is the entire model, $\hat{y}_i^{seg}$ are the logits of a predicted segmentation describing five cardiac labels with one background channel as defined in x_i^{seg}, $\hat{y}_i^{HLHS}$ are probabilities for discrete disease categories in $x_i^{HLHS} \in [0, 1]$, and y^a is an automatically optimised atlas label map. $\mathbf{M}$ consists of four modules. Image to segmentation mapping is obtained through $\hat{y}_i^{seg} = \mathbf{S}_{\theta_s}(x_i)$, which we define as a 2D UNet [18] with a SSN module [14]. The concatenation of $\hat{y}_i^{seg}$ and the atlas label map y^a is used to establish the atlas to image transformation:

$$\mathbf{d}_{b_i} = \mathbf{D}_{E,\theta_{enc}}(\hat{y}_i^{seg}, y^a),$$
$$\{v_i, T_i\} = \mathbf{D}_{D,\theta_{dec}}(\mathbf{d}_{b_i}).$$

v_i is a stationary velocity field and T_i an affine transformation matrix that is processed to a deformation field with a Transformation Computation Module $\mathbf{C}$ according to [21]. Thus, $\mathbf{C}$ yields forward and inverse transformations, Φ_i and Φ_i^{-1}. To steer the atlas generation process and emphasise disease-relevant labels,

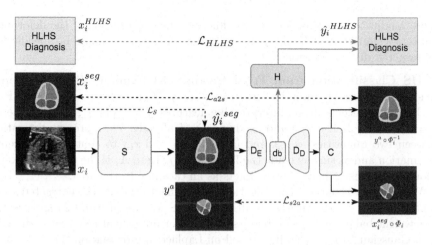

Fig. 3. Disease conditioned Atlas-ISTN network architecture. **S** the segmentation network; **D** indicates the atlas to image mapping module; **C** is the transformation computation Module; and **H** is a disease prediction branch, highlighted in red, which is the key difference to [21]. (Color figure online)

we predict $\hat{y}_i^{HLHS} = \mathbf{H}_{\theta_h}(\mathbf{d}_{b_i})$, where $\mathbf{H}_{\theta_h}$ are three fully connected layers with ReLU activations.

Additionally to the image transformer loss:

$$\mathcal{L}_S = \frac{1}{N}\left(\sum_{i=1}^{N}||x_i^{seg} - \hat{y}_i^{seg}||^2\right),$$

the atlas-to-segmentation loss:

$$\mathcal{L}_{a2s} = \frac{1}{N}\left(\sum_{i=1}^{N}\sum_{j=1}^{c}||x_{i,j}^{seg} - y_j^a \circ \Phi_i^{-1}||^2\right),$$

and the segmentation-to-atlas loss:

$$\mathcal{L}_{s2a} = \frac{1}{N}\left(\sum_{i=1}^{N}\sum_{j=1}^{c}||x_{i,j}^{seg} \circ \Phi_i - y_j^a||^2\right),$$

where j indicates the individual labels, we introduce a cross entropy loss term:

$$\mathcal{L}_{HLHS} = -(x_i^{HLHS}\log(\hat{y}_i^{HLHS}) + (1 - x_i^{HLHS})\log(1 - \hat{y}_i^{HLHS}))$$

to enforce disease-sensitive atlas generation. Thus, our final loss function is the Atlas-ISTN loss [21] with its regularisation loss term, $\mathcal{L}_{reg} = \sum_i^N||\nabla\phi||^2$, that encourages smoothness of the non-rigid deformation fields, paired with $\mathcal{L}_{HLHS}$:

$$\mathcal{L} = \mathcal{L}_S + \omega(\mathcal{L}_{a2s} + \mathcal{L}_{s2a} + \lambda\mathcal{L}_{reg}) + \gamma\mathcal{L}_{HLHS},$$

where λ adjusts smoothness of Φ, ω influences the contribution of the deformation terms similar to [21], and γ steers how much the atlas should be specific to the targeted disease category.

HLHS Classification from Fetal Cardiac 4-Chamber View Segmentations: We extract numerical features from $\hat{y}_i^{seg}$ in order to classify HLHS vs. Healthy patients from interpretable features $f = \{f_0, f_1, ..., f_N\}$ where $f_i = r_{ab} = A_a/A_b$ if $a \neq b$ and r_{ba} is not in f already. We represent the ratio between two quantities as r_{ab} and consider r_{ab} and r_{ba} to contain equivalent information and as such exclude the latter from f. Here A_a is the count of pixels belonging to class a in $\hat{y}_i^{seg}$ which acts as an estimate to the area.

We apply two common classification algorithms to classify the extracted segmentation area ratio features as healthy vs HLHS. We first use an L2 regularised, class weight balanced Logistic regression classifier implementation. Secondly we use a Gaussian Process classifier based on Laplace approximation [17].

3   Experiments and Results

Data and Pre-processing: We use a private and Ethics/IP-restricted, de-identified dataset of 1628 4CH US images (1560 healthy controls, 68 HLHS), with 1043 for training, 260 for validation, 325 for testing with equivalent class imbalance within each set (42, 10 and 16 HLHS cases respectively), acquired on Toshiba Aplio i700, i800 and Philips EPIQ V7 G devices. Class imbalance reflects the prevalence of HLHS observed in our tertiary care referral clinic ($\sim$3–4%), which is a specialised centre, thus the incident rate is relatively high. HLHS is rare, $\sim$3 in 10000 live births [2], thus this condition can be challenging to identify for primary care sonographers. Our images are taken from volunteers at 18–24 weeks gestation (Ethics: *[anonymous during review]*), acquired in a fetal cardiology clinic, where patients are given advanced screening due to their family history. Each image has been hand-picked from ultrasound videos by an expert sonographer, representing a best possible 4CH view. A fetal cardiologist and three expert sonographers delineated the images using Labelbox [1]. The images have been resampled to 288×224 pixels, centred on the heart and aligned along the cardiac main axis.

Robust Segmentation. We compare several methods for automated segmentation by average DICE score achieved for each anatomical class and summarise the results in Table 1. We show that each of the compared methods is effective for the segmentation of anatomical structures from ultrasound image views. The question remains, which method produces the most informative segmentation for downstream disease diagnostics?

Expert Derived Single Image Classification: To establish human performance on the segmentation task, heart segmentation features (area ratios of each anatomical class) are extracted from the manual ground truth segmentations. These are used to train a linear classifier and a Gaussian process classifier to predict HLHS diagnosis. We report the confusion matrices for each method using the ground truth segmentations shown in Fig. 5. Table 1 reports F1-score for positive and negative HLHS classification as well as ROC-AUC for manual as well as automated segmentation.

F1 and AUC scores in Table 1 show that our 'Area Ratios' classification method achieves state-of-the-art performance for HLHS classification over previous classification methods. Classification performance of 'area ratios' extracted from automated segmentations is on par with those extracted from expert manual segmentations. The addition of a disease-conditioned branch to the Atlas-ISTN improves the downstream 'area ratios' classification task performance over both expert segmentations and previous segmentation methods.

Figure 5 shows the performance of the 'area ratios' classification using segmentations produced by experts and by each tested segmentation method. Subfigurs. (Fig. 5e–5f) and (Fig. 5k–5i) highlight the improved sensitivity (fewer false negatives) of Atlas-ISTNs with a disease conditioning branch over expert segmentations (Fig. 5a,5g) and other segmentation methods (Fig. 5b–5d, 5h–5j). Our application is for fetal screening and as such sensitivity is the desired metric to improve, and due to the low prevalence of HLHS, F1 scores for HLHS across all methods may seem low.

Table 1 shows our diagnostic branch (**H**) is competitive with previous image classification approaches, further to this our method uses only a single 4CH image as opposed to previous methods that use multiple heart view US images or video sequences. Our method provides greater interpretability by producing a segmentation (from which 'area ratios' classification is performed) and a disease specific atlas for free. Examples for constructed atlases with different configuration are shown in Fig. 4.

Implementation: PyTorch 1.7.1+cu110 with two Nvidia Titan RTX GPUs used to train segmentation and atlas models ($\sim 10^6$ parameters) in 24–48 h; scikit-learn [16] for the LR and GP models.

(a) $Atlas_{\lambda=1}^{\gamma=0}$ (b) $Atlas_{\lambda=1}^{\gamma=1}$ (c) $Atlas_{\lambda=10^3}^{\gamma=1}$

Fig. 4. Example automatically constructed atlas images and atlas label maps in the tested configurations.

4   Discussion

Assuming a linear downstream model for clinical decision making, our results show that automated segmentation methods are en-par with human-generated annotations for the accurate identification of HLHS patients. We believe the

Table 1. DICE scores and standard deviation (Std) for all segmentation methods (left) and performance of downstream disease predictors (right). (**BG** = background; **LA** = left atrium; **RA** = right atrium; **LV** = left ventricle; **RV** = right ventricle; **WH** = whole heart; **LR** = Logistic regression; **GP** = Gaussian process; **H** = disease prediction branch; **NC** = Normal control; **HLHS** = hypo-plastic left heart syndrome.)

Method	DICE Score							F1 Score		ROC
	BG	LA	RA	LV	RV	WH		NC	HLHS	AUC
Expert	1.000	1.000	1.000	1.000	1.000	1.000	**LR:**	0.970	0.550	0.944
(Std)	–	–	–	–	–	–	**GP**	0.989	0.741	0.954
UNet [18]	0.993	0.768	0.804	0.793	0.794	0.635	**LR:**	0.972	0.585	0.922
(Std)	(0.007)	(0.192)	(0.185)	(0.184)	(0.153)	(0.105)	**GP:**	0.974	0.579	0.928
SSN [14]	0.993	0.761	0.800	0.793	0.794	0.632	**LR:**	0.955	0.471	0.883
(Std)	(0.007)	(0.196)	(0.192)	(0.194)	(0.154	(0.108)	**GP:**	0.974	0.579	0.923
$Atlas_{\lambda=1}^{\gamma=0}$	0.991	0.767	0.789	0.801	0.783	0.626	**LR:**	0.942	0.451	0.895
(Std)	(0.007)	(0.187)	(0.192)	(0.191)	(0.172)	(0.106)	**GP:**	0.981	0.625	0.970
$Atlas_{\lambda=10^3}^{\gamma=1}$	0.993	0.764	0.789	0.791	0.790	0.648	**LR:**	0.958	0.528	0.929
(Std)	(0.007)	(0.185)	(0.184)	(0.196)	(0.146)	(0.110)	**GP:**	0.974	0.619	0.973
	–	–	–	–	–	–	**H :**	0.967	0.565	0.883
$Atlas_{\lambda=1}^{\gamma=1}$	0.993	0.760	0.783	0.784	0.788	0.637	**LR:**	0.950	0.500	0.974
(Std)	(0.007)	(0.197)	(0.200)	(0.208)	(0.164)	(0.110)	**GP:**	0.974	0.636	0.978
	–	–	–	–	–	–	**H :**	0.982	0.667	0.905

reason for higher than expert performance is due to more consistent automated segmentation results that aid the following linear model in contrast to predicting from ground truth segmentations, which have been generated by different observers. An interesting observation is that a reasonable DICE score is sufficient to achieve excellent performance in diagnostic follow-up tasks.

A limitation of our study is that we require input images that resemble a 4CH acquisition orientation in a healthy subject. This can be challenging for severely affected patients. However, for cases with severely abnormal hearts, manual detection of CHD would likely be trivial at the point of care, also without segmentation analysis. Good views for borderline cases, which are in focus here, can be identified either manually or with automated view classification [5].

For this work we rigidly aligned all the data to a canonical orientation relative to the heart. This can be achieved in the clinical practice through automated localisation/segmentation/spatial transformer approaches. We observed that this data curation step has a significant impact on all models' performances compared to unaligned images, in which fetuses may present in arbitrary orientation. Accounting for flipped probe orientations paired with hyper-parameter tuning for ω, λ, γ would likely lead to further improvements.

Another limitation is that we do not consider inherent spatio-temporal information of ultrasound imaging. Experienced fetal cardiologists can derive valuable secondary information from how the heart moves. This knowledge can inform

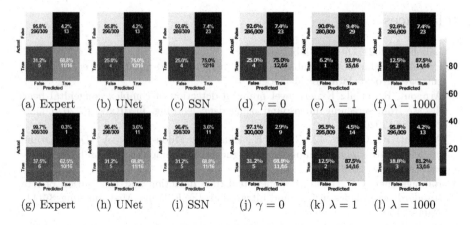

Fig. 5. Confusion matrices for expert derived classification: Top Row shows logistic regression, bottom row shows Gaussian Process. $\gamma = 1$ for (e-f)(k-l).

future work on the topic. In the clinical practice, still images, as used in our work are common practice to report and document cases, thus a direct application to retrospective quality control and diagnosis support for primary care is in reach.

5 Conclusion

We have discussed how segmentation models can be used as clinically interpretable alternative to direct image classification methods for the diagnosis of hypo-plastic left heart syndrome during routine ultrasound examinations. We test a new approach that facilitates disease-status information to bias an automatically constructed atlas label map for robust segmentation and apply Atlas-ISTNs to the problem of fetal cardiac segmentation from ultrasound images for the first time. Our analysis shows that our interpretable approach is en-par with direct image classification, for which ROC-AUC of up to 0.93 is reported [23]. Future work will investigate the true effectiveness of such methods in a prospective clinical trial, which is currently implemented in our clinic.

Acknowledgements. We thank the volunteers and sonographers at St. Thomas' Hospital London. The work of E.C.R. was supported by the Academy of Medical Sciences/the British Heart Foundation/the Government Department of Business, Energy and Industrial Strategy/the Wellcome Trust Springboard Award [SBF003/1116]. We also gratefully acknowledge financial support from the Wellcome Trust IEH 102431, EPSRC (EP/S022104/1, EP/S013687/1), EPSRC Centre for Medical Engineering [WT 203148/Z/16/Z], the National Institute for Health Research (NIHR) Biomedical Research Centre (BRC) based at Guy's and St Thomas' NHS Foundation Trust and King's College London and supported by the NIHR Clinical Research Facility (CRF) at Guy's and St Thomas', and Nvidia GPU donations.

References

1. Labelbox (2021). https://labelbox.com
2. Prevalence charts and tables (2021). https://eu-rd-platform.jrc.ec.europa.eu/eurocat/eurocat-data/prevalence_en
3. Arnaout, R., et al.: Expert-level prenatal detection of complex congenital heart disease from screening ultrasound using deep learning. medRxiv, p. 2020.06.22. 20137786 (2020)
4. Arnaout, R., Curran, L., Chinn, E., Zhao, Y., Moon-Grady, A.: Deep-learning models improve on community-level diagnosis for common congenital heart disease lesions. arXiv (2018)
5. Baumgartner, C.F., et al.: Sononet: real-time detection and localisation of fetal standard scan planes in freehand ultrasound. IEEE Trans. Med. Imaging **36**(11), 2204–2215 (2017)
6. Budd, S., et al.: Confident head circumference measurement from ultrasound with real-time feedback for sonographers. In: Shen, D., et al. (eds.) MICCAI 2019. LNCS, vol. 11767, pp. 683–691. Springer, Cham (2019). https://doi.org/10.1007/978-3-030-32251-9_75
7. Calderon, J., et al.: Impact of prenatal diagnosis on neurocognitive outcomes in children with transposition of the great arteries. J. Pediatr. **161**(1), 94–98 (2012)
8. Carneiro, G., Georgescu, B., Good, S., Comaniciu, D.: Detection and measurement of fetal anatomies from ultrasound images using a constrained probabilistic boosting tree. IEEE Trans. Med. Imaging **27**(9), 1342–1355 (2008)
9. Clough, J.R., Oksuz, I., Byrne, N., Schnabel, J.A., King, A.P.: Explicit topological priors for deep-learning based image segmentation using persistent homology. In: Chung, A.C.S., Gee, J.C., Yushkevich, P.A., Bao, S. (eds.) IPMI 2019. LNCS, vol. 11492, pp. 16–28. Springer, Cham (2019). https://doi.org/10.1007/978-3-030-20351-1_2
10. Holland, B., Myers, J., Woods, C., Jr.: Prenatal diagnosis of critical congenital heart disease reduces risk of death from cardiovascular compromise prior to planned neonatal cardiac surgery: a meta-analysis. Ultrasound Obstet. Gynecol. **45**(6), 631–638 (2015)
11. Hu, X., Li, F., Samaras, D., Chen, C.: Topology-preserving deep image segmentation. In: Wallach, H., Larochelle, H., Beygelzimer, A., d' Alché-Buc, F., Fox, E., Garnett, R. (eds.) Advances in Neural Information Processing Systems, vol. 32. Curran Associates, Inc. (2019)
12. Li, J., et al.: Automatic fetal head circumference measurement in ultrasound using random forest and fast ellipse fitting. IEEE J. Biomed. Health Inform. **22**(1), 215–223 (2018)
13. Miceli, F.: A review of the diagnostic accuracy of fetal cardiac anomalies. Australasian J. Ultrasound Med. **18**(1), 3–9 (2015)
14. Monteiro, M., et al.: Stochastic segmentation networks: modelling spatially correlated aleatoric uncertainty. In: Larochelle, H., Ranzato, M., Hadsell, R., Balcan, M.F., Lin, H. (eds.) Advances in Neural Information Processing Systems, vol. 33, pp. 12756–12767. Curran Associates, Inc. (2020)
15. NHS: NHS Fetal Anomaly Screening Programme Handbook Valid from August 2018. Technical report (2018)
16. Pedregosa, F., et al.: Scikit-learn: machine learning in Python. J. Mach. Learn. Res. **12**, 2825–2830 (2011)

17. Rasmussen, C., Nickisch, H.: Gaussian processes for machine learning (gpml) toolbox. J. Mach. Learn. Res. **11**, 3011–3015 (2010)
18. Ronneberger, O., Fischer, P., Brox, T.: U-Net: convolutional networks for biomedical image segmentation. In: Navab, N., Hornegger, J., Wells, W.M., Frangi, A.F. (eds.) MICCAI 2015. LNCS, vol. 9351, pp. 234–241. Springer, Cham (2015). https://doi.org/10.1007/978-3-319-24574-4_28
19. Rueda, S., et al.: Evaluation and comparison of current fetal ultrasound image segmentation methods for biometric measurements: a grand challenge. IEEE Trans. Med. Imaging **33**(4), 797–813 (2014)
20. Sinclair, M., et al.: Human-level performance on automatic head biometrics in fetal ultrasound using fully convolutional neural networks. In: 2018 40th Annual International Conference of the IEEE Engineering in Medicine and Biology Society (EMBC), pp. 714–717. IEEE (2018)
21. Sinclair, M., et al.: Atlas-ISTN: Joint Segmentation, Registration and Atlas Construction with Image-and-Spatial Transformer Networks. arXiv (2020)
22. Sushma, T.V., Sriraam, N., Megha Arakeri, P., Suresh, S.: Classification of fetal heart ultrasound images for the detection of CHD. In: Raj, J.S., Iliyasu, A.M., Bestak, R., Baig, Z.A. (eds.) Innovative Data Communication Technologies and Application. LNDECT, vol. 59, pp. 489–505. Springer, Singapore (2021). https://doi.org/10.1007/978-981-15-9651-3_41
23. Tan, J., et al.: Automated detection of congenital heart disease in fetal ultrasound screening. In: Hu, Y., et al. (eds.) ASMUS/PIPPI -2020. LNCS, vol. 12437, pp. 243–252. Springer, Cham (2020). https://doi.org/10.1007/978-3-030-60334-2_24
24. Wu, L., et al.: Cascaded fully convolutional networks for automatic prenatal ultrasound image segmentation. In: 2017 IEEE 14th International Symposium on Biomedical Imaging (ISBI 2017), pp. 663–666. IEEE (2017)
25. Yeo, L., Romero, R.: Fetal Intelligent Navigation Echocardiography (FINE): a novel method for rapid, simple, and automatic examination of the fetal heart. Ultrasound Obstet. Gynecol. **42**(3), 268–284 (2013)

EllipseNet: Anchor-Free Ellipse Detection for Automatic Cardiac Biometrics in Fetal Echocardiography

Jiancong Chen[1], Yingying Zhang[2,3], Jingyi Wang[4,5], Xiaoxue Zhou[4,5], Yihua He[4,5(✉)], and Tong Zhang[1(✉)]

[1] Peng Cheng Laboratory, Shenzhen, China
zhangt02@pcl.ac.cn
[2] School of Biological Science and Medical Engineering, Beihang University, Beijing, China
[3] Beijing Advanced Innovation Center for Big Data-Based Precision Medicine, Beijing, China
[4] Echocardiography Medical Center, Beijing Anzhen Hospital, Capital Medical University, Beijing, China
[5] Maternal-Fetal Medicine center in Fetal Heart Disease, Beijing Anzhen Hospital, Beijing, China

Abstract. As an important scan plane, four chamber view is routinely performed in both second trimester perinatal screening and fetal echocardiographic examinations. The biometrics in this plane including cardiothoracic ratio (CTR) and cardiac axis are usually measured by sonographers for diagnosing congenital heart disease. However, due to the commonly existing artifacts like acoustic shadowing, the traditional manual measurements not only suffer from the low efficiency, but also with the inconsistent results depending on the operators' skills. In this paper, we present an anchor-free ellipse detection network, namely EllipseNet, which detects the cardiac and thoracic regions in ellipse and automatically calculates the CTR and cardiac axis for fetal cardiac biometrics in 4-chamber view. In particular, we formulate the network that detects the center of each object as points and regresses the ellipses' parameters simultaneously. We define an intersection-over-union loss to further regulate the regression procedure. We evaluate EllipseNet on clinical echocardiogram dataset with more than 2000 subjects. Experimental results show that the proposed framework outperforms several state-of-the-art methods. Source code will be available at https://git.openi.org.cn/capepoint/EllipseNet.

Keywords: Fetal echocardiogram · Cardio-thoracic ratio · Anchor-free detection · Ellipse detection · Cardiac biometric

1 Introduction

Fetal congenital heart disease (CHD) is one of the most common forms of birth defects worldwide, which affects eight out of 1000 newborn babies worldwide

---

J. Chen and Y. Zhang—Authors contributed equally.

© Springer Nature Switzerland AG 2021
M. de Bruijne et al. (Eds.): MICCAI 2021, LNCS 12907, pp. 218–227, 2021.
https://doi.org/10.1007/978-3-030-87234-2_21

every year [1]. Despite the tremendous efforts made to improve the obstetric screening, the detection of CHD remains a relatively low yield level at 30%–60% [2,3]. Four chamber view is a standardized cardiac plane that is regularly scanned in mid-trimester antenatal screening. In 4-chamber view examinations, cardiothoracic diameter ratio (CTR) and cardiac axis are two important and commonly used metrics. The CTR estimates the heart size by measuring the maximum diameter of the cardiac and comparing it to the maximum diameter of the thorax. The cardiac axis is estimated by measuring the angle between the line bisecting the chest and the line along the cardiac interventricular septum. These cardiac biometrics are reliable perimeters in the prediction of multiple CHD [3]. Due to the commonly existing acoustic shadowing artifact in fetal echocardiogram, the traditional manual measurements are very time-consuming and suffer from large inter-operator variance depending on the sonographer's skills. Therefore, automatic and reliable fetal cardiac measurements are in high demand.

With the rapid development of deep learning, convolutional neural networks (CNN) have been applied in many medical related applications and achieved great success for a number of image analysis tasks. Baumgartner et al. [4] developed a CNN, named SonoNet, to detect and localize standardised scan planes in freehand ultrasound. Gong et al. [5] presented a one-class classification network to distinguish CHD and healthy subjects, where the improved generative adversarial networks were used for data augmentation. Sinclair et al. [6] developed a segmentation and ellipse fit network for automatic measurement of fetal head circumference and biparietal diameter. Compared to fetal head, fetal echocardiographic measurement is challenged by the moving heart and shadowing artifacts around fetal sternum.

In this work, we present a one-stage ellipse detection framework for automatic fetal cardiac biometrics in 4-chamber view scans. Unlike [6], our method directly detects the objects without the segmentation procedure. Here, we use ellipses to represent the cardiac and thoracic regions, and calculate the CTR and cardiac axis accordingly. Following the popular anchor-free detection networks [7,8], the proposed EllipseNet detects cardiac and thoracic regions and regresses the ellipses's parameters in one-stage. The main contributions of our work can be summarized as follows:

1. The proposed EllipseNet can detect cardiac and thoracic regions and calculate fetal CTR and cardiac axis in 4-chamber view. To the best of our knowledge, this is the first approach to perform cardiac biometrics automatically in fetal echocardiography.
2. EllipseNet is a one-stage ellipse detection framework. Compared to the traditional bounding box based detection, modeling the object with an ellipse better depicts a number of medical structures.
3. A rotated intersection over union (IoU) loss is introduced in ellipse regression, which significantly improves the network performance in precision of object localization.

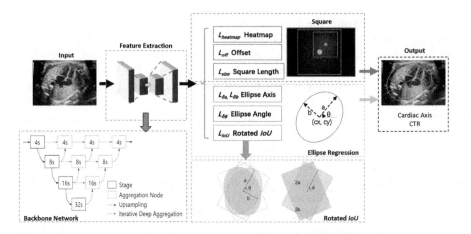

Fig. 1. Overview of the proposed ellipse detection network. EllipseNet consists of the feature extraction module, which is identical to deep layer aggregation network DLA-34 [14]; the square detection module, which detects the centers of the objects; and the ellipse regression module, where a rotated IoU loss is integrated.

1.1   Related Works

Ellipse Detection. Due to its superior presentation for objects, ellipses have been investigated in the object detection communities. Li et al. [9] replaced the Region Proposal Network (RPN) in Faster R-CNN [10] by a Gaussian proposal network (GPN), which generates elliptical proposals and represents these ellipses by 2D Gaussian distributions. Pan et al. [11] further modified the loss with Wasserstein distance in RPN to model elliptical object. Moreover, Dong et al. [12] proposed Ellipse R-CNN, a two-stage detector with ellipse regression upon the Mask R-CNN [13] for detecting elliptical objects in occluded and cluttered scenes. However, all the above ellipse detection methods are anchor-based detection frameworks, which typically introduces many hyperparameters and usually difficult to train. Further, these methods highly rely on the quality of densely predefined anchors that need sufficient overlaps with ground truth. Very few works have attempted to employ anchor-free approaches [7,8] to detect elliptical objects.

2   Method

2.1   Overview of the Framework

The overview of our proposed EllipseNet is shown in Fig. 1. It composes of a backbone network for feature extraction, an anchor-free square detection module and an ellipse regression module. All the above modules are simultaneously trained in an end-to-end manner. The backbone network and the square detection module are built upon CenterNet [7], which is a keypoint-based anchor-free detection framework with high performance on center point localization.

2.2   Square Detection

It is not stable and difficult to train if directly regressing the five parameters of ellipse. Inspired by Ellipse R-CNN [12], we extend the elliptical region to a square with length $l = 2\sqrt{a^2 + b^2}$, which shares the same center point with the ellipse. The extended square is closely related to the location and size of the ellipse but independent of the rotation angle θ. Further, the degrees of freedom of the extended square is 1 (length Q_l for shape) when its center point (Q_x, Q_y) is learning, so that it is easier to optimize by the detection network. We therefore use the square detection module for ellipse localisation.

The square detection module consists of three heads: the center point heatmap head, the center point offset head and the square length head. Given an input image $I \in R^{W \times H \times 3}$ with height H and width W. The heatmap head predict the heatmap $\hat{Y} \in [0,1]^{\frac{W}{R} \times \frac{H}{R} \times 2}$, representing the probability map of the center localization of thoracic and cardiac regions respectively. Here, we set output stride $R = 4$ as suggested in [7], where the square length head only predicts a square length but not the height and width as that within bounding box. We use pixel-wise regression with focal loss [15] to optimize the center point heatmap and L1 loss to optimize the offset and square length. The heatmap loss is presented:

$$L_{heatmap} = -\frac{1}{N} \sum_{xyc} \begin{cases} \left(1 - \hat{Y}_{xyc}\right)^{\alpha} \log\left(\hat{Y}_{xyc}\right) & \text{if } Y_{xyc} = 1 \\ (1 - Y_{xyc})^{\beta} \left(\hat{Y}_{xyc}\right)^{\alpha} \log\left(1 - \hat{Y}_{xyc}\right) & \text{otherwise} \end{cases} \tag{1}$$

where xyc represents the position in heatmap, α and β are hyper-parameters of focal loss, where $\alpha = 2$ and $\beta = 4$. The loss of square detection is formulated as:

$$L_Q = L_{heatmap} + \lambda_{size} L_{size} + \lambda_{off} L_{off}. \tag{2}$$

2.3   Ellipse Regression

When the square parameters $Q = (Q_x, Q_y, Q_l)$ are determined in the square detection module, the center points of the ellipse will be coarsely decided. Then the regression of the ellipses' parameters including semi-major axis E_a, semi-minor axis E_b, and rotation angle E_θ can be formulated as below:

$$\delta_a = (E_a/Q_l), \qquad \delta_b = (E_b/Q_l), \qquad \delta_\theta = \theta/\pi \tag{3}$$

$$\theta = \begin{cases} \text{atan2}(\sin E_\theta, \cos E_\theta) & \text{if } \cos E_\theta \geq 0 \\ \text{atan2}(-\sin E_\theta, -\cos E_\theta) & \text{if } \cos E_\theta < 0 \end{cases} \tag{4}$$

where δ_a and δ_b represent the ratios of E_a and E_b to the length of corresponding extended square, respectively. Here, we define the range of rotation angle $E_\theta \in (-\pi/2, \pi/2]$, and the difference between $-\pi/2$ and $\pi/2$ is 0 rather than π. We use separate heads to regress the parameters in Eq. 3 with L1 losses, and the ellipse regression loss L_E is defined as a weighted sum of those losses:

$$L_E = L_{\delta_a} + L_{\delta_b} + \lambda_{\delta_\theta} L_{\delta_\theta} \tag{5}$$

2.4 Rotated IoU Loss

To further constrain the ellipse regression optimisation, we introduce an IoU metric, which is very popular in object detection tasks measuring the overlap between the boxes. More generally, the IoU loss can be formulated with an additional penalty term $L_{IoU} = 1 - IoU + P$, where P further penalize the overlap between the predicted and the ground truth regions. For the original IoU loss, $P = 0$. In this work, we chose the Distance-IoU loss [16] as formulated below:

$$L_{IoU} = 1 - IoU + P_{DIoU} = 1 - \frac{R_{pred} \cap R_{gt}}{R_{pred} \cup R_{gt}} + \frac{\rho^2}{c^2}. \tag{6}$$

where a distance penalty term P_{DIoU} is added to achieve faster convergence, R_{pred} and R_{gt} represent the predicted and groud truth regions, respectively, ρ denotes the Euclidean distance between the predicted and ground truth center points, c is the diagonal length of the smallest enclosing box covering the two boxes.

To compute the IoU loss of two rotated ellipses can be very expensive. Therefore, we introduce a rotated IoU [17] to calculate the IoU of two rotated boxes tightly bounding the ellipses, as an approximate solution to the IoU of ellipses (see in Fig. 1). It is worth mentioning that the bounding box shares the same center point and rotation angle with the corresponding ellipse, while the height and width are of the same length as major and minor axes of ellipse, respectively. To this end, all the parameters for ellipse regression are defined. The overall ellipse detection loss L_{EN} can be formulated as:

$$L_{EN} = \lambda_Q L_Q + \lambda_E L_E + \lambda_{IoU} L_{IoU}. \tag{7}$$

where λ_Q, λ_E, and λ_{IoU} are the weighting parameters.

3 Experiments

3.1 Data Acquisition

The dataset consists of 2086 2D fetal echocardiographic images in 4-chamber view. The data were obtained from pregnant women aged from 19 to 45 at a gestational age of 18–39 weeks. The median age at pregnancy was 25 ± 3 years, while the median gestational age was 29.4 weeks. The machine models include the GE Voluson E8 and E10 and the Philips EPIQ 7C. All the subjects used in this study are with ethical committee approval. The cardiac and thoracic regions are labelled in ellipses using an offline ellipse annotation tool provided by VGG Image Annotator[1] [18] by two trained operators and examined by fetal echocardiographers.

---

[1] https://www.robots.ox.ac.uk/~vgg/software/via/.

3.2 Implementation Details

We randomly select 80% of the images for training and the remainder for testing. All the images are resized to 896×608 before feeding into the network. Random left-right flipping, scaling, shifting and random Gaussian noise are added for data augmentation. We use deep layer aggregation network [14] DLA-34 with output stride 4 as the backbone network, and the output heads are defined as shown in Fig. 1. Before each output head, we use a 3×3 convolutional layer with 256 channels followed by a ReLU layer. Moreover, we add a sigmoid activation function to the end of ellipse axis heads for predicting the ratios δ_a and δ_b. The network is trained by minimizing the loss function in Eq. 7 with ADAM optimizer. All loss weights λ_* are set to 1 except $\lambda_{size} = 0.1$ and $\lambda_{\delta_\theta} = 5$. We train our model with batch size 24 and initial learning rate 5e-4 for 120 epochs. All experiments were performed on a single NVIDIA Tesla V100 GPU in PyTorch.

3.3 Results

To compute the overlap between rotated ellipses, we draw the ellipses as shape masks and calculating the dice coefficient between masks to evaluate the detection performance. The CTR is formulated as $R = b_C / b_T$, where b_C represents the length of minor axis of the cardiac ellipse, b_T represents that of the thoracic ellipse. Then, the CTR precision can be defined as: $P_{avg} = 1 - \frac{|R_{true} - R_{pred}|}{R_{true}}$, where R_{pred} and R_{true} denote the predicted CTR and the ground truth CTR, respectively.

Table 1. Experimental results between different configurations and comparison results. The dice coefficient of cardiac and thoracic regions, the average of them and the precision of estimated CTR are reported here.

Methods	Setting	$Dice_T$	$Dice_C$	$Dice_{all}$	P_{avg}
Segment+ellipse-fit [6]	Residual U-Net	0.8750	0.9182	0.9112	0.8520
CircleNet [8]	Hourglass	0.8966	0.8666	0.8816	0.8555
CircleNet [8]	DLA	0.8966	0.8729	0.8847	0.8614
EllipseNet (Ours)	only IoU loss	0.8813	0.8520	0.8666	0.8855
EllipseNet (Ours)	w/o IoU loss	0.9338	0.9108	0.9224	0.8841
EllipseNet (Ours)	w/ IoU loss	**0.9430**	**0.9224**	**0.9336**	**0.8949**

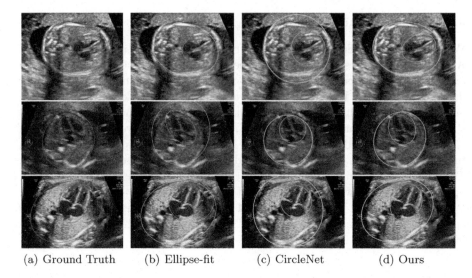

(a) Ground Truth   (b) Ellipse-fit   (c) CircleNet   (d) Ours

Fig. 2. Visualization results of different methods. The columns from left to right display the ground truth and prediction results from segment-based ellipse-fit method, CircleNet and our proposed EllipseNet with IoU loss. The red ellipses represent ground truth of cardiac and thoracic regions, while green and cyan ellipses represent the predicted cardiac and thoracic region respectively.

Comparison with Other Methods. We compare EllipseNet with several automatic estimation-related detection methods to demonstrate its effectiveness. First, we compare it with [6], which trains a Fully Convolutional Network (FCN) segmentation model to obtain the structure contour and then uses an ellipse to fit it. Instead of using the FCN in the original paper, we use MONAI[2] to train a more powerful U-Net [19] as the segmentation network for better performance. Second, we compare it with CircleNet [8], an anchor-free detection method with circle representation. The circle can be regarded as a special case of an ellipse with the same radius for all points, so the CTR can be calculated by the radii of circles. It should be noted that CircleNet cannot be used for cardiac axis measurements since it treats the objects as circles. As presented in Table 1, our proposed method outperforms the others with all the dice coefficients and CTR precision.

Although the segmentation model trained on the fetal ultrasound dataset achieves a relative high average dice scores averaging over 0.9, the fitted ellipse highly depends on the segmentation results. As shown in the middle row of Fig. 2, the segment-based ellipse-fit method performs well when the image quality is good (first row), but the performance degrades when the segmentation is affected by image artifacts such as the acoustic shadowing (second and third row). Our proposed method is more robust to image quality and shadows. We also tried to

---

[2] https://monai.io/.

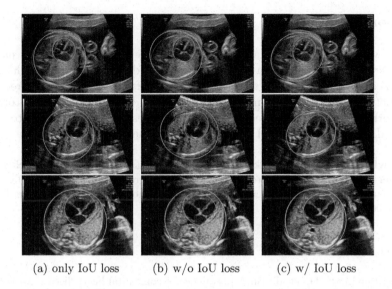

(a) only IoU loss   (b) w/o IoU loss   (c) w/ IoU loss

Fig. 3. Results of ablation study. The columns display results from different configurations of our proposed method. For better visualization, predictions and ground truth are drawn in the same image.

compare to the GPN [9] with their open source code on our dataset, however, the results are not comparable to ours and those presented in Table 1. It is difficult to conclude whether it is caused by the network itself or the training strategies. We, therefore, did not include the comparative results in this work.

Ablation Study. In this section, we evaluate the effect of different loss terms in EllipseNet by ablation study. As shown in the lower part of Table 1, IoU loss improves detection performance remarkably. EllipseNet with IoU loss achieves a gain of 0.92 and 1.16 points in cardiac and thoracic dice coefficients respectively. On the other hand, EllipseNet with only IoU loss (without ellipse regression loss defined in Eq. 5) does not achieve satisfactory performance, which demonstrate the necessarity to include the parameters' regression losses. It worth mentioning that the exact IoU of two rotated ellipses is always greater than that of corresponding rotated bounding boxes based on our observation. The IoU loss brings promotion in terms of bounding box IoU, which can be regarded as a lower bound of the ellipse IoU. This may explain why IoU loss is effective for ellipse regression in another view.

As shown in Fig. 3, both the proposed ellipse regression loss and IoU loss are necessary for ellipse detection. If the EllipseNet only supervised by IoU loss (first column), the model fails to optimize the major and minor axis separately, and the predicted ellipses degenerate into circles like the CircleNet. It is clear that the supervision of IoU loss can help to improve the prediction of location and shape (first and second row) and to correct the angle (last row).

4 Conclusion

In this work, we presented an anchor-free ellipse detection network, named EllipseNet, for automatic measurement of fetal cardiac biometrics in 4-chamber view in echocardiographic screening. The proposed framework is anchor-free one-stage network that can perform ellipse detection and CTR calculation automatically with remarkably high precision. The pipeline can be easily adapted to other elliptical object detection and biometric measurement tasks.

Acknowledgement. This work is supported by Ministry of Science and Technology of China - Peng Cheng Laboratory Special Project (grant No. PCNL2021ZDXM06), and in part by the Beijing Municipal Science and Technology Commission under Grant Z181100001918008.

References

1. Van Der Linde, D., et al.: Birth prevalence of congenital heart disease worldwide: a systematic review and meta-analysis. J. Am. Coll. Cardiol. **58**(21), 2241–2247 (2011)
2. Bakker, M.K., et al.: Prenatal diagnosis and prevalence of critical congenital heart defects: an international retrospective cohort study. BMJ Open **9**(7), e028139 (2019)
3. Sharland, G.: Fetal cardiac screening and variation in prenatal detection rates of congenital heart disease: why bother with screening at all? Future Cardiol. **8**(2), 189–202 (2012)
4. Baumgartner, C.F., et al.: SonoNet: real-time detection and localisation of fetal standard scan planes in freehand ultrasound. IEEE Trans. Med. Imaging **36**(11), 2204–2215 (2017)
5. Gong, Y., et al.: Fetal congenital heart disease echocardiogram screening based on DGACNN: adversarial one-class classification combined with video transfer learning. IEEE Trans. Med. Imaging **39**(4), 1206–1222 (2020)
6. Sinclair, M., et al.: Human-level performance on automatic head biometrics in fetal ultrasound using fully convolutional neural networks. In: 2018 40th Annual International Conference of the IEEE Engineering in Medicine and Biology Society (EMBC), pp. 714–717 (2018)
7. Zhou, X., Wang, D., Krähenbühl, P.: Objects as points. arXiv (2019)
8. Yang, H., et al.: CircleNet: anchor-free glomerulus detection with circle representation. In: Martel, A.L., et al. (eds.) MICCAI 2020. LNCS, vol. 12264, pp. 35–44. Springer, Cham (2020). https://doi.org/10.1007/978-3-030-59719-1_4
9. Li, Y.: Detecting lesion bounding ellipses with gaussian proposal networks. In: Suk, H.-I., Liu, M., Yan, P., Lian, C. (eds.) MLMI 2019. LNCS, vol. 11861, pp. 337–344. Springer, Cham (2019). https://doi.org/10.1007/978-3-030-32692-0_39
10. Ren, S., He, K., Girshick, R.B., Sun, J.: Faster R-CNN: towards real-time object detection with region proposal networks. In: NIPS, pp. 91–99 (2015)
11. Pan, S., Fan, S., Wong, S.W.K., Zidek, J.V., Rhodin, H.: Ellipse detection and localization with applications to knots in sawn lumber images. In: Proceedings of the IEEE/CVF Winter Conference on Applications of Computer Vision (WACV), pp. 3892–3901 (2021)

12. Dong, W., Roy, P., Peng, C., Isler, V.: Ellipse R-CNN: learning to infer elliptical object from clustering and occlusion. IEEE Trans. Image Process. **30**, 2193–2206 (2021)
13. He, K., Gkioxari, G., Dollar, P., Girshick, R.: Mask R-CNN. In: Proceedings of the IEEE International Conference on Computer Vision (ICCV), October 2017
14. Yu, F., Wang, D., Shelhamer, E., Darrell, T.: Deep layer aggregation. In: Proceedings of the IEEE Conference on Computer Vision and Pattern Recognition (CVPR), June 2018
15. Lin, T.Y., Goyal, P., Girshick, R., He, K., Dollár, P.: Focal loss for dense object detection. In: CVPR 2018, August 2017
16. Zheng, Z., Wang, P., Liu, W., Li, J., Ye, R., Ren, D.: Distance-IoU loss: faster and better learning for bounding box regression. Technical report (2019)
17. Zhou, D., et al.: IoU Loss for 2D/3D Object Detection. Technical report (2019)
18. Dutta, A., Zisserman, A.: The VIA annotation software for images, audio and video. In: Proceedings of the 27th ACM International Conference on Multimedia, MM 2019. ACM, New York (2019)
19. Kerfoot, E., Clough, J., Oksuz, I., Lee, J., King, A.P., Schnabel, J.A.: Left-ventricle quantification using residual U-Net. In: Pop, M., et al. (eds.) STACOM 2018. LNCS, vol. 11395, pp. 371–380. Springer, Cham (2019). https://doi.org/10.1007/978-3-030-12029-0_40

AutoFB: Automating Fetal Biometry Estimation from Standard Ultrasound Planes

Sophia Bano[1,2(✉)], Brian Dromey[1,3], Francisco Vasconcelos[1,2],
Raffaele Napolitano[3], Anna L. David[3,4], Donald M. Peebles[3,4],
and Danail Stoyanov[1,2]

[1] Wellcome/EPSRC Centre for Interventional and Surgical Sciences (WEISS),
University College London, London, UK
`sophia.bano@ucl.ac.uk`
[2] Department of Computer Science, University College London, London, UK
[3] Elizabeth Garrett Anderson Institute for Women's Health,
University College London, London, UK
[4] NIHR University College London Hospitals Biomedical Research Centre,
London, UK

Abstract. During pregnancy, ultrasound examination in the second trimester can assess fetal size according to standardized charts. To achieve a reproducible and accurate measurement, a sonographer needs to identify three standard 2D planes of the fetal anatomy (head, abdomen, femur) and manually mark the key anatomical landmarks on the image for accurate biometry and fetal weight estimation. This can be a time-consuming operator-dependent task, especially for a trainee sonographer. Computer-assisted techniques can help in automating the fetal biometry computation process. In this paper, we present a unified automated framework for estimating all measurements needed for the fetal weight assessment. The proposed framework semantically segments the key fetal anatomies using state-of-the-art segmentation models, followed by region fitting and scale recovery for the biometry estimation. We present an ablation study of segmentation algorithms to show their robustness through 4-fold cross-validation on a dataset of 349 ultrasound standard plane images from 42 pregnancies. Moreover, we show that the network with the best segmentation performance tends to be more accurate for biometry estimation. Furthermore, we demonstrate that the error between clinically measured and predicted fetal biometry is lower than the permissible error during routine clinical measurements.

Keywords: Fetal biometry estimation · Fetal ultrasound · Fetus anatomy segmentation · Computer-assisted diagnosis

Electronic supplementary material The online version of this chapter (https://doi.org/10.1007/978-3-030-87234-2_22) contains supplementary material, which is available to authorized users.

M. de Bruijne et al. (Eds.): MICCAI 2021, LNCS 12907, pp. 228–238, 2021.
https://doi.org/10.1007/978-3-030-87234-2_22

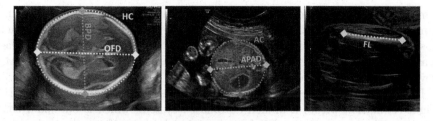

Fig. 1. Fetal biometry from transventricular plane in the head (left), transabdominal plane in the abdomen (middle) and femur plane (right).

1   Introduction

There is little global consensus on how to train, assess and evaluate skills in prenatal second trimester ultrasound (US) screening. Recommended assessment and quality control metrics vary across countries and institutions [5]. Despite this, standardized US planes and metrics to assess fetal size are well established [20]. In particular, fetal weight estimation is routinely used to assess fetal well-being, both in terms of its absolute value and its growth trajectory during pregnancy. Fetal wellbeing is considered by obstetricians for scheduling birth and by neonatologists when counselling parents on likely outcomes for their baby. There are three key structures and corresponding anatomical planes which are used for the estimation of fetal weight (Fig. 1). These are the transventricular plane to measure the head, the transabdominal plane to measure the abdomen and the femur length plane to measure the leg skeletal size. The acquisition of these standard planes is subject to intraoperator and interoperator variabilities [22] which introduces some degree of uncertainty in the clinically obtained weight measurements and consequently requires a degree of caution when clinicians are interpreting fetal growth reports. Sonography expertise has a significant impact on minimizing variability of image quality and fetal biometry [5]. Consequently, training and competence assessment are of great importance to ensure effective, reproducible and safe clinical practice. Automating fetal biometry on the standardized planes can help in minimizing the variability, specially in the case of less experienced sonographers and may also serve as expert for trainees.

There is extensive work on segmentation of anatomical structures in standard US planes, specifically those concerning second and third trimester screening [19]. These techniques can support automated fetal biometry, including measurements on the head [4,13,15,16,23,24], femur [12,15], and abdominal section [14]. These methods, however, rely on prior knowledge of which measurement to perform on a given image. A fully automated biometry system should both identify which standard plane is being imaged and whether it is of sufficient quality to perform the relevant measurements. Automatic image quality assessment has been investigated, including adequate magnification, symmetry and the visibility of relevant anatomical structures within the image [15,17]. Such methods together with classification of standard planes [1] can be used to extract appropriate planes for fetal biometry from US video or image collections [9]. Alternative approaches involve obtaining standard planes from 3D US volumes [10], in which the extracted planes approach those of an experienced sonographer but results are so far limited to the fetal head measurements. Standard plane classification has also been further developed to provide active guidance during freehand operation [6].

In this paper, we propose performing all the relevant measurements for fetal weight estimation within a unified automated system, which is our main contribution. The proposed AutoFB framework involves classifying the three standard planes and segmenting the head, abdomen and femur. This is followed by the extraction of the following measurements: biparietal diameter (BPD), occipito-frontal diameter (OFD), head circumference (HC), transverse abdominal diameter (TAD), anterior-posterior abdominal diameter (APAD), abdominal circumference (AC), and femur length (FL). We achieve this by training a multi-class segmentation neural network that automatically identifies and segments the relevant anatomy structures within any of the three standard planes. The corresponding biometry is then extracted by applying scale recovery and using ellipse fitting (head or abdomen) and bounding box fitting (femur). To the best of our knowledge, AutoFB is the first framework to automate fetal biometry estimation from all three standard planes. We demonstrate the robustness of AutoFB by experimenting using real clinical US data and validate both inferred segmentation and estimated biometry. The clinical data used for the validation contains 346 2D US planes from 42 pregnancies. AutoFB is of high clinical relevance as it will enable automating biometry, a task currently affected by high inter-operator variability [5] due to manual selection and measurement of the relevant US plane.

2   Fetal Biometry

To clinically measure fetal size and weight during a fetal US, the sonographer navigates the US probe to localize a view of each of the three standard planes. While this task is subject to operator variability, there are established guidelines on which features should be visible within each standard plane [2]. They must then lock the display and manually place calipers on key landmarks from which biometric measurements are extracted. The BPD and OFD measurements are required for the HC measurement on the transventricular plane (Fig. 1(left)). The TAD and APAD are required for the AC measurement on the transabdominal plane (Fig. 1(middle)). HC and AC are then computed using, $\pi(d_1 + d_2)/2$, where d_1 and d_2 are the BPD and OFD in the case of head and TAD and APAD in the case of abdomen measurements. Alternatively, an ellipse fitting function is available in some US machines and can be used for head and abdominal measurements, however, its usage largely depends on operator choice or established practice within a specific clinical site. This feature is not routinely used in the context of data acquired and presented within this work. To measure the femur length (FL), the extreme lateral edges including both of the epiphyses must be visualized and measured along the long axis of the femur (Fig. 1(right)).

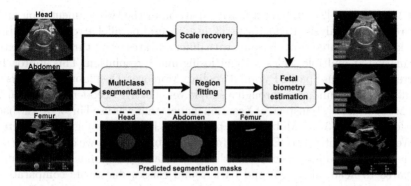

Fig. 2. Overview of the AutoFB framework. Given a US standard plane, AutoFB performs multi-class segmentation for plane detection and anatomy segmentation, followed by shape fitting and scale recovery for biometry estimation.

3 Methodology

An overview of the proposed framework is presented in Fig. 2. The framework jointly performs the 3-plane detection and anatomy segmentation by training state-of-the-art segmentation models for identifying the head, abdomen and femur anatomies and selecting the best performing architecture (Sect. 3.1). This is followed by shape fitting on the segmented regions, automated image scale retrieval and biometry estimation in millimetres units (Sect. 3.2).

3.1 Multi-class Image Segmentation

In order to build a unified system, we define our problem as semantic segmentation between 4 specific classes: head, abdomen, femur, and background. With groundtruth data, each standard plane will only contain background and one of the other 3 classes. We experimented with two state-of-the-art image segmentation models, namely, U-Net [18] and Deeplabv3+ [3]. U-Net can be regarded as the most commonly used architecture for biomedical image segmentation and is recommended when the training data is limited. Deeplabv3+ has achieved state-of-the art performance on large-scale semantic segmentation datasets (PASCAL VOC 2012). Both U-Net and Deeplabv3+ are encoder-decoder networks, where U-Net is a special case in which the decoder component is connected with the encoder through skip connections and is not decoupled from the encoder. We briefly introduce these architectures and refer the reader to [3,18] for specific details.

U-Net is a type of fully convolutional network which consists of a contraction path and an expansion path. The contraction path can be a pretrained encoder which captures the context while limiting the feature map size. The expansion path is a symmetric decoder network which also performs up-sampling to recover the segmentation map size. The encoder and decoder paths are connected through skip connections for sharing localization information. We used the ResNet50 [11] as the encoder architecture for U-Net. We also experimented with Mobilenetv2 [21] to have a fair comparison of the two segmentation architectures under analysis. Deeplabv3+ [3] uses several parallel atrous convolutions (also known as dilated convolutions) with different rates to capture the contextual information at multiple scales without losing image resolution. This approach is referred to as Atrous Spatial Pyramid Pooling. Moreover, Deeplabv3+ recovers the detailed object boundaries through a simple yet effective decoder module [3]. We used MobileNetv2 [21] instead of Xception model (that was used in [3]) as the backbone for DeeplabV3+ as MobileNetv2 backbone is both light-weight and effective.

We use cross entropy (CE) as loss function. From Table 1, we can observe that the data is highly imbalanced, with the femur class having much fewer samples compared to head, abdomen and background classes due to its comparatively small segmentation area. To handle this issue, we also use weighted CE (wCE) where given the total number of pixels per class, $[c_i]_i^4$, weight w_i for the ith class is given by, $w_i = \frac{max([c_i]_i^4)}{c_i}$. The obtained results are discussed in Sect. 5.

3.2   Fetal Biometry Estimation

Different standard planes require different biometry measurements, and therefore the first step is to detect and localize the segmented region. This is defined as the largest segmented area predicted by the networks described in the previous section. We later show experimentally that this strategy correctly identifies all planes in our test data. It is known a priori that the head and abdomen are elliptical while the femur is oblong (Fig. 1). Thus, ellipse fitting is performed on the segmented head and abdomen masks through shape contour extraction and applying direct least square to fit an ellipse [8], where the major and minor axes of the fitted ellipse represent BPD and OFD for the head and TAD and APAD for the abdomen, respectively. These are in turn used to calculate the circumference of the fitted ellipses, providing HC and AC measurements. On the femur plane, a horizontal bounding box with zero orientation is fitted on the segmented mask, where the length of its diagonal gives the FL estimate. A femur is not necessarily aligned to the horizontal/vertical axis, hence the use of bounding box diagonal as FL always holds. Finally, lengths in pixels are scaled to millimetres to obtain results that are directly comparable to clinically measured biometry.

While the metric scale of the US images (in px/mm) is usually trivial to obtain during operation, the automatic extraction of this parameter from retrospectively acquired data proved useful to fully automate the hundreds of measurements obtained in this work. Obtaining US scale is always system-dependent because it must be extracted either from the visual interface of the US machine or from the raw data, which requires access to a proprietary API. We use visual interface for scale recovery since we did not have access to the raw data. To obtain the scale, we exploit the consistent interface of the US machine used to acquire our dataset (GE Voluson), namely the caliper visible on the left-hand side of the US images. The ruler markers are detected with simple template matching and their smallest interval (can be either 5 mm or 10 mm) is determined from the relative size of the markers. The same template matching approach is easy to deploy on systems other than GE Voluson since all medical grade US machines have a similar ruler available.

Table 1. Total number of sample in each segmentation class and in each cross-validation fold and average pixels per class per frames.

	All images	Fold 1	Fold 2	Fold 3	Fold 4	Avg. pixels per class	
Total subjects	42	10	9	12	11	per frame	
Total Images	346	87	86	89	84	Background	816239
Head	135	26	44	29	36	Head	74127
Abdomen	103	32	22	26	23	Abdomen	44691
Femur	108	29	20	34	25	Femur	3833

4 Dataset and Experimental Setup

The data collection process has been reviewed and approved by the local research ethics committee under the title; "Fetal US and fetal monitoring technologies to improve prenatal diagnosis and therapy for fetal abnormality and maternal and fetal perinatal outcome", IRAS ID 230125. Patients attending University College London Hospital for US examination were enrolled and pseudo-anonomyzed by the clinical research staff. Each patient gave written consent. For the purpose of anonymizing, transferring and storing data a customized version of XNAT 1.6 was used. The complete image library from each US was transferred to the research database. The hospital protocol undertakes US screening in accordance with a National Fetal Anomaly Screening Program[1]. Each saved image represents an image of diagnostic quality. The US images saved by the operator were considered to be the optimal image for that scan given the limitations of fetal lie and stage of gestation. The measurement calipers were applied by the US operator and in most cases, the image with and without the measurement calipers were saved. A subset of images relevant to fetal biometry were extracted from the database by a clinical research fellow. A total of 346 images were included from 42 pregnancies. Each image in the set of data was classified as AC, HC or FL. The VIA annotation tool [7] was used to manually annotate the head, abdomen or femur within each image for the segmentation task. The obtained fully anonymized standard US plane images have large intra-class variability. For example, in some cases the femur is well aligned to the horizontal plane while in other cases the angle of sonnation is wider and the level of magnification is less. Although operators followed a standard protocol to capture a good quality image including all necessary anatomical details, some images have relatively poor contrast and dark patches. These are secondary to the technical limitations of US and maternal body habitus. Often unavoidable, the heterogeneity of the data set introduces challenges for the segmentation task.

The acquired data from 42 fetuses (346 US images) is divided into 4 folds, used for testing the robustness of the segmentation networks, such that each fold contains at least 80 images and all US images originating from a single fetus are only included in a particular fold. Hence, the data in a fold is unseen for all other folds

---

[1] https://tinyurl.com/NHSFetalAnomalyScreeningHB.

Table 2. Four-fold cross-validation results showing comparison of Deeplabv3+ and UNet having different configurations. Mean and standard deviation of mIoU across all folds is reported. Key: BG- background; H - head; A - abdomen; F - femur; CE - cross entropy; wCE - weighted cross entropy; MNv2 - Mobilenetv2.

Method	mIoU	mIoU-BG	mIoU-H	mIoU-A	mIoU-F
Deeplabv3+ (MNv2-CE)	0.87 ± 0.02	0.95 ± 0.02	0.93 ± 0.02	0.89 ± 0.03	0.61 ± 0.03
Deeplabv3+ (MNv2-wCE)	**0.88 ± 0.01**	**0.95 ± 0.01**	**0.93 ± 0.02**	**0.89 ± 0.02**	**0.61 ± 0.02**
UNet (MNv2-CE)	0.82 ± 0.05	0.93 ± 0.03	0.89 ± 0.05	0.85 ± 0.05	0.56 ± 0.03
UNet (MNv2-wCE)	0.86 ± 0.01	0.94 ± 0.01	0.91 ± 0.02	0.86 ± 0.02	0.58 ± 0.01
UNet (Resnet-CE)	0.75 ± 0.06	0.88 ± 0.05	0.84 ± 0.07	0.77 ± 0.05	0.53 ± 0.03
UNet (Resnet-wCE)	0.78 ± 0.04	0.87 ± 0.03	0.83 ± 0.04	0.75 ± 0.06	0.53 ± 0.02

(as mentioned in Table 1). Mean Intersection over Union (mIoU) is used for evaluating the segmentation models, and absolute error between the clinically measured and automatically predicted fetal biometry is used for evaluating the proposed AutoFB. All images are of varying sizes (resolution) as they were cropped to remove any identifiable information. Therefore, we resized all images to 1024 × 1024 pixel resolution before model training. Data augmentation is applied by introducing random scale, rotation, shift, flipping, brightness and contrast changes before obtaining an image crop of size 512 × 512 pixel at a random location which is used as the input for training the segmentation network. Data augmentation helped in avoiding model over-fitting. An initial learning rate $10e^{-3}$ with a step decay by a factor of $1/10$ at 75th and 150th is used with the ADAM optimizer. The model is trained for 600 epochs with early stopping based on the criteria of no improvement of the training set with patience of 50 epoch is used. The weights that captured the best performance on the training data are used to evaluate the segmentation model on the holdout fold. The segmentation networks are implemented in PyTorch and trained using a single Tesla V100-DGXS-32GB GPU of an NVIDIA DGX-station.

5   Results and Discussion

We perform comparison of the Deeplabv3+ and U-Net having two commonly used backbones and used both CE and wCE losses (refer to Sect. 3.1). The quantitative comparison using 4-fold cross-validation is presented in Table 2. Both configurations of Deeplabv3+ are comparable (overall mIoU = 0.88) though the standard deviation is lower when wCE is used. Deeplabv3+ also outperformed the UNet configurations. The effect of introducing wCE loss for handling class imbalance problem is more evident from the different UNet configurations. Mobilenetv2 backbone, which has significantly less number of network parameters (3.5M), showed superior performance than the Resnet50 (26M parameters) backbone. Selecting an efficient and robust backbone architecture is essential and can significantly improve the overall segmentation network performance. From

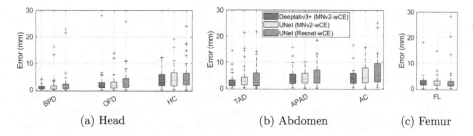

Fig. 3. Boxplots showing the comparison between the best performing models and the absolute error between the clinically measured and predicted fetal biometry.

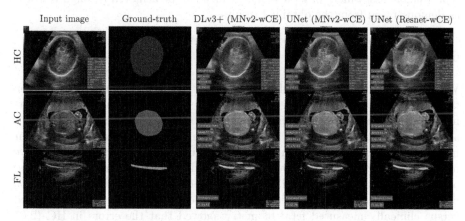

Fig. 4. Qualitative comparison of segmentation methods showing scenarios where inaccurate segmentation resulted in fetal biometry estimation failure. (Row 1 and 2) HC and AC examples where UNet resulted in inaccurate segmentation. (Row 3) FL example where all three methods failed. This image corresponds to the only outlier which is visible in Fig. 3(c) Deeplabv3+ error plot.

Table 2, we can observe that mIoU-F is particularly low compared to the mIoU-BG, mIoU-H and mIoU-A. This is because (1) the number of per-pixel samples in the femur class are very small (Table 1); (2) a small error in predicted segmentation vs the ground-truth results in a significantly low IoU value when the object size is small; (3) of large intraclass variability.

Figure 3 shows the boxplots for the absolute error between the clinically measured and predicted biometry. The error in head measurements are the lowest, with a median of 0.80 mm for BPD, 1.30 mm for OFD and 2.67 mm for HC and fewer outliers compared to other methods when segmentation masks from Deeplabv3+ (Mobilev2+wCE) are used (Fig. 3(a)). A similar trend is observed for the abdomen measurements, with a median of 2.39 mm for TAD, 3.82 mm for APAD and 3.77 mm for AC (Fig. 3(b)). FL showed comparable results with a median of 2.1 mm for Deeplabv3+ (Mobilenet-v2+wCE) but with fewer outliers (Fig. 3(c)). It is worth mentioning that the obtained error is less than the $\pm 15\%$

error permissible in the US assessment [22]. Figure 4 presents the qualitative comparison of the segmentation methods, depicting cases where either one or all methods fail in estimating the biometry due to inaccurate segmentation.

From a clinical point of view, successful interpretation of clinical US images requires an understanding that the fetus, a 3D object, fixed in neither time nor space is being represented on a 2D grey-scale. Operator experience, combined with the effects of probe motion and homogeneity of US images contributes to high inter- and intra-operator variability. US is used extensively in the assessment and management of pregnancies at high risk of fetal growth disorders. Appropriate management of these cases requires high quality assessment and reproducible assessment of fetal weight, which can be achieved through AutoFB as demonstrated from the obtained results.

6   Conclusion

We proposed AutoFB, a unified framework for estimating fetal biometry given the three standard US planes. The proposed framework exploited the existing segmentation networks for predicting the segmentation masks for the head, abdomen and femur. Head and abdomen were modelled as an ellipse with their major and minor axes and circumference providing an estimate for the respective measurements. Femur length was modelled as the diagonal on a rectangle fitted onto the segmentation mask. Through retrospective scale recovery and shape fitting, we obtained the fetal biometry estimates. Comparison of the predicted versus clinically measured fetal biometry showed that the errors in HC (2.67 mm), AC (3.77 mm) and FL (2.10 mm) were minimal and were better than the $\pm15\%$ error that is typically acceptable in fetal US assessment. Future work involves increasing the training data size for further improving the segmentation and integrating AutoFB with the standard US plane detection [1] framework. Moreover, comparing experts and novices performance with the AutoFB can provide evidence supporting its clinical translation.

Acknowledgments. The work was supported by the Wellcome/EPSRC Centre for Interventional and Surgical Sciences (WEISS) [203145Z/16/Z]; Engineering and Physical Sciences Research Council (EPSRC) [EP/P027938/1, EP/R004080/1, EP/P012841/1, NS/A000027/1]; the H2020 FET (GA 863146); Wellcome [WT101957]; and The Royal Academy of Engineering Chair in Emerging Technologies scheme (CiET1819/2/36). A. L. David and D. M. Peebles are supported by the National Institute for Health Research University College London Hospitals Biomedical Research Centre.

References

1. Baumgartner, C.F., et al.: SonoNet: Real-time detection and localisation of fetal standard scan planes in freehand ultrasound. IEEE Trans. Med. Imaging **36**(11), 2204–2215 (2017)

2. Cavallaro, A., et al.: Quality control of ultrasound for fetal biometry: results from the intergrowth-21st project. Ultrasound Obstetrics Gynecol. **52**(3), 332–339 (2018)
3. Chen, L.C., Zhu, Y., Papandreou, G., Schroff, F., Adam, H.: Encoder-decoder with atrous separable convolution for semantic image segmentation. In: Proceedings of the European Conference on Computer Vision (ECCV), pp. 801–818 (2018)
4. Chen, X., et al.: Automatic measurements of fetal lateral ventricles in 2d ultrasound images using deep learning. Front. Neurol. **11**, 526 (2020)
5. Dromey, B.P., et al.: Dimensionless squared jerk: an objective differential to assess experienced and novice probe movement in obstetric ultrasound. Prenat. Diagn. **41**(2), 271–277 (2020)
6. Droste, R., Drukker, L., Papageorghiou, A.T., Noble, J.A.: Automatic probe movement guidance for freehand obstetric ultrasound. In: Martel, A.L., et al. (eds.) MICCAI 2020. LNCS, vol. 12263, pp. 583–592. Springer, Cham (2020). https://doi.org/10.1007/978-3-030-59716-0_56
7. Dutta, A., Zisserman, A.: The VIA annotation software for images, audio and video. In: Proceedings of the 27th ACM International Conference on Multimedia. MM '19, ACM, New York, NY, USA (2019)
8. Fitzgibbon, A., Pilu, M., Fisher, R.B.: Direct least square fitting of ellipses. IEEE Trans. Pattern Anal. Mach. Intell. **21**(5), 476–480 (1999)
9. Gao, Y., Beriwal, S., Craik, R., Papageorghiou, A.T., Noble, J.A.: Label efficient localization of fetal brain biometry planes in ultrasound through metric learning. In: Hu, Y., et al. (eds.) ASMUS/PIPPI -2020. LNCS, vol. 12437, pp. 126–135. Springer, Cham (2020). https://doi.org/10.1007/978-3-030-60334-2_13
10. Grandjean, G.A., Hossu, G., Bertholdt, C., Noble, P., Morel, O., Grangé, G.: Artificial intelligence assistance for fetal head biometry: assessment of automated measurement software. Diagn. Intervent. Imaging **99**(11), 709–716 (2018)
11. He, K., Zhang, X., Ren, S., Sun, J.: Deep residual learning for image recognition. In: Proceedings of the IEEE Conference on Computer Vision and Pattern Recognition, pp. 770–778 (2016)
12. Hermawati, F., Tjandrasa, H., Sari, G.P., Azis, A., et al.: Automatic femur length measurement for fetal ultrasound image using localizing region-based active contour method. In: Journal of Physics: Conference Series, vol. 1230, p. 012002. IOP Publishing (2019)
13. van den Heuvel, T.L., de Bruijn, D., de Korte, C.L., Ginneken, B.V.: Automated measurement of fetal head circumference using 2D ultrasound images. PLOS One **13**(8), e0200412 (2018)
14. Khan, N.H., Tegnander, E., Dreier, J.M., Eik-Nes, S., Torp, H., Kiss, G.: Automatic measurement of the fetal abdominal section on a portable ultrasound machine for use in low and middle income countries. In: 2016 IEEE International Ultrasonics Symposium (IUS), pp. 1–4. IEEE (2016)
15. Khan, N.H., Tegnander, E., Dreier, J.M., Eik-Nes, S., Torp, H., Kiss, G.: Automatic detection and measurement of fetal biparietal diameter and femur length–feasibility on a portable ultrasound device. Open J. Obstet. Gynecol. **7**(3), 334–350 (2017)
16. Li, J., et al.: Automatic fetal head circumference measurement in ultrasound using random forest and fast ellipse fitting. IEEE J. Biomed. Health Inform. **22**(1), 215–223 (2017)
17. Lin, Z., et al.: Multi-task learning for quality assessment of fetal head ultrasound images. Med. Image Anal. **58**, 101548 (2019)

18. Ronneberger, O., Fischer, P., Brox, T.: U-Net: Convolutional networks for biomedical image segmentation. In: Navab, N., Hornegger, J., Wells, W.M., Frangi, A.F. (eds.) MICCAI 2015. LNCS, vol. 9351, pp. 234–241. Springer, Cham (2015). https://doi.org/10.1007/978-3-319-24574-4_28

19. Rueda, S., et al.: Evaluation and comparison of current fetal ultrasound image segmentation methods for biometric measurements: a grand challenge. IEEE Trans. Med. Imaging **33**(4), 797–813 (2013)

20. Salomon, L., et al.: ISUOG practice guidelines: ultrasound assessment of fetal biometry and growth. Ultrasound Obstetrics Gynecol. **53**(6), 715–723 (2019)

21. Sandler, M., Howard, A., Zhu, M., Zhmoginov, A., Chen, L.C.: Mobilenetv2: Inverted residuals and linear bottlenecks. In: Proceedings of the IEEE Conference on Computer Vision and Pattern Recognition, pp. 4510–4520 (2018)

22. Sarris, I., Ioannou, C., Chamberlain, P., Ohuma, E., Roseman, F., Hoch, L., Altman, D., Papageorghiou, A., International Fetal and Newborn Growth Consortium for the 21st Century (INTERGROWTH-21st): Intra-and interobserver variability in fetal ultrasound measurements. Ultrasound Obstet. Gynecol. **39**(3), 266–273 (2012)

23. Sobhaninia, Z., et al.: Fetal ultrasound image segmentation for measuring biometric parameters using multi-task deep learning. In: 2019 41st Annual International Conference of the IEEE Engineering in Medicine and Biology Society (EMBC), pp. 6545–6548. IEEE (2019)

24. Zhang, L., Dudley, N.J., Lambrou, T., Allinson, N., Ye, X.: Automatic image quality assessment and measurement of fetal head in two-dimensional ultrasound image. J. Med. Imaging **4**(2), 024001 (2017)

Learning Spatiotemporal Probabilistic Atlas of Fetal Brains with Anatomically Constrained Registration Network

Yuchen Pei[1,2], Liangjun Chen[2], Fenqiang Zhao[2], Zhengwang Wu[2], Tao Zhong[2], Ya Wang[2], Changan Chen[3], Li Wang[2], He Zhang[3], Lisheng Wang[1], and Gang Li[2(✉)]

[1] Institute of Image Processing and Pattern Recognition,
Department of Automation, Shanghai Jiao Tong University, Shanghai, China
[2] Department of Radiology and BRIC, University of North Carolina at Chapel Hill,
Chapel Hill, USA
gang_li@med.unc.edu
[3] Department of Radiology, Obstetrics and Gynecology Hospital, Fudan University,
Shanghai, China

Abstract. Brain atlases are of fundamental importance for analyzing the dynamic neurodevelopment in fetal brain studies. Since the brain size, shape, and anatomical structures change rapidly during the prenatal period, it is essential to construct a spatiotemporal (4D) atlas equipped with tissue probability maps, which can preserve sharper early brain folding patterns for accurately characterizing dynamic changes in fetal brains and provide tissue prior informations for related tasks, e.g., segmentation, registration, and parcellation. In this work, we propose a novel unsupervised age-conditional learning framework to build temporally continuous fetal brain atlases by incorporating tissue segmentation maps, which outperforms previous traditional atlas construction methods in three aspects. *First*, our framework enables learning age-conditional deformable templates by leveraging the entire collection. *Second*, we leverage reliable brain tissue segmentation maps in addition to the low-contrast noisy intensity images to enhance the alignment of individual images. *Third*, a novel loss function is designed to enforce the similarity between the learned tissue probability map on the atlas and each subject tissue segmentation map after registration, thereby providing extra anatomical consistency supervision for atlas building. Our 4D temporally-*continuous* fetal brain atlases are constructed based on 82 healthy fetuses from 22 to 32 gestational weeks. Compared with the atlases built by the state-of-the-art algorithms, our atlases preserve more structural details and sharper folding patterns. Together with the learned tissue probability maps, our 4D fetal atlases provide a valuable reference for spatial normalization and analysis of fetal brain development.

Keywords: Fetal brain · 4D atlas construction · Anatomical knowledge

© Springer Nature Switzerland AG 2021
M. de Bruijne et al. (Eds.): MICCAI 2021, LNCS 12907, pp. 239–248, 2021.
https://doi.org/10.1007/978-3-030-87234-2_23

1   Introduction

The MRI study of the developing fetal brain has been demonstrated with more challenges than the studies of aging and disease progression in adults and older children [15]. These challenges originate from the complex processes of morphological, functional, and appearance changes involved in the dynamic prenatal brain development [4,18]. To well characterize this dynamic development, a spatiotemporal (4D) rather than static fetal atlas is critically needed [17,20]. However, existing 4D fetal brain atlases [10,21] have three drawbacks. First, they are usually built at discrete time points based on subpopulations and thus are temporally discontinuous, which cannot well cover many critical ages during rapid fetal brain development. To alleviate the discrete time point issue, the kernel regression approach is proposed [14], which can compute the continuous template at any given age using weighted supports from their temporal neighbors for some populations of interest [9,11,17]. However, since the fetal brain has regionally heterogeneous and significant nonlinear development, the kernel needs to be tuned very carefully to fit for the complex development and distribution of the dataset [16]. Second, most existing atlases generally have the ambiguous appearance and low tissue contrast, which substantially degrade the spatial normalization ability of the atlas. Third, conventional atlas construction methods [1,2] require pair-wise registration from each individual to the corresponding templates in an iterative procedure, thus leading to extremely long runtime for atlas construction.

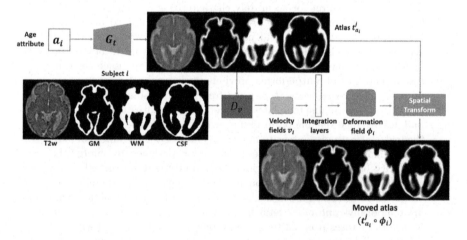

Fig. 1. The proposed age-conditional atlas construction framework. G_t represents the atlas synthesis module, and D_v represents the registration network.

Recently, deep learning-based methods are also proposed to align different subjects and synthesize templates in adult atlas buildings [6,7]. However, such methods work poorly on fetal MRI images, due to the low tissue contrast and rapid appearance changes in the fetal brain images. This makes it challenging to build a high-quality fetal brain atlas with clear tissue boundaries only based on the noisy intensity information. To solve this problem, we propose to learn the templates using both the *intensity images* and *tissue segmentation maps*, which can introduce a tissue map-based similarity loss to improve the pair-wise registration accuracy and thereby boost the quality of built atlases. Specifically, we design an unsupervised learning-based framework consisting of an atlas synthesis module and a deformable registration module, making it possible to simultaneously learn the age-conditional atlas and register the individual images to the corresponding atlases, respectively. Spatiotemporal characteristics of fetal brain development are effectively captured by our constructed atlases, which demonstrate sharper structural details, higher accuracy, and better time efficiency than conventional atlases. Besides, our learned tissue probability maps are valuable references for many atlas-based tasks, e.g., segmentation, registration, and parcellation of the fetal brain MRI.

2   Method

2.1   Dataset and Pre-processing

In this study, 4D atlases were constructed based on a fetal MRI dataset, including 82 healthy fetuses scanned at a gestational age (GA) ranging from 22 to 32 weeks. All scans were acquired by a 1.5T Siemens Avanto scanner with a resolution of $0.54 \times 0.54 \times 4.4$ mm^3. Brain localization [12], extraction [12,13], and super-resolution volume reconstruction [8] from 2D stacks were performed to generate the 3D brain volumes with an isotropic resolution of $0.8 \times 0.8 \times 0.8$ mm^3. All scans were segmented into the white matter (WM), gray matter (GM), and cerebrospinal fluid (CSF) using a learning-based method [19] and then manually corrected by experts. All scans were firstly affinely-aligned together.

2.2   Multi-channel Age-Conditional Atlas Construction Network

Figure 1 presents an overview of our framework. We design an atlas synthesis module G_t, which synthesizes atlases at given ages, and a registration network D_v, which transforms a synthesized atlas to each individual image, thus avoiding the expensive time cost of traditional multi-step registration and iterative refinement.

Let $V = \{V_1^0, V_1^1, V_1^2, V_1^3, ..., V_i^j, ..., V_N^0, V_N^1, V_N^2, V_N^3\}$ denotes a fetal volumetric dataset containing N subjects and each subject i ($i = 1, 2, ..., N$) with T2w image and three tissue label maps ($j = 0, 1, 2, 3$, representing T2w image, GM, WM, and CSF, respectively). $A = \{a_i\}$ indicates the age attributes of subject i.

The atlas generation network G_t is a decoder that takes the age attribute a_i of the data as input, and outputs the atlas t_{a_i}. The decoder consists of a fully connected layer, a reshape layer, and an upsampling layer followed by 4 convolution blocks. The registration network $D_v(t_{a_i}, V_i)$ takes the generated t_{a_i}

and input image V_i as inputs and outputs a stationary velocity field v_i. D_v is designed as a U shape with a similar architecture as in a recent registration literature [3]. Then we obtain the deformation field ϕ given the estimated velocity field v from registration network using differentiable scaling and squaring integration layers [5] to ensure diffeomorphic (invertible and topology-preserving) deformation, and the warped template $t_{a_i} \circ \phi_i$ through spatial transform layers. Then our framework can simultaneously generate an age-conditional atlas corresponding to the age attribute of the subject and provide deformation field to align the atlas to the subject image. The objective of this baseline framework is to minimize the following loss:

$$L = \sum_i NCC_{local}(V_i^0, t_{a_i}^0 \circ \phi_i) + \lambda_{AC} L_{AC} + \lambda_c \|\bar{u}\|^2 + \frac{\lambda_d}{2} \sum_i \|u_i\|^2 + \frac{\lambda_a}{2} \sum_i \|\nabla u_i\|^2 \quad (1)$$

where $t_{a_i}^0 = G(a_i)$ represents the synthesized intensity atlas at the time point a_i, $\circ$ is the spatial warp operator, and ϕ_i is the deformation field aligning the generated atlas to subject i. The first term is a localized normalized cross-correlation (NCC) loss between v_i^0 and $t_{a_i}^0 \circ \phi_i$, which enforces the similarity of the moved intensity atlas and an individual intensity image:

$$NCC_{local}(f, w) = \frac{1}{n} \sum_k \frac{\sum_{m \in \delta_k} (f_m - \overline{f_k})(w_m - \overline{w_k})}{\sqrt{\sum_{m \in \delta_k} (f_m - \overline{f_k})^2 \sum_{m \in \delta_k} (w_m - \overline{w_k})^2}} \quad (2)$$

where the size of sliding windows is n, f_m and w_m refer to the mth voxel in the subject images and warped atlases, respectively. And $\overline{f_k}$ and $\overline{w_k}$ are the average image intensity values over window δ_k, which is centered at k^{th} voxel. NCC loss is more robust to intensity variations across scans and datasets [1]. The rest terms in (1) regularize the unbiasedness, size, and smoothness of the displacement field u, where $u = \phi - Id$ [7].

2.3   Anatomical Constraint

Conventional atlas construction methods [9,16] generally take fuzzy intensity-based features into account, and ignore the informative anatomical features. However, due to a) the significant appearance and structure changes; b) low tissue contrast in the rapidly developing fetal brain, it is difficult to achieve high-quality atlases only using intensity information. To address this issue, we introduce the tissue segmentation maps to improve the registration accuracy between the constructed atlases and subject images, consequently leading to sharper image details. At the same time, the multi-channel framework can generate tissue probabilistic atlases, which can be used as a prior reference for accurate fetal brain tissue segmentation. Hence, we concatenate subjects' tissue segmentation maps with intensity images as inputs as four channels to guide the registration. In order to match all corresponding anatomical structures between the constructed atlases and subject images, we explicitly enforce the consistency of tissue segmentation maps between the subjects and the warped atlases by a devised anatomical constraint (AC), formulated as:

$$L_{AC} = \sum_i \sum_{j=1}^3 NCC_{local}(V_i^j, t_{a_i}^j \circ \phi_i) \tag{3}$$

where $t_{a_i}^j$ represents a tissue probability map from the generated atlas. We still use the localized NCC loss for the tissue maps, similarity measures.

2.4 Training Strategy

To make the training process of the proposed framework more tractable, we design a three-step training strategy. 1) We train the registration network D_v as a multi-channel registration task based on the entire training set, which computes deformation field ϕ between each pair of subject images; 2) We fix the parameters in D_v and train G_t independently until convergence. This is conducted because if we jointly train G_t and D_v from scratch, the registration module may dominate the training process due to the richer supervision information of D_v and G_t may be undertrained; 3) We jointly train the whole framework until convergence.

2.5 Implementation Details

The network is implemented using the TensorFlow with CUDA enabled on NVIDIA TITAN RTX 24GB GPU and 64-bit Linux operation system. In our experiments, all networks use Adam optimizer with a learning rate of 0.0001. λ_c for regularizing centrality is set to 1.0, λ_d for encouraging small deformation is 0.0001, λ_a for smoothing deformation is 0.5, and λ_{AC} is 1.0. We first perform affine transformation for each subject, and then the average affine parameters for each subgroup are obtained by ANTs [2]. Once the atlases are generated, we further use the above achieved affine parameters to rescale the newly constructed age-specific atlases to reflect global volume changes.

3 Experiments

In this section, we compare our framework with the state-of-the-art atlas construction methods. We show that our method efficiently synthesizes an age-conditional template, comparable to traditional methods that require significantly more executive time. Besides, we perform ablation studies to show the effectiveness of each part in our framework. Specifically, the following parts are investigated in the ablation: multi-channel architecture, anatomical constraint, and training strategy. Finally, in order to validate our method, a stratified 3-fold cross-validation strategy is employed to compare the spatial normalization ability of different atlases, and each fold consists of 64 training images, 9 validation images, and 9 testing images.

Evaluation Metrics. To quantitatively validate the atlases, we calculate the Pearson's correlation coefficient (CC) between the warped atlases and subject images. We also transform the constructed tissue probability atlases to individual space, and quantify the volume overlapping using Dice score (DSC), 95th percentile Hausdorff distance (95^{th}HD), and average surface distance (ASD). Therefore, the higher CC, DSC, and the lower 95^{th}HD, ASD indicate higher registration accuracy, and thus, a better atlas.

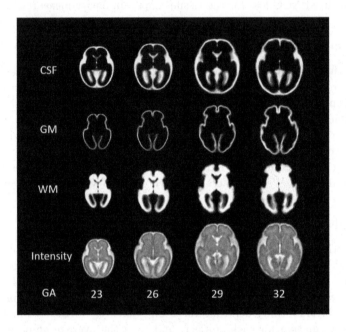

Fig. 2. Our 4D temporally continuous fetal brain atlas at typical time points from 22 to 32 gestational weeks (GA), including intensity templates and tissue probability maps.

3.1 Comparison with State-of-the-Art Method

In this section, we compare our constructed multi-channel 4D atlases with the atlases built using the state-of-the-art symmetric group-wise normalization (SyGN) method [1] provided by ANTs. In Table 1, the multi-channel features are used to construct atlases by SyGN (MC-SyGN). We can see that our method (Ours) achieves higher similarity to the ground truth and improves the segmentation accuracy between the transformed tissue probability atlases and subject's tissue segmentation maps via registration network. Moreover, our trained framework only takes 20 s for one forward inference, including atlas generation and atlas-to-individual registration, which is much faster than conventional atlas construction methods (generally taking several hours or days). We present the 4D temporally continuous fetal brain atlases constructed by our method at typical time points from 22 to 32 gestational weeks (GA) in Fig. 2.

3.2   Ablation Studies

Table 1. The results of registration accuracy based on different atlases.

Metrics	Tissue	Baseline [7]	MC-SyGN	Ours-w/o pre	Ours-w/o AC	Ours
DSC (%)	CSF	N/A	83.1 ± 2.77	80.6 ± 3.54	84.9 ± 2.72	**86.1 ± 2.69**
	GM	N/A	72.0 ± 3.81	71.2 ± 4.73	87.6 ± 3.21	**89.6 ± 2.82**
	WM	N/A	91.2 ± 2.01	89.9 ± 2.54	94.1 ± 1.98	**96.4 ± 1.12**
ASD (mm)	CSF	N/A	0.454 ± 0.076	0.543 ± 0.089	0.431 ± 0.064	**0.412 ± 0.048**
	GM	N/A	0.380 ± 0.113	0.457 ± 0.124	0.265 ± 0.098	**0.166 ± 0.056**
	WM	N/A	0.519 ± 0.13	0.879 ± 0.146	0.456 ± 0.109	**0.233 ± 0.095**
$95^{th}HD$ (mm)	CSF	N/A	1.58 ± 0.301	1.67 ± 0.342	1.47 ± 0.278	**1.39 ± 0.215**
	GM	N/A	1.31 ± 0.261	1.43 ± 0.282	1.07 ± 0.254	**0.883 ± 0.213**
	WM	N/A	1.74 ± 0.433	2.16 ± 0.467	1.43 ± 0.408	**1.07 ± 0.414**
CC (%)	N/A	96.1 ± 0.98	94.7 ± 1.0	93.7 ± 1.23	96.8 ± 0.557	**97.8 ± 0.476**

Significance of Multi-channel Based Framework. First, we trained a baseline model [7] with a single channel, which only utilized intensity images (Baseline) to construct atlas, and compared it with the proposed multi-channel framework (Ours). From Table 1, we can see that using multi-channel features to construct atlases can achieve clearly improved performance on CC, which means the informative tissue segmentation maps can help enhance the registration accuracy, and consequently sharper atlases while preserving more structural details.

Significance of Anatomical Constraint. Compared to the model without anatomical constraint (Ours-w/o AC), the proposed framework (Ours) yields largely improved performance, which implies that by leveraging the anatomical constraints, the proposed framework can capture the complex dynamically developing anatomical patterns with improved registration accuracy in synthesizing sharper 4D atlases. In Fig. 3, we also perform a qualitative comparison for atlases built with or without anatomical constraint, which indicates that the intensity atlases built with AC retain more structural details and the corresponding probability maps are clearer and sharper.

Significance of Training Strategy. To verify the effectiveness of the training strategy, we deploy the experiments by jointly training G_t and D_v without pretraining (Ours-w/o pre). From Table 1, we can see that pre-training D_v can boost the performance of our network, which is attributed to the sufficiently focused training on D_v and G_t sequentially, and then jointly training the whole network enhances the convergence of the whole network with fine-tuned G_t and D_v.

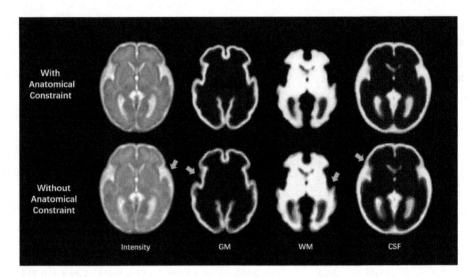

Fig. 3. Qualitative comparison of the 32-week atlases built with and without AC. Top row: the atlas constructed with AC. Bottom row: the atlas constructed without AC.

4   Conclusion

We present a novel learning-based atlas construction framework to efficiently and accurately build the spatiotemporal fetal brain atlases. The constructed 4D atlases preserve more structural details for fetal brain spatial normalization. In our method, we design an anatomical constraint to enforce multi-channel features alignment between subjects and generated atlases in order to make full use of the reliable tissue information. We characterized the performance of the proposed framework by evaluating the volumetric overlapping and tissue segmentation accuracy after registration. Both qualitative and quantitative evaluations indicate the superior performance of our framework compared to state-of-the-art methods. The tissue probability maps of the generated atlases are also provided as a valuable reference for the studies of fetal brain development. Our 4D fetal brain atlases will be released to the community soon.

Acknowledgements. Li G and Wang L were partially supported by NIH grant MH117943.

References

1. Avants, B.B., Epstein, C.L., Grossman, M., Gee, J.C.: Symmetric diffeomorphic image registration with cross-correlation: evaluating automated labeling of elderly and neurodegenerative brain. Med. Image Anal. **12**(1), 26–41 (2008)
2. Avants, B.B., Tustison, N.J., Song, G., Cook, P.A., Klein, A., Gee, J.C.: A reproducible evaluation of ants similarity metric performance in brain image registration. Neuroimage **54**(3), 2033–2044 (2011)

3. Balakrishnan, G., Zhao, A., Sabuncu, M.R., Guttag, J., Dalca, A.V.: Voxelmorph: a learning framework for deformable medical image registration. IEEE Trans. Med. Imaging **38**(8), 1788–1800 (2019)

4. Benkarim, O.M., et al.: Toward the automatic quantification of in utero brain development in 3d structural mri: a review. Hum. Brain Mapping **38**(5), 2772–2787 (2017)

5. Dalca, A.V., Balakrishnan, G., Guttag, J., Sabuncu, M.R.: Unsupervised learning for fast probabilistic diffeomorphic registration. In: Frangi, A.F., Schnabel, J.A., Davatzikos, C., Alberola-López, C., Fichtinger, G. (eds.) MICCAI 2018. LNCS, vol. 11070, pp. 729–738. Springer, Cham (2018). https://doi.org/10.1007/978-3-030-00928-1_82

6. Dalca, A.V., Balakrishnan, G., Guttag, J., Sabuncu, M.R.: Unsupervised learning of probabilistic diffeomorphic registration for images and surfaces. Med. Image Anal. **57**, 226–236 (2019)

7. Dalca, A.V., Rakic, M., Guttag, J., Sabuncu, M.R.: Learning conditional deformable templates with convolutional networks. arXiv preprint arXiv:1908.02738 (2019)

8. Ebner, M., et al.: An automated framework for localization, segmentation and super-resolution reconstruction of fetal brain MRI. NeuroImage **206**, 116324 (2020)

9. Gholipour, A., et al.: A normative spatiotemporal mri atlas of the fetal brain for automatic segmentation and analysis of early brain growth. Sci. Rep. **7**(1), 1–13 (2017)

10. Habas, P.A., et al.: A spatiotemporal atlas of mr intensity, tissue probability and shape of the fetal brain with application to segmentation. Neuroimage **53**(2), 460–470 (2010)

11. Kuklisova-Murgasova, M., et al.: A dynamic 4d probabilistic atlas of the developing brain. NeuroImage **54**(4), 2750–2763 (2011)

12. Liao, L., et al.: Joint image quality assessment and brain extraction of fetal MRI using deep learning. In: Martel, A.L., et al. (eds.) MICCAI 2020. LNCS, vol. 12266, pp. 415–424. Springer, Cham (2020). https://doi.org/10.1007/978-3-030-59725-2_40

13. Lou, J., et al.: Automatic fetal brain extraction using multi-stage U-Net with deep supervision. In: Suk, H., Liu, M., Yan, P., Lian, C. (eds.) MLMI 2019. LNCS, vol. 11861, pp. 592–600. Springer, Cham (2019). https://doi.org/10.1007/978-3-030-32692-0_68

14. Nadaraya, E.A.: On estimating regression. Theory Prob. Appl. **9**(1), 141–142 (1964)

15. Prayer, D., et al.: Mri of normal fetal brain development. Eur. J. Radiol. **57**(2), 199–216 (2006)

16. Serag, A., et al.: Construction of a consistent high-definition spatio-temporal atlas of the developing brain using adaptive kernel regression. Neuroimage **59**(3), 2255–2265 (2012)

17. Serag, A., et al.: A multi-channel 4d probabilistic atlas of the developing brain: application to fetuses and neonates. Ann. BMVA **2012**(3), 1–14 (2012)

18. Studholme, C.: Mapping fetal brain development in utero using magnetic resonance imaging: the big bang of brain mapping. Ann. Rev. Biomed. Eng. **13**, 345–368 (2011)

19. Wang, L., et al.: Volume-based analysis of 6-month-old infant brain MRI for autism biomarker identification and early diagnosis. In: Frangi, A.F., Schnabel, J.A., Davatzikos, C., Alberola-López, C., Fichtinger, G. (eds.) MICCAI 2018. LNCS, vol. 11072, pp. 411–419. Springer, Cham (2018). https://doi.org/10.1007/978-3-030-00931-1_47

20. Xia, J., et al.: Fetal cortical surface atlas parcellation based on growth patterns. Hum. Brain Mapp. **40**(13), 3881–3899 (2019)

21. Zhan, J., et al.: Spatial-temporal atlas of human fetal brain development during the early second trimester. Neuroimage **82**, 115–126 (2013)

Clinical Applications - Lung

Leveraging Auxiliary Information from EMR for Weakly Supervised Pulmonary Nodule Detection

Hao-Hsiang Yang[1](✉), Fu-En Wang[1,2], Cheng Sun[1], Kuan-Chih Huang[1],
Hung-Wei Chen[1], Yi Chen[1], Hung-Chih Chen[1], Chun-Yu Liao[1], Shih-Hsuan Kao[1],
Yu-Chiang Frank Wang[1,3], and Chou-Chin Lan[4]

[1] ASUS Intelligent Cloud Services, Asustek Computer Inc., Taipei, Taiwan
[2] National Tsing Hua University, Hsinchu, Taiwan
[3] Graduate Institute of Communication Engineering, National Taiwan University, Taipei, Taiwan
[4] Taipei Tzu Chi Hospital, Taipei, Taiwan

Abstract. Pulmonary nodule detection from lung computed tomography (CT) scans has been an active clinical research direction, benefiting the early diagnosis of lung cancer related disease. However, state-of-the-art deep learning models require instance-level annotation for the training data (i.e., a bounding box for each nodule), which require expensive costs and might not always be applicable. On the other hand, during clinical diagnosis of lung nodule detection, radiologists provide electronic medical records (EMR), which contain information such as the malignancy, number, texture of the detected nodules, and slice indices at which the nodules are located. Thus, the goal of this work is to utilize EMR information for learning pulmonary nodule detection models, without observing any nodule annotation during the training stage. To realize the above weakly supervised learning strategy, we extend multiple instance learning (MIL) and specifically take the presence and number of nodules in each CT scan, as well as the associated slice information, in our proposed deep learning framework. In our experiments, we present proper evaluation metrics for assessing and comparing the effectiveness of state-of-the-art models on multiple datasets, which verify the practicality of our proposed model.

Keywords: Pulmonary nodule detection · Weakly supervised learning · Electronic medical records

1 Introduction

Pulmonary nodules are among the important and early lesions of lung cancer. To diagnose whether a patient has pulmonary nodules, he or she needs to take computed tomography (CT) scans, which will need to be further assessed by pathological experts. Since the CT scan can be viewed as a three-dimensional volumetric space, such inspection is highly time-consuming and laboring, which makes pulmonary nodule examination inefficient. With the great achievement of deep neural networks [26–28] in recent years, automatic pulmonary nodule detection [4–6] becomes an important topic.

© Springer Nature Switzerland AG 2021
M. de Bruijne et al. (Eds.): MICCAI 2021, LNCS 12907, pp. 251–261, 2021.
https://doi.org/10.1007/978-3-030-87234-2_24

The challenges for current machine learning or deep learning based nodule detection models lie in the fact that manually labeling the ground truth bounding box for training data is expensive. Therefore, training a deep learning model in a fully supervised fashion is often not applicable. It is necessary to leverage unlabeled or partially-labeled data to improve the training of the detection network. Liu *et al.* [13] first adopt a large amount of unlabeled CT scans and develop a self-supervised framework. While impressive results were reported, a performance gap between theirs and those achieved by fully supervised models still needs to be tackled. Though some semi-supervised [22] methods are proposed, they require pre-defined instance-level annotation, which might not always be applicable in real-world scenarios. On the other hand, during clinical diagnosis of lung nodule detection, radiologists provide electronic medical records (EMR), which contain information like slice index, number, texture, and malignancy from each scanned image. These kinds of weak information can be leveraged to guide the nodule detector for better performance. Therefore, we consider a weakly-supervised framework in this paper.

Inspired by [1,24], we propose a weakly-supervised nodule detection framework based on multiple instance learning (MIL). Instead of simply utilizing the presence of nodules as prior MIL-based models do, we advance the EMR recorded by the radiologists, which reveals the information of the number of nodules and the slice index at which the nodules are located. With such auxiliary information observed from EMRs, our proposed learning framework is able to predict pseudo labels and re-assign the nodule proposals accordingly during such a weakly-supervised setting. In our evaluation, in addition to the use of the Competition Performance Metric (CPM) in fully supervised settings, we further present the evaluation metric of "Area under Hybrid Informative ROC"(AU-HIROC), which allows one to leverage the weak labels from EMR for proper evaluation.

We summarize our contributions as follows:

1. Based on CT scans and the associated EMR information, we propose a weakly supervised deep learning framework for pulmonary nodule detection.
2. Our weakly-supervised setting jointly considers EMR information, including presence, number of nodules, as well as the associated slice indices, while no nodule annotation is required.
3. Extensive experiments on both public and private datasets are conducted, which quantitatively and qualitatively support the use of our proposed model for pulmonary nodule detection.

2   Backgrounds and Related Works

2.1   Pulmonary Nodule Detection

Based on deep neural networks, recent pulmonary nodule detection works have shown promising performances [2,4–6,9,12,13,21,23,32,33]. Generally, a standard pipeline consists of two stages: *i)* nodule candidates extraction module with high recall, which is then followed by *ii)* false positive reduction module to improve the resulting precision. [4] employs Faster-RCNN [14] for candidate extraction and 3D DCNN for false positive reduction. [12] predicts a coarse-scale probability map with 3D dense CNN to indicate the existence of the nodule in a large cell. [13] builds 3D-FPN with the new self-supervised pre-trained strategy to improve the performance. [32] incorporates the methodology of radiologists by using maximum intensity projection as input CT scan pre-processing and ensemble the results from various slab thicknesses. [2] trains improved UNet [16,29] with new nodules sampling and hard example mining strategy for better candidates extraction performance; ensembling of three designed models are employed for false-positive reduction.

Despite impressive performances in literature, existing methods generally require training on large image datasets with full annotation. Their performances would drop if only a decreased amount of such labels can be utilized. Since collecting nodule annotations on CT scans is extremely expensive, a possible solution is to exploit semi-supervised learning strategies to train the detection framework. As for [22], authors leverage a semi-supervised learning framework into pulmonary nodule detection task. The labeled images and unlabeled images are used as input data. Two levels of MixUp [31] augmentations, which are image-level Mixup and object-level Mixup, are applied to each input batch. Additionally, the soft-target focal loss was used on unlabeled data to train the model. Nevertheless, semi-supervised models still require pre-defined instance-level annotation, which might not always be applicable in real-world scenarios.

2.2 Weakly Supervised Object Detection (WSOD)

Without observing instance-level annotation, weakly supervised object detection trains their models by image-level annotation. For example, [1] employs Fast R-CNN [8] and updates their detection model via MIL strategies, where an image is viewed as a bag of multiple instances (object candidates). Recently, a number of methods are proposed [15,19,20,25,30] upon the framework improving WSOD results on natural images. To use bag-level counting information as weak supervision, [7] uses a simple selection heuristic to obtain high-quality regions. There are few weakly supervised methods for medical image analysis. For example, [17] detects blood cells and classify blood cell disorders with attention-based MIL. CAMEL [24] selects the instances with the highest and lowest estimated probability for MIL semantic segmentation. [34] parses slices and coarse location information and develops an EM based WSOD to boost pulmonary nodule detection performance. However, existing WSOD works are not designed to utilize auxiliary EMR information as ours does.

3 Approach

3.1 Setting and Notations

Given a lung 3D CT scan image I with N nodules, we have $\mathbb{H} = \{H_i = (p_i, d_i)\}_{i=1}^N$, denote the set of nodules, while $p_i = (x_i, y_i, z_i) \in \mathbb{R}^3$ and d_i denote the spatial location and diameter of the ith nodule, respectively. In weakly supervised pulmonary nodule detection, $\mathbb{H}$ is *not* accessible during training. Instead, one typically observes the image label $y \in \{0, 1\}$ from electronic medical records (EMR) during the training stage, which indicates whether the CT scan contains nodules or not. In our work, we further consider auxiliary information from EMR, including the number k of nodules and the slice indices of each nodule $\{\hat{z}_1, \hat{z}_2, .., \hat{z}_n, n \leq k\}$ in CT scan.

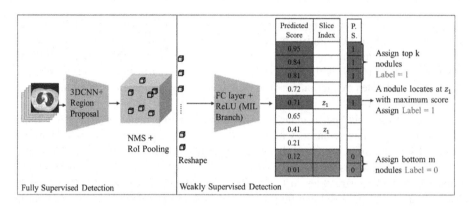

Fig. 1. Our framework for weakly supervised pulmonary nodule detection. Note that 3D CNN is a pre-trained fully supervised detector, serving as the detector backbone to extract nodule proposals and features in weakly supervised settings. In addition to image-level labels to predict the pseudo labels (denoted as "P.S.") for each proposal, our model additionally observes nodule number and slice index information from EMR for guiding the learning process.

Figure 1 shows our proposed deep learning framework for weakly supervised pulmonary nodule detection. As shown in Fig. 1, a pre-trained nodule of 3D feature pyramid network (3D-FPN) [13] is applied to extract the preliminary prediction (i.e., features, bounding boxes location) of each nodule $\hat{\mathbb{H}} = \{\hat{H}_i = (\hat{p}_i, \hat{d}_i)\}_{i=1}^M$. Such prediction outputs can be viewed as primitive nodule proposals, and the aforementioned weak EMR labels (i.e., image label y, nodule number k, and nodule slice index z) will be further utilized to guide the learning of our framework. In the following subsections, we will discuss how the above weakly supervised learning framework can be realized.

3.2 Pulmonary Nodule Detection with Weak Supervision

Image Label as Weak Supervision. Previously, multiple instance learning (MIL) [3] has been applied to address object detection in weakly supervised settings, which is realized by observing only image-level labels during training. Without the need to collect any instance-level labels, the above model aims at estimating nodule proposals $\hat{\mathbb{H}}$, which would be properly associated with the image-level label y. For each proposal $\hat{H}_i$, the pooling operation is applied to extract the corresponding feature maps from the 3D-FPN backbone detector, denoted as $\hat{\mathbb{F}} = \{\hat{f}_i\}_{i=1}^M$. In [11], fully connected layers with $ReLU$ activation function are deployed to infer the confidence score of each object proposal. Finally, to match the ground truth image-level prediction, a number of techniques have been proposed to process the predicted $\hat{y}$ from $\hat{\mathbb{H}}$ (e.g., [1, 17, 24]). In our work, we follow [24] and consider the maximum operator as the MIL pooling function:

$$\hat{y} = \max\{\hat{h}_i\}_{i=1}^M = \max\{MIL(\hat{f}_i)\}_{i=1}^M, \tag{1}$$

where MIL denotes the MIL branch and $\hat{h}_i$ is the predicted scores of proposals in the proposed learning framework. We note that, we feed the extracted visual features into our weakly supervised pulmonary nodule detection module without adjusting the weights of original ResNet-18 or FPN backbones. This allows us to focus on the network modules of predicting and re-ranking the extracted nodule proposals under different weak supervision as discussed above and in the following subsections.

Auxiliary Information from EMR as Weak Supervision. In addition to observing image labels from EMR to learn object detection models, we further advance auxiliary yet representative EMR information to guide the above weakly supervised learning process, with the goal of achieving improved detection performances without observing any instance-level ground truth labels.

In particular, besides binary image labels y, standard EMRs also contain two types of information: the number of nodules in the 3D scan and the corresponding nodule location (i.e., the slice indices at which the nodules are located). Thus, we would like to utilize these two additional EMR features as weak supervision. Together with image-level labels, we propose a unique scoring assignment for the nodule proposals, which would better reflect the nodule detection outputs while being consistent with the aforementioned weak labels. Starting from applying existing or pre-trained nodule detection models, our proposed weakly supervised learning framework first sorts the scores of nodule proposals in our MIL branch, and we assign scores of the top k proposals are 1. Next, with the slice index information z_i, we choose the detected nodule proposal $\hat{H}_j$ which at or near those slices (i.e., $z_j + 0.5d_j \geq \hat{z}_i \geq z_j - 0.5d_j$) and assign its detection label as 1. If multiple proposals are located at a certain slice, we assign the label of the proposal with the highest score as 1.

In practice, the nodules are typically sparsely located in a 3D lung CT scan. Furthermore, a large number of falsely detected nodules can be expected. Thus, we need to deal with imbalanced learning problems during training. In our proposed framework, we assign the pseudo labels of m nodules of the smallest prediction scores as 0. It is worth noting that this pseudo scoring strategy is only applicable for positive 3D CT scans (i.e., images with $y = 1$). Finally, summarizing the above learning objectives and strategies, we define and calculate the binary cross-entropy loss based function L as follows:

$$\underbrace{\sum_{\hat{h} \in P(\hat{h};1,k)} L(\hat{h},1) + \sum_{\hat{h} \in P(\hat{h};m,M)} L(\hat{h},0)}_{\text{Score based}} + \underbrace{\sum_{\hat{H} \in z_k} L(max(\hat{h}),1)}_{\text{Position based}}, \qquad (2)$$

where $P(h; p_{min}, p_{max})$ is defined as the subset of scores h, in which the predicted nodules exhibit probability scores within the minimum $p_{min}th$ and maximum $p_{max}th$ of h. As noted above, k and m denote the nodule numbers with the top largest and bottom smallest prediction scores. Calculating and minimizing the above objective would guide our proposed model without observing any instance-level ground truth data. We also note that, for negative 3D CT samples, we simply adopt Eq. (1) to calculate the predicted score. The training procedure and pseudo code of our proposed weakly supervised learning model are summarized in Algorithm 1 and Fig. 1.

Input: $\hat{h}_i$: scores of proposals from a CT scan; k: number of nodules; H_i:
corresponding bounding box; $\hat{z}_1, \hat{z}_2, ..., \hat{z}_n, n <= k$: central slice of the nodule;
m: pre-defined negative number
Output: $Loss(\hat{h}_i, k, \hat{z}_n, m)$

1 if k == 0:   Loss = $L(y, \hat{h}_i)$ # Negative samples

2 else:

3   Loss = 0

4       $\hat{h}_i$ = Sort($\hat{h}_i$)

5   For $i = 0$; $i \leq k$; $i + +$; **DO**

6   if $i < n$:   Loss += $L(\hat{h}_i, 1)$ # Equation (2), Score based

7   else if i $\geq m$:   Loss += $L(\hat{h}_i, 0)$ # Equation (2), Score based

8   For $i = 0$; $i \leq n$; $i + +$; **DO**

9   if slice $\hat{z}_i$ touches certain $\hat{H}_j$:

10   Loss += $L(\max_{\hat{H}_j \in \hat{z}_i}(\hat{h}_j), 1)$ # Equation (2), Position based

11 Return Loss

Algorithm 1: Learning of our WSOD framework using EMR information.

Quantitative Analysis in WSOD. When evaluating the performance of pulmonary nodule detection, Competition Performance Metric (CPM) [18] is among the most widely used metrics. In order to calculate CPM, it is required to have ground truth instance-level nodule information (i.e., bounding boxes) for test data, which might not be applicable in practice. To address this issue and perform the evaluation in the same weakly supervised setting, one needs to define a proper metric using only EMR but not observing the above ground truth information. Based on the EMR weak labels described above, one can simply predict whether a patient contains nodules, which is viewed as a binary classification task. From the predicted scores $\hat{y}$ and weak labels from EMR, one can apply the receiver operating characteristic curve (ROC) (or more specifically, the area under ROC (AUROC)) for evaluation. However, the nodule number and its location are not utilized when calculating the above metric. Thus, we define an extended version of AUROC, the area under hybrid informative ROC (AU-HIROC), in this work. AU-HIROC combines the aforementioned weak label EMR information. That is, for the negative samples, we take the largest score $\hat{y}$ from the 3D CT scans of interest. As for positive ones, we not only consider the top k number of nodules but also the slice indices of each detected nodule. This is to match the nodule number and slice index information from the corresponding EMR. We select the top k^{th} score and the score that a proposal matches the slice index and take a minimum of them to represent the final score. We note that, if a slice has multiple nodules, we choose the largest score, which is similar to the training phase. Besides, if the predicted nodules are not at the ground truth slices, their scores are assigned as 0. With the above definition, the predicted score of a nodule is calculated as follows:

$$Score = \begin{cases} \hat{y}, & y = 0 \\ \min(top_k(\hat{h}_i), \max(\hat{h}_i \in \hat{z}_i)) & otherwise, \end{cases} \tag{3}$$

Table 1. Detection results on our private dataset with weak supervision. Note that N and S denote the observation of nodule number and slide index, respectively.

Methods	AU-ROC	AU-HIROC	CPM	WCPM
Pre-trained	0.552	0.438	0.544	0.709
Baseline [24]	0.848	0.648	0.589	0.747
+N	0.850	0.660	0.602	0.758
+N+S	0.851	0.664	0.614	0.771

Table 2. Performance comparisons to state-of-the-art WSOD models of [1,17,24] on our private dataset and Tianchi dataset.

	Tianchi	Private Dataset			
	CPM	AU-ROC	AU-HIROC	CPM	WCPM
[1]	0.615	0.828	0.626	0.554	0.713
[17]	0.595	0.841	0.646	0.592	0.750
[24]	0.609	0.848	0.648	0.589	0.747
Ours	0.635	0.851	0.664	0.614	0.771

where top_k means the top k^{th} score in $\hat{h}_i$. With the score of each CT scan predicted, we can thus compute the above AU-HIROC based on the given EMR.

4 Experiments

4.1 Experimental Setting

We adopt three lung CT datasets for experiments: LUNA16 [18], Tianchi Lung Nodule Detection dataset[1], and an in-place (private) dataset collected by a private hospital. Following the setting of [13], the LUNA16 dataset is applied to pre-train the nodule detection model in a fully supervised fashion, with 3D-FPN as the backbone. As for Tianchi, it contains 600 weak label training data and 200 testing data with the ground truth box. The in-place dataset consists of about 15000 CT images with EMR given. For evaluation in supervised settings, we randomly select 60 CT images as the test set for doctors to label the nodule bounding boxes. For evaluation in weakly supervised settings, 1500 CT images and corresponding EMRs are adopted.

Once the pre-trained detection model with 0.89 CPM is obtained from LUNA16, it is applied to extract the nodule proposals and the corresponding features from Tianchi and the private dataset. We adopt ROI alignment [10] to extract features of their proposals. For each feature, we resize them as $5 \times 5 \times 5 \times 64$, and choose the top 20 proposals for the weakly supervised framework. We set the learning rate as 0.0003 on the SGD optimizer. For the private dataset, we apply both AUROC and AU-HIROC to evaluate the detection performances using ground truth weak labels (i.e., EMR only), while CPM is used for fully supervised evaluation (i.e., with ground truth bounding box information). We note that to alleviate the inconsistency between label annotation by different doctors, we further consider the Weighted Competition Performance Metric (WCPM), which calculates the average weighted sensitivity at predefined false-positive (FP) rates (1/8, 1/4, 1/2,1, 2, 4, and 8 FPs/case) along a free-response ROC curve. For the Tianchi dataset, we only apply CPM as the metric since all CT images in Tainchi contain nodules.

---

[1] https://tianchi.aliyun.com.

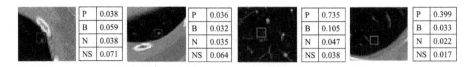

Fig. 2. Visualization of WSOD results on our private dataset with prediction scores. Positive and negative detection outputs are shown in blue and red rectangles, respectively. We have P and B denote the pre-trained and baseline method [24]. Also, N and S indicate the uses of nodule number and slice index for weak supervision.

4.2 Ablation Studies

We investigate the effectiveness and necessity of each type of weak label observed in our proposed model. Taking CAMEL [24] as the baseline weakly supervised learning network, we sequentially add weak supervision of the nodule number and their slice indices into the learning process. In Table 1, we report AUROC, AU-HIROC, CPM and WCPM for comparisons. As shown in this table, we see that such additionally observed EMR information would improve the detection performances, which cannot be achieved by pre-trained baseline models. Furthermore, we also show that AU-HIROC is more informative than AUROC during the evaluation because AU-HIROC considers much auxiliary information. Finally, we visually compare the prediction scores of proposals by leveraging different types of weak labels in Fig. 2. From this figure, it can be seen that our model effectively identified the hard positive samples while the scores of false positive samples were suppressed.

4.3 Comparison with the State-of-the-Art Methods

We select three state-of-the-art works for comparisons. They are Weakly Supervised Deep Detection Networks (WSDDN) [1], attention-based MIL [17], and CAMEL [24] as discussed in Sect. 2. Different from our works, they simply adopt the presence of nodules as weak labels. Starting from the same pre-trained models using LUNA16, we evaluate the performances of each method on both Tianchi and our private dataset. As listed and compared in Table 2, we confirm that our proposed model performed favorably against existing approaches.

5   Conclusion

In this paper, we present a novel learning framework for weakly supervised pulmonary nodule detection. In addition to image-level nodule labels, we leverage auxiliary weak labels of number and slice index of the observed nodules from electronic medical records (EMR). The above weak labels are integrated into a deep multiple instance learning based framework with the associated objectives, while no ground truth instance-level nodule information is utilized during training. We propose the pseudo scoring assignment to supervise these weak labels. In our experiments, we consider both public and private datasets with a number of evaluation metrics (in both supervised and weakly supervised settings) for evaluation. The effectiveness of our proposed framework design has been successfully confirmed, while our model is shown to perform favorably state-of-the-art WSOD methods.

Acknowledgements. This project is partly funded by Ministry of Science and Technology of Taiwan (MOST 110-2634-F-002-036).

References

1. Bilen, H., Vedaldi, A.: Weakly supervised deep detection networks. In: IEEE Conference on Computer Vision and Pattern Recognition (CVPR) (2016)
2. Cao, H., et al.: A two-stage convolutional neural networks for lung nodule detection. IEEE Journal of Biomedical and Health Informatics (2020)
3. Chikontwe, P., Kim, M., Nam, S.J., Go, H., Park, S.H.: Multiple instance learning with center embeddings for histopathology classification. In: International Conference on Medical Image Computing and Computer-Assisted Intervention (2020)
4. Ding, J., Li, A., Hu, Z., Wang, L.: Accurate pulmonary nodule detection in computed tomography images using deep convolutional neural networks. In: International Conference on Medical Image Computing and Computer-Assisted Intervention (MICCAI) (2017)
5. Dou, Q., Chen, H., Jin, Y., Lin, H., Qin, J., Heng, P.: Automated pulmonary nodule detection via 3d convnets with online sample filtering and hybrid-loss residual learning. In: International Conference on Medical Image Computing and Computer-Assisted Intervention (MICCAI) (2017)
6. Dou, Q., Chen, H., Yu, L., Qin, J., Heng, P.: Multilevel contextual 3-d cnns for false positive reduction in pulmonary nodule detection. IEEE Trans. Biomed. Eng. **64**(7), 1558–1567 (2017)
7. Gao, M., Li, A., Yu, R., Morariu, V.I., Davis, L.S.: C-wsl: count-guided weakly supervised localization. In: European Conference on Computer Vision (ECCV) (2018)
8. Girshick, R.B.: Fast R-CNN. In: IEEE International Conference on Computer Vision (ICCV) (2015)
9. Hamidian, S., Sahiner, B., Petrick, N., Pezeshk, A.: 3D convolutional neural network for automatic detection of lung nodules in chest CT. In: Medical Imaging 2017: Computer-Aided Diagnosis (2017)
10. He, K., Gkioxari, G., Dollár, P., Girshick, R.: Mask r-cnn. In: IEEE Conference on Computer Vision and Pattern Recognition (CVPR) (2017)
11. Ilse, M., Tomczak, J., Welling, M.: Attention-based deep multiple instance learning. In: International Conference on Machine Learning (ICML) (2018)
12. Khosravan, N., Bagci, U.: S4ND: single-shot single-scale lung nodule detection. In: International Conference on Medical Image Computing and Computer-Assisted Intervention (MICCAI) (2018)
13. Liu, J., Cao, L., Akin, O., Tian, Y.: Accurate and robust pulmonary nodule detection by 3D feature pyramid network with self-supervised feature learning. arXiv preprint arXiv:1907.11704 (2019)
14. Ren, S., He, K., Girshick, R.B., Sun, J.: Faster R-CNN: towards real-time object detection with region proposal networks. In: Advances in Neural Information Processing Systems (NeurIPS) (2015)

15. Ren, Z., et al.: Instance-aware, context-focused, and memory-efficient weakly supervised object detection. In: IEEE Conference on Computer Vision and Pattern Recognition (CVPR) (2020)
16. Ronneberger, O., Fischer, P., Brox, T.: U-Net: convolutional networks for biomedical image segmentation. In: Navab, N., Hornegger, J., Wells, W.M., Frangi, A.F. (eds.) MICCAI 2015. LNCS, vol. 9351, pp. 234–241. Springer, Cham (2015). https://doi.org/10.1007/978-3-319-24574-4_28
17. Sadafi, A., et al.: Attention based multiple instance learning for classification of blood cell disorders. In: International Conference on Medical Image Computing and Computer-Assisted Intervention (MICCAI) (2020)
18. Setio, A.A.A., et al.: Validation, comparison, and combination of algorithms for automatic detection of pulmonary nodules in computed tomography images: the luna16 challenge. Med. Image Anal. **42**, 1–13 (2017)
19. Tang, P., Wang, X., Bai, X., Liu, W.: Multiple instance detection network with online instance classifier refinement. In: IEEE Conference on Computer Vision and Pattern Recognition (CVPR) (2017)
20. Wan, F., Liu, C., Ke, W., Ji, X., Jiao, J., Ye, Q.: C-MIL: continuation multiple instance learning for weakly supervised object detection. In: IEEE Conference on Computer Vision and Pattern Recognition (CVPR) (2019)
21. Wang, B., Qi, G., Tang, S., Zhang, L., Deng, L., Zhang, Y.: Automated pulmonary nodule detection: high sensitivity with few candidates. In: International Conference on Medical Image Computing and Computer-Assisted Intervention (MICCAI) (2018)
22. Wang, D., Zhang, Y., Zhang, K., Wang, L.: Focalmix: semi-supervised learning for 3D medical image detection. In: CVPR (2020)
23. Wang, Q., Shen, F., Shen, L., Huang, J., Sheng, W.: Lung nodule detection in CT images using a raw patch-based convolutional neural network (2019)
24. Xu, G., et al.: Camel: a weakly supervised learning framework for histopathology image segmentation. In: IEEE International Conference on Computer Vision (ICCV) (2019)
25. Yan, G., et al.: C-MIDN: coupled multiple instance detection network with segmentation guidance for weakly supervised object detection. In: IEEE International Conference on Computer Vision (ICCV) (2019)
26. Yang, C.H., Qi, J., Chen, P.Y., Ma, X., Lee, C.H.: Characterizing speech adversarial examples using self-attention u-net enhancement. In: IEEE International Conference on Acoustics, Speech and Signal Processing (ICASSP) (2020)
27. Yang, C.H.H., Siniscalchi, S.M., Lee, C.H.: Pate-aae: Incorporating adversarial autoencoder into private aggregation of teacher ensembles for spoken command classification. arXiv preprint arXiv:2104.01271 (2021)
28. Yang, H.H., Huang, K.C., Chen, W.T.: LAFFNet: a lightweight adaptive feature fusion network for underwater image enhancement. In: IEEE International Conference on Robotics and Automation (ICRA) (2021)
29. Yang, H.H., Yang, C.H.H., Wang, Y.C.F.: Wavelet channel attention module with a fusion network for single image deraining. In: IEEE International Conference on Image Processing (ICIP) (2020)
30. Zeng, Z., Liu, B., Fu, J., Chao, H., Zhang, L.: WSOD2: learning bottom-up and top-down objectness distillation for weakly-supervised object detection. In: IEEE International Conference on Computer Vision (ICCV) (2019)
31. Zhang, H., Cisse, M., Dauphin, Y.N., Lopez-Paz, D.: mixup: Beyond empirical risk minimization. International Conference on Learning Representations (ICLR) (2018)
32. Zheng, S., Guo, J., Cui, X., Veldhuis, R.N.J., Oudkerk, M., van Ooijen, P.M.A.: Automatic pulmonary nodule detection in CT scans using convolutional neural networks based on maximum intensity projection. IEEE Transactions on Medical Imaging (2020)

33. Zhu, W., Liu, C., Fan, W., Xie, X.: Deeplung: deep 3D dual path nets for automated pulmonary nodule detection and classification. In: IEEE Winter Conference on Applications of Computer Vision (2018)
34. Zhu, W., Vang, Y.S., Huang, Y., Xie, X.: Deepem: deep 3D convnets with em for weakly supervised pulmonary nodule detection. In: International Conference on Medical Image Computing and Computer-Assisted Intervention (MICCAI) (2018)

M-SEAM-NAM: Multi-instance Self-supervised Equivalent Attention Mechanism with Neighborhood Affinity Module for Double Weakly Supervised Segmentation of COVID-19

Wen Tang[1], Han Kang[1], Ying Cao[1], Pengxin Yu[1], Hu Han[2],
Rongguo Zhang[1(✉)], and Kuan Chen[1]

[1] InferVision Medical Technology Co., Ltd., Beijing, China
zrongguo@infervision.com
[2] Institute of Computing Technology, Chinese Academy of Science, Beijing, China

Abstract. The Coronavirus Disease 2019 (COVID-19) pandemic has swept the whole world since 2019. Chest computed tomography (CT) plays an important role in clinical diagnosis, management and progression monitoring of COVID-19 patients. In order to decrease the cost of manual segmentation, weakly supervised segmentation methods, such as class activation maps (CAM) based methods, have been applied to achieve COVID-19-related lesion segmentation. Such methods could be used to localize the lesion preliminarily, but it is not precise enough to segment the lesion. In this paper, we propose a double weakly supervised segmentation method to achieve the segmentation of COVID-19 lesions on CT scans. A self-supervised equivalent attention mechanism with neighborhood affinity module is proposed for accurate segmentation. Multi-instance learning is adopted for training using annotations weaker than image-level. A simple pre-training process is also proved to be effective. We achieve a higher average Dice compared to Unet (0.782 vs 0.601) on COVID-19 lesion segmentation tasks. Codes in this paper will be available at https://github.com/TangWen920812/M-SEAM-NAM.

Keywords: Weakly supervised segmentation · Multi-instance learning · COVID-19

1 Introduction

Coronavirus Disease 2019 (COVID-19) has been announced as a global pandemic by the World Health Organization (WHO) [20]. By December 21th, over 75 million confirmed cases and 1 million deaths are reported across the globe [19]. Early detection, timely isolation and treatment of patients are advocated by WHO in order to control the disease transmission. Chest computed tomography (CT) plays an important role in the identification of suspected patients and

© Springer Nature Switzerland AG 2021
M. de Bruijne et al. (Eds.): MICCAI 2021, LNCS 12907, pp. 262–272, 2021.
https://doi.org/10.1007/978-3-030-87234-2_25

could provides quantitative evaluation of disease progression. It is listed in the diagnosis and treatment guidelines of the diagnosis and treatment of COVID-19 issued by many countries, such as China [14], Japan [8], and the United of Kingdom [3].

Deep learning methods are widely used to process medical image to assist the detection and segmentation of COVID-19. Several studies [11,17,23]have proven the effectiveness of Convolution Neural Network(CNN) to differentiate CT containing COVID-19 lesions or not. These models could help with timely patient triage but not quantitative analysis. COVID-19 related lesion segmentation could assist the localization of radiological abnormalities, and is the basis for further quantitative analysis of lesion area. Some Unet-based [16] methods [2,7,9,15] are applied to assist lesion segmentation for quantitative analysis. However, big amount of manual annotations are required for model training, which are time- and energy- consuming. Some studies adopt weakly supervised methods for COVID-19 lesion segmentation. Issam et al. [10] propose a weakly supervised method using detection or key point annotation to decrease the labeling cost. Yet the label used for weakly supervised segmentation is still not as simple as classification label. Hu et al. [6] use image-level labels and class activation maps (CAM) [24] to perform COVID-19 lesion segmentation. However, ignorance of multiple scales and details in CAM makes this method not robust enough among variable lesion sizes. Wang et al. [18] propose a new method focusing on variable sizes of targets, and achieve state-of-the-art performance with only image-level annotation on PASCAL VOC 2012 [5]. But there is no evidence shows [18] could work on COVID-19 dataset. Comparing to nature image dataset, COVID-19 data-set needs a larger number of more professional annotations. Multi-instance learning [13], a training method wherein a bag of images share one label, is a good solution requiring weaker annotations. It is usually used on huge image input which could not be put into model directly because of memory limitation. Thus, in this method, huge images are cut into several small image bags that share one label [1,22].

In this paper, we develop an end-to-end model, Multi-instance Self-supervised Equivalent Attention Mechanism with Neighborhood Affinity Module (M-SEAM-NAM), for doubly weakly supervised segmentation on COVID-19 dataset. We propose a new neighborhood affinity module on self-supervised equivalent attention mechanism to achieve better performance on lesions with variable sizes. We also adopt multi-instance learning in our model to use annotations weaker than image-level. Such designs allow us to train the model with doubly weakly classification labels while achieving better performance than fully supervised methods.

2   Method

We will introduce our method from the following three parts: the baseline method (Sect. 2.1), the improvement of our neighborhood affinity to the baseline (Sect. 2.2) and the combination of multi-instance learning with our proposed model (Sect. 2.3).

2.1   Self-supervised Equivalent Attention Mechanism (SEAM)

There is evidence [12] shows that different scales of an image would produce very different class activation maps, and thus offer different contextual information. This property has been leveraged in multi-scale CAM [12] to perform weakly supervised segmentation. However, it requires batch of network structures and complicated post-processing. Such issues are recently solved by self-supervised equivalent attention mechanism(SEAM) [18] which we will introduce in following paragraphs.

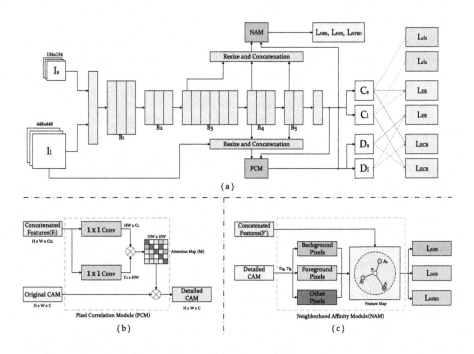

Fig. 1. (a) Overall structure of the proposed network. (b) Pixel Correlation Module (PCM). (c) Neighborhood Affinity Module (NAM).

Firstly, we define our input images and corresponding classification labels as $\{I^i\}_{i=1...n}$ and $\{y^i\}_{i=1...n}$. As shown in Fig. 1, each image I^i is sampled into two different scales: a large scale image I_l^i and a small scale image I_s^i. I_l^i and I_s^i are then put into the shared parameters backbone, ResNet38 [21], to gain two class activation maps, C_l^i and C_s^i. Then a Pixel Correlation Module (PCM) is used to produce detailed class activation maps using self-attention mechanism. I_l^i is downsampled to the size of feature maps on $B4$ and concatenated with feature maps on $B4$ and $B5$ in ResNet38 to form the concatenated feature maps: F_l^i. After that, as shown in Fig. 1 (b), F_l^i is compressed by a 1×1 convolution and flattened to calculate the non-local self-attention maps: M_l^i. Beacause F_l^i combines low-level and high-level information, M_l^i could give more reasonable details. We can get more detailed class activation maps: D_l^i and D_s^i by multiplying C_l^i and C_s^i with M_l^i.

Three losses are used in SEAM. To assist the network to gain information from different scales, the equivariant regularization loss (L_{ER}) is used between C_l^i and C_s^i, as well as D_l^i and D_s^i which should be the same regardless of scale change. To ensure the PCM could work, the equivariant cross regularization loss (L_{ECR}) is used between C_l^i and D_s^i, as well as C_s^i and D_l^i. Using L_{ECR}, D_l^i and D_s^i could locate the basic activation area, and would not make the PCM fall into local minimum. In addition, we also use soft margin loss as the classification loss (L_{cls}) to supervise the whole network training.

$$L_{cls} = log(1 + e^{-y^i * p_l^i}) + log(1 + e^{-y^i * p_s^i}) \tag{1}$$

$$L_{ER} = \frac{1}{N_s^i} \parallel Down(C_l^i) - C_s^i \parallel_1 + \frac{1}{N_s^i} \parallel Down(D_l^i) - D_s^i \parallel_1 \tag{2}$$

$$L_{ECR} = \frac{1}{N_s^i} \parallel Down(C_l^i) - D_s^i \parallel_1 + \frac{1}{N_s^i} \parallel C_s^i - Down(D_l^i) \parallel_1 \tag{3}$$

where p_l^i and p_s^i are the global average pooling result of C_l^i and C_s^i, respectively. N_s^i is the number of pixels in C_s^i. $\parallel \bullet \parallel_1$ is the L1 distance and $Down(\bullet)$ is a downsampling operation. In addition, the SEAM is pre-trained on natural images using ImageNet [4]. We use a new pre-training method for SEAM on medical images. We use CAM method to pre-train ResNet38 backbone in SEAM. In detail, we remove PCM along with L_{ER} and L_{ECR} in SEAM and only train the backbone with L_{cls}. The experiment in Sect. 3.3 shows the effectiveness of our operations.

2.2   Neighborhood Affinity Module (NAM)

Although SEAM focuses on different scales of images, it is still not good enough to solve the COVID-19 segmentation problem, as the overall changes of natural images are bigger than medical images. Such difference allows the model trained on natural images to perceive more information than the model trained on medical images, which only focuses on one significant feature. As shown in Fig. 2, SEAM mainly focuses on the edge area of large lesions and produces false positive predictions around small lesions. Based on the observation, we believe that enhancing the relevance of features from neighborhood pixels would help to improve the model performance. So we introduce a neighborhood affinity module (NAM) to the basic SEAM.

Firstly, we define the position of pixels in D_l^i as $P = \{(x^j, y^j)\}_{j=1...J_i}$, and the position of pixels around (x^j, y^j) within a radius of $r(r = 5)$ as $A_j = \{x^k, y^k\}_{k=1...K_j}$. Then, the prediction result of one pixel is defined as $\sigma(D_l^i)[x^j, y^j]$, where $\sigma(\bullet)$ is the sigmoid activation function. Considering the influence caused by uncertain pixels on network optimization, two thresholds, T_{fg}, and T_{bg}, are defined on $\sigma(D_l^i)$ to categorize P into three groups: background pixels $P_b = \{(x^j, y^j) \mid \sigma(D_l^i)[x^j, y^j] < T_{bg}\}_{j=1...J_i} = \{P_b^j\}_{j=1...J_b^i}$,

foreground pixels $P_f = \{(x^j, y^j) \mid \sigma(D_l^i)[x^j, y^j] > T_{fg}\}_{j=1...J_i} = \{P_f^j\}_{j=1...J_f^i}$
and other uncertain pixels which would not be used. We can also get $A_b^j =$
$\{(x^k, y^k) \mid \sigma(D_l^i)[x^k, y^k] < T_{bg} \cap (x^k, y^k) \in A^j\}_{k=1...K^j} = \{A_b^j(k)\}_{k=1...K_b^j}$ and
$A_f^j = \{(x^k, y^k) \mid \sigma(D_l^i)[x^k, y^k] > T_{fg} \cap (x^k, y^k) \in A^j\}_{k=1...K^j} = \{A_f^j(k)\}_{k=1...K_f^j}$
by the same way. After that, a concatenation (F'^i) of three different features on
$B3$, $B4$ and $B5$, as shown in Fig. 1, is put into NAM. Based on the definition of
P_b, P_f, A_b^j, and A_f^j, we can sample the features of each kind of pixels as $F'^i[P_b^j]$,
$F'^i[P_f^j]$, $F'^i[A_b^j(k)]$, and $F'^i[A_f^j(k)]$.

To enhance the relevance of features in neighborhood, we propose three loss,
neighborhood foreground similarity loss (L_{NFS}) measuring similarity between
foreground and foreground, neighborhood background similarity loss (L_{NBS})
measuring similarity between background and background, and neighborhood
fore-back distinctive loss (L_{NFBD}) measuring the difference between foreground
and background:

$$L_{NFS} = \frac{1}{J_f^i}\sum_j\left(\frac{1}{K_f^j}\sum_k(1 - cos(F'^i[P_f^j], F'^i[A_f^j(k)]))\right) \tag{4}$$

$$L_{NBS} = \frac{1}{J_b^i}\sum_j\left(\frac{1}{K_b^j}\sum_k(1 - cos(F'^i[P_b^j], F'^i[A_b^j(k)]))\right) \tag{5}$$

$$L_{NFBD} = \frac{1}{J_f^i}\sum_j\left(\frac{1}{K_b^j}\sum_k(cos(F'^i[P_f^j], F'^i[A_b^j(k)]))\right)$$
$$+\frac{1}{J_b^i}\sum_j\left(\frac{1}{K_f^j}\sum_k(cos(F'^i[P_b^j], F'^i[A_f^j(k)]))\right) \tag{6}$$

where $cos(\bullet, \bullet)$ is the function of cosine similarity.

2.3 Multi-instance Network with SEAM and NAM

The SEAM with NAM still requires a large amount of classification annotations.
We adopt the multi-instance training idea in our model to implement a doubly
weakly segmentation network which could use weaker classification labels to
achieve good segmentation performance. Radiologists only need to give a rough
slice range of lesion when labeling. For example, radiologists note that there are
lesions from 100 to 110 slices but do not need to record the exact slice numbers.
In this way, weaker classification labeling costs less time and we could use some
of the unknown layers in training.

The multi-instance method is usually used in pathological image. It helps to
solve the problem that a lot of images share one label. However, if the multi-
instance method could work, the positive data batch must contain at least one
positive image. In our situation, the backbone of our network is a segmentation
backbone, which means we can not use patient classification label because of
memory limitation. Thus in our experiments, we define positive batch and negative batch based on three categories of slice described in Sect. 3.1. We select

one positive layer and other seven consecutive layers (could be unknown layers or positive layers) as one positive data batch. We also randomly select eight consecutive images from negative patients as a negative batch. Because each batch has only one label, we keep other losses all the same and change the classification loss to Eq. 7, where y is the label of one batch, p^i is the prediction of one input image, and n is batch size. In addition, we update the network every eight batches to avoid jumping changes on loss and also change the CAM pre-training loss as Eq. 7.

$$L_{cls} = -log(1 + e^{-y*pb}) \, , \; pb = log(\frac{1 - \prod_i^n (1 - \sigma(p^i))}{\prod_i^n (1 - \sigma(p^i))}) \qquad (7)$$

$$L_{all} = L_{cls} + L_{ER} + L_{ECR} + L_{NBS} + L_{NFS} + 0.5 * L_{NFBD} \qquad (8)$$

3   Experiments

3.1   Data Description

The COVID-19 CT dataset used in this study is collected from two hospitals. All positive cases are confirmed by RT-PCR and show lesions related to COVID-19 on CT confirmed by radiologists, while all negatives cases are also confirmed by RT-PCR and without lesions related to COVID-19 on CT. The lesions related to COVID-19 indicate the imaging features of COVID-19 pneumonia including multiple small patchy shadows, interstitial changes appear, multiple ground-glass shadows, infiltrates shadows, and pulmonary consolidation. There are 587 positive cases and 288 negative cases from the first hospital. These cases are further divided into a training set (522 positive and 240 negative cases), and a testing set (65 positive and 48 negative cases). Cases collected from the second hospital are used for testing only, including 68 positive and 49 negative cases. Lesions related to COVID-19 of all positive cases in testing set (65 + 68 patients) are annotated by two experienced radiologists on all layers, and lesions of positive cases in training set (522 patients) are annotated every four or five layers. In all, 8309 out of 198882 CT scans layers were labeled with **Lesion** in the training set and 4725 + 5303 out of 24304 + 25447 CT scans layers are labeled with **Lesion** in the testing set. These segmentation annotations in training set are used to train fully supervised models. M-SEAM-NAM is trained using the positive/negative classification labels.

The fully supervised model used in experiments is trained using CT layers labeled with segmentation annotations and the same amount of negative layers in negative patients. The classification labels of training set are classified into 3 categories: 1) positive layers, layers with segmentation annotations; 2) negative layers, layers from negative patients; and 3) unknown layers, layers from positive patients without any annotation. The weakly supervised models we used are trained using the positive/negative classification labels. Negative layers are randomly selected to keep the sample balance.

3.2 Overall Implementations

The proposed method is implemented using Pytorch. The losses of the network are optimized by SGD, which is a method for stochastic optimization. The learning rate is 0.001 with linear decline and the model is trained for 16 epochs. Other weakly supervised segmentation methods for comparison are also trained under the same setting. The fully supervised segmentation method (Unet) is trained for 100 epochs. Common data augmentations, including shift, rotation, flip, brightness changing and center cropping are utilized during training. All CT layers are resized to 448 × 448 before inputting to the network. Additionally, we use 3 consecutive layers images as input. All the networks are trained in 2D, whereas evaluation indexes are calculated at patient level. Dice score, lesion pixel recall and lesion pixel precision are used to evaluate the models.

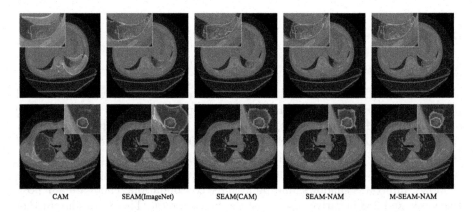

Fig. 2. Segmentation result on several weakly supervised methods. Red heatmaps are the prediction and blue contours are the ground truth. (Color figure online)

3.3 Model Comparison

We compare our method with several weakly-supervised methods including CAM, multi-scale CAM and SEAM(ImageNet pre-trained, baseline). To show the effectiveness of our proposed pre-training method, we use two different pre-trained parameters on SEAM. We also compare our model with Unet, a common fully supervised segmentation method. The dataset is split to show more detailed performance of each method, as positive cases could be used to show the model's segmentation performance and sensitivity, while negative cases could reflect the model's specificity. As shown in Table 1, SEAM pre-trained with CAM performs significantly better than SEAM pre-trained with ImageNet. Our SEAM with NAM model achieves a better result than the SEAM without NAM. It proves NAM is helpful on the segmentation of COVID-19. Our M-SEAM-NAM model outperforms all other models regarding positive, negative and all patients.

Because a big amount of unknown layers are included in the training process and this operation could increase the model's understanding of unlabeled layers, even with a weaker annotation, we still obtain the best result using the proposed model. As shown in Fig. 2, our pre-training process, NAM and multi-instance learning are all helpful for increasing true positive and decreasing false positive in lesions with variable sizes.

Table 1. Segmentation model comparison on COVID-19 testing dataset. In negative patients, if no lesions are segmented, the dice coefficient would be 1; otherwise, 0. Wilcoxon signed rank test is used to perform statistical tests. $T_{fg} = 0.1, r = 5$ is used on all method with NAM.

Method	Pre-trained	Positive patients			Negative patients	All patients
		Dice	Recall	Precision	Dice	Dice
Unet	-	0.600 ± 0.273	0.683 ± 0.280	0.574 ± 0.260	0.604 ± 0.489	0.601 ± 0.344 ($p < 0.001$)
CAM	-	0.109 ± 0.084	$\mathbf{0.974 \pm 0.04}$	0.060 ± 0.049	0.958 ± 0.200	0.336 ± 0.396 ($p < 0.001$)
Multi-CAM	-	0.509 ± 0.182	0.673 ± 0.103	0.470 ± 0.190	0.938 ± 0.242	0.637 ± 0.277 ($p < 0.001$)
SEAM (baseline)	ImageNet	0.495 ± 0.190	0.836 ± 0.127	0.383 ± 0.187	0.958 ± 0.200	0.619 ± 0.281 ($p < 0.001$)
SEAM	CAM	0.631 ± 0.139	0.834 ± 0.151	0.534 ± 0.160	0.958 ± 0.200	0.718 ± 0.214 ($p < 0.001$)
SEAM with NAM	CAM	0.683 ± 0.135	0.763 ± 0.121	0.634 ± 0.160	0.958 ± 0.200	0.757 ± 0.197 ($p < 0.001$)
multi-instance SEAM with NAM	CAM	$\mathbf{0.710 \pm 0.114}$	0.712 ± 0.124	$\mathbf{0.714 \pm 0.117}$	$\mathbf{0.979 \pm 0.143}$	$\mathbf{0.782 \pm 0.171}$

3.4 Ablation Experiment

As there are two thresholds in the neighborhood affinity module, we use ablation experiment to study their effect. Firstly, we analyse the statistical distribution of SEAM model output, i.e. the pixel probability value. Based on the statistical distribution (left histogram in Fig. 3), we set T_{bg} as 0.01 because it is confident that a pixel with prediction probability smaller than $T_{bg} = 0.01$ is background pixel. So that we only have to do ablation experiment on T_{fg}. As shown in the line chart in the right of Fig. 3, we set T_{fg} from 0.1 to 0.8. The reason we do not use 0.9 as one threshold is that when $T_{fg} = 0.9$, there is no foreground pixels at the beginning of the training process. According to the results, we choose to use $T_{fg} = 0.1$. Another hyperparameter is the neighborhood radius. As shown in Fig. 3, 5 is the best radius. According to the results, the values of two thresholds and radius have slight influence on model performance.

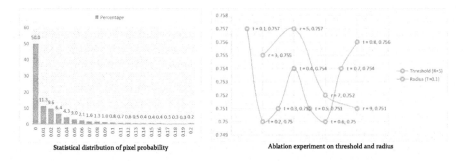

Fig. 3. Ablation experiment result

4   Conclusion

We introduce a SEAM method for COVID-19 weakly supervised segmentation and propose to use CAM model as a pre-trained model which performs better than models pre-trained with ImageNet. We also propose a NAM method to solve the problem that SEAM model performs unsatisfactorily on segmentation of lesion with variable sizes in COVID-19 datasets. The SEAM-CAM method we proposed performs best among all models. Considering the time and labor cost of annotation, we also conduct a simple labeling strategy that radiologists just label the approximate slice range of lesions other than the exact class of every single slice. By using this weaker classification labeled data, we train a doubly weakly supervised segmentation model, M-SEAM-CAM, via multi-instance learning. Our proposed method achieves an better performance because of the effective use of unlabeled data. In future study, we will try to compress our model and use patient-level classification annotation to have a much weaker supervised model.

References

1. Cances, L., Pellegrini, T., Guyot, P.: Sound event detection from weak annotations: weighted-gru versus multi-instance-learning. In: Proceedings of the Detection and Classification of Acoustic Scenes and Events 2018 Workshop (DCASE2018), pp. 64–68 (2018)
2. Cao, Y., et al.: Longitudinal assessment of covid-19 using a deep learning-based quantitative ct pipeline: illustration of two cases. Radiol. Cardiothorac. Imaging **2**(2), e200082 (2020)
3. Chua, F., et al.: The role of ct in case ascertainment and management of covid-19 pneumonia in the uk: insights from high-incidence regions. Lancet Respir. Med. **8**(5), 438–440 (2020)
4. Deng, J., Dong, W., Socher, R., Li, L.J., Li, K., Fei-Fei, L.: Imagenet: a large-scale hierarchical image database. In: 2009 IEEE Conference on Computer Vision and Pattern Recognition, pp. 248–255. IEEE (2009)

5. Everingham, M., Winn, J.: The pascal visual object classes challenge 2012 (voc2012) development kit. Pattern Analysis, Statistical Modelling and Computational Learning, Technical report 8 (2011)
6. Hu, S., et al.: Weakly supervised deep learning for covid-19 infection detection and classification from ct images. IEEE Access **8**, 118869–118883 (2020)
7. Huang, L., et al.: Serial quantitative chest ct assessment of covid-19: a deep learning approach. Radiol. Cardiothorac. Imaging **2**(2), e200075 (2020)
8. Japanese society for infection prevention and control. guide for responding to new coronavirus infections at medical institutions (ver.2.1). http://www.kankyokansen. org/uploads/uploads/files/jsipc/COVID-19_taioguide2.1.pdf. Accessed 5 Apr 2020
9. Jin, S., et al.: Ai-assisted ct imaging analysis for covid-19 screening: Building and deploying a medical ai system in four weeks. MedRxiv (2020)
10. Laradji, I., et al.: A weakly supervised region-based active learning method for covid-19 segmentation in ct images. arXiv preprint arXiv:2007.07012 (2020)
11. Li, L., et al.: Artificial intelligence distinguishes covid-19 from community acquired pneumonia on chest ct. Radiology (2020)
12. Ma, X., Ji, Z., Niu, S., Leng, T., Rubin, D.L., Chen, Q.: Ms-cam: multi-scale class activation maps for weakly-supervised segmentation of geographic atrophy lesions in sd-oct images. IEEE J. Biomed. Health Inform. **24**(12), 3443–3455 (2020)
13. Maron, O., Lozano-Pérez, T.: A framework for multiple-instance learning. Advances in Neural Information Processing Systems, pp. 570–576 (1998)
14. National health commission of the people's republic of china. chinese clinical guidance for covid-19 pneumonia diagnosis and treatment (7th edition). http:// www.nhc.gov.cn/yzygj/s7653p/202003/46c9294a7dfe4cef80dc7f5912eb1989/files/ ce3e6945832a438eaae415350a8ce964.pdf. Accessed 5 Apr 2020
15. Qi, X., et al.: Machine learning-based ct radiomics model for predicting hospital stay in patients with pneumonia associated with sars-cov-2 infection: A multicenter study. Medrxiv (2020)
16. Ronneberger, O., Fischer, P., Brox, T.: U-Net: convolutional networks for biomedical image segmentation. In: Navab, N., Hornegger, J., Wells, W.M., Frangi, A.F. (eds.) MICCAI 2015. LNCS, vol. 9351, pp. 234–241. Springer, Cham (2015). https://doi.org/10.1007/978-3-319-24574-4_28
17. Song, Y., et al.: Deep learning enables accurate diagnosis of novel coronavirus (covid-19) with ct images. MedRxiv (2020)
18. Wang, Y., Zhang, J., Kan, M., Shan, S., Chen, X.: Self-supervised equivariant attention mechanism for weakly supervised semantic segmentation. In: Proceedings of the IEEE/CVF Conference on Computer Vision and Pattern Recognition, pp. 12275–12284 (2020)
19. WHO: Coronavirus disease 2019 (covid-19) situation report. https://www.who. int/docs/default-source/coronaviruse/wou_21-dec_cleared.pdf?sfvrsn=a7575c1f_ 1&download=false. Accessed 21 Dec 2020
20. WHO: Who director-general's opening remarks at the media briefing on covid-19 - 11 March 2020. www.who.int/dg/speeches/detail/who-director-general-s-opening-remarks-at-the-media-briefing-on-covid-19--11-march-2020. Accessed 21 Dec 2020
21. Wu, Z., Shen, C., Van Den Hengel, A.: Wider or deeper: Revisiting the resnet model for visual recognition. Pattern Recogn. **90**, 119–133 (2019)
22. Yao, J., Zhu, X., Huang, J.: Deep multi-instance learning for survival prediction from whole slide images. In: Shen, D., et al. (eds.) MICCAI 2019. LNCS, vol. 11764, pp. 496–504. Springer, Cham (2019). https://doi.org/10.1007/978-3-030-32239-7_55

23. Zheng, C., et al.: Deep learning-based detection for covid-19 from chest ct using weak label. MedRxiv (2020)
24. Zhou, B., Khosla, A., Lapedriza, A., Oliva, A., Torralba, A.: Learning deep features for discriminative localization. In: Proceedings of the IEEE Conference on Computer Vision and Pattern Recognition, pp. 2921–2929 (2016)

Longitudinal Quantitative Assessment of COVID-19 Infection Progression from Chest CTs

Seong Tae Kim[2(✉)], Leili Goli[3], Magdalini Paschali[1], Ashkan Khakzar[1], Matthias Keicher[1], Tobias Czempiel[1], Egon Burian[4], Rickmer Braren[4], Nassir Navab[1,5], and Thomas Wendler[1]

[1] Computer Aided Medical Procedures, Technical University of Munich, Munich, Germany

[2] Department of Computer Science and Engineering, Kyung Hee University, Yongin-si, Korea
st.kim@khu.ac.kr

[3] Department of Computer Engineering, Sharif University of Technology, Tehran, Iran

[4] Department of Diagnostic and Interventional Radiology, Technical University of Munich, Munich, Germany

[5] Computer Aided Medical Procedures, Johns Hopkins University, Baltimore, USA

Abstract. Chest computed tomography (CT) has played an essential diagnostic role in assessing patients with COVID-19 by showing disease-specific image features such as ground-glass opacity and consolidation. Image segmentation methods have proven to help quantify the disease and even help predict the outcome. The availability of longitudinal CT series may also result in an efficient and effective method to reliably assess the progression of COVID-19, monitor the healing process and the response to different therapeutic strategies. In this paper, we propose a new framework to identify infection at a voxel level (identification of healthy lung, consolidation, and ground-glass opacity) and visualize the progression of COVID-19 using sequential low-dose non-contrast CT scans. In particular, we devise a longitudinal segmentation network that utilizes the reference scan information to improve the performance of disease identification. Experimental results on a clinical longitudinal dataset collected in our institution show the effectiveness of the proposed method compared to the static deep neural networks for disease quantification.

Keywords: Longitudinal analysis · COVID-19 · Disease progression

1 Introduction

The Coronavirus Disease 2019 (COVID-19) has infected more than 113 million people worldwide (as of February 28th, 2021[1]) and caused more than 2.52 million

---

[1] https://coronavirus.jhu.edu/map.html.

S. T. Kim and L. Goli—Equally Contributd.

M. de Bruijne et al. (Eds.): MICCAI 2021, LNCS 12907, pp. 273–282, 2021.
https://doi.org/10.1007/978-3-030-87234-2_26

deaths. Although many cases present only mild symptoms, some of them evolve into serious illnesses that require intensive medical treatment or lead to death [7].

Chest computed tomography (CT) has played an essential diagnostic role in the assessment of patients with COVID-19 by showing specific image patterns such as ground-glass opacity (GGO), crazy paving, and consolidation [22]. Several studies have proposed to automatically analyze COVID-19 infection on chest images with deep learning [5,11,12,21,26]. However, it is still challenging to automatically identify and quantify image findings associated with COVID-19 due to the subtle anatomical boundaries, pleural-based location, and variations in size, density, location, and texture [14,19,27]. Moreover, it is vital to develop an effective method to reliably assess the progression of COVID-19 and response to therapy [6,23]. Given the availability of multiple therapy options and to understand anatomical changes during the healing process, a longitudinal evaluation of CT images can be beneficial [6,9].

Only a few studies have devoted to developing a deep learning-based approach to assess the progression and response to therapy from longitudinal CT scans [17,25]. Zhang et al. propose a static segmentation model to classify the patients into mild and severe patients based on the features extracted from longitudinal CT scans [25]. Pu et al. propose a framework consisting on the following steps (1) segment the lung boundary and vessels, (2) register the boundary between serial scans using deformable registration, (3) identify regions with morphological changes due to the disease, and (4) assess disease progression [17]. The registered longitudinal CT scans are used to generate a heatmap visualizing the difference in diseased vs. healthy areas between scans. Yet, the identification of affected regions is performed in a static way, i.e., the information between serial scans is not taken into account. Besides, they do not differentiate image features of consolidation and GGO, even though these pathologies provide different information of infection in COVID-19 cases [28].

In this paper, we propose a novel framework to identify infection at a fine-grained level (identification of healthy lung, consolidation, GGO, and pleural effusion) by leveraging spatio-temporal cues between longitudinal scans and visualize the progression of COVID-19. In particular, we devise a longitudinal segmentation network that utilizes the reference scan information to improve the performance of disease segmentation. Even though longitudinal scans share structural information, differences exist due to the progression of the disease. We investigate and propose ways to use this information during the segmentation based on a dataset collected during the first COVID-19 wave of our institution. The following is the summary of the main contributions.

- To the best of our knowledge, this is the first study to explore the longitudinal segmentation of CTs of COVID-19 patients. By designing a deep network to use the information provided in the reference scan, we show that the performance of segmentation can be improved compared to the static segmentation model. Our method shows promising results with limited data.
- We propose a framework to analyze the progression of COVID-19 infection over time, which is crucial for the course of the disease and the patient's recovery.

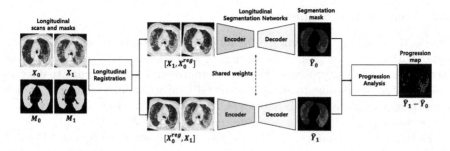

Fig. 1. Progression analysis framework: The framework is comprised of three modules: Longitudinal registration, longitudinal fine-grained segmentation to identify pathologies, and progression analysis. The inputs are two consecutive CT scans (t = 0: reference scan and t = 1: follow-up scan).

- We present comprehensive analysis and ablation studies to verify our longitudinal analysis framework's design choice.

2   Methodology

This section describes our approaches for incorporating spatio-temporal features into the framework of segmentation and progression quantification. Figure 1 shows overall framework which consists of the following three components: 1) longitudinal registration, 2) longitudinal segmentation, and 3) progression analysis. The analyses are performed based on two different time point scans.

2.1   Longitudinal Registration

Due to the nature of chest CT scans, the initial volumes of data between different time points are highly misaligned. Aspects like patient positioning, variations of the imaging parameters or devices, different phases in the breathing cycle, and the disease progression are the main reasons. This misalignment cannot be described as a linear transformation composed of translation and rotation between the time points of data and can only be expressed through non-linear transformations. The misalignment can make the network incapable of using the longitudinal information present between different time points of data.

As a solution for this problem, we utilize a deformable registration algorithm [16] where a BSpline Transform is defined using a sparse set of grid points overlaid onto the fixed domain of the image domain to deform it. Using this algorithm we register the reference scan lung mask M_0 to the follow up scan lung mask M_1 and this transform function is defined by $R_{M_0 \rightarrow M_1}(\cdot)$. Based on this function, the transformations are applied to the respective CT-scans as $X_0^{reg} = R_{M_0 \rightarrow M_1}(X_0)$. Using the lung masks, we avoid registration errors due to the pathological changes in the lung parenchyma while compensating for positioning, breathing phase, and acquisition-related differences.

2.2   Longitudinal Segmentation

Due to the challenge of training a 3D model with a limited number of training data, 2.5D approaches [2,4,18,24] have shown state-of-the-art results on various medical segmentation problems. For the 3D approach, it is challenging to directly process a full 3D volume by using current GPU memory limitations [18], which forces people to operate on 3D patches [8,20]. However, patch-wise training limits the overall spatial context for accurate semantic segmentation.

In this study, we adopt the 2.5D approach of [4] with a fully convolutional (FC) DenseNet [10] as a baseline 2D segmentation model. The FC DenseNet consists of a downsampling path (Encoder) with 5 Transitions Down blocks, each with 4 layers and an upsampling path (Decoder) with 5 Transitions Up blocks, each with 4 layers. The model is trained for all three views (coronal, sagittal, and axial view). At test time, for each given voxel, the segmentation is conducted on all three orthogonal views. Afterward, the predicted probability of a given voxel is averaged among views to assign a final predicted probability.

We extend the aforementioned 2.5D segmentation to deal with longitudinal information by modifying its architecture. To capture subtle spatio-temporal cues, we concatenate two registered scans from two different time-points as input for the longitudinal segmentation network: $[X_0^{reg}, X_1]$ for segmenting pathologies on X_1 or $[X_1, X_0^{reg}]$ for X_0. This enables the segmentation network to capture temporal changes as shown in Fig. 1. For subjects who have more than two scans, we select consecutive scans for the segmentation in each step.

2.3   Progression Analysis

To monitor the progression of the COVID-19 infection and the response to therapy, we extract the segmented pathologies from consecutive longitudinal CT scans and quantify the volume differences between them. Our approach is capable of quantifying all combinations of changes between different pathologies and healthy lung parenchyma. In this work's scope, we define two classes as consolidation and non-consolidation since consolidation has shown to be a robust biomarker for COVID-19 [15]. The progression of consolidation is computed by subtracting two registered longitudinal CT-scans. The resulting residual voxel values could be either positive or negative. Positive voxels indicate that a healthy, GGO or pleural effusion region progresses to consolidation (Progression), and negative voxels suggest that an area recovers from the severe infection of consolidation (Recovery).

2.4   Training with Progression Information

For the optimization of our model, we define an overall loss function combining a segmentation loss $\mathcal{L}_{seg}$ and a progression loss $\mathcal{L}_{prog}$. Note that as shown in Fig. 1, the outputs of our model consist of the segmentation masks for reference (t = 0) and follow-up (t = 1) scans and the subtracted volume between reference and follow-up scans for presenting progression of the COVID-19 infection. The overall loss is defined as: $\mathcal{L} = \mathcal{L}_{seg} + \mathcal{L}_{prog}$.

The segmentation loss is defined as $\mathcal{L}_{seg} = L_{MSE}(Y_0, \hat{Y}_0) + L_{MSE}(Y_1, \hat{Y}_1)$ where L_{MSE} denotes a mean squared error loss [4,24]. In our evaluation, this metric yielded better results in comparison to the dice score. Y_0 and $\hat{Y}_0$ denote a ground truth pathology segmentation map and a predicted segmentation mask for t = 0 scan, respectively. Y_1 and $\hat{Y}_1$ denote a ground truth pathology segmentation map and a predicted segmentation mask for t = 1 scan.

The progression loss is defined as $\mathcal{L}_{prog} = \mathcal{L}_{MSE}(Y_1^{con} - Y_0^{con}, \hat{Y}_1^{con} - \hat{Y}_0^{con})$ where Y^{con} denotes a ground truth consolidation map and $\hat{Y}^{con}$ denotes a predicted consolidation map. In those binary maps, consolidation is mapped to 1 and non-consolidation to 0. As explained above, the progression map is calculated by $Y_1^{con} - Y_0^{con}$. Note that, the progression loss does not minimize the distance between the two segmentations. Instead it explicitly uses the structural changes of pathologies through times as cues for modeling the optimization. In other words, if there are large changes over time, the progression loss encourages the model to predict those large changes also in the segmentation.

3   Experiment Setup

3.1   Dataset

To our knowledge there is no publicly available longitudinal CT dataset for COVID-19. Accordingly, to evaluate the proposed method, we used an in-house clinical dataset which consists of longitudinal low-dose CT-scans from 38 patients (64 ± 18 years old, 16 females, 22 males) with positive PCR from the first COVID-19 wave (March-June 2020). 28 patients had two scans and 10 had three. The CTs were separated 17 ± 10 d (1–43 days) and were taken at admission and during the hospital stay (33 ± 21 d, 0–71 days). 8 patients of the 38 died; 30 recovered from COVID-19, 20 of them needing intensive care. All scans were performed in-house using two different CT devices (IQon Spectral CT and iCT 256, Philips, Hamburg, Germany) with the same parameters (X-ray current 140–210 mA, voltage 120 kV peak, slice thickness 0.9mm, no contrast media) and covered the complete lung. The data was collected retrospectively with the approval of the institutional review board of our institution (ethics approval 111/20 S-KH).

The dataset was annotated at a voxel-level by a single expert radiologist (5 years experience), generating lung masks (lung parenchyma vs. other tissues) and pathology masks including four classes: healthy lung (HL), GGO, consolidation (CONS), and pleural effusion (PLEFF). For segmentation, the radiologist used the software ImFusion Labels (ImFusion, Munich, Germany).

The dataset was split into a training set of 16 patients (37 volumes) and an independent test set of 22 patients (49 volumes). From the training set, 12 patient scans are used for model training and 4 patient scans are used for validation. The model is finally evaluated on the unseen test set.

Table 1. Comparison of different methodologies for segmenting CoViD-19 infection on the independent test set. Dice similarity coefficient is used for metric. The average and standard error are calculated. * denotes the case that the difference with the proposed method is statistically significant ($p<0.05$). HL, CONS, GGO, PLEFF denote healthy lung, consolidation, ground-glass opacity, and pleural effusion, respsectively.

Method	HL	CONS	GGO	PLEFF
Static Network	$0.796 \pm 0.021^*$	$0.322 \pm 0.031^*$	$0.380 \pm 0.028^*$	$0.210 \pm 0.033^*$
Longitudinal Network (without progression loss)	0.835 ± 0.019	0.402 ± 0.034	$0.435 \pm 0.029^*$	$\mathbf{0.266 \pm 0.041^*}$
Proposed	$\mathbf{0.837 \pm 0.022}$	$\mathbf{0.406 \pm 0.035}$	$\mathbf{0.447 \pm 0.030}$	0.246 ± 0.040

3.2 Implementation Details

The raw CT volumes highly vary in intensity range, size, and alignment. Therefore, we perform the following pre-processing steps on the raw volumes to enable effective use of the longitudinal data:

Cropping. Since different body regions can be presented between time points and patients, we crop the volumes to the lung regions, using the manually-annotated lung masks.

Clipping and Normalization. To alleviate different intensity ranges among CT-scans, intensity values outside the range $(-1024, 600)$ are clipped and then min-max normalization is performed on each volume.

Resizing. After Cropping, resulting volumes vary in size in all three dimensions, ranging from 100 pixels to 580 pixels. Therefore all volumes are resized to a fixed size of $300 \times 300 \times 300$ with 300 being the median among the cropped-volume sizes.

Slicing and Removing Empty Slices. Finally, the volumes are sliced in each of the three dimensions to 300 slices, generating sagittal, coronal, and axial views of the lung. Slices that have a voxel-value variation smaller than 0.001% between their maximum and the minimum value are considered empty and are removed.

Model Training. For training, Adam optimizer [13] with a learning rate of 0.0001 and a decay rate of 0.1 for every 50 steps was used. The model was trained over 100 epochs with early stopping if no decrease in the validation loss was computed for 5 epochs. Our method was implemented in PyTorch 1.4 and our models were trained on an NVIDIA Titan V 12GB GPU using Polyaxon[2]. The source code is publicly available[3]. Our longitudinal model had 1.3752M parameters in comparison to its static counterpart 1.3748M.

---

[2] https://polyaxon.com/.

[3] https://github.com/lilygoli/longitudinalCOVID.

4   Results and Discussion

Effectiveness of Longitudinal Segmentation. First, we conduct comparative experiments to verify the effectiveness of our longitudinal segmentation method. In Table 1, we compare our method with a static network [24] based on FC-DenseNet [10] and with a longitudinal network without progression loss. This simpler longitudinal network has the same architecture as our proposed one, i.e., it concatenates longitudinal CT scans as an input for the segmentation model, but it is trained using only the segmentation loss. As shown in Table 1, both longitudinal networks achieved a higher Dice Similarity Coefficient (DSC) than the static network. The difference was statistically significant for all classes ($p<0.05$ by paired t-test [1]). This implies that using longitudinal information from the reference CT scan is informative to segment pathology on the target CT scan. In our longitudinal network with progression loss, the DSC was further improved for HL, CONS, and GGO. But for PLEFF, the performance slightly decreased. This can be attributed to the fact that the progression loss encourages the model to focus on CONS rather than on PLEFF. Additionally, PLEFF is a challenging, under-represented class in our dataset (only 2.17% voxels).

Table 2. Ablation study to investigate the effectiveness of the registration and using temporal information in our longitudinal segmentation model. Dice similarity coefficient is measured on the independent test set.

Method	Registration	Long. Input	HL	CONS	GGO	PLEFF
Without Registration		✓	0.761	0.311	0.388	0.146
Static Input	✓		0.774	0.327	0.224	0.160
Proposed	✓	✓	**0.837**	**0.406**	**0.447**	**0.246**

Effect of Longitudinal Registration. To showcase the importance of registration among the longitudinal scans, we report results with and without deformable registration. As seen in Table 2 the performance after registration substantially improves across the board with the increase ranging from 0.07 to 0.10.

Importance of Temporal Information in Longitudinal Network. In this experiment, we highlight the importance of the longitudinal scans for the performance of the model. Specifically, we concatenate two duplicates of the reference scan instead of the reference and follow-up scan as input to our model ('static input'). As shown in Table 2, the longitudinal input (proposed method) outperforms the static input for all classes.

Progression Analysis. Finally, we evaluate our method for progression analysis by comparing with a static network [17], a longitudinal network with multi-view approach [3] and our model trained without the progression loss. As shown in Table 3, our longitudinal architecture has a 3.4% increase compared to the static network. The proposed model with the progression loss has a 4.8% increase with respect to static network. Note that the model using the progression loss significantly outperformed the static network [17] and the longitudinal network with multi-view approach [3] ($p<0.05$). Moreover, the progression loss had statistically significant improvement for the recovery and average progression prediction compared to the longitudinal model without the progression loss.

Table 3. Comparison of different methodologies for predicting progression of CoViD-19 infection on the independent test set. Dice similarity coefficient is used for metric. The average and standard error are calculated. * denotes the case that the proposed method outperforms the baseline methods with statistical significance ($p<0.05$).

Method	Recovery	Progression	Average
Static Network [17]	$0.266 \pm 0.030^*$	$0.471 \pm 0.021^*$	$0.368 \pm 0.015^*$
Longitudinal Network (multi-view [3])	$0.287 \pm 0.031^*$	0.491 ± 0.028	$0.389 \pm 0.023^*$
Longitudinal Network (without progression loss)	$0.299 \pm 0.032^*$	0.505 ± 0.019	$0.402 \pm 0.014^*$
Proposed	$\mathbf{0.327 \pm 0.033}$	$\mathbf{0.506 \pm 0.026}$	$\mathbf{0.416 \pm 0.017}$

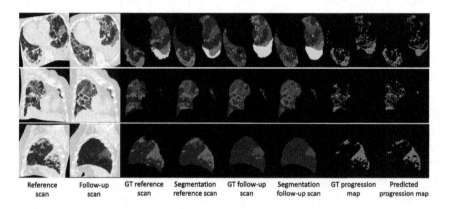

| Reference scan | Follow-up scan | GT reference scan | Segmentation reference scan | GT follow-up scan | Segmentation follow-up scan | GT progression map | Predicted progression map |

Fig. 2. Qualitative results of our method for 3 patients from different views. For the segmentation maps, blue, green, red and yellow denote healthy lung, consolidation, ground-glass opacity, and pleural effusion, respectively. For the progression map, red denotes Progression and the green Recovery.

Figure 2 showcases qualitative results of the segmentation and progression analysis of our method for all 3 different views. As shown in Fig. 2, our method successfully provides segmentation and progression maps for both reference and follow-up scans across views and patients. Even the under-represented class of PLEFF is successfully segmented. Regarding the progression, the fine-grained regions of recovered and progressed consolidation are also correctly identified.

5   Conclusion

In this work, we proposed a new longitudinal segmentation and progression analysis model for assessing COVID-19 disease over time. Comprehensive experiments were conducted to verify the effectiveness of the longitudinal model. By designing the model to exploit the reference CT scan, our method can achieve higher progression analysis performance compared to the baseline methods.

What makes our approach especially interesting is the ability to monitor the development of the infection and healing process in COVID-19, and possibly in other lung diseases. It can be used to compute the differences in disease progression of different patient subgroups, such as different COVID-19 variants under the same therapy. Moreover, it could serve as a quantitative measure to evaluate different therapy approaches. We will further investigate and improve the clinical usability of the method for a larger patient cohort.

Acknowledgement. This paper was funded by the Bavarian Research Foundation (BFS) under grant agreement AZ-1429-20C. We would like to thank NVIDIA for the GPU donation.

References

1. Altman, D.G.: Practical Statistics for Medical Research. CRC Press, Boca Raton (1990)
2. Aslani, S., Dayan, M., Storelli, L., et al.: Multi-branch convolutional neural network for multiple sclerosis lesion segmentation. NeuroImage **196**, 1–15 (2019)
3. Birenbaum, A., Greenspan, H.: Multi-view longitudinal cnn for multiple sclerosis lesion segmentation. Eng. Appl. Artif. Intell. **65**, 111–118 (2017)
4. Denner, S., et al.: Spatio-temporal learning from longitudinal data for multiple sclerosis lesion segmentation. arXiv preprint arXiv:2004.03675 (2020)
5. Fan, D.P., et al.: Inf-net: automatic covid-19 lung infection segmentation from ct images. IEEE Trans. Med. Imaging **39**(8), 2626–2637 (2020)
6. Feng, X., Ding, X., Zhang, F.: Dynamic evolution of lung abnormalities evaluated by quantitative CT techniques in patients with COVID-19 infection. Epidemiol. Infect. **148**, e136 (2020)
7. Harmon, S.A., Sanford, T.H., Xu, S., Turkbey, E.B., Roth, H., et al.: Artificial intelligence for the detection of COVID-19 pneumonia on chest CT using multinational datasets. Nat. Commun. **11**(1), 4080 (2020)
8. Hashemi, S.R., Salehi, S.S.M., Erdogmus, D., Prabhu, S.P., Warfield, S.K., Gholipour, A.: Asymmetric loss functions and deep densely-connected networks for highly-imbalanced medical image segmentation: application to multiple sclerosis lesion detection. IEEE Access **7**, 1721–1735 (2018)

9. Huang, Y., Li, Z., Guo, H., Han, D., Yuan, F., Xie, Y., et al.: Dynamic changes in chest CT findings of patients with coronavirus disease 2019 (COVID-19) in different disease stages: a multicenter study. Ann. Palliat. Med. **10**(1), 572–583 (2021)

10. Jégou, S., Drozdzal, M., Vazquez, D., Romero, A., Bengio, Y.: The one hundred layers tiramisu: fully convolutional densenets for semantic segmentation. In: CVPR Workshop, pp. 11–19 (2017)

11. Khakhar, A., et al.: Towards semantic interpretation of thoracic disease and covid-19 diagnosis models. arXiv preprint arXiv:2104.02481 (2021)

12. Khakhar, A., et al.: Explaining covid-19 and thoracic pathology model predictions by identifying informative input features. arXiv preprint arXiv:2104.00411 (2021)

13. Kingma, D.P., Ba, J.: Adam: a method for stochastic optimization. International Conference on Learning Representations (ICLR) (2014)

14. Lei, J., Li, J., Li, X., Qi, X.: Ct imaging of the 2019 novel coronavirus (2019-ncov) pneumonia. Radiology **295**(1), 18–18 (2020)

15. Li, K., Wu, J., Wu, F., Guo, D., et al.: The clinical and chest ct features associated with severe and critical covid-19 pneumonia. Investigative Radiology (2020)

16. Lowekamp, B.C., Chen, D.T., Ibáñez, L., Blezek, D.: The design of simpleitk. Front. Neuroinform. **7**, 45 (2013)

17. Pu, J., et al.: Automated quantification of covid-19 severity and progression using chest ct images. Eur. Radiol. **31**(1), 436–446 (2021)

18. Roy, A.G., Conjeti, S., Navab, N., Wachinger, C., Initiative, A.D.N., et al.: Quicknat: a fully convolutional network for quick and accurate segmentation of neuroanatomy. NeuroImage **186**, 713–727 (2019)

19. Shi, H., et al.: Radiological findings from 81 patients with covid-19 pneumonia in wuhan, china: a descriptive study. Lancet Infect. Dis. **20**(4), 425–434 (2020)

20. Wachinger, C., Reuter, M., Klein, T.: Deepnat: deep convolutional neural network for segmenting neuroanatomy. NeuroImage **170**, 434–445 (2018)

21. Wang, G., Liu, X., Li, C., et al.: A noise-robust framework for automatic segmentation of covid-19 pneumonia lesions from ct images. IEEE Trans. Med. Imaging **39**(8), 2653–2663 (2020)

22. Wong, H.Y.F., Lam, H.Y.S., et al.: Frequency and distribution of chest radiographic findings in patients positive for covid-19. Radiology **296**(2), E72–E78 (2020)

23. Wu, M.Y., et al.: Clinical evaluation of potential usefulness of serum lactate dehydrogenase (LDH) in 2019 novel coronavirus (COVID-19) pneumonia. Respir. Res. **21**(1), 171 (2020)

24. Zhang, H., et al.: Multiple sclerosis lesion segmentation with tiramisu and 2.5 d stacked slices. In: MICCAI, pp. 338–346 (2019)

25. Zhang, X., et al.: Dabc-net for robust pneumonia segmentation and prediction of covid-19 progression on chest ct scans (2020)

26. Zhou, L., Li, Z., Zhou, J., et al.: A rapid, accurate and machine-agnostic segmentation and quantification method for ct-based covid-19 diagnosis. IEEE Trans. Med. Imaging **39**(8), 2638–2652 (2020)

27. Zhou, S., Wang, Y., Zhu, T., Xia, L.: Ct features of coronavirus disease 2019 (covid-19) pneumonia in 62 patients in wuhan, china. Am. J. Roentgenol. **214**(6), 1287–1294 (2020)

28. Zhou, X., et al.: CT findings and dynamic imaging changes of COVID-19 in 2908 patients: a systematic review and meta-analysis. Acta Radiology, p. 284185121992655 (2021)

Beyond COVID-19 Diagnosis: Prognosis with Hierarchical Graph Representation Learning

Chen Liu, Jinze Cui, Dailin Gan, and Guosheng Yin[✉]

The University of Hong Kong, Pokfulam, Hong Kong
{liuchen,cjz1206,davidgan,gyin}@hku.hk

Abstract. Coronavirus disease 2019 (COVID-19), the pandemic that is spreading fast globally, has caused over 181 million confirmed cases. Apart from the reverse transcription polymerase chain reaction (RT-PCR), the chest computed tomography (CT) is viewed as a standard and effective tool for disease diagnosis and progression monitoring. We propose a diagnosis and prognosis model based on graph convolutional networks (GCNs). The chest CT scan of a patient, typically involving hundreds of sectional images in a sequential order, is formulated as a densely connected weighted graph. A novel distance aware pooling is proposed to abstract the node information hierarchically, which is robust and efficient for such densely connected graphs. Our method, combining GCNs and distance aware pooling, can integrate the information from all slices in the chest CT scans for optimal decision making, which leads to the state-of-the-art accuracy in the COVID-19 diagnosis and prognosis. With less than 1% of the total number of parameters in the baseline 3D ResNet model, our method achieves 94.8% accuracy for diagnosis, which represents a 2.4% improvement over the baseline on the same dataset. In addition, we can localize the most informative slices with disease lesions for COVID-19 within a large sequence of chest CT images. The proposed model can produce visual explanations for the diagnosis and prognosis, making the decision more transparent and explainable, while RT-PCR only leads to the test result with no prognosis information. The prognosis analysis can help hospitals or clinical centers designate medical resources more efficiently and better support clinicians to determine the proper clinical treatment.

Keywords: COVID-19 diagnosis · Lesion localization · GCN · Prognosis

1 Introduction

Coronavirus disease 2019 (COVID-19) has resulted in an ongoing pandemic in the world. To control the sources of infection and cut off the channels of transmis-

Electronic supplementary material The online version of this chapter (https://doi.org/10.1007/978-3-030-87234-2_27) contains supplementary material, which is available to authorized users.

M. de Bruijne et al. (Eds.): MICCAI 2021, LNCS 12907, pp. 283–292, 2021.
https://doi.org/10.1007/978-3-030-87234-2_27

sion, rapid testing and detection are of vital importance. The reverse transcription polymerase chain reaction (RT-PCR) is a widely-used screening technology and viewed as the standard diagnostic method for suspected cases. However, this method highly relies upon the required lab facilities and the diagnostic kits. In addition, the sensitivity of RT-PCR is not high enough for early diagnosis [1,5]. To mitigate the limitations of RT-PCR, the computed tomography (CT) has been widely used as an effective complementary tool by providing medical images of the lung to reveal the details of the disease and its prognosis [3,7], for which RT-PCR cannot. Additionally, CT is also useful in monitoring the COVID-19 disease progression and therapeutic efficacy evaluation [9,12].

The chest CT slices of a patient have a sequential and hierarchical data structure. The relationship between slices possesses more information than the order of the slices. The adjacent ones with the same abnormality could be considered as one lesion. The slices containing the same type of lesions may not be continuous as the lesions are distributed in various parts of the lung. We propose a diagnosis and prognosis system that combines graph convolutional networks (GCNs) and a distance aware pooling procedure, which integrates the information from all slices in the chest CT scans for optimal decision making. Our major contributions are three-fold: (1) Owing to the sequential structure of CT images, this is the first work to utilize GCNs to extract node information hierarchically, and conduct both diagnosis and prognosis for COVID-19. The prognosis can help facilitate medical resources, e.g., use of ventilators or admission to Intensive Care Units (ICUs), more efficiently by triaging mild or severe patients. (2) A novel pooling method, called distance aware pooling, is proposed to aggregate the graph, i.e., the patient's CT scan, effectively. The new pooling method integrated with GCNs can aggregate a densely connected graph efficiently. (3) The proposed model can localize the most informative slices within a chest CT scan, which significantly reduces the amount of work for radiologists.

2 Methodology

We propose a GCN-based diagnosis and prognosis method that models the sequential slices of CT scans hierarchically.

To downsample and learn graph-level representation from the input node features, a novel distance aware pooling method is proposed. The node features refer to the slices in a CT scan. The model gradually extracts information from the slice level to the patient level by graph convolution and pooling. Eventually, a higher-level representation is learned and further used for diagnosis, prognosis, and lesion localization. The schema of our model is illustrated in Fig. 1, which is composed of GCNs, pooling modules, a multilayer perceptron (MLP) classifier, and a one-drop localization module. The graph convolution-based method can integrate all slices in the chest CT scans for optimal decision making.

Furthermore, we propose the one-drop localization to localize the most informative slices and reduce redundancies, so that radiologists may focus on those recommended slices with the most suspected lesion areas. Consequently, the proposed model can produce visual explanations for the diagnosis and prognosis.

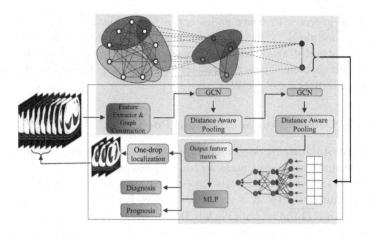

Fig. 1. Schema of our model structure. The CT scan of one patient is converted to a densely connected graph. The GCN and the Distance Aware Pooling (DAP) are integrated to learn a graph-level representation. At each of the two layers, the node embedding is learned by GCN, and the cluster membership is calculated by DAP. The aggregated graph, which is on the top right corner, is passed to an MLP. Meanwhile, the one-drop localization can localize the most informative slices in the CT scan in a weakly supervised manner.

The localized slices may help uncover how a given classification method arrives at the conclusion. Specifically, it reveals if a method is merely overfitting on the training data by examining the slices it attends to. Conversely, it may also be used to identify the intrinsic dataset bias, notably the data acquisition bias [2], and thus guide the data collection process.

2.1 Problem Statement

Let $\mathcal{G}$ ($\mathbb{V}$, $\mathbb{E}$) be a patient's CT scan graph, with $|\mathbb{V}| = N$ nodes and $|\mathbb{E}|$ edges, where $|\cdot|$ represents the cardinality of a set. For each $v_i \in \mathbb{V}$, $\boldsymbol{x}_i$ is the corresponding d-dimensional vector. Let $\boldsymbol{X} \in \mathbb{R}^{N \times d}$ be the node feature matrix, and $\boldsymbol{A}^{adj} \in \mathbb{R}^{N \times N}$ be the weighted adjacency matrix. Each entry in $\boldsymbol{A}^{adj}$ is defined using cosine similarity, which is $\boldsymbol{A}_{i,j}^{adj} = <\boldsymbol{x}_i, \boldsymbol{x}_j>/(\|\boldsymbol{x}_i\| \cdot \|\boldsymbol{x}_j\|)$.

Each graph $\mathcal{G}$ has a label y. For diagnosis, the label represents its class from COVID-19 positive, common pneumonia, or normal individuals. For prognosis, the class indicates whether a COVID-19 positive patient develops into severe/critical illness status. Thus, the diagnosis and prognosis of COVID-19 is a task of graph classification. Given a training dataset $\mathbb{T} = \{(\mathcal{G}_1, y_1), \ldots, (\mathcal{G}_M, y_M)\}$, the goal is to learn a mapping $f : \mathcal{G} \to y$, which classifies a graph $\mathcal{G}$ into the corresponding class y. Our model is composed of two modules: f_1 includes node convolution and feature pooling, and f_2 involves an MLP classifier. At each layer of f_1, the node embeddings and cluster membership are learned iteratively. The first module can be written as $f_1 : \mathcal{G} \to \mathcal{G}^p$, where $\mathcal{G}^p$

is the pooled graph with fewer nodes and a hierarchical feature representation. The second module is $f_2 : \mathcal{G}^p \rightarrow y$, which utilizes the graph-level representation learned for patient diagnosis and prognosis. The two modules are integrated in an end-to-end fashion. Thus, the cluster assignment is learnt merely based on the graph classification objective.

2.2 Node Convolution and Feature Pooling

Node Convolution. Node convolution applies the graph convolutional network to obtain a high-level node feature representation in the feature matrix X. Although several methods exist to construct the convolutional network, the method recommended by [8] is particularly effective for our case, which is given by $X^{(l+1)} = \sigma(\sqrt{D^{(l)}} A^{adj(l)} \sqrt{D^{(l)}} X^{(l)} W^{(l)})$, where $D^{(l)}$ is the diagonal degree matrix of $A^{adj(l)} - I$, and $W^{(l)} \in \mathbb{R}^{d \times k}$ is a learnable weight matrix at the l-th layer. Due to the application of feature pooling, the topology of the graph changes at each layer, and thus the dimensions of matrices involved are reduced accordingly.

Distance Aware Pooling. We propose an innovative pooling method, which includes graph-based clustering and feature pooling. Below, we outline the pooling module and illustrate how it is integrated into an end-to-end GCN-based model. Empirically, it is shown to be more robust for densely connected graphs. The overall structure of the pooling method is shown in the Supplementary Material.

- **Improved receptive field.** The concept of the receptive field, RF, used in the convolutional neural network (CNN) was extended to GNNs [11]. In particular, RF^{node} is defined as the number of hops required to cover the neighborhood of a given node, such that given a chosen node, a cluster can be obtained based on a fixed receptive field h. However, this design may not be applicable to densely connected graphs, because one node may be connected to most of the nodes in the graph even for a small value of h. Hence, we define an improved receptive field for densely connected graphs, RF^d, denoted by h^d. It is a radius centered at a given node and retains the edge weight information in the clustering process. The value of h^d is not restricted to integers. Define $\mathcal{N}(v_i)$ as the local neighborhood of the node v_i, and $\mathcal{N}_{h^d}(v_i)$ as the RF^d neighborhood of the node v_i with a radius h^d, and $\forall v_j \in \mathcal{N}(v_i)$, v_i and v_j are connected by the edge (v_i, v_j)

- **Node clustering.** Inspired by the clustering and ranking ideas in [11] and [6], we propose a local node clustering and score ranking method. Each node is considered as a center of a cluster for a given h^d. Then, we score all the clusters and choose the top k proportion of them to represent the next layer's nodes with pooled feature values, where k is a hyperparameter.

- **Cluster ranking.** Given a node v_i and a radius h^d, $\mathcal{N}_{h^d}(v_i)$ is the corresponding RF^d neighborhood. Let $\mathbb{I}_1(v_i)$ be the index set of the nodes in $\mathcal{N}_{h^d}(v_i)$. Define the distance matrix A^{dis} as $A^{dis} = \mathbb{K} - A^{adj}$, where $\mathbb{K}$ is the matrix with all entries of 1's. The cluster score is defined as $\alpha_i = \sum_{m,n \in \mathbb{I}_1(v_i)} A^{dis}_{m,n} / |\mathcal{N}_{h^d}(v_i)|$, where $m \neq n$. If α_i is small, nodes in $\mathcal{N}_{h^d}(v_i)$ are close to each other. The top k proportion of clusters form the next layer's nodes.

• **Selecting cluster centers.** We define $\mathbb{V}_j = \mathcal{N}[v_j] \cap \mathcal{N}_{h^d}(v_i), \forall v_j \in \mathcal{N}_{h^d}(v_i)$ as the set of nodes connected to v_j in $\mathcal{N}_{h^d}(v_i)$, where $\mathcal{N}[v_j]$ is the closed neighborhood of node v_j. Let $\mathbb{I}_2(v_j)$ be the index set of the nodes in $\mathbb{V}_j$. The node score is $b_j = \sum_{c \in \mathbb{I}_2(v_i)} A^{dis}_{j,c}/|\mathbb{V}_j|$. The node with the smallest value of b_j is chosen as the cluster center, and represents a new node in the next layer.

• **Center node feature pooling.** Based on the node scores, we rank the nodes in $\mathbb{V}_j$ and assign a weight w_r to $\boldsymbol{x}_r$, where $r \in \mathbb{I}_2(v_j)$. Define $\boldsymbol{b}$ as a vector containing all the weights of nodes in $\mathbb{V}_j$, and let $\tilde{\boldsymbol{b}} = \mathbf{1} - \boldsymbol{b}$, where $\mathbf{1}$ is a vector with all 1's. The weight vector $\boldsymbol{w}$ is defined as $w_r = \tilde{b}_r/(\sum_{i \in \mathbb{I}_2(v_j)} \tilde{b}_i + \epsilon)$, where ϵ is an extremely small positive value to avoid 0 in the denominator. The value of the pooled center feature is defined as $\boldsymbol{x}^p = \boldsymbol{X}_r \boldsymbol{w}$, and then $\boldsymbol{x}^p$ is used as the center feature representing $\mathcal{N}_{h^d}(v_i)$.

• **Next layer node connectivity.** Following the idea in [15], the connectivity of the nodes in the next layer is preserved as follows. According to the above ranking and pooling methods, a pooled graph $\mathcal{G}^p$ with the node set $\mathbb{V}^p$ is obtained. The next step is to determine the pooled adjacency matrix $\boldsymbol{A}^{adj,p}$. Define matrix $\boldsymbol{S}$ such that the columns of $\boldsymbol{S}$ are the top k clusters' weight vectors $\boldsymbol{w}$. Hence, the pooled adjacency matrix is defined as $\boldsymbol{A}^{adj,p} = \boldsymbol{S}^\top \boldsymbol{A}^{adj} \boldsymbol{S}$.

2.3 Classification and Localization Based on Pooled Graphs

Graph Classification. A hierarchical representation for each patient is abstracted by the above GCNs with distance aware pooling. The representation is an $N' \times D'$ feature matrix, where N' is the number of clusters and D' is the number of features in each cluster. For each feature, the mean and the maximum over the N' clusters are calculated. Subsequently, we obtain a $2D'$ feature vector, where the first and second D' elements are the mean and maximal values of each feature respectively. Then, an MLP classifies the $2D'$ feature vector into one of the three classes: COVID-19 positive, common pneumonia, or normal.

Weakly Supervised Informative CT Slices Localization. Localizing the CT slices with lesions is also vital for the diagnosis. Therefore, we propose the one-drop localization to select the most informative CT slices for the model to make a decision. This method does not require mask annotation for lesions, which can be learned in a weakly supervised manner. We are inspired by the backward elimination of stepwise regression used for knowledge discovery [4], an automatic procedure for variable selection. For each patient with the $N' \times D'$ feature matrix, we first predict the target class using the above MLP classifier. Then, we occlude one cluster each time to obtain N' new feature matrices with an equal size of $(N' - 1) \times D'$. For each of the new feature matrices, the score of the target class is calculated by the same MLP classifier. The occluded cluster with the lowest score from the N' results is chosen. Because ignoring this cluster leads to the lowest score of the target class, the cluster should contain the most crucial information. Finally, the cluster center is determined and the top k^s CT slices with the highest similarity to the center are further localized, where k^s is the number of selected slices.

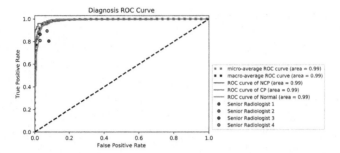

Fig. 2. The ROC curve and AUC (area under the curve) for diagnosis. 'NCP', 'CP', and 'Normal' indicate COVID-19 positive patients, common pneumonia patients, and healthy individuals respectively. The filled dots represent the performance on 'NCP' diagnosis of senior radiologists with 15 to 25 years of clinical experience [16]. It shows that our method is comparable to the senior radiologists.

3   Experiments and Results

For COVID-19 diagnosis and prognosis, we compare our model with the baseline, a 3D ResNet-18 classification network [16], and the state-of-the-art graph classification methods. Moreover, we appraise whether the proposed method can deliver meaningful and interpretable clusters on the input chest CT scans by comparing the localization results with the slices containing lesions (ground truth).

3.1   Dataset and Implementation Details

Dataset. We utilize the CT dataset from the 2019 Novel Coronavirus Resource (2019nCoVR) [16]. The dataset includes the complete chest CT scans of 929 COVID-19 positive patients, 964 common pneumonia patients, and 849 healthy individuals. The dataset also provides these patients' clinical prognosis, whether the patients developed into severe/critical illness status, referring to admission to ICUs, mechanical ventilation, or death. The prognosis analysis could support the hospitals to designate medical resources more efficiently. In addition, the dataset summarizes the slices with lesions for COVID-19 positive and common pneumonia patients, which can be used to evaluate the one-drop localization method. Each CT slice is normalized to the dimension 256×256.

Data Preprocessing. We apply the following two methods for chest CT scan image feature extraction.
• **CNN feature extraction.** We utilize Inception V3 [14] pretrained on ImageNet [13]. The feature map of the bottleneck layer, i.e., the last layer before the flatten operation, is regarded as the node representation in the graph.
• **Wavelet decomposition extraction.** Considering each slice in CT scans as a 2-dimensional signal, it can be viewed as a function with two variables, which can be reconstructed as a summation of wavelet functions multiplying their coefficients for a given resolution [10]. We choose the Haar wavelet function with

resolution 3. The flattened approximation matrix of the image signal, which is of dimension 1024, is used as the feature embedding of a slice in a CT scan.

Implementation Details. We use systematic sampling to ensure that 48 slices for each CT scan are chosen. CT scans of 60% individuals are randomly chosen as the training set, 25% as the test set, and the remaining 15% for the validation. To avoid information leakage, the dataset is split according to individuals instead of the CT slices. Similar to [17] and [15], we repeat the aforementioned data splitting on 20 random seeds. For each random seed, the model is trained from scratch. The maximum, average and standard deviation of test accuracies are reported. We use five GCN layers and a two-layer MLP classifier. The pooling proportion k is set as 0.8. The negative log-likelihood loss is used for graph classification. The Adam optimizer is applied with an initial learning rate 0.0002 and a linear decay schedule. For prognosis, the parameters of GCN and pooling are initialized using those pretrained on the diagnosis task. All models are trained for 128 epochs with possible early stopping.

3.2 Quantitative Results

Table 1. Performance evaluation of COVID-19 diagnosis and prognosis, where 'GCN-DAP' represents the proposed GCN-based method integrated with the distance aware pooling. 'ASAP', 'DiffPool', and 'HGP-SL' refer to the state-of-the-art hierarchical pooling methods. The last column is the average time in seconds to complete one training epoch for each model using a single NVIDIA V100 GPU.

(a) Performance Evaluation of COVID-19 Diagnosis				
Method	Feature Extractor	Average Accuracy (SD)	Best Accuracy	Time (s/epoch)
GCN-DAP	Inception V3	**93.93% (0.41%)**	**94.80%**	22.70
GCN-DAP	Wavelet	83.65% (1.01%)	85.16%	20.90
GCN-ASAP	Inception V3	75.20% (18.70%)	93.74%	30.00
GCN-ASAP	Wavelet	51.43% (11.90%)	81.50%	27.25
GCN-DiffPool	Inception V3	71.22% (23.73%)	94.31%	18.35
GCN-HGP-SL	Inception V3	93.89% (0.39%)	94.22%	45.60
(b) Performance Evaluation of COVID-19 Prognosis				
Method	Feature Extractor	Average Accuracy (SD)	Best Accuracy	Time (s/epoch)
GCN-DAP	Inception V3	**82.70% (3.90%)**	**91.39%**	7.67
GCN-DAP	Wavelet	78.98% (3.38%)	84.95%	1.83
GCN-ASAP	Inception V3	67.90% (11.09%)	82.80%	9.67
GCN-ASAP	Wavelet	60.22% (5.63%)	72.04%	2.30

Diagnosis and Prognosis Performance. According to the ROC (receiver operating characteristic) curve for diagnosis in Fig. 2, our method is comparable to the senior radiologists with 15 to 25 years of clinical experience. We also compare our method with the clinically applicable AI system based on 3D ResNet-18

[16], which reached 92.49% diagnosis accuracy. With less than 1% of the total number of parameters, our method achieves an improvement of 2.4% over this CNN-based state-of-the-art model.

In Table 1, we compared the performance of our model and GCN with the state-of-the-art hierarchical pooling methods, including adaptive structure aware pooling (ASAP) [11], differentiable pooling (DiffPool) [15], and hierarchical graph pooling with structure learning (HGP-SL) [17]. We observed that the Inception V3 feature extraction method constantly outperforms the wavelet decomposition method under the same model configuration. The gradient explosion occurs in around 50% of the runs under the ASAP and DiffPool, resulting in optimization failures, while this issue has not been witnessed during the training of our method and HGP-SL. Thus, the standard deviations of ASAP and DiffPool are much higher. Additionally, our method outperforms HGP-SL marginally but the training of our method is about 2 times faster. The training curves using DAP versus the aforementioned hierarchical pooling methods over 20 runs are presented in Fig. 3, which shows that our model improves the training convergence, and DAP consistently outperforms ASAP and DiffPool across almost all runs.

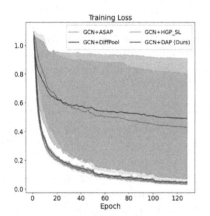 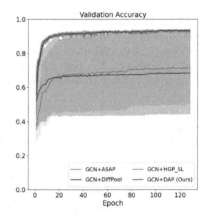

Fig. 3. Training curves of the GCN diagnosis model using DAP versus three hierarchical pooling methods over 20 runs, with varied random seeds and train-validation-test split. The solid lines represent the mean training loss and validation accuracy, and the shaded areas exhibit the intervals of one standard deviation. It shows that DAP consistently outperforms ASAP and DiffPool across almost all runs, and converges much faster.

Weakly Supervised Lesion Localization Results. In addition to graph classification, our model can also localize the most informative CT slices for each CT scan in the test set using the procedure in Sect. 2.3 with $k^s = 10$. Because the chosen CT slices are the most decisive ones for diagnosis, they contain lesions related to COVID-19 or common pneumonia. We compare our selected slices to the CT slices with labeled lesions for each CT scan. Considering the slices in the same CT scan are ordered sequentially, we compare the slices between the first and last slices localized by our model with those labeled with lesions in the dataset. Among patients in the test set with 20 random seeds, the average precision and recall are 57.39% and 79.89% with standard deviations 3.32% and 3.94%, respectively. The localization results of four patients are presented in the Supplementary Material for illustration.

4 Conclusion

This paper introduces an efficient and robust GCN-based diagnosis and prognosis system with a distance aware pooling method. It can cluster nodes and learn the patient-level representation hierarchically. Unlike previous diagnosis methods based on CT scans, our model can produce coarse localization to highlight the potential slices with lesions, making the clinical decision more interpretable and reliable. To strengthen confidence in this framework and move towards clinical use, we should ensure that the model can provide explanations for the prediction instead of merely outputting the result. The localization can help analyze prediction failure, and improve researchers' understanding of the effect of adversarial attacks in the medical imaging domain.

References

1. Ai, T., et al.: Correlation of chest CT and RT-PCR testing for coronavirus disease 2019 (COVID-19) in china: a report of 1014 cases. Radiology **296**(2), E32–E40 (2020)
2. Biondetti, G.P., Gauriau, R., Bridge, C.P., Lu, C., Andriole, K.P.: "Name that manufacturer". Relating image acquisition bias with task complexity when training deep learning models: experiments on head CT. arXiv preprint arXiv:2008.08525 (2020)
3. Chung, M., et al.: CT imaging features of 2019 novel coronavirus (2019-nCoV). Radiology **295**(1), 202–207 (2020)
4. Cios, K.J., Pedrycz, W., Swiniarski, R.W.: Data mining and knowledge discovery. In: Data Mining Methods for Knowledge Discovery, pp. 1–26. Springer (1998)
5. Fang, Y., et al.: Sensitivity of chest CT for COVID-19: Comparison to RT-PCR. Radiology **296**(2), E115–E117 (2020)
6. Gao, H., Ji, S.: Graph U-Nets. In: International Conference on Machine Learning, pp. 2083–2092 (2019)
7. Huang, C., et al.: Clinical features of patients infected with 2019 novel coronavirus in Wuhan. China Lancet **395**(10223), 497–506 (2020)
8. Kipf, T.N., Welling, M.: Semi-supervised classification with graph convolutional networks. In: International Conference on Learning Representations (2017)
9. Liechti, M.R., et al.: Manual prostate cancer segmentation in MRI: interreader agreement and volumetric correlation with transperineal template core needle biopsy. Eur. Radiol. **30**(9), 4806–4815 (2020)
10. Mallat, S.G.: A theory for multiresolution signal decomposition: the wavelet representation. IEEE Trans. Pattern Anal. Mach. Intell. **11**(7), 674–693 (1989)

11. Ranjan, E., Sanyal, S., Talukdar, P.P.: ASAP: adaptive structure aware pooling for learning hierarchical graph representations. Proc. AAAI Conf. Artif. Intell. **34**(04), 5470–5477 (2020)
12. Rodriguez-Morales, A.J., et al.: Clinical, laboratory and imaging features of COVID-19: a systematic review and meta-analysis. Travel Med. Infect. Dis. **34**, 101623 (2020)
13. Russakovsky, O., et al.: Imagenet large scale visual recognition challenge. Int. J. Comput. Vis. **115**(3), 211–252 (2015)
14. Szegedy, C., Vanhoucke, V., Ioffe, S., Shlens, J., Wojna, Z.: Rethinking the inception architecture for computer vision. In: Proceedings of the IEEE Conference on Computer Vision and Pattern Recognition, pp. 2818–2826 (2016)
15. Ying, Z., You, J., Morris, C., Ren, X., Hamilton, W., Leskovec, J.: Hierarchical graph representation learning with differentiable pooling. In: Advances in Neural Information Processing Systems, pp. 4800–4810 (2018)
16. Zhang, K., et al.: Clinically applicable AI system for accurate diagnosis, quantitative measurements, and prognosis of COVID-19 pneumonia using computed tomography. Cell **181**(6), 1423–1433 (2020)
17. Zhang, Z., et al.: Hierarchical graph pooling with structure learning. arXiv preprint arXiv:1911.05954 (2019)

RATCHET: Medical Transformer for Chest X-ray Diagnosis and Reporting

Benjamin Hou[1(✉)], Georgios Kaissis[2], Ronald M. Summers[4], and Bernhard Kainz[1,3]

[1] Imperial College London, London, UK
bh1511@imperial.ac.uk
[2] Technische Universität München, Munich, Germany
[3] FAU Erlangen–Nürnberg, Erlangen, Germany
[4] National Institutes of Health, Bethesda, USA

Abstract. Chest radiographs are one of the most common diagnostic modalities in clinical routine. It can be done cheaply, requires minimal equipment, and the image can be diagnosed by every radiologists. However, the number of chest radiographs obtained on a daily basis can easily overwhelm the available clinical capacities. We propose RATCHET: RAdiological Text Captioning for Human Examined Thoraces. RATCHET is a CNN-RNN-based medical transformer that is trained end-to-end. It is capable of extracting image features from chest radiographs, and generates medically accurate text reports that fit seamlessly into clinical work flows. The model is evaluated for its natural language generation ability using common metrics from NLP literature, as well as its medically accuracy through a surrogate report classification task. The model is available for download at: http://www.github.com/farrell236/RATCHET.

1 Introduction

Automatic report generation for clinical chest radiographs have recently attracted attention due to advances in Natural Language Processing (NLP) and the introduction of transformers. In the past decade, learning-based models for radiographic applications have evolved from single image classification [11,27], multi-label classification [2,20], to more complex systems that incorporates a mixture of modalities [10], uncertainty [28] and explainability [17].

Learning-based methods often require large and well annotated datasets, something that is available in computer vision, e.g. MS-COCO [13], but not in medical imaging. Generic large-scale annotated medical datasets are more difficult to acquire, as; (i) clinical experts need to devote extra time to annotate images, (ii) the images are likely to be bound by specific scanning parameters from the imaging device, and (iii) clinicians commonly annotate via free-text instead of single class

Electronic supplementary material The online version of this chapter (https://doi.org/10.1007/978-3-030-87234-2_28) contains supplementary material, which is available to authorized users.

© Springer Nature Switzerland AG 2021
M. de Bruijne et al. (Eds.): MICCAI 2021, LNCS 12907, pp. 293–303, 2021.
https://doi.org/10.1007/978-3-030-87234-2_28

labels during routine diagnostics. Medical reports are often preferred for diagnostics, as prose can describe a pathology in more detail compared to a single image label. However, clinical report-writing is tedious, highly variable, and a time consuming task. In countries where the population is large and radiologists are in high demand, a single radiologist may be required to read hundreds of radiology images and/or reports per day [16].

We seek to develop a model that can perform automatic report generation for chest radiographs with high *medical* accuracy. Our intention is not to replace radiologists, but instead to assist them in their day-to-day clinical duties by accelerating the process of report generation. Ultimately, the attending physician will be required to usurp responsibility of the final diagnosis, as bounded by ongoing debate in the ethics of AI, but this is beyond the scope of this paper.

Contribution: Inspired by Neural Machine Translation, we develop a transformer-based CNN-Encoder to RNN-Decoder architecture for generating chest radiograph reports. We use attention to localize regions of interest in the image and show where the network is focusing on for the corresponding generated text. This is to reinforce explainability in black-box models for clinical settings, as well as a means for extracting bounding boxes for disease localization. We evaluate the model for both, its natural language capability, and more importantly, the medical accuracy of generated reports.

Related Works: The first transformer-based network for Neural Machine Translation (NMT) was introduced by Vaswani et al. [23]. Here, the encoder and decoder LSTM networks are replaced by stacks of Masked Multi-Head Attention with Point-Wise Feed Forward Network modules that operate on whole sequences directly. The architecture decouples the time dependency of RNNs, which allows for simultaneous processing of sequential data and alleviates the problem of vanishing gradients (a well known problem with RNNs [18]). Transformer networks can be trained much faster, and are able to achieve much higher accuracy with fewer data, as the attention mechanism learns contextual relations between words and sub-words all in one sequence.

Numerous works have explored radiology report generation [16]. Wang et al. [25] introduced TieNet, a CNN-RNN architecture that combines image saliency maps with attention-encoded text embedding to jointly learn the disease class and report generation. Liu et al. [14] later proposed a Hierarchical Generation method using a CNN-RNN-RNN architecture, with a Reinforcement Learning reward signal for better readability and language coherency. At the same time, Boag et al. [1] performed several baseline experiments for chest radiograph report generation. This explored and compared three main report generation techniques: Conditional n-gram Language Model, Nearest Neighbor (similar to data retrieval, where the model recalls the caption with the largest cosine similarity to the query image), and 'Show-and-Tell' [24]. Yuan et al. [29] proposed a hierarchical approach with two nested LSTM decoders. The first decoder generates sentence hidden states from image features, whereas the second decoder generates individual words from the sentence hidden states. The image encoder also leverages additional information during the training process in the form of frontal and lateral image input, as well as image labels and Medical Text Indexer (MTI) Tags. Li et al. [12] also used a hierarchical approach for report generation, but opted for four Graph Transformer (GTR) Networks instead of CNNs and LSTMs. The input image features get encoded by a GTR to an abnormality graph, which then gets transformed by another GTR to a disease graph that yields labels specific to the input. In parallel, two more GTRs are used to convert the abnormality graph to a report template, and then subsequently to a text report. Syeda-Mahmood et al. [22] then simplified the process by training a custom deep learning network on Fine Finding Labels (FFL), which consists of four elements: the finding type, positive or negative finding, core finding, and modifiers. This constrains the network to focus only on the key and relevant pathological details.

2   Method

The process of report generation starts with tokenization of the text, which allows the machine to grasp the concept of text as numbers. There are three key aspects to training the report generation model: (i) text pre-processing, (ii) tokenization and (iii) language model formulation/training. A report consists of three parts: The *Findings* section details what the radiologist saw in each area of the body during the exam, whereas the *Impressions* is a summary of all Findings with the patients clinical history, symptoms, and reason for the exam taken into account. It is possible for the Findings section to be empty, which means that the radiologist did not find any problems to report back to the doctor in charge.

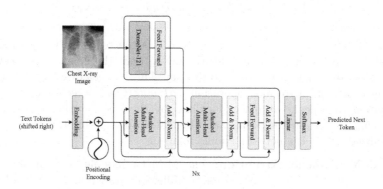

Fig. 1. Model architecture for RATCHET.

Language Pre-processing and Tokenization: Jargon heavily impacts the tokenization process, and subsequently the ability of the model to learn. Naïve tokenizers require the input sentence to be vigorously cleaned, *e.g.*, splitting of whole words, punctuation, measurements, units, abbreviations, etc. This can be especially problematic in medical texts as medical etymology involves compounding multiple Latin affixes, thus creating many unique words, as well as measurements. Too many infrequent tokens can hamper the model's ability to learn, whereas pruning too many infrequent tokens using the Out-Of-Vocabulary (OOV) token results in a stunted vocabulary.

Alternatively, a more complex tokenizer can be used, e.g., BPE [21], where words are broken down into one or more sub-words. This removes the dependency on the OOV token, and can also incorporate punctuation into the vocabulary corpus. Both sub-word tokenization methods share the same fundamental idea: more frequent words should be given unique IDs, whereas less frequent words should be decomposed into sub-words that best retain their meaning. In this work, the Huggingface 🤗 library [26] is used to create custom BPE tokenizers.

Captioning Transformer: RATCHET, as shown in Fig. 1, follows the Encoder - Decoder architecture style for NMT. Encouraged by results in CheXNet [20], the encoder module is replaced with a DenseNet-121 [6] as the primary image feature extractor. The decoder architecture remains the same as a text transformer, and is used in an auto-regressive manner. The output feature shape of DenseNet-121, for an input image of size 224×224, is $7 \times 7 \times 1024$. This is flattened to a vector of size 49×1024, which allows it to be treated the same way as a sequence of text. The mask for the self-attention sub-layer in the decoder stack is disabled for the encoder input, this ensures that the predictions for feature pixel i can depend on any or all other feature pixels.

Scaled Dot-Product Attention: [23] is a method of allowing the network to focus on a localized set of values from the input. Inspired by retrieval systems, the module features three inputs; Query, Key, Value.

$$Attention(Q, K, V) = softmax(\frac{QK^T}{\sqrt{d_k}})V, \tag{1}$$

where $Q \in \mathbb{R}^{L \times d}$, $K \in \mathbb{R}^{L \times d}$, $V \in \mathbb{R}^{L \times d_v}$, L is the sequence length and d is the feature depth. The module takes the query, finds the most similar key and returns the value that corresponds to this key. This is accelerated by two matrix multiplications and a softmax operation. $softmax(QK^T)$ creates a probability distribution that peaks at places localized by the keys for the corresponding query. This acts as a pseudo-mask, and by matrix multiplying it with V returns the localized values the network should focus on.

Masked Multi-head Attention: The M-MHA module is composed of n-stacked Scaled Dot-Product Attention modules in parallel, and is defined by:

$$h_i = Attention(QW_i^Q, KW_i^K, VW_i^V),$$
$$H = Concat(h_1, h_2, ..., h_n),$$
$$O = HW_h,$$

where $W_i^Q \in \mathbb{R}^{d_{model} \times d}$, $W_i^K \in \mathbb{R}^{d_{model} \times d}$, and $W_i^V \in \mathbb{R}^{d_{model} \times d_v}$. The attention output of each head, h_i, is concatenated together and projected to the same cardinality as the input by multiplying it with W_h, such that $W_h \in \mathbb{R}^{(n \times d_v) \times d_{model}}$ and $O \in \mathbb{R}^{L \times d_{model}}$.

The first M-MHA in the decoder is set in self-attention mode, *i.e.*, $K = Q = V =$ input. K and V behaves like a memory component of the network. It recalls the appropriate region to focus on for a given a query input based on words and sequences seen during training. The second M-MHA takes the image features as K and Q inputs, with the text features as V. The attention module then picks the necessary text features based on the image features, and simultaneously merges the image modality with the text modality. This module is used for model explainability, where highlighted features indicate the region of the image the model is particularly focusing on w.r.t. the text generated.

3 Experiment and Results

Dataset: MIMIC CXR v2.0.0 [9] is a recently released dataset of chest radiographs, with free-text radiology reports. It is publicly available and large, containing more than 300,000 images from 60,000 patients. It has been acquired at the Beth Israel Deaconess Medical Center in Boston, MA. The dataset also comes pre-annotated with disease labels, mined using either CheXpert [8] or NegBio [19] NLP tool, as well as train/validate/test splits. Since the radiology reports are free-text, they are very much prone to typos and inconsistencies that can hamper the efficacy of deep learning algorithms. The dataset has been further processed, with typos and errors corrected [5]. Only AP/PA views are used for our experiments. The gold standard ground truth for the dataset is the radiology reports. Due to the extreme time cost of labeling images by hand, the disease class labels are all mined via automatic means.

To ensure a fair testing of architectures, the input and evaluation part of the pipeline remains identical for all experiments. This includes the dataset used, image pre-processing and data augmentation methods, vocabulary construction, tokenization and NLP metrics for evaluation. We re-implemented the original LSTM-based method as presented in 'Show-and-Tell' [24] for the baseline architecture, and further modified it to create TieNet [25]. Since the implementation for TieNet has not been released, we have re-implemented it according to the descriptions provided by the original authors to the best of our abilities.

Image Input: As the images in the datasets are all non standard, each image is resized with padding to 224×224. Data augmentation includes random shifting of brightness, saturation and contrast to mimic various exposure parameters before being rescaled to an intensity range between $[0, 1]$ for network training.

Network Training: All networks are trained using Tensorflow 2.4.1 and Keras on a Nvidia Titan X, using the Adam optimiser with a learning rate of 1×10^{-4} and a batch size of 16 for 20 epochs. All weights start from random initialization, with a total parameter count of approx 51M. The hyperparameters used for all models were; $num\_layers = 6$, $n\_head = 8$, $d\_model = 512$, $dff = 2048$, $dropout = 0.2$. During training, an image I is fed into the encoder, with the output feature maps flattened into a linear sequence. The decoder is trained in a "teacher forcing" technique where the entire ground truth sequence

$S \in \mathbb{R}^{L \times n_{vocab}}$ is fed to the decoder to predict $S' \in \mathbb{R}^{L \times n_{vocab}}$. S' is the same as S except with the tokens shifted to the right by one, thus training the model to predict the next likely token of that sequence. With text represented by a one-hot vector, i.e., size of the sequence length by number of tokens in the vocabulary ($L \times n_{vocab}$), the loss is therefore defined by:

$$\log p(S|I) = \sum_{i=0}^{N} \log p(S_t|I, S_0, S_1, ...S_{t-1}, \theta), \qquad (2)$$

where (I, S) is the Image-Text pair, and θ are the network parameters. The network is trained by minimizing the cross-entropy loss to optimize for the sum of log probabilities across all N training examples. Inference takes approx. 0.12 s.

Model Inference: To generate text, RATCHET is ran in an auto-regressive fashion. For a given chest xray, it is resized to 224×224 with padding, and intensity rescaled to range of $[0, 1]$. The image is then fed into the model along with the initial token, <BOS> (Beginning Of Sentence). RATCHET then predicts the next token in accordance to the input it's received. The predicted token is then concatenated with the previous token to form a sequence, and the model is ran again (Fig. 2). This is repeated until either a <EOS> (End Of Sentence) token is predicted, or it has reached a maximum sequence length of 128 tokens, as bounded by the longest radiology report in training dataset.

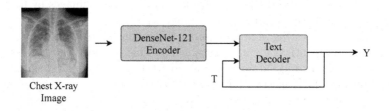

Fig. 2. Running inference using RATCHET

Results: Table 1 shows a summary of the average NLP metric scores for all predicted text compared to the ground truth radiology reports. A more detailed table containing all classes can be found in the supplementary materials. We attain better report language quality scores than both baseline method and TieNet. Interestingly, TieNet scored just sub-par compared to the baseline method. This could be attributed to the fact that TieNet was originally trained on the OpenI [3] dataset, which had 5x fewer vocabulary tokens compared to MIMIC-CXR, and non-optimal hyper-parameters as it had additional classification loss.

To assess the medical accuracy of generated reports quantitatively, discrete pathology labels of the ground truth text and generated text were obtained using Stanford's CheXpert Labeler [7]. To ensure methodological effectiveness

Table 1. NLP evaluation results on MIMIC-CXR Report Generation. A table with more detailed results can be found in the supplemental material. (B-1: BLEU-1, MET.: METEOR, R_L: ROUGE_L scores, Cm: Cardiomegaly, Ed: Edema, Co: Consolidation, At: Atelectasis, PE: Pleural Effusion. The latter are selected classes overlapping with CheXNet.) Note that the classification task has been evaluated as a surrogate to validate medical accuracy of the generated reports. We would not expect report classification to outperform direct prediction from images. Best language quality performance in bold.

Method	NLP metrics					Classification F1				
	B-1	MET.	R_L	CIDEr	SPICE	Cm	Ed	Co	At	PE
CheXNet (images)	–	–	–	–	–	0.534	0.674	0.193	0.567	0.715
TieNet (Classification)	–	–	–	–	–	0.488	0.536	0.188	0.471	0.591
Baseline (NLP)	0.208	**0.108**	0.217	0.419	0.107	–	–	–	–	–
TieNet (NLP)	0.190	0.069	0.200	0.411	0.084	0.061	0.050	0.0	0.006	0.033
RATCHET (NLP)	**0.232**	0.101	**0.240**	**0.493**	**0.127**	0.446	0.407	0.041	0.411	0.633
Class Bias	–	–	–	–	–	0.221	0.177	0.066	0.245	0.261

in radiological report generation, this was also compared against a separate direct image classifier, which was trained to perform direct multi-label classification. Following CheXNet [20], a DenseNet-121 was trained using the Adam optimizer with a learning rate of 1×10^{-4} and a batch size of 16 for 10 iterations. The epoch with the best performing weights on the validation set was used. Table 1 shows the accuracy of generated reports by the Baseline, TieNet, RATCHET and CheXNet. It can be seen that the results are on par with each other, and with related works in literature [1,14].

Of the 14 classes, CheXpert only presents the results from radiologists and original labels on ROC curves for 5 classes. Under the defined experiment setting 'LabelU' (where uncertain labels are classified as positive), the dataset is still heavily imbalanced at a ratio of 1:4 positive to negative. This is evidently seen in 'Consolidation' as the worst performing class across all models. For TieNet, the radiograph report generation head has failed to learn as shown by poor F1 scores. However, the classification head performs just under-par of CheXNet. TieNet's classification head was able to outperform 4 of 5 classes compared to the baseline model, only equaling in performance with 'Cardiomegaly'. RATCHET was able to outperform both baseline and TieNet, and also equaling in performance in 'Cardiomegaly'.

Figure 3 shows qualitative text example reports generated by RATCHET. In each case, it has picked out the correct pathological attributes, despite having very different phrasing. We observe that, at times, all models may preface generated reports with 'In comparison with the previous...'. This is a characteristic example of the network trying to imitate [4] the training set, a property that was criticized in the GPT model [15]. The network may also need to be better constrained, *e.g.* in Case1, the true text noted 'left mid-to-lower lung' whist in the generated text said 'left upper lung'. Whilst both the diagnosis of 'left' is correct, the network predicted 'lower' rather than the true text of 'upper'. Metrics from the NLP may not be the best for evaluating medical texts, due to specialized tolerances. E.g. in Case4, the true text states 'enlargement of the cardiomediastinal silhouette', which the network simplified to 'Moderate cardiomegaly'. Whilst this is medically correct, it will yield a lower NLP score.

For model explainability, the attention maps from the second Masked Multi-Head Attention module are used. The attention weights are a byproduct of the Scaled Dot-Product Attention, *i.e.* the resultant score of the softmax product between the key and query input matrices. From the entire decoder, only the attention weights from the last stacked decoder module are used, where a reduce mean operation is performed across all Attention Heads. This produces an attention map for each predicted token, Fig. 4 shows the attention of one particular test example. It can be seen that the attention module in RATCHET does indeed attend to the appropriate region of the image when referring to certain pieces of

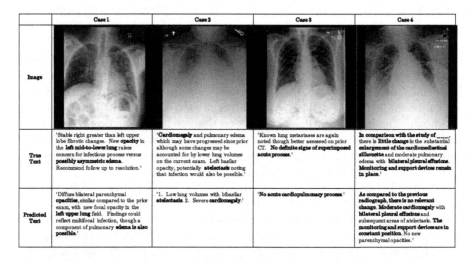

Fig. 3. Generated reports for four example test subjects by RATCHET

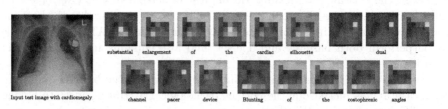

Fig. 4. A test image with generated caption and attention maps. The generated report for this case contains: *"In comparison with the study of ___, there is little overall change. Again there is substantial enlargement of the cardiac silhouette with a dual-channel pacer device in place. No evidence of vascular congestion or acute focal pneumonia. Blunting of the costophrenic angles is again seen."*

text. Notably, when describing the 'enlargement of the heart', the model focuses on the cardiac region. Similarly when 'a dual-channel pacer device' is referred to there is a peak in the region where the pacemaker device is located. 'Blunting of the costophrenic angles' is the classic sign for pleural effusion, which is a buildup of extra fluid in the space between the lungs and the chest wall (pleural space) at the base of the lungs.

4   Discussion

In our experiments, we have consistently shown that a transformer-based architecture can outperform traditional LSTM-based architectures in both NLP linguistic scores as well as medical pathology diagnosis. The attention maps from the Scaled Dot-Product Attention module highlights the region in the image that is responsible for each of the generated text tokens.

RATCHET, along with other NMT models, are trained on an individual image basis. From a medical point of view, it would be desirable to have it operating on a case by case basis, *i.e.*, to incorporate multiple images, as well as other patient data, clinical patient history and clinical biomarkers potentially including omics information. This would give a more complete picture of the diagnosis as well as constrain the network to make more accurate predictions.

5   Conclusion

Text generation is a staple aspect of NLP and remains a challenging task, especially in the realm of medical imaging. RATCHET is a transformer-based CNN-RNN founded on Neural Machine Translation for generating radiological reports. The model is prefaced with a DenseNet-121 model to perform image feature extraction, and is trained end-to-end. In this paper, we have shown that RATCHET is able to generate long medical reports.

Acknowledgement. This work is supported by the UK Research and Innovation London Medical Imaging and Artificial Intelligence Centre for Value Based Healthcare, and in part by the Intramural Research Program of the National Institutes of Health Clinical Center.

References

1. Boag, W., Hsu, T.H., McDermott, M.B.A., Berner, G., Alsentzer, E., Szolovits, P.: Baselines for chest x-ray report generation. In: ML4H@NeurIPS. Proceedings of Machine Learning Research, vol. 116, pp. 126–140. PMLR (2019)
2. Chen, H., Miao, S., Xu, D., Hager, G.D., Harrison, A.P.: Deep hierarchical multi-label classification of chest x-ray images. In: MIDL. Proceedings of Machine Learning Research, vol. 102, pp. 109–120. PMLR (2019)
3. Demner-Fushman, D., et al.: Preparing a collection of radiology examinations for distribution and retrieval. J Am. Med. Inform. Assoc. **23**(2), 304–310 (2016)

4. Foster, D.: House M.D.: Mirror Mirror. Fox Broadcasting Company, October 2007
5. Hou, B.: GitHub: farrell236/MIMIC-CXR. https://github.com/farrell236/mimic-cxr
6. Huang, G., Liu, Z., van der Maaten, L., Weinberger, K.Q.: Densely connected convolutional networks. In: CVPR, pp. 2261–2269. IEEE Computer Society (2017)
7. Irvin, J., et al.: Chexpert: a large chest radiograph dataset with uncertainty labels and expert comparison. In: Thirty-Third AAAI Conference on Artificial Intelligence (2019)
8. Irvin, J., et al.: Chexpert: a large chest radiograph dataset with uncertainty labels and expert comparison. In: AAAI, pp. 590–597. AAAI Press (2019)
9. Johnson, A.E.W., Pollard, T.J., Mark, R.G., Berkowitz, S.J., Horng, S.: MIMIC-CXR Database. In: PhysioNet (2019). https://doi.org/10.13026/C2JT1Q
10. Kraljevic, Z., et al.: Multi-domain clinical natural language processing with medcat: the medical concept annotation toolkit. CoRR abs/2010.01165 (2020)
11. Kumar, D., Wong, A., Clausi, D.A.: Lung nodule classification using deep features in CT images. In: CRV, pp. 133–138. IEEE Computer Society (2015)
12. Li, C.Y., Liang, X., Hu, Z., Xing, E.P.: Knowledge-driven encode, retrieve, paraphrase for medical image report generation. In: AAAI, pp. 6666–6673. AAAI Press (2019)
13. Lin, T.-Y., et al.: Microsoft COCO: common objects in context. In: Fleet, D., Pajdla, T., Schiele, B., Tuytelaars, T. (eds.) ECCV 2014. LNCS, vol. 8693, pp. 740–755. Springer, Cham (2014). https://doi.org/10.1007/978-3-319-10602-1_48
14. Liu, G., et al.: Clinically accurate chest x-ray report generation. In: MLHC. Proceedings of Machine Learning Research, vol. 106, pp. 249–269. PMLR (2019)
15. Marcus, G.: GPT-3, Bloviator: OpenAI's language generator has no idea what it's talking about, August 2020. https://www.technologyreview.com/2020/08/22/1007539/gpt3-openai-language-generator-artificial-intelligence-ai-opinion/
16. Monshi, M.M.A., Poon, J., Chung, V.Y.Y.: Deep learning in generating radiology reports: a survey. Artif. Intell. Med. **106**, 101878 (2020)
17. Pasa, F., Golkov, V., Pfeiffer, F., Cremers, D., Pfeiffer, D.: Efficient deep network architectures for fast chest x-ray tuberculosis screening and visualization. Sci. Rep. **9**(1), 1–9 (2019)
18. Pascanu, R., Mikolov, T., Bengio, Y.: On the difficulty of training recurrent neural networks. In: ICML (3). JMLR Workshop and Conference Proceedings, vol. 28, pp. 1310–1318. JMLR.org (2013)
19. Peng, Y., Wang, X., Lu, L., Bagheri, M., Summers, R.M., Lu, Z.: Negbio: a high-performance tool for negation and uncertainty detection in radiology reports. CoRR abs/1712.05898 (2017)
20. Rajpurkar, P., et al.: Chexnet: radiologist-level pneumonia detection on chest x-rays with deep learning. CoRR abs/1711.05225 (2017)
21. Sennrich, R., Haddow, B., Birch, A.: Neural machine translation of rare words with subword units. In: ACL (1). The Association for Computer Linguistics (2016)
22. Syeda-Mahmood, T., et al.: Chest X-ray report generation through fine-grained label learning. In: Martel, A.L., et al. (eds.) MICCAI 2020. LNCS, vol. 12262, pp. 561–571. Springer, Cham (2020). https://doi.org/10.1007/978-3-030-59713-9_54
23. Vaswani, A., et al.: Attention is all you need. In: NIPS, pp. 5998–6008 (2017)
24. Vinyals, O., Toshev, A., Bengio, S., Erhan, D.: Show and tell: a neural image caption generator. In: CVPR, pp. 3156–3164. IEEE Computer Society (2015)
25. Wang, X., Peng, Y., Lu, L., Lu, Z., Summers, R.M.: Tienet: text-image embedding network for common thorax disease classification and reporting in chest x-rays. In: CVPR, pp. 9049–9058. IEEE Computer Society (2018)

26. Wolf, T., et al.: Huggingface's transformers: State-of-the-art natural language processing. CoRR abs/1910.03771 (2019)
27. Xue, Z., et al.: Chest x-ray image view classification. In: CBMS, pp. 66–71. IEEE Computer Society (2015)
28. Yang, H., et al.: Learn to be uncertain: Leveraging uncertain labels in chest x-rays with Bayesian neural networks. In: CVPR Workshops, pp. 5–8. Computer Vision Foundation/IEEE (2019)
29. Yuan, J., Liao, H., Luo, R., Luo, J.: Automatic radiology report generation based on multi-view image fusion and medical concept enrichment. In: Shen, D., et al. (eds.) MICCAI 2019. LNCS, vol. 11769, pp. 721–729. Springer, Cham (2019). https://doi.org/10.1007/978-3-030-32226-7_80

Detecting When Pre-trained nnU-Net Models Fail Silently for Covid-19 Lung Lesion Segmentation

Camila Gonzalez[1]([envelope])[ID], Karol Gotkowski[1], Andreas Bucher[2],
Ricarda Fischbach[2], Isabel Kaltenborn[2], and Anirban Mukhopadhyay[1]

[1] Darmstadt University of Technology, Karolinenpl. 5, 64289 Darmstadt, Germany
camila.gonzalez@gris.informatik.tu-darmstadt.de
[2] University Hospital Frankfurt,
Theodor-Stern-Kai 7, 60590 Frankfurt am Main, Germany

Abstract. Automatic segmentation of lung lesions in computer tomography has the potential to ease the burden of clinicians during the Covid-19 pandemic. Yet predictive deep learning models are not trusted in the clinical routine due to *failing silently* in out-of-distribution (OOD) data. We propose a lightweight OOD detection method that exploits the Mahalanobis distance in the feature space. The proposed approach can be seamlessly integrated into state-of-the-art segmentation pipelines without requiring changes in model architecture or training procedure, and can therefore be used to assess the suitability of pre-trained models to new data. We validate our method with a patch-based nnU-Net architecture trained with a multi-institutional dataset and find that it effectively detects samples that the model segments incorrectly.

Keywords: Out-of-distribution detection · Uncertainty estimation · Distribution shift

1 Introduction

Automatic lung lesion segmentation in the clinical routine would significantly lessen the burden of radiologists, standardise quantification and staging of Covid-19 as well as open the way for a more effective utilisation of hospital resources. With this hope, several initiatives have gathered Computed Axial Tomography (CAT) scans and ground-truth annotations from expert thorax radiologists and released them to the public [6,20,23]. Experts have identified ground glass

Supported by the Bundesministerium für Gesundheit (BMG) with grant [ZMVI1-2520DAT03A].

Electronic supplementary material The online version of this chapter (https://doi.org/10.1007/978-3-030-87234-2_29) contains supplementary material, which is available to authorized users.

M. de Bruijne et al. (Eds.): MICCAI 2021, LNCS 12907, pp. 304–314, 2021.
https://doi.org/10.1007/978-3-030-87234-2_29

opacities (GGOs) and consolidations as characteristic of a pulmonary infection onset by the SARS-CoV-2 virus [24]. Deep learning models have shown good performance in segmenting these lesions. Particularly the fully-automatic *nnU-Net* framework [11] secured top spots (9 out of 10, including the first) in the leaderboard for the *Covid-19 Lung CT Lesion Segmentation Challenge* [7].

Such frameworks would ideally be utilised in the clinical practice. However, deep learning models are known to fail for data that considerably diverges from the training distribution. CAT scans are particularly prone to this *domain shift* problem [4]. The data showcased in the challenge is multi-centre and diverse in terms of patient group and acquisition protocol. A model trained with it would be presumed to produce good predictions for a wide spectrum of institutions. Yet when we evaluate a nnU-Net model on three other datasets, we notice a considerable drop in segmentation quality (see Fig. 1 (a)). Lung lesions do not manifest in large connected components (see Fig. 4), so it is not trivial for a novice radiologist to identify an incorrect segmentation.

Clinicians can still leverage models trained with large amounts of heterogeneous data, but only alongside a process that identifies when the model is unsuitable for a new data sample. Widely-used segmentation frameworks *are not designed with OOD detection in mind*, and so a method is needed that reliably identifies OOD samples post-training while requiring minimal intervention.

Several strategies have shown good OOD detection performance in classification models. Hendrycks and Gimpel [8] propose using the maximum softmax output as an OOD detection baseline. Guo et al. [5] find that replacing the regular softmax function with a *temperature-scaled* variant produces truer estimates. This can be complemented by adding perturbations to the network inputs [19]. Other methods [10,17] instead look at the KL divergence of softmaxed outputs from the uniform distribution. Some approaches use OOD data during training to explicitly train an outlier detector [1,9,17]. Bayesian-inspired techniques can also be used for outlier detection. Commonly-used are Monte Carlo Dropout [3] and Deep Ensembles [16]. These have shown promising results in the field of medical image segmentation [12,13,21]. Approaches that modify the architecture or training procedure have shown better performance in some cases, but their applicability to widely-used segmentation frameworks is limited [2,15,22].

We propose a method for OOD detection that is lightweight and seamlessly integrates into complex segmentation frameworks. Inspired by the work of Lee et al. [18], our approach estimates a multivariate Gaussian distribution from in-distribution (ID) training samples and utilises the Mahalanobis distance as a measure of uncertainty during inference. We compute the distance in a low-dimensional feature space, and down-sample it further to ensure a computationally inexpensive calculation. We validate our method on a patch-based 3D nnU-Net trained with multi-centre data from the *Covid-19 Lung CT Lesion Segmentation Challenge*. Our evaluation shows that the proposed method can effectively identify OOD samples for which the model produces faulty segmentations, and provides good model calibration estimates. Our contributions are:

- The introduction of a lightweight, flexible method for OOD detection that can be integrated into any segmentation framework.
- An extension of the nnU-Net framework to provide clinically-relevant uncertainty estimates.

2 Materials and Methods

We start by summarising the particularities of the nnU-Net framework in Sect. 2.1. In Sect. 2.2, we outline our proposed method for OOD detection, which follows a *three-step process*: (1) estimation of a Gaussian distribution from training features (2) extraction of uncertainty masks for test images and (3) calculation of subject-level uncertainty scores.

2.1 Patch-Based nnU-Net

The nnU-Net framework is a standardised baseline for medical image segmentation [11]. Without deviating from traditional U-Net architectures [26], it has won several grand challenges by automatically customising the architecture and training configuration to the data at hand [7]. The framework also performs pre- and post-processing steps, such as adapting voxel spacing and contrast normalisation, during both training and inference. In this work we utilise the patch-based full-resolution variant, which is recommended for most applications [11], but our method can be integrated into any other architecture. For the patch-based architecture, training images are first divided into overlapping patches with a sliding window approach, resulting in N patches $\{x_i\}_{i=1}^{N}$. Predictions for each patch are multiplied by a filtering operation that weights centre-voxels more heavily, and then aggregated into an output mask with the dimensions of the original image.

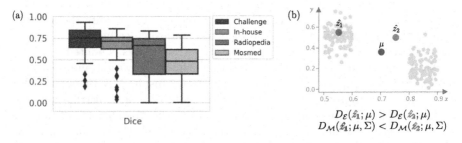

Fig. 1. (a) Dice coefficient of a model trained with *Challenge* data, evaluated with ID (*Challenge*) test data as well as on three other datasets. (b) The euclidean distance $D_{\mathcal{E}}$ does not recognize that $\hat{z}_1$ (purple marker) is closer than $\hat{z}_2$ (blue marker) to the distribution of training samples (gray markers), with mean μ (green marker) and covariance Σ. This difference intensifies in high-dimensional spaces, where it is common for regions close to the mean to be underrepresented. (Color figure online)

2.2   Estimation of a Subject-Level Uncertainty Score

We are interested in capturing *epistemic uncertainty*, which arises from a lack of knowledge about the data-generating process. Quantifying it for image *regions* instead of region boundaries is challenging, particularly for OOD data [14]. One computationally inexpensive way to assess epistemic uncertainty is to calculate the distance between training and testing activations in a low-dimensional feature space. As a model is unlikely to produce reasonable outputs for features far from any seen during training, this is a reliable signal for bad model performance [18]. Model activations have covariance and the activations of typical input images do not necessarily resemble the mean [27], so the euclidean distance is not appropriate to identify unusual activation patterns; a problem that exacerbates in high-dimensional spaces. The Mahalanobis distance $D_{\mathcal{M}}$ rescales samples into a space without covariance, supplying a more effective way to identify typical patterns in deep model features. Figure 1 (b) illustrates a situation where the euclidean distance assumes that $\hat{z}_2$ is closer to the training distribution than $\hat{z}_1$, when $\hat{z}_2$ is highly unusual and $\hat{z}_1$ is a probable sample.

In the following we describe the steps we perform to extract a subject-level uncertainty value. Note that only one forward pass is necessary for each image, keeping the computational overhead to a minimum.

Estimation of the Training Distribution: We start by estimating a multivariate Gaussian $\mathcal{N}(\mu, \Sigma)$ over model features. For all training inputs $\{x_i\}_{i=1}^N$, features $\mathcal{F}(x_i) = z_i$ are extracted from the encoder $\mathcal{F}$ of the pre-trained model. For modern segmentation networks, the dimensionality of the extracted features z_i is too large to calculate the covariance Σ in an acceptable time frame. We thus project the latent space into a lower subspace by average pooling. Finally, we flatten this subspace and estimate the empirical mean μ and covariance Σ.

$$\mu = \frac{1}{N} \sum_{i=1}^N \hat{z}_i, \quad \Sigma = \frac{1}{N} \sum_{i=1}^N (\hat{z}_i - \mu)(\hat{z}_i - \mu)^T \tag{1}$$

Extraction of Uncertainty Masks: During inference, we estimate an uncertainty mask for a subject following the process outlined in Fig. 2. For each patch x_i, features are extracted and projected into $\hat{z}_i$. Next, the Mahalanobis distance (Eq. 2) to the Gaussian distribution estimated in the previous step is calculated.

$$D_{\mathcal{M}}(\hat{z}_i; \mu, \Sigma) = (\hat{z}_i - \mu)^T \Sigma^{-1} (\hat{z}_i - \mu) \tag{2}$$

Each distance is a point estimate for the corresponding model input. These are aggregated in a similar fashion to how network outputs are combined to form a prediction mask. Following the example of the patch-based nnU-Net, a zero-filled tensor is initialised with the dimensionality of the original image. After assessing the distance for a patch, the value is replicated to the specified patch size and a filtering operation is applied to weight centre voxels more heavily. Finally, patch-level uncertainties are aggregated to an image-level mask.

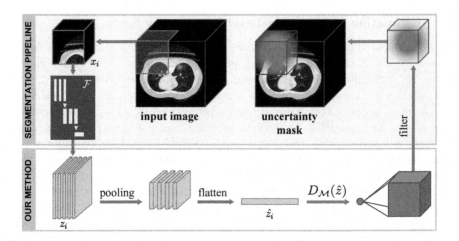

Fig. 2. Extracting an uncertainty mask based on the Mahalanobis distance $D_{\mathcal{M}}$ for an image during inference, in combination with a patch-based nnU-Net architecture.

Subject-Level Uncertainty: The process described above produces an uncertainty mask with the dimensionality of the CAT scan. In order to effectively identify highly uncertain samples, we aggregate these into a subject-level uncertainty $\mathcal{U}$ by averaging over all voxels. We then normalise uncertainties between the minimum and doubled maximum values represented in an ID validation set – which we assume to be available during training – to ensure $\mathcal{U} \in [0, 1]$.

3   Experimental Setup

We work with a total of four datasets for segmentation of Covid-19-related findings. The *Challenge* dataset [6] contains chest CAT scans for patients with a confirmed SARS-CoV-2 infection from an array of institutions. The data is heterogeneous in terms of age, gender and disease severity. We use the 199 cases made available under the *Covid Segmentation Grand Challenge*, which we randomly divide into 160 cases to train the model, 4 validation and 35 test cases.

We evaluate our method with two publicly available datasets and an in-house one. The public datasets encompass cases for patients with and without confirmed infections. *Mosmed* [23] contains fifty cases and the *Radiopedia* dataset [20], a further twenty. Finally, we utilise an in-house dataset consisting of fifty patients who were tested positive for SARS-CoV-2 with an RT PCR test. All fifty scans were reviewed for diagnostic image quality. The annotations for the in-house data were performed slice-by-slice by two independent readers trained in the delineation of GGOs and pulmonary consolidations. Central vascular structures and central bronchial structures were excluded from all segmentations. All delineations were reviewed by an expert radiologist reader. For the public datasets, the segmentation process is outlined in the corresponding publications.

With the *Challenge* data, we train a patch-based nnU-Net [11] on a *Tesla T4* GPU. Our configuration has a patch size of $[28, 256, 256]$, and adjacent patches overlap by half that size. To reduce the dimensionality of the feature space, we apply average pooling with a kernel size of $(2, 2, 2)$ and stride $(2, 2, 2)$ until the dimensionality falls below 1e4 elements. With the *Scikit Learn* library (version 0.24) [25], calculating Σ requires 85 s for 1e5 samples. Our code is available under github.com/MECLabTUDA/Lifelong-nnUNet.

We compare our approach to state-of-the-art techniques to assess uncertainty information by performing inference on a trained model. *Max. Softmax* consists of taking the maximum softmax output [8]. *Temp. Scaling* performs temperature scaling on the outputs before applying the softmax operation [5], for which we test three different temperatures $T = \{10, 100, 1000\}$. *KL from Uniform* computes the KL divergence from an uniform distribution [10]. Note that all three methods output a *confidence* score (higher is more certain), which we invert to obtain an *uncertainty* estimate (lower is more certain). Finally, *MC Dropout* consists of doing several forward passes whilst activating the Dropout layers that would usually be dormant during inference. We perform 10 forward passes and report the standard deviation between outputs as an uncertainty score. For all methods, we calculate a subject-level metric by averaging uncertainty masks, and normalise the uncertainty range between the minimum and doubled maximum uncertainty represented in ID validation data.

4   Results

We start this section by analysing the performance of the proposed method in detecting samples that vary significantly from the training distribution. We then examine how well the model estimates segmentation performance. Lastly, we qualitatively evaluate our method for ID and OOD examples.

OOD Detection: We first assess how effective our method is at identifying samples that are not ID (*Challenge* data). Due to the heterogeneity of the *Challenge* dataset, in practice data from an array of institutions would be considered ID. However, for our evaluation datasets there is a drop in performance which should manifest in higher uncertainty estimates. As is common practice in OOD detection [19], we find the uncertainty boundary that achieves a 95% true positive rate (TPR) on the ID validation set, where a *true positive* is a sample correctly identified as ID. We report for the ID test data and all OOD data the false positive rate (FPR) and *Detection Error* $= 0.5 (1 - TPR) + 0.5\ FPR$ at 95% TPR. Table 1 summarizes our findings. All methods that utilise the network outputs after one forward pass have a high detection error and FPR, while the MC Dropout approach manages to identify more OOD samples. Our proposed method displays the lowest FPR and detection error.

Segmentation Performance: While the detection of OOD samples is a first step in assessing the suitability of a model, an ideal uncertainty metric would inversely correlate with model performance, informing the user of the likely quality of a prediction without requiring manual annotations. For this we calculate the *Expected Segmentation Calibration Error* (ESCE). Inspired by Guo et al. [5], we divide the N test scans into $M = 10$ interval bins B_m according to their

Table 1. Detection Error (lower is better) and FPR (lower is better) for the boundary of 95% TPR, ESCE (lower is better) and (mean±sd) Dice (higher is better) for subjects with an uncertainty below the 95% TPR boundary. The results are reported for ID test data and all OOD samples.

Method	Det. error	FPR	ESCE	Dice
Max. Softmax [8]	0.334	0.583	0.319	0.582 ± 0.223
Temp. Scaling $T = 10$ [5]	0.508	0.758	0.407	0.601 ± 0.233
Temp. Scaling $T = 100$ [5]	0.361	0.550	0.408	0.589 ± 0.233
Temp. Scaling $T = 1000$ [5]	0.500	1.000	0.408	0.592 ± 0.233
KL from Uniform [10]	0.415	0.717	0.288	0.600 ± 0.215
MC Dropout [3]	0.177	0.183	0.215	0.614 ± 0.234
Ours	**0.082**	**0.050**	**0.125**	**0.744 ± 0.143**

normalised uncertainty. Over all bins, the absolute difference is added between average Dice $(Dice(B_m))$ and inverse average uncertainty $(1 - \mathcal{U}(B_m))$ for samples in the bin, weighted by the number of samples.

$$ESCE = \sum_{m=1}^{M} \frac{|B_m|}{N} |Dice(B_m) - (1 - \mathcal{U}(B_m))| \tag{3}$$

The results are reported in Table 1 (forth column). Our proposed approach shows the lowest $ESCE$ at 0.125. The average Dice of admitted samples (fifth column) lies at 0.744, which is consistent with the ID expected performance of the model (see Fig. 1 (a)).

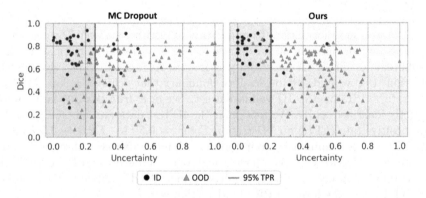

Fig. 3. Dice coefficient against normalised uncertainty for OOD (gray triangles) and test ID (black circles) samples. The vertical gray line marks the boundary of 95% TPR for ID validation data. To the right of this line, samples are classified as OOD. The lower left (red) quadrant is clinically most relevant. Unlike MC Dropout, our method does not fail silently by assigning low uncertainties to low-Dice samples. (Color figure online)

Figure 3 depicts the Dice coefficient plotted against the uncertainty for our proposed approach and MC Dropout, which has the second lowest calibration error. Relevant for a safe use of the model in clinical practice is the lower left (red) quadrant, where *silent failures* are located. Whereas MC Dropout fails to identify OOD samples with faulty predicted segmentations, our proposed method assigns these cases high uncertainty estimates. The only OOD samples that fall below the 95% TPR uncertainty boundary for our method have Dice scores over 0.6 (upper left quadrant with green background). However, our method shows room for improvement in the upper right (yellow) quadrant. Here, OOD samples for which the model produces good predictions are estimated to have a high uncertainty. An ideal calibration would place all samples in the upper left (green) and lower right (blue) quadrants.

Qualitative Evaluation: Figure 4 depicts two example images alongside corresponding ground truths and predictions. The top row shows a example from the *Challenge* dataset for which the model produces an adequate segmentation. The bottom contains a scan from the *Mosmed* dataset. The model oversegments the lesion at the middle left lobe and incorrectly marks two additional regions at the left and right superior lobes. Only our proposed method signals a possible error in the lower row with a high uncertainty, while producing a low uncertainty estimate for the upper row.

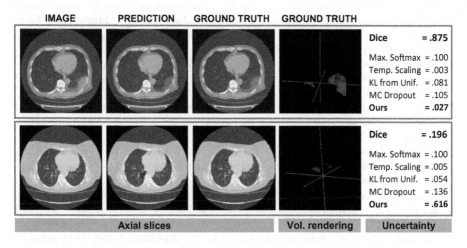

Fig. 4. Upper row: a good prediction. Lower row: a prediction for an OOD sample where two lesions are erroneously segmented in the superior lung lobes. Despite the considerable differences to the ground truth, these errors are not directly noticeable for the inexpert observer, as GGOs can manifest in superior lobes [24].

5 Conclusion

Increasingly, institutions are taking part in initiatives to gather large amounts of annotated, heterogeneous data and release it to the public. This could potentially alleviate the work burden of medical practitioners by allowing the training of robust segmentation models. Open-source end-to-end frameworks contribute to this process. But regardless of the variety of the training data, it is necessary to assess whether a model is well-suited to new samples. This is particularly true when it is not trivial to identify a faulty output, such as for the segmentation of SARS-CoV-2 lung lesions. There is currently a disconnect between methods for OOD detection, which often require special training or architectural considerations, and widely-used segmentation frameworks. We find that calculating the Mahalanobis distance to features in a low-dimensional subspace is a lightweight and flexible way to signal when a model prediction should not be trusted. Future work should explore how to better identify high-quality predictions and evaluate the methods considered in this work on other segmentation models. For now, our work increases clinicians' trust while translating trained neural networks from challenge participation to real clinics.

References

1. Bevandić, P., Krešo, I., Oršić, M., Šegvić, S.: Simultaneous semantic segmentation and outlier detection in presence of domain shift. In: Fink, G.A., Frintrop, S., Jiang, X. (eds.) DAGM GCPR 2019. LNCS, vol. 11824, pp. 33–47. Springer, Cham (2019). https://doi.org/10.1007/978-3-030-33676-9_3
2. Blundell, C., Cornebise, J., Kavukcuoglu, K., Wierstra, D.: Weight uncertainty in neural network. In: International Conference on Machine Learning, pp. 1613–1622. PMLR (2015)
3. Gal, Y., Ghahramani, Z.: Dropout as a Bayesian approximation: representing model uncertainty in deep learning. In: International Conference on Machine Learning, pp. 1050–1059. PMLR (2016)
4. Glocker, B., Robinson, R., Castro, D.C., Dou, Q., Konukoglu, E.: Machine learning with multi-site imaging data: an empirical study on the impact of scanner effects. arXiv preprint arXiv:1910.04597 (2019)
5. Guo, C., Pleiss, G., Sun, Y., Weinberger, K.Q.: On calibration of modern neural networks. In: International Conference on Machine Learning, pp. 1321–1330. PMLR (2017)
6. Harmon, S.A., et al.: Artificial intelligence for the detection of COVID-19 pneumonia on chest CT using multinational datasets. Nat. Commun. 11(1), 1–7 (2020). https://doi.org/10.1038/s41467-020-17971-2
7. Henderson, E.: Leading pediatric hospital reveals top AI models in COVID-19 grand challenge. https://www.news-medical.net/news/20210112/Leading-pediatric-hospital-reveals-top-AI-models-in-COVID-19-Grand-Challenge.aspx. Accessed 28 Feb 2021
8. Hendrycks, D., Gimpel, K.: A baseline for detecting misclassified and out-of-distribution examples in neural networks. In: International Conference on Learning Representations (2017)

9. Hendrycks, D., Mazeika, M., Dietterich, T.: Deep anomaly detection with outlier exposure. In: International Conference on Learning Representations (2018)

10. Hendrycks, D., Mazeika, M., Kadavath, S., Song, D.: Using self-supervised learning can improve model robustness and uncertainty. Adv. Neural. Inf. Process. Syst. **32**, 15663–15674 (2019)

11. Isensee, F., Jaeger, P.F., Kohl, S.A., Petersen, J., Maier-Hein, K.H.: nnU-Net: a self-configuring method for deep learning-based biomedical image segmentation. Nat. Methods **18**(2), 203–211 (2021)

12. Jungo, A., Balsiger, F., Reyes, M.: Analyzing the quality and challenges of uncertainty estimations for brain tumor segmentation. Front. Neurosci. **14**, 282 (2020)

13. Jungo, A., Reyes, M.: Assessing reliability and challenges of uncertainty estimations for medical image segmentation. In: Shen, D., et al. (eds.) MICCAI 2019. LNCS, vol. 11765, pp. 48–56. Springer, Cham (2019). https://doi.org/10.1007/978-3-030-32245-8_6

14. Kendall, A., Gal, Y.: What uncertainties do we need in Bayesian deep learning for computer vision? Adv. Neural. Inf. Process. Syst. **30**, 5574–5584 (2017)

15. Kohl, S.A., et al.: A probabilistic u-net for segmentation of ambiguous images. In: Proceedings of the 32nd International Conference on Neural Information Processing Systems, pp. 6965–6975 (2018)

16. Lakshminarayanan, B., Pritzel, A., Blundell, C.: Simple and scalable predictive uncertainty estimation using deep ensembles. Adv. Neural. Inf. Process. Syst. **30**, 6402–6413 (2017)

17. Lee, K., Lee, H., Lee, K., Shin, J.: Training confidence-calibrated classifiers for detecting out-of-distribution samples. In: International Conference on Learning Representations (2018)

18. Lee, K., Lee, K., Lee, H., Shin, J.: A simple unified framework for detecting out-of-distribution samples and adversarial attacks. In: Advances in Neural Information Processing Systems, pp. 7167–7177 (2018)

19. Liang, S., Li, Y., Srikant, R.: Enhancing the reliability of out-of-distribution image detection in neural networks. In: International Conference on Learning Representations (2018)

20. Ma, J., et al.: COVID-19 CT lung and infection segmentation dataset (2020). https://doi.org/10.5281/zenodo.3757476

21. Mehrtash, A., Wells, W.M., Tempany, C.M., Abolmaesumi, P., Kapur, T.: Confidence calibration and predictive uncertainty estimation for deep medical image segmentation. IEEE Trans. Med. Imaging **39**(12), 3868–3878 (2020)

22. Monteiro, M., et al.: Stochastic segmentation networks: modelling spatially correlated aleatoric uncertainty. In: Larochelle, H., Ranzato, M., Hadsell, R., Balcan, M.F., Lin, H. (eds.) Advances in Neural Information Processing Systems, vol. 33, pp. 12756–12767. Curran Associates, Inc. (2020)

23. Morozov, S., et al.: Mosmeddata: chest CT scans with COVID-19 related findings dataset. arXiv preprint arXiv:2005.06465 (2020)

24. Parekh, M., Donuru, A., Balasubramanya, R., Kapur, S.: Review of the chest CT differential diagnosis of ground-glass opacities in the COVID era. Radiology **297**(3), E289–E302 (2020)

25. Pedregosa, F., et al.: Scikit-learn: machine learning in python. J. Mach. Learn. Res. **12**, 2825–2830 (2012)

26. Ronneberger, O., Fischer, P., Brox, T.: U-Net: convolutional networks for biomedical image segmentation. In: Navab, N., Hornegger, J., Wells, W.M., Frangi, A.F. (eds.) MICCAI 2015. LNCS, vol. 9351, pp. 234–241. Springer, Cham (2015). https://doi.org/10.1007/978-3-319-24574-4_28
27. Wei, D., Zhou, B., Torrabla, A., Freeman, W.: Understanding intra-class knowledge inside CNN. arXiv preprint arXiv:1507.02379 (2015)

Perceptual Quality Assessment of Chest Radiograph

Mengda Guan[1], Yuanyuan Lyu[2], Wanyue Cao[3], Xingwang Wu[4], Jingjing Lu[5], and S. Kevin Zhou[6,7(✉)]

[1] University of Science and Technology of China, Heifei, China
[2] Z^2 Sky Technologies Inc., Suzhou, China
[3] Shanghai Jiao Tong University, Shanghai, China
[4] The First Affiliated Hospital of Anhui Medical University, Heifei, China
[5] Peking Union Medical College Hospital, Beijing, China
[6] Medical Imaging, Robotics and Analytic Computing Laboratory and Engineering (MIRACLE), School of Biomedical Engineering and Suzhou Institute for Advanced Research, University of Science and Technology of China, Suzhou, China
[7] Key Lab of Intelligent Information Processing of Chinese Academy of Sciences (CAS), Institute of Computing Technology, CAS, Beijing, China

Abstract. The quality of a chest X-ray image or radiograph, which is widely used in clinics, is a very important factor affects doctors' clinical decision making. Since there is no chest X-ray image quality database so far, we conduct the first study of perceptual quality assessment of chest X-ray images by introducing a Chest X-ray Image Quality Database, which contains 2,160 chest X-ray images obtained from 60 reference images. In order to simulate the real noise of X-ray images, we add different levels of Gaussian noise and Poisson noise, which are most commonly found in X-ray images. Mean opinion scores (MOS) have been collected by performing user experiments with 74 subjects (25 professional doctors and 49 non-doctors). The availability of MOS allows us to design more effective image quality metrics. We use the database to train a blind image quality assessment model based on deep neural networks, which attains better performances than conventional approaches in terms of Spearman rank-order correlation coefficient and Pearson linear correlation coefficient. The database and the deep learning models are available at https://github.com/ICT-MIRACLE-lab/CXIQ.

Keywords: Image Quality Database · Blind image quality assessment · Chest X-ray

1 Introduction

Image quality assessment (IQA) is critical in many circumstances [1]. IQA can be broadly divided into two categories: subjective assessment and objective assessment [22]. Subjective IQA evaluates image quality through psychophysical experiments [13]; although it is the best indicator, it is time-consuming, expensive,

© Springer Nature Switzerland AG 2021
M. de Bruijne et al. (Eds.): MICCAI 2021, LNCS 12907, pp. 315–324, 2021.
https://doi.org/10.1007/978-3-030-87234-2_30

and impractical [20]. The objective assessment can be divided into three categories: full reference (FR-IQA), reduced reference (RR-IQA), and no reference (NR-IQA). FR-IQA needs reference images, but in practice, pristine or perfect images as references are not easy to obtain. Although NR-IQA is a challenging method [3], it is the most appropriate and practical assessment method; yet its accuracy is not fully guaranteed [4].

With a low radiation dose and a short examination time, X-ray imaging or radiography is an imaging modality widely used in clinical diagnosis. X-ray radiography is useful in the detection of pathologies of the skeletal system as well as some disease processes in soft tissue. One notable example is chest X-ray, which can be used to identify lung diseases such as pneumonia, lung cancer, or pulmonary tuberculosis. Because of chest radiograph's wide application in clinics, its image quality is particularly important.

In the past decades, there have been a great number of studies on image quality assessment (IQA) [6,15,16,23,25,27]. Most of the existing methods are based on natural images, but medical images are inherently different from natural images [28,29]. It is questionable to apply these methods directly to medical images.

There is a lack of a dataset suitable for medical image quality assessment and a lack of a specialized medical image quality assessment method. To bridge these two gaps, we here introduce the Chest X-ray Image Quality Database (CXIQ) and propose the blind image quality assessment models, constructed by deep neural networks. Our contributions include:

- To the best of our knowledge, we are the first to conduct the study of perceptual quality assessment of chest radiograph. We introduce the Chest X-ray Image Quality Database, which contains 2,160 chest X-ray images obtained from 60 reference images. We add Poisson and Gaussian noise to the reference images to simulate the real noise in X-ray images.
- We adopt and improve the existing subjective experimental methodology to better evaluate the quality of medical images. We divide the user experiment into two parts: Part I aims to let the observer evaluate the *holistic quality* of the images, and Part II focuses on the *diagnostic value* of the images in the existence of noise. We carefully carry out the experiment to collect the mean opinion score (MOS) for each image.
- We use the database to train blind image quality assessment models, constructed by deep neural networks. We train models to approximate the MOS values of Part I and Part II, and a final aggregate of the above two MOS, respectively. These models allow us to directly score the holistic image quality and diagnostic quality of a medical image.
- We open source the entire study to benefit the whole chest X-ray imaging community.

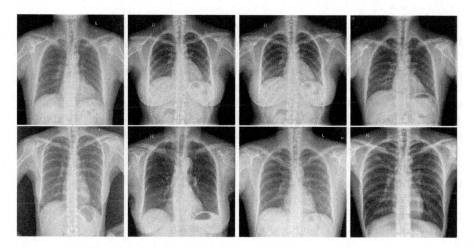

Fig. 1. The reference images in CXIQ. Images in the first row are normal cases and images in the second row are cases with manifestations of Tuberculosis.

2   CXIQ Database

In this section, we present how to construct the Chest X-ray Image Quality (CXIQ) Database from several aspects: the set of the reference images, types of distortions, levels of distortions, and details of subjective test.

2.1   Database Construction

The quality of the database strictly depends on the reference images. The reference images in the database must be in the uniform quality. We aim to select the original image with high quality and do the following to guarantee this. We first select a few public datasets with good image quality and then give these datasets to three physicians for comparison. After comparison, they choose Shenzhen chest X-ray database [11], which is known for its high image quality.

With the physicians' help, we further select 60 images from the Shenzhen chest X-ray database as the original images of our Chest X-ray Image Quality Database, half of them are collected from women and the other half from men. Also, half of the images are normal cases and half of them are cases with manifestations of tuberculosis. Due to the non-equal size of images, we resize all images to 800 × 800 pixels. Some images in the Shenzhen chest X-ray set database have black borders, thus we firstly crop the original images in order to remove the black borders, and then we scale the images to 800 × 800 pixels (Fig. 1).

There are great differences between medical images and natural images in terms of the acquisition principle. Most natural images are formed by natural light, however, for example, X-ray and CT images are obtained by attenuation of X-rays. Owing to the differences in imaging principle and acquisition method, the types of noise in images are also different. Elbakri and Fessler [5] demonstrate that the actual noise in X-ray images combines Poisson and Gaussian distribution, with the Poisson distribution arising from quantum noise and the Gaussian distribution arising from thermal noise generated by sensors and other electronic devices. So we add the Poisson noise and Gaussian noise to the reference images to simulate the real noise in X-ray images. We set five levels for all types of distortions: for five Gaussian noise levels, the corresponding PSNRs are about 38.6 dB, 37.8 dB, 36.0 dB, 34.8 dB, 33.6 dB; for five Poisson noise levels from low to high, the corresponding PSNRs are about 45.1 dB, 42.5 dB, 40.1 dB, 37.5 dB, 35.1 dB. This way, we obtain 10 noisy images per reference image. Moreover, the real noise in X-ray images is a mixture of Gaussian and Poisson, thus we combine five levels of Gaussian noise and five levels of Poisson noise and get another 25 images. To sum up, we obtain 35 noisy images associated with one reference image.

Also, each image in CXIQ is associated with:

1. The clinical variables of each X-ray image. Each reading contains the patient's age, gender, and abnormality seen in the lung, if any.
2. MOS of Part I, a continuous score in [0, 100] to represent the holistic quality of the X-ray image. A higher score indicates better holistic quality.
3. MOS of Part II, a continuous score in [0, 100] to represent the diagnostic quality of the X-ray image. A higher score indicates better diagnostic quality.
4. Aggregate MOS, a continuous score in [0, 100] to represent the aggregate quality of the X-ray image. It is weighted by the MOS scores of Part I and Part II, and both weights are 0.5 (Fig. 2).

2.2   Subjective Testing

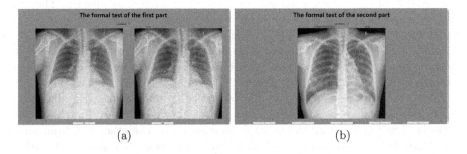

Fig. 2. Test interface used in our subjective experiments. (a) Part I. (b) Part II.

There are several frequently-used methodologies [2, 10, 12, 14, 18, 20] that can be used to evaluate the quality of an image, such as Single stimulus, Double stimuli, Forced choice, Similarity judgment, etc. However, it is inappropriate to use these methods directly due to the peculiarity of medical images. Medical images are very different from natural images, because the former are especially acquired for medical purposes, such as diagnosis, treatment, or medical research. Their image quality mainly depends on whether they can clearly and accurately identify the normal tissue or the existing lesions. For example, the Single-stimulus-based categorical rating involves displaying an image for a short and fixed duration and then asking the participant to rate it using one of the five categories: excellent, good, fair, poor or bad; but for medical images, a short and fixed duration is an excessive restriction, in general, there is no mandatory time limit for doctors to read images.

Therefore, based on the existing methods, we propose a subjective evaluation test tailored to the medical image. We divided the user test into two parts, Part I aims to let the observer evaluate the holistic quality of the images, and Part II focuses on the diagnosis details of images by, for example, clearly distinguishing normal tissues and existing lesions. In Part I, we choose an improved Forced choice as our experimental paradigm. Forced-choice pairwise comparison[2] is an ordering method, in which observers decide which of the two displayed images has higher quality, but there is no time limit. In order to make the observers pay more attention to the holistic quality, we add a time limit here. This method is very similar to the method in [17], however, in our experiment, the reference images are not used as a reference, and they also appear in the test like other distorted images. Except for the above point, the other experimental method is the same as in [17]. In Part II, we choose an improved Single stimulus as our experimental paradigm. As mentioned above, a short and fixed duration is an excessive restriction, so we don't set a time limit here. The observers are able to observe the images carefully and rate them according to whether they were confident in making a diagnostic decision. Considering that this part requires making diagnostic judgments on medical images, so only professional doctors participate in this part. Finally, we get the results of Part I from 0 to 9, and the results of Part II are from 1 to 5. We then scale the results of the two parts to 0 to 100, and weight the results of the two parts to get the aggregate MOS, and both weights are 0.5.

Totally, 74 subjects, including 25 doctors and 49 non-doctors, participate in the experiments. Each image is evaluated by a different number of subjects. In Part I, 57 groups are evaluated by 2 subjects per image, and 3 groups are evaluated by 3 subjects per image. In Part II, since only professional doctors participate in this part, each image is evaluated by 1 subject per image. Before the formal test, we give an instruction and a pilot test to facilitate the subjects to understand and adapt to the main test. Additionally, in both parts, a blank image appears for two seconds after response, in order to eliminate the visual staying phenomenon.

3   Subjective Data Analysis

In this section, we analyze the collected subjective data in CXIQ.

Figure 3 presents the MOS for each distortion level obtained from Part I and Part II, respectively. Every six is a group, the Poisson noise in each group increases sequentially, and the Gaussian noise between each group increases

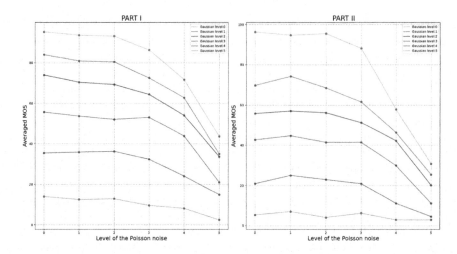

Fig. 3. Averaged MOS for Part I and Part II.

sequentially. The first group does not add any Gaussian noise. As we can see, the results of Part I and Part II are obviously correlated with the noise intensity, however, if we take a closer look at the results of Part II, in the same level of Gaussian noise, adding first-level Poisson noise to reference images may *increase* the MOS score.

We also calculate the correlation between the score obtained from different occupations and different genders for Part I. The Spearman rank-order correlation coefficient (SROCC) between MOS for groups of doctor and non-doctor is 0.91, and the SROCC between MOS for groups of male and female is 0.93. To investigate whether gender or occupation has an effect on MOS, we run a repeated measures analysis of variance (ANOVA) with gender and occupation as between-subject factors and Gaussian and Poisson noise level as within-subject factors. The results show all the main effects or interaction effects relate to gender or occupation do not reach the significance level (all $F < 1.840$, all $p > 0.102$). Based on the correlation and ANOVA results, MOS of different genders and occupations show no difference; thus, we combine all the MOS values in Part I for subsequent analysis.

4   Objective Quality Models

Based on the curated CXIQ dataset, we train deep neural networks to predict the MOS of X-ray images. We train multiple models for the MOS of Part I, Part II, and the aggregate MOS, then we compare our models with other existing BIQA models. Finally, we use the same user experiment introduced in Sect. 2 to obtain the MOS of the pediatric hand radiographs [7], used for evaluating the robustness of our image quality prediction network, which is trained based on chest radiographs.

We train several widely used deep neural networks, including VGG19 [21], ResNet50 [8], DenseNet121 [9], ResNeXt101 [24], and Wide ResNet50 [26]. The final fully connected layer in these models is changed to one output, and the softmax function is not used. We put the images directly into the network without pre-processing and we adopt L_1 norm as the loss function. In the training phase, we adopt the same training strategy, all models are initialized with the random weight, and we set the mini-batch size to 8 and the epoch number to 30. We use the stochastic gradient descent with momentum strategy, and we set the initial learning rate of 10^{-4} and momentum to 0.9.

Table 1. SROCC and PLCC results of deep learning methods with other IQA models on CXIQ.

	Model	QAC	NIQE	ILNIQE	BRISQUE	Vgg	ResNet	ResNeXt	ResNet-Wide	Dense-Net
Part I	SROCC	0.532	0.737	0.427	0.878	0.778	0.883	0.883	0.886	0.885
	PLCC	0.645	0.688	0.449	0.702	0.761	0.886	0.866	0.882	0.884
Part II	SROCC	0.564	0.790	0.482	0.905	0.848	0.896	0.909	0.917	0.921
	PLCC	0.698	0.788	0.529	0.798	0.841	0.901	0.914	0.912	0.925
Aggregate	SROCC	0.569	0.795	0.473	0.930	0.869	0.936	0.943	0.940	0.942
	PLCC	0.702	0.771	0.511	0.784	0.861	0.928	0.949	0.945	0.949

Table 2. For pediatric hand radiographs, SROCC and PLCC results of deep learning method with other IQA models on CXIQ.

Model	QAC	NIQE	ILNIQE	BRISQUE	OURS
SROCC	0.661	0.753	0.770	0.735	0.935
PLCC	0.722	0.786	0.791	0.681	0.943

4.1 Performance Evaluation

We select several image quality models designed based on different methods to compare with our model. Among them are NSS-based model-BRISQUE [15], NIQE [16], ILNIQE [27], and codebook-based model-QAC [25]. We test these models on our CXIQ using their officially released codes.

We randomly sample 80% of the images in CXIQ for training and the rest are for testing. We train and test MOS of Part I, MOS of Part II, and MOS separately. SROCC and Pearson linear correlation coefficient (PLCC) are used to compare which model has a better correspondence to image quality assessment by humans. The results are shown in Table 1, we get some observations. First, QAC does not work well for X-ray images, due to the huge distribution difference between the medical images and natural images. NIQE and ILNIQE use the same method, on which they extract a set of local features from an image, then fit the feature vectors to a multivariate Gaussian (MVG) model, and these two models do not require MOSs. ILNIQE does not work well, while NIQE delivers superior performance on CXIQ, this may be due to the fact that many features extracted by NIQE are not applicable to medical images. Third,

BRISQUE obtains a comparable performance, suggesting supervised learning with MOS has a great influence. Fourth, models based on deep neural network achieve very good results, and this seems to suggest that deep neural network has a good potential of learning *perceptual characteristics* of X-ray images.

In order to verify the robustness of the model, we request five observers to score five sets of pediatric hand radiographs. The processing and evaluation methods of these five sets of pediatric hand radiographs are consistent with the methods we introduce in Sect. 2. We select our model based on ResNet-50, and test our model together with other models for the aggregate MOS. The results are shown in Table 2, our model trained on Chest X-ray image achieves better prediction on the MOS of pediatric hand radiographs compared with other models. It proves the generalizability of the proposed model.

4.2 Visualization

In this subsection, we utilize NormGrad [19] to understand our model's behavior. We select our model based on ResNet-50, and Fig. 4 shows the heat maps of the layer before the Average Pooling layer, by using the virtual con 1×1 and conv 3×3 layer. Red and blue colors within the heat maps point to regions with high and low values. It can be seen from the results that the model mainly focuses on the central part of the image, which is also in line with the characteristics of people when observing the images. At the same time, unlike classification tasks and target detection tasks, high values are concentrated in a very small area. We think this is reasonable, although medical images have many details to pay more attention to, in most cases, it is not common that partial details affect the quality of the images.

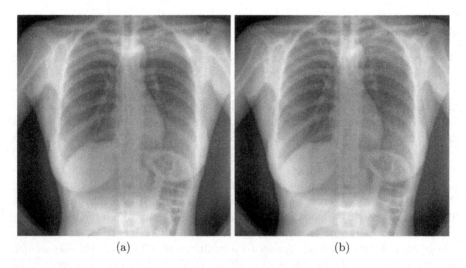

(a)   (b)

Fig. 4. Heat maps at the layer before the Average Pooling layer. (a) conv 1×1. (b) conv 3×3. (Color figure online)

5   Conclusion

We firstly build Chest X-ray Image Quality Database, which contains 2,160 chest X-ray images, attached with clinical variable, MOS of Part I and Part II, and MOS. We also construct several BIQA models using deep neural networks, which achieve a better result. We visualize the area which the model focuses on and test our model on the pediatric hand radiographs. It demonstrates the proposed models are very robust. We believe that our curated database and proposed models lay the foundation for the development of medical images quality assessment, which will impact the development of Medical Imaging.

References

1. Bovik, A.C.: Handbook of Image and Video Processing. Academic Press, Cambridge (2010)
2. ITU-R Rec. BT: Methodology for the subjective assessment of the quality of television pictures. International Telecommunication Union (2002)
3. Chandler, D.M.: Seven challenges in image quality assessment: past, present, and future research. In: International Scholarly Research Notices 2013 (2013)
4. Chow, L.S., Paramesran, R.: Review of medical image quality assessment. Biomed. Signal Process. Control **27**, 145–154 (2016)
5. Elbakri, I.A., Fessler, J.A.: Statistical image reconstruction for polyenergetic x-ray computed tomography. IEEE Trans. Med. Imaging **21**(2), 89–99 (2002)
6. Fang, Y., Zhu, H., Zeng, Y., Ma, K., Wang, Z.: Perceptual quality assessment of smartphone photography. In: Proceedings of the IEEE/CVF Conference on Computer Vision and Pattern Recognition, pp. 3677–3686 (2020)
7. Halabi, S.S., et al.: The RSNA pediatric bone age machine learning challenge. Radiology **290**(2), 498–503 (2019)
8. He, K., Zhang, X., Ren, S., Sun, J.: Deep residual learning for image recognition. In: Proceedings of the IEEE Conference on Computer Vision and Pattern Recognition, pp. 770–778 (2016)
9. Huang, G., Liu, Z., Van Der Maaten, L., Weinberger, K.Q.: Densely connected convolutional networks. In: Proceedings of the IEEE Conference on Computer Vision and Pattern Recognition, pp. 4700–4708 (2017)
10. ITU-T Recommendation P.910: Subjective audiovisual quality assessment methods for multimedia applications (1998)
11. Jaeger, S., Candemir, S., Antani, S., Wáng, Y.X.J., Lu, P.X., Thoma, G.: Two public chest x-ray datasets for computer-aided screening of pulmonary diseases. Quant. Imaging Med. Surg. **4**(6), 475 (2014)
12. Leveque, L., et al.: On the subjective assessment of the perceived quality of medical images and videos. In: 2018 Tenth International Conference on Quality of Multimedia Experience (QoMEX), pp. 1–6. IEEE (2018)
13. Ma, K., Liu, W., Liu, T., Wang, Z., Tao, D.: dipIQ: blind image quality assessment by learning-to-rank discriminable image pairs. IEEE Trans. Image Process. **26**(8), 3951–3964 (2017)
14. Mantiuk, R.K., Tomaszewska, A., Mantiuk, R.: Comparison of four subjective methods for image quality assessment. In: Computer Graphics Forum, vol. 31, pp. 2478–2491. Wiley Online Library (2012)

15. Mittal, A., Moorthy, A.K., Bovik, A.C.: No-reference image quality assessment in the spatial domain. IEEE Trans. Image Process. **21**(12), 4695–4708 (2012)
16. Mittal, A., Soundararajan, R., Bovik, A.C.: Making a "completely blind" image quality analyzer. IEEE Signal Process. Lett. **20**(3), 209–212 (2012)
17. Ponomarenko, N., Lukin, V., Zelensky, A., Egiazarian, K., Carli, M., Battisti, F.: TID 2008-a database for evaluation of full-reference visual quality assessment metrics. Adv. Mod. Radioelectron. **10**(4), 30–45 (2009)
18. Ponomarenko, N., Silvestri, F., Egiazarian, K., Carli, M., Astola, J., Lukin, V.: On between-coefficient contrast masking of DCT basis functions. In: Proceedings of the Third International Workshop on Video Processing and Quality Metrics, vol. 4 (2007)
19. Rebuffi, S.A., Fong, R., Ji, X., Vedaldi, A.: There and back again: revisiting back-propagation saliency methods. In: Proceedings of the IEEE/CVF Conference on Computer Vision and Pattern Recognition, pp. 8839–8848 (2020)
20. Sheikh, H.R., Sabir, M.F., Bovik, A.C.: A statistical evaluation of recent full reference image quality assessment algorithms. IEEE Trans. Image Process. **15**(11), 3440–3451 (2006)
21. Simonyan, K., Zisserman, A.: Very deep convolutional networks for large-scale image recognition. arXiv preprint arXiv:1409.1556 (2014)
22. Wang, Z., Bovik, A.C.: Modern image quality assessment. Synth. Lect. Image Video Multimed. Process. **2**(1), 1–156 (2006)
23. Wang, Z., Bovik, A.C., Sheikh, H.R., Simoncelli, E.P.: Image quality assessment: from error visibility to structural similarity. IEEE Trans. Image Process. **13**(4), 600–612 (2004)
24. Xie, S., Girshick, R., Dollár, P., Tu, Z., He, K.: Aggregated residual transformations for deep neural networks. In: Proceedings of the IEEE Conference on Computer Vision and Pattern Recognition, pp. 1492–1500 (2017)
25. Xue, W., Zhang, L., Mou, X.: Learning without human scores for blind image quality assessment. In: Proceedings of the IEEE Conference on Computer Vision and Pattern Recognition, pp. 995–1002 (2013)
26. Zagoruyko, S., Komodakis, N.: Wide residual networks. arXiv preprint arXiv:1605.07146 (2016)
27. Zhang, L., Zhang, L., Bovik, A.C.: A feature-enriched completely blind image quality evaluator. IEEE Trans. Image Process. **24**(8), 2579–2591 (2015)
28. Zhou, S.K., et al.: A review of deep learning in medical imaging: image traits, technology trends, case studies with progress highlights, and future promises. arXiv preprint arXiv:2008.09104 (2020)
29. Zhou, S.K., Rueckert, D., Fichtinger, G.: Handbook of Medical Image Computing and Computer Assisted Intervention. Academic Press, Cambridge (2019)

Pristine Annotations-Based Multi-modal Trained Artificial Intelligence Solution to Triage Chest X-Ray for COVID-19

Tao Tan[1]([⊠]), Bipul Das[2], Ravi Soni[3], Mate Fejes[4], Sohan Ranjan[2],
Daniel Attila Szabo[4], Vikram Melapudi[2], K. S. Shriram[2], Utkarsh Agrawal[2],
Laszlo Rusko[4], Zita Herczeg[4], Barbara Darazs[4], Pal Tegzes[4], Lehel Ferenczi[4],
Rakesh Mullick[2], and Gopal Avinash[3]

[1] GE Healthcare, Hoevelaken, The Netherlands
tao.tan@ge.com
[2] GE Healthcare, Bangalore, India
[3] GE Healthcare, San Ramon, USA
[4] GE Healthcare, Budapest, Hungary

Abstract. The COVID-19 pandemic continues to spread and impact the well-being of the global population. The front-line imaging modalities computed tomography (CT) and X-ray play an important role for triaging COVID-19 patients. Considering the limited access to resources (both hardware and trained personnel) and decontamination, CT may not be ideal for triaging suspected subjects. Artificial intelligence (AI) assisted X-ray based applications for triaging and monitoring COVID-19 patients in a timely manner with the additional ability to delineate the disease region boundary are seen as a promising solution. Our proposed solution differs from existing solutions by industry and academic communities. We demonstrates a functional AI model to triage by inferencing using a single x-ray image, while the AI model is trained using both X-ray and CT data. We report on how such a multi-modal training improves the solution compared to X-ray only training. The multi-modal solution increases the AUC (area under the receiver operating characteristic curve) from 0.89 to 0.93 for the classification between COVID-19 and non-COVID-19 cases. It also positively impacts the Dice coefficient (0.59 to 0.62) for segmenting the COVID-19 pathology.

Keywords: Multi-modal · COVID-19 · Artificial intelligence

1 Introduction

Coronavirus disease 2019 (COVID-19) is extremely contagious and has become a pandemic [3,17]. It has spread inter-continentally in the first wave [15] and suspected to be currently entering the third wave [4] in various countries, having infected more than 30 million people and caused nearly 3M deaths till March 2021 [11]. From the imaging domain, chest CT may be considered as a primary tool for COVID-19 detection [2] and the sensitivity of chest CT can be greater than that

© Springer Nature Switzerland AG 2021
M. de Bruijne et al. (Eds.): MICCAI 2021, LNCS 12907, pp. 325–334, 2021.
https://doi.org/10.1007/978-3-030-87234-2_31

of RTPCR (98% vs 71%) [8], but the cost, time and risks of imaging including dose and the need for system decontamination can be prohibitive. Portable X-ray (XR) units are cost and time effective, and thus is the primary imaging modality used in diagnosis and management. Since it has limited capability to provide detailed 3D structure of anatomy or pathology of chest cavity, it is not regarded as an optimum tool for quantitative analysis [6]. Also due to the imaging apparatus and nature of the x-ray projection, it is challenging for the radiologists to identify relevant disease regions for accurate interpretation [10].

Although the performance of a host of artificial intelligence solutions approaches the level of radiologists for chest X-rays classification, very limited studies have verified the detection and segmentation of the disease regions compared to human annotation on X-rays. The contribution of our work builds on establishing a multi-modal protocol for our analysis and downstream classification of X-ray images. Barbosa Jr. et al. [5] have leveraged CT ground-truth for generating synthetic X-rays. The approach takes ratios of disease region to lung region using both synthetic X-rays (SXR) and regular X-ray images and force them to be equal to the ratio of disease volume to lung volume in CT. This may be challenging to be incorporated into clinical practice. Our approach also trains the AI model using data from CT, by generating synthetic X-ray images to complement the original X-ray images. At the inference end, only X-ray images are used, not requiring any CT data. In our differentiated approach, we have established a comprehensive synthetic X-ray generation schema to create a multitude of realistic synthetic X-ray to significantly augment X-ray images. Second, we use synthetic X-ray as a bridge to transfer the ground-truth from CT to the original X-ray geometry. Another contribution to highlight is that we show the serious domain shift issue when collecting images from multiple data sources for training and testing.

2   Method

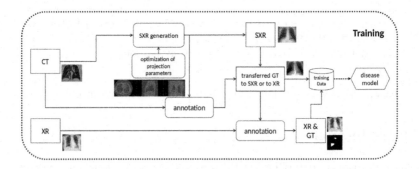

Fig. 1. The training scheme design

The key idea is to learn the disease patterns jointly using multi-modal (CT and XR) data but to inference the solution using only XR. In this study, we are not particularly focusing on deep-learning network design, but rather on improving data sufficiency and enhancing ground-truth quality to improve the visibility of abnormal tissue on a lower dimensional modality. In order to leverage available X-ray data, CT scans and paired X-ray and CT images, we have designed a pipeline as illustrated in Fig. 1. Disease regions in X-ray and CT are annotated independently by trained staff with varying levels of experience. Using CT, we generate multiple synthetic X-ray images (SXR) per CT volume and the corresponding projected 2D masks of the diseased region (see Sect. 2.1). For patients where X-rays and CTs (paired XR and CT data) are acquired within a small time-window (48 h given that COVID-19 is a fast evolving disease and pathology regions in the lung may change dramatically between scans), we automatically transfer the masks from the SXR to the corresponding XR. The transferred annotations on XR are further adjusted manually by trained staff to build our pristine mask ground-truth/annotations (see Sect. 2.2). For patients where the time-window of CT and X-ray is larger than 48 h, no ground-truth transfer is performed but are still added in the training pool. With all available XR and SXR together with corresponding disease masks, we train a deep-learning convolutional neural network for diagnosing and segmenting COVID-19 disease regions on X-ray.

2.1 Synthetic X-Ray Generation

The attenuation of X-ray beams in matter follows the BeerLambert law, stating that the decrease in the beam's intensity is proportional to the intensity itself (I) and the linear attenuation coefficient value (μ) of the material being traversed (1).

$$\frac{dI}{dx} = -\mu(x)I(x) \tag{1}$$

An inverse process can help create X-ray images from CT images by projecting the constructed 3D volume back into a 2D plane along virtual rays originating from a virtual source. Since the X-ray image is an integral/summation of attenuation values along the projection rays, established methods create projection images by converting the CT voxel values from Hounsfield unit (HU) to linear attenuation coefficient and simply summing the pixel values along the virtual rays (weighted by the travelled path-length by the ray withing each pixel). To achieve realistic pseudo X-ray we used a point source for the projection which accurately models the source of the X-ray. For the projection itself we used the ASTRA-toolbox [1], which allowed parameterization of the beam's angle (vertical and horizontal) as well as the distances between the virtual point source, the origin of the subject and the virtual detector). To generate realistic synthetic X-ray, the range of angles between the center of the CT image and the virtual X-ray point source fall between -10 and $10°$ around the longitudinal axis, and -5 and $5°$ around the mediolateral axis of the patient. We used 2 distance settings between the center of the CT volume and the virtual X-ray source: 1 m and 1.8 m, while the distance between the CT volume center and the virtual detector is set to 0.25 m. It should be noted that synthetic X-rays have less spatial resolutions compared to real X-rays as the original slice thickness of CT ranges from 3 mm to 6 mm.

During our work we also projected the segmentation ground truth masks for CT images to 2D format. The same transformations were applied to these binary masks as to the corresponding CT volumes.

2.2 Pristine Annotation Generation

For paired XR and CT of same patients, our aim was to transfer the pixel-wise ground-truth from CT to SXR and from SXR to XR. SXR serves as a bridge between CT and XR.

For each CT volume, we generate a number of SXRs and corresponding disease masks by varying the imaging parameters as mentioned in Sect. 2.1. To register SXR to XR, one challenge is that the fields of body view are different between the two. Further more, SXR is imaged usually with hands and arms up while for XR the hands are hanging downward in the normal position of the body. To alleviate this problem during image registration, instead of using original images, we create lung ROI image by applying lung segmentation on synthetic X-ray and X-ray. The segmentation solution is an in-house developed method using U-Net [16] trained from a different dataset outside the scope of this study. We use affine registration to maximize the mutal information between lung ROI from SXR and that from XR. Once registration error converges or registration step limit is reached, the obtained transformation is applied to the annotations corresponding to the SXR to generate transferred annotations. Although we have an approximate spatial match between SXR and XR, we cannot blindly use the transferred groundtruth for training. The disease could have rapidly changed between the first and the second exams. In this case the transferred groundtruth can only be used as a directional guidance which a human expert can use to annotate the 2D X-rays. The general instruction to annotators is to disallow the transferred regions where lesions are invisible on X-rays and to keep minimally visible regions even if they appear in the heart/diaphragm region.

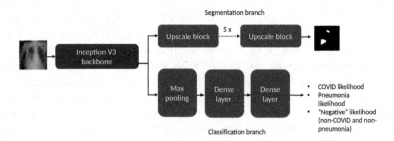

Fig. 2. The schematic overview of our proposed classification and segmentation deep-learning model

2.3   COVID-19 Modeling and Evaluation Strategy

For this extra-ordinary COVID-19 pandemic, AI systems have been investigated to identify anomalies in the lungs and assist in the detection, triage, quantification and stratification (e.g. mild, moderate and severe) of COVID-19 stages. To help radiologists do the triage, our deep-learning model takes frontal (anterial-posterior view or posterior-antierial view) X-ray as input and output two types of information (see Fig. 2): location of the disease regions and classification. For location, the network generates a segmentation mask to identify disease pixels on X-ray related to COVID-19 or other pneumonia. For classification, the fully connected neural network (FCN) outputs the likelihood of COVID-19 pneumonia, other pneumonia or a negative finding for each given input AP/PA X-ray image. The classification branch consists of one maxpooling player, a dense layer of 10 nodes and a dense layer of 3 nodes. The segmentation branch consists of 5 upscale bocks and each block consists of one residual block and one transpose convolution layer. Each upscale block doubles the dimensions of each feature channel. All images are resized to 1024×1024 pixels and normalized by Z-score method before being input to the model. For training, we use the combination of cross-entropy loss for the classification branch and Dice loss for the segmentation branch. The model is trained for 50 epochs with batch size 3 and Adam optimizer [14].

Our deep learning model was trained with and without the multi-modal data. We evaluated our approach in three separate aspects. First, AI predictions were compared with the subject classification labels (COVID-19, other pneumonia or negative). Second, the model segmentation of disease regions for the COVID-19 class was evaluated against direct X-ray pathology masks. Third, the model segmentation of disease regions was evaluated against pristine pixel-wise annotations.

3   Data, Groundtruth and Experiment Setting

In this study, we formed a large experimental dataset consisting of real X-ray images and synthetic X-ray images originating from CT volumes. The data was sourced from in-house/internal collections as well as publicly available data sources including Kaggle Pneumonia RSNA [13], Kaggle Chest Dataset [12], PadChest Dataset [7], IEEE github dataset [9], NIH dataset [18]. Any image databases with limiting non-commercial use licenses were excluded from our train/test cohorts. Representative paired and unpaired CT and XR datasets from US, Africa, and European population were also leveraged and sourced through our data partnerships. Outcomes were derived from information aggregated from radiological and laboratory reports. A summary of the database and selected categories used for our experiments is summarized in Table 1 where the in-house test dataset is a subset of the complete test dataset. As the trained model will be inferenced on real X-ray images, we remove synthetic X-rays from validation and testing cohorts. The reason that we have a dedicated in-house testing dataset is that this data source provide all COVID-19, other pneumonia and negative images while in the general testing datset, some data source does not contain all three-class images which may cause domain shift bias. Another important reason to form an in-house testset is to leverage the paired CT data for the evaluation, something lacking in the public data.

Table 1. Dataset breakdown for our experiments

Dataset	X-rays (# XMA, # PMA)	Synthetic X-rays (# SMA)
train COVID-19	974 (247, 77)	21487 (8322)
train pneumonia	10175 (6108, 17)	11312 (5380)
train negative	14859 (NA, NA)	12542 (NA)
val COVID-19	113 (37, 2)	NA
val pneumonia	531 (473, 8)	NA
val negative	3301 (NA, NA)	NA
test COVID-19	307 (68, 52)	NA
test pneumonia	1006 (345, 33)	NA
test negative	2271 (NA, NA)	NA
in-house test COVID-19	266 (68, 52)	NA
in-house test pneumonia	116 (45, 33)	NA
in-house test negative	37 (NA, NA)	NA

For the pixel-wise/segmentation groundtruth, we have four types of annotations. X-ray manual mask annotations (XMA) are made by annotators purely based on X-rays without any information from CTs. Synthetic mask annotations (SMA) are generated by the projection algorithms based on CT annotations for synthetic X-rays. Transferred CT mask annotations (TMA) are automatically generated by registration algorithms which transfer annotations from CT to X-ray using SXR as a bridge. Pristine mask annotations (PMA) which are adjusted annotations by annotators from TMA. The voxel/pixel-wise annotations from CT and X-ray except for RSNA dataset are performed by internal annotators. The RSNA pixel annotations were generated by fitting ellipses to the bounding boxes provided from the data source.

To show the benefits of multi-modal training for developing COVID-19 model, we have conducted training with 4 different training datasets summarized in Table 2 where T3 contains paired images used twice with XMA and PMA.

Table 2. Different training sets

Training set	Description
T1	X-ray images with XMA
T2	T1 + synthetic X-ray images with SMA
T3	T1 + X-ray images with PMA
T4	T1 + synthetic X-ray images with SMA + X-ray images with PMA

To evaluate the performance of our inferred classification on the test subjects, we used the area under the receiver operating characteristic curve (AUC) between different combinations of positive and negative classes including COVID-19 vs pneumonia, COVID-19 vs pneumonia+negative and COVID-19 vs negative for different usage scenarios.

Dice coefficient was used to evaluate the exactness of our pathology localization.

4   Results

4.1   Pristine Annotation Creation

With three different pixel-wise annotations on X-ray, we evaluated overlap between XMA, PMA and TMA.

After using TMA (transferred CT annotations), the consistency of human annotations to CT annotations is largely improved from 0.28 to 0.47 in terms of *Dice* coefficient. The *Dice* coefficient between XMA (X-ray manual mask annotations) and PMA (pristine mask annotations) are also moderate which means PMA is somewhere between X-ray direct annotations and CT transferred annotations. Figure 3 shows a number of examples with TMA, XMA and PMA. The X-ray annotations show large inconsistency with automated CT transferred annotations.

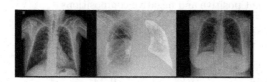

Fig. 3. Examples with TMA as red contour, XMA as blue regions and PMA as green regions (Color figure online)

4.2   Model Evaluation

The evaluation of the model is performed in terms of both classification and segmentation of COVID-19 disease regions. To test the effect of domain shift, we show the evaluation results on both all X-ray test dataset and in-house test dataset where COVID-19 other pneumonia and normal cases are all available.

Regarding classification, Table 4 shows AUC based on different combination for positive and negative classes. By adding synthetic X-rays, the AUC increases for COVID-19 vs pneumonia + negative increase from 0.89 to 0.93. The addition of adding pristine groundtruth does not further increase the AUC. The same increase is observed if AUC is computed using COVID-19 as positive and pneumonia as negative class. We form triaging of COVID-19 patients as three-classification problem to cope with different application situations. When inferencing, clinicians can adjust AI outputs depending on the different scenarios. For example, if clinicians want minimum other pneumonia and negative patients in the recall, the likelihood from COVID-19 class is a sufficient indicator. If clinicians prefer high sensitivity, and recalling other pneumonia patients is not a burden, the maximum among likelihood of COVID-19 class and pneumonia class from our solution can be an indicator.

We measure the Dice coefficients to estimate the segmentation accuracy. As PMA were only obtained for in-house testing images, we have measured Dice for that cohort and shown that in Table 3. Adding PMA can largely improve the *Dice* measures across different test settings, pushing it up to 0.70.

Table 3. Segmentation results: dice measures of different training schemes on different datasets

Train dataset vs test Dice	XR with XMA	in-house XR with XMA	in-house XR with PMA
T1	0.58	0.59	0.59
T2	0.57	0.56	0.58
T3	**0.60**	**0.62**	**0.70**
T4	0.57	0.58	0.62

Table 4. Classification results: AUC measures of different training schemes on different datasets with different positive and negative compositions

Training vs test AUC	AUC C vs P+N on XR	AUC C vs P on XR	AUC C vs P+N on in-house XR	AUC C vs P on in-house XR
T1	0.98	0.98	0.89	0.87
T2	**0.99**	0.98	**0.93**	**0.92**
T3	**0.99**	0.98	0.91	0.90
T4	**0.99**	**0.99**	**0.93**	**0.92**

5   Conclusion and Discussion

In this study, from multi-modal perspective, we have developed an artificial intelligence system which learns from a mix of high and low dimensional modality data but inferencing is only using low dimensional mono-modality X-ray for COVID-19 diagnosis and segmentation. The system classifies a given image into three categories: COVID-19, pneumonia and negative. We show that by learning from CT, the performance of the AI system seems to improve both classification and segmentation. Our AI system achieves a classification AUC of 0.99 and 0.93 between COVID-19 and pneumonia plus negative on full testing dataset and the subset in-house dataset, respectively. The Dice of 0.57 and 0.58 are obtained for COVID-19 disease regions on full testing dataset and the subset in-house dataset using X-ray direct annotations, respectively. The Dice measure is increased to 0.62 when pristine ground-truth transferred from CT is used for training and testing.

Learning from a second modality (CT) in our multi-modal approach has two main implications. One impact is to add synthetic X-ray and corresponding disease masks with different projection parameters to significantly augment training image pool and ensure data diversity. Another benefit related to additional pathology evidence from a higher-sensitivity modality when we transfer the CT annotations via synthetic X-ray, to original X-ray, thereby allowing manual fine-tuning of the annotations used for training. The first addition contributes mostly towards gain in the classification accuracy and the second addition contributes substantially to the gain in disease localization.

One important observation we want to point out is that domain shift issue can occur when developing an AI model using data collected from different data sources. Excellent performance is achieved in a general testing dataset but prominent performance drop is observed in the results of in-house testing set. On one hand, the COVID-19 and the pneumonia cases are confirmed with RTPCR tests, while RTPCR tests have lower sensitivity making the ground-truth less accurate. On the other hand the training dataset includes pubic datasets which do not contain all three-class images confounding the trained model to recognize both disease and perhaps the data source at the same time for the classification task. Although both normalization and strong augmentations are applied during training, when this model is tested on the general testing dataset, the recognized data-source may help to boost classification results due to the data-source bias.

References

1. van Aarle, W., et al.: Fast and flexible X-ray tomography using the ASTRA toolbox. Opt. Express **24**, 25129–25147 (2016). http://www.opticsexpress.org/abstract.cfm?URI=oe-24-22-25129
2. Ai, T., et al.: Correlation of chest CT and RT-PCR testing in coronavirus disease 2019 (COVID-19) in China: a report of 1014 cases. Radiology 200642 (2020)
3. Al-Awadhi, A.M., Al-Saifi, K., Al-Awadhi, A., Alhamadi, S.: Death and contagious infectious diseases: impact of the COVID-19 virus on stock market returns. J. Behav. Exp. Finance 100326 (2020)
4. Ali, I.: COVID-19: are we ready for the second wave? Disaster Med. Public Health Prep. **14**, 1–3 (2020)
5. Mortani Barbosa, Jr, E., et al.: Automated detection and quantification of COVID-19 airspace disease on chest radiographs: a novel approach achieving radiologist-level performance using a CNN trained on digital reconstructed radiographs (DRRS) from CT-based ground-truth (2020)
6. Blanchon, T., et al.: Baseline results of the depiscan study: a French randomized pilot trial of lung cancer screening comparing low dose CT scan (LDCT) and chest X-ray (CXR). Lung Cancer **58**(1), 50–58 (2007)
7. Bustos, A., Pertusa, A., Salinas, J.M., de la Iglesia-Vayá, M.: Padchest: a large chest x-ray image dataset with multi-label annotated reports. arXiv preprint arXiv:1901.07441 (2019)
8. Fang, Y., et al.: Sensitivity of chest CT for COVID-19: comparison to RT-PCR. Radiology 200432 (2020)
9. IEEE-Github: COVID-19 image data collection (2020). https://github.com/ieee8023/covid-chestxray-dataset/

10. Jacobi, A., Chung, M., Bernheim, A., Eber, C.: Portable chest x-ray in coronavirus disease-19 (COVID-19): a pictorial review. Clin. Imaging **64**, 35–42 (2020)
11. JHU (2020). https://coronavirus.jhu.edu
12. Kaggle: Kaggle chest (2017). https://www.kaggle.com/paultimothymooney/chest-xray-pneumonia. Accessed 30 June 2020
13. Kaggle: Kaggle RSNA pneumonia (2018). https://www.kaggle.com/c/rsna-pneumonia-detection-challenge. Accessed 30 June 2020
14. Kingma, D.P., Ba, J.: Adam: a method for stochastic optimization (2017)
15. Leung, K., Wu, J.T., Liu, D., Leung, G.M.: First-wave COVID-19 transmissibility and severity in china outside hubei after control measures, and second-wave scenario planning: a modelling impact assessment. The Lancet **395**(10233), 1382–1393 (2020)
16. Ronneberger, O., Fischer, P., Brox, T.: U-Net: convolutional networks for biomedical image segmentation. In: Navab, N., Hornegger, J., Wells, W.M., Frangi, A.F. (eds.) MICCAI 2015. LNCS, vol. 9351, pp. 234–241. Springer, Cham (2015). https://doi.org/10.1007/978-3-319-24574-4_28
17. Shaker, M.S., et al.: COVID-19: pandemic contingency planning for the allergy and immunology clinic. J. Allergy Clin. Immunol. Pract. **8**(5), 1477–1488 (2020)
18. Wang, X., Peng, Y., Lu, L., Lu, Z., Bagheri, M., Summers, R.M.: Chestx-ray8: hospital-scale chest x-ray database and benchmarks on weakly-supervised classification and localization of common thorax diseases. In: Proceedings of the IEEE Conference on Computer Vision and Pattern Recognition, pp. 2097–2106 (2017)

Determination of Error in 3D CT to 2D Fluoroscopy Image Registration for Endobronchial Guidance

Nicole Varble[1(✉)], Alvin Chen[1], Ayushi Sinha[1], Brian Lee[1],
Quirina de Ruiter[2], Bradford Wood[2], and Torre Bydlon[1]

[1] Philips Research North America, Cambridge, USA
nicole.varble@philips.com
[2] National Institutes of Health, Center for Interventional Oncology, Bethesda, USA

Abstract. Endobronchial biopsy is the preferred method for assessing lung lesions. However, navigation to pulmonary lesions and obtaining adequate tissue samples for diagnosis remains challenging. Utilizing information from high-resolution pre-procedural CT scans intra-procedurally could provide real-time guidance and confirmation during biopsy. An image registration algorithm was developed to automatically fuse thoracic 3D pre-operative CT images to 2D intra-procedural fluoroscopic images with a single 2D image or a limited C-arm sweep. A rigid intensity-based technique was applied and the CT image was iteratively transformed to minimize the sum of squared error between intraoperative fluoroscopy and closest forward projections. The registration errors were measured by computing the sum of squared difference and manually identified fiducial markers. In a swine model, error was minimized when using a CT with an inhalation breath hold (7.7 ± 4.4 mm) and when using an anterior-posterior positioning of the C-arm (3.7 ± 2.4 mm). Error increased marginally when the FOV was decreased (10.9 ± 5.9 mm) and was larger in peripheral (9.7 ± 5.7 mm) and distal (9.2 ± 3.2 mm) lung, compared to central (6.2 ± 4.5 mm) and proximal (7.6 ± 5.9 mm) lung. To determine the features that contribute most to registration, features were systematically masked and registration was performed. The largest error was seen when the spine was masked (52.5 ± 27.6 mm). When multiple images were used for registration, error converges ($<5\%$ change) when 50 images acquired in a $100°$ sweep were used. This work establishes a protocol and identifies sources of registration error for a reliable and automatic 2D-3D registration method that requires minimal changes to procedural workflow and equipment in the endobronchial suite.

Keywords: Endobronchial biopsy · Image registration · Augmented visualization · Device navigation

1 Introduction

Lung cancer is the leading cause of cancer-related death. The high mortality rate is due, in part, to the discovery of lung cancer in the late stages of the disease.

© Springer Nature Switzerland AG 2021
M. de Bruijne et al. (Eds.): MICCAI 2021, LNCS 12907, pp. 335–344, 2021.
https://doi.org/10.1007/978-3-030-87234-2_32

This had led to the emergence and implementation of low-dose CT screening trials for high-risk individuals [13]. These screening trials have led to decreased lung cancer mortality and, concurrently, has increased the discovery of smaller and more peripheral lung lesions [13, 14]. To adequately identify, diagnose, and screen these lesions a biopsy must be performed. As opposed to percutaneous biopsy, where an interventional radiologist obtains access to the lesion by puncturing the skin, an endobronchial biopsy is preferred due to the lower overall complication rates [7, 8, 19]. However, endobronchial biopsies suffer from low diagnostic yields [1], as navigation to peripheral lung lesions with traditional endobronchial tools and imaging lacks accuracy [2].

New approaches to endobronchial biopsies include the use of robotics systems [4, 17], electromagnetic navigation [5], virtual bronchoscopy [12], ultrasound [15] or augmented visualization and navigation using cone-beam computed tomography (CBCT) [16, 18]. Although the reported diagnostic yield for these advanced imaging and robotic technologies has shown promise in preliminary trials, access to high-cost imaging equipment can be challenging. Enhancement of currently used, portable, and accessible imaging platforms could provide an opportunity to increase diagnostic yield without drastic change to overall workflow.

To better visualize anatomical lung features or assist in intra-procedural navigation, one proposed solution is to utilize information from high resolution pre-procedural CT scans. Pre-procedural images can be analyzed and features such as airways and lesions can be automatically segmented and extracted. Then, intra-procedurally, 3D features can be registered and visualized on real-time intra-procedural fluoroscopic image. As the definition of a pathway to lesions is challenging on typical 2D fluoroscopic images and lesions are often fluoroscopically obscured, intra-procedural augmented fluoroscopy could greatly enhance the procedure. An augmented fluoroscopic image has the potential to provide real-time guidance to a lesion as well as confirmation when a biopsy tool is within the lesion. Since lesions that are targeted during endobronchial biopsies may be up to a few centimeters in diameter, successful augmented fluoroscopy can be achieved using registrations with sub-centimeter errors. Additionally, a successful image registration algorithm is not time consuming to the operator or require drastic changes to the procedural workflow to produce reliable registration results.

Prior registration algorithms has utilized gradient difference, pattern intensity, and distance metrics to register 3D CT images to 2D fluoroscopic tissue such as liver [6, 10, 11]. Intensity-based registration methods can be automated, are robust, and fast since they do not require the time-consuming step of segmentation. This technique does, however, require that the images to have equivalent spatial dimensions and suffers from additional errors when applied to thoracic imaging, where attenuation of features is less dramatic than other body imaging. As such, a comprehensive analysis of how an intensity-based 3D to 2D registration method in thoracic imaging is impacted by various imaging protocols and features has yet to be explored and warrants further exploration.

This study presents an automatic method to register 3D pre-operative CT scans to 2D intra-procedural fluoroscopic images, with the goal of providing a real-time augmented fluoroscopic image guidance. The algorithm accuracy was evaluated on retrospective images from an endobronchial in vivo swine study where respiratory and cardiac motion were present. The registration error was assessed based on the respiratory phase of the pre-operative scan, final positioning of the C-arm, image resolution and field of view (FOV). In addition, error at different anatomical locations in the lung, how different thoracic imaging features contribute to registration error, and the recommended number of 2D images required for a reliable registration were examined.

2   Methods

2.1   Image Registration Workflow

To perform registration, a rigid intensity-based method was utilized as outlined in Fig. 1 in MATLAB 2019a (mathworks.com). During preprocessing, 2D fluoroscopy and 3D CT images were resized, masked based on intensity or using a manually generated mask, the user estimated the approximate C-arm angle, and both image intensities were normalized. Masks were generated automatically based on image intensity to disregard collimated image boarders or were manually generated to mask out specific image features. Next, the user estimated the approximate C-arm angle to initialize the registration. For registration, the 3D image was transformed and a forward projection was generation. Particle swarm optimization (PSO) was used to iteratively (n = 50 with 60 particles) minimize the sum of the squared difference (SSD) between 2D and projected 3D image intensities. Masked areas were not included in the optimization and, therefore, SSD was not computed in these areas. A final local optimization was performed using a quasi-Newton method (Broyden-Fletcher-Goldfarb-Shanno or BFGS) with a cubic line search to minimize the objective function. When multiple images were used, information from each single image registration informed

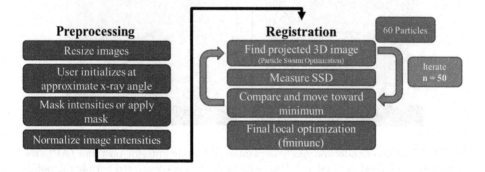

Fig. 1. 2D-3D image registration workflow

and led to the overall adjustment of the final registration. Since each image had an unspecified transformation error and position of the 2D images were known relative to each other, the final transformation matrix was based on the average transformations (specifically, rotation) of all images used in the simulation. Error was evaluated based on SSD or manual identification of the location of fiducial markers (coils), which were implanted endobronchially.

2.2   Subject and Image Acquisitions

Under a protocol approved by the Institutional Animal Care and Use Committee, images were collected from a 47 kg castrated male Yorkshire domestic swine (Oak Hill Genetics, Ewing, Illinois). The animal was sedated with intramuscular ketamine (25 mg/kg), midazolam (0.5 mg/kg), and glycopyrrolate (0.01 mg/kg) and was mechanically ventilated during the procedure. 3D images were taken under an inhalation breath hold (20 mmHg) or under passive exhalation. Four coils were endobronchially implanted in the lungs via a catheter endobronchially (2 per lung). The coils (fiducial markers) were stainless steel embolization coils (Cook, USA), size $0.035'' \times 5$ cm $\times$ 5 mm (wire diameter, deployed coil mass diameter, length). Coils were evenly spaced laterally (2 on right side, 2 on left side) and coronally (2 proximal, 2 distal) and is shown in Fig. 2. Due to the size of the coils ($<< 1\%$ of total volume), we do not expect that they play a role in the overall registration.

CT images were taken with Brilliance-16 CT scanner (Philips, Best, The Netherlands) and fluoroscopic and CBCT images were acquired on an Allura Xper FD20 angiographic fixed C-arm system (Philips, Best, The Netherlands). Both an inhalation and exhalation CT were acquired. In the same session, the swine was transferred to the C-arm table. CBCT images were acquired using a

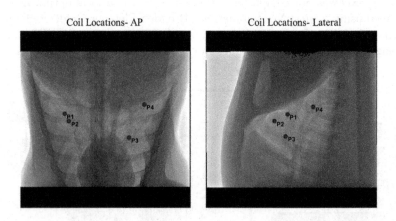

Fig. 2. Locations of endobronchially implanted coils in an anterior-posterior (AP) view (left) and in a lateral view (right). Points P1 and P4 were considered peripheral, points P2 and P3 were considered central. Points P1, P2 and P4 were considered distal and point P3 was considered proximal.

thoracic-imaging roll protocol (120 kV, 60 frames per second, 8-second acquisition time, 480 images). Two fluoroscopic imaging protocols were used, (1) a thorax single shot, which produced high-quality chest images, and (2) a lung fluoroscopic imaging run. The lung fluoroscopic images were acquired at 2 frames per second (fps) and a single frame was extracted. Breath holds were not performed when 2D images were acquired.

2.3   Registration Performance Analysis

The registration error was analyzed in various scenarios and the sources of inaccuracies were determined. Registration was first performed with a single 2D image. To determine the optimal pre-operative image protocol, the error was assessed when an inhalation or exhalation 3D image was used. To examine the impact on image quality, two different fluoroscopic imaging protocols were used (either a high quality thorax single shot, or a lower quality single frame lung fluoroscopic run). To determine the impact on the final C-arm position and the registration error, single image registration was assessed for 2D images that were acquired between the anterior-posterior to lateral positions. The accuracy to which the user has to estimate the final C-arm position and the error at different anatomical positions for single-image registration was analyzed. Next, various thoracic imaging features (heart, diaphragm, ribs, spine, and lung) were systematically and manually masked and registration was performed on the masked images to determine the features that contribute most to registration or caused the highest registration error.

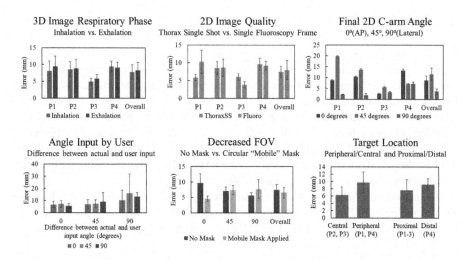

Fig. 3. 2D-3D registration results and determination of and impact of error from a single 2D image. P1–P4 are locations of the implanted coils as explained in Fig. 2.

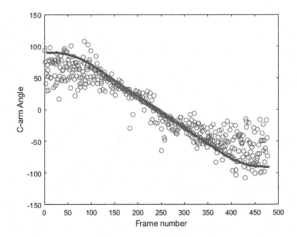

Fig. 4. The C-arm position (solid blue line), with registration results (orange circles) along a single axis. Error is lowest as the AP position (0°) and increases as the C-arm moves to more lateral positions (±90°).

Next, the impact of increasing number of 2D images was assessed by extracting 2D images from an inhalation CBCT scan. Error was assessed as increasing number of images were used over 4 different C-arm sweep angles (40°, 60°, 80°, and 100°). Error was measured using SSD and was considered to be converged when the change in SSD was <5%.

3   Results

Registration results when a single 2D image is used to register to the 3D preoperative scan are shown in Fig. 3. Registration with a single image takes less than 1 min. The registration error was lower when using an inhalation breath hold was used when taking the 3D image (error = 7.7 ± 4.4 mm) compared to when an exhalation breath hold was done (8.3 ± 4.7 mm). Error was lower when anterior-posterior positioning of the C-arm (3.7 ± 2.4 mm) was used rather than an oblique (11.5 ± 5.9 mm) or lateral (8.8 ± 4.1 mm) C-arm position was used. This is also illustrated in Fig. 4, which shows the 1D error from a single 2D image registration across an entire C-arm sweep.

From baseline images, error increased marginally (10.9 ± 5.9 mm) when the FOV was decreased to mimic a mobile C-arm FOV. Error was larger in peripheral (9.7±5.7 mm) and distal (9.2±3.2 mm) lung, compared to central (6.2±4.5 mm) and proximal (7.6 ± 5.9 mm) lung.

When manually masking features, masking any feature increased the error by a minimum of 21.3 ± 6.2 mm (Fig. 5). The largest error was seen when the spine was masked from images (52.5±27.6 mm), suggesting the highest relative importance. Further investigation into the resulting transformation matrix revealed that when the heart and diaphragm were masked, a large translation error occurred. When the ribs and spine were masked, a large rotation error occurred.

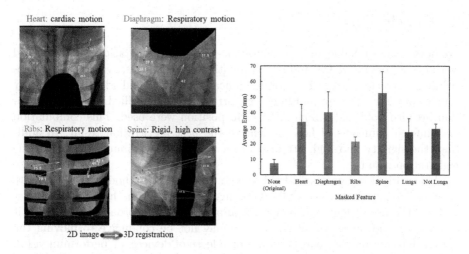

Fig. 5. The error was determined by assessing the change in coil location (left) from the 2D image (blue dot) to the registered 3D image (red dot). Quantified errors when different features were masked is shown on the right. (Color figure online)

When multiple images are used for registration, error converges when 50 to 80 images were used and when a sweep is performed over 100° (Fig. 6). However, error convergence was still achieved when a 40° sweep is used.

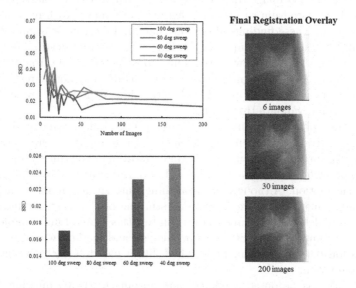

Fig. 6. SSD compared to the number of images used over a limited sweep from 40° to 100° (top left). Convergence (change in SSD <5%) was seen between 50 to 80 images. When a maximum number of images are used (bottom left) a 100° sweep has a minimum SSD (error). The final registration error for 6, 30, and 200 images is shown (right) with the 3D image in green and 2D in magenta, which both (green and magenta) become less visible as alignment improves. (Color figure online)

4   Conclusions

We developed a reliable and automatic 2D-3D image registration algorithm and workflow for operative CT and intra-operative X-ray images, evaluated the algorithm registration errors in various clinical scenarios, and identified primary sources of error. The optimal image registration was identified when an inhalation breadth hold and a final AP C-arm position were used. This combination may inform clinicians of how to achieve the best registration result. In addition, the rigidity and high attenuation of the spine contributes greatly to the registration result.

This work also suggests that this image registration method is robust despite decreased image quality and FOV. This suggests that this method may be robust without the use of high-end imaging equipment. During bronchoscopy, patient positioning and operation of equipment may not be conducive to moving the patient into an imaging bore, removal or release of devices, or performing a full 180° CBCT sweep. This work shows, if using a burst of multiple images, a sweep as small as 40° could result in adequate registration.

Once registration is achieved, this work also demonstrates that, with a static registration, a clinician could expect the largest error in the peripheral and distal regions of the lung, a region which is often difficult to navigate to without imaging assistance. These findings also suggest that the spine contributes the most to image registration. We hypothesize that the rigidity of the spine during respiration and the high radio-density make it the most relevant feature and necessary to capture in the 2D image. In contrast, the bronchial tree is relatively opaque, hard to differentiate from surrounding features, and moves with respiratory and cardiac motion. These findings also show that the lowest error occurs when the 2D image is taken at an oblique angle. This result was interesting and may be due to the unobstructed view of the spine (obstructed in the frontal plane by ribs/heart/lungs). It is recognized that a swine model may not be representative as human arms are typically positioned at the side of the body. Further studies on humans could validate this finding.

Limitations of this study includes the use of a preclinical swine model for data acquisition. Advantages of this model includes the ability to precisely control breath holds, for image acquisition beyond traditional treatment, and give the ability to plant fiducial markers while still performing analysis with respiratory and cardiac motion. Although the branching pattern and number of lobes in a swine are somewhat different, swine is still one of the most commonly used pulmonary models [3,9]. Future studies will test this method in available human data. Further improvement of the error analysis and method will also take into consideration diseased human lung, which likely has different compliance and X-ray attenuation.

This work has proposed a reliable and automatic 2D-3D imaging registration method that requires minimal changes to the procedural workflow and equipment. This technique could provide detailed anatomic views of airways and tumors in the lung or could provide navigation assistance toward small and peripheral tumors. With additional navigation and imaging assistance, interventional pulmonologist could increase the diagnostic success for endobronchial biopsies.

Acknowledgement. Data was collected at the National Institutes of Health Center for Interventional Oncology with Animal Care and Use Committee approval and under a Cooperative Research and Development Agreement. The authors would like to thank William Pritchard, John Karanian, Juan Esparza-Trujillo, Ivane Bakhutashvili, and Ming Li for assistance in collection of preclinical data and stimulating conversation that has enhanced the work.

References

1. Bhatt, K.M., et al.: Electromagnetic navigational bronchoscopy versus CT-guided percutaneous sampling of peripheral indeterminate pulmonary nodules: a cohort study. Radiology **286**(3), 1052–1061 (2018). https://doi.org/10.1148/radiol.2017170893

2. Chen, Y., Khemasuwan, D., Simoff, M.J.: Lung cancer screening: detected nodules, what next? Lung Cancer Manag. **5**(4), 173–184 (2016). https://doi.org/10.2217/lmt-2016-0008

3. Dondelinger, R.F., et al.: Relevant radiological anatomy of the pig as a training model in interventional radiology. Eur. Radiol. **8**(7), 1254–1273 (1998). https://doi.org/10.1007/s003300050545

4. Fielding, D.I.K., et al.: First human use of a new robotic-assisted fiber optic sensing navigation system for small peripheral pulmonary nodules. Respiration **98**(2), 142–150 (2019). https://doi.org/10.1159/000498951

5. Folch, E.E., Pritchett, M.A., Nead, M.A., et al.: Electromagnetic navigation bronchoscopy for peripheral pulmonary lesions: one-year results of the prospective, multicenter NAVIGATE study. J. Thorac. Oncol. **14**(3), 445–458 (2019). https://doi.org/10.1016/j.jtho.2018.11.013

6. Gilhuijs, K.G., van de Ven, P.J., van Herk, M.: Automatic three-dimensional inspection of patient setup in radiation therapy using portal images, simulator images, and computed tomography data. Med. Phys. **23**(3), 389–399 (1996). https://doi.org/10.1118/1.597801

7. Gould, M.K., et al.: Evaluation of individuals with pulmonary nodules: when is it lung cancer? Diagnosis and management of lung cancer, 3rd ed: American college of chest physicians evidence-based clinical practice guidelines. Chest **143**(5), 93–120 (2013). https://doi.org/10.1378/chest.12-2351

8. Heerink, W.J., de Bock, G.H., de Jonge, G.J., Groen, H.J.M., Vliegenthart, R., Oudkerk, M.: Complication rates of CT-guided transthoracic lung biopsy: meta-analysis. Eur. Radiol. **27**(1), 138–148 (2016). https://doi.org/10.1007/s00330-016-4357-8

9. Judge, E.P., Hughes, J.M.L., Egan, J.J., Maguire, M., Molloy, E.L., O'Dea, S.: Anatomy and bronchoscopy of the porcine lung. a model for translational respiratory medicine. Am. J. Respir. Cell Mol. Biol. **51**(3), 334–343 (2014). https://doi.org/10.1165/rcmb.2013-0453TR

10. Khamene, A., Bloch, P., Wein, W., Svatos, M., Sauer, F.: Automatic registration of portal images and volumetric CT for patient positioning in radiation therapy. Med. Image Anal. **10**(1), 96–112 (2006). https://doi.org/10.1016/j.media.2005.06.002

11. Lemieux, L., Jagoe, R., Fish, D.R., Kitchen, N.D., Thomas, D.G.: A patient-to-computed-tomography image registration method based on digitally reconstructed radiographs. Med. Phys. **21**(11), 1749–1760 (1994). https://doi.org/10.1118/1.597276

12. Memoli, J.S.W., Nietert, P.J., Silvestri, G.A.: Meta-analysis of guided bronchoscopy for the evaluation of the pulmonary nodule. Chest **142**(2), 385–393 (2012). https://doi.org/10.1378/chest.11-1764

13. National Lung Screening Trial Research Team: The National Lung Screening Trial: overview and study design. Radiology **258**(1), 243–253 (2011). https://doi.org/10.1148/radiol.10091808

14. National Lung Screening Trial Research Team: Reduced lung-cancer mortality with low-dose computed tomographic screening. N. Engl. J. Med. **365**(5), 395–409 (2011). https://doi.org/10.1056/NEJMoa1102873

15. Ost, D.E., Ernst, A., Lei, X., et al.: Diagnostic yield and complications of bronchoscopy for peripheral lung lesions: results of the AQuIRE registry. Am. J. Respir. Crit. Care **193**(1), 68–77 (2016). https://doi.org/10.1164/rccm.201507-1332OC

16. Pritchett, M.A., Schampaert, S., de Groot, J.A.H., Schirmer, C.C., van der Bom, I.: Cone-beam CT with augmented fluoroscopy combined with electromagnetic navigation bronchoscopy for biopsy of pulmonary nodules. J. Bronchology Interv. Pulmonol. **25**(4), 274–282 (2018). https://doi.org/10.1097/LBR.0000000000000536

17. Rojas-Solano, J.R., Ugalde-Gamboa, L., Machuzak, M.: Robotic bronchoscopy for diagnosis of suspected lung cancer: A feasibility study. J. Bronchol. Interv. Pulmonol. **25**(3), 168–175 (2018). https://doi.org/10.1097/LBR.0000000000000499

18. de Ruiter, Q.M.B., Karanian, J.W., Bakhutashvili, I., et al.: Endobronchial navigation guided by cone-beam CT-based augmented fluoroscopy without a bronchoscope: feasibility study in phantom and swine. J. Vasc. Interv. Radiol. **31**(12), 2122–2131 (2020). https://doi.org/10.1016/j.jvir.2020.04.036

19. Wiener, R.S., Schwartz, L.M., Woloshin, S., Welch, H.G.: Population-based risk for complications after transthoracic needle lung biopsy of a pulmonary nodule: an analysis of discharge records. Ann. Intern. Med. **155**(3), 137–144 (2011). https://doi.org/10.7326/0003-4819-155-3-201108020-00003

Chest Radiograph Disentanglement for COVID-19 Outcome Prediction

Lei Zhou[1]($\boxtimes$), Joseph Bae[2], Huidong Liu[1], Gagandeep Singh[3], Jeremy Green[3], Dimitris Samaras[1], and Prateek Prasanna[2]

[1] Department of Computer Science, Stony Brook University, Stony Brook, NY, USA
lezzhou@cs.stonybrook.edu
[2] Department of Biomedical Informatics, Stony Brook University,
Stony Brook, NY, USA
[3] Department of Radiology, Newark Beth Israel Medical Center, Newark, NJ, USA

Abstract. Chest radiographs (CXRs) are often the primary front-line diagnostic imaging modality. Pulmonary diseases manifest as characteristic changes in lung tissue texture rather than anatomical structure. Hence, we expect that studying changes in only lung tissue texture without the influence of possible structure variations would be advantageous for downstream prognostic and predictive modeling tasks. In this paper, we propose a generative framework, Lung Swapping Autoencoder (LSAE), that learns a factorized representation of a CXR to *disentangle* the tissue texture representation from the anatomic structure representation. Upon learning the disentanglement, we leverage LSAE in two applications. 1) After adapting the texture encoder in LSAE to a thoracic disease classification task on the large-scale ChestX-ray14 database (N = 112,120), we achieve a competitive result (mAUC: 79.0%) with unsupervised pre-training. Moreover, when compared with Inception v3 on our multi-institutional COVID-19 dataset, COVOC (N = 340), for a COVID-19 outcome prediction task (estimating need for ventilation), the texture encoder achieves 13% less error with a 77% smaller model size, further demonstrating the efficacy of texture representation for lung diseases. 2) We leverage the LSAE for data augmentation by generating hybrid lung images with textures and labels from the COVOC training data and lung structures from ChestX-ray14. This further improves ventilation outcome prediction on COVOC.

The code is available here: https://github.com/cvlab-stonybrook/LSAE.

Research reported in this publication was enabled by the Renaissance School of Medicine at Stony Brook University's "COVID-19 Data Commons and Analytic Environment", a data quality initiative instituted by the Office of the Dean, and supported by the Department of Biomedical Informatics. Research was supported by SBU OVPR and IEDM seed grant 2019 (P.P, D.S), NIGMS T32GM008444 (J.B). D.S was partially supported by the Partner University Fund, the SUNY2020 Infrastructure Transportation Security Center, and a gift from Adobe.

Electronic supplementary material The online version of this chapter (https://doi.org/10.1007/978-3-030-87234-2_33) contains supplementary material, which is available to authorized users.

© Springer Nature Switzerland AG 2021
M. de Bruijne et al. (Eds.): MICCAI 2021, LNCS 12907, pp. 345–355, 2021.
https://doi.org/10.1007/978-3-030-87234-2_33

Keywords: Chest radiographs · Disentanglement · Lung swapping autoencoder · Unsupervised learning

1   Introduction

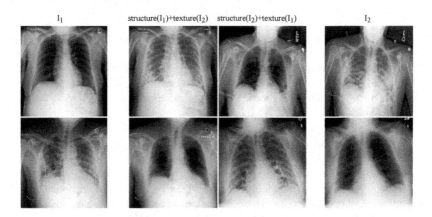

I_1 structure(I_1)+texture(I_2)   structure(I_2)+texture(I_1) I_2

Fig. 1. Lung Swapping Result. Two examples of lung swapping between images in column I_1 and images in column I_2. The Lung Swapping Autoencoder (LSAE) is able to successfully transfer target lung textures without affecting the lung shape. The swapping results are shown in the second and the third columns.

Chest radiographs (CXRs) are the primary diagnostic tool for COVID-19 pneumonia because they are widely available, and have lower risk of cross infection compared to Computed Tomography (CT) scans. Despite these advantages, CXRs are less sensitive to subtle disease changes compared to CT scans [12].

In COVID-19 CXRs, we observe that lung tissue texture may change drastically during hospitalization due to varying infiltrate levels. However, chest anatomy remains mostly unchanged. Therefore, we hypothesize that *disease information is more related to lung tissue texture rather than the anatomical structure of the lung*. To be concise, we use the terms *texture* and *structure* in the rest of the article. Our hypothesis is also supported by recent findings that COVID-19 on CXR is observed as opacities within lung regions, and their extent and location is associated with disease severity and progression [22,25].

Previous medical image analysis in COVID-19 has mostly focused on diagnosis [7,13]. CT-based models are better at predicting COVID-19 outcomes, compared to CXR-based models [1,11,13]. This is largely due to the lack of COVID-19 CXR datasets with relevant endpoints and the limited information that CXRs contain relative to CT scans. Several data augmentation techniques have been proposed for CT scans in the COVID-19 setting [11], but CXR approaches have continued to rely on publicly sourced datasets [2,13]. These datasets tend to be homogeneous, and often lack disease outcome labels (hospitalization, mechanical ventilation requirement, etc.) [2,13]. Generative Adversarial Networks (GANs) [3] and

Autoencoders [10] have been widely used for data augmentation, including in medical image applications [6,19,20]. However, standard GAN-based methods [8,26] are not suitable for CXR generation due to the lack of explicit structure supervision, which can lead to generating distorted shapes.

In this paper, we propose the Lung Swapping AutoEncoder (LSAE), which learns a factorized representation of a CXR to *disentangle* the texture factor from the structure factor. LSAE shares the same core idea as the recently-proposed Swapping AutoEncoder (SAE) [15], namely that a successful disentanglement model should be able to generate a realistic hybrid image that merges the structure of one image and the texture of another. To achieve this, images are encoded as a combination of two latent codes representing structure and texture respectively. The SAE is trained to generate realistic images from the swapped codes of arbitrary image pairs. Moreover, the SAE is also forced to synthesize the target texture supervised by patches sampled from the target image. However, this vanilla SAE does not work well for CXR disentanglement. First, by sampling texture patches from the whole target image, irrelevant out-of-lung textures diminish the effect of the in-lung texture transfer. More importantly, because texture supervision is derived from image patches, irrelevant structure clues may leak into the hybrid image, resulting in undesired lung shape distortion and interference with successful disentanglement.

We address these problems by: 1) Sampling patches from the lung area instead of the whole image for texture supervision (as the infiltrates of interest are located within lung zones), and 2) Adding a patch contrastive loss to explicitly force out-of-lung local patches in the hybrid image to mimic the corresponding patch in the structure image (to prevent structure information in texture patches from leaking into the hybrid image). LSAE, trained on a large public CXR dataset, ChestX-ray14, can generate realistic and plausible hybrid CXRs with one patient's lung structure and another patient's disease texture (see Fig. 1). We further provide quantitative results for disentanglement in the experiment section.

The trained disentanglement model is used in two applications: 1) If textures represent disease infiltrates, the texture encoder in LSAE, Enc^t, should be discriminative in downstream CXR semantic tasks. To prove this, we finetune Enc^t in LSAE on both the large ChestX-ray14 database (N = 112,120) and our multi-institutional COVID-19 outcome prediction dataset, COVOC (N = 340). For thoracic disease classification on ChestX-ray14, Enc^t achieves competitive results without supervised pre-training. On COVOC, compared with a strong baseline Inception v3 [21], Enc^t reduces error by 13% with a much smaller model size. Results show that Enc^t learns effective and transferable representations of lung diseases. 2) To exploit the generative potential of LSAE, we generate hybrid images with textures and labels from COVOC training data and lung structures from ChestX-ray14. Augmenting with these hybrid images further improves ventilation prediction on COVOC.

In summary: 1) LSAE is the first approach to disentangle chest CXRs into structure and texture representations. LSAE succeeds by explicit in-lung texture supervision and out-of-lung distortion suppression, 2) We achieve superior performance for COVID-19 outcome prediction with an efficient model, and 3) We propose a hybrid image augmentation technique, to further improve ventilation prediction.

2   Methodology

2.1   Swapping AutoEncoder (SAE)

The recent Swapping AutoEncoder (SAE) [15] consists of an encoder Enc, and a decoder Dec, where Enc is composed of a structure branch Enc^s and a texture branch Enc^t to encode the input into structure and texture codes z^s and z^t.

Latent Code Swapping. SAE aims to generate realistic images from swapped latent codes of arbitrary image pairs. Two sampled images I_1 and I_2 are first encoded as (z_1^s, z_1^t) and (z_2^s, z_2^t). Then, the latent codes are swapped to get a hybrid code (z_1^s, z_2^t). Finally, the hybrid code is decoded to yield a hybrid image I_{hybrid} which is expected to maintain the structure of I_1 but present the texture of I_2. We use G to denote the composite generation process as $G(I_1, I_2) \stackrel{\text{def}}{=} Dec(Enc^s(I_1), Enc^t(I_2))$. The objective has reconstruction and GAN loss, i.e., $\mathcal{L}_{\text{recon}} = \mathbb{E}_{I_1 \sim \mathbf{X}} \| G(I_1, I_1) - I_1 \|_1$, and $\mathcal{L}_{\text{G}} = \mathbb{E}_{I_1, I_2 \sim \mathbf{X}} - \log\big(D\big(G(I_1, I_1)\big)\big) - \log\big(D\big(G(I_1, I_2)\big)\big)$, where $\mathbf{X}$ denotes the image training set, and D is a discriminator [3]. Note that the complete GAN loss also includes a discriminator part. To be concise, we only show the generator part in the paper.

Texture Supervision. To ensure I_{hybrid} texture matches that of I_2, SAE samples patches from I_2 to supervise the I_{hybrid} texture. Patch discriminator D_{patch} is trained to distinguish patches in I_{hybrid} from patches in I_2. SAE is trained adversarially to generate a I_{hybrid} whose patches can confuse D_{patch} by mimicking the texture of I_2. The texture loss is formulated as $\mathcal{L}_{\text{tex}} = \mathbb{E}_{\substack{\tau_1, \tau_2 \sim \mathcal{T} \\ I_1, I_2 \sim \mathbf{X}}} \big[- \log\big(D_{\text{patch}}\big(\tau_1(I_2), \tau_2(G(I_1, I_2))\big)\big)\big]$ where $\mathcal{T}$ is the distribution of multi-scale random cropping, and τ_1 and τ_2 are two random operations sampled from $\mathcal{T}$.

2.2   Lung Swapping AutoEncoder (LSAE)

Using SAE to generate a hybrid CXR from an image pair does not work as desired. First, the disease level in I_{hybrid} is usually diminished when compared with I_2. We hypothesize that the out-of-lung irrelevant texture patterns may hinder target texture synthesis in I_{hybrid}. Results in Table 1 (left) support our analysis. Second, since there is no structure supervision in SAE, I_{hybrid} shows undesired lung shape distortion towards I_2. This leads us to design two new features, in-lung texture supervision and out-of-lung structural distortion suppression. We name the new model as *Lung Swapping AutoEncoder* (LSAE) (Fig. 2).

In-lung Texture Supervision. As infiltrates on CXRs are observed inside lung regions, we should only sample texture patches from within the lung zone. Guided by a lung segmentation mask, we rewrite the texture supervision loss as,

$$\mathcal{L}_{\text{inTex}} = \mathbb{E}_{\substack{\tau_1, \tau_2 \sim \mathcal{T}_{\text{LungMask}}^{\text{in}} \\ I_1, I_2 \sim \mathbf{X}}} \left[- \log\left(D_{\text{patch}}\left(\tau_1(I_2), \tau_2\big(G(I_1, I_2)\big)\right)\right)\right] \qquad (1)$$

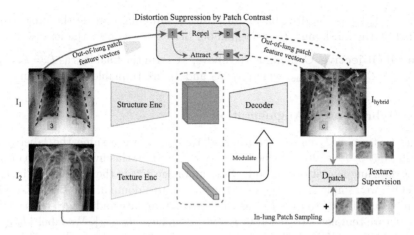

Fig. 2. Lung Swapping AutoEncoder (LSAE) consists of a structure encoder, a texture encoder, and a decoder. After inputing images (I_1, I_2), the LSAE generates a hyrbid image with factorized representations from I_1 and I_2 respectively. To ensure the texture in I_{hybrid} matches I_2, we adversarially train a patch discriminator D_{patch} to supervise the texture synthesis within lungs. To ensure I_{hybrid} maintains the structure of I_1, we apply a patch contrastive loss outside the lungs to minimize structural distortion.

where $\mathcal{T}$ in $\mathcal{L}_{\text{tex}}$ is replaced by $\mathcal{T}_{\text{LungMask}}^{\text{in}}$, indicating that we only sample patches from in-lung regions to train the generator G.

Out-of-Lung Structural Distortion Suppression. To suppress structural distortion, we introduce a patch contrastive loss [14] in LSAE. The patch contrastive loss, based on Noise Contrastive Estimation [4], preserves image content in image-to-image translation tasks. We first encode I_{hybrid} with structure encoder Enc^s. Then, we randomly sample a feature vector q_i from the h_{th} layer's output $Enc_h^s(I_{\text{hybrid}})$, and a bag of feature vectors $\{p^+, p_1^-, p_2^-, \cdots, p_{N-1}^-\}$ from $Enc_h^s(I_1)$, where p^+ is from the corresponding position with q_i, and $\{p^-\}$ from other positions. The loss objective is to force q_i from I_{hybrid} to attract to the corresponding feature vector p^+ but repel the others $\{p^-\}_{N-1}$ from I_1. Therefore, we write it as a cross-entropy loss by regarding the process as an N-way classification problem where q_i is the query and p^+ is the target:

$$l_{\text{NCE}}(q_i, \{p\}_N) = -\log\left(\frac{\exp(q_i \cdot p^+/\alpha)}{\exp(q_i \cdot p^+/\alpha) + \sum_{j=1}^{N-1}\exp(q_i \cdot p_j^-/\alpha)}\right) \quad (2)$$

where α is the temperature, and $\{p\}_N = \{p^+\} \cup \{p^-\}_{N-1}$. To prevent l_{NCE} from affecting the in-lung texture transfer, we only apply it outside the lung region. Thus, the structural distortion suppression loss is expressed as

$$\mathcal{L}_{\text{sup}} = \mathop{\mathbb{E}}_{\substack{\{\tau\}_N \sim \mathcal{T}_{\text{LungMask}}^{\text{out}} \\ I_1, I_2 \in \mathbf{X}}} \sum_{h}^{H} \sum_{i}^{N} l_{\text{NCE}}\Big(\tau_i\big(Enc_h^s(I_{\text{hybrid}})\big), \{\tau\big(Enc_h^s(I_1)\big)\}_N\Big) \quad (3)$$

where $\mathcal{T}_{\text{LungMask}}^{\text{out}}$ is the distribution of position sampling outside the lung region guided by the mask, and H is the number of layers we apply the loss to.

Overall Objective. Combining the above, we term the final loss as $\mathcal{L} = \mathcal{L}_{\text{recon}} + \lambda_1 \mathcal{L}_{\text{G}} + \lambda_2 \mathcal{L}_{\text{inTex}} + \lambda_3 \mathcal{L}_{\text{sup}}$, where λ_1, λ_2 and λ_3 are tunnable hyper-parameters.

2.3 Hybrid Image Augmentation

If the texture in I_2 can be transferred to I_{hybrid}, we assume the label (e.g., ventilation) of I_2 is also attached to I_{hybrid}. Based on this hypothesis, we design a new data augmentation method: Given an image I^{dst} in the target domain, e.g., COVOC, we sample K images $\{I^{\text{src}}\}$ from a source domain, e.g., ChestX-ray14. Then, we use LSAE to take I^{dst} as the texture template and images from $\{I^{\text{src}}\}$ as structure templates to generate K hybrid images $\{I_{\text{hybrid}}\}$. We label $\{I_{\text{hybrid}}\}$ with the label of I^{dst}. Following this protocol, we can enlarge the training set in target domain K times. Experimental results show the augmentation method can improve performance further on the COVID-19 ventilation prediction task.

3 Experimental Design and Results

3.1 Dataset Description

ChestX-ray14 [24] is a large-scale CXR database consisting of 112,120 frontal-view CXRs from 32,717 patients. We report all the results based on the official split which consists of training ($\sim$70%), validation ($\sim$10%), and testing ($\sim$20%) sets. Images from the same patient will only appear in one of the sets.

COVID-19 Outcome (COVOC) is a COVID-19 CXR dataset curated from two institutions. It consists of 340 CXRs from 327 COVID-19 patients acquired upon disease presentation [9]. Each CXR in COVOC is labeled based upon whether the patient required mechanical ventilation (henceforth ventilation) or not. We separate COVOC randomly into 3 splits. Each split has 250 samples for training, 30 for validation, and 60 for testing.

3.2 Implementation

For LSAE, both decoder and discriminator architectures follow StyleGAN2. Encoders are built with residual blocks [5]. The texture encoder outputs a flattened vector while the structure encoder's output preserves spatial dimension. The input image size is of 256×256. Patch sizes sampled for texture supervision vary from 16×16 to 64×64. We set $\lambda_1 = 0.5$, $\lambda_2 = 1$, and $\lambda_3 = 1$. The temperature α is set to 0.07. LSAE is optimized by Adam with learning rate 1e-3. Our code is based on PyTorch 1.7. We preprocess all CXRs with histogram equalization.

3.3   Hybrid CXR Generation

We start with a larger dataset, i.e., ChestX-ray14, to pre-train LSAE by learning to generate hybrid CXRs from swapped latent code.

Experimental Settings. We train LSAE on the training set of ChestX-ray14 with a batch of 16 images for 150K iterations. To evaluate performance, we define a lung swapping task by creating two sets of 9,000 images, $\{I^+\}$ and $\{I^-\}$, from the test set of ChestX-ray14. $\{I^+\}$ are sampled from the images diagnosed with at least one of the diseases: infiltration, pneumonia, and fibrosis, which are tightly related with COVID-19 [18]. In contrast, $\{I^-\}$ are sampled from healthy lungs. We generate hybrid image sets by mixing texture and structure from $\{I^+\}$ and $\{I^-\}$ in both directions. We measure the distance of hybrid image set between both the target texture image set and the structure image set.

Evaluation Protocol. First, we propose a new metric, Masked SIFID, to measure the disease level distance between I_{hybrid} and I_2. Masked SIFID is based on the SIFID [17] which calculates the FID distance between two images in the Inception v3 feature space. To customize it to disease level distance, we only consider features within the lung region. Additionally, we use the ChestX-ray14 pre-trained Inception v3 (mAUC: 79.56%) to infer features. Second, to quantify structural distortion, we use lung segmentation metrics as surrogates. Given a lung segmentation model Seg, we can compute segmentation metrics by treating $Seg(G(I_1, I_2))$ as the query and $Seg(I_1)$ as the ground truth. Third, we solicit feedback from 5 radiologists through a 4-question survey. Please refer to the supplementary material for details. In the first two questions regarding image quality, radiologists misconstrue 56% of the generated images as real, and 74% of the hybrid image patches as real. The third question is to verify the correlation between Masked SIFID and disease level distance. When picking which of two query images is closer to the reference, 78.67% of the radiologists' answers match with Masked SIFID. The fourth question is to ascertain whether our method can transfer disease correctly. When the radiologists were shown the original and the hybrid image with the same texture, 60% of the images passed the test.

Results. To get reference values, we first report initial results by directly comparing $\{I^-\}$ with $\{I^+\}$. In Table 1 (left), LSAE achieves lower Masked SIFID when compared with SAE, which demonstrates that our design of in-lung texture supervision works as expected. LSAE also outperforms SAE by a large margin in the segmentation metrics, and achieves over 90% in all segmentation metrics, which proves out-of-lung patch contrastive loss can effectively suppress structural distortion. Please refer to the supplementary material for details.

3.4   Semantic Prediction in CXRs by Texture Encoder

Based on hypothesis that pulmonary diseases are tightly related to CXR texture, the texture encoder in a well-trained LSAE should be discriminative on CXR semantic tasks. To verify this hypothesis, we evaluate texture encoder, Enc^t, on both lung disease classification and COVID-19 outcome prediction tasks.

Table 1. Left. On the hybrid image generation task, LSAE surpasses SAE in both texture synthesis and structure maintenance. We report the average of two directions' texture transfer. **Right.** For the 14 pulmonary diseases classification task on ChestX-ray14, the texture encoder in LSAE achieves competitive results with a smaller model size. *Note that the data split in [16] is not released. We reimplemented CheXNet on the official split. The Inception v3 model is also self-implemented.

Method	Masked SIFID ↓	mIoU ↑	Pixel Acc ↑	Dice ↑	Pre-train	Method	Params	mAUC ↑
Init	0.0335	0.60	0.82	0.72		CXR14-R50[24]	23M	0.745
SAE	0.0257	0.76	0.91	0.85		ChestNet [23]	60M	0.781
LSAE	**0.0245**	**0.91**	**0.97**	**0.95**	Sup	CheXNet* [16]	7M	0.789
						Inception v3	22M	**0.796**
					Unsup	Enc^t	**5M**	0.790

Table 2. COVOC Outcome Prediction. We evaluate models with Balanced Error Rate (BER) and average AUC (mAUC). The texture encoder surpasses Inception v3 by a large margin. It also outperforms the texture encoder in a baseline model SAE, which demonstrates that better disentanglement does lead to better discrimination. We report mean and std of 5 random runs.

BER(%)↓	Inception v3	Enc^t in SAE	Enc^t in LSAE
split 1	20.25 ± 1.46	20.25 ± 1.63	19.00 ± 1.84
split 2	19.25 ± 3.67	20.50 ± 1.12	17.75 ± 1.66
split 3	17.75 ± 3.48	14.00 ± 1.85	12.75 ± 2.15
Avg	19.08	18.25	16.50

mAUC(%)↑	Inception v3	Enc^t in SAE	Enc^t in LSAE
split 1	85.45 ± 1.89	89.03 ± 2.16	89.17 ± 0.68
split 2	86.02 ± 1.27	85.63 ± 1.77	87.07 ± 1.91
split 3	89.12 ± 1.38	92.60 ± 1.25	95.00 ± 0.29
Avg	86.86	89.09	90.41

Disease Classification on ChestX-ray14. We finetune Enc^t on the training set of ChestX-ray14 with 14 disease labels, and report mean AUC in Table 1 (right). We achieve competitive results, despite having a smaller model size.

Outcome Prediction on COVOC. Considering the possible domain discrepancy between ChestX-ray14 and COVOC, we first adapt LSAE to the new domain by training to generate hybrid images on COVOC for 10K iterations. Then, we evaluate the texture encoder Enc^t on the outcome prediction task by further finetuning. As COVOC is imbalanced, we report the ventilation prediction Balanced Error Rate (BER) together with mAUC in Table 2. Compared with Inception v3, Enc^t reduces BER by 13.5%, and improves mAUC by 4.1%. When comparing with the texture encoder in baseline model SAE, LSAE also performs better. It demonstrates that better disentanglement leads to better discrimination. We also report the prediction of mortality in Table 2 in SM.

3.5 Data Augmentation with Hybrid Images

As described in Sect. 2.3, we generate hybrid images to augment the training data in COVOC. To control the training budget, we set $K = 2$, i.e., the training data of COVOC is augmented 2 times. To avoid introducing irrelevant diseases, we only sample structure images from the healthy lungs in ChestX-ray14. With the same experimental setup with Table 2, the augmentation method can reduce error rate further from 16.50% to 15.67%, and improve mAUC from 90.41% to 92.04% on the ventilation prediction task. Moreover, we implement Mixup independently for training on COVOC which achieves 16.41% BER/90.82% mAUC. Our method still shows superior performance.

4   Conclusion

We propose LSAE to disentangle texture from structure in CXR images, enabling analysis of disease-associated textural changes in COVID-19 and other pulmonary diseases. We also create a data augmentation technique which synthesizes images with structural and textural information from two CXRs. This technique can be used to augment data for machine learning applications. Our resulting predictive model for mechanical ventilation in COVID-19 patients outperforms conventional methods and may have clinical significance in enabling improved decision making. We will apply our texture disentanglement and data augmentation methods to study other pulmonary diseases in the future.

Acknowledgement. We thank Amit Gupta, Nicole Sakla, and Rishabh Gattu for their expert evaluation of the generated radiographs.

References

1. Bae, J., et al.: Predicting mechanical ventilation requirement and mortality in COVID-19 using radiomics and deep learning on chest radiographs: a multi-institutional study. arXiv preprint arXiv:2007.08028 (2020)
2. Cohen, J.P., Morrison, P., Dao, L., Roth, K., Duong, T.Q., Ghassemi, M.: COVID-19 image data collection: Prospective predictions are the future. arXiv:2006.11988 (2020)
3. Goodfellow, I., et al.: Generative adversarial nets. In: Advances in Neural Information Processing Systems (2014)
4. Gutmann, M., Hyvärinen, A.: Noise-contrastive estimation: a new estimation principle for unnormalized statistical models. In: Proceedings of the Thirteenth International Conference on Artificial Intelligence and Statistics, pp. 297–304. JMLR Workshop and Conference Proceedings (2010)
5. He, K., Zhang, X., Ren, S., Sun, J.: Deep residual learning for image recognition. In: Proceedings of the IEEE Conference on Computer Vision and Pattern Recognition, pp. 770–778 (2016)

6. Hou, L., Samaras, D., Kurc, T.M., Gao, Y., Davis, J.E., Saltz, J.H.: Patch-based convolutional neural network for whole slide tissue image classification. In: Proceedings of the IEEE Conference on Computer Vision and Pattern Recognition, pp. 2424–2433 (2016)

7. Hu, Q., Drukker, K., Giger, M.L.: Role of standard and soft tissue chest radiography images in COVID-19 diagnosis using deep learning. In: Medical Imaging 2021: Computer-Aided Diagnosis, vol. 11597, p. 1159704. International Society for Optics and Photonics, February 2021

8. Isola, P., Zhu, J.Y., Zhou, T., Efros, A.A.: Image-to-image translation with conditional adversarial networks. In: Proceedings of the IEEE Conference on Computer Vision and Pattern Recognition (CVPR), July 2017

9. Konwer, A., et al.: Predicting COVID-19 lung infiltrate progression on chest radiographs using spatio-temporal LSTM based encoder-decoder network. In: Medical Imaging with Deep Learning (2021)

10. Li, Y., Liu, S., Yang, J., Yang, M.H.: Generative face completion. In: Proceedings of the IEEE Conference on Computer Vision and Pattern Recognition, pp. 3911–3919 (2017)

11. Li, Z., et al.: A novel multiple instance learning framework for COVID-19 severity assessment via data augmentation and self-supervised learning. arXiv:2102.03837 [cs, eess], February 2021

12. Litmanovich, D.E., Chung, M., Kirkbride, R.R., Kicska, G., Kanne, J.P.: Review of chest radiograph findings of COVID-19 pneumonia and suggested reporting language. J. Thorac. Imaging 35(6), 354–360 (2020)

13. López-Cabrera, J.D., Orozco-Morales, R., Portal-Diaz, J.A., Lovelle-Enríquez, O., Pérez-Díaz, M.: Current limitations to identify COVID-19 using artificial intelligence with chest X-ray imaging. Heal. Technol. 11(2), 411–424 (2021). https://doi.org/10.1007/s12553-021-00520-2

14. Park, T., Efros, A.A., Zhang, R., Zhu, J.-Y.: Contrastive learning for unpaired image-to-image translation. In: Vedaldi, A., Bischof, H., Brox, T., Frahm, J.-M. (eds.) ECCV 2020. LNCS, vol. 12354, pp. 319–345. Springer, Cham (2020). https://doi.org/10.1007/978-3-030-58545-7_19

15. Park, T., Zhu, J.Y., Wang, O., Lu, J., Shechtman, E., Efros, A.A., Zhang, R.: Swapping autoencoder for deep image manipulation. arXiv preprint arXiv:2007.00653 (2020)

16. Rajpurkar, P., et al.: Chexnet: radiologist-level pneumonia detection on chest x-rays with deep learning. arXiv preprint arXiv:1711.05225 (2017)

17. Rott Shaham, T., Dekel, T., Michaeli, T.: Singan: learning a generative model from a single natural image. In: IEEE International Conference on Computer Vision (ICCV) (2019)

18. Salehi, S., Abedi, A., Balakrishnan, S., Gholamrezanezhad, A.: Coronavirus disease 2019 (COVID-19): a systematic review of imaging findings in 919 patients. Am. J. Roentgenol. 215(1), 87–93 (2020)

19. Sandfort, V., Yan, K., Pickhardt, P.J., Summers, R.M.: Data augmentation using generative adversarial networks (CycleGAN) to improve generalizability in CT segmentation tasks. Sci. Rep. 9(1), 1–9 (2019)

20. Shin, H., et al.: Medical image synthesis for data augmentation and anonymization using generative adversarial networks. CoRR abs/1807.10225 (2018)

21. Szegedy, C., Vanhoucke, V., Ioffe, S., Shlens, J., Wojna, Z.: Rethinking the inception architecture for computer vision. In: Proceedings of the IEEE Conference on Computer Vision and Pattern Recognition, pp. 2818–2826 (2016)

22. Toussie, D., et al.: Clinical and chest radiography features determine patient outcomes in young and middle age adults with COVID-19. Radiology 201754 (2020)
23. Wang, H., Xia, Y.: Chestnet: a deep neural network for classification of thoracic diseases on chest radiography. arXiv preprint arXiv:1807.03058 (2018)
24. Wang, X., Peng, Y., Lu, L., Lu, Z., Bagheri, M., Summers, R.M.: Chestx-ray8: hospital-scale chest x-ray database and benchmarks on weakly-supervised classification and localization of common thorax diseases. In: Proceedings of the IEEE Conference on Computer Vision and Pattern Recognition, pp. 2097–2106 (2017)
25. Wong, H.Y.F., et al.: Frequency and distribution of chest radiographic findings in COVID-19 positive patients. Radiology 201160 (2020)
26. Zhu, J.Y., Park, T., Isola, P., Efros, A.A.: Unpaired image-to-image translation using cycle-consistent adversarial networks. In: Proceedings of the IEEE International Conference on Computer Vision, pp. 2223–2232 (2017)

Attention Based CNN-LSTM Network for Pulmonary Embolism Prediction on Chest Computed Tomography Pulmonary Angiograms

Sudhir Suman[1], Gagandeep Singh[2], Nicole Sakla[2], Rishabh Gattu[2], Jeremy Green[2], Tej Phatak[2], Dimitris Samaras[3], and Prateek Prasanna[4(✉)]

[1] Department of Electrical Engineering, Indian Institute of Technology, Bombay, Mumbai, India
[2] Department of Radiology, Newark Beth Israel Medical Center, Newark, NJ, USA
[3] Department of Computer Science, Stony Brook University, Stony Brook, NY, USA
[4] Department of Biomedical Informatics, Stony Brook University, Stony Brook, NY, USA
prateek.prasanna@stonybrook.edu

Abstract. With more than 60,000 deaths annually in the United States, Pulmonary Embolism (PE) is among the most fatal cardiovascular diseases. It is caused by an artery blockage in the lung; confirming its presence is time-consuming and is prone to over-diagnosis. The utilization of automated PE detection systems is critical for diagnostic accuracy and efficiency. In this study we propose a two-stage attention-based CNN-LSTM network for predicting PE, its associated type (chronic, acute) and corresponding location (leftsided, rightsided or central) on computed tomography (CT) examinations. We trained our model on the largest available public Computed Tomography Pulmonary Angiogram PE dataset (RSNA-STR Pulmonary Embolism CT (RSPECT) Dataset, $N = 7279$ CT studies) and tested it on an in-house curated dataset of $N = 106$ studies. Our framework mirrors the radiologic diagnostic process via a multi-slice approach so that the accuracy and pathologic sequela of true pulmonary emboli may be meticulously assessed, enabling physicians to better appraise the morbidity of a PE when present. Our proposed method outperformed a baseline CNN classifier and a single-stage CNN-LSTM network, achieving an AUC of 0.95 on the test set for detecting the presence of PE in the study.

Keywords: Computer-aided diagnosis · CNN · LSTM · Pulmonary embolism

1 Introduction

Clinical Motivation. Pulmonary embolism (PE) is the most common preventable cause of hospital death in the United States with diagnostic delay and

© Springer Nature Switzerland AG 2021
M. de Bruijne et al. (Eds.): MICCAI 2021, LNCS 12907, pp. 356–366, 2021.
https://doi.org/10.1007/978-3-030-87234-2_34

diagnostic challenges among the most common etiologies for resultant mortality [6]. The time dependent nature of PE diagnosis is especially critical given that the treatment involves anti-coagulant therapy which must be administered in an attempt to halt thrombus growth. Delayed diagnosis and treatment enables undiagnosed clots to increase in size. The patient's mortality therefore becomes compounded by not only the resultant ischemic insult to the pulmonary parenchyma but also by the potential development of decompensated right heart failure. Currently, attempts to reduce diagnostic delays have predominantly focused on the speed of PE symptom identification in the emergency setting and little to no changes made to the imaging diagnosis paradigm through which treatment is ultimately decided [6]. Through the application of machine learning techniques, the diagnostic delays and errors in PE identification may be mitigated, thereby decreasing patient mortality.

Technical Motivation. Similar to the neoplastic applications of deep learning paradigms [3,5,22], the utilization of automated systems for PE detection is critical when diagnostic accuracy and efficiency are considered. Early works in automated PE diagnosis have mostly relied on traditional feature engineering and pulmonary vessel segmentation to reduce the search space which is quite computationally intensive [14,16–18,27]. Furthermore, most of these studies have reported results on small datasets. Recent Convolutional Neural Network-Long Short-Term memory (CNN-LSTM) methods [19,20] and an end-to-end deep learning model, PENet [8], have demonstrated promise when diagnosing PE on Computed Tomography Pulmonary Angiogram (CTPA) examinations, but are overall limited in their binary classification of whether a PE is present or not. Methods based on blood clot segmentation [19,25] for detecting PE are quite computationally-intensive, due to their use of only segmented blood clot regions; other regions which may show pulmonary abnormalities suggestive of clots get ignored.

In this work, we present a two-stage attention-based network consisting of a CNN and a Sequence model (LSTM + Dense) for prediction of PE and its associated characteristics. Current detection algorithms are largely modeled on the premise that every single CT slice is evaluated as an independent unit versus analyzing the relationship between successive CT slices. Our model extrapolates information from this 3D relationship in an attempt to better mimic the human cognitive process when examining a cross-sectional image. Due to the fact that PE is often observed in a small subset of 2D slices of a CT volume, the proposed attention module assists the network in locating insightful slices and assigns them higher weights in a bag-level feature aggregation mechanism. The slices with a higher attention are more likely to contain important information for PE identification thereby aiding in subsequent PE prediction. Besides, unlike other methods, our framework is trained to identify forms of PE on each CTPA examination slice and other PE attributes such as laterality, chronicity, and the RV/LV (Right Ventricular to Left Ventricular) ratio which can prove clinically useful in determining patients at risk [7,13]. Our contributions can be summarized as follows:

1. Our attention based network provides image-level (ψ_{image}) PE prediction for each of the CTPA examination slices, study-level (ψ_{study}) prediction of PE in the CTPA volume and other associated PE characteristics.
2. Multiple Instance Learning (MIL) pooling [10] is used as an attention mechanism that provides insight into the contribution of each CTPA examination slice and the aggregate feature representation for study-level prediction.
3. Our network is trained on the largest publicly available PE dataset (RSPECT N > 7000 studies) [4] and its performance has been evaluated on an external test dataset (N = 106 studies).
4. Our network is designed and trained to explicitly obey clinically-defined label hierarchy and avoid making conflicting label predictions.

2   Proposed Methodology

Our network architecture (Fig. 1) consists of two stages: 1) a CNN classifier to capture the image properties and study labels and 2) a sequence model for learning inter-slice dependencies. The CNN is used to extract features from every slice from the study and the sequence model combines these *spatially dependent* features and captures long-range dependencies to give the network information about the global change around each CT slice.

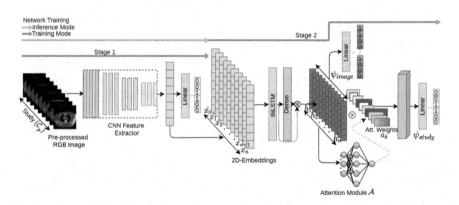

Fig. 1. Overview of our two-stage attention based PE detection model. The first stage involves training with all slices in study $\mathcal{C}_p$ for all image and study labels. The 2D-embedding $\{\hat{x}_1, \hat{x}_2, .., \hat{x}_3\}$ from stage 1 after the pooling layer is used for the stage 2 LSTM-based training with an attention module for generating final predictions

2.1   Pre-processing

CT scans capture tissue radiodensity information. Voxel intensities range from −1000 to 3000 Hounsfield Units. Different window levels are used by radiologists to evaluate lung pathologies. We first convert the single channel images to a 3-channel RGB format [2], where each channel is a distinct window [24] corresponding to the lung (window level $= -600$, width $= 1500$), PE (window level $= 100$, width $= 700$), and mediastinal ranges (window level $= 40$, width $= 400$). The PE and mediastinal windows enable detection of blood clots. The lung window, on the other hand, may show pulmonary abnormalities suggestive of a clot. The constructed 3-channel RGB images were used to train our network.

2.2 *Stage 1:* CNN Classifier to Capture Different Label Properties

We used a Convolutional Neural Network (CNN) Efficient-Net [23] architecture as a backbone to extract features, followed by average pooling and a linear layer to classify each image and study. The pooling layer reduces the size of features maps after the convolutional layer so that we can make use of all the slices present in the study at once while training the Sequence model in Stage 2. This CNN classifier is trained for multi-label prediction on the RSPECT [4] dataset to capture the properties for different study labels. We make use of spatial feature maps generated in Stage 1 to train our Sequence model in Stage 2.

2.3 *Stage 2:* Sequence Model for Learning Inter-slice Dependencies

The Sequence model, consisting of a bidirectional long short-term memory (BiL-STM) [9] and a dense layer, is used to capture long-range dependencies in CT scans. It makes use of the extracted features from study slices using the trained CNN classifier from Stage 1 and passes them through a Bi-LSTM and a dense layer to provide the network additional contextual information about the global changes around the slices. The features extracted using CNN-LSTM network capture spatio-temporal information in the CT scan volume. The 'temporal' aspect refers to the global relationship between successive slices. These features are subsequently used for the final prediction. The Sequence model has two heads - one for image (ψ_{image}) and another for study (ψ_{study}) level predictions. The ψ_{image} head is used for detecting the presence of PE on each slice in the study and the ψ_{study} head provides predictions regarding different characteristics of PE in the given study consisting of hundreds of slices.

Attention Mechanism. Features from the Sequence Model's Bi-LSTM and dense layers, $(\hat{h}_1, \hat{h}_2, .., \hat{h}_n)$, corresponding to slices $\mathcal{C}_{pq}, q \in [1, 2, ...n]$ in study volume $\mathcal{C}_p$ are fed into the attention module $\mathcal{A}$ to obtain bag level features for $\mathcal{C}_p$. The aggregated bag-level features are used for global prediction by passing it through the ψ_{study} classifier. Attention pooling is the weighted average of the slice-wise features present in the study and these attention weights a_k are learned from the neural network as $a_k = \frac{\exp\{w^T tanh(V\hat{h}_k^T)\}}{\sum_{j=1}^n \exp\{w^T tanh(V\hat{h}_j^T)\}}$, where w and V are learnable parameters and a_k denotes the learned attention weight distribution of $\mathcal{A}$. Features from each slice have a corresponding weight, and the bag-level feature set is obtained as $z = \sum_{k=1}^n a_k\hat{h}_k$. The attention module enables the network to locate informative slices present in the study. During feature aggregation, it assigns higher weight to the more informative slices for final study-level prediction.

Loss Function. We use a custom loss function for training, similar to the performance metrics for the RSNA STR Pulmonary Embolism Detection challenge [1]. The study-level weighted log loss is defined as

$$L_{ij} = -w_j * [y_{ij} * log(p_{ij}) + (1 - y_{ij}) * log(1 - p_{ij})],$$

where i and j denote the study and label respectively, y_{ij} is ground truth label $\{0, 1\}$ and p_{ij} is prediction probability of label j for study i that $y_{ij} = 1$. w_j signifies the weight of label j. Similarly, the image-level weighted log loss is defined as

$$L_{ik} = -[w * \frac{\sum_{k=1}^{n_i} y_{ik}}{n_i}] * [y_{ik} * log(p_{ik}) + (1 - y_{ik}) * log(1 - p_{ik})]$$

Here i is the study and $k = 1, 2, 3 ..., n_i$, where n_i is the total number of images in study i, y_{ik} is ground truth $\{0, 1\}$ and p_{ik} is prediction probability (for presence of PE) on image k in study i. The total loss of study i is the sum of the image-level and study-level loss divided by the total weights, given by

$$L_i = \frac{\sum_{k=1}^{n_i} L_{ik} + \sum_{j=1}^{n} L_{ij}}{\sum_{j=1}^{n} w_j + w * \sum_{k=1}^{n_i} y_{ik}}.$$

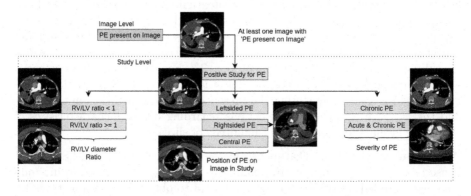

Fig. 2. Flowchart outlining relationships between labels

Label Consistency. Our model ensures logical consistency of the predicted labels using constraint-based modified activation function. According to the rules in [1], a label is considered as predicted if it is assigned a probability (p) of more than 0.5. Label consistency rules of PE are as follows:

– If any image in a study is predicted positive for the presence of PE

- At least one of the three labels, i.e., *Right Sided, Left Sided or Central PE* must be assigned a probability greater than 0.5.
- Only one of *RV/LV < 1 or RV/LV ≥ 1* will have $p > 0.5$ and one of it must be present if at least one image is found positive for PE in study.
- Both *Chronic & Acute* and *Chronic* cannot be predicted positive at the same time, i.e., only one can have $p > 0.5$.

− If no image in the study volume is found positive for presence of PE, then
  - Either *indeterminate* or *negative for PE* labels must have a $p > 0.5$; both cannot have a $p > 0.5$.
  - All positive study related labels, i.e., *right sided PE, left sided PE, RV/LV < 1, RV/LV ≥ 1, central PE* must have a $p < 0.5$.

Table 1. Data distribution in training, validation and test sets.

Labels	Train	Val	Test
# studies	5824	1455	106
# studies with positive PE	1878	490	56
# study with $RV/LV \geq 1$	754	186	43
# study with $RV/LV < 1$	1878	490	63
# study with leftsided PE	1242	302	51
# study with rightsided PE	1486	389	53
# study with central PE	318	83	43
# study with Chronic PE	220	72	2
# study with Acute&Chronic PE	116	29	0
# slice	1431318	359276	87764
# slice with PE	77453	19087	−

3 Experimental Design

3.1 Dataset Description

Training Dataset. We use the largest publicly available annotated PE dataset (the RSNA Pulmonary Embolism CT, RSPECT dataset), comprising more than 12,000 CT studies [4]. This dataset, composed of CTPA scans and associated ground truth annotations has been contributed to by five international research centers and labeled by a group of more than 80 expert thoracic radiologists. There are multiple labels for each study and one label for each slice present in the study. The flowchart outlining the relationships between different labels is shown in Fig. 2. 1,790,594 slices corresponding to 7279 studies have a complete set of ground truth labels available. 5824 of these studies were randomly selected as the training (D_1) and 1455 studies as the validation (D_2) cohorts.

Independent Test Set. The datasets are summarized in Table 1. We retrospectively collected data from 106 patients (D_3) who received CTPA studies under the PE protocol (GE Revolution CT, GE Lightspeed VCT, GE Lightspeed 16, Phillips Brilliance 64) at Newark Beth Israel Medical Center. The images were anonymized in accordance with Health Insurance Portability and Accountability Act and our Institutional Review Board guidelines. Both male and female adult patients (over 18 years) were selected with random sampling utilized in order to obtain similar numbers of positive versus negative PE cases (56 positive and 50 negative). CTPA studies with extensive artifacts (i.e. beam hardening, motion artifact, over/under exposure) were excluded. The ground truth study-level labels were provided by three expert readers working in consensus. Two readers (>15 yr and >10 yr exp) are board certified radiologists and the third reader (>3 yr exp) is a radiology resident.

3.2   Framework Implementation

We trained our PE network in two stages. First, we trained the CNN classifier using all pre-processed 3-channel slices in D_1, with both image and study-level labels as targets. Each batch consists of slices from the same study and the final study-level prediction was based on the prediction with maximum confidence. This multitask learning stage not only captures the presence of PE in slices but also learns other PE properties based on different study-level targets. We extracted the features of each slice from study C_p using the trained CNN network and map it to a 2-D embedding $\{\hat{x}_1, \hat{x}_2, .., \hat{x}_3\}$ using global average pooling. To exploit the sequential inter-slice spatial relationship and learn global intensity variations, we passed the embedding in order of their captured time points to the LSTM layer. The extracted features from the LSTM module passes through the attention module to yield bag-level features for the study volume C_p, which is subsequently fed to the final study-level classifier (ψ_{study}); features from the LSTM are directly provided to the image-level classifier(ψ_{image}) for detecting the presence of PE on individual slices. Both stages were trained with the Adam optimizer [12] and the custom loss function discussed in Sect. 2.3. The model with the lowest loss value on D_2 was chosen for evaluation on D_3. Both stages were implemented in PyTorch 1.4 and trained using one NVIDIA Tesla T4 Google Cloud Instance GPU.

4   Results

We used area under the receiver operating characteristic curve (AUC) as our performance metric. Our proposed method was compared with a CNN classifier (Baseline 1) and a CNN-LSTM network without the attention module (Baseline 2). The AUCs of Baseline 1, Baseline 2, and our method were 0.5, 0.94, and 0.95, respectively. The corresponding accuracies were 0.74, 0.65, and 0.88, respectively. The performance curves of our best model for different images and study labels on D_2 and D_3 are shown in Fig. 3. Our proposed network achieved an AUC of 0.82 for detecting PE on D_2 and 0.95 on the D_3. CT volumes in D_3 had different

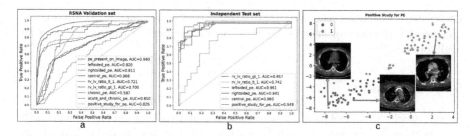

Fig. 3. (a, b) ROC curves for different image and study-level labels on the validation and independent test set. (c) 2D scatter plot representation using t-SNE of bag level features from Attention Module corresponding to studies in $D3$

slice thickness and higher number of slices per study as compared to those in D_1 and D_2, which potentially provides more contextual information to the model.

To provide an intuitive understanding into the features and the context learnt by the network, class activation maps (CAMs) [26] were computed by averaging the feature maps from the final convolutional layer of CNN network. As may be observed on a few representative slices in Fig. 4, there is a strong activation over the regions surrounding the clots, as verified by our collaborating radiologists (see interpretation in Fig. 4 caption). This corroborates the clinical characteristics of PE diagnosis. A t-Distributed Stochastic Neighbor Embedding (t-SNE) [15] was applied to the attention module's bag level features. The 2D scatter plot visualization is shown in Fig. 3.

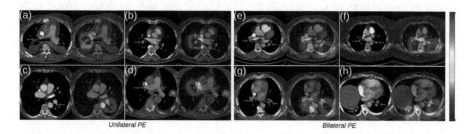

Fig. 4. Visualization of class activation maps. CTPA slices (a-d) demonstrate unilateral PE marked by red arrows. High intensity regions in the CAMs represent the precise location of PE. (a, b, d) show right-sided lobar/segmental PE. (c) shows lower segmental PE. Slices (e-h) demonstrate bilateral PE. (f, h) show lobar/segmental PE and CAMs depict the exact location of PE. In (e, g), the CAMs do not accurately define the maximal intensity on the right-sided central/lobar PE; however, they depict left-sided PE accurately. (Color figure online)

As may be observed in Fig. 3, our two-stage CNN-LSTM network with an attention mechanism is able to learn and predict not only presence of PE, but also its associated attributes with a high AUROC. Note that the number of ROC curves are fewer in D_3 as compared to D_2 in Fig. 3 since image-level annotations and enough number of studies for *chronic* and *acute & chronic PE* were not available for D_3.

5   Conclusion

In this work, we developed an attention based two-stage network for CTPA classification on the largest publicly available PE dataset [4] and tested its performance on a held out validation and an independent test set from a different clinical site. We believe the contribution is significant not only due to the large-scale multi-institutional validation, but also because of how the framework is designed to explicitly obey a clinically-defined label hierarchy. Our framework is among the first to identify forms of PE on each CTPA slice as well as other attributes such as laterality, chronicity, and the RV/LV ratio which can prove clinically useful in determining patient risk. Our model is also designed to mimic the human cognitive process when examining cross-sectional scans. Furthermore, the generated CAMs have been validated by radiologists. Our motivation behind a two-stage 2D CNN-LSTM training approach stems from the GPU memory limitations in the PE challenge [1]. The results of the challenge are also suggestive of the fact that a 2D-CNN with a Sequence model works better than a 3D-CNN for this task. We achieved a loss of < 0.180 on the Kaggle test set in the final leaderboard [1]. The state-of-the-art 3D-CNN, PENet [8], achieved a mean AUC of around 0.85 for detecting PE. Our study-level PE prediction performance is similar to PENet; however, unlike PENet, our network provides slice-level predictions as well as predictions for associated PE attributes. There are a few limitations to our approach. The CAMs are able to detect blood clots as informative regions with high certainty in unilateral PE. However, in bilateral PE, we can observe that the some discriminative regions are not precisely from the PE regions. This will be investigated in our future work. Additionally, we will evaluate the efficacy of an end-to-end network and radiomic features [11,21] for the prediction tasks. Refining the radiologic PE diagnostic paradigm through the use of a two-stage network may enable radiologists to reduce inherent human error and also decrease patient morbidity and mortality.

Acknowledgment. Dimitris Samaras was partially supported by the Partner University Fund, the SUNY2020 Infrastructure Transportation Security Center, and a gift from Adobe.

References

1. RSNA-STR pulmonary embolism CT (RSPECT) dataset, copyright RSNA. https://www.rsna.org/education/ai-resources-and-training/ai-image-challenge/rsna-pe-detection-challenge-2020
2. Anam, C., Budi, W., Haryanto, F., Fujibuchi, T., Dougherty, G.: A novel multiple-windows blending of CT images in red-green-blue (RGB) color space: phantoms study. Sci. Vis. **11**(5) (2019)

3. Bejnordi, B.E., et al.: Diagnostic assessment of deep learning algorithms for detection of lymph node metastases in women with breast cancer. JAMA **318**(22), 2199–2210 (2017)

4. Colak, E., et al.: The RSNA pulmonary embolism CT dataset. Radiol.: Artif. Intell. **3**(2), e200254 (2021)

5. Coudray, N., et al.: Classification and mutation prediction from non-small cell lung cancer histopathology images using deep learning. Nat. Med. **24**(10), 1559–1567 (2018)

6. Friedman, T., Winokur, R.S., Quencer, K.B., Madoff, D.C.: Patient assessment: clinical presentation, imaging diagnosis, risk stratification, and the role of pulmonary embolism response team. In: Seminars in Interventional Radiology, vol. 35, pp. 116–121. Thieme Medical Publishers (2018)

7. Ghaye, B., Ghuysen, A., Bruyere, P.J., D'Orio, V., Dondelinger, R.F.: Can CT pulmonary angiography allow assessment of severity and prognosis in patients presenting with pulmonary embolism? What the radiologist needs to know. Radiographics **26**(1), 23–39 (2006)

8. Huang, S.C., et al.: PENet-a scalable deep-learning model for automated diagnosis of pulmonary embolism using volumetric CT imaging. NPJ Digit. Med. **3**(1), 1–9 (2020)

9. Huang, Z., Xu, W., Yu, K.: Bidirectional LSTM-CRF models for sequence tagging. arXiv preprint arXiv:1508.01991 (2015)

10. Ilse, M., Tomczak, J., Welling, M.: Attention-based deep multiple instance learning. In: International Conference on Machine Learning, pp. 2127–2136. PMLR (2018)

11. Khorrami, M., et al.: Changes in CT radiomic features associated with lymphocyte distribution predict overall survival and response to immunotherapy in non-small cell lung cancer. Cancer Immunol. Res. **8**(1), 108–119 (2020)

12. Kingma, D.P., Ba, J.: Adam: a method for stochastic optimization. arXiv preprint arXiv:1412.6980 (2014)

13. Lankeit, M.: Always think of the right ventricle, even in "low-risk" pulmonary embolism (2017)

14. Liang, J., Bi, J.: Computer aided detection of pulmonary embolism with tobogganing and mutiple instance classification in CT pulmonary angiography. In: Karssemeijer, N., Lelieveldt, B. (eds.) IPMI 2007. LNCS, vol. 4584, pp. 630–641. Springer, Heidelberg (2007). https://doi.org/10.1007/978-3-540-73273-0_52

15. Van der Maaten, L., Hinton, G.: Visualizing data using t-SNE. J. Mach. Learn. Res. **9**(11) (2008)

16. Masutani, Y., MacMahon, H., Doi, K.: Computerized detection of pulmonary embolism in spiral CT angiography based on volumetric image analysis. IEEE Trans. Med. Imaging **21**(12), 1517–1523 (2002)

17. Özkan, H., Osman, O., Şahin, S., Boz, A.F.: A novel method for pulmonary embolism detection in CTA images. Comput. Methods Programs Biomed. **113**(3), 757–766 (2014)

18. Park, S.C., Chapman, B.E., Zheng, B.: A multistage approach to improve performance of computer-aided detection of pulmonary embolisms depicted on CT images: preliminary investigation. IEEE Trans. Biomed. Eng. **58**(6), 1519–1527 (2010)

19. Rajan, D., Beymer, D., Abedin, S., Dehghan, E.: Pi-PE: a pipeline for pulmonary embolism detection using sparsely annotated 3D CT images. In: Machine Learning for Health Workshop, pp. 220–232. PMLR (2020)

20. Shi, L., Rajan, D., Abedin, S., Yellapragada, M.S., Beymer, D., Dehghan, E.: Automatic diagnosis of pulmonary embolism using an attention-guided framework: a large-scale study. In: Medical Imaging with Deep Learning, pp. 743–754. PMLR (2020)
21. Singh, G., et al.: Radiomics and radiogenomics in gliomas: a contemporary update. Br. J. Cancer 1–17 (2021)
22. Sirinukunwattana, K., Raza, S.E.A., Tsang, Y.W., Snead, D.R., Cree, I.A., Rajpoot, N.M.: Locality sensitive deep learning for detection and classification of nuclei in routine colon cancer histology images. IEEE Trans. Med. Imaging **35**(5), 1196–1206 (2016)
23. Tan, M., Le, Q.: Efficientnet: rethinking model scaling for convolutional neural networks. In: International Conference on Machine Learning, pp. 6105–6114. PMLR (2019)
24. Wittram, C., Maher, M.M., Yoo, A.J., Kalra, M.K., Shepard, J.A.O., McLoud, T.C.: CT angiography of pulmonary embolism: diagnostic criteria and causes of misdiagnosis. Radiographics **24**(5), 1219–1238 (2004)
25. Yang, X., Lin, Y., Su, J., Wang, X., Li, X., Lin, J., Cheng, K.T.: A two-stage convolutional neural network for pulmonary embolism detection from CTPA images. IEEE Access **7**, 84849–84857 (2019)
26. Zhou, B., Khosla, A., Lapedriza, A., Oliva, A., Torralba, A.: Learning deep features for discriminative localization. In: Proceedings of the IEEE Conference on Computer Vision and Pattern Recognition, pp. 2921–2929 (2016)
27. Zhou, C., et al.: Preliminary investigation of computer-aided detection of pulmonary embolism in threedimensional computed tomography pulmonary angiography images. Acad. Radiol. **12**(6), 782 (2005)

LuMiRa: An Integrated Lung Deformation Atlas and 3D-CNN Model of Infiltrates for COVID-19 Prognosis

Amogh Hiremath[1], Lei Yuan[2], Rakesh Shiradkar[1], Kaustav Bera[1],
Vidya Sankar Viswanathan[1], Pranjal Vaidya[1], Jennifer Furin[3],
Keith Armitage[3], Robert Gilkeson[4], Mengyao Ji[5], Pingfu Fu[6], Amit Gupta[4],
Cheng Lu[1], and Anant Madabhushi[1,7(✉)]

[1] Department of Biomedical Engineering, Case Western Reserve University,
Cleveland, USA
axm788@case.edu
[2] Department of Information Center, Renmin Hospital of Wuhan University,
Wuhan, China
[3] Department of Infectious Diseases, University Hospitals Cleveland Medical Center,
Cleveland, USA
[4] Department of Radiology, University Hospitals Cleveland Medical Center,
Cleveland, USA
[5] Department of Gastroenterology, Renmin Hospital of Wuhan University,
Wuhan, China
[6] Department of Population and Quantitative Health Sciences,
Case Western Reserve University, Cleveland, USA
[7] Louis Stokes Cleveland Veterans Administration Medical Center, Cleveland, USA

Abstract. Although, recently convolutional neural networks (CNNs) based prognostic models have been developed for COVID-19 severity prediction, most of these studies have analyzed characteristics of lung infiltrates (ground-glass opacities and consolidations) on chest radiographs or CT. However, none of the studies have explored the possible lung deformations due to the disease. Our hypothesis is that more severe disease results in more pronounced deformation. The key contributions of this work are three-fold: (1) A new lung deformation based biomarker analyzing regions of differential distensions between COVID-19 patients with mild and severe disease. (2) Integrating 3D-CNN characterization of lung deformation regions and lung infiltrates on lung CT into a novel framework (LuMiRa) for prognosticating COVID-19 severity. (3) Validating LuMiRa on one of the largest multi-institutional cohort till date (N = 948 patients). We found that majority of the shape deformations were observed in the mediastinal surface of both the lungs and in left interior lobe. On a testing cohort based on two institutions, $\mathbf{A_v}$ (N = 419) and $\mathbf{B_v}$ (N = 113), LuMiRa yielded an area under the receiver operating characteristic curve (AUC) of 0.89 and 0.77 respectively showing

Electronic supplementary material The online version of this chapter (https://doi.org/10.1007/978-3-030-87234-2_35) contains supplementary material, which is available to authorized users.

M. de Bruijne et al. (Eds.): MICCAI 2021, LNCS 12907, pp. 367–377, 2021.
https://doi.org/10.1007/978-3-030-87234-2_35

significant improvement over a 3D-CNN trained over just lung infiltrates (AUC = 0.85 (p < 0.001), AUC = 0.75 (p = 0.01)). Additionally, LuMiRa performed significantly better than machine learning models trained on clinical and radiomic features (0.82, 0.78 and 0.72, 0.72 on $\mathbf{A_v}$ and $\mathbf{B_v}$ respectively).

1 Introduction

Artificial intelligence (AI) based approaches with the use of convolutional neural networks (CNNs) have been extensively used to detect and diagnose COVID-19 [12,13] using chest radiographs or CT. Majority of these studies are focused on diagnosis, specifically distinguishing COVID-19 cases from other pneumonias [15,16]. However, there is a parallel and unmet clinical need in developing AI based prognostic models predicting the severity of COVID-19 disease.

Recently, prognostic models using CNNs and radiomic based approaches [14,22] have been developed for COVID-19 severity prediction. Additionally, some of the approaches have even combined clinical features such as albumin, creatinine, neutrophil count etc. with radiomic signatures [3,20] to predict COVID-19 patient outcomes. However, most of these imaging studies have mostly focused on characteristics of lung infiltrates (ground-glass opacities and consolidation regions)[2,23]. On the imaging aspect that has not been considered to date and that could have a bearing on severity of COVID-19 is the distension induced in the lungs on account of more severe disease.

A number of recent studies have shown that severe COVID-19 disease causes lung damage. Bussani et al. [1] on postmortem samples of 41 patients showed extensive damage, persistent distortion of normal lung structure. Tonelli et al. [21] and Dimbath et al. [5] have shown the link between increasing strain and mechanical deformation of lungs in COVID-19 patients with severe disease. These findings can be explained by severe lung damage that causes intense disruption of the normal lung parenchyma and interstitium, with fibrosis as a sequela. Our hypothesis is that distensions induced in the lung by COVID-19 are a reflection of more advanced lung fibrosis or interstitial lung changes induced by more severe disease. Early characterization of these physiological based changes can portend the possible long-term sequela of COVID-19 and help in better prognosis and treatment planning for COVID-19 patients.

We present a new approach, LuMiRa, that combines a population atlas based lung distension biomarker with 3D-CNN based features of the infiltrates on CT scans to prognosticate COVID-19 disease severity. Key contributions of our study are as follows;

- A new imaging biomarker using population atlas-based approach to analyze regions of lung shape distension differences between COVID-19 patients with mild and severe disease.
- Integrating 3D-CNN characterization of lung deformation regions and lung infiltrate regions on lung CT into a novel framework (LuMiRa) for better identifying patients with severe disease requiring mechanical ventilation.

- Using a large multi-institutional cohort of $N = 948$ patients, we show that the signatures encoded by LuMiRa generalize well across multiple sites. We also demonstrate that LuMiRa outperforms a 3D-CNN, radiomics and clinical based machine learning models respectively.

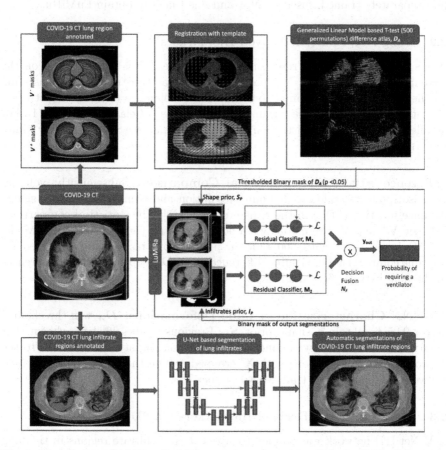

Fig. 1. Flowchart of LuMiRa. LuMiRa consists of two parallel 3D CNNs (M_1, M_2). While M_1 analyzes regions of shape differences of the lung between $\mathbf{V}^+$ and $\mathbf{V}^-$ patients via a shape prior S_p (top row), M_2 encodes spatial information of automatically segmented lung infiltrates, I_p^i (bottom row). The decisions from M_1 and M_2 are fused at the decision fusion node (N_F) to obtain the final predictions.

2   Methodology

2.1   Notations and Brief Overview

We denote a CT scan of a patient as $\mathbf{X}^i$ where $i = (1, ..., n)$ with n being the total number of scans, and severity of COVID-19 disease as mild ($\mathbf{V}^-$: patients who did not require a ventilator) and severe ($\mathbf{V}^+$: patients who required a ventilator). The target outcome of a patient ($\mathbf{V}^-$ or $\mathbf{V}^+$) is represented by $\mathbf{V}_T$.

LuMiRa consists of two parallel 3D CNNs (M_1, M_2). While M_1 analyzes regions of shape deformations of the lung between $\mathbf{V}^+$ and $\mathbf{V}^-$ patients via a shape prior S_p derived from difference atlas, D_A (Fig. 1 top row), M_2 characterizes regions of lung infiltrates, I_p^i (Fig. 1 bottom row). Both M_1 and M_2 are first separately trained, fused at N_F, and fine-tuned to obtain LuMiRa.

2.2   Lung Deformation Derived Biomarkers

Atlas Construction: A representative template T was chosen based on the median lung volume from $\mathbf{V}^-$ patients. Atlas was constructed by registering all the images to a common canonical frame of reference (T). The registration algorithm was based on the parameter file database of elastix toolbox [11]. For detailed information of registration parameters, please see Supplementary Materials (pp 1).

Difference Atlas Using Statistical Comparison: Subsequently, volumes were isotropically scaled with $1\,\mathrm{mm}^3$ resolution and transformed into a signed distance function (SDF) in the registered space [7]. The statistical comparisons between $\mathbf{V}^+$ and $\mathbf{V}^-$ were done via a non-parametric Generalized Linear Model (GLM) based t-test with 500 random permutation testing and corrected for multiple comparison [7], yielding difference atlas, D_A depicting shape differences of the lung between $\mathbf{V}^-$ and $\mathbf{V}^+$.

3D-CNN Characterization of Lung Deformations: D_A was thresholded at $p < 0.05$ to obtain a shape prior S_p. Previous studies [6] have shown that binary masks as auxiliary channels can help in setting attention regions to the network. Therefore S_p along with $\mathbf{X}^i$ cropped around the lung region were used as two input channels to train M_1 to predict $\mathbf{V}_T$.

2.3   Lung Infiltrates Derived Biomarkers via 3D-CNN

A U-Net [17] network was trained to segment the infiltrate regions in the lung chest CTs. First, the lung region on CT was segmented using a previously used automatic lung segmentation method utilizing watershed transform [19]. Next, 2D segmentation of infiltrate regions (I_p^i) was performed by U-Net by providing patches of right and left lung separately, in turn reducing the huge imbalance between the positive and negative class pixels in the image. Subsequently, $\mathbf{X}^i$ and I_p^i cropped around the lung region were used as two input channels to train M_2 to predict $\mathbf{V}_T$.

2.4   Model Fusion

Pre-trained models, M_1 and M_2 were integrated at decision fusion node N_F (Fig. 1). The output, y_{out} from LuMiRa is given by Eq. 1;

$$y_{out} = M_1(X_i, S_p, \theta_{c1}, \theta_{d1}) + M_2(X_i, I_p^i, \theta_{c2}, \theta_{d2}) \qquad (1)$$

where $\boldsymbol{X}_i$ is the input CT volume, $\boldsymbol{S}_p$ is the shape prior, $\boldsymbol{I}_p^i$ is the infiltrates' prior, $\boldsymbol{\theta}_{c1}$, $\boldsymbol{\theta}_{c2}$ are the parameters of convolutional layers of M_1 and M_2 respectively and, $\boldsymbol{\theta}_{d1}$ and $\boldsymbol{\theta}_{d2}$ are parameters of dense layers of M_1 and M_2 respectively. Subsequently, the training of LuMiRa was performed by fine-tuning only the dense layers $\boldsymbol{\theta}_{d1}$ and $\boldsymbol{\theta}_{d2}$.

3   Experimental Results and Discussion

3.1   Data Description

The dataset used consists of $N = 948$ chest CT scans of patients from two institutions ($\mathbf{A}$, $\mathbf{B}$) with Reverse transcription polymerase chain reaction (RT-PCR) positivity. Additional details of the dataset including the inclusion and exclusion criteria and CT acquisition parameters are presented in Supplementary materials (pp 1-2). The dataset $\mathbf{A}$ was stratified into 50% training ($\mathbf{A}_t$, $N = 416$) and 50% testing $\mathbf{A}_v$: $N = 419$) and the scans from $\mathbf{B}$ ($\mathbf{B}_v$, $N = 113$) was used as an external test set. 42.5%, 40.57% and 69.9% had a mild disease in $\mathbf{A}_t$, $\mathbf{A}_v$ and $\mathbf{B}_v$ respectively which resolved while 57.5%, 59.43% and 30.1% had severe disease $\mathbf{A}_t$, $\mathbf{A}_v$ and $\mathbf{B}_v$ respectively, needing invasive mechanical ventilation. An expert radiologist with 14 years of experience delineated the lung region and lung infiltrates on a stratified randomly selected subset, $\mathbf{A}_s$ ($N = 181$ patients) of training set $\mathbf{A}_t$ using the editor module of 3D Slicer [10].

3.2   Data Preprocessing and Implementation

All CT volumes were first pre-processed by converting them from Hounsfield units to image intensities by ensuring that air inside the lung as having zero intensity value. $\mathbf{A}_s$ was used for construction of the atlas ($\boldsymbol{D}_A$) and segmentation of infiltrate regions (I_p^i). To train the 3D-CNNs to predict $\mathbf{V}_T$ the CT scans were further rescaled to a resolution of $2\,\mathrm{mm}^3$. The input size to all the 3D-CNNs were based on the lung size of the template, T whose dimensions covered 96×144 voxels in the axial plane spanning over 128 slices. For each of the experiments a 3-fold cross validation was performed on $\mathbf{A}_t$ and the ensemble of the predictions trained on 3-fold cross validation (average predictions of the predictors) was used to evaluate the performance on $\mathbf{A}_v$ and $\mathbf{B}_v$.

To determine the effect of the classification architecture, three different base architectures, 3D-SEResNet (Squeeze and Excitation ResNet) [8,9], and 3D-DenseNet [9] and 3D-ResNeXt [9], were compared in terms of their performance. All the CNNs were initialized with manual seeding. Following parameters were used to train the networks; a) Dice loss (1 - Dice Similarity Coefficient (DSC)) for segmentation while binary cross entropy loss function for classification tasks, b) Adam optimizer with weight decay of 1e-3; c) learning rate of 1e-5, d) batch size of 2. An early stopping criteria (patience = 10) with respect to the cross-validation loss was used to stop the network training.

3.3   Experiment 1: COVID-19 Prognosis via LuMiRa

The registration accuracy for atlas creation, D_A was observed to be high with a DSC of 0.91 ± 0.08 between the template, T and the registered lung masks. Figure 2 illustrates the regions in the lung where significant deformations were observed between the V^+ and V^- populations in D_A. Majority of the shape differences were observed in the mediastinal surface of both the lungs and in the left interior lobe. Additionally, Gradient-weighted Class Activation Maps (Grad-CAM) [18] (Fig. 3 (1a-1f)) illustrate that M_1 (ResNeXt) primarily focuses on the lung deformations.

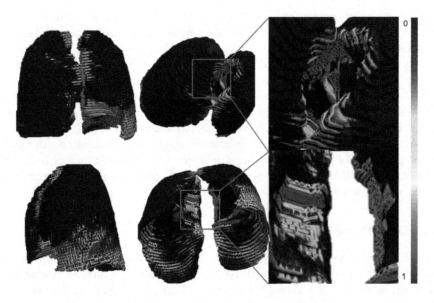

Fig. 2. Shape differences illustrated through different views of the lung based on statistical comparisons between V^+ and V^- done via a non-parametric Generalized Linear Model (GLM) based t-test with 500 random permutation testing. All p-values are mapped to atlases and thresholded with $p < 0.05$. Blue value signifies low difference and red value signifies a bigger difference. Majority of the shape differences were observed in the mediastinal surface of both the lungs and in left interior lobe. (Color figure online)

On the other hand, the infiltrate regions were defined as being detected if ≥ 0.2 DSC overlap existed between the network output and the ground-truth delineation of the corresponding region. U-Net [17] achieved a DSC of 0.59 ± 0.02 on the detected regions. Consequently, Grad-CAM activations [18] (Fig. 3 (2a-2f)) illustrate that M_2 (ResNeXt) which was used to predict V_T primarily focuses on the lung infiltrate regions as indicated by I_p^i.

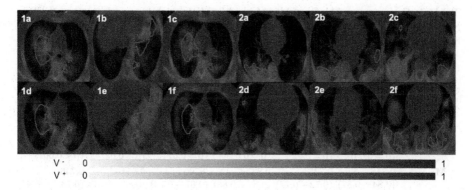

Fig. 3. Regions (1a-1f) are based on the shape prior (S_p) derived from difference atlas (D_A), while regions (2a-2f) are based on infiltrates prior (I_p^i). Grad-CAM interpretability results illustrate that the binary masks, S_p and I_p^i encoded into M_1 (1a-1f) and M_2 (2a-2f) respectively as auxiliary channels to the network can aid the network in setting an attention region helping the network to focus on these regions, while at the same time, providing the context of the whole lung region. The color "blue" indicates pixels contributing towards V^- while "red" corresponds V^+. The color bar gradient corresponds to the strength of the contribution. (Color figure online)

3.4   Experiment 2: Sensitivity Analysis

Architecture Sensitivity: No statistically significant difference was found between performances of different architectures (SEResNet, DenseNet, ResNeXt). However, since, ResNeXt architecture performed better than the other architectures (Table 1) yielding an AUC = 0.887; 95% CI [0.87 − 0.90] and AUC = 0.773; 95% CI [0.79 − 0.79] on A_v and B_v respectively, we chose ResNeXt as the base architecture for LuMiRa. The corresponding sensitivity, specificity for LuMiRa was found to be 81.52%, 79.41% and 79.41%, 67.08% on A_v and B_v respectively.

Effect of Template (T) Selection: To study the effect of template selection, we constructed a difference shape atlas D_A^r and subsequently S_p^r by choosing a random template (T^r) from V^- patients in A_s instead of T. Although, DSC = 0.64 was observed between S_p and S_p^r when T^r was further aligned with T, the trained LuMiRa yielded an AUC = 0.873; 95% CI[0.84 − 0.90] and AUC = 0.756; 95%CI [0.69 − 0.82] on A_v and B_v respectively with S_p^r having no statistically significant (p > 0.08) difference in the performance.

3.5   Experiment 3: Comparative Strategies

LuMiRa Versus Individual Prognosis Models (M_1 and M_2):   From Table 1 we can observe that LuMiRa outperformed both M_1 and M_2 in terms of AUC. For the ResNeXt architecture, the increase in the AUC compared to M_1, M_2 was found to be 1.5%, 3.3% on A_v respectively and 4.7%, 2.7% on B_v

Table 1. Performance (AUC) of M_1, M_2 and LuMiRa with base architectures (SERes-Net, DenseNet, ResNeXt) on $\mathbf{A}_v$ and $\mathbf{B}_v$ in predicting $\mathbf{V}_T$. All p-values indicated are based on DeLong's test with comparisons made against LuMiRa

Model	SEResNet		DenseNet		ResNeXt	
	$\mathbf{A}_v$ (95% CI) p-value	$\mathbf{B}_v$ (95% CI) p-value	$\mathbf{A}_v$ (95% CI) p-value	$\mathbf{B}_v$ (95% CI) p-value	$\mathbf{A}_v$ (95% CI) p-value	$\mathbf{B}_v$ (95% CI) p-value
M_1	0.871 (0.86 − 0.88) (p = 0.03)	0.704 (0.68 − 0.72) (p = 0.02)	0.860 (0.84 − 0.88) (p = 0.04)	0.720 (0.69 − 0.75) (p = 0.07)	0.872 (0.86 − 0.88) (p = 0.02)	0.726 (0.69 − 0.75) (p < 0.001)
M_2	0.856 (0.84 − 0.87) (p = 0.006)	0.735 (0.72 − 0.77) (p = 0.14)	0.837 (0.81 − 0.85) (p < 0.001)	0.700 (0.67 − 0.73) (p = 0.02)	0.854 (0.83 − 0.87) (p < 0.001)	0.746 (0.73 − 0.76) (p = 0.01)
LuMiRa	0.885 (0.87 − 0.91)	0.748 (0.72 − 0.77)	0.885 (0.86 − 0.91)	0.749 (0.73 − 0.77)	0.887 (0.87 − 0.90)	0.773 (0.76 − 0.79)

Table 2. Other performance metrics (AUC, sensitivity, specificity) of LuMiRa with base architecture (ResNeXt) on $\mathbf{A}_v$ and $\mathbf{B}_v$

ResNeXt	$\mathbf{A}_v$			$\mathbf{B}_v$		
	AUC (95% CI) p-value	Sensitivity (%)	Specificity (%)	AUC (95% CI) p-value	Sensitivity (%)	Specificity (%)
M_1	0.872 (0.86 − 0.88) (p = 0.02)	81.45	75.88	0.726 (0.69 − 0.75) (p < 0.001)	73.52	68.35
M_2	0.854 (0.83 − 0.87) (p < 0.001)	75.40	78.23	0.746 (0.73 − 0.76) (p = 0.01)	70.58	65.82
LuMiRa	0.887 (0.87 − 0.90)	81.52	79.41	0.773 (0.76 − 0.79)	79.41	67.08

respectively with improvement being statistically significant using DeLong's test [4]. Furthermore, from Table 2 we can also notice that LuMiRa achieves a higher sensitivity and specificity compared to M_1 and M_2. This shows that coupling the deformation measurements via lung atlas with characterization of lung infiltrates results in a better prognosis prediction.

LuMiRa Versus 3D-CNN Without Shape and Infiltrate Priors: When ResNeXt was trained with just lung CT without including shape (S_p) and infiltrate priors (I_p^i), an AUC = 0.8; 95% CI [0.78 − 0.82], AUC = 0.645; 95% CI[0.60 − 0.69] on $\mathbf{A}_v$ and $\mathbf{B}_v$ respectively was observed. This accounted to a statistically significant (p < 0.001) decrease of 7.2%, 5.4% and 8.7% AUC compared to M_1, M_2 and LuMiRa respectively on $\mathbf{A}_v$. Similarly, a statistically significant (p < 0.001) decrease of 8.1%, 10.1% and 12.8% AUC compared to M_1, M_2 and LuMiRa respectively was observed on $\mathbf{B}_v$.

LuMiRa Versus Clinical Model: A univariate analysis was conducted to identify top 5 features from clinical and laboratory factors (For the full list of clinical features analyzed, please refer to the Supplementary materials (pp 1)). Subsequently, a logistic regression classifier was trained with a 3-fold cross validation to compare the performance against LuMiRa. Lactate dehydrogenase, albumin, absolute neutrophils count, % lymphocyte and prothrombin time were identified as the most prognostic clinical features. For Table 3, we can observe that LuMiRa was found to outperform the clinical model and the difference was found to be statistically significant ($p < 0.001$).

Table 3. Comparison of LuMiRa with clinical and radiomics based model

	LuMiRa (95% CI)	Clinical (95% CI)	Radiomics (95% CI)
A_v	0.89 (0.87-0.90)	0.82 (0.79, 0.85)	0.78 (0.73, 0.83)
B_v	0.77 (0.76-0.79)	0.72 (0.61, 0.82)	0.72 (0.61, 0.83)

LuMiRa Versus Radiomics Model: A LASSO logistic regression-based model was used to train a radiomics based model to predict V_T (For the full list of radiomic features extracted, please refer to the Supplementary materials (pp 1,2)). By excluding cases with very small infiltrate regions on which radiomic analysis could not be performed, we extracted radiomic features from infiltrate regions on $N = 928$ cases. From Table 3, we observe that LuMiRa outperformed radiomics model with the difference being statistically significant ($p < 0.001$).

4   Conclusion

In this work we presented a novel imaging biomarker using a population atlas based approach capturing differential distensions in lung shape between patients with mild and severe COVID-19 disease. A new prognostic model, LuMiRa, a 3D-CNN was presented analyzing regions of lung distensions along with regions of lung infiltrates on CT scans to identify COVID-19 patients with severe disease requiring a ventilator. Using a large multi-institutional cohort we demonstrated that LuMiRa was able to outperform other state-of-the-art approaches including clinical and radiomics based machine learning models. Future work will entail comparison of LuMiRa's performance with expert radiologist interpretation of severity. Additionally, evaluating LuMiRa in a prospective setting by following up patients till discharge will be part of the future work.

Acknowledgements. Research reported in this publication was supported by the National Cancer Institute under award numbers (1U24CA199374-01, R01CA249992-01A1, R01CA202752-01A1, R01CA208236-01A1, R01CA216579-01A1, R01CA220581-01A1), National Heart, Lung and Blood Institute 1R01HL15127701A1, National Institute of Biomedical Imaging and Bioengineering 1R43EB028736-01, National Center for Research Resources under award number 1 C06 RR12463-01, VA Merit Review Award

IBX004121A from the United States Department of Veterans Affairs Biomedical Laboratory Research and Development Service, the Office of the Assistant Secretary of Defense for Health Affairs, through the Breast Cancer Research Program (W81XWH-19-1-0668), the Prostate Cancer Research Program (W81XWH-15-1-0558, W81XWH-20-1-0851), the Lung Cancer Research Program (W81XWH-18-1-0440, W81XWH-20-1-0595), the Peer Reviewed Cancer Research Program (W81XWH-18-1-0404), the Kidney Precision Medicine Project (KPMP) Glue Grant, the Ohio Third Frontier Technology Validation Fund, the Clinical and Translational Science Collaborative of Cleveland (UL1TR0002548) from the National Center for Advancing Translational Sciences (NCATS) component of the National Institutes of Health and NIH roadmap for Medical Research, The Wallace H. Coulter Foundation Program in the Department of Biomedical Engineering at Case Western Reserve University. Sponsored research agreements from Bristol Myers-Squibb, Boehringer-Ingelheim, and Astrazeneca. The content is solely the responsibility of the authors and does not necessarily represent the official views of the National Institutes of Health, the U.S. Department of Veterans Affairs, the Department of Defense, or the United States Government.

References

1. Bussani, R., Schneider, E., et al.: Persistence of viral RNA, pneumocyte syncytia and thrombosis are hallmarks of advanced COVID-19 pathology. EBioMedicine **61**, 103104 (2020)
2. Cai, W., Liu, T., et al.: CT quantification and machine-learning models for assessment of disease severity and prognosis of COVID-19 patients. Acad. Radiol. **27**(12), 1665–1678 (2020)
3. Chao, H., Fang, X., et al.: Integrative analysis for COVID-19 patient outcome prediction. Med. Image Anal. **67**, 101844 (2021)
4. DeLong, E.R., DeLong, D.M., Clarke-Pearson, D.L.: Comparing the areas under two or more correlated receiver operating characteristic curves: a nonparametric approach. Biometrics 837–845 (1988)
5. Dimbath, E., Maddipati, V., et al.: Implications of microscale lung damage for COVID-19 pulmonary ventilation dynamics: a narrative review. Life Sci. 119341 (2021)
6. Eppel, S.: Setting an attention region for convolutional neural networks using region selective features, for recognition of materials within glass vessels. arXiv preprint arXiv:1708.08711 (2017)
7. Ghose, S., Shiradkar, R., et al.: Prostate shapes on pre-treatment MRI between prostate cancer patients who do and do not undergo biochemical recurrence are different: preliminary findings. Sci. Rep. **7**(1), 1–8 (2017)
8. Hu, J., Shen, L., Sun, G.: Squeeze-and-excitation networks. CoRR (2017)
9. Kataoka, H., Wakamiya, T., Hara, K., Satoh, Y.: Would mega-scale datasets further enhance spatiotemporal 3D CNNs? arXiv preprint arXiv:2004.04968 (2020)
10. Kikinis, R., Pieper, S.D., Vosburgh, K.G.: 3D slicer: a platform for subject-specific image analysis, visualization, and clinical support. In: Jolesz, F.A. (ed.) Intraoperative Imaging and Image-Guided Therapy, pp. 277–289. Springer, New York (2014). https://doi.org/10.1007/978-1-4614-7657-3_19
11. Klein, S., Staring, M., et al.: Elastix: a toolbox for intensity-based medical image registration. IEEE Trans. Med. Imaging **29**(1), 196–205 (2009)

12. Lessmann, N., Sánchez, C.I., et al.: Automated assessment of CO-RADS and chest CT severity scores in patients with suspected COVID-19 using artificial intelligence. Radiology (2020)

13. Mei, X., Lee, H.C., et al.: Artificial intelligence-enabled rapid diagnosis of patients with COVID-19. Nat. Med. **26**(8), 1224–1228 (2020)

14. Meng, L., et al.: A deep learning prognosis model help alert for COVID-19 patients at high-risk of death: a multi-center study. IEEE J. Biomed. Health Inform. **24**(12), 3576–3584 (2020)

15. Murphy, K., Smits, H., et al.: COVID-19 on chest radiographs: a multireader evaluation of an artificial intelligence system. Radiology **296**(3), E166–E172 (2020)

16. Oh, Y., Park, S., Ye, J.C.: Deep learning COVID-19 features on CXR using limited training data sets. IEEE Trans. Med. Imaging **39**(8), 2688–2700 (2020)

17. Ronneberger, O., Fischer, P., Brox, T.: U-net: convolutional networks for biomedical image segmentation. CoRR (2015)

18. Selvaraju, R.R., Cogswell, M., et al.: Grad-cam: visual explanations from deep networks via gradient-based localization. In: Proceedings of the IEEE International Conference on Computer Vision, pp. 618–626 (2017)

19. Shojaii, R., Alirezaie, J., Babyn, P.: Automatic lung segmentation in CT images using watershed transform. In: IEEE International Conference on Image Processing 2005. IEEE (2005)

20. Tang, Z., Zhao, W., et al.: Severity assessment of COVID-19 using CT image features and laboratory indices. Phys. Med. Biol. **66**(3), 035015 (2021)

21. Tonelli, R., Marchioni, A., et al.: Spontaneous breathing and evolving phenotypes of lung damage in patients with COVID-19: review of current evidence and forecast of a new scenario. J. Clin. Med. **10**(5), 975 (2021)

22. Wu, Q., et al.: Radiomics analysis of computed tomography helps predict poor prognostic outcome in COVID-19. Theranostics **10**(16), 7231 (2020)

23. Yue, H., Yu, Q., et al.: Machine learning-based CT radiomics method for predicting hospital stay in patients with pneumonia associated with SARS-COV-2 infection: a multicenter study. Ann. Transl. Med. **8**(14) (2020)

Clinical Applications - Neuroimaging - Brain Development

Multi-site Incremental Image Quality Assessment of Structural MRI via Consensus Adversarial Representation Adaptation

Siyuan Liu, Kim-Han Thung, Weili Lin, and Pew-Thian Yap[✉]

Department of Radiology and Biomedical Research Imaging Center (BRIC),
University of North Carolina at Chapel Hill, Chapel Hill, NC, USA
ptyap@med.unc.edu

Abstract. Deep learning based image quality assessment (IQA) is useful for automatic quality control of medical images but requires a large number of training data. Though using multi-site data can significantly increase the training sample size and improve the performance of the IQA model, there are technical and legal issues involved in the sharing of patient data across different sites. When data are not sharable, devising a single IQA model that is applicable to all sites is challenging. To overcome this problem, we introduce a multi-site incremental IQA (MSI-IQA) method for structural MRI, which first trains an IQA model from one site, and then sequentially and incrementally improves the IQA model in other sites using transfer learning and consensus adversarial representation adaptation (CARA) without explicit data sharing between sites.

Keywords: Image quality assessment · Deep learning · Semi-supervised learning · Multi-site learning

1 Introduction

Image quality assessment (IQA) is a crucial step in determining the usability and the re-scan necessity of acquired structural magnetic resonance (MR) images, so that problematic images can be identified and excluded in subsequent processing and analysis. IQA is typically performed by a human rater [1,2]. However, subjective IQA is time-consuming, costly, and error-prone [3]. In contrast, objective IQA carried out using a computer algorithm can be fully automated, a trait that is desirable for large-scale studies.

Recently, deep neural networks (DNNs), particularly convolutional neural networks (CNNs) have demonstrated great potential for objective IQA due to its automatic learning of pertinent features for IQA. To sufficiently train a CNN,

This work was supported in part by United States National Institutes of Health (NIH) grant EB006733 and the efforts of the UNC/UMN Baby Connectome Project Consortium.

© Springer Nature Switzerland AG 2021
M. de Bruijne et al. (Eds.): MICCAI 2021, LNCS 12907, pp. 381–389, 2021.
https://doi.org/10.1007/978-3-030-87234-2_36

a large amount of data is typically required. However, medical data samples are often scarce, especially for rarer diseases [4] or unique populations such as babies [5]. Small sample sizes may result in neural network models with low generalizability.

Collaborating with multiple institutions to increase the amount and diversity of data is essential to achieve high algorithm performance. Ideally, this can be done by depositing patient data from multiple sites into a central location to continuously train a neural network model. However, this is highly impractical and is challenging to implement in reality. First, it may be cumbersome to share a large amount data (e.g., high-resolution images that are increasingly common). Second, there are often technical, legal, or ethical barriers to sharing patient data [4]. Third, institutions might simply prefer not to share valuable patient data [6]. Existing methods for IQA lack the mechanism to take into account inter-site differences [2]. Even if they do, they require imaging data to be shared across sites for training purposes. Data sharing agreements are required between institutions for this to happen. Adapting these methods to large-scale multi-site studies is hence challenging.

In this paper, we address the above issues by introducing a multi-site incremental IQA (MSI-IQA) framework, where the IQA neural network is first learned using images from one site (source site), and then incrementally updated using images from the other sites (target sites) using transfer learning and consensus adversarial representation adaptation (CARA), without involving any data sharing between sites. MSI-IQA requires only a small amount of annotated images for training at each site and is robust to potential annotation errors. Our MSI-IQA consists of four sequential stages, i.e., source-site slice training, optimal transfer learning, CARA, and target-site volume training. In source-site slice training, a slice quality assessment network (SQA-Net), which consists of an encoder and a classifier, is first pre-trained using a small set of labeled data and then iteratively refined using a slice self-training algorithm, which removes or relabels unreliable labels and retrains the network. In optimal transfer learning, the target-site SQA-Net is first initialized with the weights of the trained source-site SQA-Net, then pre-trained with target-site data, and eventually iteratively refined using a slice self-training algorithm. In CARA, we employ adversarial learning to incrementally ensure the consistency of feature representations learned for existing labeled and new unlabeled data. During the target-site volume training stage, by agglomerating the quality ratings of slices belonging to one volume, we train a random forest on all target-site data using a volume self-training algorithm.

2   Method

Figure 1 shows the architecture of MSI-IQA, consisting of four sequential stages, i.e., source-site slice training, optimal transfer learning, consensus adversarial representation adaptation (CARA), and target-site volume training. The first three stages are for slice quality assessment and the fourth stage is for volume quality assessment. The source-site slice training stage is designed to predict

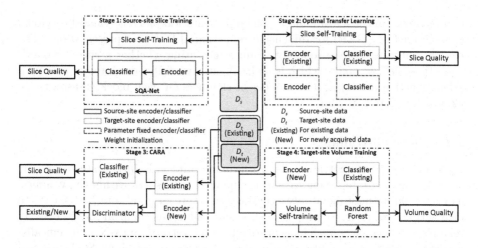

Fig. 1. Architecture of MSI-IQA. Stage 1: Source-site encoder and classifier are trained using source-site data and refined via slice self-training. Stage 2: Target-site encoder and classifier are initialized with the source-site network weights, and refined via slice self-training using target-site data. Stage 3: If exists new data with a distribution that differs from the existing data, feature discrepancy is resolved via adversarial learning. Stage 4: Random forest and volume self-training are used to aggregate the QA results of all slices belonging to a single volume to output a volume quality score.

the quality of each source-site slice via iterative slice self-training. The optimal transfer learning is designed to transfer source-site SQA-Net to target site and then optimize the target-site SQA-Net via slice self-training. The CARA stage is designed to learn a consensus feature representation of existing and incremental new data and to refine the classifier. By assembling the quality ratings of all slices belonging to a volume, the fourth stage employs a random forest to evaluate the quality of each volume with volume self-training. We introduce the details of the proposed MSI-IQA method in the following sections.

2.1 Slice Quality Assessment Network (SQA-Net)

Our SQA-Net is designed with both accuracy and speed in mind. It involves two parts, i.e., an encoder and a classifier. The encoder consists of six sequential blocks, i.e., two convolution blocks, two residual blocks, a nonlocal block [7], and a representation block. The convolution block consists of a convolution layer followed by batch normalization (BN) and exponential linear unit (ELU) activation [8]. The residual block is constructed by integrating 3×3 convolution with unit stride into a residual block, where each convolution is followed by BN and ELU activation, and a 2×2 max-pooling layer is adopted to down-sample the intermediate features. The shortcut branch of the residual block is implemented by a 1×1 convolution with a 2×2 stride. The nonlocal block is utilized to capture long-range dependencies between features regardless of their positional distances and to compute the response function at each position as a weighted summation of all features in the feature maps. The representation block is realized by global average pooling (GAP) for dimension reduction of feature representations. The classifier outputs three probability values indicating whether a slice is "pass", "questionable", or "fail", which is realized with two fully-connected layers and a softmax activation layer. The number of channels for each block is respectively 32, 64, 128, 256, 256, and 3.

2.2   Slice Self-training

Typically, only volumetric labels are available and it is inaccurate to assume that the labels are representative of the quality of all slices within the volumes, as each volume may inevitably contains a mixture of good and bad slices. To deal with noisy labels, we propose a slice self-training mechanism that iteratively relabels/removes slices with inaccurate labels and retrains SQA-Net. Specifically, we first predict the quality of each slice using the pre-trained SQA-Net and then keep slices that meet one of the following criteria: 1) the predicted label is identical to the prediction in the previous iteration; 2) the label is predicted with high-confidence, i.e., the maximum probability exceeds a threshold. After removing slices that do not meet the above criteria, we relabel the remaining slices with high-confidence labels, and retrain the SQA-Net. The above slice self-training process is repeated until improvement in accuracy is negligible.

2.3   Optimal Transfer Learning

After the source-site SQA-Net is completely trained, it will not necessarily perform satisfactorily at the target site due to data discrepancy between the two sites. Instead of training the target-site SQA-Net from scratch, which is time-consuming, we initialize the target-site SQA-Net using weights from the source-site SQA-Net. Considering that low-level features typically vary very little across different tasks [9], we freeze the weights of the first two convolution blocks of the target-site SQA-Net and then adapt it to the target site via transfer learning [10], as illustrated in Stage 2 of Fig. 1. We also utilize the slice self-training mechanism to iteratively remove/relabel inaccurate slice labels and retrain the network using the remaining relabeled dataset, until convergence.

2.4   Consensus Adversarial Representation Adaptation (CARA)

The target-site SQA-Net mentioned above is trained using existing annotated target-site data. When newly acquired but unannotated MRI data become available, CARA utilizes two encoders to respectively embed the existing and new data in a common feature space. The feature consensus between the encoders is achieved with the help of adversarial learning. The feature discrepancy between the existing and new data is minimized in the common space, yielding generalizable representations that avoid over-fitting to either existing or new data, as illustrated in Stage 3 of Fig. 1. Specifically, we initialize these two encoders with the weights of the SQA-Net trained in Stage 2, and then employ adversarial learning to fine-tune these two encoders until their outputs are indistinguishable by a discriminator for existing and new data. This is akin to the GAN mechanism used in training generators so that they produce images that are indistinguishable from real ones. After updating the encoders and discriminator in each iteration, we also fine-tune the classifier trained on existing data while freezing the encoders to achieve consistent prediction of slice quality.

2.5   Volume Self-training

We employ random forest, which is effective even with small training datasets [11], to predict volumetric quality of a volume based on the quality ratings of its constituent slices. Both existing and new volumes are utilized to pre-train the random forest, where the initial quality ratings of the unannotated new volumes are determined based on the following rules: 1) "Pass" if more than 80% of slices within a volume is labeled as "pass"; 2) "Fail" if more slices are labeled as "fail" than "pass" or "questionable"; 3) "Questionable" if otherwise. Similar to the slice self-training algorithm, we utilize volume self-training to reduce the influence of volumetric label noise. The volume self-training also involves iterative volume relabeling and removal, and retraining of random forest, as illustrated in Stage 4 of Fig. 1.

2.6   Loss Function

SQA-Net Training. To deal with data imbalance, a multi-class balanced focal loss [12] with L_2 regularization is used, i.e.,

$$\mathcal{L}(p_t) = -\alpha_t(1 - p_t)^\kappa \log(p_t) + \frac{\lambda}{2n_w} \sum_w \|w\|_2^2, \tag{1}$$

where p_t, $t = 1, 2, 3$, are the predicted probabilities for "pass", "questionable", and "fail", $\kappa \geq 0$ is a focusing parameter, w's are the weight matrices of the SQA-Net, $\lambda = 0.01$ is a tuning parameter for L_2 regularization, and n_w is the number of weight matrices. Here, the class weights $\alpha_t = \max(N_1, N_2, N_3)/N_t$ are used for balancing the contributions of imbalanced datasets, where N_t $(t = 1, 2, 3)$ is the number of slices associated with the t-th class.

CARA. An entropy based adversarial loss is used for CARA as follows:

$$\mathcal{L}_D(\theta_{E_e}, \theta_{E_n}, \theta_D) = -\mathbb{E}_{i_e \sim \mathcal{E}} \left(\log\left(D(f_e, \theta_D)\right)\right) - \mathbb{E}_{i_n \sim \mathcal{N}} \left(\log\left(1 - D(f_n, \theta_D)\right)\right), \tag{2}$$

where θ_{E_e}, θ_{E_n} and θ_D, respectively, are the weights of the encoder E_e for existing data $\mathcal{E}$, the encoder E_n of new data $\mathcal{N}$ in target site and the discriminator D. f_e and f_n, respectively, are the output feature representations of E_e and E_n. The GAN loss of the encoder for existing data is

$$\mathcal{L}_{E_e}(\theta_{E_e}, \theta_D) = -\mathbb{E}_{i_e \sim \mathcal{E}} \left(\log(D(f_e, \theta_D))\right) \tag{3}$$

Table 1. Dataset

Data Site		Training Data (Unit: Slice)			Testing Data (Unit: Slice)			Unlabeled Data (Unit: Slice)
		Pass	Ques	Fail	Pass	Ques	Fail	
Source Site		1740	1500	420	–	–	–	–
Target Site	Existing	1680	1680	360	305	115	60	–
	New	–	–	–	725	115	120	3780

and the inverted label GAN loss is employed to train the encoder for new data as follows:

$$\mathcal{L}_{E_n}(\theta_{E_n}, \theta_D) = -\mathbb{E}_{i_n \sim \mathcal{N}} \left(\log(D(f_e, \theta_D)) \right). \tag{4}$$

The loss used for classifier fine-tuning is the same as Eq. (1).

3 Experimental Results

3.1 Dataset

We evaluated our MSI-IQA framework using T1-weighted MR images of pediatric subjects from birth to six years of age [5]. As shown in Table 1, the images were separated into three datasets: 1) Training dataset with noisy labels; 2) Testing dataset with reliable labels at the target site; 3) Unlabeled new dataset at the target site. We also categorized the images into three datasets according to imaging sites and ages, as follows, 1) Source-site dataset with 8- to 10-month-old subjects; 2) Target-site existing dataset with 10 to 13-month-old subjects; 3) Target-site new dataset with 13- to 24-month-old subjects. Note that due to site and age differences, the data from these three datasets may have different distributions, which we resolve using transfer learning and adversarial learning in MSI-IQA. According to the artifact level, the slices are labeled as "pass" (no/minor artifacts), "questionable" (moderate artifacts), and "fail" (heavy artifacts). Each slice was uniformly padded to 256×256, min-max intensity normalized, and labeled according to the volume it belongs to.

3.2 Implementation Details

MSI-IQA was implemented using Keras with Tensorflow backend. During training, the RMSprop optimizer was utilized with initial learning rate 1×10^{-5} and decay rate 5×10^{-8}. To avoid overfitting, the training data were augmented via rotation and horizontal flipping. The slice/volume self-training algorithm was repeated twice with the same confidence threshold of 0.9.

3.3 Evaluation Metrics

We evaluate MSI-IQA using three performance metrics, including classification accuracy (ACC), sensitivity (SEN), and specificity (SPE), which are respectively defined as ACC= (TP+TN)/(TP+TN+FP+FN), SEN=TP/(TP+FN),

Table 2. Confusion matrix, sensitivity, specificity, and accuracy for IQA using existing data at the target-site.

(a) Source-Site Training

Image Quality		Predicted			SEN	SPE	ACC
		Pass	Ques	Fail			
Actual	Pass	322	28	0	0.9200	0.9421	
	Ques	11	119	0	0.9154	0.7976	0.8259
	Fail	0	55	5	0.0833	1.0000	

(b) Individual Target-Site Training

Image Quality		Predicted			SEN	SPE	ACC
		Pass	Ques	Fail			
Actual	Pass	328	22	0	0.9371	0.9789	
	Ques	4	126	0	0.9692	0.9415	0.9481
	Fail	0	2	58	0.9667	1.0000	

(c) MSI-IQA

Image Quality		Predicted			SEN	SPE	ACC
		Pass	Ques	Fail			
Actual	Pass	347	3	0	0.9914	0.9842	
	Ques	3	127	0	0.9769	0.9902	0.9870
	Fail	0	1	59	0.9833	1.0000	

and SPE=TN/(FP+TN), where TP, TN, FP, and FN respectively denote the number of true positive, true negative, false positive, and false negative cases.

3.4 Performance Analysis

The confusion matrix, sensitivity, specificity, and accuracy of slice QA using existing and new testing slices are shown in Tables 2 and 3. In both tables, (a) and (b) respectively show the results given by networks that are trained using only the source-site and target-site data and (c) show the results of MSI-IQA, which incrementally uses data from both sites. Both tables show that IQA networks trained at the source site and target site alone are not sufficient to yield satisfactory slice-wise QA results at the target site. A similar conclusion can be drawn for volume-wise QA, where the accuracy values for the three methods are respectively 0.5556, 0.8889 and 1.0000 when using existing volumes for testing, and 0.6923, 0.9231, and 1.0000 when using new volumes for testing. These results demonstrate the effectiveness of MSI-IQA in leveraging multi-site data for improving IQA without explicit data sharing between sites.

Table 3. Confusion matrix, sensitivity, specificity, and accuracy for IQA using new data at the target-site.

(a) Source-site Training

Image Quality		Predicted			SEN	SPE	ACC
		Pass	Ques	Fail			
Actual	Pass	563	162	0	0.7766	0.9702	
	Ques	7	108	0	0.9391	0.6911	0.7208
	Fail	0	99	21	0.175	1.0000	

(b) Individual Target-site Training

Image Quality		Predicted			SEN	SPE	ACC
		Pass	Ques	Fail			
Actual	Pass	708	17	0	0.9821	0.9745	
	Ques	6	109	0	0.9478	0.9846	0.9760
	Fail	0	0	120	1.0000	1.0000	

(c) MSI-IQA

Image Quality		Predicted			SEN	SPE	ACC
		Pass	Ques	Fail			
Actual	Pass	723	3	0	0.9959	0.9830	
	Ques	4	111	0	0.9652	0.9965	0.9938
	Fail	0	0	120	1.0000	1.0000	

4　Conclusion

In this paper, we have proposed a deep learning based incremental multi-site IQA method for pediatric structural MR images without the need of inter-site data sharing. Our method first employs quality-annotated images at the source site to train a network for slice-level quality assessment and then adapt the network to a target site via transfer learning. Our method incorporates self-training algorithms to remove/relabel noisy labels both at the slice and volume level. Our method also include a mechanism to resolve feature discrepancy between existing data used to train the network and newly acquired but unannotated data. Experimental results validated that our method yields high accuracy for mutli-site IQA without the need for explicit data sharing.

References

1. Sheikh, H., Bovik, A.: Image information and visual quality. IEEE Trans. Image Process. **15**(2), 430–444 (2006)
2. Chow, L.S., Paramesran, R.: Review of medical image quality assessment. Biomed. Signal Process. Control **27**, 145–154 (2016)
3. Gedamu, E.L., Collins, D., Arnold, D.L.: Automated quality control of brain MR images. J. Magn. Reson. Imaging **28**(2), 308–319 (2008)

4. Dluhoš, P., et al.: Multi-center machine learning in imaging psychiatry: a meta-model approach. Neuroimage **155**, 10–24 (2017)
5. Howell, B.R., et al.: The UNC/UMN baby connectome project (BCP): an overview of the study design and protocol development. Neuroimage **185**, 891–905 (2019)
6. Xia, W., et al.: It's all in the timing: calibrating temporal penalties for biomedical data sharing. J. Am. Med. Inf. Assoc. **25**(1), 25–31 (2017)
7. Wang, X., Girshick, R., Gupta, A., He, K.: Non-local neural networks. In: Proceedings of IEEE Conference on Computer Vision and Pattern Recognition (CVPR), Salt Lake City, USA, pp. 7794–7803, June 2018
8. Clevert, D.A., Unterthiner, T., Hochreiter, S.: Fast and accurate deep network learning by exponential linear units (ELUS). In: Proceedings of International Conference on Learning Representations (ICLR) (2016)
9. Oquab, M., Bottou, L., Laptev, I., Sivic, J.: Learning and transferring mid-level image representations using convolutional neural networks. In: Proceedings of IEEE Conference on Computer Vision and Pattern Recognition (CVPR), June 2014
10. Shin, H.C., et al.: Deep convolutional neural networks for computer-aided detection: CNN architectures, dataset characteristics and transfer learning. IEEE Trans. Med. Imaging **35**(5), 1285–1298 (2016)
11. Breiman, L.: Random forest. Mach. Learn. **45**(1), 5–32 (2001)
12. Lin, T.Y., Goyal, P., Girshick, R., He, K., Dollar, P.: Focal loss for dense object detection. In: Proceedings of IEEE International Conference on Computer Vision (ICCV), Venice, Italy, pp. 2980–2988, October 2017

Surface-Guided Image Fusion for Preserving Cortical Details in Human Brain Templates

Sahar Ahmad, Ye Wu, and Pew-Thian Yap[✉]

Department of Radiology and Biomedical Research Imaging Center (BRIC),
The University of North Carolina at Chapel Hill, Chapel Hill, NC, USA
ptyap@med.unc.edu

Abstract. Human brain templates are a basis for comparison of brain features across individuals. They should ideally capture anatomical details at both coarse and fine scales to facilitate comparison at varying granularity. Brain template construction typically involves spatial normalization and image fusion. While significant efforts have been dedicated to improving brain templates with sophisticated spatial normalization algorithms, image fusion is typically carried out using intensity-based averaging, causing blurring of anatomical structures. Here, we present an image fusion method that exploits cortical surfaces as guidance to help preserve details in brain templates. Our method encodes cortical boundary information given by a cortical surface mesh in a signed distance function (SDF) map. We use the SDF map to help determine localized contributions of the individual images, especially at cortical boundaries, in image fusion. Experimental results demonstrate that our method significantly improves the preservation of fine gyral and sulcal details, resulting in detailed brain templates with good surface-volume agreement.

Keywords: Brain templates · Cortical surface · Image fusion

1 Introduction

The increasing availability of imaging data collected at multiple sites empowers brain researchers with the ability to precisely quantify brain growth and degeneration [8,9]. This is realized using computational anatomy (CA) methods [2] that allow numerical comparison of geometrically complex and convoluted brain structures across individuals [19]. Differences in brain geometry and appearance are typically investigated by mapping the images to a reference coordinate space defined by a brain template [5]. For greater sensitivity to subtle changes, brain templates should ideally preserve fine anatomical details and faithfully capture gyral and sulcal patterns [21].

Brain templates are typically created by (i) spatial normalization of individual brain scans to a common space and (ii) fusion of the spatially normalized brain scans. Abundant efforts have been dedicated to advancing spatial normalization

© Springer Nature Switzerland AG 2021
M. de Bruijne et al. (Eds.): MICCAI 2021, LNCS 12907, pp. 390–399, 2021.
https://doi.org/10.1007/978-3-030-87234-2_37

techniques [12]. In [16], the brain templates for older adults were obtained by iteratively moving the initial template to a common space with rigid, affine, and nonlinear deformations [3]. Schuh *et al.* [17] constructed the spatio-temporal templates
for neonates using unbiased stationary velocity free-form deformation (SVFFD)
algorithm and adaptive kernel regression (AKR). Luo *et al.* [13] used intensity and
sulcal landmark information in a groupwise registration framework to construct a
pediatric brain template. While establishing good correspondences, these methods fall short in preserving subtle details during image fusion. The often employed
AKR technique [14,18] weighs the whole image equally without sufficient consideration of spatially-varying contributions of local details in arriving at the template,
often leading to the blurring of cortical details.

In this paper, we propose an image fusion technique that leverages cortical surfaces to weigh the contributions of local details in images involved in template construction. This is achieved by representing cortical surfaces using signed distance
function (SDF) maps and then using these maps for voxel-wise weighting to accen-
tuate the white-matter/gray-matter (WM/GM) and gray-matter/cerebrospinal
fluid (GM/CSF) interfaces. Our results indicate that the proposed method for
surface-guided image fusion (SGIF) produces brain templates with sharp and distinct anatomical details and good surface-volume agreement.

2   Materials and Methods

2.1   Structural MRI Data

The dataset for building the brain templates consisted of MRI data from three
lifespan human connectome projects (HCPs) [20]: (i) 47 subjects from HCP
Development (HCP-D; 14.58–21.92 years of age); (ii) 67 subjects from HCP
Young Adult (HCP-YA; 22–35 years of age); and (iii) 89 subjects from HCP
Aging (HCP-A; 36.25–70 years of age). The minimally preprocessed data [6,7,
10,11] were used, including T1-weighted (T1w) images, WM and cortical ribbon
parcellation maps, and pial and WM cortical surfaces. The labels defined in
the parcellation maps were grouped into WM, GM, and CSF to create tissue
segmentation maps.

2.2   Construction of Brain Templates

The proposed method involves the following steps: (i) Construction of reference
cortical surface and tissue segmentation templates at reference time points; (ii)
Centralization of reference templates; (iii) Template propagation; (iv) Computation of T1w brain templates via SGIF; and (v) Inter-cohort intensity harmonization of the T1w brain templates.

Construction of Reference Brain Templates. Using subjects scanned within
time interval $\mathcal{W}_r = [(\tau_r - \delta\tau_r), (\tau_r + \delta\tau_r)]$, where $(\tau_r, \delta\tau_r) \in \{(16, 2), (30, 5),$
$(60, 10)\}$ (unit: years), we construct reference templates $\{\mathcal{P}_{\tau_r}^\kappa : \kappa = \text{tissue/surf}\}$
for the HCP-D, HCP-YA, and HCP-A cohorts. For each reference time point, the

tissue segmentation map $T_{(j,\tau)}$ and cortical surfaces $S_{(j,\tau)}$ (pial and WM surfaces for both hemispheres) of each subject $\{\mathcal{D}_{(j,\tau)} : \tau \in \mathcal{W}_r\}$ are registered respectively to the tissue segmentation map and cortical surfaces of a reference subject $\mathcal{R}_\tau{}^1$ using the surface-constrained dynamic elasticity model (SC-DEM) [1]. The aligned tissue segmentation maps are fused by majority voting to construct the reference tissue segmentation template $\mathcal{P}_{\tau_r}^{\text{tissue}}$, which is further iteratively updated by repeating the registration step described above. The corresponding reference cortical surface template $\mathcal{P}_{\tau_r}^{\text{surf}}$ is constructed and updated by weighted averaging of the registered cortical surfaces $\{\hat{S}_{(j,\tau)}\}$:

$$\mathcal{P}_{\tau_r}^{\text{surf}} = \frac{\sum_{(j,\tau)} w(\tau, \tau_r)\hat{S}_{(j,\tau)}}{\sum_{(j,\tau)} w(\tau, \tau_r)} \tag{1}$$

with weights given by

$$w(\tau, \tau_r) = \frac{1}{\sigma_{\tau_r}\sqrt{2\pi}}\exp\left(\frac{-(\tau - \tau_r)^2}{2\sigma_{\tau_r}^2}\right), \tag{2}$$

where temporal smoothness is controlled by σ_{τ_r}, which is set to $\sqrt{2\delta\tau_r}$.

Centralization of Brain Templates. We shift each reference template so that it is at the center of the population of images scanned in the time interval $\mathcal{W}_r$. This reduces bias associated with the selection of the reference subject. We register all subjects $\mathcal{D}_{(j,\tau)}$ to the reference template $\mathcal{P}_{\tau_r}^\kappa$ and average the resulting displacement fields $\{\phi_{(j,\tau)}\}$ using the weights computed in (2) to obtain displacement field

$$\phi_{\tau_r} = \frac{\sum_{(j,\tau)} w(\tau, \tau_r)\phi_{(j,\tau)}}{\sum_{(j,\tau)} w(\tau, \tau_r)}. \tag{3}$$

$\mathcal{P}_{\tau_r}^{\text{surf}}$ and $\mathcal{P}_{\tau_r}^{\text{tissue}}$ are warped with ϕ_{τ_r} and $\phi_{\tau_r}^{-1}$, respectively, to obtain the final templates, denoted as $\mathcal{A}_{\tau_r}^{\text{tissue}}$ and $\mathcal{A}_{\tau_r}^{\text{surf}}$.

Propagation of Brain Templates. For each cohort, we propagate the reference templates to non-reference time points. We register the subjects $\{\mathcal{D}_{(j,\tau)} : \tau \in \mathcal{W}_{nr}\}$ scanned within time interval $\mathcal{W}_{nr} = [(\tau_{nr} - \delta\tau_{nr}), (\tau_{nr} + \delta\tau_{nr})]$, where $(\tau_{nr}, \delta\tau_{nr}) = \{(20, 2), (25, 5), (40, 10)\}$ (unit: years), to the corresponding $\mathcal{A}_{\tau_r}^\kappa$ using SC-DEM. The subjects displacement fields $\{\phi_{(j,\tau)}\}$ are averaged to obtain the final displacement field $\phi_{\tau_{nr}}$ for propagating the brain template:

$$\phi_{\tau_{nr}} = \frac{\sum_{(j,\tau)} w(\tau, \tau_{nr})\phi_{(j,\tau)}}{\sum_{(j,\tau)} w(\tau, \tau_{nr})}, \tag{4}$$

with the weight associated with each subject computed using (2). $\mathcal{A}_{\tau_r}^{\text{surf}}$ and $\mathcal{A}_{\tau_r}^{\text{tissue}}$ are warped with $\phi_{\tau_{nr}}$ and $\phi_{\tau_{nr}}^{-1}$, respectively, to obtain brain templates $\mathcal{A}_{\tau_{nr}}^{\text{tissue}}$ and $\mathcal{A}_{\tau_{nr}}^{\text{surf}}$.

---

^{1} An image closest in appearance to all other images.

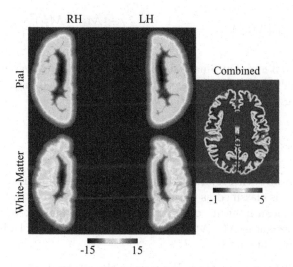

Fig. 1. SDF maps of the 16-year-old pial and WM cortical surface templates combined into a single SDF map $\mathcal{M}_{16}$.

Construction of T1w Brain Templates via SGIF. Next, we construct the T1w brain templates at all time points using surface-guided image fusion (SGIF). The four SDF maps[2] associated with the pial and WM cortical surface templates of both hemispheres, i.e., $\mathcal{M}_{\tau'}^{(h,c)}$, where $h \in \{L, R\}$ and $c \in \{pial, WM\}$, are combined to yield a single SDF map $\mathcal{M}_{\tau'}$ (Fig. 1):

$$\mathcal{M}_{\tau'}^h = \begin{cases} \sum_c \mathcal{M}_{\tau'}^{(h,c)} & \text{if } |\mathcal{M}_{\tau'}^{(h,pial)}| < 0.8 \wedge |\mathcal{M}_{\tau'}^{(h,WM)}| < 5.5, \\ \mathcal{M}_{\tau'}^{(h,pial)} & \text{if } |\mathcal{M}_{\tau'}^{(h,pial)}| < 0.8 \wedge |\mathcal{M}_{\tau'}^{(h,WM)}| > 5.5, \\ \mathcal{M}_{\tau'}^{(h,WM)} & \text{if } |\mathcal{M}_{\tau'}^{(h,pial)}| > 0.8 \wedge |\mathcal{M}_{\tau'}^{(h,WM)}| < 5.5, \\ 0; & \text{otherwise} \end{cases} \tag{5}$$

and

$$\mathcal{M}_{\tau'} = \begin{cases} \sum_h \mathcal{M}_{\tau'}^h & \text{if } \sum_h \mathcal{M}_{\tau'}^h > -0.5, \\ 0; & \text{otherwise.} \end{cases} \tag{6}$$

The three thresholds in (5) and (6) are set empirically.

Next, we use FreeSurfer [6] to reconstruct the cortical surfaces of the warped T1w images $\hat{I}_{(j,\tau)} = I_{(j,\tau)}(\Phi_{(j,\tau)} \circ \Phi_{\tau'}^{-1})$, where the deformation fields $\Phi_{(j,\tau)}$ and $\Phi_{\tau'}^{-1}$ are computed using SC-DEM and (4), respectively. The cortical surfaces are subsequently used to compute SDF maps of each subject $\mathcal{M}_{(j,\tau)}^{(h,c)}$. These hemispheric SDF maps associated with the subject pial and WM cortical surfaces are combined using (5)–(6) to yield a single SDF map $\mathcal{M}_{(j,\tau)}$. $\mathcal{M}_{\tau'}$ and $\mathcal{M}_{(j,\tau)}$ help determine the voxel-wise contribution of the candidate images by computing the weights $w(\mathcal{M}_\tau, \mathcal{M}_{\tau'})$ as

---

[2] https://www.humanconnectome.org/software/connectome-workbench.

$$w(\mathcal{M}_\tau, \mathcal{M}_{\tau'}) = \frac{1}{\sigma_m \sqrt{2\pi}} \exp\left(\frac{-(\mathcal{M}_\tau - \mathcal{M}_{\tau'})^2}{2\sigma_m^2}\right), \tag{7}$$

where $\sigma_m = \sqrt{0.3}$ mm. Our image fusion method takes into account both global and local contributions of the images; thus, intensity-based brain templates are computed using temporal and spatial weights:

$$\mathcal{A}_{\tau'}^{\mathrm{T1w}} = \frac{\sum_{(j,\tau)} w(\tau, \tau') w(\mathcal{M}_\tau, \mathcal{M}_{\tau'}) \hat{I}_{(j,\tau)}}{\sum_{(j,\tau)} w(\tau, \tau') w(\mathcal{M}_\tau, \mathcal{M}_{\tau'})}. \tag{8}$$

Next, we employ an outlier rejection technique to remove spurious voxel intensities. A voxel in $\hat{I}_{(j,\tau)}$ is classified as outlier if its intensity is two standard deviations away from the mean intensity of $\mathcal{A}_{\tau'}^{\mathrm{T1w}}$. For all the outlier voxels, we reduce the corresponding weight $w(\mathcal{M}_\tau, \mathcal{M}_{\tau'})$ by half and recompute the T1w brain template with modified weights in (8).

Inter-Cohort Harmonization. We remove the inter-cohort intensity variability of the brain templates using ComBat-GAM [15]. We fit a generalized additive model (GAM) to capture the nonlinear age effects on image intensity and contrast, captured using intensity mean and standard deviation. The model adjusts the location and scale differences in image intensity and contrast across cohorts. The brain templates are updated according to the harmonized intensity means and standard deviations.

Construction of Brain Templates with Existing Methods. For evaluation, we construct T1w brain templates with (i) adaptive kernel regression (AKR); and (ii) tissue-guided image fusion (TGIF). AKR uses only temporal weighting given as

$$\mathcal{B}_{\tau'}^{\mathrm{T1w}} = \frac{\sum_{(j,\tau)} w(\tau, \tau') \hat{I}_{(j,\tau)}}{\sum_{(j,\tau)} w(\tau, \tau')}. \tag{9}$$

For TGIF-based brain templates, we compute voxel-wise weights using the difference between the tissue segmentation template and the tissue segmentation map of each subject:

$$w(\mathcal{A}_{\tau'}^{\mathrm{tissue}}, T_\tau) = \frac{1}{\sigma_t \sqrt{2\pi}} \exp\left(\frac{-(\mathcal{A}_{\tau'}^{\mathrm{tissue}} - T_\tau)^2}{2\sigma_t^2}\right), \tag{10}$$

where $\sigma_t = \sqrt{8000}$. The T1w brain templates for TGIF ($\mathcal{C}_{\tau'}^{\mathrm{T1w}}$) are constructed using both the temporal and spatial weights, similar to SGIF:

$$\mathcal{C}_{\tau'}^{\mathrm{T1w}} = \frac{\sum_{(j,\tau)} w(\tau, \tau') w(\mathcal{A}_{\tau'}^{\mathrm{tissue}}, T_\tau) \hat{I}_{(j,\tau)}}{\sum_{(j,\tau)} w(\tau, \tau') w(\mathcal{A}_{\tau'}^{\mathrm{tissue}}, T_\tau)}. \tag{11}$$

Inter-cohort harmonization was performed on brain templates constructed with AKR and TGIF.

3   Results and Discussion

We performed qualitative and quantitative evaluation of the brain templates constructed with AKR, TGIF, and SGIF. The SGIF-based brain templates are sharper with distinct structural boundaries compared with templates obtained using AKF and TGIF (Fig. 2). These differences are more noticeable in the close-ups of the brain templates, as shown for selected time points in Fig. 3.

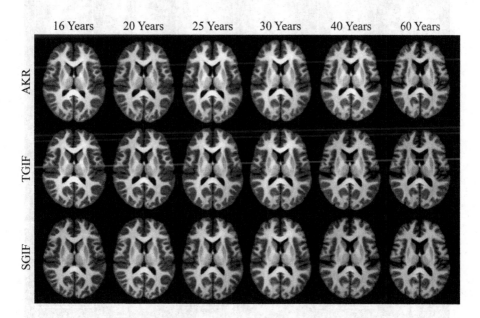

Fig. 2. Brain templates constructed by AKR, TGIF, and SGIF.

Most apparent are the clearer cortical boundaries in the parietal and occipital lobes of the brain templates constructed with SGIF. AKR and TGIF result in brain templates with blurred cortical details; adjacent gyri merge and cerebral sulci fade. SGIF also yields better alignment of the WM/GM and GM/CSF interfaces with the cortical surface templates. Overlaying the pial surface templates onto the volumetric templates, shown in Fig. 4 for the 40-year-old brain template, indicates good alignment at the GM/CSF interface.

We quantified the sharpness of the three sets of brain templates using a frequency domain image blur measure [4] and the standard deviation of GM intensities (Fig. 5). SGIF scores higher values for both measures at all time points, indicating brain templates with more high-frequency contents capturing finer details and better contrast. Furthermore, we evaluated the alignment accuracy of the volumetric and cortical surface templates by computing the modified Hausdorff distance (MHD) [1] between the cortical surface templates and the cortical surfaces reconstructed from the T1w brain templates (Fig. 6(a)). The MHD values for SGIF

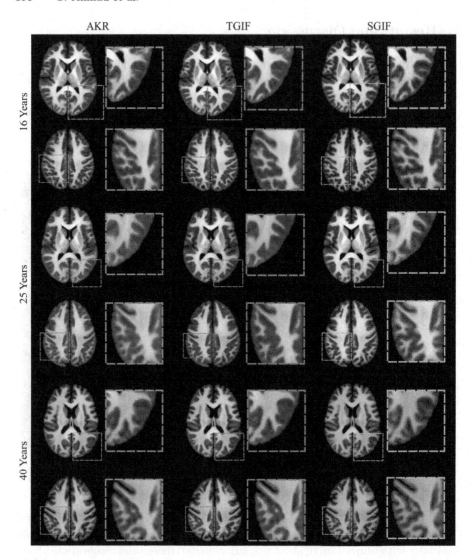

Fig. 3. Axial views (*left*) and the corresponding closeups (*right*) of the brain templates constructed by the three methods.

are lower (mean $\pm$ std $= 0.56 \pm 0.04$) than AKR (mean $\pm$ std $= 0.68 \pm 0.05$) and TGIF (mean $\pm$ std $= 0.66 \pm 0.05$). To complement our evaluation, we computed the mean absolute distance (MAD) between the SDF maps of the cortical surface templates and the T1w brain templates (Fig. 6(b)). The mean $\pm$ std of MAD are 0.54 ± 0.06, 0.58 ± 0.07, and 0.37 ± 0.05 for AKR, TGIF, and SGIF, respectively. The improvements given by SGIF are statistically significant for both distance measures ($p < 0.01$).

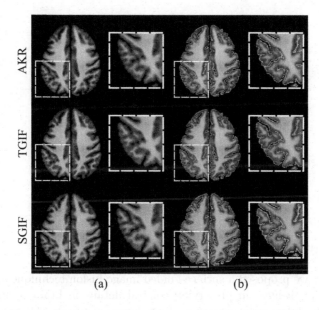

Fig. 4. The 40-year-old brain template (a) overlaid with pial surface template in (b).

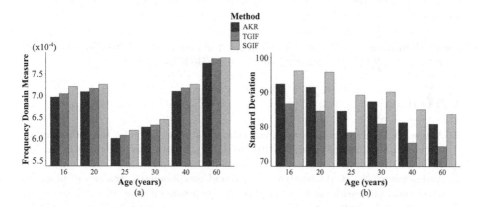

Fig. 5. (a) Frequency domain image blur measure and (b) standard deviation of GM intensities of the brain templates constructed by the three methods.

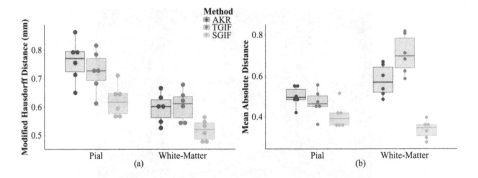

Fig. 6. Boxplots for MHD and MAD.

4   Conclusion

In this paper, we proposed a surface-guided image fusion technique and demonstrated its effectiveness in preserving cortical details in brain templates. Our method utilizes information on cortical boundaries encoded in SDF maps to determine the voxel-wise contributions of images involved in constructing the templates. Experimental results indicate that our method yields templates that preserve significantly more details than conventional kernel regression and local fusion methods.

Acknowledgments. This work was supported in part by the United States National Institutes of Health (NIH) grants EB008374 and EB006733.

References

1. Ahmad, S., et al.: Surface-constrained volumetric registration for the early developing brain. Med. Image Anal. **58**, 101540 (2019). https://doi.org/10.1016/j.media.2019.101540
2. Andescavage, N., et al.: In vivo textural and morphometric analysis of placental development in healthy and growth-restricted pregnancies using magnetic resonance imaging. Pediatr. Res. **85**, 974–981 (2019). https://doi.org/10.1038/s41390-019-0311-1
3. Avants, B.B., et al.: The optimal template effect in hippocampus studies of diseased populations. NeuroImage **49**(3), 2457–2466 (2010). https://doi.org/10.1016/j.neuroimage.2009.09.062
4. De, K., Masilamani, V.: Image sharpness measure for blurred images in frequency domain. Procedia Eng. **64**, 149–158 (2013). https://doi.org/10.1016/j.proeng.2013.09.086
5. Evans, A.C., et al.: Brain templates and atlases. Neuroimage **62**(2), 911–922 (2012). https://doi.org/10.1016/j.neuroimage.2012.01.024
6. Fischl, B.: Freesurfer. NeuroImage **62**(2), 774–781 (2012). https://doi.org/10.1016/j.neuroimage.2012.01.021

7. Glasser, M.F., et al.: The minimal preprocessing pipelines for the human connectome project. NeuroImage **80**, 105–124 (2013). https://doi.org/10.1016/j.neuroimage.2013.04.127

8. He, C., et al.: Structure-function connectomics reveals aberrant developmental trajectory occurring at preadolescence in the autistic brain. Cereb. Cortex **30**(9), 5028–5037 (2020). https://doi.org/10.1093/cercor/bhaa098

9. Herting, M.M., et al.: Development of subcortical volumes across adolescence in males and females: a multisample study of longitudinal changes. NeuroImage **172**, 194–205 (2018). https://doi.org/10.1016/j.neuroimage.2018.01.020

10. Jenkinson, M., et al.: Improved optimization for the robust and accurate linear registration and motion correction of brain images. NeuroImage **17**(2), 825–841 (2002). https://doi.org/10.1006/nimg.2002.1132

11. Jenkinson, M., et al.: FSL. NeuroImage **62**(2), 782–790 (2012). https://doi.org/10.1016/j.neuroimage.2011.09.015

12. Kuklisova-Murgasova, M., et al.: A dynamic 4D probabilistic atlas of the developing brain. NeuroImage **54**(4), 2750–2763 (2011). https://doi.org/10.1016/j.neuroimage.2010.10.019

13. Luo, Y., et al.: Intensity and sulci landmark combined brain atlas construction for Chinese pediatric population. Hum. Brain Mapp. **35**(8), 3880–3892 (2014). https://doi.org/10.1002/hbm.22444

14. Makropoulos, A., et al.: Regional growth and atlasing of the developing human brain. NeuroImage **125**, 456–478 (2016). https://doi.org/10.1016/j.neuroimage.2015.10.047

15. Pomponio, R., et al.: Harmonization of large MRI datasets for the analysis of brain imaging patterns throughout the lifespan. NeuroImage **208**, 116450 (2020). https://doi.org/10.1016/j.neuroimage.2019.116450

16. Ridwan, A.R., et al.: Development and evaluation of a high performance T1-weighted brain template for use in studies on older adults. Hum. Brain Mapp., 1–19 (2021). https://doi.org/10.1002/hbm.25327

17. Schuh, A., et al.: Unbiased construction of a temporally consistent morphological atlas of neonatal brain development. bioRxiv (2018). https://doi.org/10.1101/251512

18. Serag, A., et al.: A multi-channel 4D probabilistic atlas of the developing brain: application to fetuses and neonates. Spec. Issue Ann. Br. Mach. Vis. Assoc. (2011)

19. Valk, S.L., et al.: Shaping brain structure: genetic and phylogenetic axes of macroscale organization of cortical thickness. Sci. Adv. **6**(39) (2020). https://doi.org/10.1126/sciadv.abb3417

20. Van Essen, D.: The human connectome project: a data acquisition perspective. NeuroImage **62**(4), 2222–2231 (2012). https://doi.org/10.1016/j.neuroimage.2012.02.018

21. Zhang, Y., et al.: Detail-preserving construction of neonatal brain atlases in space-frequency domain. Hum. Brain Mapp. **37**(6), 2133–2150 (2016). https://doi.org/10.1002/hbm.23160

Longitudinal Correlation Analysis for Decoding Multi-modal Brain Development

Qingyu Zhao[1(✉)], Ehsan Adeli[1,2], and Kilian M. Pohl[1,3]

[1] School of Medicine, Stanford University, Stanford, USA
qingyuz@stanford.edu
[2] Computer Science Department, Stanford University, Stanford, USA
[3] Center of Health Sciences, SRI International, Menlo Park, USA

Abstract. Starting from childhood, the human brain restructures and rewires throughout life. Characterizing such complex brain development requires effective analysis of longitudinal and multi-modal neuroimaging data. Here, we propose such an analysis approach named Longitudinal Correlation Analysis (LCA). LCA couples the data of two modalities by first reducing the input from each modality to a latent representation based on autoencoders. A self-supervised strategy then relates the two latent spaces by jointly disentangling two directions, one in each space, such that the longitudinal changes in latent representations along those directions are maximally correlated between modalities. We applied LCA to analyze the longitudinal T1-weighted and diffusion-weighted MRIs of 679 youths from the National Consortium on Alcohol and Neurodevelopment in Adolescence. Unlike existing approaches that focus on either cross-sectional or single-modal modeling, LCA successfully unraveled coupled macrostructural and microstructural brain development from morphological and diffusivity features extracted from the data. A retesting of LCA on raw 3D image volumes of those subjects successfully replicated the findings from the feature-based analysis. Lastly, the developmental effects revealed by LCA were inline with the current understanding of maturational patterns of the adolescent brain.

1 Introduction

The human brain undergoes profound changes over the entire life span [1]. Starting from childhood, gray matter volume declines [2], in part reflecting synaptic pruning [3], while white matter volume increases, contributing to efficient neural signaling and transmission [2]. The marked morphological development, along with the maturation in microstructural fiber pathways [4], gives rise to the remodeling of functional brain circuits, supporting the capacity for high-level cognition [5]. To advance the understanding of such multifaceted neurodevelopment, studies have increasingly relied on analyzing *longitudinal* and *multi-modal* MRI data collected over specific age ranges [6].

© Springer Nature Switzerland AG 2021
M. de Bruijne et al. (Eds.): MICCAI 2021, LNCS 12907, pp. 400–409, 2021.
https://doi.org/10.1007/978-3-030-87234-2_38

While deep learning techniques have led to tremendous progress in the analysis of brain MRIs [7], the focus has been on either cross-sectional or single modality data. For example, existing longitudinal models for assessing neurodevelopment are primarily designed for structural MRIs only [8]. Multi-modal analysis [9–12], on the other hand, explores common factors across modalities using cross-sectional data, so the identified factors are not guaranteed to characterize development. Another limitation of deep learning in analyzing development is that they are often formulated as supervised learning models with respect to pseudo-marker of developmental stages, such as age. As suggested in [8], such age-based supervision is sub-optimal as brain development is highly heterogeneous in populations with similar chronological ages.

To identify developmental effects from two longitudinal MRI modalities without the guidance of age, we propose a longitudinally self-supervised deep learning approach, named Longitudinal Correlation Analysis (LCA). LCA consists of an autoencoding structure that separately reduces the data of either modality to a latent representation. Motivated by Canonical Correlation Analysis (CCA) [13], LCA then relates the two latent spaces by jointly estimating one direction in each space, such that the longitudinal changes (of latent representations) along those directions are maximally correlated between modalities. Training of LCA is self-supervised as it only requires grouping the repeated measures of each individual but does not rely on chronological age (or any other supervisory signal).

We applied LCA to analyze the longitudinal multi-modal data of 679 subjects from the National Consortium on Alcohol and Neurodevelopment in Adolescence (NCANDA) [14] to investigate the macrostructural and microstructural brain development of adolescents. Based on features extracted from longitudinal T1 and diffusion-weighted MRIs, LCA successfully revealed coupled developmental effects from the two modalities while baseline methods failed to do so. We then successfully replicated the findings of this feature-based analysis by retesting LCA on the raw longitudinal T1 images and Fractional Anisotropy maps. Lastly, we show that the developmental effects revealed by LCA were inline with the current understanding of maturational patterns of the adolescent brain.

2   Method

We first review the Canonical Correlation Analysis (CCA) [13] for identifying association across two modalities. We then discuss the difficulties of linking the CCA solution to interpretable factors. This limitation motivates our design of self-supervised learning based on the repeated measures of longitudinal data to jointly disentangle developmental effects across modalities.

Canonical Correlation Analysis. Let $X := [\mathbf{x}^1, \cdots, \mathbf{x}^S] \in \mathbb{R}^{n \times S}$ be the data of the first modality from S subjects, where $\mathbf{x}^s \in \mathbb{R}^n$ is the data of subject 's'. Likewise, let $Y := [\mathbf{y}^1, \cdots, \mathbf{y}^S] \in \mathbb{R}^{m \times S}$ be the data of the second modality. CCA aims to find two vectors $\boldsymbol{\tau}_x \in \mathbb{R}^n$ and $\boldsymbol{\tau}_y \in \mathbb{R}^m$, such that the projections of X and Y onto these vectors are maximally correlated, i.e.,

$$(\boldsymbol{\tau}_x, \boldsymbol{\tau}_y) := \arg\max_{\boldsymbol{\tau}'_x, \boldsymbol{\tau}'_y} \ \text{corr}(\boldsymbol{\tau}'_x{}^\top X, \boldsymbol{\tau}'_y{}^\top Y). \tag{1}$$

We call the optimal τ_x and τ_y the canonical vectors and the resulting projections $p_x := \tau_x^\top X$ and $p_y := \tau_y^\top Y$ the *canonical variables*.

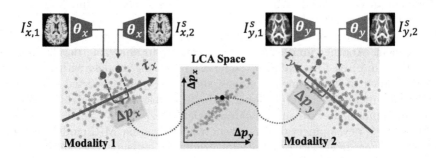

Fig. 1. LCA finds two directions τ_x and τ_y in two latent spaces, one for each modality, such that the projections $(\Delta p_x, \Delta p_y)$ of longitudinal changes are maximally correlated.

While these canonical variables characterize the strongest association across modalities, their meanings are not easily interpretable as they cannot be exactly linked to any single real-life factor underlying the data (such as age or gender). To conceptually illustrate this ambiguity problem, we assume that the data are generated by K hidden factors $\boldsymbol{\alpha} := [\alpha^1, \cdots, \alpha^K]^\top$ through linear functions. In other words, $\mathbf{x} = U\boldsymbol{\alpha}$ and $\mathbf{y} = V\boldsymbol{\alpha}$ with $U := [\mathbf{u}^1, \cdots, \mathbf{u}^K] \in \mathbb{R}^{n \times K}$ and $V := [\mathbf{v}^1, \cdots, \mathbf{v}^K] \in \mathbb{R}^{m \times K}$ consisting of K components. Now, we define the canonical variables p_x and p_y as being linked to a single factor α_k if τ_x and τ_y are orthogonal to all components except for $\mathbf{u}^k$ and $\mathbf{v}^k$, such that

$$p_x = \tau_x^\top \mathbf{x} = \tau_x^\top U\boldsymbol{\alpha} = \tau_x^\top \mathbf{u}^k \alpha^k \propto \alpha^k, \text{ and likewise, } p_y \propto \alpha^k. \qquad (2)$$

In practice, the above condition cannot be guaranteed in the CCA setting as $\boldsymbol{\alpha}$ cannot be uniquely determined. In fact, for any orthogonal transformation R, one can re-define the generative procedure with respect to $\boldsymbol{\alpha}' = R\boldsymbol{\alpha}$ as

$$\mathbf{x} = U\boldsymbol{\alpha} = UR^\top R\boldsymbol{\alpha} = U'\boldsymbol{\alpha}', \text{ and likewise, } \mathbf{y} = V'\boldsymbol{\alpha}'. \qquad (3)$$

This indicates the same p_x and p_y are linked to a family of possible factors defined by the quotient group of orthogonal transformation. In fact, such ambiguity is not limited to CCA but generally exists in probabilistic formulations of independent component analysis (ICA) and principal component analysis (PCA), where the generative factors can only be determined up to rotation [15].

Longitudinal Correlation Analysis (LCA). We now show that we can leverage longitudinal data to relate canonical variables to the single factor of brain development without ambiguity by assessing the 'differential effects' between repeated measures. To do this, we first observe that while $\mathbf{x}$ is a composite of multiple factors, the partial derivative $\partial \mathbf{x}/\partial \alpha^k$ is a single component $\mathbf{u}^k$. In other words, if we perturb α^k by $\Delta \alpha^k$ while fixing the value of other factors, the resulting difference in $\mathbf{x}$ is simply $\Delta\mathbf{x} = \mathbf{u}^k \Delta\alpha^k$, which is solely dependent on the kth component. Moreover, the resulting difference in canonical variables are also guaranteed to link to one component because

$$\Delta p_x = \boldsymbol{\tau}_x^\top \Delta\mathbf{x} = (\boldsymbol{\tau}_x^\top \mathbf{u}^k)\Delta\alpha^k \propto \Delta\alpha^k, \text{ and likewise, } \Delta p_y \propto \Delta\alpha^k. \tag{4}$$

This observation implies that we can transfer the CCA setup on $\mathbf{x}$ and $\mathbf{y}$ to a setup on the differential vectors $\Delta\mathbf{x}$ and $\Delta\mathbf{y}$ to focus on finding cross-modal information related to one single factor. A definition of $\Delta\mathbf{x}$ and $\Delta\mathbf{y}$ that well fits with longitudinal studies is the difference between the repeated measures of an individual. Unlike the unknown factors underlying $\mathbf{x}$ and $\mathbf{y}$, these differential vectors solely encode a known factor of brain development as all other factors such as race and gender are static across visits (assuming no brain damaging events occur between visits). LCA then maximizes the correlation between Δp_x and Δp_y to unravel coupled developmental effects across modalities.

We embed LCA in an autoencoder setting (Fig. 1). Let $I_{x,1}^s$ and $I_{x,2}^s$ be two images (or feature vectors) of the first modality from subject s with $I_{x,2}^s$ scanned after $I_{x,1}^s$. We assume the images can be reduced to low-dimensional latent representations by an encoder network with parameters θ_x, i.e., $\mathbf{x}_1^s := f(I_{x,1}^s; \theta_x)$ and $\mathbf{x}_2^s := f(I_{x,2}^s; \theta_x)$. The decoder network g with parameters ϕ_x can then reconstruct the latent representations. Let $\Delta\mathbf{x}^s = \mathbf{x}_2^s - \mathbf{x}_1^s$, $\Delta\mathbf{y}^s = \mathbf{y}_2^s - \mathbf{y}_1^s$ and let $\Delta X(\theta_x) := [\Delta\mathbf{x}^1, \cdots, \Delta\mathbf{x}^s]$, $\Delta Y(\theta_y) := [\Delta\mathbf{y}^1, \cdots, \Delta\mathbf{y}^s]$. We now propose to couple the two developmental effects by finding $\boldsymbol{\tau}_x$ and $\boldsymbol{\tau}_y$ such that projections of the longitudinal changes are maximally correlated across the group while reducing the reconstruction loss based on the mean-squared error $|| \cdot ||^2$:

$$\arg\max_{\boldsymbol{\tau}_x, \boldsymbol{\tau}_y, \theta_x, \theta_y, \phi_x, \phi_y} \text{corr}(\boldsymbol{\tau}_x^\top \Delta X(\theta_x), \boldsymbol{\tau}_y^\top \Delta Y(\theta_y)) \tag{5}$$

$$- \lambda \sum_{s=1}^{S} \sum_{t=1}^{2} \left(||g(f(I_{x,t}^s; \theta_x); \phi_x), I_{x,t}^s||^2 + ||g(f(I_{y,t}^s; \theta_y); \phi_y), I_{y,t}^s||^2 \right)$$

This optimization is fully self-supervised as we only need to pair the repeated measures within each individual. Although the training only correlates Δp_x and Δp_y, the actual values of p_x and p_y can be directly computed by projecting data onto $\boldsymbol{\tau}_x$ and $\boldsymbol{\tau}_x$ during validation.

3 Experiments

3.1 Setup

We investigate adolescent brain development by applying LCA to feature vectors or 3D volumes derived from T1-weighted (T1) MRIs and from 3D diffusivity maps associated with diffusion weighted MRIs (DWI).

Data. The testing data set consisted of the T1 MRI and DWI of 2169 visits that were from 679 healthy youths recruited by NCANDA (no more than 6 annual visits and in average 3.2 visits per subject; age at baseline visits: 15.7 ± 2.4; 340

boys/339 girls) who met the no-to-low drinking criteria based on the adjusted Cahalan score [16]. For each subject, we constructed pairs from two different visits and randomly selected 3 visit pairs for subjects with more than 3 visits to avoid over-emphasizing those subjects. This procedure resulted in 1054 visit pairs with the average age difference between paired visits being 2.7 ± 0.9 years.

Preprocessing. T1 MRI were preprocessed by denoising, bias-field correction, and skull striping. Cortical thickness and surface area of 68 bilateral cortical regions and volume of 31 sub-cortical regions were extracted by FreeSurfer 5.3 [17] resulting in a 99 dimensional feature for each image. Then the skull-stripped images were further affinely registered to the SRI24 atlas [18] and reduced to $64 \times 64 \times 64$ resolution. DWIs were first registered to the structural images and skull stripped. Preprocessing continued with bad single shots removal, echo-planar, and Eddy-current distortion correction. Camino [19] computed diffusivity maps including fractional anisotropy (FA), medium diffusivity (MD), L1, and LT maps. Average regional measures were extracted from each map by reducing the map to a skeleton via Tract-Based Spatial Statistics (TBSS) [20], dividing the skeleton into 27 regions according to the Johns Hopkins University (JHU) DTI atlas [21], and averaging the skeleton values for each region. This resulted in a 108 dimensional feature. In parallel, the 3D FA maps were reduced to $64 \times 64 \times 64$ resolution.

Implementation. For feature-based analysis, the encoder was a two-layer perceptron with dimension (64, 32) with tanh activation. The decoder adopted the inverse structure. For 3D-image-based analysis, the encoder was an encoder composed of 4 stacks of $3 \times 3 \times 3$ convolution/ReLU/max-pooling layers with feature channel (16, 32, 64, 16). Then, a dense layer reduced the output to a 512 dimensional representation. We set a non-informative $\lambda = 1/(2S)$. Models were implemented in Keras 2.2.2 (see code at https://github.com/QingyuZhao/LCA) and run on an Nvidia Quadro P6000 GPU. Training was confined to the 1054 visit pairs and performed by the Adam optimizer with a 0.0002 learning rate for 200 epochs.

Evaluation. We applied the trained model to derive the canonical variables p_{T1} and p_{DWI} of all 2169 visits (i.e., including also visits that were not paired during training). We then used two types of statistical analysis to examine the relationship between the canonical variables with age (we regard age only as an approximate marker for quantifying brain development but not the ground-truth for p_{T1} and p_{DWI}). To remove the impact of repeated measures, we first computed the Pearson's correlation r between p_{T1} at the baseline visit of each subject and their age at baseline. To account for the consistency of p_{T1} across the repeated measures, we then parameterized a linear mixed effect (LME) model, which estimated a cubic group-level trajectory of p_{T1} over age with each subject having a random intercept. The goodness of fit of LME was examined by the Akaike information criterion (AIC) [22]. Standard deviation of these statistics was generated based on a bootstrapping procedure, which repeated the training and validation 10 times based on resampling for each subject the visit pairs used for training. Lastly, the age-related statistics were also generated for p_{DWI}.

Baselines. While many existing models are designed to leverage association between multi-view data [10–12], to the best of our knowledge, there are no unsupervised or self-supervised multi-view methods that also disentangle latent directions from longitudinal data. We, therefore, compared LCA to multi-modal cross-sectional methods, including Deep CCA (DCCA) [23] and Deep Canonical Correlation Autoencoders (DCCAE) [24], which were extensions of CCA in deep learning settings. All 2169 T1 and DWI images were used for training these two methods as they did not rely on repeated measures. We then further compared LCA to Longitudinal Self-Supervised Learning (LSSL) [8], a longitudinal method that estimates a brain aging direction in the latent space of a single modality. Training LSSL was based on the same set of visit pairs but was independently performed on either modality.

3.2   Results

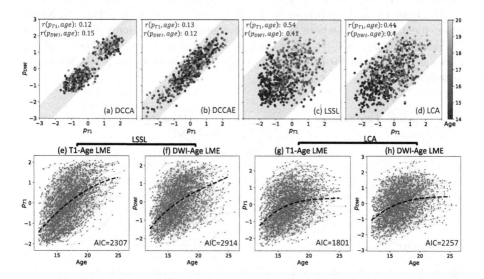

Fig. 2. Feature-based analysis. (a)–(d): canonical variables at the baseline visits of subjects for the 4 comparison methods. Each point is color-coded by age. Width of the gray bands corresponds to Pearson's correlation $r(p_{T1}, p_{DWI})$; (e)–(f): LME fitting between age and p_{T1}, p_{DWI} for LSSL and LCA. Black curves are the average estimation of the cubic fixed effects by bootstrapping.

Feature-Based Analysis. Figures 2(a)–(d) display the canonical variables associated with the two modalities of the 679 subjects at their baseline visits. The gray band in each figure corresponds to the Pearson's correlation $r(p_{T1}, p_{DWI})$. Unlike the baselines, LCA resulted in highly correlated p_{T1} and p_{DWI}, both of which also significantly correlated with age (see also Fig. 3(a)). Note, the significant group-level correlation between p_{T1} and p_{DWI} ($r = 0.61$, Fig. 2d) were the result of training LCA solely on longitudinal differences Δp_{T1} and Δp_{DWI}. Moreover, both canonical variables highly correlated with age ($r \geq 0.4$), indicated by the smooth color transition along the diagonal direction (older youths were color-coded as yellow). Again, this group-level correlation within the whole age span from 12 to 25 years was learned from pairs of visits, whose average time between visits was 2.7 years. This indicates that LCA successfully disentangled the factor linked to brain development within the multi-modal data. In comparison, the first pair of canonical variables identified by DCCA and DCCAE had higher correlation ($r(p_{T1}, p_{DWI}) = 0.87$), but neither variable was related to brain development, indicated by the substantially lower correlation with age ($r \leq 0.15$, Fig. 3a) and the less pronounced color transition in Figs. 2(a)+(b). In fact, age correlation for all 32 pairs of canonical variables of DCCA and DCCAE were relatively low ($r = 0.08 \pm 0.06$), which supports our speculation that each canonical variable of conventional CCA is a composite of multiple factors. While the single-modal analysis of LSSL could separately reveal developmental effect in each modality, the identified p_{T1} and p_{DWI} showed lower correlation ($r = 0.39$, Fig. 2c) than LCA. The differences between modality-specific LSSL results are also evident in Fig. 3a, where the developmental effect was more pronounced in T1 features ($r(p_{T1}, age) = 0.54$) than in DWI features ($r(p_{DWI}, age) = 0.40$). This was not the case for LCA, which resulted in balanced age correlation across modalities (Fig. 3a).

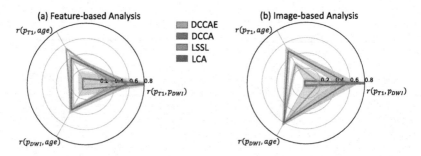

Fig. 3. Mutual correlation between age and the canonical variables p_{T1}, p_{DWI} for feature-based and image-based experiments. Width of triangle edges correspond to standard deviation of the correlation derived by bootstrapping.

Figure 2(e)–(h) shows individual developmental trajectories (blue) estimated by LSSL and LCA and the group-level trajectories derived by the post-hoc LME fitting. Despite both methods producing strong age correlation, the developmental trajectories of p_{T1} and p_{DWI} revealed by LCA were more robust and realistic than those by LSSL. According to the LME fitting (Figs. 2(g)+(h)), LSSL resulted in lower AIC metrics than LCA, suggesting the individual trajectories estimated by LCA were more consistent with the group-level trajectory.

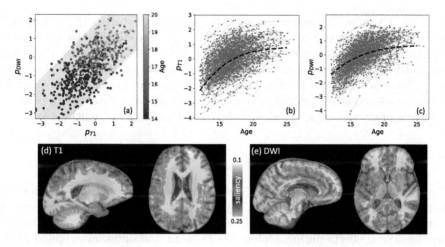

Fig. 4. Image-based analysis by LCA: (a) p_{T1}, p_{DWI} at the baseline visits of subjects; (b)-(c): LME fitting on age; (d)-(e): Saliency maps overlaid on the structural atlas indicating regions extracted from T1 and DWI that show coupled development.

Moreover, LCA revealed that both macrostructural and microstructural development was fast in the teen years but plateaued during adulthood (after age 20), which comported with maturational patterns reported for the NCANDA cohort [16, 25]. In comparison, p_{T1} and p_{DWI} estimated by LSSL showed significant increase even after reaching adulthood (Figs. 2(e)+(f)), which might be due to overfitting. With respect to LCA, the variance in p_{T1} or p_{DWI} at any given age potentially indicates that the brain developmental stage is highly heterogeneous during adolescence, which supports our claim that self-supervised learning is potentially a better approach for quantifying that stage compared to age-guided supervision.

Image-Based Analysis. Similar to the previous experiment, DCCA and DCCAE revealed strongly coupled factors that did not reflect brain development (Fig. 3b). Unlike the previous experiment, LSSL revealed pronounced developmental effects only in DWI but resulted in significantly lower correlation (than the feature-based analysis) between age and p_{T1}. As a result, we observed dissociation between p_{T1} and p_{DWI} by LSSL ($r(p_{T1}, p_{DWI}) = 0.13$, Fig. 3b) compared to the feature-based result. On the other hand, the results of LCA were inline with the findings from the previous experiment. The mutual correlation among p_{T1}, p_{DWI}, and age all increased compared to the feature-based results, suggesting that the raw imaging data contained more comprehensive information compared to low-dimensional features. The three correlation values were well balanced (Fig. 3b), suggesting that joint self-supervised learning might regularize the disentanglement across modalities.

Lastly, we visualized regions that were associated with development in either modality. We did so by quantifying the voxelwise importance in driving the change in p_{T1} and p_{DWI} via guided saliency map [26] being applied to each T1

MRI and FA map. The average saliency maps across the cohort (Figs. 4(d)–(e)) highlighted gray matter areas from the T1 images and highlighted white matter pathways (including the corpus callosum and brainstem) in the FA maps. Such distinction in regions was to be expected as gray matter morphology was only visible in structural images, whereas FA maps were specific to the quantification of white matter integrity. This supports that LCA successfully revealed complementary aspects for adolescent brain maturation from multi-modal data.

4   Conclusion

We have proposed a self-supervised multi-modal data analysis approach named Longitudinal Correlation Analysis. By coupling the developmental effects of repeated measures in latent spaces, LCA successfully uncovered the multifaceted developmental effects during adolescence from T1 and diffusion-weighted MRI. Reproducibility between feature-based and image-based analyses highlighted the robustness of the method. The correlation-based self-supervision may have a broader impact for multi-view and contrastive learning, applications of which remained to be explored in the future.

Acknowledgment. This research was supported in part by NIH U24 AA021697 and Stanford HAI AWS Cloud Credit. The data were part of the public NCANDA data release NCANDA_PUBLIC_5Y_REDCAP_V01, NCANDA_PUBLIC_5Y_STRUCTU-RAL_V01, NCANDA_PUBLIC_5Y_DIFFUSION_V01 [27], whose collection and distribution were supported by NIH funding AA021697, AA021695, AA021692, AA021696, AA021681, AA021690, and AA02169.

References

1. Petrican, R., Taylor, M., Grady, C.: Trajectories of brain system maturation from childhood to older adulthood: implications for lifespan cognitive functioning. Neuroimage **163**, 125–149 (2017)
2. Giedd, J.: Geidd JN. structural magnetic resonance imaging of the adolescent brain. Ann. N. Y. Acad. Sci. **1021**, 77–85 (2004)
3. Natu, V., et al.: Apparent thinning of human visual cortex during childhood is associated with myelination. Proc. Natl. Acad. Sci. **116**, 20750–20759 (2019)
4. Lebel, C., Walker, L., Leemans, A., Phillips, L., Beaulieu, C.: Microstructural maturation of the human brain from childhood to adulthood. Neuroimage **40**, 1044–1055 (2008)
5. Simmonds, D., Hallquist, M., Asato, M., Luna, B.: Developmental stages and sex differences of white matter and behavioral development through adolescence: a longitudinal diffusion tensor imaging (DTI) study. Neuroimage **92**, 356–68 (2013)
6. Sadeghi, N., Prastawa, M., Gilmore, J., Lin, W., Gerig, G.: Towards analysis of growth trajectory through multi-modal longitudinal MR imaging. Proc. Soc. Photo Opt. Instrum. Eng. **7623** (2010)
7. Zhu, G., Jiang, B., Tong, L., Xie, Y., Zaharchuk, G., Wintermark, M.: Applications of deep learning to neuro-imaging techniques. Front. Neurol. **10**, 1–13 (2019)

8. Zhao, Q., Liu, Z., Adeli, E., Pohl, K.M.: Longitudinal self-supervised learning. Med. Image Anal. **7**, 1–11 (2021)
9. Chen, L., Bentley, P., Mori, K., Misawa, K., Fujiwara, M., Rueckert, D.: Self-supervised learning for medical image analysis using image context restoration. Med. Image Anal. **58**, 3–12 (2019)
10. Han, T., Xie, W., Zisserman, A.: Self-supervised co-training for video representation learning. In: NeurIPS (2020)
11. He, K., Fan, H., Wu, Y., Xie, S., Girshick, R.: Momentum contrast for unsupervised visual representation learning. In: Proceedings of the IEEE/CVF Conference on Computer Vision and Pattern Recognition, pp. 9729–9738 (2020)
12. Tsai, Y.H.H., Wu, Y., Salakhutdinov, R., Morency, L.P.: Self-supervised learning from a multi-view perspective. In: International Conference on Learning Representations (ICLR) (2020)
13. Härdle, W.K., Simar, L.: Applied Multivariate Statistical Analysis, 2nd edn. (2007)
14. Brown, S.A., et al.: The national consortium on alcohol and neurodevelopment in adolescence (NCANDA): a multisite study of adolescent development and substance use. J. Stud. Alcohol Drugs **76**(6), 895–908 (2015)
15. Locatello, F., et al.: Challenging common assumptions in the unsupervised learning of disentangled representations. In: ICML (2018)
16. Pfefferbaum, A., et al.: Altered brain developmental trajectories in adolescents after initiating drinking. Am. J. Psychiatry **175**(4), 370–380 (2018)
17. Fischl, B.: Freesurfer. Neuroimage **62**, 774–81 (2012)
18. Rohlfing, T., Zahr, N., Sullivan, E., Pfefferbaum, A.: The SRI24 multichannel atlas of normal adult human brain structure. Hum. Brain Mapp. **31**, 798–819 (2009)
19. Cook, P., et al.: Open-source diffusion-MRI reconstruction and processing. Proc. Intl. Soc. Magn. Reson. Med. **14**, 2759 (2005)
20. Smith, S., et al.: Tract-based spatial statistics: voxelwise analysis of multi-subject diffusion data. NeuroImage **31**, 1487–505 (2006)
21. Mori, S., Wakana, S., Nagae, L.: MRI Atlas of the Human White Matter, vol. 27 (2005)
22. Aho, K., Derryberry, D., Peterson, T.: Model selection for ecologists: the world-views of AIC and BIC. Ecology **95**, 631–6 (2014)
23. Andrew, G., Arora, R., Bilmes, J., Livescu, K.: Deep canonical correlation analysis. In: Proceedings of the 30th International Conference on Machine Learning (2013)
24. Wang, W., Arora, R., Livescu, K., Bilmes, J.: On deep multi-view representation learning. In: ICML (2015)
25. Pohl, K., et al.: Harmonizing DTI measurements across scanners to examine the development of white matter microstructure in 803 adolescents of the NCANDA study. NeuroImage **130**, 194–213 (2016)
26. Simonyan, K., Vedaldi, A., Zisserman, A.: Deep inside convolutional networks: visualising image classification models and saliency maps (2014)
27. Pohl, K., et al.: The NCANDA_PUBLIC_5Y_REDCAP_V01, NCANDA_PUBLIC_5Y_STRUCTURAL_V01, NCANDA_PUBLIC_5Y_DIFFUSION_V01 data releases of the national consortium on alcohol and neurodevelopment in adolescence (ncanda). Sage Bionetworks Synapse. https://doi.org/10.7303/syn25955956. https://doi.org/10.7303/syn25955955. https://doi.org/10.7303/syn24240020

ACN: Adversarial Co-training Network for Brain Tumor Segmentation with Missing Modalities

Yixin Wang[1,2], Yang Zhang[3(✉)], Yang Liu[1,2], Zihao Lin[4], Jiang Tian[5], Cheng Zhong[5], Zhongchao Shi[5], Jianping Fan[5], and Zhiqiang He[1,2,3(✉)]

[1] Institute of Computing Technology, Chinese Academy of Sciences, Beijing, China
wangyixin19@mails.ucas.ac.cn, hezq@lenovo.com
[2] University of Chinese Academy of Sciences, Beijing, China
[3] Lenovo Corporate Research and Development, Lenovo Ltd., Beijing, China
zhangyang20@lenovo.com
[4] Department of Electronic and Computer Engineering,
Duke University, Durham, NC, USA
[5] AI Lab, Lenovo Research, Beijing, China

Abstract. Accurate segmentation of brain tumors from magnetic resonance imaging (MRI) is clinically relevant in diagnoses, prognoses and surgery treatment, which requires multiple modalities to provide complementary morphological and physiopathologic information. However, missing modality commonly occurs due to image corruption, artifacts, different acquisition protocols or allergies to certain contrast agents in clinical practice. Though existing efforts demonstrate the possibility of a unified model for all missing situations, most of them perform poorly when more than one modality is missing. In this paper, we propose a novel Adversarial Co-training Network (ACN) to solve this issue, in which a series of independent yet related models are trained dedicated to each missing situation with significantly better results. Specifically, ACN adopts a novel co-training network, which enables a coupled learning process for both full modality and missing modality to supplement each other's domain and feature representations, and more importantly, to recover the 'missing' information of absent modalities. Then, two unsupervised modules, i.e., entropy and knowledge adversarial learning modules are proposed to minimize the domain gap while enhancing prediction reliability and encouraging the alignment of latent representations, respectively. We also adapt modality-mutual information knowledge transfer learning to ACN to retain the rich mutual information among modalities. Extensive experiments on BraTS2018 dataset show that our proposed method significantly outperforms all state-of-the-art methods under any missing situation.

Keywords: Co-training · Missing modalities · Brain tumor segmentation

---

This work was done at AI Lab, Lenovo Research.

M. de Bruijne et al. (Eds.): MICCAI 2021, LNCS 12907, pp. 410–420, 2021.
https://doi.org/10.1007/978-3-030-87234-2_39

1   Introduction

Malignant brain tumors have become an aggressive and dangerous disease that leads to death worldwide. Accurate segmentation of brain tumor is crucial to quantitative assessment of tumor progression and surgery treatment planning. Magnetic resonance imaging (MRI) provides various tissue contrast views and spatial resolutions for brain examination [2,22], by which tumor regions can be manually segmented into heterogeneous subregions (i.e., GD-enhancing tumor, peritumoral edema, and the necrotic and non-enhancing tumor core) by comparing MRI modalities with different contrast levels (i.e., T1, T1ce, T2 and Flair). These modalities are essential as they provide each other with complementary information about brain structure and physiopathology. In recent years, deep learning has been widely adopted in medical image analysis due to its promising performance as well as the high cost of manual diagnosis. Joint learning from multiple modalities significantly boosts the segmentation or registration accuracy, and thus has been successfully applied in previous works [9,12,19–21,23]. In clinical settings, however, MRI sequences acquired may be partially unusable or missing due to corruption, artifacts, incorrect machine settings, allergies to certain contrast agents or limited scan time. In such a situation, in only a subset of the full treatment modalities is available.

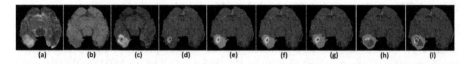

Fig. 1. Example images. (a)–(d) Four modalities in brain MRI images; (e) Ground-truth of three brain tumors: enhancing tumor (yellow), edema (green), necrotic and non-enhancing tumor (red); (f) Our proposed ACN method; (g) KD-Net [11]; (h) HeMIS[10]; (i) U-HVED [8]. (Color figure online)

Despite efforts made for missing modalities, existing works usually learn common representations from full modality, and build a uniform model for all possible missing situations during inference [6–8,10,16,17]. However, they perform much worse when more than one modality is missing, which is quite common in clinical practice. Therefore, instead of a 'catch-all' general model for all missing situations with low accuracies, it is more valuable to train a series of independent yet related models dedicated to each missing situation with good accuracy. Hu et al. [11] propose a 'dedicated' knowledge distillation method from multi-modal to mono-modal segmentation. However, bias easily occurs to training a single model for a single modality. Moreover, their models are unable to learn domain knowledge from full modality and rich features from different levels. What's more, the teacher model may contain information irrelevant to the mono-modal model, which wastes the network's learning capacity.

To address the obstacles mentioned above, we propose a novel Adversarial Co-training Network (ACN) for missing modalities, which enables a coupled learning process from both full modality and missing modality to supplement each other's domain and feature representations, and more importantly, to recover the 'missing' information. Specifically, our model consists of two learning paths: a multimodal path to derive rich modality information and a unimodal path to generate modality-specific representations. Then, a co-training approach is introduced to establish a coupled learning process between them, which is achieved by three major components, namely, 1) an entropy adversarial learning module (EnA) to bridge the domain gap between multimodal and unimodal path; 2) a knowledge adversarial learning module (KnA) to encourage the alignment of latent representations during multimodal and unimodal learning; 3) a modality-mutual information knowledge transfer module (MMI) to recover the most relevant knowledge for incomplete modalities from different feature levels via variational information maximization. The proposed ACN outperforms all state-of-the-arts on a popular brain tumor dataset. Figure 1 highlights our superior results under the missing condition that only one modality is avaiable. To the best of our knowledge, this is the **first** attempt to introduce the concept of unsupervised domain adaptation (UDA) to missing modalities and the **first** exploration of transferring knowledge from full modality to missing modality from the perspective of information theory.

2   Method

2.1   ACN: Adversarial Co-training Network

Co-training approach [5] has been applied to tasks with multiple feature views. Each view of the data is independently trained and provides independent and complementary information. Similarly, the fusion of all brain MRI modalities provides complete anatomical and functional information about brain structure and physiopathology. Models trained on incomplete modalities can still focus on the specific tumor or structure of brain MRI and provide important complementary information. Therefore, we propose ACN (see Fig. 2) to enable a coupled learning process to enhance the learning ability of both multimodal and unimodal training. It consists of two learning paths, i.e., the multimodal path and the unimodal path, which are responsible for deriving generic features from full modality and the most relevant features from available incomplete modalities, respectively.

Concretely, patches from full modality are concatenated together to generate an input $X_{multi} \in \mathbb{R}^{H \times W \times M}$ to the multimodal path, where $M = 4$ (Flair, T2, T1 and T1ce) in our task. The unimodal path receives patches from N available incomplete modalities, denoted by $X_{uni} \in \mathbb{R}^{H \times W \times N}$. The two paths share the same U-Net architecture and are trained independently and simultaneously. We distill semantic knowledge from the output distributions of both paths and design a consistency loss term $\mathcal{L}_{con}$ to minimize their Kullback-Leibler (KL) divergence, which is defined as:

$$\mathcal{L}_{con} = \frac{1}{C} \sum_c \left(\mathcal{D}_{KL}\left(s_c^m \| s_c^u\right) + \mathcal{D}_{KL}\left(s_c^u \| s_c^m\right) \right) = \frac{1}{C} \sum_c \left(\sum s_c^m \log \frac{s_c^m}{s_c^u} + \sum s_c^u \log \frac{s_c^u}{s_c^m} \right), \quad (1)$$

where s_c^m and s_c^u denote the softened logits of multimodal and unimodal path. C is the total number of classes. Such a coupled learning process not only promotes the unimodal path to learn from multimodal path but also adds necessary

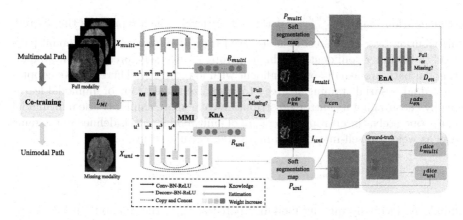

Fig. 2. Overview of our proposed ACN, consisting of a multimodal path, a unimodal path and three modules: (a) an entropy adversarial learning module (EnA); (b) a knowledge adversarial learning module (KnA); (c) a modality-mutual information knowledge transfer module (MMI). In (a), given full modality and missing modality, D_{en} is trained to predict domain label based on entropy map $\{I_{multi}, I_{uni}\}$. In (b), D_{kn} is trained to discriminate high-level features $\{R_{multi}, R_{uni}\}$ to encourage soft-alignment of knowledge. In (c), given K pairs of representations $\left\{\left(m^{(k)}, u^{(k)}\right)\right\}_{k=1}^{K}$ from both paths' encoder layers, 'missing' knowledge from multiple levels is recovered via variational information maximization.

guidance and regularization to multimodal path to further replenish the learned generic features.

This consistency exploits the correlations between full modality and missing modality in the output level. However, the two paths learn from different modalities and domains. It is significant to explicitly enhance the alignment of different domains and high-level knowledge. Therefore, we propose two adversarial learning modules to minimize the distribution distance between the two paths and utilize knowledge transfer to retain rich mutual information from a perspective of information theory.

2.2 Entropy Adversarial Learning

The prediction of multimodal path is not necessarily more accurate than unimodal path. However, models trained on multiple modalities tend to be more reliable than those on incomplete modalities. Therefore, we adopt the principle of entropy maps as a confidence measurement from unsupervised domain adaptation (UDA) tasks [15,18] and design an entropy adversarial module to match the distributions between two paths.

In detail, the multimodal and unimodal path receive X_{multi} and X_{uni} separately, along with their corresponding pixel-level C-class ground-truth $Y^{(h,w)} = \left[Y^{(h,w,c)}\right]_c$. The multimodal path takes X_{multi} as input and generates the soft segmentation map $\left[P_{multi}^{(h,w,c)}\right]_c$ at the output level, which represents the discrete distribution over C classes. It is observed that predictions from unimodal path tend to be under-confident with high-entropy. Conversely, predictions from multimodal path are usually over-confident with low-entropy. Therefore, we adopt a unified adversarial training strategy to minimize the entropy of unimodal learning by encouraging its entropy distribution to be more similar to the multimodal one. Concretely, the entropy map I_{multi} of multimodal path is defined in the same way as weighted self-information maps [18], calculated by:

$$I_{multi}^{(h,w)} = \sum_c -P_{multi}^{(h,w,c)} \cdot \log P_{multi}^{(h,w,c)}. \tag{2}$$

Similarly, receiving missing modality patches X_{uni}, the unimodal path can be seen as a generator G_{en}, which generates the soft segmentation map $\left[P_{uni}^{(h,w,c)}\right]_c$ and entropy map I_{uni}. These pixel-level vectors can be seen as the disentanglement of the Shannon Entropy, which reveals the prediction confidence. We then introduce a fully-convolutional network as a discriminator D_{en}. The generator G_{en} tries to generate an unimodal entropy map I_{uni} and fools D_{en}, while the discriminator D_{en} aims to distinguish I_{uni} from the multimodal entropy map I_{multi}. Accordingly, the optimization of G_{en} and D_{en} is achieved by the following objective function:

$$\mathcal{L}_{en}^{adv}\left(X_{multi}, X_{uni}\right) = \sum_{h,w} \log\left(1 - D_{en}\left(I_{uni}^{(h,w)}\right)\right) + \log\left(D_{en}\left(I_{multi}^{(h,w)}\right)\right). \tag{3}$$

2.3   Knowledge Adversarial Learning

Considering that the high-level representations contain richer information, we also need to encourage features distribution alignment in latent space. Simply minimizing the KL divergence between the features of two paths' bottlenecks $\{R_{multi}, R_{uni}\}$ may easily disturb the underlying learning of unimodal path in the deep layers. Therefore, we encourage the high-level representations of both paths to be aligned using a knowledge adversarial module. Similar to EnA, the generator G_{kn} tries to generate high-level features to mislead another discriminator D_{kn}. The objective function of this process is formulated as:

$$\mathcal{L}_{kn}^{adv}\left(X_{multi}, X_{uni}\right) = \log\left(1 - D_{kn}\left(R_{uni}\right)\right) + \log\left(D_{kn}\left(R_{multi}\right)\right). \tag{4}$$

KnA serves as a soft-alignment to encourage unimodal path to learn abundant and 'missing' knowledge from full modality.

2.4   Modality-Mutual Information Knowledge Transfer Learning

Multimodal path may contain information irrelevant to the task, which requires superfluous effort for alignment by unimodal path. To address this issue, we introduce the modality-mutual information knowledge transfer learning to retain high mutual information between two paths. Specifically, K pairs of representations $\left\{\left(m^{(k)}, u^{(k)}\right)\right\}_{k=1}^{K}$ can be obtained from K encoder layers of the multimodal and unimodal path separately. Given the entropy $H(m)$ and conditional entropy $H(m \mid u)$, the mutual information MI between each pair (m, u) can be defined by: $MI(m, u) = H(m) - H(m \mid u)$, which measures a reduction in uncertainty in the knowledge of the multimodal learning encoded in its layers when the unimodal knowledge is known. Following [1] to measure the exact values, we use variational information maximization [4] for each $MI(m, u)$:

$$
\begin{aligned}
MI(m, u) &= H(m) - H(m \mid u) = H(m) + \mathbb{E}_{m, u \sim p(m, u)}[\log p(m \mid u)] \\
&= H(m) + \mathbb{E}_{u \sim p(u)}\left[D_{\mathrm{KL}}(p(m \mid u) \| q(m \mid u))\right] + \mathbb{E}_{m, u \sim p(m, u)}\left[\log q(m \mid u)\right] \quad (5) \\
&\geq H(m) + \mathbb{E}_{m, u \sim p(m, u)}[\log q(m \mid u)].
\end{aligned}
$$

The distribution $p(m \mid u)$ is approximated by a variational distribution $q(m \mid u)$. Accordingly, the optimization of hierarchical mutual information can be formulated with the following loss function $\mathcal{L}_{\mathcal{MI}}$:

$$
\mathcal{L}_{\mathcal{MI}} = - \sum_{k=1}^{K} \gamma_k MI\left(m^{(k)}, u^{(k)}\right) = - \sum_{k=1}^{K} \gamma_k \mathbb{E}_{m^{(k)}, u^{(k)} \sim p\left(m^{(k)}, u^{(k)}\right)}\left[\log q\left(m^{(k)} \mid u^{(k)}\right)\right],
$$
$$(6)$$

where γ_k increases with k, indicating that higher layers contain more semantic information which should be assigned with larger weights for guidance. For specific implementation of our co-training network, the variation distribution can be realized by:

$$
- \log q(m \mid u) = \sum_{c=1}^{C} \sum_{h=1}^{H} \sum_{w=1}^{W} \log \sigma_c + \frac{\left(m_{c,h,w} - \mu_{c,h,w}(u)\right)^2}{2\sigma_c^2} + Z, \quad (7)
$$

where $\mu(\cdot)$ and σ denote the heteroscedastic mean and homoscedastic variance function of a Gaussian distribution. W and H denote width and height, C is the channel numbers of the corresponding layers and Z is a constant value.

2.5   Overall Loss and Training Procedure

The overall loss function $\mathcal{L}$ is formulated by:

$$
\mathcal{L} = \lambda_{multi} \mathcal{L}_{multi}^{dice} + \lambda_{uni} \mathcal{L}_{uni}^{dice} + \omega(t)\mathcal{L}_{con} + \lambda_0 \mathcal{L}_{en}^{adv} + \lambda_1 \mathcal{L}_{kn}^{adv} + \lambda_2 \mathcal{L}_{MI}. \quad (8)
$$

where $\mathcal{L}_{multi}^{dice}$ and $\mathcal{L}_{uni}^{dice}$ are commonly used segmentation Dice loss in medical tasks for multimodal and unimodal path respectively. $\omega(t) = 0.1 * e^{\left(-5(1-S/L)^2\right)}$ is a time-dependent Gaussian weighting function, where S and L represent the current training step and ramp-up length separately. λ_{multi}, λ_{uni}, λ_0, λ_1 and λ_2 are trade-off parameters, which are set as 0.2, 0.8, 0.001, 0.0002, and 0.5 in our model. They are chosen according to the performance under the circumstance that only T1ce modality is available since it's the major modality for tumor diagnosis. The ultimate goal of the overall co-training procedure is the following optimization: $\min_{G_{en}, G_{kn}} \max_{D_{en}, D_{kn}} \mathcal{L}$.

3   Experimental Results

3.1   Experimental Setup

Dataset. BraTS2018 training dataset [2,3,13] consists of 285 multi-contrast MRI scans with four modalities: a) native (T1), b) post-contrast T1-weighted (T1ce), c) T2-weighted (T2), and d) T2 Fluid Attenuated Inversion Recovery (Flair) volumes. Each of these modalities captures different properties of brain tumor subregions: GD-enhancing tumor (ET), the peritumoral edema (ED), and the necrotic and non-enhancing tumor core (NCR/NET). These subregions are combined into three nested subregions: whole tumor (WT), tumor core (TC) and enhancing tumor (ET). All the volumes have been co-registered to the same anatomical template and interpolated to the same resolution.

Implementation and Evaluation. Experiments are implemented in Pytorch and performed on NVIDIA Tesla V100 with 32GB Ram. For fair comparison, we follow the experimental settings of the winner of BraTS2018 [14]. Both paths share the same U-Net backbone as [14] and the discriminators D_{en} and D_{kn} are two fully-convolutional networks. The input patch size is set as $160 \times 192 \times 128$ and batch size as 1. The adam optimizer is applied, and the initial learning rate is set as $1e-4$ and progressively decreases according to a poly policy $(1 - \text{epoch}/\text{epoch}_{\max})^{0.9}$, where $\text{epoch}_{\max}$ is the total number of epochs (300). We randomly split the dataset into training and validation sets by a ratio of 2:1. The segmentation performance is evaluated on each nested subregion of brain tumors using 'Dice similarity coefficient (DSC)' and 'Hausdorff distance (HD95)'. A higher DSC and a lower HD95 indicate a better segmentation performance. Codes are available at https://github.com/Wangyixinxin/ACN.

3.2   Results and Analysis

Comparisons with State-of-the-Art Methods. We compare the proposed ACN with three state-of-the-art methods: two 'catch-all' models HeMIS [10] and U-HVED [8] and a 'dedicated' model KD-Net [11]. For fair comparison, U-Net is applied as the benchmark to all methods.

Table 1 shows that ACN significantly outperforms both HeMIS and U-HVED for all the 15 possible combinations of missing modality, especially when only one or two modalities are available. It is noted that missing T1ce leads to a severe decreasing on both ET and TC while missing Flair causes a significant drop on WT. Both HeMIS and U-HVED perform poorly under the above circumstances, with unacceptable DSC scores of only about 10%–20% on the ET and TC. In contrast, our method achieves promising DSC scores of above 40% in such hard situations, which is much more valuable in clinical practice. Table 2 shows the DSC comparison with the 'dedicated' KD-Net. Since KD-Net aims to deal with mono-modal segmentation, we set the comparison under the condition that only one modality is available.

Table 1. Comparison with state-of-the-art 'catch-all' methods (DSC %) on three nested subregions (ET, TC and WT). Modalities present are denoted by ●, the missing ones by ○.

Modalities				ET						TC						WT					
				U-HeMIS		U-HVED		ACN		U-HeMIS		U-HVED		ACN		U-HeMIS		U-HVED		ACN	
Flair	T1	T1ce	T2	DSC	HD95	DSC	HD95	DSC	HD95	DSC	HD95	DSC	HD95	DSC	HD95	DSC	HD95	DSC	HD95	DSC	HD95
○	○	○	●	25.63	14.42	22.82	14.28	**42.98**	**10.66**	57.20	17.88	54.67	15.38	**67.94**	**10.07**	80.96	12.53	79.83	14.63	**85.55**	**7.24**
○	○	●	○	62.02	22.87	57.64	31.90	**78.07**	**3.57**	65.29	28.21	59.59	38.01	**84.18**	**5.04**	61.53	28.23	53.62	34.14	**80.52**	**8.42**
○	●	○	○	10.16	26.06	8.60	28.80	**41.52**	**10.68**	37.39	30.70	33.90	32.29	**71.18**	**10.46**	57.62	27.40	49.51	31.87	**79.34**	**10.22**
●	○	○	○	11.78	25.85	23.80	14.39	**42.77**	**11.44**	26.06	30.69	57.90	15.16	**67.72**	**11.75**	52.48	28.21	84.39	12.40	**87.30**	**7.81**
○	○	●	●	67.83	9.06	67.83	10.70	**75.65**	**4.36**	76.64	11.16	73.92	14.56	**84.41**	**6.41**	82.48	10.23	81.32	12.24	**86.41**	**7.41**
○	●	●	○	66.22	14.89	61.11	26.92	**75.21**	**3.77**	72.46	16.86	67.55	29.65	**84.59**	**5.76**	68.47	19.55	64.22	28.46	**80.05**	**9.27**
●	○	●	○	10.71	25.71	27.96	15.09	**43.71**	**11.38**	41.12	25.06	61.14	14.70	**71.30**	**11.87**	64.62	20.69	85.71	12.20	**87.49**	**8.88**
○	●	○	●	32.39	13.60	24.29	14.33	**47.39**	**9.10**	60.92	15.18	56.26	14.66	**73.28**	**8.72**	82.41	11.90	81.56	11.82	**85.50**	**7.96**
●	●	○	○	30.22	13.44	32.31	12.84	**45.96**	**10.45**	57.68	15.73	62.70	12.76	**71.61**	**10.31**	82.95	11.51	87.58	7.89	**87.75**	**6.65**
●	○	●	●	66.10	15.33	68.36	9.66	**77.46**	**4.22**	71.49	19.34	75.07	11.77	**83.35**	**5.83**	68.99	19.77	85.93	11.57	**88.28**	**7.47**
●	●	●	○	68.54	12.32	68.60	9.66	**76.16**	**5.62**	76.01	12.43	77.05	10.23	**84.25**	**6.75**	72.31	14.29	86.72	11.11	**88.96**	**6.93**
●	●	○	●	31.07	13.94	32.34	11.79	**42.09**	**10.81**	60.32	14.22	63.14	11.58	**67.86**	**10.69**	83.43	11.58	88.07	7.88	**88.35**	**6.14**
●	○	●	●	68.72	8.03	68.93	7.72	**75.97**	**4.34**	77.53	9.02	76.75	9.21	**82.85**	**6.48**	83.85	9.26	88.09	8.00	**88.34**	**7.03**
○	●	●	●	69.92	7.81	67.75	10.19	**76.10**	**5.01**	78.96	8.93	75.28	10.69	**84.67**	**5.86**	83.94	9.09	82.32	11.08	**86.90**	**6.24**
●	●	●	●	70.24	7.43	69.03	8.37	**77.06**	**5.09**	79.48	8.95	77.71	8.63	**85.18**	**5.94**	84.74	8.99	88.46	7.80	**89.22**	**6.71**
Average				46.10	15.38	46.76	15.11	**61.21**	**7.37**	62.57	17.62	64.84	16.62	**77.62**	**8.13**	74.05	16.22	79.16	14.87	**85.92**	**7.62**

Table 2. Comparison with state-of-the-art 'dedicated' method KD-Net [11] (DSC %).

Modalities				ET		TC		WT		Average	
Flair	T1	T1ce	T2	KD-Net	ACN	KD-Net	ACN	KD-Net	ACN	KD-Net	ACN
○	○	○	●	39.04	**42.98**	66.01	**67.94**	82.32	**85.55**	62.46	**65.49**
○	○	●	○	75.32	**78.07**	81.89	**84.18**	76.79	**80.52**	78.00	**80.92**
○	●	○	○	39.87	**41.52**	70.02	**71.18**	77.28	**79.34**	62.39	**64.01**
●	○	○	○	40.99	**42.77**	65.97	**67.72**	85.14	**87.30**	64.03	**65.93**

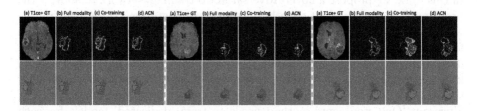

Fig. 3. Qualitative results on BraTS2018 dataset. Column (a) shows one of the input modalities (T1ce) and the corresponding segmentation ground-truth. Column (b) (c) (d) show the prediction entropy maps from multimodal path (full modality), unimodal path (only T1ce modality) without ACN modules and unimodal path with ACN modules, along with their segmentation results.

Ablation Study. In this section, we investigate the effectiveness of each proposed module, i.e., EnA, KnA and MMI. We choose a major modality T1ce as the single available modality. First we build a co-training baseline network which only uses consistency loss $\mathcal{L}_{con}$. Then we add the three modules gradually. It is observed in Table 3 that the unimodal path for missing modality is improved simultaneously through adding these modules, which proves the mutual benefit of proposed co-training network and the cooperative work of the three modules. Figure 3 further verifies the superiority and rationality of the proposed modules. The entropy maps indicate the prediction of missing modality is less reliable than full modality, while our proposed ACN produces significantly better segmentation for missing modality with reduced uncertainty.

Table 3. Effectiveness of each module on unimodal (T1ce modality) paths (DSC %).

	$\mathcal{L}_{con}$	EnA	KnA	MMI	ET	TC	WT	Average
	√				74.57	82.08	77.72	78.12
	√	√			76.40	83.78	79.78	79.98
T1ce modality	√	√	√		77.14	83.85	78.82	79.94
	√	√		√	76.12	84.24	79.89	80.09
	√	√	√	√	**78.07**	**84.18**	**80.52**	**80.92**

4   Conclusion

In this work, we propose a novel Adversarial Co-training Network to address the problem of missing modalities in brain tumor segmentation. More importantly, we present two unsupervised adversarial learning modules to align domain and feature distributions between full modality and missing modality. We also introduce a modality-mutual information module to recover 'missing' knowledge via knowledge transfer. Our model outperforms all existing methods on the multimodal BraTS2018 dataset in all missing situations by a considerable margin. Since our 'One stop shop' method needs to train 'dedicated' models for each missing situation, it may bring training cost, but it brings large improvement during inference without extra cost, which is of great value to clinical application and can be also generalized to other incomplete data domains.

References

1. Ahn, S., Hu, S.X., Damianou, A., Lawrence, N.D., Dai, Z.: Variational information distillation for knowledge transfer. In: Proceedings of the IEEE/CVF Conference on Computer Vision and Pattern Recognition (CVPR), June 2019
2. Bakas, S., et al.: Advancing the cancer genome atlas glioma MRI collections with expert segmentation labels and radiomic features. Scientific Data 4 (2017). https://doi.org/10.1038/sdata.2017.117

3. Bakas, S., et al.: Identifying the best machine learning algorithms for brain tumor segmentation, progression assessment, and overall survival prediction in the brats challenge (2018)
4. Barber, D., Agakov, F.: The IM algorithm: a variational approach to information maximization (2003)
5. Blum, A., Mitchell, T.: Combining labeled and unlabeled data with co-training. In: Proceedings of the Annual ACM Conference on Computational Learning Theory (2000). https://doi.org/10.1145/279943.279962
6. Chartsias, A., Joyce, T., Giuffrida, M.V., Tsaftaris, S.A.: Multimodal MR synthesis via modality-invariant latent representation. IEEE Trans. Med. Imaging **37**(3), 803–814 (2018). https://doi.org/10.1109/TMI.2017.2764326
7. Chen, C., Dou, Q., Jin, Y., Chen, H., Qin, J., Heng, P.-A.: Robust multimodal brain tumor segmentation via feature disentanglement and gated fusion. In: Shen, D., et al. (eds.) MICCAI 2019. LNCS, vol. 11766, pp. 447–456. Springer, Cham (2019). https://doi.org/10.1007/978-3-030-32248-9_50
8. Dorent, R., Joutard, S., Modat, M., Ourselin, S., Vercauteren, T.: Hetero-modal variational encoder-decoder for joint modality completion and segmentation. In: Shen, D., et al. (eds.) MICCAI 2019. LNCS, vol. 11765, pp. 74–82. Springer, Cham (2019). https://doi.org/10.1007/978-3-030-32245-8_9
9. Havaei, M., et al.: Brain tumor segmentation with deep neural networks. Med. Image Anal. **35**, 18–31 (2017). http://www.sciencedirect.com/science/article/pii/S1361841516300330
10. Havaei, M., Guizard, N., Chapados, N., Bengio, Y.: HeMIS: hetero-modal image segmentation. In: Ourselin, S., Joskowicz, L., Sabuncu, M.R., Unal, G., Wells, W. (eds.) MICCAI 2016. LNCS, vol. 9901, pp. 469–477. Springer, Cham (2016). https://doi.org/10.1007/978-3-319-46723-8_54
11. Hu, M., et al.: Knowledge distillation from multi-modal to mono-modal segmentation networks. In: Martel, A.L., et al. (eds.) MICCAI 2020. LNCS, vol. 12261, pp. 772–781. Springer, Cham (2020). https://doi.org/10.1007/978-3-030-59710-8_75
12. Kamnitsas, K., et al.: Efficient multi-scale 3D CNN with fully connected CRF for accurate brain lesion segmentation. Med. Image Anal. **36** (2016). https://doi.org/10.1016/j.media.2016.10.004
13. Menze, B.H., et al.: The multimodal brain tumor image segmentation benchmark (BRATS). IEEE Trans. Med. Imaging **34**(10), 1993–2024 (2015)
14. Myronenko, A.: 3D MRI brain tumor segmentation using autoencoder regularization (2018)
15. Pan, F., Shin, I., Rameau, F., Lee, S., Kweon, I.S.: Unsupervised intra-domain adaptation for semantic segmentation through self-supervision. In: Proceedings of the IEEE/CVF Conference on Computer Vision and Pattern Recognition (CVPR), June 2020
16. Sharma, A., Hamarneh, G.: Missing MRI pulse sequence synthesis using multi-modal generative adversarial network. IEEE Trans. Med. Imaging **39**(4), 1170–1183 (2020)
17. van Tulder, G., de Bruijne, M.: Why does synthesized data improve multi-sequence classification? In: Navab, N., Hornegger, J., Wells, W.M., Frangi, A.F. (eds.) MICCAI 2015. LNCS, vol. 9349, pp. 531–538. Springer, Cham (2015). https://doi.org/10.1007/978-3-319-24553-9_65
18. Vu, T.H., Jain, H., Bucher, M., Cord, M., Perez, P.: ADVENT: adversarial entropy minimization for domain adaptation in semantic segmentation. In: Proceedings of the IEEE/CVF Conference on Computer Vision and Pattern Recognition (CVPR), June 2019

19. Wang, Y., et al.: Modality-pairing learning for brain tumor segmentation. In: Crimi, A., Bakas, S. (eds.) BrainLes 2020. LNCS, vol. 12658, pp. 230–240. Springer, Cham (2021). https://doi.org/10.1007/978-3-030-72084-1_21

20. Xu, Z., et al.: Adversarial uni- and multi-modal stream networks for multimodal image registration. In: Martel, A.L., et al. (eds.) MICCAI 2020. LNCS, vol. 12263, pp. 222–232. Springer, Cham (2020). https://doi.org/10.1007/978-3-030-59716-0_22

21. Xu, Z., Yan, J., Luo, J., Li, X., Jagadeesan, J.: Unsupervised multimodal image registration with adaptative gradient guidance. In: ICASSP 2021–2021 IEEE International Conference on Acoustics, Speech and Signal Processing (ICASSP), pp. 1225–1229 (2021). https://doi.org/10.1109/ICASSP39728.2021.9414320

22. Yan, J., Chen, S., Zhang, Y., Li, X.: Neural architecture search for compressed sensing magnetic resonance image reconstruction. Comput. Med. Imaging Graph. (2020)

23. Zhou, T., Ruan, S., Canu, S.: A review: deep learning for medical image segmentation using multi-modality fusion. Array **3–4**, 100004 (2019)

Covariate Correcting Networks for Identifying Associations Between Socioeconomic Factors and Brain Outcomes in Children

Hyuna Cho[1], Gunwoong Park[2], Amal Isaiah[3], and Won Hwa Kim[1,4(✉)]

[1] POSTECH, Pohang, South Korea
{hyunacho,wonhwa}@postech.ac.kr
[2] University of Seoul, Seoul, South Korea
[3] University of Maryland School of Medicine, Baltimore, USA
[4] University of Texas at Arlington, Arlington, USA

Abstract. Brain development in adolescence is synthetically influenced by various factors such as age, education, and socioeconomic conditions. To identify an independent effect from a variable of interest (e.g., socioeconomic conditions), statistical models such as General Linear Model (GLM) are typically adopted to account for covariates (e.g., age and gender). However, statistical models may be vulnerable with insufficient sample size and outliers, and multiple tests for a whole brain analysis lead to inevitable false-positives without sufficient sensitivity. Hence, it is necessary to develop a unified framework for multiple tests that robustly fits the observation and increases sensitivity. We therefore propose a unified flexible neural network that optimizes on the contribution from the main variable of interest as introduced in original GLM, which leads to improved statistical outcomes. The results on group analysis with fractional anisotropy (FA) from Diffusion Tensor Images from Adolescent Brain Cognitive Development (ABCD) study demonstrate that the proposed method provides much more selective and meaningful detection of ROIs related to socioeconomic status over conventional methods.

1 Introduction

Experiences consistent with different levels of socioeconomic status (SES) are key to brain development [4,15]. Brain structure and functional connectivity were thought to be explained solely by their heritable aspects, consolidated by magnetic resonance imaging (MRI) based studies of twins with shared characteristics for brain volume and structure [3,16,31]. High correlation in genetically similar regions of interest (ROIs) support genetic aspects of intelligence [28]. However, recent evidence points to SES being a principal determinant of brain structural characteristics such as Pediatric Imaging, Neurocognition and Genetics (PING) study [17] that offer unique insights into the associations of SES and children's brain structure. However, most studies remain cross-sectional and are susceptible to confounds that impact delineation of associations between SES and brain

© Springer Nature Switzerland AG 2021
M. de Bruijne et al. (Eds.): MICCAI 2021, LNCS 12907, pp. 421–431, 2021.
https://doi.org/10.1007/978-3-030-87234-2_40

development [1,5]. Regardless of the method, most conventional approaches do not effectively differentiate meaningful associations between environmental exposures and brain morphometric or diffusion characteristics with MRIs.

In conventional approaches, statistical tests with neuroimaging measures under Gaussian assumption are independently performed at each ROI to obtain evaluative measures (e.g., p-value), and correcting for multiple comparisons delineates ROIs that are statistically associated with a variable of interest X (e.g., diseased vs. healthy) [8,9]. As several factors affect the brain synthetically, statistical models such as General Linear Model (GLM) are utilized to control for effects from nuisance covariates Z and identify the true effect from X [14,18,24]. GLM achieves this by comparing a pair of models, i.e., Full and Reduced Models, where the Full Model considers X and Z together while the Reduced Model takes only Z to fit given observations. Investigating the difference of errors explained by the two models, i.e., analysis of variance (ANOVA), yields the true effect from the variable of interest X corrected for covariates Z [26].

Notice that, while GLM offers a satisfactory formulation to control for covariates to describe marginal effects, it does not necessarily increase sensitivity of the method. It uses Ordinary Least Square (OLS) to fit a pair of linear models [24], and performs F-test based on residuals from OLS, which can often be misleading owing to multicollinearity and outliers [32]. Moreover, a whole brain analysis on several voxels or ROIs requires multiple tests which ends up with inevitable false-positives [2,27]. In machine learning, Domain Adaptation (DA) provides a way to control for an unwanted variable, i.e., domain. DA operates with a source and a target domain where separate neural networks for each domain are compared for a predictive task to transfer knowledge from source to target [7,25]. While the architectures of GLM and DA are similar, the objectives are quite different as the target variable in DA is a covariate in GLM. Also, DA may become very complex to control multiple covariates.

To address the issues above, we propose to develop a novel Artificial Neural Network (ANN) architecture that constructs an ensemble of multiple pairs of linear models that correct for covariates and optimize on statistical sensitivity. The framework shares the same hypothesis space bias with GLM with a loss inspired by F-test; we choose to maximize the overall F-statistics with constraints from domain knowledge. Optimizing such an ANN model in a constrained space improves sensitivity corrected for covariates together with selective identification of task-specific ROIs. Our contributions here are summarized as: **1)** Devising a novel ANN that marginalizes effects from covariates, **2)** The framework can simultaneously optimize multiple models for whole brain analyses, **3)** Various experiments identify associations between socioeconomic status and brain outcomes, which were not available with conventional approaches.

2   Preliminary: General Linear Model

GLM is a generalization of multiple linear regression to the case of more than one target variable. Hence, it can be written as

$$Y_j = \beta_{1j}X_1 + \cdots + \beta_{pj}X_p + \epsilon_j$$

Y_j $(j = 1, \cdots, m)$ is target variable (e.g., ROI measures), $\{X_k\}_{k=1}^{p}$ is a set of independent variables (e.g., group), and ϵ_j is a Gaussian distributed error.

GLM provides a general statistical framework for testing various associations and hypotheses such as ANOVA, [26], analysis of covariance (ANCOVA) [26] and the multivariate analysis of covariance (MANCOVA) [30]. Hence, GLM is an appropriate method of determining the effect of variables of interest $X = (X_1, ..., X_p)$ on target variables Y_j even with nuisance covariates $Z = (Z_1, \cdots, Z_q)$ (e.g., age, gender, and so on). Specifically, GLM compares a Full Model (with X and Z) and a Reduced Model (with Z), which perform linear operations on Y_j:

$$\text{Full Model: } Y_j = Z\lambda_j^F + X\beta_j^F + \epsilon_j, \quad \text{Reduced Model: } Y_j = Z\lambda_j^R + \epsilon_j$$

where coefficients $\lambda_j^F, \lambda_j^R \in \mathbb{R}^q$ and $\beta_j^F \in \mathbb{R}^p$. In the finite sample setting, the influence of X for Y_j can be measured by comparing the sum of squared of errors (SSE) for the Full and Reduced Models:

$$SSE_j^R := \sum_{i=1}^{n} \left(y_{ij} - z_i\lambda_j^R\right)^2 \quad \text{and} \quad SSE_j^F := \sum_{i=1}^{n} \left(y_{ij} - z_i\lambda_j^F - x_i\beta_j^F\right)^2 \quad (1)$$

where $(y_{ij})_{i=1}^{n}$, $(x_i)_{i=1}^{n}$, and $(z_i)_{i=1}^{n}$ are set of observed values of Y_j, X, and Z for the i-th subject, respectively.

F-test is one of the popular statistical approaches to determine whether the influence of X is significant. More precisely, it considers the following hypotheses:

$$H_{0j} : \beta_j^F = \mathbf{0} \quad \text{vs.} \quad H_{1j} : \beta_j^F \neq \mathbf{0} \quad (2)$$

where $\mathbf{0} = (0, \cdots, 0)' \in \mathbb{R}^p$ and resultant statistics F_j^* is defined as

$$F_j^* := \left(\frac{SSE_j^R - SSE_j^F}{df_R - df_F}\right)\left(\frac{SSE_j^F}{df_F}\right)^{-1} \quad (3)$$

where $df_R = n-q$ and $df_F = n-p-q$ are the degrees of freedom from the Reduced and Full Model, respectively. The F-test rejects H_{0j} if F_j^* is sufficiently large. This is feasible because a larger F_j^* comes from a larger $SSE_j^R - SSE_j^F$ which means that X is providing auxiliary information on Y_j that covariates Z cannot explain. In general, coefficients β_j^F, λ_j^F and λ_j^R are estimated by OLS which can be vulnerable to outliers and lack of sample size. Also, m-multiple hypothesis tests often suffer from a multicollinearity and lack of sensitivity problems. These issues motivate to develop a new robust framework, to be described shortly.

3   CoCoNet: Covariate Correcting Network

Consider the same setting as in GLM (under Gaussain assumption) where a set of measurements Y_j at location j, a main variable of interest X (e.g., group), and covariates Z (e.g., age and gender) across n samples are given. The aim here is to quantify the true effect from X on Y_j with *high sensitivity* across all j, while correcting for effects from confounding Z. For this, we design a novel framework, Covariate Correcting Network (CoCoNet), whose architecture is motivated from performing independent hypothesis tests across all j using GLM. Unlike conventional approaches, CoCoNet optimizes the multiple GLM at the same time using specialized ANN architecture with regularizers rather than performing multiple independent ANOVA.

From Sect. 2, as larger F_j^* is desired for those j's rejecting the H_{0j} showing *significant group differences*, the optimization should maximize the sum of F_j^* across j (i.e., minimizing its inverse). One can easily see from (3) that increasing SSE_j^R will increase F_j^*, hence naively optimizing on (3) will lead to loose fitting of the Reduced Model. To avoid this, Mean Squared Error (MSE) of predicting Y with X, Z and their coefficients from both models, i.e., MSE^F and MSE^R, are considered such that the regressions for both models become tight. Lastly, from a neuroscience perspective, it is intuitive to include a sparsity constraint as changes in the brain due to certain conditions may manifest on selective regions. We therefore include ℓ_1-norm penalty term on the $F^* = (F_1^*, \cdots, F_m^*)$ indicating sparse detection of ROIs that reject the null. Formulating the ideas discussed above, the initial loss to minimize is defined as

$$\frac{1}{\sum_j^m F_j^*} + \gamma_{\ell_1}\|F^*\|_{\ell_1} + \sum_j^m (\gamma_R MSE_j^R + \gamma_F MSE_j^F) \tag{4}$$

where γ_{ℓ_1}, γ_R and γ_F are hyperparameters to balance individual terms.

Carefully observing (3), maximizing F_j^* is equivalent to maximizing the ratio of SSE^R and SSE^F (as in the natural ANOVA) as

$$F_j^* = \left(\frac{SSE_j^R - SSE_j^F}{SSE_j^F}\right)\left(\frac{df_R - df_F}{df_F}\right)^{-1} \propto \left(\frac{SSE_j^R - SSE_j^F}{SSE_j^F}\right) \propto \left(\frac{SSE_j^R}{SSE_j^F}\right) \tag{5}$$

as the degrees of freedoms are constants. Moreover, applying ℓ_1 penalty on (3) is equivalent to imposing the ℓ_1 constraint on $SSE_j^R - SSE_j^F = 0$, denoting $\beta_j^F = \mathbf{0}$. In the end, the loss function (4) is revised as

$$L(\lambda, \beta) = \frac{1}{\sum_j^m \frac{SSE_j^R}{SSE_j^F}} + \gamma_{\ell_1}\|\beta^F\|_{\ell_1} + \sum_j^m \left(\frac{\gamma_R}{n - df_R}SSE_j^R + \frac{\gamma_F}{n - df_F}SSE_j^F\right) \tag{6}$$

where γ_{ℓ_1}, γ_R, and γ_F are hyperparameters that balance contributions from each term. It is important to set γ_R and γ_F properly to avoid underfitting of individual models, as the Full and Reduced Models converge at different speeds.

The overall architecture of CoCoNet is illustrated in Fig. 1, that consists of m pairs of Full and Reduced Models. Reduced Model takes q covariates, i.e., Z,

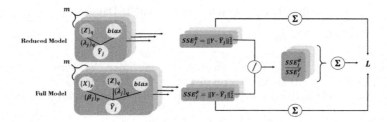

Fig. 1. Overview of CoCoNet structure, consisting of m Full Models and m Reduced Models for all target variables. The residuals from each model pairs are diversely used to define the loss function, which lead to maximizing statistical outcomes.

with a bias term. The input of the Full Model includes p variables of interest (i.e., X) along with Z and bias. The outputs from a single Reduced and Full Model pair are estimations $\hat{Y}_j^R$ and $\hat{Y}_j^F$ of the same target variable Y_j computed from pertinently trained weights and inputs so that the model ultimately minimizes (6). The loss L given in (6) to be minimized is computed from $\hat{Y}$'s based on SSE_j^R's and SSE_j^F's together with a ℓ_1 penalty on β.

4   Experimental Result

We performed several group analyses based on household income criteria; from high to subtle effect-size groups. The results show quantitative improvements with CoCoNet over baselines, which yield clinically meaningful results (Figs. 2 and 3).

4.1   Experiment on ABCD Dataset

Dataset. The Adolescent Brain Cognitive Development (ABCD) study [11] is an ongoing observational assessment of brain development in children recruited from 21 sites in the U.S. at 9–10 years of age. The rationale for study are provided by [6,29]. Enrolled children underwent cognitive, behavioral, demographic, health, and sleep assessments annually and brain imaging every two years. DTI from individuals were parcellated by co-registration with the Destrieux Atlas [10] comprising 148 regions, and average fractional anisotropy (FA) was computed at each ROI. The current study used the baseline dataset (v2.0.1) approved by institutional review boards. This fully processed data for our experiment is hosted by the National Institutes of Health Data Archive (NDA), and was downloaded and analyzed following a data use agreement.

Setup. Baselines are as follows: 1) conventional GLM, 2) GLM with Neural Network for regression (GLM_{NN}), 3) GLM with Lasso for the Full Model (GLM_{Lasso}). All of these methods analyze variance at each ROI independently and yield corresponding F-statistics and p-values. For GLM_{Lasso}, sparsity coefficient was set to 0.01. For CoCoNet, hyperparameters were set as: $m = 148$

Table 1. Demographics of the ABCD Dataset.

Category	BP	NP	High	Mid	Low
# of Subjects	952	8876	4205	2792	2831
Gender (M/F)	487/465	4608/4268	2207/1998	1448/1344	1440/1391
Age (Mean ± std)	118.5 ± 7.3	119.1 ± 7.5	119.4 ± 7.5	118.8 ± 7.5	118.7 ± 7.5
Scanner (Siemens/GE/Philips)	623/215/114	5719/2024/1133	2717/891/597	1866/591/335	1759/757/315

Table 2. ROIs identified using GLM from BP vs. NP groups and comparisons of statistics (p-values with Bonferroni correction at $\alpha = 0.01$ and F-statistics).

idx	ROI	p-value				F-statistic			
		GLM	GLM_{NN}	GLM_{Lasso}	CoCoNet	GLM	GLM_{NN}	GLM_{Lasso}	CoCoNet
1	left g.temp. sup.plan.polar	5.45E-05	5.45E-05	5.45E-05	1.34E-11	16.30	16.30	16.30	45.86
2	right g.and.s. cingul.ant	3.92E-05	3.92E-05	3.92E-05	1.11E-16	16.92	16.92	16.92	94.82
3	right g.front. sup	3.06E-05	3.06E-05	5.05E-05	6.66E-16	17.39	17.39	16.44	65.42
4	right g.oc.temp. med.parahip	2.15E-05	2.15E-05	2.21E-05	1.11E-16	18.06	18.07	18.02	111.76
5	right g.temp.sup. plan.polar	1.86E-06	1.86E-06	1.86E-06	1.11E-16	22.76	22.76	22.76	92.41
6	right s.front. sup	3.46E-06	3.46E-06	5.98E-06	1.18E-08	21.57	21.57	20.52	32.58

(i.e., # of ROIs), $p = 1$ (i.e., group), $q = 4$ (i.e., age, gender, scanner$_{1,2}$), $\gamma_{stat} = 1$, $\gamma_{\ell_1} = 10$, $\gamma_R = 0.1$, $\gamma_F = 0.1$, and learning rate was 0.01. CoCoNet was implemented with Pytorch and optimized using Adam optimizer [19]. All the p-values were corrected for multiple comparisons using Bonferroni correction at $\alpha = 0.01$.

Two sets of experiments based on several group criteria were performed. In the first experiment, the subjects were divided into two groups, i.e., Below Poverty and Non-Poverty, using the poverty criteria from U.S. Census Bureau ($16,910) [21]. In the latter, the subjects were categorized into three groups: 1) Low (<$50,000), 2) Mid (between $50,000 and $100,000), and High (≥$100,000) income groups based on [22] to look at more detailed development differences between Low vs. Middle and Middle vs. High income groups.

Below Poverty vs. Non-Poverty. We first performed a group analysis on Below Poverty (BP) vs. Non-Poverty (NP) groups. All models yielded several ROIs that showed statistically significant group differences between the BP and NP (6, 7, 6, and 52 ROIs for GLM, GLM_{NN}, GLM_{Lasso}, and CoCoNet respectively) and CoCoNet showed decrease in the p-values.

Quantitative results with F-statistics and p-values on the 6 ROIs that were detected with conventional GLM are compared in Table 2. All these ROIs were commonly detected across all models tested on the ABCD data. We first used a Linear NN, i.e., GLM_{NN}, to fit the same data with the same objective function of OLS, and used its regression result for F^* in (3) and p-values. This baseline was designed to confirm that NN can replace OLS in GLM. As expected, we observed that both GLM with OLS and NN performed almost the same with very marginal difference as seen in the F^*'s and p-values. Lasso with penalty on β (i.e., GLM_{Lasso}) yielded increase in SSE^F as the parameter space was restricted by the regularizer. However, it still resulted in very similar results with GLM and GLM_{NN} in their F^* and p-values on the ROIs detected with GLM. On the other hand, notice that both F^* and resultant p-values from CoCoNet got significantly better over the three baseline approaches. This result was obtained by leveraging a slightly different model fitting; we observed $\sim$0.01 decrease on average in R^2 for the Reduced Model but overall statistical outcomes have improved. Some of the F^* went negative for unimportant ROIs with GLM_{Lasso} and CoCoNet due to regularizers, but it was not a problem for the ROIs with β's close to 0.

Low vs. Middle vs. High Income Group Analyses. The Low vs. Middle/ Middle vs. High group comparisons are more challenging than the BP vs. NP analysis as their effect sizes are more subtle in the income spectrum. The results are summarized in Table 3 with the identified ROI labels and their p-values.

Table 3. Identified ROIs and p-values from Low vs. Mid (Left) and Mid vs. High income (Right) analyses. (detected ROIs surviving Bonferroni at α=0.01 in bold)

(a) Low vs. Middle income

idx	ROI	GLM	GLM_{NN}	GLM_{Lasso}	CoCoNet
1	left g.and.s.transv.frontopol	(4.45E-01)	(4.45E-01)	(1.00)	**8.03E-13**
2	left s.front.sup	(4.22E-01)	(4.22E-01)	(4.41E-01)	**1.11E-16**
3	left s.interm.prim.jensen	(2.59E-03)	**4.26E-05**	(2.59E-03)	**1.11E-16**
4	left s.postcentral	(2.30E-01)	(2.30E-01)	(1.00)	**5.45E-14**
5	left s.precentral.inf.part	(2.00E-01)	(2.00E-01)	(2.00E-01)	**1.11E-16**
6	right g.oc.temp.med.parahip	**3.63E-05**	**3.63E-05**	**3.63E-05**	**1.11E-16**
7	right g.temp.sup.plan.polar	(7.81E-04)	**3.30E-05**	(7.81E-04)	**8.68E-10**
8	right s.calcarine	(1.18E-01)	(1.18E-01)	(1.00)	**6.99E-15**
9	right s.front.middle	(1.68E-01)	(1.68E-01)	(1.00)	**1.10E-05**
10	right s.orbital.h.shaped	(1.55E-01)	(1.55E-01)	(1.00)	**1.11E-16**
11	right s.parieto.occipital	(1.78E-03)	(1.78E-03)	(1.78E-03)	**1.11E-16**

(b) Middle vs. High income

idx	ROI	GLM_{NN}	CoCoNet
1	left g.front.inf.triangul	(1.00)	**1.10E-08**
2	left g.oc.temp.med.lingual	(1.02E-01)	**1.11E-16**
3	left g.pariet.inf.supramar	(1.31E-01)	**1.85E-06**
4	left g.parietal.sup	(1.35E-01)	**2.59E-08**
5	left g.postcentral	**3.65E-05**	**1.11E-16**
6	left g.temp.sup.lateral	(1.95E-01)	**1.02E-11**
7	left s.orbital.med.olfact	(7.96E-01)	**5.62E-08**
8	right g.front.middle	(9.58E-02)	**4.82E-13**
9	right g.front.sup	**6.38E-05**	**1.11E-16**
10	right g.temp.sup.plan.tempo	(9.70E-01)	**1.11E-16**
11	right s.occipital.ant	(4.98E-01)	**2.13E-09**
12	right s.oc.temp.lat	(7.38E-01)	**1.11E-16**
13	right s.temporal.transverse	(6.53E-01)	**1.11E-16**

In the comparison of Low vs. Middle income groups, CoCoNet identified 11 ROIs of which p-values survive Bonferroni correction at $\alpha = 0.01$, while GLM and GLM_{Lasso} yielding only 1 and GLM_{NN} detecting 3 ROIs subsumed by the 11 ROIs from CoCoNet. These results are visually compared in Fig. 2. Notably, the overall p-values from CoCoNet are distributed at smaller values than those from baselines, meaning that CoCoNet was able to intensively detect valid ROIs.

In the Middle vs. High group analysis, GLM and GLM_{Lasso} did not yield any ROIs surviving Bonferroni correction even after hyperparameter tuning. However, GLM_{NN} and CoCoNet detected 2 and 13 regions respectively. Comparing the p-values of the detected ROIs, the associations between socioeconomic characteristics and brain outcomes were much more sensitively identified with CoCoNet. These descriptions can be visually seen in Fig. 3 whose first row is the p-value map from GLM_{NN} and the second row is that from CoCoNet.

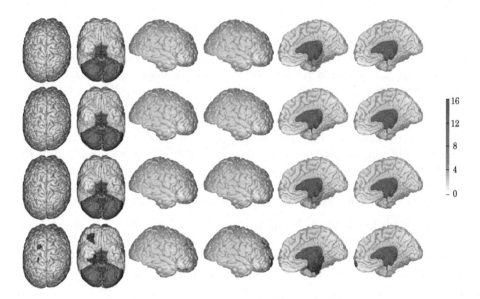

Fig. 2. p-value maps in $-\log_{10}$ scale from Low vs. Mid income group analysis. First row: GLM, Second row: GLM$_{NN}$, Third row: GLM$_{Lasso}$, Fourth row: CoCoNet. CoCoNet detects more ROIs with lower p-values than the baselines.

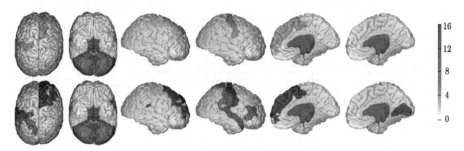

Fig. 3. Middle vs. High income group analysis and resultant p-value maps in $-\log_{10}$ scale. First row: GLM$_{NN}$, Second row: CoCoNet. GLM and GLM$_{Lasso}$ yielded no ROIs.

4.2 Discussions on the Results

Statistical Parametric Mapping (SPM) is a conventional approach. While SPM performs multiple ROI/voxel-wise independent regressions, we proposed to solve them simultaneously as a unified framework. Almost identical results between GLM and GLM$_{NN}$ show this is doable, which share the same inductive bias with CoCoNet without ℓ_1-penalty. With ℓ_1-penalty in CoCoNet, we observed marginal decrease in R^2 and the residuals should be very close to F-distribution.

The results demonstrate substantially improved performance over conventional regression-based tools in isolating distinct ROIs with structural alternations. Notably, these results support the refinement of current methods used to identify meaningful effects associated with environmental exposures that impact brain development. As effect sizes are tied to sample size in imaging-based datasets such as the ABCD study [11], there is a need to extend conventional approaches prone to errors tied to differences detected by chance alone.

Our results show that the right parahippocampal gyrus has the highest F-statistic with CoCoNet when compared to GLM-based methods. This is an ROI implicated in memory encoding and retrieval, which are key to overall cognition in children. Other meaningful ROIs include the anterior cingulate and the planum temporale, which are regions critical to motivational learning and executive function, as well as language development. Our method consistently identified key ROIs such as the post-central sulcus [33] and the calcarine sulcus [20] in comparisons of children from low and middle income groups. These ROIs, responsible for the somatosensory and visual attributes of cortical processing, appear to be consistent with regional substrates for specific aspects of sensory processing thought to be impacted by environmental exposures [12]. Fusiform gyrus [23] and parts of the parietal cortex [13] (e.g. supramarginal gyrus) are detected from middle and high income comparison using CoCoNet. Biological specificity of structural alterations in the brain is emphasized from the results.

5   Conclusion

In this paper, we developed a novel framework, i.e., CoCoNet, which is an ANN that finds appropriate regressions to enhance multiple ANOVA. The results demonstrate substantially improved performance over conventional tools in isolating brain regions with structural alternations potentially impacted by environmental factors. We see very high potential of CoCoNet being adopted in various studies that require higher sensitivity accounting for confounding effects.

Acknowledgement. This research was supported by NSF IIS CRII 1948510, NSF IIS Core 2008602, NIH R03 AG070701, NIH RF1 AG059312, and IITP-2019- 0-01906 funded by MSIT (AI Graduate School Program at POSTECH), and partially supported by NRF-2021R1C1C1004562 funded by MSIT. The ABCD Study is supported by NIH and federal partners under various awards: U01DA041048, U01DA050989, U01DA051016, U01DA041022, U01DA051018, U01DA051037, U01DA050987, U01DA041174, U01DA 041106, U01DA041117, U01DA041028, U01DA041134, U01DA050988, U01DA051039, U01DA041156, U01DA041025, U01DA041120, U01DA051038, U01DA041148, U01D A041093, U01DA041089, U24DA041123, U24DA041147. A full list of supporters at https://abcdstudy.org/federal-partners, A listing of participating sites and study investigators at https://abcdstudy.org/consortium_members. ABCD consortium investigators designed and implemented the study and/or provided data but did not necessarily participate in analysis or writing of this report. This manuscript reflects the views of the authors and may not reflect the opinions or views of the NIH or ABCD consortium investigators.

References

1. Ahirwar, A.: Study of techniques used for medical image segmentation and computation of statistical test for region classification of brain MRI. IJ Inf. Technol. Comput. Sci. **5**(5), 44–53 (2013)
2. Barnes, G.R., Litvak, V., Brookes, M.J., et al.: Controlling false positive rates in mass-multivariate tests for electromagnetic responses. Neuroimage **56**(3), 1072–1081 (2011)
3. Bowyer, R.C., Jackson, M.A., Le Roy, C.I., et al.: Socioeconomic status and the gut microbiome: a Twinsuk cohort study. Microorganisms **7**(1), 17 (2019)
4. Brito, N.H., Noble, K.G.: Socioeconomic status and structural brain development. Front. Neurosci. **8**, 276 (2014)
5. Bullmore, E.T., Suckling, J., Overmeyer, S., et al.: Global, voxel, and cluster tests, by theory and permutation, for a difference between two groups of structural MR images of the brain. IEEE Trans. Med. Imaging **18**(1), 32–42 (1999)
6. Casey, B., Cannonier, T., Conley, M.I., et al.: The adolescent brain cognitive development (ABCD) study: imaging acquisition across 21 sites. Dev. Cogn. Neurosci. **32**, 43–54 (2018)
7. Daumé III, H.: Frustratingly easy domain adaptation. arXiv preprint arXiv:0907. 1815 (2009)
8. Della Nave, R., Ginestroni, A., Tessa, C., et al.: Brain white matter damage in sca1 and sca2. an in vivo study using voxel-based morphometry, histogram analysis of mean diffusivity and tract-based spatial statistics. Neuroimage **43**(1), 10–19 (2008)
9. Della Nave, R., Ginestroni, A., Tessa, C., et al.: Brain white matter tracts degeneration in Friedreich ataxia. An in vivo MRI study using tract-based spatial statistics and voxel-based morphometry. Neuroimage **40**(1), 19–25 (2008)
10. Destrieux, C., Fischl, B., Dale, A., et al.: Automatic parcellation of human cortical gyri and sulci using standard anatomical nomenclature. Neuroimage **53**(1), 1–15 (2010)
11. Dick, A.S., Lopez, D.A., Watts, A.L., et al.: Meaningful associations in the adolescent brain cognitive development study. BioRxiv (2021). https://doi.org/10.1101/2020.09.01.276451
12. Farah, M.J., Shera, D.M., Savage, J.H., et al.: Childhood poverty: specific associations with neurocognitive development. Brain Res. **1110**(1), 166–174 (2006)
13. Fogassi, L., Ferrari, P.F., Gesierich, B., et al.: Parietal lobe: from action organization to intention understanding. Science **308**(5722), 662–667 (2005)
14. Glueck, D.H., Muller, K.E.: Adjusting power for a baseline covariate in linear models. Stat. Med. **22**(16), 2535–2551 (2003)
15. Hackman, D.A., Farah, M.J.: Socioeconomic status and the developing brain. Trends Cogn. Sci. **13**(2), 65–73 (2009)
16. Ivanovic, D.M., Leiva, B.P., Pérez, H.T., et al.: Nutritional status, brain development and scholastic achievement of Chilean high-school graduates from high and low intellectual quotient and socio-economic status. Br. J. Nutr. **87**(1), 81–92 (2002)
17. Jernigan, T.L., Brown, T.T., Hagler Jr., D.J., et al.: The pediatric imaging, neurocognition, and genetics (ping) data repository. Neuroimage **124**, 1149–1154 (2016)
18. Kim, W.H., Adluru, N., Chung, M.K., et al.: Multi-resolution statistical analysis of brain connectivity graphs in preclinical Alzheimer's disease. Neuroimage **118**, 103–117 (2015)

19. Kingma, D.P., Ba, J.: Adam: a method for stochastic optimization. arXiv preprint arXiv:1412.6980 (2014)
20. Lambert, S., Sampaio, E., Scheiber, C., et al.: Neural substrates of animal mental imagery: calcarine sulcus and dorsal pathway involvement-an fMRI study. Brain Res. **924**(2), 176–183 (2002)
21. Lee, A.: Us poverty thresholds and poverty guidelines: What's the difference. Population Reference Bureau (2018)
22. Marshall, A.T., Betts, S., Kan, E.C., et al.: Association of lead-exposure risk and family income with childhood brain outcomes. Nat. Med. **26**(1), 91–97 (2020)
23. McCarthy, G., Puce, A., Gore, J.C., et al.: Face-specific processing in the human fusiform gyrus. J. Cogn. Neurosci. **9**(5), 605–610 (1997)
24. Oakes, T.R., Fox, A.S., Johnstone, T., et al.: Integrating VBM into the general linear model with voxelwise anatomical covariates. Neuroimage **34**(2), 500–508 (2007)
25. Pan, S.J., Tsang, I.W., Kwok, J.T., et al.: Domain adaptation via transfer component analysis. IEEE Trans. Neural Networks **22**(2), 199–210 (2010)
26. Rutherford, A.: ANOVA and ANCOVA: A GLM Approach. Wiley (2011)
27. Scarpazza, C., Tognin, S., Frisciata, S., et al.: False positive rates in voxel-based morphometry studies of the human brain: should we be worried? Neurosci. Biobehav. Rev. **52**, 49–55 (2015)
28. Thompson, P.M., Cannon, T.D., Narr, K.L., et al.: Genetic influences on brain structure. Nat. Neurosci. **4**(12), 1253–1258 (2001)
29. Volkow, N.D., Koob, G.F., Croyle, R.T., et al.: The conception of the ABCD study: From substance use to a broad NIH collaboration. Dev. Cogn. Neurosci. **32**, 4–7 (2018)
30. Weerahandi, S.: Generalized Inference in Repeated Measures: Exact Methods in MANOVA and Mixed Models, vol. 500. Wiley (2004)
31. Yang, F., Isaiah, A., Kim, W.H.: COVLET: covariance-based wavelet-like transform for statistical analysis of brain characteristics in children. In: Martel, A.L., et al. (eds.) MICCAI 2020. LNCS, vol. 12267, pp. 83–93. Springer, Cham (2020). https://doi.org/10.1007/978-3-030-59728-3_9
32. Zimmerman, D.W.: Increasing the power of the ANOVA F test for outlier-prone distributions by modified ranking methods. J. Gen. Psychol. **122**(1), 83–94 (1995)
33. Zlatkina, V., Amiez, C., Petrides, M.: The postcentral SULCAL complex and the transverse postcentral sulcus and their relation to sensorimotor functional organization. Eur. J. Neurosci. **43**(10), 1268–1283 (2016)

Symmetry-Enhanced Attention Network for Acute Ischemic Infarct Segmentation with Non-contrast CT Images

Kongming Liang[1], Kai Han[2], Xiuli Li[2], Xiaoqing Cheng[3], Yiming Li[2], Yizhou Wang[4], and Yizhou Yu[2,5(✉)]

[1] Pattern Recognition and Intelligent System Laboratory, School of Artificial Intelligence, Beijing University of Posts and Telecommunications, Beijing, China
[2] Deepwise AI Lab, Beijing, China
[3] Department of Medical Imaging, Jinling Hospital, Nanjing University School of Medicine, Nanjing, Jiangsu, China
[4] Department of Computer Science and Technology, Peking University, Beijing, China
[5] The University of Hong Kong, Pokfulam, Hong Kong
yizhouy@acm.org

Abstract. Quantitative estimation of the acute ischemic infarct is crucial to improve neurological outcomes of the patients with stroke symptoms. Since the density of lesions is subtle and can be confounded by normal physiologic changes, anatomical asymmetry provides useful information to differentiate the ischemic and healthy brain tissue. In this paper, we propose a symmetry enhanced attention network (SEAN) for acute ischemic infarct segmentation. Our proposed network automatically transforms an input CT image into the standard space where the brain tissue is bilaterally symmetric. The transformed image is further processed by a U-shape network integrated with the proposed symmetry enhanced attention for pixel-wise labelling. The symmetry enhanced attention can efficiently capture context information from the opposite side of the image by estimating long-range dependencies. Experimental results show that the proposed SEAN outperforms some symmetry-based state-of-the-art methods in terms of both dice coefficient and infarct localization.

Keywords: Computer aided diagnosis · Infarct segmentation · Acute ischemic stroke · Attention mechanism · Deep learning

1 Introduction

Acute ischemic stroke is one of the leading causes of death and disability worldwide and imposes an enormous burden for the health care system [8]. The use

---

Electronic supplementary material The online version of this chapter (https://doi.org/10.1007/978-3-030-87234-2_41) contains supplementary material, which is available to authorized users.

© Springer Nature Switzerland AG 2021
M. de Bruijne et al. (Eds.): MICCAI 2021, LNCS 12907, pp. 432–441, 2021.
https://doi.org/10.1007/978-3-030-87234-2_41

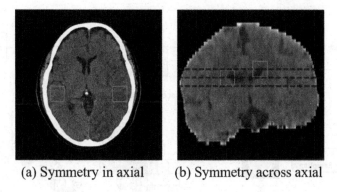

(a) Symmetry in axial   (b) Symmetry across axial

Fig. 1. The anatomical symmetry of the brain CT images. (a) and (b) show the two symmetrical patches of the image in axial view and across axial view. The dotted red line in (b) denotes the slice from axial view. Due to the rotation of patient's head, the symmetrical landmarks may appear on different images in the axial view. (Color figure online)

of pretreatment neuroimaging is critical to improve neurological outcomes of patients with stroke symptoms. Compared to MRI, non-contrast head CT scan is commonly used as the initial imaging because of its wide availability and low acquisition time. To interpret early infarct signs in CT, the Alberta Stroke Program Early CT Score (ASPECTS) evaluation was proposed at the beginning in the 2000s [2] and has found increasing acceptance in clinical practice. However, ASPECTS evaluation is only an approximation of the assessment of early ischemic changes. Since the density of lesions is subtle and can be confounded by normal physiologic changes, quantitative estimation of acute ischemic infarct is challenging. In clinical practice [9], bilaterally symmetric (illustrated in Fig. 1) provides useful information for the identification of acute ischemic infarct.

Anatomical asymmetry has been utilized in previous works to localize and segment the abnormal regions for neuroimaging analysis. [12,15] leverage the symmetry by adding extra information beyond the input image. [12] calculates the differences of each voxel by subtracting the original brain from the mirrored brain. The difference map is further used as the input to train a random forest classifier to yield lesion segmentation. [15] extracts both the original patch and its symmetric patch, and feeds them into the network simultaneously. Except for calculating the asymmetry on image-level, [3,11,16] propose to explore feature-level fusion of the two symmetry regions. For instance, two-branch networks (e.g. siamese network) can learn the features of left and right hemispheres and measure the difference between the features of two hemispheres to analyze abnormalities such as Alzheimer's disease [11], ischemic stroke [3,10] and brain tumors [16]. Even though the pixel-wise difference is widely used in previous methods, it can not efficiently exploit the bilaterally symmetric information due to the limitation of context modeling. In addition, all the above methods need the input images to be already calibrated which cannot be guaranteed in practice.

In this paper, a symmetry enhanced attention network (SEAN) is proposed for acute ischemic infarct segmentation. The proposed SEAN can automatically

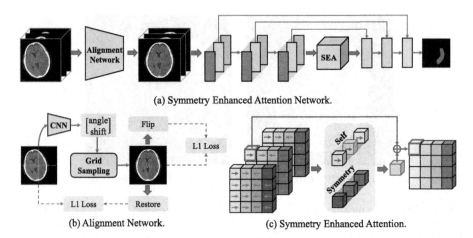

(a) Symmetry Enhanced Attention Network.

(b) Alignment Network.　　　　　(c) Symmetry Enhanced Attention.

Fig. 2. Overview of our proposed network architecture.

transform an input image into the standard space without any human super-vision. The transformed image is further processed by a U-shape network that contains encoding and decoding stages. Different from the original design [13], the encoder performs 3d convolution to leverage context information of adjacent images in axial. Then, a symmetry enhanced attention module is integrated between the encoding and decoding stages to efficiently model the anatomical symmetry. In summary, the main contributions of our paper are as follows.

1. A symmetry enhanced attention is proposed to capture both in-axial and cross-axial symmetry information by explicitly estimating the long-range dependencies.
2. A symmetry-based alignment network is proposed to transform an input image as bilaterally symmetric in axial without any human supervision.
3. We release the dataset at https://github.com/GriffinLiang/AISD.

2 Method

In this section, we first introduce the symmetry based alignment network in Sect. 2.1 and give the detail information of how to make the input image bilater-ally symmetric in axial. Then we define the structure of the proposed symmetry enhanced attention network which can capture both the in-planar and across-planar symmetry information for ischemic infarct segmentation in Sect. 2.2. The whole pipeline is shown in Fig. 2.

2.1 Symmetry Based Alignment

Since the poses of patients are arbitrary when they perform CT scans, the brain images are usually not in standard space. In order to effectively use the sym-metry of the brain, we attempt to align the image to keep the region of brain in the center of the image and horizontally symmetrical. However, traditional registration based method can not be applied in clinical practice due to the high time complexity. Therefore, we proposed a Symmetry based Alignment Network as show in Fig. 2 which can automatically align the brain images with only the information of images itself. Inspired by Spatial Transformer Networks [7], we design the symmetry based alignment network as: c2d[32, 7, 7]-relu-max2d[2, 2]-c2d[32, 5, 5]-relu-max2d[2, 2]-fc[3] where c2d[n, c_w, c_h] denotes a 2d convolutional layer with n filters of size $c_w \times c_h$, max2d[s_h, s_w] is a 2d max-pooling layer with the kernel size and the stride as $s_h \times s_w$, fc[n] is a fully connected layer with n units. The output of the network is interpreted as the parameters α (rotation, horizontal shift and vertical shift) of rigid transformation matrix.

Given an input volume, we define $\mathbf{A}_i$ as the i-th slice in the axial view. During training, the output parameters α is applied to the input slice $\mathbf{A}_i^t = f_\alpha(\mathbf{A}_i)$ using parameterised sampling grid. Then we generate the flipped version of $\mathbf{A}_i^t$ as $\tilde{\mathbf{A}}_i^t$. The total loss is designed as the following:

$$L_a = L_1(\tilde{\mathbf{A}}_i^t, \mathbf{A}_i^t) + L_1(f_\alpha^{-1}(\mathbf{A}_i^t), \mathbf{A}_i), \tag{1}$$

where $f_\alpha^{-1}(\cdot)$ denotes the inverse transformation function and L_1 denotes the L1 loss. We define the first term of Eq. (1) as symmetry loss which is the L1 distance between the aligned image and its horizontally flipped image. The symmetry loss is based on the assumption that the difference between the image of brain and its horizontally flipped version will be minimized when the brain is perfectly aligned. In addition, we need to add constraints to the symmetry loss to avoid trivial solution where the alignment network can simply transform the brain region out of the input image. Therefore, a restoration loss is defined as the L1 distance between the restored image and the original image to learn useful transformation parameters.

2.2 Symmetry Based Segmentation

The proposed segmentation network adopts the structure of UNet [13] which is mainly composed of two parts: the encoder stage and the decoder stage. Inspired by [5], we use 3D convolutions as the basic encoding block to keep the context information from adjacent images in axial view. For the decoding stage, the middle plane of input volume is retained as the target image and upsampled to the original resolution for pixel-wise labelling. We name the above network as HybridUnet. Finally, we cascade the last encoding block with the symmetry attention module. In this way, the feature representation can be enhanced by its symmetry information to efficiently assess the presence and extent of ischemic infarct.

To exploit the context information of the i-th axial image, HybridUnet takes its adjacent images $\{\mathbf{A}_{i+t} | t = -T, \cdots, T\}$ as the input. The input images are firstly processed by the 3d encoder. We design the encoder block as: c3d-bn-relu-c3d-bn-relu-max3d where c3d denotes 3d convolutional layer, bn denotes 3d batchnorm layer and max denotes a 3d max-pooling layer. The encoder stage contains five encoder blocks. The output feature from the last encoder block is denoted as $\mathbf{X}_i \in \mathbb{R}^{C \times H \times W}$ for the input image $\mathbf{A}_i$ where H and W represent the height and width of the output feature respectively and C is the number of the output channels. The output feature $\mathbf{X}_i$ is further processed by symmetry attention module in Sect. 2.2 and further processed by four decoding blocks with the same structure as the original UNet. We train the proposed SEAN by minimizing the combination of the generalized Dice loss and the cross-entropy loss in an end-to-end manner.

Symmetry Enhanced Attention. The symmetry information is hard to explore by the conventional operation which only processes a local neighborhood in space. Since the input position and its symmetrical position usually have long spatial interval distances, local operations need to be applied repeatedly to capture such long-range dependencies. As mentioned in [6], stacking multiple convolution operations is computationally inefficient and increases the difficulty of optimization. To compensate for the drawback of convolution operation, we propose to model the relationships between symmetrical position with attention mechanism [14]. Given an input feature map $\mathbf{X}_i \in \mathbb{R}^{C \times H \times W}$, we first divide it into $P \times Q$ partitions as below,

$$
\mathbf{X}_i = \begin{bmatrix} \mathbf{X}_{i,1,1} & \mathbf{X}_{i,1,2} & \cdots & \mathbf{X}_{i,1,Q} \\ \mathbf{X}_{i,2,1} & \mathbf{X}_{i,2,2} & \cdots & \mathbf{X}_{i,2,Q} \\ \vdots & \vdots & \ddots & \vdots \\ \mathbf{X}_{i,P,1} & \mathbf{X}_{i,P,2} & \cdots & \mathbf{X}_{i,P,Q} \end{bmatrix}, \tag{2}
$$

where $\mathbf{X}_{i,j,k} \in \mathbb{R}^{C \times H' \times W'}$ is a subset of $\mathbf{X}_i$ ($H = H' \times P$ and $W = W' \times Q$). Then we flip $\mathbf{X}_i$ horizontally to generate its mirrored feature map $\tilde{\mathbf{X}}_i$. Therefore, the symmetrical partition of $\mathbf{X}_{i,j,k}$ can be denoted as $\tilde{\mathbf{X}}_{i,j,k}$.

For the input partition $\mathbf{X}_{i,j,k}$, its symmetry enhanced attention is composed of the self-attention module and the symmetry-attention module:

$$
\mathbf{S}^t_{i,j,k} = Softmax(\frac{\theta(\mathbf{X}_{i,j,k})^\top \phi(\mathbf{X}_{i+t,j,k})}{\sqrt{d}}),
$$

$$
\tilde{\mathbf{S}}^t_{i,j,k} = Softmax(\frac{\theta(\mathbf{X}_{i,j,k})^\top \phi(\tilde{\mathbf{X}}_{i+t,j,k})}{\sqrt{d}}), \tag{3}
$$

$$
\mathbf{Y}_{i,j,k} = \left(\sum_{t=-T}^{T} \mathbf{S}^t_{i,j,k} \cdot g(\mathbf{X}_{i+t,j,k})^\top + \sum_{t=-T}^{T} \tilde{\mathbf{S}}^t_{i,j,k} \cdot h(\tilde{\mathbf{X}}_{i+t,j,k})^\top \right)^\top,
$$

where $\mathbf{S}^t_{i,j,k}, \tilde{\mathbf{S}}^t_{i,j,k} \in \mathbb{R}^{N' \times N'}$ ($N' = H' \times W'$) are the similarity matrix of the self attention and symmetry attention respectively. $\theta(\cdot)$ and $\phi(\cdot)$ perform convolution operations to reduce the number of input channels to d (e.g. $d = \frac{C}{2}$) and reshape the output to $\mathbb{R}^{d \times H'W'}$. We use $\sqrt{d}$ as a scaling factor for the inner product to solve the small gradient problem of softmax function. $\mathbf{X}_{i+t,j,k}$ and $\tilde{\mathbf{X}}_{i+t,j,k}$ are also fed into $g(\cdot)$ and $h(\cdot)$ to compute the new representation by convolution operations and reshape the output feature map to $\mathbb{R}^{\frac{C}{2} \times H'W'}$. The symmetry enhance feature $\mathbf{Y}_{i,j,k} \in \mathbb{R}^{C \times H'W'}$ is reshaped to $C \times H' \times W'$ and further considered as the residual mapping of $\mathbf{X}_{i,j,k}$ to acquire the final output of the symmetry enhanced attention.

3 Experiments

In this section, we first introduce the data acquisition and evaluation indicators of our model in Sect. 3.1. Then we show the detail information of implementation in Sect. 3.2. Finally, we compare our approach with the state-of-the-art methods and conduct extensive ablation studies in Sect. 3.3.

3.1 Experiment Setup

Data Acquisition. We obtain 397 Non–Contrast-enhanced CT (NCCT) scans of acute ischemic stroke with the interval from symptom onset to CT less than 24 h. The patients underwent diffusion-weighted MRI (DWI) within 24 h after taking the CT. The slice thickness of NCCT is 5mm. We name the above CT-MRI pairs as acute ischemic stroke dataset (AISD). 345 scans are used to train and validate the model, and the remaining 52 scans are used for testing. Ischemic lesions are manually contoured on NCCT by a doctor using MRI scans as the reference standard. Then a senior doctor double-reviews the labels.

Evaluation Metrics. To quantitatively evaluate the result of our proposed symmetry based alignment network, we compare the output transformation parameters with the human annotated rotation and offset. Specifically, a doctor annotates the beginning and end point of cerebral falx on the middle slice of each CT scan. The offset angle and the center of brain region are further calculated as ground truth. We use the average difference of the rotation angle and the offset distance in the horizontal direction between the model's output and the ground truth for each data. As for the segmentation results, we utilize Dice coefficient to quantitatively evaluate the performance. In addition, we also calculate infarct-level evaluation metrics such as recall, precision and F1 score. To evaluate the clinical value, Pearson correlation between the estimated ASPECTS and the ground-truth is also performed.

Comparison Methods. We compare the proposed SEAN with following methods: 1) No symmetry modelling method (Unet) which is a Vanilla Unet without considering the symmetry information; 2) Symmetry modelling on image-level (Unet-IM-L1) [12] which calculates the bilateral density L_1 difference between symmetric brain regions as one of the input; 3) Symmetry modelling on feature-level by concatenation (Unet-FT-CC) which takes the original image and its flipped image as the input of Unet and concatenates the two output features from the last encoder; 4) Symmetry on feature-level with L1 distance (Unet-FT-L1) [16] which takes the distance map between the feature from the last encoder

Table 1. Quantitative results on the AIS dataset.

Network	Fusion-level	Fusion-Type	Dice	F1	Recall	Precision
Unet	N/A	N/A	0.4588	0.5105	0.5019	0.5196
Unet-IM-L1	Image	L1	0.5035	0.5457	0.5318	0.5603
Unet-FT-L1	Feature	L1	0.5121	0.567	0.5468	0.5888
Unet-FT-CC	Feature	Concat	0.5354	0.572	0.5655	0.5786
HybridUnet	N/A	N/A	0.4952	0.5433	0.6105	0.4895
HybridUnet-IM-L1	Image	L1	0.5437	0.5992	0.5581	0.6471
HybridUnet-FT-L1	Feature	L1	0.5445	0.5982	0.5543	0.6497
HybridUnet-FT-CC	Feature	Concat	0.5577	0.6015	0.5431	0.6742
SEAN	Feature	Attention	0.5784	0.6218	0.5880	0.6597

and its flipped version as an extra information. We also implement the above methods using HybridUnet [5] as the backbone for further comparison.

3.2 Implementation Details

Our implementation is based on Pytorch framework. For data pre-processing, we truncate the raw intensity values to the range [40, 100] HU and normalize each raw CT case to have zero mean and unit variance. The Adam optimizer is used to train the model with parameters $\beta_1 = 0.9, \beta_2 = 0.99$ for 150 epochs. And we set the base learning rate as 1×10^{-4} and deploy a poly learning rate policy where the initial learning rate is multiplied by $(1 - \frac{iter}{total\_iter})^{power}$ and $power = 0.9$ after each iteration.

3.3 Results and Discussions

Efficacy of Alignment Network. We evaluate the similarity of the rotation angle and the offset distance in the horizontal direction between the model's output and the ground truth on test data (52 patients). The error of rotation angle is 3 degrees and the error of shift distance is 5 pixels. From the results, we can see that the output of the symmetry based alignment network is very close to the ground truth even though the proposed method is fully unsupervised. Besides, we register the NCCT images to the standard brain template to align the images as described in [12]. This operation cost 134 s per patient on average, while the time consumption of our proposed symmetry based alignment network is only 0.46 s per patient on average. The registration method is based on Advanced Normalization Tools(ANTS) [1], and all experiments are performed on a Linux server with Intel(R) Xeon(R) CPU E5-2697 v4 @ 2.30GHz and a NVIDIA 2080ti GPU.

Efficacy of SEAN. As shown in Table 1, the proposed SEAN achieves the highest performance (Dice: 0.5784, F1: 0.6218) compared to the other methods. Since the proposed SEAN benefits from leveraging both the in-planar and

Table 2. Ablation study of SEAN on AISD.

Method	Dice	F1	Recall	Precision
Baseline	0.4952	0.5433	0.6105	0.4895
+ Ours (Align)	0.5281	0.5767	0.5506	0.6056
+ Ours (Align+Self)	0.5635	0.5834	0.5506	0.6205
+ Ours (Align+Self+Sym)	0.5784	0.6218	0.5880	0.6597

across-planar symmetry information, it can differentiate the infarct and the normal physiologic change more efficiently. In general, symmetry based methods observed significant improvements according to both Dice and F1. This is also consistent with doctors' habit in clinical practice. For the effectiveness of the backbone network, HybridUnet achieves better performance than the original Unet, which demonstrates the importance of context information from adjacent images. Feature-level method outperforms image-level method, since the feature-level method is robust to the misalignment of the input image. We conduct ablation studies of SEAN and show the results in Table 2. The influence of the proposed alignment network and the two type of attention mechanism: self-attention and symmetry-attention is investigated. According to the results, it can be seen that the proposed alignment network improves over the baseline for a large margin. We can also see that symmetry enhanced attention yields a higher increase in both Dice and F1 comparing to only using self-attention. We also show some qualitative examples in Fig. 3.

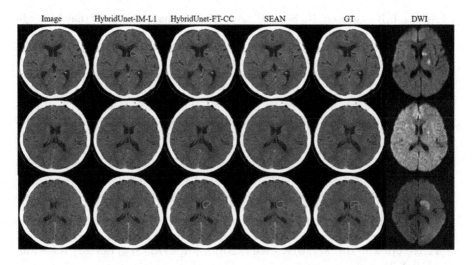

Fig. 3. Qualitative comparison of different methods.

Efficacy of Clinical Usage. To validate the clinical efficacy of the proposed SEAN, Pearson correlation between the estimated ASPECTS and the ground-truth is performed. A standard template with ASPECTS regions in the Montreal Neurologic Institute space [4] is registered to all NCCT images by performing affine transformation using ANTS. The Pearson correlation between the scores estimated by SEAN and doctor is 0.75, which further indicates the efficiency of the proposed method.

4   Conclusion

In this paper, we propose a symmetry enhanced attention network (SEAN) for acute ischemic infarct segmentation. The proposed network calibrates an input CT image and capture bilateral symmetry information by explicitly estimating the long-range dependencies. With the seamless integration, the proposed symmetry enhanced attention can be applied to any lesion segmentation task. Experimental results on acute ischemic stroke dataset (AISD) show that the proposed SEAN outperforms some symmetry-based state-of-the-art methods in terms of both dice coefficient and infarct localization. The acute ischemic stroke dataset (AISD) is published for future study.

Acknowledgments. This work was supported in part by following grants, MOST-2018AAA0102004, NSFC-62061136001, the Key Program of Beijing Municipal Natural Science Foundation (7191003), and the Key Projects of the National Natural Science Foundation of China (81830057).

References

1. Avants, B.B., Tustison, N., Song, G.: Advanced normalization tools (ants). Insight J **2**(365), 1–35 (2009)
2. Barber, P.A., Demchuk, A.M., Zhang, J., Buchan, A.M., Group, A.S., et al.: Validity and reliability of a quantitative computed tomography score in predicting outcome of hyperacute stroke before thrombolytic therapy. Lancet **355**(9216), 1670–1674 (2000)
3. Barman, A., Inam, M.E., Lee, S., Savitz, S.I., Sheth, S.A., Giancardo, L.: Determining ischemic stroke from CT-angiography imaging using symmetry-sensitive convolutional networks. In: International Symposium on Biomedical Imaging, pp. 1873–1877 (2019)
4. Evans, A.C., Collins, D.L., Mills, S., Brown, E.D., Kelly, R.L., Peters, T.M.: 3D statistical neuroanatomical models from 305 MRI volumes. In: 1993 IEEE Conference Record Nuclear Science Symposium and Medical Imaging Conference, pp. 1813–1817. IEEE (1993)
5. Fang, C., Li, G., Pan, C., Li, Y., Yu, Y.: Globally guided progressive fusion network for 3d pancreas segmentation. In: Shen, D., et al. (eds.) MICCAI 2019. LNCS, vol. 11765, pp. 210–218. Springer, Cham (2019). https://doi.org/10.1007/978-3-030-32245-8_24
6. He, K., Zhang, X., Ren, S., Sun, J.: Deep residual learning for image recognition. In: Computer Vision and Pattern Recognition (2015)

7. Jaderberg, M., Simonyan, K., Zisserman, A., et al.: Spatial transformer networks. In: Advances in Neural Information Processing Systems, pp. 2017–2025 (2015)

8. Katan, M., Luft, A.: Global burden of stroke. In: Seminars in Neurology, vol. 38, pp. 208–211. Georg Thieme Verlag (2018)

9. Khan Academy: Diagnosing strokes with imaging CT, MRI, and angiography. https://www.khanacademy.org

10. Kuang, H., Menon, B.K., Qiu, W.: Automated infarct segmentation from follow-up non-contrast CT scans in patients with acute ischemic stroke using dense multi-path contextual generative adversarial network. In: Shen, D., et al. (eds.) MICCAI 2019. LNCS, vol. 11766, pp. 856–863. Springer, Cham (2019). https://doi.org/10.1007/978-3-030-32248-9_95

11. Liu, C.F., et al.: Using deep Siamese neural networks for detection of brain asymmetries associated with Alzheimer's disease and mild cognitive impairment. Magn. Reson. Imaging **64**, 190–199 (2019)

12. Qiu, W., et al.: Machine learning for detecting early infarction in acute stroke with non-contrast-enhanced CT. Radiology **294**(3), 638–644 (2020)

13. Ronneberger, O., Fischer, P., Brox, T.: U-net: convolutional networks for biomedical image segmentation. In: Navab, N., Hornegger, J., Wells, W.M., Frangi, A.F. (eds.) MICCAI 2015. LNCS, vol. 9351, pp. 234–241. Springer, Cham (2015). https://doi.org/10.1007/978-3-319-24574-4_28

14. Wang, X., Girshick, R., Gupta, A., He, K.: Non-local neural networks. In: Proceedings of the IEEE Conference on Computer Vision and Pattern Recognition, pp. 7794–7803 (2018)

15. Wang, Y., Katsaggelos, A.K., Xue, W., Parrish, T.B.: A deep symmetry convnet for stroke lesion segmentation. In: IEEE International Conference on Image Processing (ICIP) (2016)

16. Zhang, H., Zhu, X., Willke, T.L.: Segmenting brain tumors with symmetry. In: Proceedings of NIPS Workshop (2017)

Modality Completion via Gaussian Process Prior Variational Autoencoders for Multi-modal Glioma Segmentation

Mohammad Hamghalam[1,2]($\boxtimes$), Alejandro F. Frangi[4,5,6], Baiying Lei[7], and Amber L. Simpson[1,3]

[1] School of Computing, Queen's University, Kingston, ON, Canada
[2] Faculty of Electrical, Biomedical and Mechatronics Engineering, Qazvin Branch, Islamic Azad University, Qazvin, Iran
[3] Department of Biomedical and Molecular Sciences, Queen's University, Kingston, ON, Canada
[4] CISTIB Centre for Computational Imaging and Simulation Technologies in Biomedicine, School of Computing, University of Leeds, Leeds LS2 9LU, UK
[5] LICAMM Leeds Institute of Cardiovascular and Metabolic Medicine, School of Medicine, Leeds LS2 9LU, UK
[6] Medical Imaging Research Center (MIRC) – University Hospital Gasthuisberg, KU Leuven, Herestraat 49, 3000 Leuven, Belgium
[7] National-Regional Key Technology Engineering Laboratory for Medical Ultrasound, Guangdong Key Laboratory for Biomedical Measurements and Ultrasound Imaging, School of Biomedical Engineering, Health Science Center, Shenzhen University, Shenzhen, China

Abstract. In large studies involving multi protocol Magnetic Resonance Imaging (MRI), it can occur to miss one or more sub-modalities for a given patient owing to poor quality (e.g. imaging artifacts), failed acquisitions, or hallway interrupted imaging examinations. In some cases, certain protocols are unavailable due to limited scan time or to retrospectively harmonise the imaging protocols of two independent studies. Missing image modalities pose a challenge to segmentation frameworks as complementary information contributed by the missing scans is then lost. In this paper, we propose a novel model, Multi-modal Gaussian Process Prior Variational Autoencoder (MGP-VAE), to impute one or more missing sub-modalities for a patient scan. MGP-VAE can leverage the Gaussian Process (GP) prior on the Variational Autoencoder (VAE) to utilize the subjects/patients and sub-modalities correlations. Instead of designing one network for each possible subset of present sub-modalities or using frameworks to mix feature maps, missing data can be generated from a single model based on all the available samples. We show the applicability of MGP-VAE on brain tumor segmentation where either, two, or three of four sub-modalities may be missing. Our experiments against competitive segmentation baselines with missing sub-modality on BraTS'19 dataset indicate the effectiveness of the MGP-VAE model for segmentation tasks.

© Springer Nature Switzerland AG 2021
M. de Bruijne et al. (Eds.): MICCAI 2021, LNCS 12907, pp. 442–452, 2021.
https://doi.org/10.1007/978-3-030-87234-2_42

Keywords: Missing modality · Gaussian process · Variational autoencoder · Glioma segmentation · MRI

1   Introduction

Glioma tumor segmentation in MR scans plays a crucial role during the diagnosis, survival prediction, and brain tumor surgical planning. Multiple MRI submodalities, FLAIR (F), T1, T1c, and T2, are regularly utilized to detect and evaluate the brain tumor subregions such as the whole tumor (WT), tumor core (TC), and the enhancing tumor (ET) region. These sequences provide comprehensive information regarding tumor brain tissues. In clinical settings, it is common for physicians to have one or more sub-modalities to be missing for a patient due to patient artifacts, acquisition problems, and other clinical reasons.

Segmentation with missing modalities techniques can be categorized into three approaches: 1) training a segmentation model for any subset of input sub-modalities; 2) training a synthesis model to impute missing sequence from input sub-modalities [13]; 3) instead of designing different models for every potential missing modality combination, designing a single model that operates based on the shared feature space through all input sub-modalities (such as taking the mean) [5,11,19]. The two first solutions associate with the training and handling of a different network for $2^{(\#of\ sub-modalities)} - 1$ combinations. The third group extracts a shared feature space from the sub-modalities, which is independent of the number of sub-modalities, to provide a unique model that shares its extracted features.

The current methods, which work based on the common representation to address missing modality, are Hetero-modal Image Segmentation (HeMIS) [11] and its relevant extension Permutation Invariant Multi-Modal Segmentation (PIMMS) technique [19]. They computed first and second moments of extracted feature maps across available sub-modalities to combine them separately. Although using these statistics are independent of the number of sub-modalities, they do not compel their convolutional model to learn a shared latent representation. For this aim, Dorent *et al.* [5] introduced Hetero-Modal Variational Encoder-Decoder (HVED) based on the Variational Autoencoder (VAE) [12] to provide the common latent variable z. Furthermore, conditional VAE (CVAE) [17] includes extra data in both the decoder and the encoder to produce samples with specific properties. Based on this method, some models used the auxiliary information to synthesis missing sub-modalities [16]. However, the prior assumption that latent representations are identically and independently distributed (*i.i.d.*) is one of the VAE model's limitations. In contrast, the Gaussian Process Prior Variational Autoencoder (GPPVAE) [4] models correlation between input sub-modalities through a Gaussian Process (GP) prior on the latent representations.

In this paper, we extend the GPPVAE for missing MRI sub-modalities imputation in a 3D framework for brain glioma tumor segmentation. The contribution of this work is three-fold. First, we extend the GPPVAE for 3D MRI modalities imputation for any scenario of missing sequences as input in one model. Second, we adapt a kernel function to capture multiple levels of correlation between sub-modalities and subjects in our Multi-modal Gaussian Process Prior Variational Autoencoder (MGP-VAE). Finally, we show that our model outperforms HVED and HeMIS in terms of DSC from multi-modal brain tumor MRI scans with any configuration of the available sub-modalities.

2 Multi-modal Gaussian Process Prior Variational Autoencoder

Assume we have a collection of 3D MRI scans to visualize brain and tumor tissue in different contrasts in MR pulse sequences, each scan coupled with auxiliary data: patient/subject IDs and sub-modality entities. Individually, we consider BraTS'19 datasets with MRI scans of the brain with glioma tumors in four sub-modalities per patient: F, T1, T1c, and T2, each of which contributes differing tissue contrast views and spatial resolutions. Each unique patient and sub-modality is assigned to a feature vector, which we define as a subject feature vector and a modality feature vector, respectively. Subject and modality features refer to elements of kernel covariance function calculated based on present data. Subject feature vectors might contain brain features such as brain tissue, tumor location, or dimension, while modality feature vectors may contain contrast information or tumor sub-region features of each modality. Figure 1 shows an overview of the MGP-VAE for synthesizing missing modalities. We assume that at least one sub-modality of each test subject is present during training.

Formulation. Let P denote the number of subjects, M the number of sub-modalities, $N = P \times M$ denote the number of all samples, and consider $\mathbf{Y} = \{y_n\}_{n=1}^N \in \mathbb{R}^{N \times K}$ for all N input samples and K denotes k-dimensional representation for N samples; let $\mathbf{X} = \{x_p\}_{p=1}^P \in \mathbb{R}^{P \times Q}$ denote Q-dimensional patient feature vectors for the P patients, and let $\mathbf{W} = \{w_m\}_{m=1}^M \in \mathbb{R}^{M \times R}$ denote R-dimensional modality feature vectors for the M sub-modalities. Four sub-modalities provide complementary information about the brain tissue, $\mathbf{W} = \{w_1 = F, w_2 = T1, w_3 = T1c, w_4 = T2\}$. Finally, let $\mathbf{Z} = \{z_n\}_{n=1}^N \in \mathbb{R}^{N \times L}$ denote the L-dimensional latent representations which abstract input samples through GP, $f_{\mathcal{GP}}$. We examine the following process for the available input data:

- The latent representation of MRI scan p_n in sub-modality m_n is generated from the subject feature vector x_{p_n} and modality feature vector w_{m_n} as: $z_n = f_{\mathcal{GP}}(x_{p_n}, w_{m_n})$, where $f_{\mathcal{GP}}$ is a GP prior to compute sample covariances as a function of subject and modality feature vectors in the latent space, z.
- Reconstructed output $\hat{y}_n$ is created from its latent representation z_n as: $\hat{y}_n = f_d(z_n)$, where f_d is a convolutional neural network with decoder architecture to map latent representation, z, into the reconstruction space, $\hat{y}$.

The marginal likelihood of the MGP-VAE model is:

$$p(\mathbf{Y}|\mathbf{X}, \mathbf{W}, \theta_d, \theta_{\mathcal{GP}}) = \int p(\mathbf{Y}|\mathbf{Z}, \theta_d) \, p(\mathbf{Z}|\mathbf{X}, \mathbf{W}, \theta_{\mathcal{GP}}) \, d\mathbf{Z} \qquad (1)$$

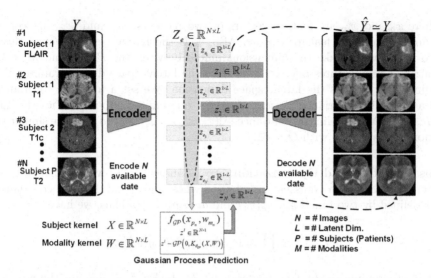

Fig. 1. Overview of the proposed MGP-VAE. Each sub-modality volume is mapped to a 1024-dimensional (L) space and decoded to the initial space. Covariances among input volumes are formed through a GP Prior to each column of the latent representation matrix $\mathbf{Z}_e$. The subject and modality correlations are modeled in the latent space due to its compact superiority.

where θ_d denotes the parameters of the decoder and $\theta_{\mathcal{GP}}$ indicates the parameters of GP's kernels. Equation 1 cannot be optimized straightforwardly as the integral is not tractable. Thus we resort to variational inference, which requires introducing an approximate posterior distribution.

2.1 Proposed Kernel Functions for MGP-VAE

The GP defines a set of random variables on z^l, l–th column of $\mathbf{Z}$, so any finite number of them have a multivariate Gaussian distribution. In case $z^l = f_{\mathcal{GP}}(\mathbf{X}, \mathbf{W})$ is a GP, then given L observations, the joint distribution of the random variables, $z^1 = f_{\mathcal{GP}}(\mathbf{X}, \mathbf{W}), z^2 = f_{\mathcal{GP}}(\mathbf{X}, \mathbf{W}), ..., z^L = f_{\mathcal{GP}}(\mathbf{X}, \mathbf{W})$, is Gaussian. For our L–dimensional latent representation, we have:

$$p(\mathbf{Z}|\mathbf{X}, \mathbf{W}, \theta_{\mathcal{GP}}) = \prod_{l=1}^{L} \mathcal{GP}(z^l|0, K_{\theta_{\mathcal{GP}}}(\mathbf{X}, \mathbf{W})) \tag{2}$$

where $K_{\theta_{\mathcal{GP}}}(\mathbf{X}, \mathbf{W})$ is the covariance function with the kernel parameters, $\theta_{\mathcal{GP}}$, which comprises a modality kernel and a patient kernel. The former models covariance among sub-modalities, while the latter models covariance between patients. $K_{\theta_{\mathcal{GP}}}(\mathbf{X}, \mathbf{W})$ can be factorized into [3]:

$$K_{\theta_{\mathcal{GP}}}(\mathbf{X}, \mathbf{W}) = \underbrace{\mathcal{K}(x_p, x'_p)}_{patient\ kernel} \otimes \underbrace{\mathcal{K}(w_m, w'_m)}_{modality\ kernel} \tag{3}$$

where x_p and x'_p are feature vectors of two patients, w_m and w'_m are corresponding modality feature vectors. Also, $\otimes$ is the Kronecker product of these two matrices to make the dimensions match between the $P \times P$ and $M \times M$ matrix. These features are extracted from the latent space during training. We define $L = 1024$ as the latent space dimension. We set a full-rank covariance as a modality covariance ($\mathcal{K}(w_m, w'_m)$) for our limited sub-modalities (F, T1, T1c, and T2) and a linear covariance ($\mathcal{K}(x_p, x'_p) = x_p^T . x'_p$) to measure similarity among the subjects with $Q = 64$.

Loss Function and Optimization. As a standard VAE, we approximate the latent variables by a Gaussian distribution whose mean and diagonal covariance are defined by two functions, $\mu_e(y_n)$ and $diag(\sigma_e^2(y_n))$. Thus, we have:

$$q(\mathbf{Z}_e|\mathbf{Y}) = \prod_{n=1}^{N} \mathcal{N}\left(z_{e_n}|\mu_e(y_n), diag(\sigma_e^2(y_n))\right), \qquad (4)$$

which approximates the true posterior on $\mathbf{Z}_e$. In Eq. 4, θ_e denotes the weights of the encoder in auto-encoder neural network architecture. Latent representations $\mathbf{Z}_e = [z_{e_1}, z_{e_2}, ..., z_{e_N}] \in \mathbb{R}^{N \times L}$ are also sampled employing the reparameterization method [12], $z_{e_n} = \mu_e(y_n) + v_n \odot \sigma_e(y_n)$, where $\odot$ denotes the element-wise product and v is a random number drawn from a normal distribution. We compute the resulting evidence lower bound (ELBO) as:

$$\log p(\mathbf{Y}|\mathbf{X}, \mathbf{W}, \theta_d, \theta_{\mathcal{GP}}) \geq \mathbb{E}_{Z \sim q_{\theta_e}}\left[\sum_n \log \mathcal{N}(y_n|f_d(z_n)) + \log p(\mathbf{Z}_e|\mathbf{X}, \mathbf{W}, \theta_{\mathcal{GP}})\right]$$

$$+ \frac{1}{2}\sum_{nl} \log(\sigma_q^2(y_n)_l) + const.$$

$$(5)$$

To increase the ELBO as much as possible, we apply stochastic backpropagation [12]. Individually, we approximate the expectation by sampling from a reparameterized variational posterior over the latent representations, achieving the resulting loss function:

$$\mathcal{L}(\theta_d, \theta_e, \theta_{\mathcal{GP}})$$

$$= \sum_n \underbrace{\frac{\|y_n - f_d(z_{e_n})\|^2}{2\sigma_y^2}}_{L2 \ reconstruction \ loss} - \underbrace{\log p(\mathbf{Z}_e|\mathbf{X}, \mathbf{W}, \theta_{\mathcal{GP}})}_{\mathcal{GP}} + \underbrace{\frac{1}{2}\sum_{nl} \log(\sigma_{z_e}^2(y_n)_l)}_{regularization} \qquad (6)$$

$$+ NK \log \sigma_y^2$$

where we optimize regarding θ_d, θ_e, and $\theta_{\mathcal{GP}}$. We optimize loss function through Adam optimizer with a learning rate of 0.001. We experimentally noted that minimizing loss function was developed by first training the encoder and the decoder within the VAE, next optimizing the GP weights by frozen encoder and decoder for 100 epochs (the learning rate of 0.01), last, optimizing all parameters jointly in our MGP-VAE model.

2.2   Missing Modality Imputation

We derive an approximate predictive posterior for MGP-VAE that enables missing modality predictions of high-dimensional samples. Specifically, given training samples Y, subject feature vectors X, and modality feature vectors W, the prediction for the missing data y_t of subject p_t in modality m_t is given by:

$$p(y_t|x_t, w_t, \mathbf{Y}, \mathbf{X}, \mathbf{W})$$
$$= \int \underbrace{p(y_t|z_t)}_{decoding\ missing\ data}\ \underbrace{p(z_t|x_t, w_t, \mathbf{Z}_e, \mathbf{X}, \mathbf{W})}_{\mathcal{GP}\ prediction\ of\ z_t}\ \underbrace{q(\mathbf{Z}_e|\mathbf{Y})}_{encoding\ all\ training\ data}\ dz_t d\mathbf{Z}_e$$

(7)

where x_t and w_t are feature vectors of subject p_t and sub-modality m_t, respectively. The approximation in Eq. 7 is achieved by substituting the exact posterior on $\mathbf{Z}_e$ with the variational distribution $q(\mathbf{Z}_e|\mathbf{Y})$ (see [4]). According to Eq. 7, the missing sub-modality can be computed by the three steps. First, we encode all training image data in the latent space by the encoder, $\mathbf{Z}_e$. Next, predict latent representation z_t of image y_t through the GP model using m, $\mathbf{X}$, and $\mathbf{W}$. Lastly, latent representation z_t is decoded to the high-dimensional image space through the decoder as missing modality imputation.

3D Variational Encoder-Decoder Network Architecture. The encoder part employs four spatial levels, where each level consists of two convolution layers with $3 \times 3 \times 3$ kernel and ELU. The first convolution layer is without downsampling (stride = 1), while the second one applies strode convolution for downsizing. We follow a typical VAE approach to downsize image dimensions by two progressively, but with fixed feature size equal to 32 except encoder endpoint with 16 feature maps. The encoder endpoint has size $16 \times 4 \times 4 \times 4$, followed by a fully connected layer, and is 16 times spatially smaller than the input volume of $64 \times 64 \times 64$. The decoder structure is similar to the encoder one, but each level begins with volumetric upsampling using the nearest neighbor algorithm.

3   Experiments and Results

Data. We assess our method on the training set of BRATS'19 [1,2,14], which includes the scans of 335 patients. Each subject is scanned with four T1, T1c, T2, and F sequences. All scans are skull-striped and re-sampled to an isotropic 1mm resolution, and four sequences of each patient have been co-registered. Radiologists provided the ground truth labels. The segmentation comprise the following tumor tissue labels: 1) non-enhancing tumor, 2) edema, 3) enhancing core. Implementation of the MGP-VAE is available[1].

3.1   Missing Modality Imputation

Figure 2(a) illustrates a qualitative evaluation of each sub-modality reconstruction with one missing sub-modality (second column) and two missing sub-modalities (third column). Our model proposes to reconstruct the brain and

---

[1] https://github.com/hamghalam/MGP-VAE.

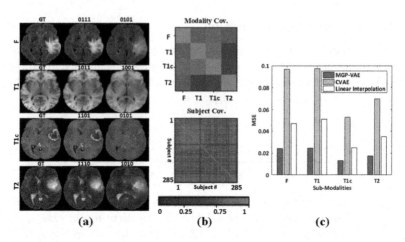

Fig. 2. (a) Example of modality completion (each row corresponds to a particular sub-modality) given a subset of sub-modalities as input. The 4-bit strings on top of each slice determine present and absent sub-modality with 1 and 0, respectively (bit order from left to right F, T1, T1c, and T2). (b) Covariances between sub-modalities and subjects are modeled through a GP prior model. (c) We compared our method with CVAE and VAE, the baseline of HVED and other well-known imputation methods.

tumor tissue even when the tumor information is missing or not clear by coupling information from available samples. Comparing reconstructions using two sub-modalities and three sub-modalities confirms that the reconstructed volumes preserve high-frequency details. This suggests that the MGP-VAE model can effectively learn relations between available sub-modalities in different subjects (Fig. 2(b)). The PSNR values for imputation are F = 27.95, T1 = 27.80, T1c = 29.43, and T2 = 27.99 based on three available sub-modalities. Similarly, we have F = 22.36, T1 = 22.56, T1c = 24.86, and T2 = 22.66 with two available sub-modalities. Besides, Mean Squared Error (MSE) for each sub-modality is considered as an evaluation metric to compare MGP-VAE with CVAE and linear interpolation. The latter applies linear interpolation between available sub-modalities of a subject in the latent space learned through VAE to predict the missing sequence (Fig. 2(c)). The CVAE indeed improves VAE in image generation by conditioning the encoder and decoder to the desired input. However, when we have missing input data, CVAE has a confined ability to create the latent variable for unseen data compared to VAE. This might be because CVAE is more restricted to learn particular data features (latent representation) from observed input data. Therefore it has dedicated latent variables with limited features from missing data. The latent space of VAE contains more general characteristics which can be used to predict missing data.

3.2   Glioma Segmentation

To assess the MGP-VAE, we examine it on the brain tumor segmentation framework and compare it with two state-of-the-art methods for all the possible subset of sub-modalities. The first, HVED [5] is the state-of-the-art method based on VAE for brain tumor segmentation with missing sub-modalities. The second approach, HeMIS [11], combines the available sub-modalities based on feature maps moments. We adopt the 3D U-Net architecture? [15] to segment glioma where the available sub-modalities and imputed ones are concatenated as input multimodal MRI scans. We use the Dice score to measure segmentation accuracy in clinically significant glioma subregions: WT, TC, and ET in Table 1. We have almost the same performance in Table 1 if all sub-modalities are available (without imputation). Our method is designed and optimized to address problems where either one, two, or three of four sub-modalities may be missing. When all the sub-modalities are available, this is a different scenario [6–10,18]. Moreover, Fig. 3 shows comparative segmentation results of the BraTS'19 dataset without (first row) and with (second row) imputation through MGP-VAE. The last column (first row) is the ground truth.

Table 1. Comparison of MGP-VAE model with HeMIS and HVED model (Dice %) for all subset of available sub-modalities. Sub-modalities present are denoted by 1, the missing ones by 0. The IQR is the interquartile range (IQR) and * indicates significant improvement by a Wilcoxon test ($p < 0.05$).

Modalities				WT			TC			ET		
F	T1	T1c	T2	HeMIS	HVED	Ours	HeMIS	HVED	Ours	HeMIS	HVED	Ours
0	0	0	1	78.0	79.9	**81.1***	49.3	52.8	**56.2***	22.1	29.9	**30.8**
0	0	1	0	57.6	61.7	**63.2***	57.7	65.8	**68.1***	59.5	65.1	**66.4***
0	1	0	0	**53.3**	51.5	53.2	37.0	36.2	**39.9***	11.3	13.2	**14.2**
1	0	0	0	78.9	81.0	**83.3***	48.8	49.9	**52.7***	24.2	24.1	**25.2**
0	0	1	1	80.1	81.8	**83.1***	68.3	73.1	**75.7***	67.5	69.2	**70.5***
0	1	1	0	62.6	66.0	**67.6***	63.4	69.1	**71.8***	64.3	66.6	**68.0***
1	1	0	0	82.9	83.2	**84.7***	**56.0**	54.4	55.3	**28.2**	24.1	25.1
0	1	0	1	79.8	81.1	**82.0**	52.5	56.6	**58.4***	27.1	30.3	**31.8***
1	0	0	1	84.8	86.5	**87.9***	58.0	58.9	**61.0**	26.9	33.7	**35.0***
1	0	1	0	82.2	84.8	**85.8**	66.7	72.6	**74.6***	67.0	**70.4**	70.2
1	1	1	0	83.9	85.6	**87.4***	69.9	73.6	**75.8***	68.8	70.2	**71.5***
1	1	0	1	85.9	87.1	**88.3**	60.1	61.0	**63.0***	32.3	33.3	**34.5***
1	0	1	1	85.9	87.5	**89.0***	71.6	75.2	**77.4***	68.9	70.3	**71.4***
0	1	1	1	81.3	82.3	**83.6**	69.9	74.5	**76.4***	68.7	70.4	**71.6***
Median				80.7	82.05	83.45	59.05	63.4	65.55	45.9	49.4	50.7
IQR				9.98	9.40	9.80	17.0	19.23	19.75	41.58	41.78	41.33

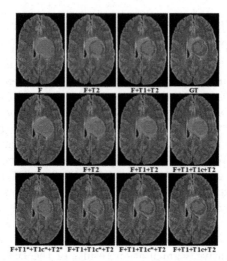

Fig. 3. The first row explains the effects of different combinations of input sub-modalities overlaid on the FLAIR(F) slice without the imputation. The second row illustrates the HVED results. The last row represents the segmentation with imputation based on MGP-VAE method. The segmentation colors describe edema (green), non-enhancing (blue), and enhancing (red). * indicates imputed sub-modalities. (Color figure online)

4   Conclusion

We have introduced MGP-VAE to predict missing data of subjects in specified MRI sub-modalities using a specialized VAE. Our model incorporates a GP prior over the encoded representation of available volumes to compute correlations among available sub-modalities and subjects. We also validated the robustness of the method with all possible missing sub-modality scenarios on glioma tumors and achieved state-of-the-art segmentation results. Finally, our method offers promising insight for leveraging large but incomplete data sets through one single model. Possible future work focuses on extending the method to various imaging modalities (MRI, PET, and CT) as well as genetic data.

Acknowledgements. This work was funded in part by National Institutes of Health R01CA233888.

References

1. Bakas, S., et al.: Advancing the cancer genome atlas glioma MRI collections with expert segmentation labels and radiomic features. Sci. Data **4**(1), 1–13 (2017)
2. Bakas, S., et al.: Identifying the best machine learning algorithms for brain tumor segmentation, progression assessment, and overall survival prediction in the brats challenge. arXiv preprint arXiv:1811.02629 (2018)

3. Bonilla, E.V., Agakov, F.V., Williams, C.K.: Kernel multi-task learning using task-specific features. In: Artificial Intelligence and Statistics, pp. 43–50 (2007)
4. Casale, F.P., Dalca, A.V., Saglietti, L., Listgarten, J., Fusi, N.: Gaussian process prior variational autoencoders. In: Advances in Neural Information Processing Systems, pp. 10369–10380 (2018)
5. Dorent, R., Joutard, S., Modat, M., Ourselin, S., Vercauteren, T.: Hetero-modal variational encoder-decoder for joint modality completion and segmentation. In: Shen, D., et al. (eds.) MICCAI 2019. LNCS, vol. 11765, pp. 74–82. Springer, Cham (2019). https://doi.org/10.1007/978-3-030-32245-8_9
6. Hamghalam, M., Lei, B., Wang, T.: Convolutional 3D to 2D patch conversion for pixel-wise glioma segmentation in MRI scans. In: Crimi, A., Bakas, S. (eds.) BrainLes 2019. LNCS, vol. 11992, pp. 3–12. Springer, Cham (2020). https://doi.org/10.1007/978-3-030-46640-4_1
7. Hamghalam, M., Lei, B., Wang, T.: High tissue contrast MRI synthesis using multi-stage attention-GAN for segmentation **34**, 4067–4074 (2020). https://doi.org/10.1609/aaai.v34i04.5825
8. Hamghalam, M., Wang, T., Lei, B.: High tissue contrast image synthesis via multistage attention-GAN: application to segmenting brain MR scans. Neural Netw. **132**, 43–52 (2020)
9. Hamghalam, M., Wang, T., Qin, J., Lei, B.: Transforming intensity distribution of brain lesions via conditional GANs for segmentation. In: 2020 IEEE 17th International Symposium on Biomedical Imaging (ISBI), pp. 1–4. IEEE (2020)
10. Hatami, T., Hamghalam, M., Reyhani-Galangashi, O., Mirzakuchaki, S.: A machine learning approach to brain tumors segmentation using adaptive random forest algorithm. In: 2019 5th Conference on Knowledge Based Engineering and Innovation (KBEI), pp. 076–082 (2019). https://doi.org/10.1109/KBEI.2019.8735072
11. Havaei, M., Guizard, N., Chapados, N., Bengio, Y.: HeMIS: hetero-modal image segmentation. In: Ourselin, S., Joskowicz, L., Sabuncu, M.R., Unal, G., Wells, W. (eds.) MICCAI 2016. LNCS, vol. 9901, pp. 469–477. Springer, Cham (2016). https://doi.org/10.1007/978-3-319-46723-8_54
12. Kingma, D.P., Welling, M.: Auto-encoding variational bayes. arXiv preprint arXiv:1312.6114 (2013)
13. Li, R., et al.: Deep learning based imaging data completion for improved brain disease diagnosis. In: Golland, P., Hata, N., Barillot, C., Hornegger, J., Howe, R. (eds.) MICCAI 2014. LNCS, vol. 8675, pp. 305–312. Springer, Cham (2014). https://doi.org/10.1007/978-3-319-10443-0_39
14. Menze, B.H., et al.: The multimodal brain tumor image segmentation benchmark (BraTS). IEEE Trans. Med. Imaging **34**(10), 1993–2024 (2015). https://doi.org/10.1109/TMI.2014.2377694
15. Ronneberger, O., Fischer, P., Brox, T.: U-net: convolutional networks for biomedical image segmentation. In: Navab, N., Hornegger, J., Wells, W.M., Frangi, A.F. (eds.) MICCAI 2015. LNCS, vol. 9351, pp. 234–241. Springer, Cham (2015). https://doi.org/10.1007/978-3-319-24574-4_28
16. Sharma, A., Hamarneh, G.: Missing MRI pulse sequence synthesis using multimodal generative adversarial network. IEEE Trans. Med. Imaging **39**(4), 1170–1183 (2019)
17. Sohn, K., Lee, H., Yan, X.: Learning structured output representation using deep conditional generative models. Adv. Neural. Inf. Process. Syst. **28**, 3483–3491 (2015)

18. Soleymanifard, M., Hamghalam, M.: Segmentation of whole tumor using localized active contour and trained neural network in boundaries. In: 2019 5th Conference on Knowledge Based Engineering and Innovation (KBEI), pp. 739–744. IEEE (2019)

19. Varsavsky, T., Eaton-Rosen, Z., Sudre, C.H., Nachev, P., Cardoso, M.J.: PIMMS: permutation invariant multi-modal segmentation. In: Stoyanov, D., et al. (eds.) DLMIA/ML-CDS -2018. LNCS, vol. 11045, pp. 201–209. Springer, Cham (2018). https://doi.org/10.1007/978-3-030-00889-5_23

Joint PVL Detection and Manual Ability Classification Using Semi-supervised Multi-task Learning

Jingyun Yang[1], Jie Hu[2], Yicong Li[1], Heng Liu[2], and Yang Li[1(✉)]

[1] Tsinghua-Berkeley Shenzhen Institute, Tsinghua University, Shenzhen, China
yangli@sz.tsinghua.edu.cn
[2] Department of Radiology, Affiliated Hospital of Zunyi Medical University,
Guizhou, China

Abstract. Among symptoms of cerebral palsy (CP), the degree of hand function impairment in young children is hard to assess due to large inter-personal variability and differences in evaluators' experience. To help design better treatment strategies, accurate identification and delineation of manual ability injury level is a major clinical concern. Periventricular leukomalacia (PVL), a form of brain lesion in periventriular white matter in premature infants, is a leading cause of CP and have clinical associations with motor function injuries. In this paper, we exploit the correlation between PVL lesion segmentation and manual ability classification (MAC) to improve the identification performance of both tasks for T2 FLAIR MRI scans. Particularly, we propose a semi-supervised multi-task learning framework to jointly learn from heterogeneous datasets. Two clinically related auxiliary tasks, lesion localization and ventricle segmentation, are also incorporated to improve the classification accuracy while requiring only a small amount of manual annotations. Using two datasets containing 24 labeled PVL samples and 87 labeled MAC samples, the proposed model significantly outperforms single-task methods, achieving a dice score of 0.607 for PVL lesion segmentation and 84.3% accuracy for manual ability classification.

Keywords: Heterogeneous multi-task learning · Semi-supervised learning · PVL lesion segmentation · Manual ability classification · Computer-aided diagnosis

1 Introduction

Cerebral palsy (CP) is a non-progressive interference to the developing brain, causing a range of motor function disorder among young children [11]. Among various CP symptoms, hand function impairment have large inter-personal variations and would adversely affect patients' self-care ability in their daily life [16,21]. To help develop better treatment strategy, accurate identification and delineation of injury level in early childhood is a major clinical concern [21]. Several examination scales,

© Springer Nature Switzerland AG 2021
M. de Bruijne et al. (Eds.): MICCAI 2021, LNCS 12907, pp. 453–463, 2021.
https://doi.org/10.1007/978-3-030-87234-2_43

such as the Manual Ability Classification System (MACS) and the Assisting Hand Assessment (AHA) have been frequently used to assess the degree of hand function impairment in children with CP [1]. However, an individual's performance may vary across different scales and the accuracy of assessment relies on the experience of the evaluator [15]. Furthermore, it is difficult to perform detailed assessments during infancy due to incomplete development.

Recent clinical studies have attempted to use MRI, a more objective and quantitative measurement tool to reflect motor function. For instance, [20] uses diffusion weighted MRI to explore the correlation between damaged white matter pathways and specific motor functions such as upper limbs functions. Contrary to clinical observations, the reported correlations were mostly low to moderate. On the other hand, periventricular leukomalacia (PVL), a form of ischemic brain white matter injury in prematured infants have been shown to be frequently associated with motor function disorder in children with CP [9,12]. Hence we aim to utilize the relationship between PVL and manual ability in children with CP to improve the identification of PVL and the classification of manual ability level based on T2-Weighted MRI data.

Previous works on brain MRI lesion segmentation generally focused on extracting handcrafted image features, including modal intensities, tissue probabilities, multi-scale annular filter for blobness detection [10] or patient features like age and gender for computer-aided disease diagnosis [19]. Research that focus on PVL segmentation and manual ability classification, however, seem rare except for [25], which only uses handcrafted features that is hard to generalize to new datasets. Recently, deep learning based methods that can automatically learn informative features are extensively used to tackle certain brain MRI lesion segmentation tasks [27]. *Multi-task learning*, a learning paradigm that trains two or more related tasks with shared parameters or features, is a common approach for automatically incorporating auxiliary information in neural networks. Several studies have adopted multi-task learning for MRI image processing tasks [3–5]. However, most of the deep learning methods rely on having large amounts of annotated training data, which is not a practical solution due to the high cost obtaining expert annotations.

To solve the aforementioned challenges, we propose a semi-supervised multi-task framework for jointly segmenting the PVL lesions and classifying manual ability in a small data setting, as shown in Fig. 1. In addition to the two target tasks, we consider two medically related auxiliary tasks to help improve the performance of target tasks: *lesion localization* (LL) and *ventricle segmentation* (VS). The former identifies the coordinate of each lesion center, as clinical evidence shows that PVL lesions are always detected regularly in specific locations like the thalamus and basal ganglia. The latter is also related to PVL as the volume of the ventricles changes due to white matter injuries. One advantage of this framework is that it only requires the data for each task to be partially labeled. Information learned from known labels will propagate to unlabeled data during training by using pseudo-labeling. We show in the experiment that, having only 10 out of 80 samples with ventricle annotations is sufficient to improve the target task performance.

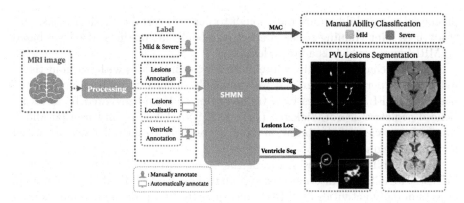

Fig. 1. Illustration of SHMN framework. Tasks outlined in red are the target tasks, those in green are the auxiliary tasks. (Color figure online)

The main contributions of this work are as follows:

1. We show that PVL lesions on MRI images does correlate with the manual ability level of CP patients.
2. We propose a novel deep learning framework, named Semi-supervised Heterogeneous Multi-task Network (SHMN), that can jointly learn to perform heterogeneous medical imaging tasks including segmentation, localization and classification. It effectively uses the correlation among PVL lesion condition, ventricle shape and MA level to improve the performance of manual ability classification and PVL segmentation based on MRI data.
3. We adopted an effective semi-supervised algorithm that utilizes unlabeled data in auxiliary tasks to reduce dependency on manually annotated data.

2   Materials and Method

2.1   Data and Preprocessing

Two private datasets are used in our research. `Dataset1` contains 24 T2 FLAIR MRI images with PVL lesion annotation and `Dataset2` contains 87 T2 FLIAIR MRI scans with manual ability labels. For all MRI data, participants were identified through a medical record management system of the First Affiliated Hospital of Xi'an Jiaotong University. Brain images were captured using GE SignaHDxt 3.0T magnetic resonance scanner (GE Healthcare, Milwaukee, Wisconsin, USA) and 8-channel head coil.

`Dataset1` is captured from PVL patients between 19 to 28 months. Although PVL mostly occurs in newborns, premature infants younger than 1 year old often cannot tolerate MRI examinations, therefore we use MRI images of infants between 1–2 years old who are still in the relatively early stage of PVL. `Dataset2` is captured from CP patients with MA tests between 1–12 years old. These children are older since CP is commonly developed from PVL and can only be diagnosed when children are old enough to perform certain physical tasks. Our study uses joint training to learn a better model that predicts MA level based on MRI images, which is a novel design.

PVL was characterized by white matter hyperintensities with/without tissue reduction in periventricular and was manually delineated on each slice of the patient's T2-FLAIR images by a trained rater (HJ) and reviewed by an experienced pediatric neuroradiologist (LH). ITK-SNAP [26], a software that allows simultaneous view of the brain on axial, coronal, and sagittal planes was used for manual segmentation.

Manual ability was classified using the Manual Ability Classification System (MACS). MACS is an international system to classify a child's ability of handling objects in daily activities [8]. MACS measures five levels of the manual abilities of children with CP. Level I indicates the best while level V indicates the worst level of manual ability. The specific assessment steps were described as the prior protocol [18]. All the participants were classified into two groups according to MACS levels, level I–II as mild injury group and level III–V as severe injury group.

We automatically create labels for Lesion Localization as follows: extract the topological structure of the lesion shape from the binary segmentation mask by outermost border following [23] and compute the center of mass using spatial moments.

All MRI images used in experiments are preprocessed via a standard pipeline: (i) Images are transformed into the same coordinate system by FSL FLIRT [22]. (ii) Skull stripping is performed to remove the redundant parts. (iii) Bias field correction is conducted using ANTs [2].

2.2   Network Architecture

Since medical images commonly encompass three dimensions, we choose 3D U-Net [6] for as our base model. The framework resides on a modified version of U-Net [14], which deviates from the original architecture in that it replaces ReLU activation functions with leaky ReLUs and uses instance normalization [24] instead of the more popular batch normalization [13]. Figure 2 shows the architecture of the proposed SHMN model which jointly learns a common encoder for all target and auxiliary tasks. The correlation among these tasks enables us to assume that their discriminative features lie in a common multi-scale feature space, represented by the encoder network. The model takes input data through one common encoder and then branch out to perform different tasks through corresponding decoders. Meanwhile, our data-driven framework incorporates semi-supervision into the learning procedures to make it few-shot friendly and reduce cost of expert annotation.

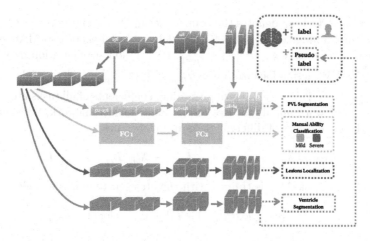

Fig. 2. Architecture of the multi-task semi-supervised model. The purple blocks represent the shared backbone for extracting features, and the orange and green branches are the task-specific decoders for the target tasks and auxiliary tasks, respectively. (Color figure online)

As shown in Fig. 2, the input of the network contains $182 \times 218 \times 182$ MRI image patches with M_{cls} label (class of manual ability), P_{seg} label (PVL lesions annotation), L_{loc} label (density map) and V_{seg} label (ventricle annotation) while the target output includes the classification of manual ability and segmentation mask of PVL lesions. The common encoder uses two plain convolutional layers between pooling in each block to extract the latent features from the given images and down-samples the resolution by using the pooling technique and the respective branches receive the feature maps from the bottom. In the M_{cls} branch there is a sequence of two fully connected (FC) layers. We take the feature maps as the input of the FC1 followed by a rectified linear unit (ReLU) activation function and the output of FC1 as the input of the FC2 followed by a sigmoid activation function to predict the class probability. When the probability is greater than 0.5, we classify the sample into severe injury group. For other branches, decoders take transposed convolution operations and recombines the semantic information with higher resolution feature maps obtained directly from the encoder through skip connections. As V_{seg} aims to make full use of those unlabeled MRI data, which can be acquired easily in practice, we therefore modify the regular decoder to predict the pseudo-label for each unlabeled sample in the fine-tuning phases [17]. Due to the large image size and memory constraints we set a batch size of 2. We use the Adam optimizer with an initial learning rate of 3e-4 and set it to decrease periodically if the losses do not improve enough. The network is trained in a semi-supervised fashion with labeled and unlabeled data simultaneously. For unlabeled samples, pseudo-Labels recalculated after every weights update are used for the modified loss function of supervised learning task. The training procedure can be mathematically described as follows:

Let $\mathcal{X}^1 = \{\mathbf{X}_n^1\}_{n=1}^{N_1}$ denotes the training set in Dataset1 and $\mathcal{X}^2 = \{\mathbf{X}_n^2\}_{n=1}^{N_2}$ denotes the training set in Dataset2. We define the labels of two MAC groups (mild and severe) as $\mathbf{y}_{m_{cls}} = \{y_n^{m_{cls}}\}_{n=1}^{N_1}$, PVL lesion segmentation as $\mathbf{y}_{p_{seg}} = \{y_n^{p_{seg}}\}_{n=1}^{N_2}$, PVL lesion localization as $\mathbf{y}_{p_{loc}} = \{y_n^{p_{loc}}\}_{n=1}^{N_2}$, and ventricle segmentation as $\mathbf{y}_{v_{seg}} = \{y_n^{v_{seg}}\}_{n=1}^{N_1}$. During training, coefficients of all tasks are equally set to 1 as our work mainly focuses on the multi-task framework, however we'll explore the coefficients of different tasks in future work.

$$L_{SHMN} = \lambda_{m_{cls}} L_{m_{cls}} + \lambda_{p_{seg}} L_{p_{seg}} + \lambda_{p_{loc}} L_{p_{loc}} + \lambda_{v_{seg}} L_{v_{seg}} \tag{1}$$

Specifically, task M_{cls} is trained with the cross-entropy loss as follows:

$$L_{m_{cls}} = -[y_{m_{cls}} \log(\hat{y}_{m_{cls}}) + \varepsilon(1 - y_{m_{cls}}) \log(1 - \hat{y}_{m_{cls}})] \tag{2}$$

where ε is the bias of positive examples to negative examples, which can increase the penalty for misclassification of positive examples.

Task P_{seg} is trained with a combination of dice and cross-entropy loss, $L_{p_{seg}} = L_{CE} + L_{dice}$. The dice loss function here is an adaptation of the variant proposed in [7]:

$$L_{dice} = -2 \frac{\sum_{i \in I} o_i v_i}{\sum_{i \in I} u_i + \sum_{i \in I} v_i} \tag{3}$$

where o is the softmax output of the network and v is a one hot encoding of the ground truth segmentation map $y_{p_{seg}}$ and $i \in I$ is the number of pixels in the training batch.

As for task V_{seg}, in order to provide pseudo-labels for unlabeled samples, we select the class with maximum predicted probability for each unlabeled sample.

$$y_{v_{seg}}^{c'} = \begin{cases} 1 & \text{if } c = \text{argmax}_{c'} \hat{y}_{v_{seg}}^{c'} \\ 0 & \text{otherwise} \end{cases} \tag{4}$$

and we can add the pseudo-label to the unlabeled data and update our dataset to train with the loss function:

$$L_{v_{seg}^*} = \alpha(t) \sum_{c=1}^{C} L_{v_{seg}} \left(y_{v_{seg}}^{c'}, \hat{y}_{v_{seg}}^{c} \right) \tag{5}$$

where $\alpha(t)$ is a coefficient balancing labeled data and pseudo-labeled data to maximize benefit of the unlabeled data.

$$\alpha(t) = \begin{cases} 0, & t < T_1; \\ \frac{t - T_1}{T_2 - T_1} \alpha_f, & T_1 \leq t < T_2; \\ \alpha_f, & T_2 \leq t. \end{cases} \tag{6}$$

Task V_{seg} is trained with standard binary cross-entropy loss while task P_{loc} is trained with MSE loss for its simplicity and effectiveness.

3   Experiments and Results

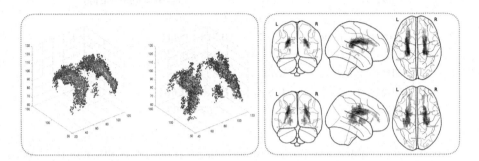

Fig. 3. Lesions distribution of mild (left) and severe (right) MA groups on the left and distributions of total lesion volume in mild (above) and severe (below) MA groups on the right.

First, we present the empirical evidence of the correlation between PVL lesions and MAC levels. In Fig. 3, we visualize the PVL lesion distributions in the mild injury group and the severe injury group based on 16 lesions labeled samples annotated by expert from Dataset2. Compared with the mild group, the distribution map of the CP patients with severe impairment demonstrates that the occurrence of injure follows a characteristic spatial pattern with more predilection for thalamus, basal ganglia and regions around the central sulcus which are indeed related to manual ability. The result is consistent with the clinical research.

Model training is done on two datasets containing 24 images with PVL segmentation mask and 87 images with manual ability classification 68 mild, 29 severe respectively. We divide both dataset1 and dataset2 into 70% training, 10% validation and 20% testing for two target tasks. For auxiliary tasks, Lesion Localization is trained on Dataset1 and Ventricle Segmentation is trained on Dataset2, both with a 9:1 training-validation split. The performance of manual ability classification is evaluated by accuracy and F1 score while the performance of PVL lesion segmentation is evaluated by dice. To further show the validity of the task setting, we train with different combinations of tasks via SHMN. The setting includes (1) joint MAC and PVL Lesion Segmentation (SHMN-2), (2) joint MAC and PVL Lesion Segmentation with Lesion Localization (SHMN-3), (3) joint MAC and PVL Lesion segmentation with Lesion Localization and Ventricle Segmentation (SHMN-4), (4) joint MAC and PVL Lesion Segmentation with Lesion Localization and Ventricle Segmentation using less labeled data (SHMN-4*). Specifically we use 25 ventricle labeled MRI data in SHMN-4 while pseudo-labeling 70 more ventricle unlabeled data with only 10 labeled data in SHMN-4*.

Table 1. Results of manual ability classification and PVL lesion segmentation

Method	MAC		Lesion segmentation
	Accuracy	F1	Dice
U-Net set single task	0.627	0.513	0.441
SHMN-2	0.667	0.583	0.532
SHMN-3	0.745	0.649	0.583
SHMN-4	**0.843**	**0.789**	**0.607**
SHMN-4*	0.824	0.757	0.600

As shown in Table 1, the proposed methods yield better results in both MAC and lesion segmentation compared to the conventional approach of single-task training with 3D U-Net [6]. The standard errors of MAC and of PVL segmentation are 1.2e-3 and 1.6458e-2 respectively in the SHMN-4 setting. This experiment implies that our heterogeneous multi-task model is superior to models which learn from different tasks separately. Note that every auxiliary task does help improving target task performance, proving that these medically related tasks can indeed assist in the diagnosis. Specifically, even we just use 10 ventricle labeled data in SHMN-4*, the model can still achieve quite satisfying results. Thus we can achieve high segmentation and classification accuracy under the condition of few annotation cost.

We have also compared our model with a SOTA multi-task learning architecture [3] by augmenting the 2D U-Net baseline model with an image classification subnet. The final loss is a combination of categorical cross-entropy for image classification, and the dice loss and cross-entropy loss for PVL segmentation. Trained with the same setting as SHMN-2, the accuracy of MAC and dice score of PVL segmentation are 0.627 and 0.531 respectively, compared to 0.667 and 0.532 for our result with SHMN-2. This demonstrates that our architecture choice for multi-task learning is more effective than the previous work.

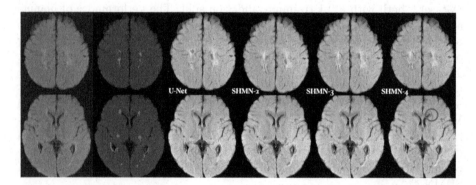

Fig. 4. A visual demonstration of the performance of the proposed SHMN. From left to right: sample T2 FLAIR slices; expert annotations of PVL lesions in green; segmentation of PVL lesions generated by U-Net and SHMN with different task setting in red. Two rows represent different axial slices of the same instance. (Color figure online)

Moreover, Fig. 4 demonstrates the segmentation results compared with the expert annotation for the visual assessment of SHMN. Task setting is the same as above. It is worth noting that after adding the ventricle segmentation task, the model become more ventricle-sensitive in the PVL lesions segmentation observed from the circled area.

To further evaluate SHMN's effectiveness in the semi-supervised scenario, we train our model with 5, 10, 15 and 20 ventricle segmentation labels and show their MAC results in Fig. 5. Good classification accuracy can be achieved with as few as 10 ventricle labels using our semi-supervised model.

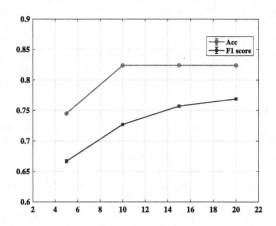

Fig. 5. Accuracy and F1 score of M_{cls} with different sizes of labeled data used in V_{seg}

4   Discussion

In this paper an automatic and simple framework is presented for the medical domain that can precisely detect small periventricular white matter lesions and assist in the diagnosis of manual ability injury. Our work provides this novel direction that explores auxiliary tasks and make full use of them which can inspire more medical research. In our future research, we want to extend our framework with even less annotation requirements and automatic auxiliary task selection. We will consider unsupervised lesion detection by using anomaly detection and develop an algorithm that picks the most useful auxiliary tasks to train based on automatically computed task relevance.

Acknowledgements. This work was supported by the National Natural Science Foundation of China (Grant No. 81901732) and the Science and Technology Supporting Program of Guizhou Province (Grant No. qiankehezhicheng S[2020]2359).

References

1. Arner, M., Eliasson, A.C., Nicklasson, S., Sommerstein, K., Hägglund, G.: Hand function in cerebral palsy. Report of 367 children in a population-based longitudinal health care program. J. Hand Surg. **33**(8), 1337–1347 (2008)

2. Avants, B.B., Tustison, N.J., Stauffer, M., Song, G., Wu, B., Gee, J.C.: The insight toolkit image registration framework. Front. Neuroinform. **8**, 44 (2014)

3. Sainz de Cea, M.V., Diedrich, K., Bakalo, R., Ness, L., Richmond, D.: Multi-task learning for detection and classification of cancer in screening mammography. In: Martel, A.L., et al. (eds.) MICCAI 2020. LNCS, vol. 12266, pp. 241–250. Springer, Cham (2020). https://doi.org/10.1007/978-3-030-59725-2_24

4. Chen, C., Bai, W., Rueckert, D.: Multi-task learning for left atrial segmentation on GE-MRI. In: Pop, M., et al. (eds.) STACOM 2018. LNCS, vol. 11395, pp. 292–301. Springer, Cham (2019). https://doi.org/10.1007/978-3-030-12029-0_32

5. Cheng, G., Cheng, J., Luo, M., He, L., Tian, Y., Wang, R.: Effective and efficient multitask learning for brain tumor segmentation. J. Real-Time Image Proc. **17**(6), 1951–1960 (2020)

6. Çiçek, Ö., Abdulkadir, A., Lienkamp, S.S., Brox, T., Ronneberger, O.: 3D U-Net: learning dense volumetric segmentation from sparse annotation. In: Ourselin, S., Joskowicz, L., Sabuncu, M.R., Unal, G., Wells, W. (eds.) MICCAI 2016. LNCS, vol. 9901, pp. 424–432. Springer, Cham (2016). https://doi.org/10.1007/978-3-319-46723-8_49

7. Drozdzal, M., Vorontsov, E., Chartrand, G., Kadoury, S., Pal, C.: The importance of skip connections in biomedical image segmentation. In: Carneiro, G., et al. (eds.) LABELS/DLMIA -2016. LNCS, vol. 10008, pp. 179–187. Springer, Cham (2016). https://doi.org/10.1007/978-3-319-46976-8_19

8. Eliasson, A.C., et al.: The manual ability classification system (MACS) for children with cerebral palsy: scale development and evidence of validity and reliability. Dev. Med. Child Neurol. **48**(7), 549–554 (2006)

9. Franki, I., et al.: The relationship between neuroimaging and motor outcome in children with cerebral palsy: a systematic review-part a. Structural imaging. Res. Dev. Disabil. **100**, 103606 (2020)

10. Ghafoorian, M., et al.: Small white matter lesion detection in cerebral small vessel disease. In: Medical Imaging 2015: Computer-Aided Diagnosis, vol. 9414, p. 941411. International Society for Optics and Photonics (2015)

11. Graham, H.K., et al.: Erratum: cerebral palsy. Nat. Rev. Dis. Primers. **2**(1), 15082 (2016)

12. Holmefur, M., et al.: Neuroradiology can predict the development of hand function in children with unilateral cerebral palsy. Neurorehabil. Neural Repair **27**(1), 72–78 (2013)

13. Ioffe, S., Szegedy, C.: Batch normalization: accelerating deep network training by reducing internal covariate shift. In: International Conference on Machine Learning, pp. 448–456. PMLR (2015)

14. Isensee, F., et al.: nnU-Net: self-adapting framework for u-net-based medical image segmentation. arXiv preprint arXiv:1809.10486 (2018)

15. Jiang, H., et al.: Specific white matter lesions related to motor dysfunction in spastic cerebral palsy: a meta-analysis of diffusion tensor imaging studies. J. Child Neurol. **35**(2), 146–154 (2020)

16. Klingels, K., et al.: Upper limb impairments and their impact on activity measures in children with unilateral cerebral palsy. Eur. J. Paediatr. Neurol. **16**(5), 475–484 (2012)

17. Lee, D.H., et al.: Pseudo-label: the simple and efficient semi-supervised learning method for deep neural networks. In: Workshop on Challenges in Representation Learning, ICML, vol. 3 (2013)

18. Liu, H., et al.: Treatment response prediction of rehabilitation program in children with cerebral palsy using radiomics strategy: protocol for a multicenter prospective cohort study in west China. Quant. Imaging Med. Surg. **9**(8), 1402 (2019)

19. Liu, M., Zhang, J., Adeli, E., Shen, D.: Deep multi-task multi-channel learning for joint classification and regression of brain status. In: Descoteaux, M., Maier-Hein, L., Franz, A., Jannin, P., Collins, D.L., Duchesne, S. (eds.) MICCAI 2017. LNCS, vol. 10435, pp. 3–11. Springer, Cham (2017). https://doi.org/10.1007/978-3-319-66179-7_1

20. Mailleux, L., Franki, I., Emsell, L., Peedima, M.L., Fehrenbach, A., Feys, H., Ortibus, E.: The relationship between neuroimaging and motor outcome in children with cerebral palsy: a systematic review-part b diffusion imaging and tractography. Res. Dev. Disabil. **97**, 103569 (2020)

21. Mailleux, L., et al.: White matter characteristics of motor, sensory and interhemispheric tracts underlying impaired upper limb function in children with unilateral cerebral palsy. Brain Struct. Funct. **225**(5), 1495–1509 (2020)

22. Smith, S.M., et al.: Advances in functional and structural MR image analysis and implementation as FSL. Neuroimage **23**, S208–S219 (2004)

23. Suzuki, S., et al.: Topological structural analysis of digitized binary images by border following. Comput. Vis. Graph. Image Process. **30**(1), 32–46 (1985)

24. Ulyanov, D., Vedaldi, A., Lempitsky, V.: Instance normalization: the missing ingredient for fast stylization. arXiv preprint arXiv:1607.08022 (2016)

25. Wang, J., Huang, C., Wang, Z., Zhang, Y., Ding, Y., Xiu, J.: Anchor-based segmentation of periventricular white matter for neonatal HIE. IEEE Access **8**, 73547–73557 (2020)

26. Yushkevich, P.A., et al.: User-guided 3D active contour segmentation of anatomical structures: significantly improved efficiency and reliability. Neuroimage **31**(3), 1116–1128 (2006)

27. Zeng, C., Gu, L., Liu, Z., Zhao, S.: Review of deep learning approaches for the segmentation of multiple sclerosis lesions on brain MRI. Front. Neuroinform. **14**, 55 (2020)

Clinical Applications - Neuroimaging - DWI and Tractography

Active Cortex Tractography

Ye Wu, Yoonmi Hong, Sahar Ahmad, and Pew-Thian Yap[✉]

Department of Radiology and Biomedical Research Imaging Center (BRIC),
University of North Carolina, Chapel Hill, USA
ptyap@med.unc.edu

Abstract. Most existing diffusion tractography algorithms are affected by gyral bias, causing the termination of streamlines at gyral crowns instead of sulcal banks. In this paper, we propose a tractography technique, called active cortex tractography (ACT), to overcome gyral bias by enabling fiber streamlines to curve naturally into the cortex. We show that the cortex can play an active role in cortical tractography by providing anatomical information to overcome orientation ambiguities as the streamlines enter the superficial white matter in gyral blades and approach the cortex. This is achieved by devising a direction scouting mechanism that takes into account the white matter surface normal vectors. The scouting mechanism allows probing of directions further in space to prepare the streamlines to turn at appropriate angles. The surface normal vectors guide the streamlines to turn into the cortex, perpendicular to the white-gray matter interface. Evaluation using synthetic, macaque and human data with different streamline seeding schemes demonstrates that ACT improves cortical tractography.

Keywords: Diffusion MRI · Tractography · Structural connectivity · Gyral bias

1 Introduction

Diffusion tractography is commonly used to infer the human brain connectome via tracing of axonal trajectories [1–4]. These tracts are often long-range and the underlying axons transverse cortical gray matter (GM) sites through superficial white matter (WM), then the deep WM, and into a distant cortical or subcortical region. Technical difficulties in tracing substantially curved axonal trajectories across WM-GM boundaries in gyral blades can result in gyral bias, with streamlines terminating predominately in gyral crowns instead of sulcal banks. Gyral bias caused by existing tractography algorithms can bias connectivity analysis [5–8].

Gyral bias can be attributed to the complexity of fiber arrangement in superficial WM underneath the cortical sheet [5,9]. It has been shown that up to 70%

This work was supported in part by United States National Institutes of Health (NIH) grants MH125479 and EB006733.

of the streamlines produced by state-of-the-art tractography algorithms fail to reach the GM, even with high angular resolution data [6,7,10,11]. The typical symmetric representations of fiber orientation distributions are not suitable for complex fiber configurations such as fanning and bending, which are ubiquitous in gyral blades. Gyral bias can be mitigated by (i) increasing spatial resolution [12,13], (ii) modeling sub-voxel geometry [7,14], (iii) utilizing anatomical constraints [15–17], and (iv) improving tractography algorithms [8,18]. The first option generally requires time-consuming data acquisition and possibly expensive scanner upgrades. We will therefore focus on a combination of the last three options and show that gyral bias can be reasonably controlled with consistent results across spatial resolutions.

In this paper, we introduce a novel method, called active cortex tractography (ACT), to overcome gyral bias by enabling fiber streamlines to enter the cortex terminating at sulcal banks instead of gyral crowns and nearly perpendicular to the WM-GM interface. The key features of ACT include (i) allowing the cortex to play an active role in overcoming orientation ambiguities; (ii) better characterization of complex fiber configurations in sub-voxel using an asymmetric representation of the fiber orientation distribution; (iii) incorporating a direction scouting mechanism that allows probing of distant directions to help the streamlines turn at appropriate angles. Unlike existing tractography methods, ACT is designed with cortical tractography in mind, aiming to improve cortico-cortical connectivity by alleviating gyral bias.

2   Methods

2.1   Deterministic Tractography

Deterministic tractography methods [19–21] typically reconstruct fiber streamlines by successively following local directions using the Euler forward method,

$$\mathbf{r}_{i+1} = \mathbf{r}_i + \rho \mathbf{t}_i, \tag{1}$$

where $\mathbf{r}_0$ is a given seed point, and the streamline $\mathcal{X} = [\mathbf{r}_0, \cdots, \mathbf{r}_n]$ is estimated using a fixed step size ρ. The forward direction $\mathbf{t}_i$ is typically chosen from M directions $\{\mathbf{d}_1, \cdots, \mathbf{d}_M\}$ that correspond to the local maxima of the fiber orientation distribution function (FODF) [2] in the voxel containing $\mathbf{r}_i$, minimizing the angular deviation with respect to the previous direction $\mathbf{t}_{i-1}$:

$$\mathbf{t}_i = \operatorname*{argmin}_{\mathbf{d}} \arccos(\mathbf{t}_{i-1} \cdot \mathbf{d}), \ \mathbf{d} \in \{\mathbf{d}_1, \cdots, \mathbf{d}_M\},$$
$$\text{s.t.} \ \arccos(\mathbf{t}_{i-1} \cdot \mathbf{d}) \le \theta, \tag{2}$$

where the maximum allowed angular deviation is θ. FODFs are typically assumed to be antipodally symmetric; hence the configurations, such as fanning and bending, cannot be represented correctly.

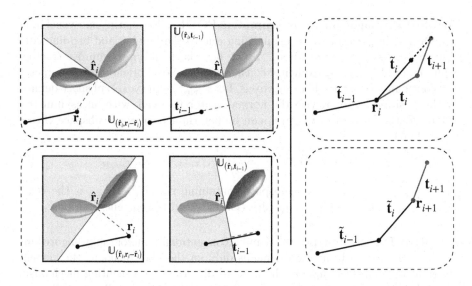

Fig. 1. Left: Direction searching in voxel subspace (non-shaded area). **Right**: Updating forward direction.

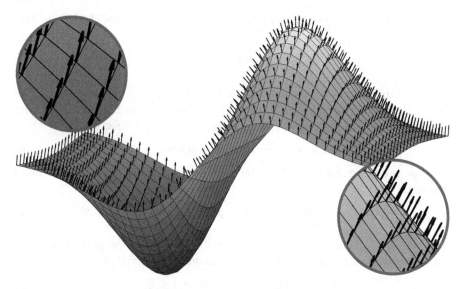

Fig. 2. Normal vectors at surface vertices.

2.2 Active Cortex Tractography

We propose a tractography method to adaptively update the forward direction and the step size based on the asymmetric fiber orientation distribution (AFODF). We further incorporate surface prior to help the streamlines bend more naturally in gyral blades into the cortex.

Forward Propagation via Asymmetric FODFs. Estimating sub-voxel fiber orientations in gyral blades is challenging due to the fanning and bending configurations. Asymmetric FODFs [7,14,22] have been shown to improve tractography in gyral blades by explicitly considering sub-voxel orientation asymmetry and continuity across neighboring voxels. Let $\mathbb{U}_{(\mathbf{r},\mathbf{n})}$ be the voxel subspace defined by the position vector $\mathbf{r}$ with the normal vector $\mathbf{n}$. Then, with an asymmetric FODF $\mathcal{F}$, the forward direction $\mathbf{t}_i$ can be determined in voxel subspace $\mathbb{U}$ (see Fig. 1, Left):

$$\mathbf{t}_i = \operatorname*{argmax}_{\mathbf{t}^* \in \mathbb{U}} \mathcal{F}_{\mathbf{r}_i}(\mathbf{t}^*), \quad \mathbb{U} := \mathbb{U}_{(\hat{\mathbf{r}}_i, \mathbf{r}_i - \hat{\mathbf{r}}_i)} \cap \mathbb{U}_{(\hat{\mathbf{r}}_i, \mathbf{t}_{i-1})}, \tag{3}$$

where $\mathbb{U}_{(\hat{\mathbf{r}}_i, \mathbf{r}_i - \hat{\mathbf{r}}_i)}$ is the voxel subspace containing $\mathbf{r}_i$ and $\mathbb{U}_{(\hat{\mathbf{r}}_i, \mathbf{t}_{i-1})}$ is the voxel subspace defined by the previous direction $\mathbf{t}_{i-1}$ and the current position $\hat{\mathbf{r}}_i$.

Adaptive Direction Update. Unlike conventional tractography algorithms in which the forward direction depends only on the fiber peaks at the current position, we adaptively update the forward direction via a scouting mechanism (see Fig. 1, Right) based on the asymmetric FODF [7,22] as follows:

$$\tilde{\mathbf{t}}_i = \delta_i \mathbf{t}_i + \delta_{i+1} \mathbf{t}_{i+1}, \tag{4}$$

where $\delta_i = \rho \mathcal{F}_{\mathbf{r}_i}(\mathbf{t}_i)$. Then, the next position $\mathbf{r}_{i+1}$ is updated with the adaptive step size:

$$\mathbf{r}_{i+1} = \mathbf{r}_i + \delta_i \tilde{\mathbf{t}}_i. \tag{5}$$

Surface Guidance. We harness the anatomical information provided by the cortex to guide the tracking of the streamline when it enters the gyral blade and into the cortex. This is achieved by updating the forward direction by

$$\tilde{\mathbf{t}}_i \leftarrow \tilde{\mathbf{t}}_i + \exp(-\|\mathbf{r}_i - \mathbf{p}_i\|)\mathbf{v}_i, \tag{6}$$

where $\mathbf{p}_i$ is a vertex on the WM surface with normal vector $\mathbf{v}_i$, which is chosen such that

$$\{\mathbf{p}_i, \mathbf{v}_i\} = \operatorname*{argmin}_{\mathbf{p}, \mathbf{v}} \exp(-\|\mathbf{r}_i - \mathbf{p}\|) \arccos(\mathbf{v} \cdot \tilde{\mathbf{t}}_i), \quad \mathbf{v} \in \mathbb{V}, \tag{7}$$

and $\mathbb{V}$ is a field of surface normals (see Fig. 2). The sign of the normal vector is flipped when the streamline leaves the cortex into superficial WM:

$$\mathbf{v}_i = \begin{cases} \mathbf{v}_i, & \text{if } \arccos(\mathbf{v}_i \cdot \tilde{\mathbf{t}}_i) \leq \frac{\pi}{2}, \\ -\mathbf{v}_i, & \text{if } \arccos(\mathbf{v}_i \cdot \tilde{\mathbf{t}}_i) > \frac{\pi}{2}. \end{cases} \tag{8}$$

3   Results

Qualitative and quantitative comparisons were performed with respect to two common algorithms, i.e., deterministic tractography (DT) based on spherical deconvolution [23] and tractography based on first-order integration over FODFs (iFOD2) [23], with the same seeding strategies and the same number of streamlines.

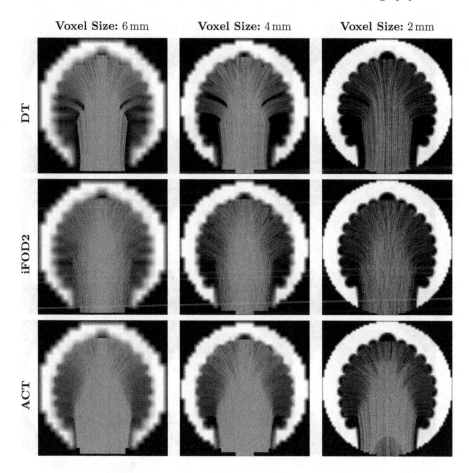

Fig. 3. Streamlines generated by various methods.

3.1   Phantom Data

We used Phantomas [24] to generate an *in silico* gyral blade with a fanning configuration (SNR = 30). The gradient table from the Human Connectome Project (HCP) diffusion MRI data was used. We used a fixed axial diffusivity of $1.7 \times 10^{-3}\,\mathrm{mm^2/s}$, a fixed radial diffusivity of $0.2 \times 10^{-3}\,\mathrm{mm^2/s}$, and different spatial resolutions with voxel sizes from 6 mm to 2 mm in steps of 2 mm. Tractography was performed with seeding at the entrance to the gyral blade. We placed seeds on a 3D mesh grid, resulting in 27 seeds per voxel. Figure 3 indicates that the ACT produces streamlines across WM-GM boundaries with greater cortical coverage and with reduced gyral bias. The results remain consistent despite changes in spatial resolution. The average curvature maps of the streamlines, shown in Fig. 4, confirm that ACT is able to generate streamlines with higher curvatures when entering the cortex in comparison with DT and iFOD2.

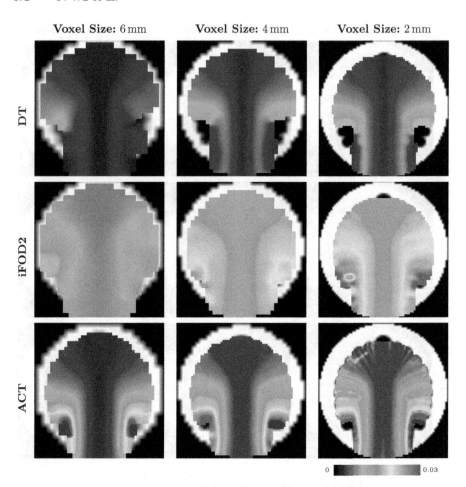

Fig. 4. Voxel-wise average curvature maps of tractograms generated by various methods.

3.2 Macaque and HCP Data

We also utilized the diffusion MRI datasets of a macaque brain and a human brain (from the HCP; ID: 105923) for validation. The acquisition details of the macaque data and the HCP data can be found in [25] and [6], respectively. For both datasets, We placed seeds on a 3D mesh grid, resulting in 27 seeds per voxel. Figure 5 shows the spatial distribution of the average curvatures of the streamlines at WM-GM boundary voxels of the macaque brain using WM, GM, and WM-GM boundary seeding. The results confirm that ACT consistently improves cortical tractography by tracing highly-curved axonal trajectories across WM-GM boundaries in gyral blades, irrespective of the seeding scheme. Figure 6 shows the average curvature distribution for the human brain using WM seeding. The results confirm that ACT improves cortical tractography by mitigating the effects

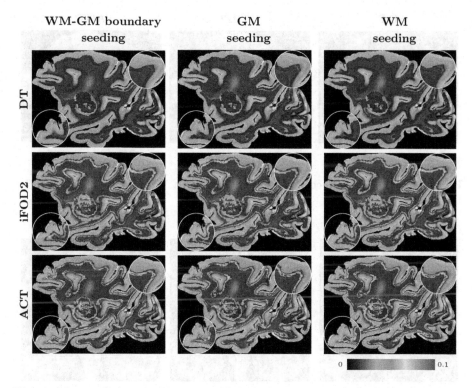

Fig. 5. (Macaque) Average curvatures of streamlines in WM-GM boundary given by DT, iFOD2, and ACT using different seeding strategies.

of gyral bias and allows fiber streamlines at gyral blades to make sharper turns into the cortex. Table 1 further indicates that ACT enables fiber streamlines to enter the cortex at angles that are more perpendicular to the WM-GM interface.

We further validated the effectiveness of ACT in tracking the corticospinal tract (CST), which is a tract connecting the postcentral gyrus and the brain stem. Seeding is performed in these two regions with the help of FreeSurfer cortical parcellation. The results, shown in Fig. 7, indicate that ACT yields fiber streamlines that cover most of the postcentral gyri (white arrows), indicating better cortical connectivity.

Table 1. Average angle between normal vector and streamlines entering WM-GM boundaries at sulcal walls and gyral crowns.

Angle	Macaque			HCP		
	DT	iFOD2	ACT	DT	iFOD2	ACT
Sulcal walls	57.32°	50.03°	**35.71°**	64.55°	56.93°	**38.18°**
Gyral crowns	26.84°	**25.78°**	26.12°	31.25°	30.66°	**29.07°**

DT   iFOD2   ACT

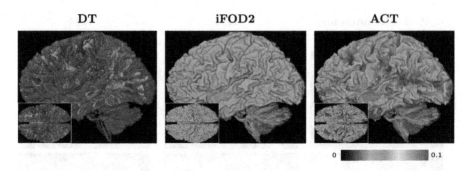

Fig. 6. (Human) Average curvatures of streamlines in WM-GM boundary given by DT, iFOD2 and ACT.

DT   iFOD2   ACT

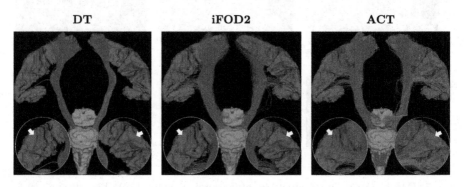

Fig. 7. The part of the CST that connects the postcentral gyrus and the brainstem.

4   Conclusion

In this work, we presented a tractography method for tracking high-curvature WM pathways in the gyral blade with cortical prior information to mitigate gyral bias. Our method allows adaptive estimation of the forward direction in fiber tracking based on asymmetric FODFs as well as direction priors at the WM-GM boundary. We showed that, with ACT, fiber streamlines in gyral blades are able to make sharper turns into the cortex.

References

1. Johansen-Berg, H., Behrens, T.E.J.: Diffusion MRI: From Quantitative Measurement to in Vivo Neuroanatomy. Academic Press, Cambridge (2013)
2. Jeurissen, B., Descoteaux, M., Mori, S., Leemans, A.: Diffusion MRI fiber tractography of the brain. NMR Biomed. **32**(4), e3785 (2019)
3. Bastiani, M., Shah, N.J., Goebel, R., Roebroeck, A.: Human cortical connectome reconstruction from diffusion weighted MRI: the effect of tractography algorithm. NueroImage **62**(3), 1732–1749 (2012)

4. Jbabdi, S., Johansen-Berg, H.: Tractography: where do we go from here? Brain Connect. **1**(3), 169–183 (2011)
5. Reveley, C., et al.: Superficial white matter fiber systems impede detection of long-range cortical connections in diffusion MR tractography. Proc. Natl. Acad. Sci. **112**(21), E2820–E2828 (2015)
6. Schilling, K., Gao, Y., Janve, V., Stepniewska, I., Landman, B.A., Anderson, A.W.: Confirmation of a gyral bias in diffusion MRI fiber tractography. Hum. Brain Mapp. **39**(3), 1449–1466 (2018)
7. Wu, Y., Hong, Y., Feng, Y., Shen, D., Yap, P.T.: Mitigating gyral bias in cortical tractography via asymmetric fiber orientation distributions. Med. Image Anal. **59**, 101543 (2020)
8. Wu, Y., Feng, Y., Shen, D., Yap, P.-T.: Penalized geodesic tractography for mitigating gyral bias. In: Frangi, A.F., Schnabel, J.A., Davatzikos, C., Alberola-López, C., Fichtinger, G. (eds.) MICCAI 2018. LNCS, vol. 11072, pp. 12–19. Springer, Cham (2018). https://doi.org/10.1007/978-3-030-00931-1_2
9. Thompson, E., Robinson, E., Bozek, J., Jbabdi, S., Bastiani, M., Sotiropoulos, S.N.: Exploring the gyral bias on white matter tractography in neonates. In: Annual Meeting of the Organization for Human Brain Mapping, Rome (2019)
10. Girard, G., Whittingstall, K., Deriche, R., Descoteaux, M.: Towards quantitative connectivity analysis: reducing tractography biases. NueroImage **98**, 266–278 (2014)
11. Maier-Hein, K.H., et al.: The challenge of mapping the human connectome based on diffusion tractography. Nat. Commun. **8**(1), 1349 (2017)
12. Sotiropoulos, S.N., et al.: Fusion in diffusion MRI for improved fibre orientation estimation: an application to the 3T and 7T data of the human connectome project. NueroImage **134**, 396–409 (2016)
13. Cottaar, M., et al.: A gyral coordinate system predictive of fibre orientations. NueroImage **176**, 417–430 (2018)
14. Bastiani, M., et al.: Improved tractography using asymmetric fibre orientation distributions. NueroImage **158**, 205–218 (2017)
15. St-Onge, E., Daducci, A., Girard, G., Descoteaux, M.: Surface-enhanced tractography (SET). NueroImage **169**, 524–539 (2018)
16. Frank, L.R., Zahneisen, B., Galinsky, V.L.: JEDI: joint estimation diffusion imaging of macroscopic and microscopic tissue properties. Magn. Reson. Med. **84**(2), 966–990 (2020)
17. Teillac, A., Beaujoin, J., Poupon, F., Mangin, J.-F., Poupon, C.: A novel anatomically-constrained global tractography approach to monitor sharp turns in gyri. In: Descoteaux, M., Maier-Hein, L., Franz, A., Jannin, P., Collins, D.L., Duchesne, S. (eds.) MICCAI 2017. LNCS, vol. 10433, pp. 532–539. Springer, Cham (2017). https://doi.org/10.1007/978-3-319-66182-7_61
18. Donahue, C.J., et al.: Using diffusion tractography to predict cortical connection strength and distance: a quantitative comparison with tracers in the monkey. J. Neurosci. **36**(25), 6758–6770 (2016)
19. Basser, P.J., Pajevic, S., Pierpaoli, C., Duda, J., Aldroubi, A.: In vivo fiber tractography using DT-MRI data. Magn. Reson. Med. **44**(4), 625–632 (2000)
20. Tournier, J.D., Calamante, F., Connelly, A.: MRtrix: diffusion tractography in crossing fiber regions. Int. J. Imaging Syst. Technol. **22**(1), 53–66 (2012)
21. Mori, S., Crain, B.J., Chacko, V.P., van Zijl, P.C.: Three-dimensional tracking of axonal projections in the brain by magnetic resonance imaging. Ann. Neurol. **47**(2), 265–269 (1999)

22. Wu, Y., Feng, Y., Shen, D., Yap, P.-T.: A multi-tissue global estimation framework for asymmetric fiber orientation distributions. In: Frangi, A.F., Schnabel, J.A., Davatzikos, C., Alberola-López, C., Fichtinger, G. (eds.) MICCAI 2018. LNCS, vol. 11072, pp. 45–52. Springer, Cham (2018). https://doi.org/10.1007/978-3-030-00931-1_6

23. Tournier, J.D., Calamante, F., Connelly, A.: Improved probabilistic streamlines tractography by 2nd order integration over fibre orientation distributions. In: Proceedings of the International Society for Magnetic Resonance in Medicine (2010)

24. Caruyer, E., Daducci, A., Descoteaux, M., Houde, J.C., Thiran, J.P., Verma, R.: Phantomas: a flexible software library to simulate diffusion MR phantoms. In: Proceedings of Joint Annual Meeting ISMRM-ESMRMB, volume 17, p. 20013 (2014)

25. Van Essen, D.C., Smith, S.M., Barch, D.M., Behrens, T.E., Yacoub, E., Ugurbil, K.: The WU-Minn human connectome project: an overview. NeuroImage **80**, 62–79 (2013)

Highly Reproducible Whole Brain Parcellation in Individuals via Voxel Annotation with Fiber Clusters

Ye Wu, Sahar Ahmad, and Pew-Thian Yap$^{(\boxtimes)}$

Department of Radiology and Biomedical Research Imaging Center (BRIC), The University of North Carolina at Chapel Hill, Chapel Hill, NC, USA
ptyap@med.unc.edu

Abstract. A central goal in systems neuroscience is to parcellate the brain into discrete units that are neurobiologically coherent. Here, we propose a strategy for consistent whole-brain parcellation of white matter (WM) and gray matter (GM) in individuals. We parcellate the brain into coherent parcels using non-negative matrix factorization based on voxel annotation using fiber clusters. Tractography is performed using an algorithm that mitigates gyral bias, allowing full gyral and sulcal coverage for reliable parcellation of the cortical ribbon. Experimental results indicate that parcellation using our approach is highly reproducible with 100% test-retest parcel identification rate and is highly consistent with significantly lower inter-subject variability than FreeSurfer parcellation. This implies that reproducible parcellation can be obtained for subject-specific investigation of brain structure and function.

Keywords: Brain parcellation · Tractography · Fiber clustering

1 Introduction

Parcellation of gray matter (GM) and white matter (WM) in the human brain into functional units is fundamental to neuroscience [1–3]. While automatic parcellation of the cortex into functionally differentiable areas can be performed based on the geometries of cortical gyri and sulci, parcellation of WM in association with cortically defined function [1], structure [2,3], and cytoarchitecture is less straightforward.

Demarcation of WM regions poses interesting challenges due to the need to integrate both local and global information. A typical approach is to first parcellate the cortex, and then propagate the cortical labels to annotate WM using a predefined distance function [4,5]. Several WM atlases [4,6–8] classify WM voxels based on their distances to cortical regions. WM parcellation can also be performed by label transfer from a WM atlas. However, this approach can

This work was supported in part by United States National Institutes of Health (NIH) grants EB008374, MH125479, and EB006733.

The original version of this chapter was revised: Grant number has been added. The correction to this chapter is available at
https://doi.org/10.1007/978-3-030-87234-2_75

M. de Bruijne et al. (Eds.): MICCAI 2021, LNCS 12907, pp. 477–486, 2021.
https://doi.org/10.1007/978-3-030-87234-2_45

be error-prone due to the reliance on registration accuracy [2,9]. Tractography-based parcellation (TBP) leverages structural connectivity for brain parcellation [4,10–12] by determining regions with common connectivity patterns [13]. TBP typically involves high computation complexity due to the massiveness of connectivity data and therefore dimension reduction is often required [11,14,15]. TBP-driven atlases [16,17] are often population-based and are unable to reflect individual variations, especially in superficial WM near the cortex [18].

In this paper, we propose a data-driven approach for subject-specific whole brain parcellation without relying on the annotations of cortical regions or fiber tracts. Our method uses a robust unsupervised algorithm for groupwise bilateral clustering of fiber streamlines in both cerebral hemispheres. Each voxel is labeled according to the streamline clusters that traverse it. Non-negative matrix factorization (NMF) is then applied to simultaneously parcellate the GM and WM into distinct regions. Experimental results indicate that our technique yields parcellation of the GM and WM that is highly consistent across subjects, test-retest scans, and hemispheres.

2   Materials and Methods

2.1   Dataset and Preprocessing

We utilized the test-retest anatomical MRI and diffusion MRI data of 22 subjects from the Human Connectome Project (HCP) [19]. The complete description of the HCP acquisition protocol can be found in [19]. The data were minimally preprocessed with the standard HCP preprocessing pipeline [20,21]. For reproducibility analysis, we applied our algorithm to the test and retest data separately.

We performed whole brain tractography using asymmetric fiber orientation distribution functions (AFODFs) to better capture complex axonal configurations, such as bending, fanning and crossing, and to mitigate gyral bias for better cortico-cortical connectivity [22]. Fiber streamlines were generated by successively following the local directions determined from the AFODFs [22]. Whole brain tractography performed with 64 random seeds per voxel resulted in approximately 100 million streamlines.

2.2   Whole-Brain Parcellation

Our method involves (i) population and subject-specific streamline clustering (Fig. 1 (a–b)); (ii) voxel annotation based on the streamline clusters (Fig. 1(c)); and (iii) NMF for joint WM and GM parcellation (Fig. 1(d)).

Unsupervised Bilateral Fiber Clustering. We warp a random subset of the fiber streamlines of each subject to the MNI atlas space with affine and non-linear transforms estimated between the subject anatomical scan and the atlas.

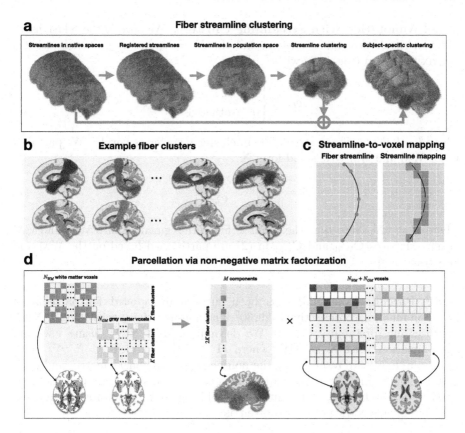

Fig. 1. Overview of the proposed whole-brain parcellation method. (a) Fiber clusters obtained from streamlines generated using whole-brain tractography. (b) Example fiber clusters. (c) Inter-voxel connectivity. (d) Parcellation obtained via NMF.

The spatially-normalized streamlines of all subjects are combined for population-level fiber clustering using an automated unsupervised fiber clustering method. The streamlines are mapped to a Hilbert space via parameterization as coefficients of cosine series [23] and then grouped into K clusters via K-medoids clustering [24,25]. Streamlines in both cerebral hemispheres are clustered bilaterally for consistency and robustness [25]. Streamlines with distances greater than 2.5 standard deviations [26,27] from their cluster centers are removed as outliers. Common non-anatomical false-positive (FP) clusters, as defined in [28], are removed. All streamlines for each subject are warped to the atlas space and are consistently clustered based on the population fiber clusters [27] by assigning each streamline to the closest population cluster. Outlier streamlines are removed using the same criterion described above.

Voxel Annotation with Streamline Clusters. We encode the relationship between the K streamline clusters and the $N = N_{\text{WM}} + N_{\text{GM}}$ voxels using a matrix $\mathbf{C} = (c_{ki})$ (Fig. 1(c)) with

$$c_{ki} = \begin{cases} 1 & \text{if } i \sim k, \\ 0 & \text{otherwise} \end{cases}, \tag{1}$$

where $i \sim k$ is true if any streamline in cluster k traverses voxel i. We partition $\mathbf{C}$ for WM and GM voxels and form $\mathbf{X}$:

$$\mathbf{X} = \begin{bmatrix} \mathbf{C}_{\text{WM}} & \mathbf{0} \\ \mathbf{0} & \mathbf{C}_{\text{GM}} \end{bmatrix}_{P \times N}, \tag{2}$$

where $P = 2K$. Note that the brain is first pre-segmented into WM and GM voxels. Matrices $\mathbf{C}_{\text{WM}}$ and $\mathbf{C}_{\text{GM}}$ in (2) are partitioned based on the WM and GM tissue maps.

NMF Parcellation. NMF has been successfully applied for parcellation based on functional connectivity [29,30]. The non-negativity constraint in NMF improves interpretability [31,32]. We decompose $\mathbf{X} \in \mathbb{R}^{P \times N}$ using low-rank approximation with $\mathbf{X} \approx \mathbf{HF}$, where $\mathbf{H} \in \mathbb{R}^{P \times M}$ and $\mathbf{F} \in \mathbb{R}^{M \times N}$ with $M < \min\{P, N\}$ are respectively the non-negative component matrix and coefficient matrix (Fig. 1(d)). We require $\mathbf{H}$ and $\mathbf{F}$ to be sparse for parsimonious voxel representation and for limiting the solution space to improve robustness [33]. The sparse NMF problem is given as

$$\min_{\mathbf{H} \succeq 0, \ \mathbf{F} \succeq 0} \frac{1}{2} \|\mathbf{X} - \mathbf{HF}\|_{\text{F}} + \alpha \|\mathbf{H}\|_1 + \beta \|\mathbf{F}\|_1, \tag{3}$$

where $\|\cdot\|_{\text{F}}$ is the Frobenius norm, $\|\cdot\|_1$ is the matrix ℓ_1-norm defined as $\|\mathbf{P}\|_1 = \max_j \sum_i |p_{i,j}|$, and α and β are the regularization parameters. We solve (3) with the algorithm detailed in [34]. Non-negative singular value decomposition (SVD) is employed for initialization to improve accuracy [31,35].

Each column of $\mathbf{F}$ encodes the membership of each voxel $n \in \{1, \dots, N\}$ with respect to each component $m \in \{1, \dots, M\}$. By determining the largest value in each column of $\mathbf{F}$, each voxel is assigned to the most likely component, generating an initial parcellation map $\mathbf{S}$ with M parcels. We correct for isolated voxels by incorporating information from neighboring voxels by updating $\mathbf{S}$ using

$$\min_{\hat{\mathbf{S}} \succeq 0} \frac{1}{2} \left\| \hat{\mathbf{S}} - \mathbf{AS} \right\|_{\text{F}} + \gamma \left\| \hat{\mathbf{S}} \right\|_1, \quad \mathbf{A} = \begin{bmatrix} \mathbf{A}_{\text{WM}} & \mathbf{0} \\ \mathbf{0} & \mathbf{A}_{\text{GM}} \end{bmatrix}, \tag{4}$$

where $\mathbf{A}_{\text{WM}}$ and $\mathbf{A}_{\text{GM}}$ are the adjacency matrices [36,37] that encode spatial voxel relationships in WM and GM, respectively, and γ is the regularization parameter. (4) can be efficiently solved using the SPAMS toolbox[1].

---

[1] http://spams-devel.gforge.inria.fr.

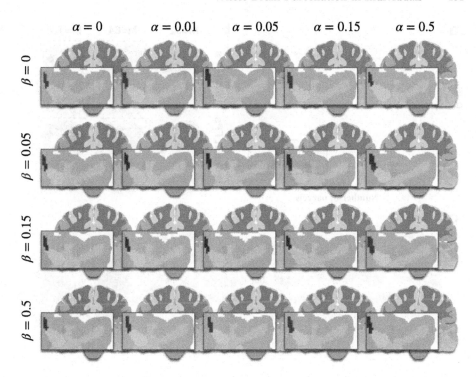

Fig. 2. The effect of the regularization parameters on population parcellation with $M = 16$.

3   Results and Discussion

We evaluated the test-retest reproducibility of streamline clustering with different numbers of clusters. We found that $K = 800$ gives the best consistency across scans. This setting results in low variation in streamline count per cluster and across subjects. Increasing K further does not improve cluster consistency but increases computation time.

We evaluated parcellation test-retest reproducibility and consistency using the following metrics:

– *Dice coefficient* for each corresponding parcel P_m in the test and retest scans:

$$\text{Dice}_m = \frac{P_m^{\text{Test}} \bigcap P_m^{\text{Retest}}}{P_m^{\text{Test}} \bigcup P_m^{\text{Retest}}}, \quad m = 1, \cdots, M. \tag{5}$$

A higher Dice coefficient indicates higher reproducibility.

– *Relative difference* (RD) of the voxel-averaged T1w/T2w ratio [38] in each corresponding parcel in the test and retest scans:

$$\text{RD} = \frac{\left| \text{Ratio}_m^{\text{Test}} - \text{Ratio}_m^{\text{Retest}} \right|}{\left| \text{Ratio}_m^{\text{Test}} + \text{Ratio}_m^{\text{Retest}} \right|}, \quad m = 1, \cdots, M. \tag{6}$$

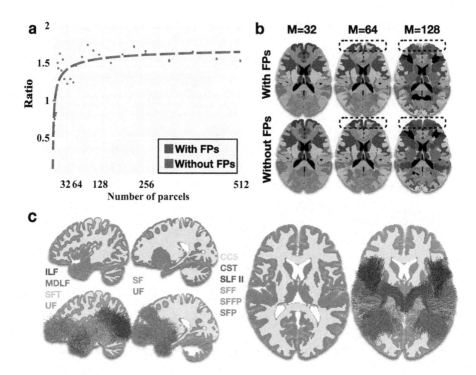

Fig. 3. Population parcellation. (a) The ratio of the number of WM parcels to the number of GM parcels converges with increasing M, and is insensitive to FP fiber clusters. (b) Parcellation examples for different values of M. (c) Fiber clusters corresponding to selected parcels.

A lower RD value indicates higher reproducibility.

- *Coefficient of variation* (CV) of the volume of a corresponding parcel across subjects. The CV of the volume of a parcel v_m is computed across subjects as

$$CV_m = \frac{\text{std}(v_m)}{\text{mean}(v_m)}, \quad m = 1, \cdots, M. \tag{7}$$

A lower CV indicates higher inter-subject consistency.

3.1 Population Parcellation

We evaluated the impact of the regularization parameters in (3) on population parcellation for $M = 16$ (Fig. 2). We found that $\alpha = \beta = 0.15$ is sufficient for capturing fine structures near the cortex.

We analyzed the impact of M on parcellation. The ratio of the number of WM parcels to the number of GM parcels converges with increasing M to around 1.6, consistent with FreeSurfer parcellation (called 'wmparc') based on the Desikan-Killiany atlas [16], and is insensitive to FP fiber clusters (Fig. 3(a)).

Table 1. Parcel identification rates across 22 subjects.

	NMF		wmparc [16]		aparc (2006) [16]		aparc (2009) [17]	
	Test	Retest	Test	Retest	Test	Retest	Test	Retest
IR	100%	100%	98.9%	99.45%	99.12%	99.12%	99.48%	99.48%
Lowest IR	100%	100%	63.64%	63.64%	63.64%	63.64%	63.64%	63.64%
Overall IR	100%		98.9%		99.12%		99.48%	

Results for different values of M indicate that parcellation is affected by FP fiber clusters, as marked in Fig. 3(b). Therefore, we remove the common FP fiber clusters in the subsequent experiments.

Our method allows the identification of fiber clusters (out of K fiber clusters) passing through a parcel (Fig. 3(c)) using the component matrix $\mathbf{H}$. We identified the fiber clusters corresponding to selected parcels. For each fiber cluster, we anatomically annotated a fiber cluster using the white matter query language (WMQL) [39]. For each parcel, the corresponding fiber clusters are consistent across subjects.

3.2 Subject-Specific Parcellation

We evaluated whether the parcels can be successfully identified for $M = 3, \ldots, 512$. Each parcel is deemed as identified if there is at least a voxel that belongs to the parcel. For each parcel, the identification rate (IR) is defined as the percentage of subjects with the parcel identified over the total number of subjects. The overall IR is defined as the percentage of identified parcels over the total number of parcels. Table 1 indicates that (i) our method consistently identifies all M parcels from all subjects from either the test or retest dataset; (ii) the lowest IRs associated with least reproducible parcels are much lower for FreeSurfer; (iii) our method yields perfect overall IRs.

Figure 4(a) indicates that subject-specific parcellation is highly reproducible for different values of M. Figure 4(b) shows the distributions of Dice and RD of each parcel with respect to its mean volume for the NMF method (for $M = 3, \ldots, 512$) and all FreeSurfer parcellation schemes (wmparc, aparc 2006, and aparc 2009). It is apparent that the Dice score is significantly higher ($p < 0.001$) for our method compared with Freesurfer. Our method results in low RD values, indicating test-retest reliability is high in terms of the T1/T2 ratio, comparable to FreeSurfer.

The test-retest reproducibility of the parcels is highly related to parcel volumes. Parcels with larger volumes tend to be more reproducible. Our method yields high reproducibility for parcels of different sizes. Particularly, small-size parcels given by our method are more reproducible than FreeSurfer. The consistently lower CV values (Fig. 4(c)) confirm that parcellation using our method is less variable across subjects.

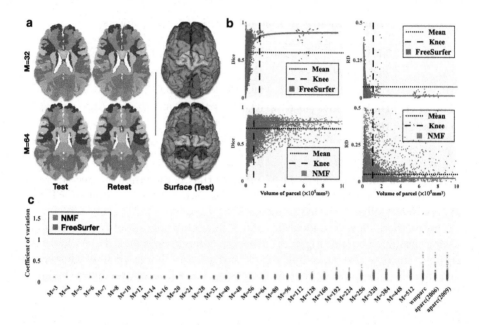

Fig. 4. Evaluation of subject-specific parcellation. (a) Individual parcellation examples for different values of M. (b–c) Quantitative comparison of Dice, RD, and CV between NMF and FreeSurfer.

4   Conclusion

In this paper, we presented a method for subject-specific whole-brain parcellation. Joint parcellation of WM and GM is performed using NMF based on clustered fiber streamlines. Our results demonstrate that the method is highly reproducible between test and retest datasets, and is highly consistent between subjects, providing a reliable tool for large-scale parcellation at the individual level.

References

1. Arslan, S., Rueckert, D.: Multi-level parcellation of the cerebral cortex using resting-state fMRI. In: Navab, N., Hornegger, J., Wells, W.M., Frangi, A.F. (eds.) MICCAI 2015. LNCS, vol. 9351, pp. 47–54. Springer, Cham (2015). https://doi.org/10.1007/978-3-319-24574-4_6
2. Bloy, L., Ingalhalikar, M., Eavani, H., Schultz, R.T., Roberts, T.P.L., Verma, R.: White matter atlas generation using HARDI based automated parcellation. NeuroImage **59**(4), 4055–4063 (2012)
3. Destrieux, C., Fischl, B., Dale, A.M., Halgren, E.: A Sulcal depth-based anatomical parcellation of the cerebral cortex. NeuroImage **47**, S151 (2009)
4. Schiffler, P., Tenberge, J.G., Wiendl, H., Meuth, S.G.: Cortex parcellation associated whole white matter parcellation in individual subjects. Front. Hum. Neurosci. **11**, 352 (2017)

5. Fischl, B., et al.: Whole brain segmentation: automated labeling of neuroanatomical structures in the human brain. Neuron **33**(3), 341–355 (2002)
6. Wakana, S., Jiang, H., Nagae-Poetscher, L.M., van Zijl, P.C.M., Mori, S.: Fiber tract-based atlas of human white matter anatomy. Radiology **230**(1), 77–87 (2004)
7. Mori, S., et al.: Stereotaxic white matter atlas based on diffusion tensor imaging in an ICBM template. NeuroImage **40**(2), 570–582 (2008)
8. Atkinson, A.P., Vuong, Q.C., Smithson, H.E.: Modulation of the face- and body-selective visual regions by the motion and emotion of point-light face and body stimuli. NeuroImage **59**(2), 1700–1712 (2012)
9. Rohlfing, T.: Incorrect ICBM-DTI-81 atlas orientation and white matter labels. Front. Neurosci. **7**, 4 (2013)
10. Xu, J., et al.: Tractography-based parcellation of the human middle temporal gyrus. Sci. Rep. **5**, 18883 (2015)
11. Li, H., et al.: ATPP: a pipeline for automatic Tractography-Based brain parcellation. Front. Neuroinform. **11**, 35 (2017)
12. López-López, N., Vázquez, A., Poupon, C., Mangin, J., Guevara, P.: Cortical surface parcellation based on intra-subject white matter fiber clustering. In: 2019 IEEE CHILEAN Conference on Electrical, Electronics Engineering, Information and Communication Technologies (CHILECON), pp. 1–6, November 2019
13. López-López, N., et al.: From coarse to Fine-Grained parcellation of the cortical surface using a Fiber-Bundle atlas. Front. Neuroinform. **14**, 32 (2020)
14. Moreno-Dominguez, D., Anwander, A., Knösche, T.R.: A hierarchical method for whole-brain connectivity-based parcellation: whole-Brain Connectivity-Based parcellation. Hum. Brain Mapp. **35**(10), 5000–5025 (2014)
15. Lefranc, S., et al.: Groupwise connectivity-based parcellation of the whole human cortical surface using watershed-driven dimension reduction. Med. Image Anal. **30**, 11–29 (2016)
16. Desikan, R.S., et al.: An automated labeling system for subdividing the human cerebral cortex on MRI scans into gyral based regions of interest. NeuroImage **31**(3), 968–980 (2006)
17. Destrieux, C., Fischl, B., Dale, A., Halgren, E.: Automatic parcellation of human cortical gyri and sulci using standard anatomical nomenclature. NeuroImage **53**(1), 1–15 (2010)
18. Sotiropoulos, S.N., Zalesky, A.: Building connectomes using diffusion MRI: why, how and but. NMR Biomed. **32**(4), e3752 (2019)
19. Van Essen, D.C., Smith, S.M., Barch, D.M., Behrens, T.E.J., Yacoub, E., Ugurbil, K.: WU-Minn HCP consortium: The WU-Minn human connectome project: an overview. NeuroImage **80**, 62–79 (2013)
20. Sotiropoulos, S.N., Jbabdi, S., Andersson, J.L., Woolrich, M.W., Ugurbil, K., Behrens, T.E.J.: RubiX: combining spatial resolutions for Bayesian inference of crossing fibers in diffusion MRI. IEEE Trans. Med. Imaging **32**(6), 969–982 (2013)
21. Glasser, M.F., et al.: The minimal preprocessing pipelines for the Human Connectome Project. NeuroImage **80**, 105–124 (2013)
22. Wu, Y., Hong, Y., Feng, Y., Shen, D., Yap, P.T.: Mitigating gyral bias in cortical tractography via asymmetric fiber orientation distributions. Med. Image Anal. **59**, 101543 (2020)
23. Chung, M.K., Adluru, N., Lee, J.E., Lazar, M., Lainhart, J.E., Alexander, A.L.: Cosine series representation of 3D curves and its application to white matter fiber bundles in diffusion tensor imaging. Stat. Interface **3**(1), 69–80 (2010)

24. Park, H.S., Jun, C.H.: A simple and fast algorithm for k-medoids clustering. Expert Syst. Appl. **36**(2, Part 2), 3336–3341 (2009)
25. Wu, Y., Hong, Y., Ahmad, S., Lin, W., Shen, D., Yap, P.-T.: Tract dictionary learning for fast and robust recognition of fiber bundles. In: Martel, A.L., et al. (eds.) MICCAI 2020. LNCS, vol. 12267, pp. 251–259. Springer, Cham (2020). https://doi.org/10.1007/978-3-030-59728-3_25
26. Kriegel, H.P., Kroger, P., Schubert, E., Zimek, A.: Interpreting and unifying outlier scores. In: Proceedings of the 2011 SIAM International Conference on Data Mining. Proceedings, pp. 13–24. Society for Industrial and Applied Mathematics, April 2011
27. O'Donnell, L.J., et al.: Automated white matter fiber tract identification in patients with brain tumors. NeuroImage. Clinical **13**, 138–153 (2017)
28. Zhang, F., Wu, Y., Norton, I., Rigolo, L., Rathi, Y., Makris, N., O'Donnell, L.J.: An anatomically curated fiber clustering white matter atlas for consistent white matter tract parcellation across the lifespan. NeuroImage **179**, 429–447 (2018)
29. Fürböck, C.: Non-negative matrix factorization as a tool for fMRI analysis of dynamicity and individuality of functional networks (2020)
30. Varikuti, D.P., et al.: Evaluation of non-negative matrix factorization of grey matter in age prediction. NeuroImage **173**, 394–410 (2018)
31. Thompson, E., et al.: Non-negative data-driven mapping of structural connections with application to the neonatal brain. NeuroImage **222**, 117273 (2020)
32. Eickhoff, S.B., Yeo, B.T.T., Genon, S.: Imaging-based parcellations of the human brain. Nature reviews. Neuroscience **19**(11), 672–686 (2018)
33. Hoyer, P.O.: Non-negative matrix factorization with sparseness constraints. J. Mach. Learn. Res. JMLR 5, 1457–1469 (2004)
34. Mairal, J., Bach, F., Ponce, J., Sapiro, G.: Online learning for matrix factorization and sparse coding. J. Mach. Learn. Res. JMLR **11**(1) (2010)
35. Boutsidis, C., Gallopoulos, E.: SVD based initialization: a head start for nonnegative matrix factorization. Pattern Recogn. **41**(4), 1350–1362 (2008)
36. Arslan, S., Ktena, S.I., Makropoulos, A., Robinson, E.C., Rueckert, D., Parisot, S.: Human brain mapping: a systematic comparison of parcellation methods for the human cerebral cortex. NeuroImage **170**, 5–30 (2018)
37. Bastiani, M., Shah, N.J., Goebel, R., Roebroeck, A.: Human cortical connectome reconstruction from diffusion weighted MRI: the effect of tractography algorithm. NeuroImage **62**(3), 1732–1749 (2012)
38. Glasser, M.F., Van Essen, D.C.: Mapping human cortical areas in vivo based on myelin content as revealed by T1- and T2-weighted MRI. J. Neurosci. Official J. Soc. Neurosci. **31**(32), 11597–11616 (2011)
39. Wassermann, D., et al.: The white matter query language: a novel approach for describing human white matter anatomy. Brain Struct. Funct. **221**(9), 4705–4721 (2016). https://doi.org/10.1007/s00429-015-1179-4

Accurate Parameter Estimation in Fetal Diffusion-Weighted MRI - Learning from Fetal and Newborn Data

Davood Karimi[1]([✉]), Lana Vasung[2], Fedel Machado-Rivas[1], Camilo Jaimes[1], Shadab Khan[1], and Ali Gholipour[1]

[1] Computational Radiology Laboratory (CRL), Department of Radiology, Boston Children's Hospital, and Harvard Medical School, Boston, USA
davood.karimi@childrens.harvard.edu
[2] Department of Pediatrics at Boston Children's Hospital, and Harvard Medical School, Boston, MA, USA

Abstract. Recent works have used deep learning for accurate parameter estimation in diffusion-weighted magnetic resonance imaging (DW-MRI). However, no prior study has addressed the fetal brain, mainly because obtaining reliable fetal DW-MRI data with accurate ground truth parameters is very challenging. To overcome this obstacle, we present a novel method that uses both fetal scans as well as high-quality pre-term newborn scans. We use the newborn scans to estimate accurate parameter maps. We then use these parameter maps to generate DW-MRI data that match the measurement scheme and noise distributions that are characteristic of fetal scans. To demonstrate the effectiveness and reliability of the proposed data generation pipeline, we use the generated data to train a convolutional neural network for estimating color fractional anisotropy. We show that the proposed machine learning pipeline is significantly superior to standard estimation methods in terms of accuracy and expert assessment of reconstruction quality. Our proposed methods can be adapted for estimating other diffusion parameters for fetal brain.

Keywords: Fetal MRI · Diffusion weighted imaging · Deep learning

1 Introduction

Diffusion weighted magnetic resonance imaging (DW-MRI) is one of the most promising tools for non-invasive probing of brain micro-architecture and microstructure in vivo [2,12]. DW-MRI has dramatically enhanced our knowledge of the human brain in the past three decades. However, most of the research in this field has studied the brain after birth. In utero fetal brain imaging, on the other hand, is much more challenging because 1) the fetus moves constantly and unpredictably, 2) the signal to noise ratio (SNR) is very low and there can be strong artifacts, and 3) scan times should be kept short to minimize maternal

© Springer Nature Switzerland AG 2021
M. de Bruijne et al. (Eds.): MICCAI 2021, LNCS 12907, pp. 487–496, 2021.
https://doi.org/10.1007/978-3-030-87234-2_46

discomfort and, hence, the number of measurements is small [6,8]. As a result, accurate and robust estimation of parameters of interest in fetal brain DW-MRI remains an open problem.

There have been remarkable strides in developing more accurate fetal head motion tracking and slice-to-volume registration (SVR) methods [11,24,28]. However, these methods fall far short of fully compensating for the fetal motion, which can be very fast and unpredictable. Furthermore, after motion compensation, most previous studies have used classical least squares (LS)-based methods to estimate the model parameters [25,32]. Those estimation methods have been originally developed for non-fetal DW-MRI, and they can be highly sub-optimal for fetal data that is marred by excessive motion, very strong noise, and artifacts. Furthermore, almost all these estimation methods work on a voxel-wise basis, without exploiting the spatial regularity of the parameters of interest. This is a major shortcoming because while the signal in each voxel is usually too noisy for accurate parameter estimation, there is considerable spatial regularity that can be exploited to improve the accuracy and robustness of estimation.

Given the above limitations, we expect that machine learning methods can have significant advantages over classical methods for parameter estimation in fetal DW-MRI. Unlike classical techniques, machine learning methods do not need to assume a model for the diffusion signal or the noise and motion artifacts. Rather, they can directly learn a mapping from the noisy and imperfect measurements to the parameters of interest from training data. Furthermore, many machine learning models can effectively learn the spatial correlations. Despite these important potential advantages, there is one major challenge facing the application of machine learning methods in fetal DW-MRI: *lack of reliable training data with accurate ground-truth.* To the best of our knowledge, no prior work has successfully addressed this challenge to enable application of machine learning methods for parameter estimation in fetal DW-MRI. The goal of this work is to propose a solution to this problem.

To overcome this challenge, we propose a novel strategy that involves the use of fetal scans as well as high-quality pre-term newborn scans. We show how a combination of fetal and pre-term newborn scans can be used to generate large amounts of reliable training data. We demonstrate the effectiveness of the proposed approach by using the generated data in reconstruction of color fractional anisotropy (CFA), which is one of the most widely used parameters in studying fetal brain development, micro-architecture, and organization [20,29]. Note that the choice of CFA as the parameter of interest is to illustrate the effectiveness of the proposed methodology. Our proposed methods should be applicable to other parameters such as FA and ADC. We propose to use a convolutional neural network (CNN) to learn to estimate CFA from the generated data. We compare our method with a standard reconstruction method, both quantitatively and also in terms of expert evaluations.

1.1 Related Works

Data-driven and machine learning-based parameter estimation methods for DW-MRI have been explored by several studies in recent years. For estimation of certain diffusion parameters, it was reported that deep learning could reduce the required number of measurements by a factor of 12 [10]. Subsequent studies used deep learning to estimate diffusion kurtosis and generalized fractional anisotropy [1,9,31], number of diffusion compartments [16], fiber orientation distribution function [18,23,27], and for tractography [14,30]. However, none of these studies have addressed the fetal DW-MRI, which has its own very specific challenges and is the focus of this paper.

2 Materials and Methods

2.1 Description of the Problem

Figure 1 shows example DW-MRI scans from a fetal subject and a pre-term newborn subject of the same gestational age (34 weeks). It also shows the corresponding CFA images, reconstructed using the commonly-used weighted linear least squares (WLLS) method [22]. The newborn scan is from the developing Human Connectome Project (dHCP) dataset [3]. Newborn subjects can be imaged with minimal motion in long acquisition sessions. The dHCP subjects were imaged in natural sleep at three different shells ($b = 400, 1000, 2600$) for a total of 280 gradient directions. For this dataset, we have estimated the SNR to be in the range 15–20 dB. The fetal DWI volume was reconstructed from a scan acquired in our institution and processed with motion correction and SVR pipelines such as [15,24]. For fetal scans, to minimize the fetal and maternal discomfort we are usually restricted to approximately 24–48 gradient directions in one shell. Furthermore, the SNR in these volumes is in the range 5–16 dB and, as we show below, the noise distribution is very complex. Therefore, as can be seen in the example shown in Fig. 1, the reconstructed fetal CFA has a very low quality and lacks much of the architectural details that can be seen in the newborn CFA image of the same age.

We expect that more accurate parameter estimation in fetal DW-MRI should be possible by using machine learning methods. Machine learning techniques do not require prior knowledge of the noise distribution. Moreover, unlike the LS-based methods that estimate the CFA one voxel at a time, machine learning methods can effectively use the measurements in the neighboring voxels and learn the spatial patterns as well. However, as we mentioned above, the major challenge in developing a machine learning model in fetal DW-MRI is the lack of reliable training data. One could use a set of fetal DWI scans and corresponding LS-based CFA reconstructions; however, as shown in Fig. 1, CFA reconstructions are inaccurate and noisy. Alternatively, one could use newborn measurements and CFA reconstructions; however, although in this case reconstructions would have high quality, the measurement scheme and the noise distribution are very

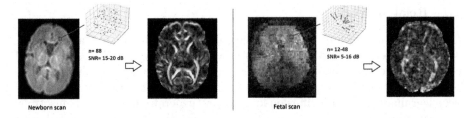

Fig. 1. Example CFA images reconstructed from typical newborn and fetal scans. Note that the number of diffusion-sensitized measurements in this fetal scan was 36; the number of data points in each voxel will be larger than the number of measurements because of the point-spread function used in common SVR methods used to build the data volume [15, 24].

different between newborn and fetal data. Furthermore, usually a much lower diffusion strength is used in fetal imaging than in newborn imaging, which renders newborn data useless for fetal applications.

2.2   Proposed Solution

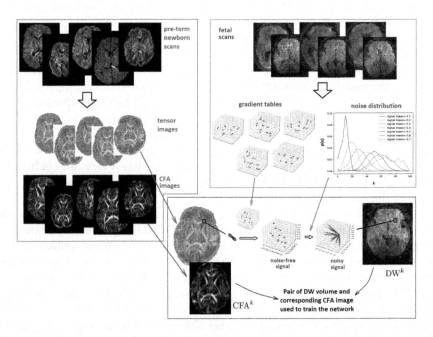

Fig. 2. A schematic representation of our data generation pipeline. Left pane: high-quality pre-term newborn scans are used to estimate accurate CFA and tensor images. Top right pane: fetal scans are used to estimate the noise probability distribution and to store gradient tables that are typical of fetal DW-MRI acquisitions. Lower right pane: accurate tensor images from the newborn scans are used to synthesize DW-MRI data that follow the acquisition scheme and noise distribution of fetal scans.

Since neither newborn data nor fetal data are adequate on their own, we propose a method that uses the best of the two worlds: 1) parameter maps estimated from newborn scans, and 2) measurement scheme and noise distribution from fetal scans. Figure 2 shows our data generation pipeline, which we explain below.

We process a set of $n = 20$ fetal DW scans using our motion correction and SVR pipeline. From the reconstructed DW volumes, we estimate the noise probability distribution function (PDF). The noise distribution in DW-MRI is very complex. In non-fetal applications, it is usually modeled with Rician or non–central chi–square PDFs [4,7]. Not only the noise in fetal scans is much more strong, it also involves residual (uncorrected) motion errors. Since we are unaware of the true distribution of noise in the reconstructed fetal volumes, we use kernel density estimation with a Parzen window [26] to estimate the noise PDF from data. Since the noise is known to be signal-dependent, we estimate the noise PDF for signal levels from 0.01 to 0.99 in steps of 0.01. Example estimated PDFs are shown in Fig. 2. Furthermore, we store the gradient tables of these 20 fetal scans.

We estimate the diffusion tensor and CFA images for all pre-term newborn scans ($n = 82$, gestational age < 38 weeks) from the dHCP dataset [3]. For this estimation, we use the 88 measurements in the $b = 1000$ shell from this dataset. The b-value used in this estimation will not pose a restriction on the b values of the fetal scans used at test time because we use only the estimated parameters from the dHCP data, and not the measurements. We used the $b = 1000$ shell from the dHCP dataset in this stage based on the recommendations in [13]. For fetal DW-MRI, on the other hand, lower b values in the range $[500, 700]$ are commonly used [19,20]. We used $b = 500$ for acquiring all fetal scans in this work.

Now, to generate a training data sample (i.e., DW-MRI volume and corresponding CFA image) we follow these steps:

1. Randomly select one of the fetal DW volumes, which we denote with V^k.
2. Randomly select one of the newborn subjects and pick the reconstructed tensor and CFA images for that subject, which we denote as Tensor^k and CFA^k, respectively.
3. CFA^k is the high-quality target CFA image. Initialize an empty image, DW^k, with the same size as CFA^k in the voxel space to hold the generated diffusion data. In order to generate fetal DW data that match this target, for each voxel in CFA^k perform these steps:
   (a) Select the tensor, D, from the same voxel in Tensor^k and the gradient table, G, from a voxel in V^k.
   (b) Estimate the mean signal in each direction $g \in G$ using the standard diffusion tensor model: $m = S/S_0 = \exp(-bg^T Dg)$, where b is the diffusion strength ($b = 500$ in this work), S is the diffusion-weighted signal, and S_0 is the baseline non-diffusion-weighted signal. Then, "add" noise to m.

This is done by sampling the kernel density-estimated PDF corresponding to m using standard PDF sampling methods; in this work we use the inversion method [5].

(c) Repeat the above step for all $g \in G$ and place the generated noisy diffusion signal in the corresponding voxel in DW^k.

The above steps can be repeated K times to generate a set of diffusion-weighted images and corresponding CFA images $\mathcal{D}_{train} = \{DW^k, CFA^k\}_{k=1}^{K}$. Because of the construction of our training data generation pipeline described above: 1) the diffusion images $DW^k \in \mathcal{D}_{train}$ follow the noise distribution and measurement scheme (i.e., gradient table) of fetal DWI volumes, and 2) the corresponding CFA images $CFA^k \in \mathcal{D}_{train}$ are the most accurate practically achievable targets obtained from high-quality in-vivo pre-term newborn scans. We generated a total of 600 images using the above method. We randomly divided this dataset into subsets of 400, 100, and 100 for training, validation, and test. This division was done on a patient-wise basis: newborn and fetal scans used in the generation of training, validation, and test sets had no overlap.

2.3   Model Architecture and Training

In DW-MRI, the gradient tables are in general different between subjects/scans. In fetal scans, even the gradient tables between adjacent voxels are generally different due to the SVR process used to generate the DW volumes. For a machine learning method to generalize to all subjects/scans, the measurements obtained with different gradient tables need to be transformed into the same basis/grid in q space. This is typically performed using either spherical harmonics representation [23] or interpolation [17]. In this work we used the interpolation approach by considering a uniform spherical grid of size 200 as proposed in [17]. We generated a uniform spherical grid following an approach similar to [13]. Therefore, the input to our CNN has 200 channels and the CNN output has three channels, for the three (RGB) components of the CFA image.

The CNN that we use follows 3D UNet++ [33] with 3D input patches of size 48^3 voxels. Since the network architecture design is not the focus of this work, we refer the reader to [33] for details. We set the number of feature maps in the first stage of the network to be 12, which was the largest possible on our GPU memory. During training, we sample blocks from random locations in the training images and use them to update the network weights. At test time, we use a sliding window with a stride of 16 voxels in each dimension to estimate CFA for an entire DW image of arbitrary size. We train the CNN by minimizing the ℓ_2 norm between the predicted and ground truth CFA using Adam optimizer [21], a batch size of 10, and an initial learning rate of 10^{-3} that is reduced by 0.5 every time the validation loss does not decrease after a training epoch.

2.4   Evaluation Strategy

We compared the trained model with standard WLLS reconstruction [22,24] in three different ways:

1. For the 100 test DW volumes from Sect. 2.2 we have high-quality target CFA images. Therefore, on these images we compare the methods in terms of reconstruction accuracy by computing the root mean square of the reconstruction error (RMSE).

2. We selected scans of 20 separate fetuses (gestational age 31.0 ± 5.3) that had not been used in any way in the generation of the training data above. We presented the CFA reconstructions to a neuroanatomist with extensive experience in fetal neuroanatomy and asked her to rate the fidelity of the reconstructions based on her knowledge of fetal neuroanatomy on a 1–5 scale with 5 being the best quality. The images reconstructed with the proposed method and WLLS were randomly named "CFA 01" and "CFA 02". The neuroanatomist was told that the names were randomized, but she was blind to the order.

3   Results and Discussion

3.1   Reconstruction Accuracy on Test Images with Known Ground Truth

Figure 3 shows CFA images reconstructed with WLLS and our proposed method on a test volume from the data generated in Sect. 2.2. The RMSE for WLLS and the proposed method were 0.0807 ± 0.0034 and 0.0379 ± 0.0030, respectively. Paired t-test showed that the difference was statistically significant ($p < 0.001$).

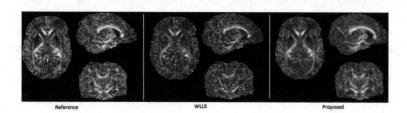

Fig. 3. Comparison of the proposed method and WLLS on a test DW volume from Sect. 2.2.

3.2   Expert Assessment of Reconstruction Accuracy

Example images reconstructed by WLLS and the proposed method and the scores assigned by our expert neuroanatomist are shown in Fig. 4. The mean and standard deviation of the score assigned to the CFA images reconstructed with WLLS and the proposed method were, respectively, 1.30 ± 0.46 and 3.85 ± 0.79. On all 20 fetuses, the score received by our proposed method was higher than WLLS. Wilcoxon signed-rank test showed that the difference between our method and WLLS was statistically significant ($p < 0.001$).

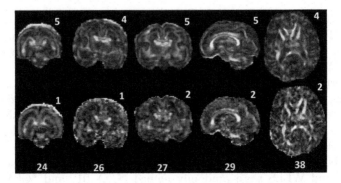

Fig. 4. Slices from images reconstructed by the proposed method (top) and WLLS (bottom). The scores assigned by an expert neuroanatomist are shown on the top right of each image. The number below each image pair shows the gestational age in weeks.

4   Conclusions

The fetal period represents a critical stage in brain development. Accurate and robust parameter estimation in fetal DW-MRI can greatly enhance our understanding of fetal brain development and disorders and their impact on neurological disorders and cognitive disabilities later in life. In this work, we presented the first successful application of deep learning for parameter estimation in fetal DWI. Our main contribution was a methodology that enabled us to generate massive amounts of reliable training data. While we focused on CFA estimation to demonstrate the effectiveness of the proposed methods, our methods can be adapted for other parameters. For other diffusion tensor imaging parameters, for example, this would be as simple as replacing CFA with the parameter of interest (e.g., fractional anisotropy or mean diffusivity) in the methods described above and other obvious necessary changes such as changing the number of CNN's output channels.

Acknowledgement. This work was supported in part by the National Institute of Neurological Disorders and Stroke, the National Institute of Biomedical Imaging and Bioengineering, and the National Library of Medicine of the National Institutes of Health (NIH) award numbers R01NS106030, R01EB031849, R01EB019483, and R01LM013608; by the Office of the Director of the NIH Award Number S10OD0250111; and by a Technological Innovations in Neuroscience Award from the McKnight Foundation. The content is solely the responsibility of authors and does not necessarily represent the official views of the NIH. Data were provided by the developing Human Connectome Project, KCL-Imperial-Oxford Consortium funded by the European Research Council under the European Union Seventh Framework Programme (FP/2007-2013)/ERC Grant Agreement no. [319456]. We are grateful to the families who generously supported this trial.

References

1. Aliotta, E., Nourzadeh, H., Sanders, J., Muller, D., Ennis, D.B.: Highly accelerated, model-free diffusion tensor MRI reconstruction using neural networks. Med. Phys. **46**(4), 1581–1591 (2019)
2. Basser, P.J., Pierpaoli, C.: Microstructural and physiological features of tissues elucidated by quantitative-diffusion-tensor MRI. J. Magn. Reson. **213**(2), 560–570 (2011)
3. Bastiani, M., et al.: Automated processing pipeline for neonatal diffusion MRI in the developing human connectome project. Neuroimage **185**, 750–763 (2019)
4. Canales-Rodríguez, E.J., Daducci, A., Sotiropoulos, S.N., et al.: Spherical deconvolution of multichannel diffusion MRI data with non-gaussian noise models and spatial regularization. PloS one **10**(10), e0138910 (2015)
5. Casella, G., Berger, R.L.: Statistical inference. Cengage Learning (2021)
6. Deprez, M., Price, A., Christiaens, D., Estrin, G.L., et al.: Higher order spherical harmonics reconstruction of fetal diffusion MRI with intensity correction. IEEE Trans. Med. Imaging **39**(4), 1104–1113 (2019)
7. Dietrich, O., Raya, J.G., Reeder, S.B., Ingrisch, M., et al.: Influence of multichannel combination, parallel imaging and other reconstruction techniques on MRI noise characteristics. Magn. Reson. Imaging **26**(6), 754–762 (2008)
8. Gholipour, A., et al.: Fetal MRI: a technical update with educational aspirations. Concepts Magn. Resonance Part A **43**(6), 237–266 (2014)
9. Gibbons, E.K., et al.: Simultaneous NODDI and GFA parameter map generation from subsampled q-space imaging using deep learning. Magn. Reson. Med. **81**(4), 2399–2411 (2019)
10. Golkov, V., et al.: q-space deep learning: twelve-fold shorter and model-free diffusion MRI scans. IEEE Trans. Med. Imaging **35**(5), 1344–1351 (2016)
11. Jiang, S., et al.: Diffusion tensor imaging (DTI) of the brain in moving subjects: application to in-utero fetal and ex-utero studies. Magn. Reson. Med. Official J. Int. Soc. Magn. Reson. Med. **62**(3), 645–655 (2009)
12. Johansen-Berg, H., Behrens, T.E.: Diffusion MRI: From Quantitative Measurement to in Vivo Neuroanatomy. Academic Press (2013)
13. Jones, D.K., Horsfield, M.A., Simmons, A.: Optimal strategies for measuring diffusion in anisotropic systems by magnetic resonance imaging. Magn. Reson. Med. Official J. Int. Soc. Magn. Reson. Med. **42**(3), 515–525 (1999)
14. Jörgens, D., Smedby, Ö., Moreno, R.: Learning a single step of streamline tractography based on neural networks. In: Kaden, E., Grussu, F., Ning, L., Tax, C.M.W., Veraart, J. (eds.) Computational Diffusion MRI. MV, pp. 103–116. Springer, Cham (2018). https://doi.org/10.1007/978-3-319-73839-0_8
15. Kainz, B., et al.: Fast volume reconstruction from motion corrupted stacks of 2D slices. IEEE Trans. Med. Imaging **34**(9), 1901–1913 (2015)
16. Karimi, D., et al.: A machine learning-based method for estimating the number and orientations of major fascicles in diffusion-weighted magnetic resonance imaging. Med. Image Anal., 102129 (2021)
17. Karimi, D., et al.: A machine learning-based method for estimating the number and orientations of major fascicles in diffusion-weighted magnetic resonance imaging. arXiv preprint arXiv:2006.11117 (2020)
18. Karimi, D., Vasung, L., Jaimes, C., Machado-Rivas, F., Warfield, S.K., Gholipour, A.: Learning to estimate the fiber orientation distribution function from diffusion-weighted MRI. NeuroImage, 118316 (2021)

19. Kasprian, G., et al.: In utero tractography of fetal white matter development. Neuroimage **43**(2), 213–224 (2008)
20. Khan, S., et al.: Fetal brain growth portrayed by a spatiotemporal diffusion tensor MRI atlas computed from in utero images. Neuroimage **185**, 593–608 (2019)
21. Kingma, D.P., Ba, J.: Adam: a method for stochastic optimization. In: Proceedings of the 3rd International Conference on Learning Representations (ICLR) (2014)
22. Koay, C.G., Chang, L.C., Carew, J.D., Pierpaoli, C., Basser, P.J.: A unifying theoretical and algorithmic framework for least squares methods of estimation in diffusion tensor imaging. J. Magn. Reson. **182**(1), 115–125 (2006)
23. Lin, Z., et al.: Fast learning of fiber orientation distribution function for MR tractography using convolutional neural network. Med. Phys. **46**(7), 3101–3116 (2019)
24. Marami, B., et al.: Temporal slice registration and robust diffusion-tensor reconstruction for improved fetal brain structural connectivity analysis. Neuroimage **156**, 475–488 (2017)
25. Marami, B., Scherrer, B., Afacan, O., Erem, B., Warfield, S.K., Gholipour, A.: Motion-robust diffusion-weighted brain MRI reconstruction through slice-level registration-based motion tracking. IEEE Trans. Med. Imaging **35**(10), 2258–2269 (2016)
26. Murphy, K.P.: Machine Learning: A Probabilistic Perspective (2012)
27. Nat, V., et al.: Deep learning reveals untapped information for local white-matter fiber reconstruction in diffusion-weighted MRI. Magn. Reson. Imaging **62**, 220–227 (2019)
28. Oubel, E., Koob, M., Studholme, C., Dietemann, J.L., Rousseau, F.: Reconstruction of scattered data in fetal diffusion MRI. Med. Image Anal. **16**(1), 28–37 (2012)
29. Wakana, S., Jiang, H., Nagae-Poetscher, L.M., Van Zijl, P.C., Mori, S.: Fiber tract-based atlas of human white matter anatomy. Radiology **230**(1), 77–87 (2004)
30. Wasserthal, J., Neher, P.F., Maier-Hein, K.H.: Tract orientation mapping for bundle-specific tractography. In: Frangi, A.F., Schnabel, J.A., Davatzikos, C., Alberola-López, C., Fichtinger, G. (eds.) MICCAI 2018. LNCS, vol. 11072, pp. 36–44. Springer, Cham (2018). https://doi.org/10.1007/978-3-030-00931-1_5
31. Ye, C., Li, X., Chen, J.: A deep network for tissue microstructure estimation using modified LSTM units. Med. Image Anal. **55**, 49–64 (2019)
32. Zanin, E., et al.: White matter maturation of normal human fetal brain. An in vivo diffusion tensor tractography study. Brain Behav. **1**(2), 95–108 (2011)
33. Zhou, Z., Rahman Siddiquee, M.M., Tajbakhsh, N., Liang, J.: UNet++: a nested U-net architecture for medical image segmentation. In: Stoyanov, D., et al. (eds.) DLMIA/ML-CDS -2018. LNCS, vol. 11045, pp. 3–11. Springer, Cham (2018). https://doi.org/10.1007/978-3-030-00889-5_1

Deep Fiber Clustering: Anatomically Informed Unsupervised Deep Learning for Fast and Effective White Matter Parcellation

Yuqian Chen[1,2], Chaoyi Zhang[2], Yang Song[3], Nikos Makris[1], Yogesh Rathi[1], Weidong Cai[2], Fan Zhang[1(✉)], and Lauren J. O'Donnell[1]

[1] Harvard Medical School, Boston, MA, USA
fzhang@bwh.harvard.edu
[2] The University of Sydney, Sydney, NSW, Australia
[3] The University of New South Wales, Sydney, NSW, Australia

Abstract. White matter fiber clustering (WMFC) enables parcellation of white matter tractography for applications such as disease classification and anatomical tract segmentation. However, the lack of ground truth and the ambiguity of fiber data (the points along a fiber can equivalently be represented in forward or reverse order) pose challenges to this task. We propose a novel WMFC framework based on unsupervised deep learning. We solve the unsupervised clustering problem as a self-supervised learning task. Specifically, we use a convolutional neural network to learn embeddings of input fibers, using pairwise fiber distances as pseudo annotations. This enables WMFC that is insensitive to fiber point ordering. In addition, anatomical coherence of fiber clusters is improved by incorporating brain anatomical segmentation data. The proposed framework enables outlier removal in a natural way by rejecting fibers with low cluster assignment probability. We train and evaluate our method using 200 datasets from the Human Connectome Project. Results demonstrate superior performance and efficiency of the proposed approach.

Keywords: Diffusion MRI · Tractography · Fiber clustering · Deep embedding · Self-supervised learning

1 Introduction

Diffusion magnetic resonance imaging (dMRI) [1] uniquely enables mapping of the brain's white matter fiber tracts via tractography [2], to study the brain's connections in health and disease [9]. Tractography of a single brain can generate hundreds of thousands of streamlines (fibers), which are not immediately useful to clinicians or researchers. Therefore, tractography parcellation, i.e. dividing the massive number of tractography fibers into multiple subdivisions, is needed.

We acknowledge funding provided by the following National Institutes of Health (NIH) grants: R01MH125860, R01MH119222, R01MH074794, and P41EB015902.

ⓒ Springer Nature Switzerland AG 2021
M. de Bruijne et al. (Eds.): MICCAI 2021, LNCS 12907, pp. 497–507, 2021.
https://doi.org/10.1007/978-3-030-87234-2_47

One widely used tractography parcellation strategy, white matter fiber clustering (WMFC), groups fiber streamlines with similar geometric trajectory into clusters [22]. WMFC is useful in applications such as disease classification [42], anatomical tract identification [33] and neurosurgical brain mapping [29]. In general, WMFC first computes pairwise fiber geometric similarities, then applies a computational clustering method to group similar fibers into clusters [7,31,43]. Existing WMFC methods show good performance, but key challenges remain. First, it is computationally expensive to compute pairwise fiber geometric similarities. Second, the computation of fiber similarity is sensitive to the order of points along the fibers, even though a fiber can equivalently start from either end [7]. Third, false positive fibers are prevalent in tractography; thus outlier fiber removal is needed to filter undesired fibers from the clustering result [16,18]. Fourth, it is a challenge for WMFC to use all available information to improve cluster anatomical quality: most methods use either fiber spatial coordinate information [7,43] or anatomical information about brain regions that fibers pass through [27]. Fifth, WMFC methods should ideally consider inter-subject correspondence of fiber clusters, which is essential for group-wise analysis [21]. To achieve this goal, some studies perform WMFC across subjects (to form an atlas) and predict clusters of new subjects with correspondence to the atlas [23,38,39], while other approaches first perform within-subject WMFC then match (or cluster) the fiber clusters across subjects [7,10,13,27].

In computer vision, clustering has been extensively studied as an unsupervised learning task [3,11,28,34,37], which requires a data feature representation and similarity computation between the features for cluster assignment. Autoencoder-based approaches are popularly used for unsupervised clustering [11,28,34]. The Deep Embedding Clustering (DEC) framework performs simultaneous embedding of input data and cluster assignments in an end-to-end way [34]. Deep Convolutional Embedded Clustering (DCEC) is an extension of DEC to the image clustering task [11]. In addition to autoencoder approaches, [3] and [37] also realized joint embedding learning and cluster assignments by alternative feature learning and traditional clustering, which is time consuming.

Self-supervised learning is a promising subclass of unsupervised learning that shows advanced performance in many applications [15,25]. It aims to learn high-level features without requiring manual annotations. This is achieved by designing pretext tasks, such as predicting context [5] or image rotation [8], and giving the network pseudo annotations generated from the input itself. The high-level representations learned from the pretext task can then be transferred to downstream tasks such as clustering. Therefore, besides the classical autoencoder network, the self-supervised learning framework can also be a promising approach to learn deep embeddings of inputs.

Considering the advances of deep neural networks in feature extraction, deep learning is a promising direction for WMFC. In related work, multiple deep learning methods have been proposed for white matter tractography segmentation [12,32,35,40]. In [12,32,40], known fiber labels are provided for training. One proposed method [35] has shown the potential of unsupervised deep learning for

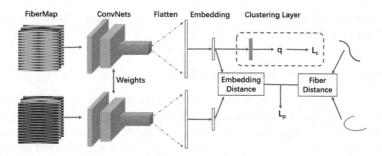

Fig. 1. Overview of our DFC framework. A self-supervised learning strategy is adopted with the pretext task of pairwise fiber distance prediction. In the pretraining stage, a pair of FiberMaps are encoded as embeddings with Siamese Networks, and prediction loss (L_p) is calculated based on the difference between embedding distance and fiber distance. In the clustering stage, a clustering layer is connected to the embedding layer and generates soft label assignment (as shown in the dashed box). A KL divergence loss (L_c) and the prediction loss are combined to optimize the neural network.

fiber clustering; however, the anatomical utility of this approach was not tested as results were limited to a maximum of 11 clusters in the whole brain. The goal of our study is to propose an anatomically meaningful unsupervised deep learning framework, Deep Fiber Clustering (DFC), for fast and effective white matter fiber clustering. The paper has four contributions. First, we propose a novel deep learning pipeline that adopts self-supervised learning for deep embedding and achieves joint representation learning and cluster assignment. Second, anatomical information is incorporated into the neural network to improve cluster anatomical coherence. Third, outliers are removed by rejecting fibers with low soft label assignment probabilities. Our approach automatically creates a multi-subject fiber cluster atlas that is applied for white matter parcellation of new subjects. Finally, our approach has demonstrated superior performance and efficiency via evaluations on a large scale dataset.

2   Methods

As shown in Fig. 1, our training pipeline includes two stages, *pretraining* and *clustering*. In the pretraining stage (Sect. 2.1), a CNN is trained in a self-supervised way with a designed pretext task to obtain deep embeddings. After that, k-means clustering is performed on the embeddings to get initial clusters, which is performed only once during training. In the clustering stage (Sect. 2.2), the clustering results are fine-tuned in a self-learning manner and cluster centroids are automatically optimized as parameters of the network. During network inference, when the model is applied to a new subject, cluster assignments are obtained from the network directly in an end-to-end way without any k-means clustering.

In this work, we adopt the FiberMap fiber representation [40], which was found to be effective for tractography segmentation in supervised learning. One benefit of using FiberMap is that it is a 2D multi-channel feature descriptor (analogous to a RGB image); thus it can be effectively processed by CNNs.

2.1 Self-supervised Deep Embedding

We propose a novel self-supervised learning strategy for learning deep fiber embeddings. The goal is to obtain embeddings with similar distances to fiber distances in the brain space, enabling subsequent WMFC in the embedding space. (We note that a DCEC model with a convolutional autoencoder could be adopted here for unsupervised WMFC, but as we show in the Results, this straightforward approach is sensitive to fiber point ordering.) To learn the embeddings, a pretext task is first designed to predict the distance between a pair of input fibers. Specifically, the input to the network is the FiberMaps of a fiber pair and a pseudo annotation of the fiber pair distance. For the pairwise fiber distance, we use the minimum average direct-flip (MDF) distance which is widely successful in WMFC [7,43]. The computation of fiber distance considers the order of points along the fibers; thus, fiber distance is not affected if a fiber point sequence is flipped. A Siamese Network [4], a neural network that encodes different inputs and computes comparable outputs with shared weights, is then adopted to learn embeddings of an input FiberMap pair and output Euclidean distance between the embeddings. The distance prediction loss L_p is the mean squared error between embedding distance and fiber-distance pseudo annotations. By using fiber distances as pseudo annotations, the network is guided to generate similar embeddings for close fibers, even those with flipped point orders.

2.2 Clustering Layer and Clustering Loss

Here we adopt the DCEC model design [11]. In the clustering stage, a clustering layer is designed to encapsulate cluster centroids as its trainable weights and compute a soft assignment label q_{ij} using Student's t-distribution [17,34]:

$$q_{ij} = (1 + \|z_i - \mu_j\|^2)^{-1} / (\textstyle\sum_{j'}(1 + \|z_i - \mu'_j\|^2)^{-1}) \tag{1}$$

where z_i is the embedding of fiber i and μ_j is the centroid of cluster j. q_{ij} is the probability of assigning fiber i to cluster j. The network is trained in a self-training manner and its clustering loss L_c is defined as a KL divergence loss [34]: $L_c = KL(P\|Q) = \sum_i \sum_j p_{ij} log \frac{p_{ij}}{q_{ij}}$, where $p_{ij} = (q_{ij}^2 / \sum_i q_{ij}) / (\sum_{j'} (q_{ij'}^2 / \sum_i q_{ij'}))$. The distance prediction loss is retained in this stage, and the total loss is $L = L_p + \lambda L_c$, where λ is the weight of L_c. During inference, a fiber i is assigned to the cluster with the maximum q_{ij}.

2.3 Incorporation of Anatomical Information and Outlier Removal

We extend our proposed self-learning framework described above to enable two important tasks in WMFC, i.e., inclusion of additional anatomical information for anatomical coherence and removal or filtering of false positive outlier fibers. For the first task, we propose to incorporate Freesurfer parcellation [6] information during the clustering stage. We design a new soft label assignment probability definition which is used to calculate loss and extends Eq.(1) to further regularize that fibers within a cluster pass through the same brain regions:

$$q_{ij} = (1 + \|z_i - \mu_j\|^2 * (1 - D_{ij}))^{-1} / (\textstyle\sum_{j'} (1 + \|z_i - \mu_{j'}\|^2 * (1 - D_{ij'}))^{-1}) \quad (2)$$

where D_{ij} is the Dice score between the set of Freesurfer regions of fiber i and the set of Freesurfer regions of cluster j. We use the Tract Anatomical Profile (TAP) proposed in [43] to define the set of Freesurfer regions commonly intersected by the fibers in a cluster (at least 40% of fibers, as in [43]). During training, the TAP is initially calculated from the clusters generated by k-means and is updated iteratively with new predictions during the training process. During inference, soft label assignments are calculated with Eq. (2) and fibers are assigned to the cluster with maximum q_{ij}, referred to as q_m.

For outlier removal, we remove fibers using the maximum label assignment probability q_m, considering that fibers with higher q_m tend to have more confidence of belonging to the corresponding cluster and are less likely to be outliers. Therefore, we remove outliers by setting a threshold h on the q_m values of fibers, meaning that fibers with $q_m < h$ will be rejected from the final clusters.

2.4   Implementation Details

As shown in Fig. 1, our model architecture includes three convolutional layers of sizes $5 \times 5 \times 32$, $5 \times 5 \times 64$ and $3 \times 3 \times 128$, respectively, to extract feature maps. These feature maps are flattened to a vector, followed by a fully connected layer to compute embeddings with a dimension of 10 (suggested in [11]). In the pretraining and clustering stages, the network is trained for 25000 iterations with a learning rate of 0.0001 and another 4000 iterations with a learning rate of 0.00001, which are sufficient to achieve training convergence. Admax [14] is used for optimization in both stages. All experiments are performed on an NVIDIA RTX 2080Ti GPU using Pytorch (v1.7.1) [26]. The weight of clustering loss λ is set to be 0.1, as suggested in [11]. We set the threshold h for outlier removal to be 0.015 to reject fibers with extremely low cluster assignment probabilities.

3   Experiments and Results

3.1   Dataset

In our experiments, we used a dataset of 200 healthy adults from the Human Connectome Project [30]. 100 subjects were used for training, 50 for validation and 50 for testing. Tractography data were generated using a two-tensor unscented Kalman filter (UKF) method [19], and tractography co-registration was performed using an affine followed by a nonrigid registration [24]. Fibers longer than 40 mm were retained to avoid any bias towards implausible short fibers. For each training subject, 10,000 fibers were randomly selected, generating a training dataset of

1 million samples. For testing and validation, all whole-brain tractography fibers were used (around 500,000 per subject). Fibers were downsampled to 14 points [40] to obtain the FiberMap input to neural network. We performed diffusion MRI tractography and visualization in 3D Slicer (www.slicer.org) via the SlicerDMRI project (http://dmri.slicer.org) [20,41].

3.2   Evaluation Metrics

Three evaluation metrics were adopted to quantify performance of our proposed method and enable comparisons among approaches. The first one is the Davies–Bouldin (DB) index [36], which is computed as:

$$DB = (1/n)\sum_{k=1}^{n} max_{x \neq y}(\frac{\alpha_i + \alpha_j}{d(c_i, c_j)}) \tag{3}$$

where n is the number of clusters, α_i and α_j are mean pairwise intra-cluster fiber distances, and $d(c_i, c_j)$ is the inter-cluster fiber distance between centroids c_i and c_j of cluster i and j [31]. A smaller DB score indicates a better separation between clusters. The second metric is White Matter Parcellation Generalization (WMPG) [43], which is used to represent the percentage of clusters successfully detected across the testing subjects. In our work, clusters with a over 10 fibers are considered to be successfully detected [43]. The last metric is Tract Anatomical Profile Coherence (TAPC) [43], which measures if the fibers within a cluster c commonly pass through the same brain anatomical regions:

$$TAPC(c) = (\sum_{f=1}^{NF(c)} Dice(TAP(f), TAP_{atlas}(c)))/NF(c) \tag{4}$$

Higher TAPC scores indicate better anatomical coherence.

3.3   Evaluation Results

Comparison with State-of-the-Art Methods. We compare our proposed approach with two open-source state-of-the-art WMFC algorithms, WhiteMatterAnalysis (WMA)[43] and QuickBundles (QB) [7]. WMA is an atlas-based WMFC method that shows high performance and strong correspondence across subjects. QB is a widely used WMFC method that performs clustering within each subject and achieves group correspondence with post-processing steps. We use the open-source software packages WMA v0.3.0 and Dipy v1.3.0 with their default settings. For all experiments, we perform WMFC into 800 clusters (which has been suggested to be a good whole brain tractography parcellation scale [43]). Dipy does not accept an input number of clusters; therefore, we tuned parameters in each subject to obtain a number as close as possible to 800 clusters (greater than or equal to 800). All results are reported using data from the 50 test subjects. The WMPG and TAPC metrics require corresponding clusters across all subjects; these are automatically generated by WMA and our proposed DFC method. For QB, correspondence is achieved by matching cluster centroids from all subjects to those of one selected subject (with exactly 800 clusters) according to the fiber distances between centroids, as suggested by the QB developers [7].

Table 1. Quantitative comparison results. SOTA: state of the art.

	Methods	DB index	WMPG	TAPC	Time (s)
SOTA comparison	WMA	3.231 ± 0.153	99.22% ± 0.79%	0.802 ± 0.006	3210
	QB	**2.419 ± 0.096**	81.14% ± 2.64%	0.690 ± 0.015	240
	DFC	2.661 ± 0.107	**99.35% ± 0.54%**	**0.836 ± 0.006**	**205**
Baseline comparison	DCEC	15.661 ± 4.390	**99.87% ± 0.35%**	0.755 ± 0.009	–
	DFC	**2.661 ± 0.107**	99.35% ± 0.54%	**0.836 ± 0.006**	–
Ablation study	DFC$_{no-fs-ro}$	3.095 ± 0.156	99.80% ± 0.48%	0.773 ± 0.009	–
	DFC$_{no-ro}$	3.152 ± 0.139	**99.82% ± 0.32%**	0.816 ± 0.007	–
	DFC	**2.661 ± 0.107**	99.35% ± 0.54%	**0.836 ± 0.006**	–

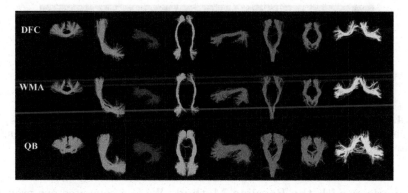

Fig. 2. Visualization of example clusters generated from DFC, WMA, and QB in one subject. Similar clusters were identified across methods for visualization.

As shown in Table 1, our DFC method exhibits the best performance in general. For the DB index metric, QB obtained a slightly lower value than DFC, likely because intra-cluster distances are lower when performing within-subject clustering since the obtained clusters do not describe anatomical variability across subjects. When compared to the atlas-based WMA, the DB index of our method is obviously smaller, indicating more compact and/or better separated clusters. As for WMPG, both our method and WMA successfully detected over 99% of clusters while the WMPG score of QB is around 80% indicating poor correspondence across subjects. The TAPC metric of DFC obtained the highest value among the three methods owing to the incorporation of anatomical information, indicating the best anatomical coherence of clusters. Figure 2 gives a visual illustration of obtained clusters for each method.

To evaluate efficiency of approaches, inference time of one subject is also recorded and shown in Table 1. All methods were tested on a computer equipped with a 2.1 GHz Intel Xeon E5 CPU (8 DIMMs; 32 GB Memory). For fair comparison, DFC was set to run on CPU instead of GPU. The results show that our method is much faster than WMA and slightly better than QB.

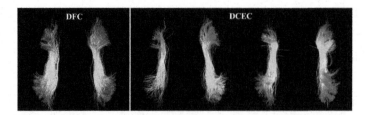

Fig. 3. Illustration of corresponding clusters from DFC and DCEC. Colors represent order of points along a fiber with red for starting point and blue for ending point. (Color figure online)

Fig. 4. Illustration of outlier removal process. Left: cluster before outlier removal; Middle: fiber soft label assignment probability (rainbow coloring with red representing 0); Right: cluster after outlier removal.

Comparison with the Baseline Method. We also compared our proposed method with the DCEC baseline model. The results in Table 1 show a large improvement of the DB index of our method compared to DCEC, because DCEC separately clusters fibers with close positions but flipped point orders. As shown in Fig. 3, spatially close fibers with different point orders are split into two clusters in DCEC, while our proposed DFC method groups them together.

Ablation Study. We performed an ablation study to investigate how different factors influence performance of our method. Evaluation of three models was performed, including $DFC_{no-fs-ro}$ (DFC without FreeSurfer information and outlier removal), DFC_{no-ro} (DFC without outlier removal but with FreeSurfer information) and $DFC_{proposed}$, as shown in Table 1. By adding FreeSurfer information into the model, the DB index and WMPG metrics do not show much difference, while the TAPC score exhibits obvious improvement. With implementation of outlier removal, the DB index and TAPC improve obviously, while WMPG shows slight decrease, which is inevitable due to the decreased number of fibers (but it still remains a high percentage). These results demonstrate effectiveness of our designed modules. As shown in Fig. 4, outlier fibers have apparently low values of soft label assignment probabilities and are then removed.

4   Conclusion

In this paper, we present a novel unsupervised deep learning framework for dMRI tractography WMFC. We adopt the self-supervised learning strategy to enable joint deep embedding and cluster assignment. Our method can handle several key challenges in WMFC methods, including handling flipped order of points along fibers, incorporating anatomical brain segmentation information, false positive fiber filtering and inter-subject correspondence of fiber clusters. Our results show advantages over clustering performance as well as efficiency compared to the state-of-art algorithms. Further research could be conducted to improve the framework, such as designing more complex network architectures, incorporating additional sources of anatomical information and balancing anatomical and fiber geometry information for clustering.

References

1. Basser, P.J., Mattiello, J., LeBihan, D.: MR diffusion tensor spectroscopy and imaging. Biophys. J. **66**(1), 259–267 (1994)
2. Basser, P.J., Pajevic, S., Pierpaoli, C., Duda, J., Aldroubi, A.: In vivo fiber tractography using DT-MRI data. Mag. Res. Med. **44**(4), 625–632 (2000)
3. Caron, M., Bojanowski, P., Joulin, A., Douze, M.: Deep clustering for unsupervised learning of visual features. In: Ferrari, V., Hebert, M., Sminchisescu, C., Weiss, Y. (eds.) Computer Vision – ECCV 2018. LNCS, vol. 11218, pp. 139–156. Springer, Cham (2018). https://doi.org/10.1007/978-3-030-01264-9_9
4. Chopra, S., Hadsell, R., LeCun, Y.: Learning a similarity metric discriminatively, with application to face verification. In: CVPR, vol. 1, pp. 539–546 (2005)
5. Doersch, C., Gupta, A., Efros, A.A.: Unsupervised visual representation learning by context prediction. In: ICCV, pp. 1422–1430 (2015)
6. Fischl, B.: Freesurfer. Neuroimage **62**(2), 774–781 (2012)
7. Garyfallidis, E., Brett, M., Correia, M.M., et al.: QuickBundles, a method for tractography simplification. Front. Neurosci. **6**, 175 (2012)
8. Gidaris, S., Singh, P., Komodakis, N.: Unsupervised representation learning by predicting image rotations. arXiv:1803.07728 (2018)
9. Griffa, A., Baumann, P.S., Thiran, J.P., Hagmann, P.: Structural connectomics in brain diseases. Neuroimage **80**, 515–526 (2013)
10. Guevara, P., et al.: Automatic fiber bundle segmentation in massive tractography datasets using a multi-subject bundle atlas. Neuroimage **61**(4), 1083–1099 (2012)
11. Guo, X., Liu, X., Zhu, E., Yin, J.: Deep clustering with convolutional autoencoders. In: ICNIP, pp. 373–382 (2017). https://doi.org/10.1007/978-3-319-70096-0_39
12. Gupta, V., Thomopoulos, S.I., Rashid, F.M., Thompson, P.M.: FiberNET: an ensemble deep learning framework for clustering white matter fibers. In: Descoteaux, M., Maier-Hein, L., Franz, A., Jannin, P., Collins, D.L., Duchesne, S. (eds.) MICCAI 2017. LNCS, vol. 10433, pp. 548–555. Springer, Cham (2017). https://doi.org/10.1007/978-3-319-66182-7_63
13. Huerta, I., et al.: Inter-subject clustering of brain fibers from whole-brain tractography. In: EMBC, pp. 1687–1691. IEEE (2020)
14. Kingma, D.P., Ba, J.: Adam: a method for stochastic optimization. arXiv:1412.6980 (2014)

15. Kolesnikov, A., Zhai, X., Beyer, L.: Revisiting self-supervised visual representation learning. In: CVPR, pp. 1920–1929 (2019)
16. Legarreta, J.H., et al.: Tractography filtering using autoencoders. arXiv:2010.04007 (2020)
17. Van der Maaten, L., Hinton, G.: Visualizing data using t-SNE. J. Mach. Learn. Res. **9**(11) (2008)
18. Maier-Hein, K.H., Neher, P.F., et al.: The challenge of mapping the human connectome based on diffusion tractography. Nat. Commun. **8**(1), 1–13 (2017)
19. Malcolm, J.G., Shenton, M.E., Rathi, Y.: Filtered multitensor tractography. IEEE Trans. Med. Imaging **29**(9), 1664–1675 (2010)
20. Norton, I., Essayed, W.I., Zhang, F., Pujol, S., Yarmarkovich, A., et al.: SlicerDMRI: open source diffusion MRI software for brain cancer research. Cancer Res. **77**(21), e101–e103 (2017)
21. O'Donnell, L., Westin, C.-F.: White matter tract clustering and correspondence in populations. In: Duncan, J.S., Gerig, G. (eds.) MICCAI 2005. LNCS, vol. 3749, pp. 140–147. Springer, Heidelberg (2005). https://doi.org/10.1007/11566465_18
22. O'Donnell, L.J., Golby, A.J., Westin, C.F.: Fiber clustering versus the parcellation-based connectome. NeuroImage **80**, 283–289 (2013)
23. O'Donnell, L.J., et al.: Automated white matter fiber tract identification in patients with brain tumors. NeuroImage: Clin. **13**, 138–153 (2017)
24. O'Donnell, L.J., Wells, W.M., Golby, A.J., Westin, C.-F.: Unbiased groupwise registration of white matter tractography. In: Ayache, N., Delingette, H., Golland, P., Mori, K. (eds.) MICCAI 2012. LNCS, vol. 7512, pp. 123–130. Springer, Heidelberg (2012). https://doi.org/10.1007/978-3-642-33454-2_16
25. Oord, A.V.D., Li, Y., Vinyals, O.: Representation learning with contrastive predictive coding. arXiv:1807.03748 (2018)
26. Paszke, A., Gross, S., Massa, F., Lerer, A., et al.: PyTorch: an imperative style, high-performance deep learning library. arXiv:1912.01703 (2019)
27. Siless, V., Chang, K., Fischl, B., Yendiki, A.: AnatomiCuts: hierarchical clustering of tractography streamlines based on anatomical similarity. NeuroImage **166**, 32–45 (2018)
28. Tian, F., Gao, B., Cui, Q., Chen, E., Liu, T.Y.: Learning deep representations for graph clustering. In: AAAI, vol. 28 (2014)
29. Tunç, B., et al.: Individualized map of white matter pathways: connectivity-based paradigm for neurosurgical planning. Neurosurgery **79**(4), 568–577 (2016)
30. Van Essen, D.C., Smith, S.M., Barch, D.M., et al.: The WU-Minn human connectome project: an overview. Neuroimage **80**, 62–79 (2013)
31. Vázquez, A., et al.: FFClust: fast fiber clustering for large tractography datasets for a detailed study of brain connectivity. NeuroImage **220**, 117070 (2020)
32. Wasserthal, J., Neher, P., Maier-Hein, K.H.: Tractseg-fast and accurate white matter tract segmentation. NeuroImage **183**, 239–253 (2018)
33. Wu, Y., Hong, Y., Ahmad, S., Lin, W., Shen, D., Yap, P.-T.: Tract dictionary learning for fast and robust recognition of fiber bundles. In: Martel, A.L., et al. (eds.) MICCAI 2020. LNCS, vol. 12267, pp. 251–259. Springer, Cham (2020). https://doi.org/10.1007/978-3-030-59728-3_25
34. Xie, J., Girshick, R., Farhadi, A.: Unsupervised deep embedding for clustering analysis. In: ICML, pp. 478–487. PMLR (2016)
35. Xu, C., Sun, G., Liang, R., Xu, X.: Vector field streamline clustering framework for brain fiber tract segmentation. arXiv preprint arXiv:2011.01795 (2020)
36. Xu, D., Tian, Y.: A comprehensive survey of clustering algorithms. Ann. Data Sci. **2**(2), 165–193 (2015)

37. Yang, J., Parikh, D., Batra, D.: Joint unsupervised learning of deep representations and image clusters. In: CVPR, pp. 5147–5156 (2016)
38. Yeh, F.C., et al.: Population-averaged atlas of the macroscale human structural connectome and its network topology. NeuroImage **178**, 57–68 (2018)
39. Yoo, S.W., et al.: An example-based multi-atlas approach to automatic labeling of white matter tracts. PLoS ONE **10**(7), e0133337 (2015)
40. Zhang, F., Karayumak, S.C., Hoffmann, N., Rathi, Y., Golby, A.J., O'Donnell, L.J.: Deep white matter analysis (DeepWMA): fast and consistent tractography segmentation. Med. Image Anal. **65**, 101761 (2020)
41. Zhang, F., et al.: SlicerDMRI: diffusion MRI and tractography research software for brain cancer surgery planning and visualization. JCO Clin. Cancer Inform. **4**, 299–309 (2020)
42. Zhang, F., et al.: Whole brain white matter connectivity analysis using machine learning: an application to autism. NeuroImage **172**, 826–837 (2018)
43. Zhang, F., et al.: An anatomically curated fiber clustering white matter atlas for consistent white matter tract parcellation across the lifespan. NeuroImage **179**, 429–447 (2018)

Disentangled and Proportional Representation Learning for Multi-view Brain Connectomes

Yanfu Zhang[1], Liang Zhan[1], Shandong Wu[2], Paul Thompson[3],
and Heng Huang[1,4(✉)]

[1] Department of Electrical and Computer Engineering, University of Pittsburgh,
Pittsburgh, PA 15260, USA
[2] Department of Radiology, University of Pittsburgh, Pittsburgh, PA 15260, USA
[3] Imaging Genetics Center, Institute for Neuroimaging and Informatics, University of
Southern California, Los Angeles, CA 90032, USA
[4] JD Finance America Corporation, Mountain View, CA 94043, USA
heng.huang@pitt.edu

Abstract. Diffusion MRI-derived brain structural connectomes or brain networks are widely used in the brain research. However, constructing brain networks is highly dependent on various tractography algorithms, which leads to difficulties in deciding the optimal view concerning the downstream analysis. In this paper, we propose to learn a unified representation from multi-view brain networks. Particularly, we expect the learned representations to convey the information from different views fairly and in a disentangled sense. We achieve the disentanglement via an approach using unsupervised variational graph autoencoders. We achieve the view-wise fairness, *i.e.* proportionality, via an alternative training routine. More specifically, we construct an analogy between training the deep network and the network flow problem. Based on the analogy, the fair representations learning is attained via a network scheduling algorithm aware of proportionality. The experimental results demonstrate that the learned representations fit various downstream tasks well. They also show that the proposed approach effectively preserves the proportionality.

Keywords: Brain connectome · Alzheimer's disease · Multi-view · Prediction

1 Introduction

Human brain connectomes [6] are models of complex brain networks and can be derived from diverse experimental modalities and tractography algorithms. Large-scale brains connections convey important insights for understanding the underlying yet largely unknown mechanisms of many mental disorders [7,11,15,26]. Nevertheless, the apparent characteristics of brain networks

© Springer Nature Switzerland AG 2021
M. de Bruijne et al. (Eds.): MICCAI 2021, LNCS 12907, pp. 508–518, 2021.
https://doi.org/10.1007/978-3-030-87234-2_48

are profoundly influenced by the tractography algorithms. The designs of tractography algorithms, including tensor-based deterministic algorithms [2], probabilistic approaches [18], random forest [17] and Deep Neural Network (DNN) [20], and regularized methods guided by biologically plausible fascicle structures [3], are inspired by specific experimental questions [5], *e.g.*, different tractography algorithms are used for predicting or classifying neurodegenerative or neurodevelopmental conditions based on various brain abnormalities. For example, the selection and accuracy of the extracted fibers are different for different tractography algorithms, and the relevance of the extracted fiber bundles depend on the different tasks and questions being addressed. Therefore, it is elusive to decide a universally optimal modality of brain networks and associated processing pipeline for distinct diagnostic tasks [5,23].

Multi-view methods can leverage the available information from diverse tractography algorithms simultaneously, and tentative studies have demonstrated that multi-modal brain networks can provide complementary viewpoints for the classification tasks, *e.g.*, multi-view graph convolutional network [25] is found to have state-of-the-art performance in classifying Parkinson's disease (PD) status. However, previous multi-view methods have two restrictions regarding general prediction tasks of neurodegenerative conditions. First, many methods are designed for some specific tasks. If one want to tailor these methods to other tasks, it is necessary to carefully tune the hyperparameters. Second, though some methods learn representations from multi-view brain networks, the learning is guided by some predefined prediction tasks, which may introduce bias to overemphasize a particular modality. As such, the learned embeddings cannot represent multi-modal brain networks comprehensively, and their application to the related analysis in a broader scope is potentially constrained.

To address these problem, we propose to learn unified representations from multi-modal brain networks via unsupervised learning techniques. To extend the generalization ability of the learned representations to different downstream analysis, the representations shall be of *disentanglement* and *proportionality* concerning different modalities. Here, disentanglement refers to the representations encoding salient attributes of data explicitly, which can help the analysis of the prediction tasks and the modalities. Proportionality refers to a balanced contribution to the representations of each modality, which avoids the potential bias on specific modalities. In other words, in our approach the learned representations can fairly convey the information from different modalities and can be exploited by various downstream analysis. More specifically, in this paper we propose a multi-view graph auto-encoder to learn the disentangled graph embeddings from brain networks. We formulate the proportionality-awareness in multi-view representation learning as a network scheduling problem via an analogy between training deep networks and the graph flow problems. The experimental results demonstrate the effectiveness of the proposed method.

2   Methodology

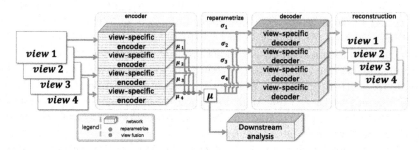

Fig. 1. The structure of the proposed method. Each view uses an independent VGAE to learn a unified μ, while the σ is different.

The proposed method is illustrated in Fig. 1. For each view, a Variational Graph Auto-encoder (VGAE) [10] is exploited. Let $G^{(v)}$ denote the brain networks of the vth view, $f^{(v)}$ and $g^{(v)}$ the corresponding encoder and decoder, $[\mu^{(v)}|\sigma^{(v)}] = f(G^{(v)})$ is the estimated mean and variance of the encoder. The unified representations are computed by max-out the stacked $\mu^{(v)}$ by the position, which can be denoted as $\mu = maxpool1d([\mu^{(v)}])$. The reparameterization for the vth view is then computed using μ and $\sigma^{(v)}$. $\mu \in \mathbb{R}^k$ is also used as the embeddings. According to the structure of VGAE, $\sigma^{(v)} \in \mathbb{R}^k$. Besides the view-wise VGAE loss, we push μ and $\mu^{(v)}$ to be close so that the learned embeddings for different views are consistent. The disentanglement of the representations is acquired via introduce the β-VAE loss [8]. Disentangled representations are compact and interpretable [4]. The objective for our multi-view GVAE is:

$$\mathcal{L} = \sum_{v \in \mathcal{V}} \mathbb{B}\left(log\left(P(\tilde{G}^{(v)})\right)\right) + \beta KL\left(P(z^{(v)}|\mathcal{N}(0,1))\right) + \lambda(\mu^{(v)} - \mu)^2 \quad (1)$$

here the first term is the reconstruction loss, the second is the Kullback-Leibler divergence, and the last is the multi-view consistency.

As aforementioned, the representations shall also be fair to different views. In the above auto-encoder framework, the decoder is used for evaluating the vividness of the learned representations. However, for multi-view data, the reconstruction for different views is not necessarily equally accurate. When the imbalance occurs, some views are less included in the learned representations. To address this problem, we consider to learn fair representations regarding different views, which indicates *the view-wise loss in* (1) *is close to each other*. Such fairness, referred to as *proportionality*, can be achieved via an alternative training routine of the above model. We will formulate an analog between flow network problem and the training of multi-view model in the following. Based on the formulation, we design a scheduling algorithm to satisfy the proportionality requirement.

Training Multi-view Network: A Flow Network Perspective. Directed Acyclic Graph (DAG) is an important tool in graphical models [9]. It is also exploited to express network structures by many popular deep learning frameworks [1,19]. Inspired by this idea, we make an analogy between training the deep network and the flow network problems.

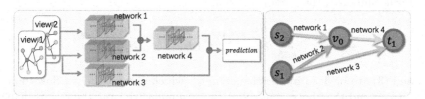

Fig. 2. Left: a simple DNN. Right: the corresponding DAG. Each edge represents a network, and each node denote an intermediate representation.

In Fig. 2, we illustrate an example for multi-view learning. To simplify the elaboration, we consider a structure taking two views s_1 and s_2, as inputs. The network consists of four sub-networks, each corresponding to one edge in the DAG. v_0 is a fused hidden representation, and t_1 is the prediction. For multiple inputs, $\oplus$ denotes the fusion operation for the outputs of multiple sub-networks, and it can either a weighted summation or concatenation. Consider a network trained after t steps using gradient based method. In the $t + 1$ step, we can define the flow $d_{i,j}$ from predecessor i to successor j as $d_{i,j} = \triangle \mathcal{L} \left(f_j^{(t+1)}(h_i^{(t+1)}, \mathcal{H}_{j \backslash i}^{(t)}) \right)$, here $\mathcal{L}$ is an objective defined on the targets, and $\triangle \mathcal{L}$ denotes the loss difference between step $t + 1$ and t. Let $\mathcal{P}_{ij}$ represent the set of all paths from sources to targets containing $e_{i,j}$. $f_j^{(t+1)}$ refers to the network to compute the final outputs with all paths in $\mathcal{P}_{ij}$ updated. $\mathcal{P}_{ij}$ can be defined on the node i and a set $\mathcal{H}_{j \backslash i}$. Here $\mathcal{H}_{j \backslash i}$ denotes any cut set containing node j that separate sources and targets, and $\mathcal{H}_{j \backslash i}$ does not include any node in $\mathcal{P}_{ij}$ except j.

Our definition satisfies the flow conservation, which states that if a node is neither a source or a target, its net flow shall be 0. For a node j with multiple incoming flow, the fusion operation is defined as $h_j = \sum_{i \in \mathcal{P}_j} P_{ij} W_{ij} f_{ij}(h_i)$, here $\mathcal{P}_j$ is the predecessor set of node j, f_{ij} is the sub-network between node i and j. For different fusion operations, P_i and W_i take different forms. For example, when both P_i and W_i are the identity matrices, the fusion is by summation; if W_i is the augmented matrix $(I_i | 0)$, fusion by concatenation is feasible by setting P_i as the corresponding permutation matrix. For a node with multiple outgoing flow, the output is equally distributed. We abuse the notation $\mathcal{H}_j \equiv \mathcal{H}_{j \backslash i} \cup \{i\}$. Consider a fixed given cut $\mathcal{H}_j$ for node j, we can induce two additional cuts: $\mathcal{H}_{\mathcal{P}_j}$, which excludes j and include all its predecessors; and $\mathcal{H}_{\mathcal{S}_j}$, which excludes j and include all its successors. Under the updating rule of backward propagation, the incoming flow with respect to node j is,

$$\sum_{i \in \mathcal{P}_j} d_{i,j} \approx \frac{\partial \mathcal{L}}{\partial f_j} \sum_{i \in \mathcal{P}_j} P_{ij} W_{ij} \frac{\partial f_j}{\partial h_i} dh_i = \frac{\partial \mathcal{L}}{\partial f_j} \frac{\partial f_j}{\partial h_j} dh_j, \tag{2}$$

the above equation follows because the partial differential is 0 except dh_i and dh_j term. Similarly, the outgoing flow is,

$$\sum_{k \in \mathcal{S}_j} d_{j,k} \approx \sum_{k \in \mathcal{S}_j} \frac{\partial \mathcal{L}}{\partial f_k} P_{jk} W_{jk} \frac{\partial f_k}{\partial h_j} dh_j = \frac{\partial \mathcal{L}}{\partial f_j} \frac{\partial f_j}{\partial h_j} dh_j, \tag{3}$$

(2) and (3) are bridged by the change in h_j, which ensures the net flow to be 0.

If we extend the above analogy to the accumulative case, the flow is defined to be the loss decrease with respect to the particular structure represented by $i \rightarrow j$. Noteworthy, it is not the pure contribution of $i \rightarrow j$. Rather, it is more of the quantification of the total loss decrease of the particular structure, as the definition considers both the upstream and downstream computation of the entire network. The empirical loss is related to the generalization bound of the learned representations concerning downstream tasks. As such, the accumulated flow can be interpreted as the amount of information learned from each view informally. Based on this analogy, we define that the proportionality is achieved if the view-wise flow, *i.e.* the accumulated $\sum_{k \in \mathcal{S}_j} d_{j,k}$ for some view j, is balanced.

Alternative Training Routine with Proportionality Awareness. Conventionally, the proportionality concerning different views can be written as a constrained optimization problem, and a standard training routine is based on SGD. From the flow perspective, the proportional training can be interpreted as multiple views competing for the updating resources in the backward propagation, which is a network scheduling algorithm. More specifically, during the training, the accumulated flow is continuously updated, which reflexes the dynamic of loss decrease and the generalization ability. A proportional representation is then equivalent to a balanced flow avoiding the overload of some specific path.

In detail, we define the total flow as the loss decrease. When the learning rate is small enough, the summation of view-wise SGD update is equivalent to a *round-robin* update with respect to each view. Here, the objective associated with each view is optimized in a predefined turn. To avoid a specific view taking up too much updating resources, we can maximize the total flow of the network while allowing minimal level of service for all views via introducing a competing mechanism for each view to occupy the update based on the estimated flow. We refer to this method as *proportionality*. The updating priority of each view is based on the current loss decrease and the historical cumulative loss decrease. Assume the loss decrease of view i at update t can be foreseen as $r_{i,t}$. The throughput of view i is defined as historical cumulative loss decrease at step t:

$$\theta_{i,t} = \theta_{i,0} + \sum_{l=1}^{t} \frac{r_{i,l} I_{i,l}}{t} = \frac{n-1}{n} \theta_{i,t-1} + \frac{1}{n} r_{i,t-1} I_{i,t-1}, \tag{4}$$

where $I_{i,l}$ is an indicator. $I_{i,l} = 1$ if the lth update is conducted on view i, and 0 otherwise. Based on (4), the priority $p_{i,t}$ for view i can be defined, and the $t+1$ update is then applied to the view with the highest priority:

$$\arg\max_{i \in \mathcal{V}} \{p_{i,t}\}, \qquad p_{i,t} = \frac{r_{i,t+1}}{\epsilon + \theta_{i,t}} \tag{5}$$

where ϵ is a small positive number for computational stability. Notably, the above algorithms is not immediately applicable to our formulation, as that $r_{i,t}$ is not pre-assigned as in standard proportionally fairness algorithms. Instead, the values are only known after the update is finished. Thus, we propose a compensation update method: at the beginning, we use one round robin update and compute initial $r_{i,0}$. In the following steps we use proportionally fairness algorithm, but computing the priority using the loss decrease from the last applied update:

$$\arg\max_{i \leq v} \left\{ \frac{r_{i,t_i}}{d_i + \theta_{i,t}} \right\}, \qquad t_i = \max l, \quad s.t. \quad l \leq t, \quad I_{i,l} = 1, \tag{6}$$

The proportionality and convergence of our scheduling algorithm are guaranteed under some weak conditions, and the analysis can be found in [13].

3 Experimental Results

In this experiments we use three datasets, including the data from the Alzheimer's Disease Neuroimaging Initiative (ADNI) and National Alzheimer's Coordinating Center (NACC), and the Parkinson Progression Marker Initiative (PPMI). The preprocessed ADNI brain networks [22] include 51 healthy controls (HC) (mean age = 69.69 ± 10.27, 29 males), 112 people with Mild Cognitive Impairment (MCI) (mean age = 71.68 ± 9.89, 41 males) and 39 individuals with AD (mean age = 75.56 ± 8.99, 14 males). The similarly preprocessed NACC brain networks [21] include 329 HCs (mean age = 60.96 ± 8.96, 107 males), 57 with MCI (mean age = 73.60 ± 7.93, 38 males), and 54 AD patients (mean age = 72.02 ± 10.41, 32 males). The similarly preprocessed PPMI brain networks [27, 28] includes 145 HC (mean age = 66.70 ± 10.95, 96 males) and 474 subjects with PD (mean age = 67.33 ± 9.33, 318 males). Nine different views are reconstructed using T-FACT, T-RK2, T-TL, T-SL, O-FACT and O-RK2, Probt, Hough, and PICo (Please refer to [24] for more details on the brain network reconstruction). We use a modified network structure based on graph variational auto-encoder. The view-wise graph is the averaged brain connectome, and the node features are the corresponding row for each brain connectome. We set $\beta = 4$ recommended by β-VAE [8]. The performance is not sensitive to λ, and we set it to 0.001. In the encoder, we use three graph convolutional layers for μ and σ respectively. The first two layers are shared, both with 64 hidden units. The embedding length is 32. The encoder are limited in layers due to the potential over-smoothing for graph convolutional layers. Our model is trained 100 epochs using ADAM with batch size 32 and learning rate 0.0001.

Table 1. The comparison on classification tasks.

Sparse logistic regression

		ADNI	NACC	PPMI
Single view	FSL	0.7786 ± 0.0976	0.7669 ± 0.0799	$\mathbf{0.6597 \pm 0.0584}$
	PICo	0.7615 ± 0.1408	0.7119 ± 0.1103	0.6065 ± 0.0486
	T-FACT	0.7451 ± 0.0379	0.6581 ± 0.0411	0.5850 ± 0.0433
	O-FACT	0.7278 ± 0.1066	0.7094 ± 0.0866	0.5921 ± 0.0353
	ODF-Rk2	0.7568 ± 0.0821	0.6890 ± 0.0366	0.5942 ± 0.0331
	T-RK2	0.7276 ± 0.0797	0.7281 ± 0.0674	0.5921 ± 0.0353
	T-SL	0.7402 ± 0.1371	0.6582 ± 0.0785	0.5884 ± 0.0389
	T-TL	0.6875 ± 0.0682	0.7358 ± 0.0799	0.5851 ± 0.0423
	Hough	0.7559 ± 0.0780	0.7271 ± 0.0549	0.5536 ± 0.0391
Multi View	All views	0.7966 ± 0.0904	0.7301 ± 0.1325	0.5716 ± 0.0378
	MVNMF	0.8149 ± 0.0550	0.7685 ± 0.0958	0.6104 ± 0.0332
	MVSC	0.8203 ± 0.0791	0.7595 ± 0.1013	0.6205 ± 0.0373
	DMGCN	0.8058 ± 0.1006	0.7557 ± 0.0898	0.6141 ± 0.0707
	Proposed-I	0.8074 ± 0.0493	0.7491 ± 0.0897	0.6122 ± 0.0442
	Proposed-II	0.8185 ± 0.0770	0.7549 ± 0.0790	0.6240 ± 0.0234
	Proposed*	$\mathbf{0.8278 \pm 0.1537}$	$\mathbf{0.8090 \pm 0.1472}$	0.6250 ± 0.0472

Random forest

		ADNI	NACC	PPMI
Single view	FSL	0.8124 ± 0.0455	0.3737 ± 0.7065	0.5753 ± 0.0255
	PICo	0.7838 ± 0.1067	0.1588 ± 0.9463	0.5475 ± 0.0244
	T-FACT	0.8383 ± 0.0483	0.7029 ± 0.1184	0.5654 ± 0.0331
	O-fact	0.7817 ± 0.1512	0.3789 ± 0.6903	0.5478 ± 0.0228
	O-RK2	0.7617 ± 0.1087	0.7879 ± 0.1333	0.5566 ± 0.0382
	T-RK2	0.7764 ± 0.1275	0.7029 ± 0.1333	0.5486 ± 0.0361
	T-SL	0.8148 ± 0.0587	0.7235 ± 0.1163	0.5386 ± 0.0347
	T-TL	0.7695 ± 0.0862	0.7009 ± 0.1164	0.5411 ± 0.0410
	Hough	0.8368 ± 0.0671	0.7011 ± 0.1797	0.5276 ± 0.0344
Multi view	All views	0.8560 ± 0.0574	0.7615 ± 0.1053	0.5743 ± 0.0464
	MVNMF	0.8826 ± 0.0830	0.8317 ± 0.1561	0.5659 ± 0.0528
	MVSC	0.8827 ± 0.0457	0.7997 ± 0.1435	0.5753 ± 0.0348
	DMGCN	0.8862 ± 0.0503	0.8307 ± 0.1493	0.5683 ± 0.0323
	Proposed-I	0.8578 ± 0.0516	0.7919 ± 0.0725	0.5590 ± 0.0250
	Proposed-II	0.8678 ± 0.0573	0.8327 ± 0.0988	0.5699 ± 0.0382
	Proposed*	$\mathbf{0.8946 \pm 0.0510}$	$\mathbf{0.8359 \pm 0.1321}$	$\mathbf{0.5814 \pm 0.0274}$

Evaluating the Proposed Method in Down-Streaming Analysis: We compare our approach with related baselines on several classification and regression tasks. The ablation study is also included.

Table 1 summarizes the classification results. For ADNI and NACC, we predict the HC and AD. For PPMI we predict HC and PD. For multi-view predictions, we include principal component analysis (PCA), multi-view non-negative matrix factorization (MVNMF) [14], co-regularized spectral clustering (MVSC) [12] and Deep Metric Graph Convolutional Network (DMGCN) [11]. We use the aforementioned methods to learn the representations, and then exploit two off-the-shelves methods, sparse logistic regression, and random forest to make the final prediction. We report AUC on 5-fold cross-validation. To make the comparison self-contained, single view results are also included. For the ablation study, in *propose-I* neither disentanglement nor proportionality is considered, and in *proposed-II* the disentanglement is considered. The full approach is *proposed**. We omit more single-view ablation study in the experiments because our objective is designed for multi-view data. Of note, integrating multi-view data is also shown to be beneficial for brain network analysis [27]. From the results, we find that the prediction ability of different views with respect to different tasks are complicated, and heavily coupled with the algorithms. Multi-view methods, generally, can improve the prediction ability. However, the advantage of multi-view data is intriguing and needs careful examination. The proposed method have good performances and are robust with respect to different tasks. And at last, the ablation study demonstrates that the performance can be improved through considering disentanglement and proportionality.

Table 2 summarizes the regression results. We use the learned representation to predict several clinical scores, including the Tremor Dominant scores (TD), the University of Pennsylvania Smell Identification Test (UPSIT), and the Montreal Cognitive Assessment Test (MoCA). Mean squared error (MSE) is used as the metric evaluating the prediction. All scores are normalized to $[0, 1]$. The results show that the prediction is more complicated with respect to the particular medical scores and views. Similarly, we can observe the advantage of utilizing multi-view data and the robust and superior prediction abilities of our approach.

Evaluating the Proportionality During Training: In this section, we demonstrate the proposed method can achieve proportionality using the proposed training scheduling method. Figure 3 illustrates the training loss of the proposed deep network against epochs, and the shaded area represents the variance regarding different views. From the results, we can observe that the proposed method effectively reduces the variance during training, which indicates the learned representations proportionally represent different modalities of brain networks. The results also show the training routine aware of proportionality converges slightly slower than the standard training routine. However, with moderate epochs their performance difference is negligible.

Discussions: There some works applying the fairness principle on brain analysis [16]. Our method is designed for representation learning for multi-view brain connectomes, particularly focusing on the disentangled and proportional prop-

Table 2. The comparison on regression tasks.

		TD	UPSIT	MoCA
Single view	FSL	0.0749 ± 0.0167	0.0794 ± 0.0054	0.0381 ± 0.0033
	PICo	0.0714 ± 0.0078	0.0983 ± 0.0101	0.0394 ± 0.0027
	T-FACT	0.0404 ± 0.0037	0.0545 ± 0.0043	0.0210 ± 0.0021
	O-FACT	0.0410 ± 0.0018	0.0508 ± 0.0066	0.0208 ± 0.0036
	O-RK2	0.0428 ± 0.0079	0.0503 ± 0.0015	0.0208 ± 0.0027
	T-RK2	0.0441 ± 0.0034	0.0500 ± 0.0051	0.0212 ± 0.0025
	T-SL	0.0427 ± 0.0012	0.0512 ± 0.0058	0.0212 ± 0.0017
	T-TL	0.0406 ± 0.0059	0.0517 ± 0.0019	0.0210 ± 0.0027
	Hough	0.0434 ± 0.0036	0.0495 ± 0.0058	0.0225 ± 0.0044
Multi view	All views	0.0414 ± 0.0045	0.0524 ± 0.0034	0.0227 ± 0.0074
	MVNMF	0.0378 ± 0.0122	0.0507 ± 0.0041	0.0207 ± 0.0030
	MVSC	0.0355 ± 0.0047	0.0499 ± 0.0046	$\mathbf{0.0199 \pm 0.0013}$
	DMGCN	0.0365 ± 0.0071	0.0487 ± 0.0085	0.0202 ± 0.0015
	Proposed-I	0.0358 ± 0.0024	0.0501 ± 0.0022	0.0209 ± 0.0019
	Proposed-II	0.0361 ± 0.0035	0.0492 ± 0.0033	0.0200 ± 0.0022
	Proposed*	$\mathbf{0.0351 \pm 0.0059}$	$\mathbf{0.0484 \pm 0.0044}$	$\mathbf{0.0199 \pm 0.0022}$

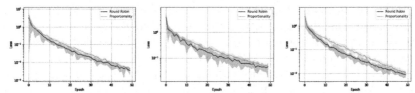

Fig. 3. Left to right: ADNI, NACC, PPMI.

erty (which is related to algorithmic fairness) for the learned embeddings. Our experimental results demonstrate that the proposed method can be applied to various downstream works. As such, it is of potential to apply our method to broader applications, including generating a refined connectome matrix,

4   Conclusion

In this paper, we propose an unsupervised method to learn unified graph embeddings for multi-view brain networks. We design a multi-view graph variational auto-encoder to learn the representations with disentanglement and proportionality. The experimental results demonstrate that the learned representations can be effectively used by various downstream tasks.

Acknowledgements. This work was partially supported by NSF IIS 1845666, 1852606, 1838627, 1837956, 1956002, 2045848, IIA 2040588, and NIH U01AG068057, R01AG049371, R01AG071243, RF1MH125928.

The NACC database was funded by NIA U01AG016976. The ADNI data were funded by the Alzheimer's Disease Metabolomics Consortium (NIA R01AG046171, RF1AG051550 and 3U01AG024904-09S4). The PPMI data were obtained from the Parkinson's Progression Markers Initiative (PPMI) database.

References

1. Abadi, M., et al.: TensorFlow: a system for large-scale machine learning. In: OSDI, pp. 265–283 (2016)
2. Aganj, I., et al.: A Hough transform global probabilistic approach to multiple-subject diffusion MRI tractography. Med. Image Anal. **15**(4), 414–425 (2011)
3. Aminmansour, F. et al.: Learning macroscopic brain connectomes via group-sparse factorization. In: Advances in Neural Information Processing Systems, pp. 8847–8857 (2019)
4. Bengio, Y., et al.: Representation learning: a review and new perspectives. IEEE Trans. Pattern Anal. Mach. Intell. **35**(8), 1798–1828 (2013)
5. Bullmore, E., et al.: Complex brain networks: graph theoretical analysis of structural and functional systems. Nat. Rev. Neurosci. **10**(3), 186–198 (2009)
6. Bullmore, E.T., Bassett, D.S.: Brain graphs: graphical models of the human brain connectome. Ann. Rev. Clin. Psychol. **7**, 113–140 (2011)
7. Caspell-Garcia, C., et al.: Multiple modality biomarker prediction of cognitive impairment in prospectively followed de novo Parkinson disease. PLoS ONE **12**(5), e0175674 (2017)
8. Higgins, I., et al.: beta-VAE: learning basic visual concepts with a constrained variational framework. ICLR **2**(5), 6 (2017)
9. Huang, F., Chen, S.: Learning dynamic conditional gaussian graphical models. IEEE Trans. Knowl. Data Eng. **30**(4), 703–716 (2017)
10. Kipf, T.N., Welling, M.: Variational graph auto-encoders. arXiv preprint arXiv:1611.07308 (2016)
11. Ktena, S.I., et al.: Distance metric learning using graph convolutional networks: application to functional brain networks. In: Descoteaux, M., Maier-Hein, L., Franz, A., Jannin, P., Collins, D.L., Duchesne, S. (eds.) MICCAI 2017. LNCS, vol. 10433, pp. 469–477. Springer, Cham (2017). https://doi.org/10.1007/978-3-319-66182-7_54
12. Kumar, A., Rai, P., Daume, H.: Co-regularized multi-view spectral clustering. In: Advances in Neural Information Processing Systems, pp. 1413–1421 (2011)
13. Kushner, H.J., et al.: Convergence of proportional-fair sharing algorithms under general conditions. IEEE Trans. Wirel. Commun. **3**(4), 1250–1259 (2004)
14. Liu, J., Wang, C., Gao, J., Han, J.: Multi-view clustering via joint nonnegative matrix factorization. In: SIAM International Conference on Data Mining, pp. 252–260. SIAM (2013)
15. Luo, L., Xu, J., Deng, C., Huang, H.: Robust metric learning on Grassmann manifolds with generalization guarantees. In: Proceedings of the AAAI Conference on Artificial Intelligence, vol. 33, pp. 4480–4487 (2019)
16. Moyer, D., Ver Steeg, G., Tax, C.M., Thompson, P.M.: Scanner invariant representations for diffusion MRI harmonization. Magn. Reson. Med. **84**(4), 2174–2189 (2020)

17. Neher, P.F., Götz, M., Norajitra, T., Weber, C., Maier-Hein, K.H.: A machine learning based approach to fiber tractography using classifier voting. In: Navab, N., Hornegger, J., Wells, W.M., Frangi, A.F. (eds.) MICCAI 2015. LNCS, vol. 9349, pp. 45–52. Springer, Cham (2015). https://doi.org/10.1007/978-3-319-24553-9_6

18. Parker, G.J., et al.: A framework for a streamline-based probabilistic index of connectivity (PICo) using a structural interpretation of MRI diffusion measurements. J. Magn. Reson. Imaging **18**(2), 242–254 (2003)

19. Paszke, A. et al.: Automatic differentiation in PyTorch (2017)

20. Poulin, P., et al.: Learn to track: deep learning for tractography. In: Descoteaux, M., Maier-Hein, L., Franz, A., Jannin, P., Collins, D.L., Duchesne, S. (eds.) MICCAI 2017. LNCS, vol. 10433, pp. 540–547. Springer, Cham (2017). https://doi.org/10.1007/978-3-319-66182-7_62

21. Wang, Q., et al.: The added value of diffusion-weighted MRI-derived structural connectome in evaluating mild cognitive impairment: a multi-cohort validation. J. Alzheimer's Dis. **64**(1), 149–169 (2018)

22. Wang, Q., Sun, M., Zhan, L., Thompson, P., Ji, S., Zhou, J.: Multi-modality disease modeling via collective deep matrix factorization. In: Proceedings of the 23rd ACM SIGKDD International Conference on Knowledge Discovery and Data Mining, pp. 1155–1164 (2017)

23. Wang, Q., Zhan, L., Thompson, P.M., Dodge, H.H., Zhou, J.: Discriminative fusion of multiple brain networks for early mild cognitive impairment detection. In: 2016 IEEE 13th International Symposium on Biomedical Imaging (ISBI), pp. 568–572. IEEE (2016)

24. Zhan, L., et al.: Comparison of nine tractography algorithms for detecting abnormal structural brain networks in Alzheimer's disease. Front. Aging Neurosci. **7**, 48 (2015)

25. Zhang, X., et al.: Multi-view graph convolutional network and its applications on neuroimage analysis for Parkinson's disease. In: AMIA Annual Symposium Proceedings, vol. 2018, p. 1147 (2018)

26. Zhang, Y., Huang, H.: New graph-blind convolutional network for brain connectome data analysis. In: Chung, A.C.S., Gee, J.C., Yushkevich, P.A., Bao, S. (eds.) IPMI 2019. LNCS, vol. 11492, pp. 669–681. Springer, Cham (2019). https://doi.org/10.1007/978-3-030-20351-1_52

27. Zhang, Y., Zhan, L., Cai, W., Thompson, P., Huang, H.: Integrating heterogeneous brain networks for predicting brain disease conditions. In: Shen, D., et al. (eds.) MICCAI 2019. LNCS, vol. 11767, pp. 214–222. Springer, Cham (2019). https://doi.org/10.1007/978-3-030-32251-9_24

28. Zhang, Y., Zhan, L., Thompson, P.M., Huang, H.: Biological knowledge guided deep neural network for brain genotype-phenotype association study. In: Zhu, D., et al. (eds.) MBIA/MFCA -2019. LNCS, vol. 11846, pp. 84–92. Springer, Cham (2019). https://doi.org/10.1007/978-3-030-33226-6_10

Quantifying Structural Connectivity
in Brain Tumor Patients

Yiran Wei[1], Chao Li[1,2(✉)], and Stephen John Price[1]

[1] Department of Clinical Neurosciences, University of Cambridge, Cambridge, UK
cl647@cam.ac.uk
[2] Department of Neurosurgery, Shanghai General Hospital, Shanghai Jiao Tong
University School of Medicine, Shanghai, China

Abstract. Brain tumors are characterised by infiltration along the
white matter tracts, posing significant challenges to precise treatment.
Mounting evidence shows that an infiltrative tumor can interfere with the
brain network diffusely. Therefore, quantifying structural connectivity
has potential to identify tumor invasion and stratify patients more accurately. The tract-based statistics (TBSS) is widely used to measure the
white matter integrity. This voxel-wise method, however, cannot directly
quantify the connectivity of brain regions. Tractography is a fiber tracking approach, which has been widely used to quantify brain connectivity.
However, the performance of tractography on the brain with tumors is
biased by the tumor mass effect. A robust method of quantifying the
structural connectivity in brain tumor patients is still lacking.

Here we propose a method which could provide robust estimation
of tract strength for brain tumor patients. Specifically, we firstly construct an unbiased tract template in healthy subjects using tractography. The voxel projection procedure of TBSS is employed to quantify
the tract connectivity in patients, based on the location of each tract
fiber from the template. To further improve the standard TBSS, we propose an approach of iterative projection of tract voxels, under the guidance of tract orientation measured by voxel-wise eigenvectors. Compared
to the conventional tractography methods, our approach is more sensitive in reflecting functional relevance. Further, the different extent of
network disruption revealed by our approach correspond to the clinical
prior knowledge of tumor histology. The proposed method could provide a robust estimation of the structural connectivity for brain tumor
patients.

Keywords: Brain networks · Structural connectivity · Brain tumor ·
Tractography · Diffusion MRI

1 Introduction

1.1 Brain Tumors and White Matter Tracts

A brain tumor refers to a mass lesion identified within the brain or related structures. Among them, gliomas and meningiomas are the most common primary

© Springer Nature Switzerland AG 2021
M. de Bruijne et al. (Eds.): MICCAI 2021, LNCS 12907, pp. 519–529, 2021.
https://doi.org/10.1007/978-3-030-87234-2_49

tumor types in adults. Impacts of tumors are characterized by diffuse infiltration and interference with the white matter tracts [24,26], the neuronal fiber bundles forming a complex network connecting cortical regions. As a result, the tumor infiltration along white matter tracts may lead to structural disturbance of the brain [13].

Brain tumor patients frequently demonstrate neurological deficits that are not directly explained by the focal lesion [8]. Therefore, it is increasingly accepted that brain tumors may cause broader impacts to the global brain beyond the focal site through the white matter tracts [1]. Therefore, accurately measuring structural connectivity strength offers a promising imaging surrogate to detect the tumor-related brain alterations. Further, the pre-treatment mapping of the structural connectivity provides significance for the planning of both surgery and radiotherapy [18].

1.2   Connectivity Strength Measurement

Recent development of neuroimaging techniques have facilitated characterization of brain connectivity at the whole-brain level, based on diffusion MRI (dMRI). The derived fractional anisotropy (FA) map is commonly used to measure the tract strength. The tract-based spatial statistics (TBSS) is a method to estimate the strength of specific tracts using a FA skeleton derived by mapping the local maxima FA voxels to the template [19]. As a voxel-based method, however, TBSS fails to consider the spatial continuity of the fiber pathway, which has difficulties in multiple testing and tracking the fibers that tumor infiltrates along [4]. Further, as the projection procedure in TBSS is purely based on the FA values, the traditional TBSS does not consider the orientation of the tract fibers that are frequently affected in brain tumor patients.

Tractography is a widely-used fiber tracking method to measure the tract connectivity. This method has the advantage of detecting the fiber pathway with spatial consistency. Performing tractography on brain tumor patients, however, has the below challenges: 1) The tract pathways in vicinity to the tumor are often anatomically deviated, which may cause errors in fiber tracking. 2) Many brain tumors are remarkably heterogeneous [10,11,27,28]. Particularly, the edema region surrounding the tumor may cause artefacts in fiber tracking, as the FA value is commonly affected in these regions [17].

1.3   Related Work

The conventional tractography methods include Fiber Assignment by Continuous Tracking (FACT) and Unscented Kalman Filter (UKF) tractography.

FACT Tractography. FACT [14] is a deterministic method that tracks fiber streamlines from a seed region by following the primary eigenvector from one voxel to the next. FACT is highly sensitive to the changes of FA and tensor values in the white matter. As a result, the tracking frequently fails when encountering peritumoral edema, leading to overestimated disruption of the structural networks. In some other cases, it may produce spurious tract rings around the tumor, which may underestimate the connectivity reduction [6].

UKF Tractography. UKF is shown to be able to track inside the peritumoral edema using two tensor models [12]. However, in order to achieve this, the model weakly controls the false positive fibers compared to healthy subjects [6]. The trade-off between false positive and false negative rates is challenging to be optimized among patients, which could particularly cause bias in the group analysis when individual tumors have heterogeneous extent of peritumoral edema.

1.4 Our Contributions

We propose a method of estimating structural connectivity for network construction in brain tumor patients, without directly fiber tracking on patients (Fig. 1). Inspired by a method for traumatic brain injury [20], we firstly performed tractography on healthy controls and generated an unbiased tract template, with the location of each tract fiber derived. The TBSS approach was then employed to derive the skeletonized FA maps from patients, for estimating the connectivity strength of each specific tract in patients. To mitigate the bias in TBSS voxel projection caused by tumor mass effect, we further proposed an improved TBSS tailored to brain tumors, using an iterative projection of FA voxels guided by the tract orientation based on the voxel-wise eigenvector.

We compared the connectivity strength estimated by the proposed approach and the conventional tractography methods in multiple datasets of brain tumor patients. The results show that our approach is reflecting more functional relevance. Further, the different extent of network disruption revealed by our approach correspond to the clinical prior knowledge of tumor histology.

2   Methods

2.1   Datasets

Three datasets of glioma or meningioma patients were included, with both diffusion MRI (dMRI) and resting-state functional MRI (rs-fMRI) available: 1) 4 low-grade gliomas (LGGs), 5 high-grade gliomas (HGGs) and 2 meningiomas (dMRI: 60 directions, b = $1000 \, s/mm^2$; rs-fMRI : TR = 2.5 s) were obtained from [16]; 2) OpenNeuro database: 7 LGGs, 4 HGGs, and 14 meningiomas (dMRI: 101 directions, b = 0, 700, 1200, $2800 \, s/mm^2$; rs-fMRI : TR = 2.4 s) [2]. 3) an in-house dataset: 12 HGGs (dMRI: 12 directions for each b values, b = 0, 350, 650, 1000, 1300, $1600 \, s/mm^2$; rs-fMRI : TR = 2.43 s). In total, 11 LGGs, 21 HGGs and 16 meningiomas were included.

2.2   Template-Based Brain Networks

The proposed method includes three main steps:

- Producing an unbiased tractography template (Fig. 1A),
- Generating individualized FA skeleton (Fig. 1B),
- Combining tractography template with FA skeleton to estimate the connectivity strength between each pair of brain regions in patients

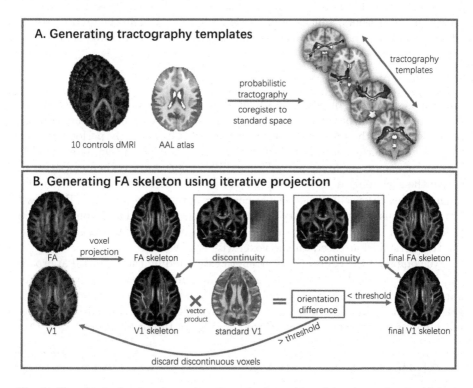

Fig. 1. Flowchart of network construction. A. An unbiased tractography template is generated using probabilistic tractography in ten healthy subjects to indicate the location of the tracts between brain regions parcellated using the brain atlas. B. By using an iterative projection approach that considers the direction of voxels (V1), a skeletonized FA map with maximum tract continuity can be produced. The red arrows indicate the discontinuity in the FA skeleton from the traditional TBSS that is corrected with our proposed method. (Color figure online)

Tractography Template. An unbiased tract template was generated in four steps using ten age-matched elderly healthy controls obtained from the Alzheimer's disease Neuroimaging Initiative (ADNI, http://adni.loni.usc.edu/).

1) A diffusion model was fitted at each voxel of dMRI using Bedpostx of FMRIB software library (FSL, https://fsl.fmrib.ox.ac.uk/fsl/). Pairwise probabilistic tractography was performed between 90 regions of Automated Anatomical Labelling atlas (AAL) atlas [22] using Probtrackx2 to generate a path distribution map in healthy controls [5].
2) The path distribution map was nonlinearly transformed to standard MNI-152 space [7] using the Advanced Normalization Tools (ANTs) [3] and finally averaged, to produce a group distribution map for each tract. The maps were thresholded and binarized to preserve the top 5% strongest connections and generate a conservative tract atlas in the standard space.

Individual FA Skeleton from Iterative Projection. An FA skeleton of each patient was generated at the individualized native space using an improved TBSS approach.

1) FA maps of all patients were first non-linearly co-registered to the standard space using ANTs. In addition, the corresponding principal eigenvector V1 maps were also transformed to the standard space. Instead of using the patient group thinned FA skeleton, we used the standard FA skeleton of FSL as the target for projection to reduce bias introduced by tumor.
2) We further improved TBSS that only projects the voxels with local maximum value at the tract center on the FA skeleton. Specifically, we compared the principal eigenvector (V1) of the projected voxel to the standard V1 and calculated the orientation difference using vector product. The voxels with the orientation difference over 90° were discarded, where neighbor voxels were selected instead. The projection iteration continued until all voxels on the skeleton converged to minimum direction difference with the standard FA skeleton. Using this procedure, the continuity of the voxel directions on each tract could be improved. An individualized FA skeleton was generated in each patient.

Structural Connectivity Strength Estimation. The segments of the individualized FA skeleton within each tract of the template was extracted and averaged, representing the connectivity strengths of the major tracts.

2.3 Baseline Methods

FACT was performed using diffusion toolkits [25], while UKF was performed using the UKFTratography packages in 3D Slicer (https://www.slicer.org/) via the SlicerDMRI project (http://dmri.slicer.org) [15,30]. For both methods, the mean FA value of the tracts connecting two ROIs of AAL atlas was calculated using MRTrix3 by sampling and averaging the FA values along the streamlines generated by the tractography [21].

2.4   Functional Brain Networks

We constructed functional networks from rs-fMRI using GRETNA[23]. Firstly, the rs-fMRI signals were regressed, bandpass-filtered, smoothed with a kernel with a full width at half maximum of 6mm and co-registered to the standard space. Secondly, the brain regions of each patient were parcellated using the AAL atlas. Finally, pairwise Pearson correlation was performed between the mean rs-fMRI signals in 90 brain regions.

2.5   Functional Relevance of the Structural Networks

Whole-Brain Structural and Functional Connectivity Coupling. To measure the individualized functional relevance of a structural network, we calculated the whole brain structural connectivity-functional connectivity (SC-FC) coupling. Spearman rank correlation was performed between the non-zero elements of the structural connections with their corresponding functional connections to produce correlation coefficient for each patient. To compare the group difference, the correlation coefficients were transformed using Fisher Z-Transformation.

Group-Wise Structural-Functional Connection Correlation. The strength of functional connections varies due to the distinct extent of disruptions in corresponding structural connections. By testing the correlation of each functional and structural connection, the functional connections sensitive to structural damage can be identified. To weakly control the family-wise error in multiple comparisons, we used the network-based statistics (NBS) [29], providing higher sensitivity in controlling false discovery rate.

2.6   Clinical Validation

The clinical prior knowledge establishes that meningiomas normally cause less disturbance to the brain than gliomas. A robust network construction method should be sensitive to the difference between meningioma and glioma. We used the global efficiency, calculated according to graph theory, to characterize the brain networks, which were compared in meningioma and glioma patients [9].

3   Results

3.1   Iterative TBSS Projection

The proposed approach showed smaller orientation differences between patient FA skeleton and standard FA, compared to the baseline methods (Table 1). One example is illustrated on Fig. 2. Tracts that are displaced by the tumor have voxels from the wrong directions projected onto the FA skeleton. In comparison, our approach could ensure the maximum orientation continuity of the tracts.

Table 1. Mean orientation differences between patient FA skeleton and standard FA

Traditional TBSS projection (radians)	Iterative TBSS projection (radians)	P-value
0.424 ± 0.057	0.388 ± 0.035	3.8e−4

Table 2. SC-FC coupling coefficient (z-transformed)

Methods	Mean ± SD	P-value (vs FACT)	P-value (vs UKF)
Proposed	0.25 ± 0.07	3.3e−26	3.5e−8
UKF	0.16 ± 0.08	3.8e−10	–
FACT	0.07 ± 0.05	–	–

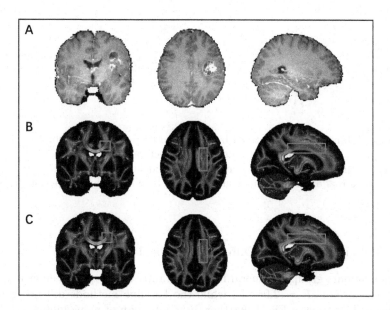

Fig. 2. Example of voxel projection in coronal (left), axial (middle) and sagittal (right) views. A. T1 contrast MRI indicating tumor location. B. Traditional TBSS voxel projection: the tracts surrounding the tumor display different direction (green) from the contralesional tract (blue). C. The proposed method improves the direction continuity of the tract (Color figure online).

3.2   Whole Brain Structural and Functional Connectivity Coupling

The proposed method achieved higher SC-FC coupling over the baseline methods (Table 2), suggesting the robustness of the brain network constructed using the proposed method.

3.3   Group-Wise Structural-Functional Connection Correlation

Only the proposed method identified significantly correlated functional and structural sub-networks across the patient group, suggesting its high sensitivity to functional related structural disruptions (Fig. 3).

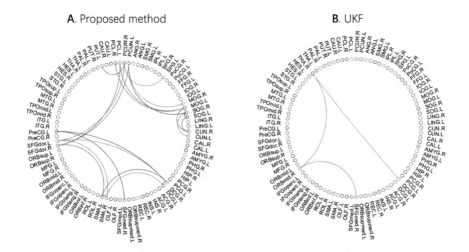

Fig. 3. A. Sub-network is identified by the proposed method, with clusters of significantly correlated structural and functional connections ($P = 0.0014$). B. Fewer correlated structural and functional connections are identified by the UKF ($P > 0.05$). Family wise errors of both methods are corrected by NBS with 5000 permuations.

3.4   Global Efficiency of Different Tumor Types

The meningioma group in general has significantly higher global efficiency than the glioma group ($P = 2.9\mathrm{e}{-}4$), while the baseline methods did not capture significant difference in Fig. 4. Further, in a subgroup of meningioma and high-grade glioma with comparable size, meningioma patients still have significant higher global efficiency comparing to high-grade glioma group ($P = 9.3\mathrm{e}{-}5$), while baseline methods did not show significant difference.

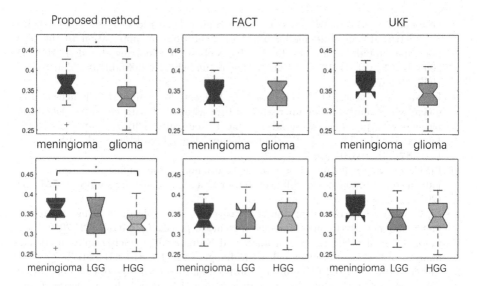

Fig. 4. Global efficiency of the structural networks constructed using the proposed method, FACT, and UKF. Compared to glioma, meningioma patients display significantly higher global efficiency in the network constructed using the proposed approach. $* : P < 0.05$

4   Conclusion

This study proposes an approach to construct structural network and estimate the connectivity, which employs an improved TBSS approach and the tractography template from healthy controls. This method is shown to be robust compared to the conventional tractography, with higher functional and clinical relevance. The proposed method shows promise to aid treatment planning and patient risk assessment.

References

1. Aerts, H., Fias, W., Caeyenberghs, K., Marinazzo, D.: Brain networks under attack: robustness properties and the impact of lesions. Brain **139**(12), 3063–3083 (2016)
2. Aerts, H., Marinazzo, D.: Brain tumor connectomics data (2019). https://doi.org/10.18112/openneuro.ds001226.v1.0.0
3. Avants, B.B., Tustison, N., Song, G.: Advanced normalization tools (ANTS). Insight J. **2**(365), 1–35 (2009)
4. Bach, M., et al.: Methodological considerations on tract-based spatial statistics (TBSS). Neuroimage **100**, 358–369 (2014)
5. Behrens, T.E., Berg, H.J., Jbabdi, S., Rushworth, M.F., Woolrich, M.W.: Probabilistic diffusion tractography with multiple fibre orientations: what can we gain? Neuroimage **34**(1), 144–155 (2007)
6. Chen, Z., et al.: Reconstruction of the arcuate fasciculus for surgical planning in the setting of peritumoral edema using two-tensor unscented Kalman filter tractography. NeuroImage: Clin. **7**, 815–822 (2015)

7. Grabner, G., Janke, A.L., Budge, M.M., Smith, D., Pruessner, J., Collins, D.L.: Symmetric atlasing and model based segmentation: an application to the hippocampus in older adults. In: Larsen, R., Nielsen, M., Sporring, J. (eds.) MICCAI 2006. LNCS, vol. 4191, pp. 58–66. Springer, Heidelberg (2006). https://doi.org/10.1007/11866763_8

8. Incekara, F., Satoer, D., Visch-Brink, E., Vincent, A., Smits, M.: Changes in language white matter tract microarchitecture associated with cognitive deficits in patients with presumed low-grade glioma. J. Neurosurg. **130**(5), 1538–1546 (2018)

9. Latora, V., Marchiori, M.: Efficient behavior of small-world networks. Phys. Rev. Lett. **87**(19), 198701 (2001)

10. Li, C., et al.: Intratumoral heterogeneity of glioblastoma infiltration revealed by joint histogram analysis of diffusion tensor imaging. Neurosurgery **85**(4), 524–534 (2019)

11. Li, C., et al.: Characterizing tumor invasiveness of glioblastoma using multiparametric magnetic resonance imaging. J. Neurosurg. **132**(5), 1465–1472 (2019)

12. Liao, R., et al.: Performance of unscented Kalman filter tractography in edema: analysis of the two-tensor model. NeuroImage: Clin. **15**, 819–831 (2017)

13. Liu, Y., et al.: Altered rich-club organization and regional topology are associated with cognitive decline in patients with frontal and temporal gliomas. Front. Hum. Neurosci. **14**, 23 (2020)

14. Mori, S., Crain, B.J., Chacko, V.P., Van Zijl, P.C.: Three-dimensional tracking of axonal projections in the brain by magnetic resonance imaging. Ann. Neurol. Off. J. Am. Neurol. Assoc. Child Neurol. Soc. **45**(2), 265–269 (1999)

15. Norton, I.: SlicerDMRI: open source diffusion MRI software for brain cancer research. Cancer Res. **77**(21), e101–e103 (2017)

16. Pernet, C.R., Gorgolewski, K.J., Job, D., Rodriguez, D., Whittle, I., Wardlaw, J.: A structural and functional magnetic resonance imaging dataset of brain tumour patients. Sci. Data **3**(1), 1–6 (2016)

17. Schult, T., Hauser, T.K., Klose, U., Hurth, H., Ehricke, H.H.: Fiber visualization for preoperative glioma assessment: tractography versus local connectivity mapping. PLoS ONE **14**(12), e0226153 (2019)

18. Sinha, R., Dijkshoorn, A.B., Li, C., Manly, T., Price, S.J.: Glioblastoma surgery related emotion recognition deficits are associated with right cerebral hemisphere tract changes. Brain Commun. **2**(2), fcaa169 (2020)

19. Smith, S.M., et al.: Tract-based spatial statistics: voxelwise analysis of multi-subject diffusion data. Neuroimage **31**(4), 1487–1505 (2006)

20. Squarcina, L., Bertoldo, A., Ham, T.E., Heckemann, R., Sharp, D.J.: A robust method for investigating thalamic white matter tracts after traumatic brain injury. Neuroimage **63**(2), 779–788 (2012)

21. Tournier, J.D., et al.: MRtrix3: a fast, flexible and open software framework for medical image processing and visualisation. NeuroImage **202**, 116137 (2019)

22. Tzourio-Mazoyer, N., et al.: Automated anatomical labeling of activations in SPM using a macroscopic anatomical parcellation of the MNI MRI single-subject brain. Neuroimage **15**(1), 273–289 (2002)

23. Wang, J., Wang, X., Xia, M., Liao, X., Evans, A., He, Y.: GRETNA: a graph theoretical network analysis toolbox for imaging connectomics. Front. Hum. Neurosci. **9**, 386 (2015)

24. Wang, J., et al.: Invasion of white matter tracts by glioma stem cells is regulated by a NOTCH1-SOX2 positive-feedback loop. Nat. Neurosci. **22**(1), 91–105 (2019)

25. Wang, R., Benner, T., Sorensen, A.G., Wedeen, V.J.: Diffusion toolkit: a software package for diffusion imaging data processing and tractography. In: Proceedings of the International Society for Magnetic Resonance in Medicine, Berlin, vol. 15 (2007)
26. Wei, Y., et al.: Structural connectome quantifies tumor invasion and predicts survival in glioblastoma patients. bioRxiv (2021)
27. Yan, J.L., et al.: Multimodal MRI characteristics of the glioblastoma infiltration beyond contrast enhancement. Thera. Adv. Neurol. Disord. **12**, 1756286419844664 (2019)
28. Yan, J.L., Li, C., van der Hoorn, A., Boonzaier, N.R., Matys, T., Price, S.J.: A neural network approach to identify the peritumoral invasive areas in glioblastoma patients by using MR radiomics. Sci. Rep. **10**(1), 1–10 (2020)
29. Zalesky, A., Fornito, A., Bullmore, E.T.: Network-based statistic: identifying differences in brain networks. Neuroimage **53**(4), 1197–1207 (2010)
30. Zhang, F., et al.: SlicerDMRI: Diffusion MRI and tractography research software for brain cancer surgery planning and visualization. JCO Clin. Cancer Inform. **4**, 299–309 (2020)

Q-space Conditioned Translation Networks for Directional Synthesis of Diffusion Weighted Images from Multi-modal Structural MRI

Mengwei Ren, Heejong Kim$^{(\boxtimes)}$, Neel Dey, and Guido Gerig

Department of Computer Science and Engineering, New York University, New York, NY, USA
heejong.kim@nyu.edu

Abstract. Current deep learning approaches for diffusion MRI modeling circumvent the need for densely-sampled diffusion-weighted images (DWIs) by directly predicting microstructural indices from sparsely-sampled DWIs. However, they implicitly make unrealistic assumptions of static q-space sampling during training and reconstruction. Further, such approaches can restrict downstream usage of variably sampled DWIs for usages including the estimation of microstructural indices or tractography. We propose a generative adversarial translation framework for high-quality DWI synthesis with arbitrary q-space sampling given commonly acquired structural images (e.g., B0, T1, T2). Our translation network linearly modulates its internal representations conditioned on continuous q-space information, thus removing the need for fixed sampling schemes. Moreover, this approach enables downstream estimation of high-quality microstructural maps from arbitrarily subsampled DWIs, which may be particularly important in cases with sparsely sampled DWIs. Across several recent methodologies, the proposed approach yields improved DWI synthesis accuracy and fidelity with enhanced downstream utility as quantified by the accuracy of scalar microstructure indices estimated from the synthesized images. Code is available at https://github.com/mengweiren/q-space-conditioned-dwi-synthesis.

1 Introduction

Diffusion MRI is fundamental to *in vivo* tissue microstructure characterization. Recent dMRI models such as neurite orientation dispersion and density imaging (NODDI) [33] and diffusion kurtosis imaging (DKI) [14] give deeper insight into tissue configurations than traditional indices such as fractional anisotropy.

M. Ren and H. Kim—These authors contributed equally.

Electronic supplementary material The online version of this chapter (https://doi.org/10.1007/978-3-030-87234-2_50) contains supplementary material, which is available to authorized users.

© Springer Nature Switzerland AG 2021
M. de Bruijne et al. (Eds.): MICCAI 2021, LNCS 12907, pp. 530–540, 2021.
https://doi.org/10.1007/978-3-030-87234-2_50

However, advanced dMRI models require dense q-space sampling and prolonged acquisitions, leading to increases in motion corruption and eddy-current and susceptibility artifacts and necessitate outlier DWI exclusion and/or correction.

Several deep network approaches propose to predict high-quality microstructure models such as NODDI/DKI from undersampled DWIs given a training set of fully sampled images [4,5,7,9]. [7,9] use deep networks to estimate DKI and NODDI parameters from only 12 and 8 DWIs, respectively. [5] applies graph convolutional networks for microstructure estimation from 36 DWIs. [20] uses fiber orientation distribution function coefficients from fully sampled images to obtain high-quality tractography from subsampled DWIs. Image translation methods [12,16,34] have also been applied towards prediction of diffusion indices from other commonly co-acquired modalities, including structural/functional MRI to DTI translation [2,29,30] and translation to DTI-derived scalars [1,10]. While promising, current deep network approaches require *static* q-space sub-sampling schemes. Practical DWI processing pipelines [8,24] create unpredictable sampling patterns during motion & eddy-current correction which involves the reorientation of gradients and/or their complete exclusion. As corruptions cannot be anticipated, DWI gradients can be degraded unpredictably and require methods for arbitrary directional restoration.

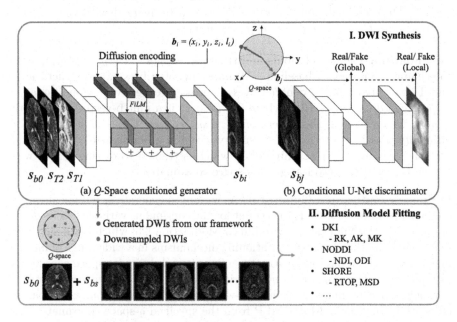

Fig. 1. Framework Overview. (Top) Our q-space conditioned translation framework. **(Bottom)** Once trained, the generator is able to synthesize DWIs along gradients in q-space (blue dots). Merged with arbitrarily downsampled DWIs, various microstructural indices can be calculated with diffusion model fitting. (Color figure online)

We propose a DWI synthesis framework compatible with arbitrary q-space sampling by constructing a generative adversarial translation network to learn mappings between commonly co-acquired structural MRI (e.g., B0, T1, T2) to DWI with user-specified b-values and b-vectors. We incorporate diffusion sampling schemes into the translation by linearly modulating internal network representations with q-space coordinate information. The synthesized DWIs can be merged with sparsely-sampled DWIs to simulate dense q-space acquisition to enable the usage of diffusion models which require high angular resolution. Our contributions include: (1) An adversarial learning framework for structural-to-diffusion MRI translation *independent* of any dMRI model assumption and q-space sampling scheme; (2) Q-space conditioned generator and discriminator architectures which encode diffusion priors into the spatial and spectral translation framework leading to improved visual and structural fidelity over current methodologies; (3) Flexible downstream utility enabling the usage of any microstructural index, tractography strategy, and dODF estimation method.

2   Methods

Figure 1 gives an overview of our framework (yellow outline) alongside its potential downstream utility (gray outline) towards model fitting in downsampled q-space. Once trained, the generated DWIs can be merged with the sparsely-sampled data to simulate dense q-space sampling.

Formulation. The model inputs are structural images $s = (s_{b_0}, s_{T_2}, s_{T_1})$, where the channels represent a baseline ($B0$) image denoted by s_{b_0}, and structural T2 and T1-weighted images, denoted by s_{T_2} and s_{T_1}, respectively. Conditioned on q-space coordinates $\mathbf{b} = (\theta, l)$ where $\theta = (x, y, z)$ is a unit b-vector, and l is the b-value for each shell, the generator $G : \{s_{b_0}, s_{T_2}, s_{T_1}, \mathbf{b}\} \rightarrow s_b$ learns to conditionally translate s to a DWI s_b supervised by a groundtruth DWI alongside a q-space conditioned discriminator D which is trained to distinguish between synthesized and real DWI at given q-space coordinates.

Q-Space Conditioned Generator. As we aim to generate s_b with arbitrary $\mathbf{b}$, our generator network needs to account for the continuous variable b in its generation. We do this via Feature-wise Linear Modulation [25] which has been shown to outperform other network conditioning mechanisms in a wide range of medical imaging tasks [13,26]. Taking structural images and desired q-space coordinates as input, the generator injects q-space information into pre-activation instance-normalized feature maps via a channel-wise linear transformation, whose parameters are learned by a 3-layer MLP from the specified q-space coordinates. The transformation is defined as $\gamma_c^k(\mathbf{b}) \left(\frac{h_c^k - \mu_c^k}{\sigma_c^k} \right) + \beta_c^k(\mathbf{b})$, where h_c^k is the feature map from the cth channel of kth convolutional layer before activation, μ_c^k and $\sigma_{n,c}^k$ are the mean and standard deviation of channel c, and γ_c^k and β_c^k are the scale and shift parameters conditionally learned with fully connected layers from $\mathbf{b}$. Given both structural information and gradient configurations, the generator decodes and upsamples the deep features to a translated dMRI.

Conditional U-Net Discriminator. Existing translation GANs typically use unconditional PatchGAN discriminators [12,34], providing only patch level feedback to the generator. We make two important changes: (1) to provide assessments of both local and *global* image realism for improved synthesis, we use a U-Net discriminator [28] D whose bottleneck discerns global realism and whose output feature determines pixel-level realism; (2) Following recent work in conditional synthesis [22], our discriminator assesses both whether the synthesized image was sampled from the real distribution and whether the synthesized image corresponds to the desired conditioning based on the q-space diffusion vector $\mathbf{b}$. D takes $x = (s_{b0}, s_{T_2}, s_{T_1}, s_{b_i})$ as input, where s_{b_i} is either sampled from a real dMRI or a generated image $G(s_{b0}, s_{T_2}, s_{T_1}, b_i)$, and extracts a global representation $\phi(x)$ via the encoding path D_{enc} to assess global realism. Additionally, a decoder D_{dec} expands $\phi(x)$ to the input size and outputs per-pixel realism feedback. To incorporate q-space coordinates into the discriminator, a conditional projection via inner product is inserted before the last layer of both global and local branches following [22]. The final layer of D is defined as $f(x, \mathbf{b}) := \mathbf{b}^T V \boldsymbol{\phi}(\boldsymbol{x}) + \psi(\boldsymbol{\phi}(\boldsymbol{x}))$, where V is a learnable embedding of condition $\mathbf{b}$; $\phi(x)$ is the output before conditioning, and $\psi(\cdot)$ is a scalar function of $\phi(x)$.

Q-Space Data Augmentation. For improved generalization, we develop two forms of DWI-physics informed augmentation: (1) With input $l = 0$ s/mm^2, the model should produce the baseline image ($B0$). We zero-out the b-value input with probability 0.1 and replace the target image with the $B0$ image; (2) Diffusion signals from spherically antipodal directions are ideally identical. Therefore, we replace θ with $-\theta$ with probability 0.1 and train the generator and discriminator to produce the same output.

Learning Objectives. We use the least squares adversarial objective [21] for D computed from both a local and global perspective as $\mathcal{L}_{GAN}(D) = \mathcal{L}_{GAN}(D_{enc}) + \mathcal{L}_{GAN}(D_{dec})$, where $\mathcal{L}_{GAN}(D_{enc})$ and $\mathcal{L}_{GAN}(D_{dec})$ are defined as,

$$\mathcal{L}_{GAN}(D_{enc}) = \frac{1}{2}\mathbb{E}_y\left[||1 - D_{enc}(y, \mathbf{b}_j)||_2^2\right] + \frac{1}{2}\mathbb{E}_x\left[||D_{enc}(x, \mathbf{b}_i)||_2^2\right],$$

$$\mathcal{L}_{GAN}(D_{dec}) = \frac{1}{2}\mathbb{E}_y\left[\sum_{u,v}||1 - D_{dec}(y, \mathbf{b}_j)_{u,v}||_2^2\right] + \frac{1}{2}\mathbb{E}_x\left[\sum_{u,v}||D_{dec}(x, \mathbf{b}_i)_{u,v}||_2^2\right],$$

where $x = (s_i, G(s_i, b_i)), s_i = (s_{b_{0i}}, s_{T_{2i}}, s_{T_{1i}})$ is the concatenated synthesized tuple of structural inputs and generated dMRI, $y = (s_j, s_{b_j})$ is the real tuple with s_{b_j} sampled from the training data and (u, v) is the image-plane indices of each pixel location. The least-squares generator loss is defined as follows:

$$\mathcal{L}_{GAN}(G) = \frac{1}{2}\mathbb{E}_x\left[||D_{enc}(x, \mathbf{b}_i) - 1||_2^2\right] + \frac{1}{2}\mathbb{E}_x\left[\sum_{u,v}||D_{dec}(x, \mathbf{b}_i)_{u,v} - 1||_2^2\right],$$

To better match low-frequency details and maintain consistency with the input, we further use an intensity-based translation term as below,

$$\mathcal{L}_{L_1}(G) = \begin{cases} \mathbb{E}_{s,b}\left[\|s_b - G(s, \mathbf{b})\|_1\right], & \text{if } l > 0 \\ \mathbb{E}_{s,b}\left[\|s_{b_0} - G(s, \mathbf{b})\|_1\right], & \text{if } l = 0 \end{cases}$$

where $s = (s_{b0}, s_{T2}, s_{T1})$ is the two-channel structural input, s_b is the reference DWI with diffusion $\mathbf{b} = (\theta, l)$. Our complete objective function is summarized as $\mathcal{L}(G, D) = \lambda_{GAN}\mathcal{L}_{GAN}(G, D) + \lambda_{L_1}\mathcal{L}_1(G)$, where λ_{GAN} and λ_{L_1} represent the weights applied to adversarial terms and translation terms, respectively.

3   Experiments

Data and Preprocessing. We experiment on the HCP 500 release [31] including preprocessed images from 19 unrelated subjects. The DWIs are obtained at 3T with b-values of 1000, 2000, and 3000 s/mm^2 covering 270 gradient directions with a voxel size of $1.25 \times 1.25 \times 1.25$ mm^3. The prealigned T1 and T2 images are resampled to match the image resolution of B0 images. We exclude the dataset-provided corrected DWI gradients [3] such that the gradients used for training are raw acquisitions. Consequently, each subject may have a different number of DWIs (ranging from 253 to 269). We use 9 subjects for training; 1 subject for validation and model selection; and 9 held-out subjects for final testing.

Implementation Details. The generator and discriminator learning rates are 10^{-4} and 5×10^{-5}, respectively, using Adam [17] optimization ($\beta_1 = 0.5$, $\beta_2 = 0.999$) and batch size 12. Loss function weights are set to $\lambda_{GAN} = 1$ and $\lambda_{L1} = 100$. During training, the discriminator is updated once for every two generator updates. We train on 2D axial slices sampled from random gradient directions. DWI intensities are rescaled voxel-wise by their corresponding B0 image and b-values are normalized by their maximum value. All models are implemented in PyTorch 1.7.1 and trained on an Nvidia P100 GPU (12 GB vRAM). Inference time is 0.01 s/slice. Further details are elaborated in the appendix.

Diffusion-Weighted Image Synthesis. We compare four models (configurations A, B, C, and D in Table 3) under varying loss functions and input modalities to gauge the benefits of the adversarial loss and multi-modal inputs in both isolation and combination, with a summary of test set metrics calculated in 3D and averaged across subjects given in Table 3. Figure 2 visualizes two example images generated from the four evaluated settings. To assess straightforward translations of B0 images to DWI, Model A is constructed as a single-modality non-adversarial network. As shown in Fig. 2A, Model A can capture overall DWI structure in its translation, but loses significant high-frequency details associated with DWI due to suboptimal objectives [12,19,23]. On top of Model A, we add an adversarial objective for Models B, C, and D with all settings leading to improved visual fidelity as compared to Model A. Compared to Model B (which uses only $B0$ images), Models C and D show progressively enhanced structure by concatenating additional T1 and T2 channels (see yellow insets in Fig. 2) alongside improved PSNR, SSIM and MAE (Table 3). Following this investigation of model configurations, we use Models C and D for further experiments.

Simulated Dense q-Space for Downstream Diffusion Model Fitting.
Here, we evaluate simulated dense q-space sampling by merging synthesized DWIs with the original sparsely-sampled DWIs for the downstream estimation of microstructural indices, tractograms, and dODFs. We randomly downsample in q-space at rate r and evaluate the accuracy of downstream estimations. On held-out test subjects, we downsample in q-space such that $k = (1 - r) * N$ DWIs are retained. As each subject has different numbers of gradient volumes, we assume that the gradient tables are independent across subjects and thus downsample q-space subject-wise with $r = 30\%, 50\%, 70\%, 90\%$, and 95%.

We benchmark our methods against 3 model-free learning frameworks for the estimation of seven different microstructural indices: NDI, ODI (NODDI [11, 33]); AK, MK, RK (DKI [14,32]); and RTOP, MSD (SHORE [6,18]). The derived diffusion maps from both the original DWIs and the synthesized DWIs are calculated and compared with the scalar maps calculated from full DWIs as reference. Compared methods include: (1) q-DL (MLP) [9], (2) q-DL (2D CNN) [7], and (3) a network trained to regress scalar maps from structural MRI inputs following [10,27,29]. (1) and (2) learn scalar maps from the downsampled DWIs, while (3) is implemented as a benchmark to verify our claim that synthesizing DWIs prior to estimating microstructural maps is beneficial both in terms of flexibility and fidelity. To implement (3), we remove all conditioning mechanisms in our

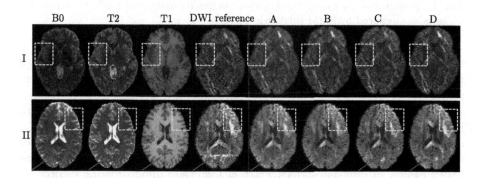

Fig. 2. DWI synthesis results under various model configurations detailed in Table 3. All potential input channels (B0, T1, T2) are visualized on the left. I. Shows translation results with a standard DWI slice in the test set as the reference; II. Visualizes a test slice where artifacts (red arrows) appear in the reference DWI, which are corrected by the generated DWIs from our methods. Readers are encouraged to zoom in. (Color figure online)

Model	Input	Loss	PSNR (↑)	SSIM (↑)	MAE (↓)
A	b0	L1	29.58(±7.31e-1)	0.939(±1.12e-2)	0.0563(±2.62e-3)
B	b0	L1+GAN	29.65(±7.61e-1)	0.939(±1.11e-2)	0.0559(±2.80e-3)
C	b0+T2	L1+GAN	29.64(±7.38e-1)	0.942(±1.04e-2)	0.0558(±2.64e-3)
D	b0+T2+T1	L1+GAN	**29.77(±7.18e-1)**	**0.944(±1.02e-2)**	**0.0550(±2.28e-3)**

Fig. 3. Quantitative DWI synthesis quality on held-out test subjects as measured by PSNR (higher is better), SSIM (higher is better), and MAE (lower is better).

generator and experiment with three input settings (B0 only; B0 and T2; B0, T2, and T1) for generating the desired 7-channel scalar maps.

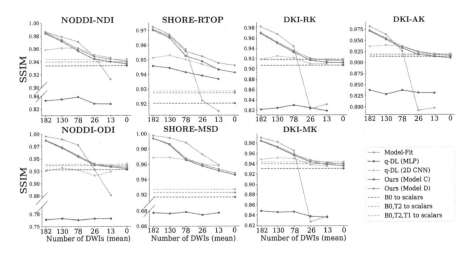

Fig. 4. Quantitative microstructural estimation quality under varying DWI acquisition budgets as measured by SSIM (higher is better).

Figure 4 presents SSIM between scalar maps estimated by the compared methods and the fully-sampled ground truth. As expected, under lower down-sampling rates, standard diffusion model-fits show strong similarity to their fully-sampled references as they are calculated by the same methods. However, with increased downsampling, model-fitting performance drops dramatically, as also qualitatively shown in Fig. 5. q-DL (MLP) [9] does not yield feasible results for this use-case due to its reliance on a predefined downsampling scheme and as MLPs cannot fit diffusion signals which change their input ordering subject-to-subject. Significantly improved performance was obtained with q-DL (CNN) [7]. However, q-DL (CNN) still inaccurately predicts structural details (see yellow box in Fig. 5). The three dashed lines in Fig. 4 present the direct structural MRI to scalar map translation results, with performance gradually increasing with more input modalities. Qualitatively, the produced images lose high-frequency detail as compared to the reference, whereas ours does not. Further, our method

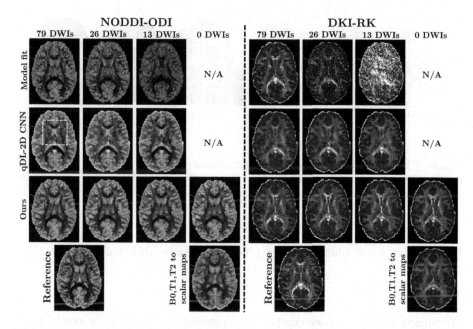

Fig. 5. NODDI (ODI) and DKI (RK) scalar maps estimated by different methods from arbitrarily downsampled DWIs. As the number of DWIs decreases, model-fitting results show strong quality degradation and qDL-2D produces inaccurate structural details (see yellow insets), whereas our method preserves microstructure with high fidelity (see external capsule/red arrows). Under fully synthetic settings (0 DWIs), our result displays improved fidelity compared to direct B0, T2, T1 to scalar translation. (Color figure online)

possesses increased generalizability across arbitrary downsampling schemes and enables enhanced scalar map estimation with standard model fits.

To summarize, at low downsampling rates, our method achieves competitive results with model-fitting and as the downsampling rate increases, our method generally outperforms other methods. When fully synthetic DWIs are used for generating scalar maps (the 0 DWI setting), our method achieves competitive or better SSIM as compared to the scalar translation framework, alongside significantly improved qualitative reconstruction as shown in Fig. 5. We note that while we use SSIM to quantify image similarity, it may not correlate exactly with human perception [23].

To demonstrate further downstream utility beyond microstructural indices, we calculate tensor-based fiber tractography [15] and SHORE-based diffusion orientation distribution functions (dODF) [18] for a randomly selected test subject. Figure 6 presents qualitative results from downsampled DWIs and the q-space restored DWIs in comparison to the fully-sampled reference. While the downsampled results are noisy and inaccurate, the results from our method yield higher fidelity and coherence with respect to the reference.

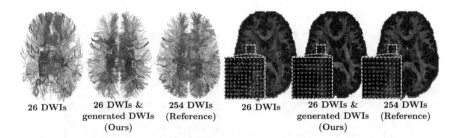

| 26 DWIs | 26 DWIs & generated DWIs (Ours) | 254 DWIs (Reference) | 26 DWIs | 26 DWIs & generated DWIs (Ours) | 254 DWIs (Reference) |

Fig. 6. Whole brain fiber-tractography and estimated dODFs overlaid on generalized fractional anisotropy maps calculated from 26 sparsely-sampled DWIs (left), the results from our framework (center), and results from fully-sampled DWIs (right).

Supplemental animations including synthetic DWIs on continuously interpolated q-space are available at https://heejongkim.com/dwi-synthesis.

4 Conclusion

We present a novel q-space conditioned translation framework for synthesizing high-fidelity DWIs at arbitrary q-space coordinates from multi-modal structural MRIs. The model outputs can be merged with downsampled DWIs to simulate dense q-space sampling, enabling the usage of advanced diffusion models which require high angular resolution. Some open questions exist, as we train and evaluate on a healthy population under relatively controlled imaging settings. As different imaging modalities may reveal disparate anatomical structure and diffusion properties (e.g., via lesions), the proposed framework requires further validation on larger and more neuroanatomically diverse cohorts. Additionally, real-world considerations such as inter-modality misregistration, resolution differences, and skull stripping discrepancies in multi-modal structural inputs may further impede translation performance. Fortunately, reasonable performance is achieved even when only a single structural modality is available.

Acknowledgements. Work supported by NIH grants 1R01DA038215-01A1, R01-HD 055741-12, 1R01HD088125-01A1, 1R01MH118362-01, 1R34DA050287, R01MH122447, and R01ES032294.

References

1. Alexander, D.C., et al.: Image quality transfer and applications in diffusion MRI. NeuroImage **152**, 283–298 (2017)
2. Anctil-Robitaille, B., Desrosiers, C., Lombaert, H.: Manifold-aware cycleGAN for high resolution structural-to-DTI synthesis. arXiv preprint arXiv:2004.00173 (2020)
3. Andersson, J.L., Sotiropoulos, S.N.: Non-parametric representation and prediction of single-and multi-shell diffusion-weighted MRI data using gaussian processes. Neuroimage **122**, 166–176 (2015)

4. Chen, G., Dong, B., Zhang, Y., Lin, W., Shen, D., Yap, P.T.: XQ-SR: Joint XQ space super-resolution with application to infant diffusion MRI. Med. Image Anal. **57**, 44–55 (2019)

5. Chen, G., et al.: Estimating tissue microstructure with undersampled diffusion data via graph convolutional neural networks. In: Martel, A.L., et al. (eds.) MICCAI 2020. LNCS, vol. 12267, pp. 280–290. Springer, Cham (2020). https://doi.org/10.1007/978-3-030-59728-3_28

6. Garyfallidis, E., et al.: Dipy, a library for the analysis of diffusion MRI data. Front. Neuroinform. **8**, 8 (2014)

7. Gibbons, E.K., et al.: Simultaneous NODDI and GFA parameter map generation from subsampled q-space imaging using deep learning. Magn. Reson. Med. **81**(4), 2399–2411 (2019)

8. Glasser, M.F., et al.: The minimal preprocessing pipelines for the human connectome project. Neuroimage **80**, 105–124 (2013)

9. Golkov, V., et al.: q-space deep learning: twelve-fold shorter and model-free diffusion MRI scans. IEEE Trans. Med. Imaging **35**(5), 1344–1351 (2016)

10. Gu, X., Knutsson, H., Nilsson, M., Eklund, A.: Generating diffusion MRI scalar maps from T1 weighted images using generative adversarial networks. In: Felsberg, M., Forssén, P.-E., Sintorn, I.-M., Unger, J. (eds.) SCIA 2019. LNCS, vol. 11482, pp. 489–498. Springer, Cham (2019). https://doi.org/10.1007/978-3-030-20205-7_40

11. Harms, R.L., Fritz, F., Tobisch, A., Goebel, R., Roebroeck, A.: Robust and fast nonlinear optimization of diffusion MRI microstructure models. Neuroimage **155**, 82–96 (2017)

12. Isola, P., Zhu, J.Y., Zhou, T., Efros, A.A.: Image-to-image translation with conditional adversarial networks. In: CVPR (2017)

13. Jacenków, G., O'Neil, A.Q., Mohr, B., Tsaftaris, S.A.: INSIDE: steering spatial attention with non-imaging information in CNNs. In: Martel, A.L., et al. (eds.) MICCAI 2020. LNCS, vol. 12264, pp. 385–395. Springer, Cham (2020). https://doi.org/10.1007/978-3-030-59719-1_38

14. Jensen, J.H., Helpern, J.A., Ramani, A., Lu, H., Kaczynski, K.: Diffusional kurtosis imaging: the quantification of non-gaussian water diffusion by means of magnetic resonance imaging. Magn. Reson. Med. Off. J. Int. Soc. Magn. Reson. Med. **53**(6), 1432–1440 (2005)

15. Jiang, H., Van Zijl, P.C., Kim, J., Pearlson, G.D., Mori, S.: DtiStudio: resource program for diffusion tensor computation and fiber bundle tracking. Comput. Meth. Programs Biomed. **81**(2), 106–116 (2006)

16. Karras, T., Laine, S., Aila, T.: A style-based generator architecture for generative adversarial networks. In: Proceedings of the IEEE/CVF Conference on Computer Vision and Pattern Recognition, pp. 4401–4410 (2019)

17. Kingma, D.P., Ba, J.: Adam: a method for stochastic optimization. arXiv preprint arXiv:1412.6980 (2014)

18. Koay, C., Basser, P.: Simple harmonic oscillator based estimation and reconstruction for one-dimensional q-space MR. In: Proceedings of the ISMRM, vol. 16, p. 35 (2008)

19. Larsen, A.B.L., Sønderby, S.K., Larochelle, H., Winther, O.: Autoencoding beyond pixels using a learned similarity metric. In: International Conference on Machine Learning, pp. 1558–1566. PMLR (2016)

20. Lin, Z., et al.: Fast learning of fiber orientation distribution function for MR tractography using convolutional neural network. Med. Phys. **46**(7), 3101–3116 (2019)

21. Mao, X., Li, Q., Xie, H., Lau, R.Y.K., Wang, Z.: Multi-class generative adversarial networks with the L2 loss function. CoRR arXiv:1611.04076 (2016)
22. Miyato, T., Koyama, M.: cGANs with projection discriminator. CoRR arXiv:1802.05637 (2018)
23. Nilsson, J., Akenine-Möller, T.: Understanding SSIM. arXiv preprint arXiv:2006.13846 (2020)
24. Nir, T.M., et al.: Effectiveness of regional DTI measures in distinguishing Alzheimer's disease, MCI, and normal aging. NeuroImage: Clin. **3**, 180–195 (2013)
25. Perez, E., Strub, F., De Vries, H., Dumoulin, V., Courville, A.: FiLM: visual reasoning with a general conditioning layer. In: Proceedings of the AAAI Conference on Artificial Intelligence, vol. 32 (2018)
26. Ren, M., Dey, N., Fishbaugh, J., Gerig, G.: Segmentation-renormalized deep feature modulation for unpaired image harmonization. IEEE Trans. Med. Imaging **40**(6), 1519–1530 (2021)
27. Schilling, K.G., et al.: Synthesized b0 for diffusion distortion correction (Synb0-DisCo). Magn. Reson. Imaging **64**, 62–70 (2019)
28. Schonfeld, E., Schiele, B., Khoreva, A.: A U-Net based discriminator for generative adversarial networks. In: IEEE Computer Vision and Pattern Recognition (2020)
29. Son, S.J., Park, B.Y., Byeon, K., Park, H.: Synthesizing diffusion tensor imaging from functional MRI using fully convolutional networks. Comput. Biol. Med. **115**, 103528 (2019)
30. Tian, Q., et al.: DeepDTI: high-fidelity six-direction diffusion tensor imaging using deep learning. NeuroImage **219**, 117017 (2020)
31. Van Essen, D.C., et al.: The WU-Minn human connectome project: an overview. Neuroimage **80**, 62–79 (2013)
32. Veraart, J., Sijbers, J., Sunaert, S., Leemans, A., Jeurissen, B.: Weighted linear least squares estimation of diffusion MRI parameters: strengths, limitations, and pitfalls. Neuroimage **81**, 335–346 (2013)
33. Zhang, H., Schneider, T., Wheeler-Kingshott, C.A., Alexander, D.C.: NODDI: practical in vivo neurite orientation dispersion and density imaging of the human brain. NeuroImage **61**(4), 1000–1016 (2012)
34. Zhu, J.Y., Park, T., Isola, P., Efros, A.A.: Unpaired image-to-image translation using cycle-consistent adversarial networks. In: The IEEE International Conference on Computer Vision, October 2017

Clinical Applications - Neuroimaging - Functional Brain Networks

Detecting Brain State Changes by Geometric Deep Learning of Functional Dynamics on Riemannian Manifold

Zhuobin Huang[1], Hongmin Cai[1], Tingting Dan[1(✉)], Yi Lin[2], Paul Laurienti[3], and Guorong Wu[2,4]

[1] School of Computer Science and Engineering, South China University of Technology, Guangzhou 510006, China
[2] Department of Psychiatry, University of North Carolina at Chapel Hill, Chapel Hill, NC 27599, USA
[3] Department of Radiology, Wake Forest School of Medicine, Winston Salem, NC 27157, USA
[4] Department of Computer Science, University of North Carolina at Chapel Hill, NC 27599, USA

Abstract. Functional resonance magnetic imaging (fMRI) technology has been widely used in understanding cognition and behavior by characterizing the functional interaction between distant brain regions. Since the network topology of functional connectivity (FC) often dynamically shifts along with the change of brain states, it is challenging to identify the change point (transition between tasks) of functional connectivity without requiring prior knowledge of experiment settings. Although striking efforts have been made to detect changes on BOLD (blood-oxygen-level-dependent) signals, little attention has been paid to characterize the trajectory of whole-brain functional connectivity, which is more closely correlated to brain state change. Since FC is essentially a symmetric positive definite (SPD) correlation matrix, we present a change point detection network (CPD-Net) tailored to (1) learn the low-dimensional geometric feature representations of whole-brain functional connectivity on the Riemannian manifold of SPD matrices, and (2) automatically detect the brain state changes on the unseen functional neuroimages. It is worth noting that our CPD-Net is a manifold-based neural network to the extent that we leverage the alignment between the known functional tasks and the stratification underlying the learned low-dimensional FC feature representation on the Riemannian manifold of SPD matrices to steer the learning of geometric patterns from functional brain networks. We have evaluated the accuracy and replicability of our CPD-Net on task-based fMRI data from HCP (human connectome project) database, where our manifold-based CPD-Net achieves more accurate and consistent results than current learning-based CPD methods.

Keywords: Deep neural network · Symmetric positive definite matrix · Riemannian manifold · Change detection · Functional MRI

© Springer Nature Switzerland AG 2021
M. de Bruijne et al. (Eds.): MICCAI 2021, LNCS 12907, pp. 543–552, 2021.
https://doi.org/10.1007/978-3-030-87234-2_51

1 Introduction

Understanding how cognition and behavior emerge from brain functions is a fundamental scientific problem in the field of neuroscience. In the last couple of decades, striking efforts have been made to deliver structural and functional brain mapping *in-vivo* using cutting-edge neuroimaging technology [1]. Recently, the research focus of fMRI studies has been shifted to functional dynamics since the cognitive states frequently oscillate even in a short period of resting time [2]. In light of this, characterizing functional dynamics from the observed BOLD (blood-oxygen-level-dependent) signals becomes an important step to understand the psychophysiological mechanism of cognition and discover disease-relevant neuroimaging biomarkers. Due to the practical limitations in fMRI, such as low signal-to-noise level and the indirect proxy of neural activity, it is difficult to detect the change of cognitive states without knowing the setting of tasks in fMRI experiments.

To address this challenge, various change point detection (CPD) approaches have been proposed in the literature. Generally speaking, current state-of-the-art change detection methods can be categorized into either supervised or unsupervised scenarios. Regarding the supervised CPD methods, the learning component is often trained to establish the mapping between the functional neuroimaging data and the cognitive states. For example, the Bayesian-based statistical inference has been used to partition the time course into segments by modeling the statistics of BOLD signals and the temporal transition probability [3, 4]. A recurrent neural network (RNN) has been proposed in [5] by considering the brain state change detection as a classification problem where the RNN is trained to predict the task label based on the functional connectivity (FC) signatures vectors generated through non-negative matrix decomposition [6].

Regarding the unsupervised CPD approach, it is a common practice to cast the change detection into a data clustering process. The working premise is that time points associated with the same functional tasks are supposed to fall into the same cluster. Although most of the current unsupervised CPD methods adapt the clustering technique in computer vision and machine learning areas, many efforts have been made to contrive the putative and low-dimensional feature representation for each time point, which describes the functional connectivity profile within the underlying sliding window. For example, the FC matrix in each sliding window is often vectorized into a data array and then stratified into several groups using classic clustering techniques, such as k-means and locally linear embedding (LLE) [7]. By doing so, the whole scanning period can be partitioned into several segments, where each segment is supposed to correspond to a specific cognitive task during fMRI scan. Recently, the graph learning technique has been proposed to construct the subspace of individual fMRI data, which is jointly spanned by Fourier bases and the Eigen bases of graph Laplacian [8]. After that, the spectrum coefficients after projecting BOLD time course into the learned subspace are used to stratify time points into different clusters.

The functional brain network is fundamentally a covariance matrix often assessed by statistical measures such as correlation [9]. In this context, FC matrix essentially resides on a Riemannian manifold of SPD matrix [10]. Despite the rich mathematical insight of SPD manifold, such intrinsic geometry of functional network data is not well investigated in functional brain network analysis, including the state change point detection. In this paper, we present a manifold-based change point detection network (called CPD-Net) to (1) learn the low-dimensional geometric feature representation (called FC fingerprint) on the SPD manifold and (2) stratify the time points into different functional tasks through a sequence of learned mean-shift kernels. Specifically, the input to our CPD-Net is a set of FC matrices (Fig. 1(c)), which is constructed using the sliding window technique (centered at the underlying time point), based on the mean time course of BOLD signal (Fig. 1(b)) from task fMRI scans (Fig. 1(a)). To perverse the intrinsic geometry in FC matrices, we adapt a manifold-based deep neural network (DNN) to boil down the FC matrices from high-dimensional SPD manifold to the FC fingerprints in a low-dimensional vector space (displayed in Fig. 1(d) and called SPD-DNN). Furthermore, we integrate a recurrent neural network (RNN) of mean-shift to find the modes of leaned FC fingerprints (Fig. 1(e)), where the difference between identified modes and brain states (ground truth) is used to steer the feature representation learning and kernel density estimation in the iterative process of mean-shift (called MS-RNN, Fig. 1(f)). We have evaluated our manifold-based CPD-Net on working memory fMRI data from HCP database. Experiments and comparisons with the current learning-based CPD methods demonstrate that our manifold-based CPD-Net achieves more accurate detection results with the enhanced replicability on the test-retest experiments, which indicates the great applicability of our CPD-Net in neuroscience studies.

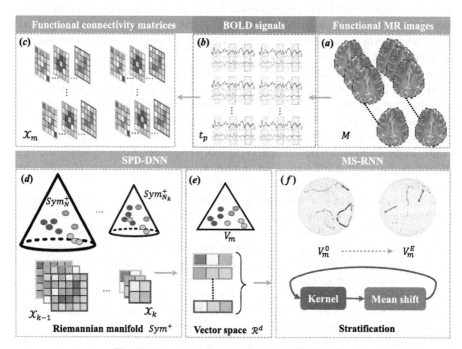

Fig. 1. Overview of our manifold-based CPD-Net.

2 Method

Suppose the training data consist of M task fMRI scans (Fig. 1 (a)). Each scan has been processed into N mean time courses of BOLD signal with P time points (Fig. 1 (b)), where N denotes the number of brain parcellations. We use the sliding window technique to capture functional dynamics. Specifically, we construct the $N \times N$ FC matrix at the underlying time point t_p ($p = 1, \ldots, P$) based on the BOLD signal within the time window w_p, where the sliding window center is t_p. Thus, the functional dynamics for each scan can be encoded in a sequence of FC matrices $\mathbb{X}_m = \{ \mathcal{X}_m(t_p) | p = 1, \ldots, P \}$ ($m = 1, \ldots, M$), as shown in Fig. 1(c).

There are Q cognitive tasks $\Gamma = \{ \gamma_q | q = 1, \ldots, Q \}$ in each scan. Since we know the functional task schedule of fMRI scan for training data, we can associate each $\mathcal{X}_m(t_p)$ with a task label $y_m(t_p) \in \Gamma$. In this paper, we present our manifold-based CPD-Net to predict the stratification result based on the temporal-evolving FC matrices of the new instance of fMRI scan. The overall network architecture is shown in Fig. 1(d)-(f), which consists of two major learning components described below.

2.1 Deep Neural Network of SPD Manifold for Feature Representation Learning

To preserve the intrinsic geometric properties of FC matrix, we exploit a Riemannian network inspired by [11] to (1) learn the low-dimensional geometric feature representation on the Riemannian manifold of SPD matrix, and (2) project each FC matrix from the Riemannian manifold into a vector space which allows us to stratify functional tasks using the classic Euclidean operations.

In a nutshell, the Riemannian network for SPD matrix works the sense of convolutional neural network (CNN), except the convolution and ReLU (rectified linear units) operations in CNN have been replaced with manifold algebra. In what follows, we use $\mathcal{F}_\Theta(\mathcal{X})$ to denote the SPD-DNN (with the network parameters Θ) for inferring the task-specific feature vector $v \in \mathcal{R}^d$ from the high dimensional input $\mathcal{X} \in Sym_N^+$, where Sym_N^+ stands for the Riemannian manifold of $N \times N$ SPD matrices. Since we apply the same SPD-DNN $\mathcal{F}_\Theta$ to all $\mathcal{X}_m(t_p)$ one after another, we drop the variable t_p (for time) and subscript m (for subject) in Sect. 2.1, for clarity.

As shown in Fig. 1(d-e), the non-linear dimension reduction in SPD-DNN is achieved by alternating the following two steps. Suppose the SPD matrix has been reduced from Sym_N^+ to $Sym_{N_{k-1}}^+$ before starting the kth ($k = 1, \ldots, K$) iteration. We first find a bilinear mapping W_k to further reduce the dimensionality of SPD matrix from $N_{k-1} \times N_{k-1}$ to $N_k \times N_k$ by $\mathcal{X}_k = f_b(\mathcal{X}_{k-1}) = W_k \mathcal{X}_{k-1} W_k^T$, where $W_k \in \mathcal{R}^{N_k \times N_{k-1}}$ ($N_k < N_{k-1}$). Similar to the ReLU layer in CNN, the non-linearity is obtained by a hard-thresholding smoothing process in the spectrum domain of the underlying $\mathcal{X}_k$ by $\mathcal{X}_k = f_r(X_k) = U_k max(\varepsilon I, \Lambda_k) U_k^T$, where U_k and Λ_k are eigenvectors and the diagonal matrix of corresponding eigenvalues, respectively. ε is a pre-defined scalar, which controls the regularization of rectifying smaller eigenvalues. At the end of Kth iteration, the Log-Euclidean operator [12] is adapted to project $\mathcal{X}_K \in Sym_{N_K}^+$ from the Riemannian manifold of SPD matrix to the d-length vector $v \in \mathcal{R}^d$ by $v = f_g(\mathcal{X}_K) = U_K log(\Lambda_K) U_K^T$, where $log(\Lambda_K)$ is the diagonal matrix of eigenvalue logarithms.

It is clear that the network hyper-parameters Θ of SPD-DNN consists of a set of mapping matrices, i.e., $\Theta = \{W_k | k = 1, \ldots, K\}$. Given Θ, we can project the sequence of subject-specific FC matrices $\mathbb{X}_m$ from Sym_N^+ manifold to the stack of vectors $V_m = [v_m(t_p)]_{p=1}^P$, where each column $v_m(t_p)$ represents the fingerprint of functional connectivity at the time t_p for m^{th} subject. Since our goal is to stratify Q cognitive tasks based on V_m for the new fMRI scan, we expect the FC fingerprint $v_m(t_p)$ has high similarity to other $v_m(t_{p'})$ ($p' \neq p$) if t_p and $t_{p'}$ are associated with the same cognitive task, i.e., $y_m(p) = y_m\left(p'\right)$. Otherwise, their similarity $s_m^{pp'} = s(v_m(t_p), v_m(t_{p'}))$ ($0 \leq s_m^{pp'} \leq 1$) is required to be as small as possible. In light of this, the loss function ℓ for learning task-specific FC fingerprint is given by:

$$\ell = \sum_{m=1}^M \sum_{p,p'=1}^P \left(1_{\{y_m(p)=y_m(p')\}}\left(1 - s_m^{pp'}\right) + 1_{\{y_m(p)\neq y_m(p')\}}\left[s_m^{pp'} - \alpha\right]_+ \right) \quad (1)$$

where the scalar α controls the maximum margin for negative pairs of time points that bear with different functional tasks.

Although we can learn the task-specific FC fingerprints by finding the best network parameter Θ that minimizes the loss function ℓ, a post hoc process (such as clustering) is required to stratify the functional task based on the learned FC fingerprints. In doing so, the gap between feature representation learning and task stratification often yield a sub-optimal result. To solve this issue, we present an integrated solution by concatenating SPD-DNN with a recurrent neural network of mean-shift (MS-RNN), which iteratively groups the learned FC fingerprints V_m into a number of meaningful functional tasks on the vector space $\mathcal{R}^d$.

2.2   Recurrent Neural Network of Mean-Shift for Functional Task Stratification

Mean-shift is a non-parametric density estimation technique. The backbone of mean-shift is a set of kernels $h_\delta\left(v_m(t), v_m(t_p)\right) = \frac{1}{\sqrt{(2\pi)^d}}exp\left(-\frac{\delta^2 \cdot \|v_m(t)-v_m(t_p)\|_2^2}{2}\right)$, which allows us to delineate the density function of V_m on the vector space $\mathcal{R}^d$ as $\varphi_m(t) = \frac{1}{P}\sum_{p=1}^P h_\delta(v_m(t), v_m(t_p))$. The kernel width δ is the network parameter that controls the smoothness of the density function $\varphi_m(t)$. Given a set of FC fingerprints V_m of mth subject, the discretized process of mean-shift iteratively repeat the following three steps: (1) construct subject-specific kernel matrix $H_m = \psi_h(V_m) = \exp(\delta^2 V_m^T V_m)$; (2) calculate the mean-shift $Z_m = (1 - \eta)I + \eta H_m D^{-1}$, where η is the step size and $D = diag(H_m^T 1)$ is the diagonal matrix of the row-wise sum of H_m; and (3) update V_m by $V_m = \psi_z(V_m) = V_m Z_m$.

It is clear that the iterative process of mean-shift can be formulated into a recurrent neural network (called MS-RNN in short) Ψ_Δ, where network hyper-parameters include the kernel width $\{\delta_e | e = 1, \ldots, E\}$ in total E iterations. As shown in Fig. 1(f), the distribution of FC fingerprints V_m^E upon the last iteration of mean-shift becomes more and more concentrated at the latent modes than V_m^0 (input to MS-RNN) right after the SPD-DNN.

Furthermore, we require the stratification of FC fingerprints to be matched with the functional tasks. Thus, the loss function of our manifold-based CPD-Net is shown in Eq. (1), where the pair-wise similarity matrix is measured by $s_m^{pp'} = \frac{1}{2}\left(1 + \frac{[v_m(t_p)]^T v_m(t_{p'})}{\|v_m(t_p)_2\|\|v_m(t_{p'})_2\|}\right)$.

2.3 Back-Propagation to Solve the Network Minimization

Given $\mathbb{X}_m$, the output to CPD-Net is $V_m^E = \Psi_\Delta(\mathcal{F}_\Theta(\mathbb{X}_m))$. Specifically, the function of SPD-DNN can be formulated as $\mathcal{F} = f_g \circ \underbrace{(f_r \circ f_b) \circ \ldots \circ (f_r \circ f_b)}_{K}$. Likewise, the function of MS-RNN is boiled down to $\Psi = \underbrace{(\psi_z \circ \psi_h) \circ \ldots \circ (\psi_z \circ \psi_h)}_{E}$. Since the loss function (in Eq. 1) eventually examines the mis-stratification based on V_m^E, the back-propagation process starts from the gradient $\frac{\partial l}{\partial V_m^E} = \frac{\partial l}{\partial S} V_m^E$, where $\frac{\partial l}{\partial S} = \frac{1}{P^2} 1_{P \times P} \odot (1_{P \times P} - 2G)$, where l is the average loss of each subject, G denotes binary indicator matrix, if $y_m(p) = y_m(p')$, $G_{pp'} = 1$, otherwise, $G_{pp'} = 0$. Following the gradient chain in [13], it is straightforward to obtain the gradient $\frac{\partial l}{\partial V_m^0}$ at the beginning of MS-RNN. The network flow is $\mathbb{X}_m \to f_b \to f_r \to f_b \to f_r \to f_b \to f_g \to V_m^0 \to \psi_h \to \psi_z \to \psi_h \to \psi_z \to \psi_h \to \psi_z \to V_m^E$. Given $\frac{\partial l}{\partial V_m^0}$, we follow the manifold-based operations in [11] to continue the back-propagation in SPD-DNN until we reach $\frac{\partial l}{\partial W_1}$. We use stochastic gradient descent algorithm to optimize network hyperparameters Δ and Θ sequentially.

In general, we consider each functional brain network as a data instance in the Riemannian manifold of SPD matrix. Furthermore, we cast the change detection problem as a clustering process on the Riemannian manifold. In a nutshell, we can adapt the iterative mean-shift procedure from Euclidean space to Riemannian manifold using the well-studied Lie-group algebra. By doing so, our CPD-Net is able to associate the functional brain networks with the pre-scheduled cognitive tasks on the Riemannian manifold.

3   Experiments

In this section, we evaluated the performance of the proposed CPD-Net for detecting change points on real task-based fMRI data from the HCP (Human Connectome Project) [14]. We select a total of 743 subjects (splitting 378 training set, 47 validation set, 318 testing set) from the HCP database, each with test and retest fMRI scans of the working memory, which includes 2-back and 0-back task events for body, place, face, and tools, as well as fixation periods. Each task fMRI scan has 393 scanning time points and 268 brain regions parcellated by Shen 268 brain region atlas [15]. Since the working memory task is highly related to the attention and default mode network areas of the brain [8], we pick up 58 regions from the two sub-networks.

We compare our CPD-Net with three methods, including (1) the spectral clustering method (SC); (2) the recent clustering method by seeking for density peaks (DP) of the data distribution [16]; (3) the dynamic graph embedding (dGE) method [8]. In contrast to our method, three counterpart methods first vectorize the functional brain networks and then perform change detection. We use a purity score between the ground truth (pre-defined task schedules) and stratified brain states to evaluate the detection accuracy [8]. The kernel width δ is initially set to 0.1, the weights W are initialized as random semi-orthogonal matrices and the rectifying threshold ε is set as 10^{-4}. Regarding the spec of PC, we use an Intel (R) Core (TM) i7–8700 CPU @ 3.20 GHz PC without graphic card.

3.1 Ablation Study

We conduct the ablation study on the collected fMRI data by testing all combinations of turning on/off SPD-DNN and MS-RNN (as shown in Table 1). By shutting down SPD-DNN (denoted by ×), the input to MS-RNN is the vectorized FC matrices. By shutting down MS-RNN, we use the learned FC fingerprints for clustering. By shutting down both, we simply use the vectorized FC matrix to stratify the time points by data clustering. It is clear that each component plays a vital role in change detection, as evidenced by the significant difference after turning off either component.

Table 1. Ablation study of the SPD-DNN and MS-RNN in our CPD-Net.

SPD-DNN	×	×	√	√
MS-RNN	×	√	×	√
Purity (mean±std)	0.674±0.065	0.701±0.041	0.729±0.059	**0.745±0.040**

3.2 Change Detection on Real Task-Based fMRI Data

Evaluating the Accuracy of Change Detection. In this experiment, we train our manifold-based CPD-Net for detecting state changes, where the training and testing data are mixed with test/retest fMRI data. In Fig. 2(a), we show the change detection result of one test data (in the green dash box) and one retest data (in the blue dash box), where the bar plots with different colors and heights denote the pre-defined time schedule of functional tasks. The automatic detection results by our CPD-Net, SC, DP and dGE are displayed at the bottom of Fig. 2(a) with the matched color of ground truth. By examining the temporal alignment between the pre-defined functional tasks and the automatic detection result, our CPD-Net yields more accurate prediction results than the other three methods. In addition, we show the distribution of FC fingerprints on the vector space $\mathcal{R}^d$ (in the sphere) before and after stratification by MS-RNN in Fig. 2(c)-(d), respectively. It is clear that MS-RNN is quite effective in grouping the timepoints into isolated modes, which makes change detection much easier.

Next, we evaluate the accuracy of brain state change detection under different window sizes (from 10 to 60 time points). The mean and standard deviation of purity scores are shown in Fig. 2 (b) for test and retest data separately since they have different task schedules. Our CPD-Net shows significant improvement ($p < 10^{-3}$) of detection accuracy using t-test over SC and DP in all setting windows, dGE except $w = 15 - 25$.

Evaluating the Replicability of Change Detection. In this experiment, we opt for the window size of 30 and evaluate the replicability of the brain state change detections between test and retest data where the task schedules are different. Specifically, we first train one CPD-Net using the training set of the test data only (called CPD-Net-1) and another CPD-Net using the training set of retest data only (called CPD-Net-2), separately. Next, we apply the trained CPD-Net-1 to the testing set of the test data and the testing data of the retest data, where CPD-Net-1 has not seen any instance of retest data in the training stage, and vice versa. We evaluate such test/retest replicability in Table 2 by showing the change detection accuracy by the scenario of training/testing on the same task schedule versus the scenario of training/testing on different task schedules. By running a two-sample t-test between the purity scores of two scenarios, we have not found significant differences by CPD-Net. It implies that CPD-Net has high replicability performance when the task schedules are completely different between test and retest data.

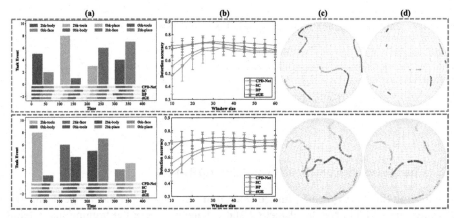

Fig. 2. (a) The typical change detection result by our CPD-Net, SC, DP and dGE methods on the test (in the green box) and retest data (in the blue box). (b) The accuracy of change detection by CPD-Net (in red), SC (in blue), DP (in brown) and dGE (in green). (c)-(d) The distribution of FC fingerprints before and after stratification by MS-RNN, where the different colors represent the different functional tasks (corresponding to the color of the barplot in (a)). (Color figure online)

Table 2. The replicability test results of change detection, where CPD-Net-1 and CPD-Net-2 denote the neural network trained exclusively using test and retest data, respectively.

		Same task schedule vs. Different task schedule
CPD-Net-1	mean±std	0.7467±0.0443 vs. 0.7425±0.0402
	t-test	No significant difference found ($p = 0.7459$)
CPD-Net-2	mean±std	0.7471±0.0438 vs. 0.7440±0.0457
	t-test	No significant difference found ($p = 0.8249$)

4 Conclusion

In this study, we devised a novel deep manifold-based change point detection network (CPD-Net) by learning task-specific functional connectivity fingerprints to characterize the dynamic functional fluctuations. Our CPD-Net enjoys two major merits. (1) We presented a novel integrated solution to simultaneously infer task-specific geometric feature representations and detect state changes, compared to the current approaches where feature learning and change detection are often performed in two steps separately. (2) To our knowledge, our work is the first attempt to detect the change of brain states using the manifold-based neural network. Compared to the conventional methods, which often break down the network topological information by vectorizing the brain network into a data array, the learned geometric patterns by our CPD-Net allow us to track the trajectory of the whole-brain functional network more accurately along with the brain stage changes. Furthermore, our method probably provides the basis for exploring the geometry insight of the whole. functional brain network. Extensive experiments demonstrate that our CPD-Net can be effectively detecting the change of functional brain states with preferable accuracy and replicability change detection.

Acknowledgment. This work was partially supported by the National Natural Science Foundation of China (61472145, 61771007).

References

1. Buckner, R.L., Krienen, F.M., Yeo, B.T.T.: Opportunities and limitations of intrinsic functional connectivity MRI. Nat. Neurosci. **16**(7), 832–837 (2013)
2. Filippi, M., Spinelli, E.G., Cividini, C., Agosta, F.: Resting state dynamic functional connectivity in neurodegenerative conditions: a review of magnetic resonance imaging findings. Front. Neurosci. **13**, 657 (2019)
3. Xu, Y., Lindquist, M.: Dynamic connectivity detection: an algorithm for determining functional connectivity change points in fMRI data. Front. Neurosci. **9**, 285 (2015)
4. Cribben, I., Haraldsdottir, R., Atlas, L.Y., Wager, T.D., Lindquist, M.A.: Dynamic connectivity regression: determining state-related changes in brain connectivity. Neuroimage **61**(4), 907–920 (2012)

5. Li, H., Fan, Y.: Identification of temporal transition of functional states using recurrent neural networks from functional MRI. In: Frangi, A.F., Schnabel, J.A., Davatzikos, C., Alberola-López, C., Fichtinger, G. (eds.) MICCAI 2018. LNCS, vol. 11072, pp. 232–239. Springer, Cham (2018). https://doi.org/10.1007/978-3-030-00931-1_27

6. Li, H., Satterthwaite, T.D., Fan, Y.: Large-scale sparse functional networks from resting state fMRI. Neuroimage **156**, 1–13 (2017)

7. Feldt, S., Waddell, J., Hetrick, V.L., Berke, J.D., Zochowski, M.: Functional clustering algorithm for the analysis of dynamic network data. Phys. Rev. E Stat. Nonlinear Soft Matter Phys. **79**(5), 056104 (2009)

8. Lin, Y., Hou, J., Laurienti, P.J., Wu, G.: Detecting changes of functional connectivity by dynamic graph embedding learning. In: Martel, A.L., et al. (eds.) MICCAI 2020. LNCS, vol. 12267, pp. 489–497. Springer, Cham (2020). https://doi.org/10.1007/978-3-030-59728-3_48

9. Biswal, B., Yetkin, F.Z., Haughton, V.M., Hyde, J.S.: Functional connectivity in the motor cortex of resting human brain using echo-planar MRI. Magn. Reson. Med. **34**(4), 537–541 (1995)

10. Dai, M., Zhang, Z., Srivastava, A.: Analyzing dynamical brain functional connectivity as trajectories on space of covariance matrices. IEEE Trans. Med. Imaging **39**(3), 611–620 (2020)

11. Huang, Z., Gool, L.V.: A riemannian network for SPD matrix learning. In: Proceedings of the Thirty-First AAAI Conference on Artificial Intelligence, pp. 2036–2042. AAAI Press, San Francisco (2017)

12. Arsigny, V., Fillard, P., Pennec, X., Ayache, N.: Geometric means in a novel vector space structure on symmetric positive-definite matrices. SIAM J. Matrix Anal. Appl. **29**(1), 328–347 (2007)

13. Kong, S., Fowlkes, C.: Recurrent pixel embedding for instance grouping. In: 2018 IEEE/CVF Conference on Computer Vision and Pattern Recognition, pp. 9018–9028 (2018)

14. Barch, D.M., et al.: Function in the human connectome: task-fMRI and individual differences in behavior. Neuroimage **80**, 169–189 (2013)

15. Shen, X., Tokoglu, F., Papademetris, X., Constable, R.T.: Groupwise whole-brain parcellation from resting-state fMRI data for network node identification. Neuroimage **82**, 403–415 (2013)

16. Rodriguez, A., Laio, A.: Clustering by fast search and find of density peaks. Science **344**(6191), 1492–1496 (2014)

From Brain to Body: Learning Low-Frequency Respiration and Cardiac Signals from fMRI Dynamics

Roza G. Bayrak[1(✉)], Colin B. Hansen[1], Jorge A. Salas[1], Nafis Ahmed[1], Ilwoo Lyu[2], Yuankai Huo[1], and Catie Chang[1,3,4]

[1] Computer Science, Vanderbilt University, Nashville, TN 37235, USA
roza.g.bayrak@vanderbilt.edu
[2] Computer Science and Engineering, UNIST, Ulsan 44919, Korea
[3] Biomedical Engineering, Vanderbilt University, Nashville, TN 37235, USA
[4] Vanderbilt University Institute of Imaging Science, Nashville, TN 37235, USA

Abstract. Functional magnetic resonance imaging (fMRI) is a powerful technique for studying human brain activity and large-scale neural circuits. However, fMRI signals can be strongly modulated by slow changes in respiration volume (RV) and heart rate (HR). Monitoring cardiac and respiratory signals during fMRI enables modeling and/or reducing such effects; yet, physiological measurements are often unavailable in practice, and are missing from a large number of fMRI datasets. Very recent work has demonstrated the ability to reconstruct RV signals from resting-state fMRI data, but it is currently unclear whether such an approach generalizes to other physiological signals (such as HR) or across fMRI task conditions. Here, we propose a joint learning approach for inferring RV and HR signals directly from fMRI time-series dynamics. Our models are trained on resting-state fMRI data using the largest dataset employed for the problem, and tested both on resting-state fMRI and on separate fMRI paradigms that were acquired during three task conditions: emotion processing, social cognition, and working memory. We demonstrate that our deep LSTM model successfully captures both RV and HR patterns, outperforming existing approaches, and translates to scans of variable lengths and different experimental conditions. Source code is available at: https://github.com/neurdylab/deep-physio-recon.

Keywords: fMRI time series · Physiological signals · Heart rate · Respiration

1 Introduction

Functional magnetic resonance imaging (fMRI) measures indirect, hemodynamic (blood oxygen) consequences of neural activity. As a result, blood oxygen level dependent (BOLD) fMRI signal can be influenced by other physiological processes that modulate blood oxygenation [16], in addition to neural activity. Two

© Springer Nature Switzerland AG 2021
M. de Bruijne et al. (Eds.): MICCAI 2021, LNCS 12907, pp. 553–563, 2021.
https://doi.org/10.1007/978-3-030-87234-2_52

major physiological drivers of fMRI signal changes arise from natural, slowly varying ($< 0.15\,\mathrm{Hz}$) fluctuations in respiration volume (RV) [2,3,27] and heart rate (HR) [6,23]. Low-frequency RV ad HR changes modulate fMRI by altering blood oxygen concentrations [29], and are shown to widely correlate with fMRI signal variation across the brain [10,14,19]. Hence, having access to RV and HR signals collected during fMRI is critical for interpreting fMRI data, improving the delineation of large-scale networks and studying brain-body interactions.

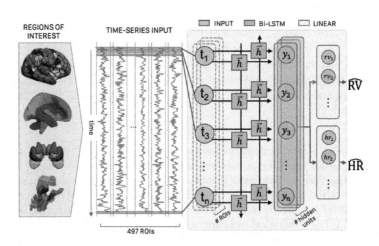

Fig. 1. Framework for estimating respiration volume (RV) and heart rate (HR) signals from fMRI time-series dynamics. Regions of interest were defined using 4 published atlases that had been constructed from different imaging modalities, comprising areas in cerebral cortex, white matter, subcortex, and the ascending arousal network. Time courses are extracted from each ROI, and the preprocessed signals are provided to a candidate network as input channels. A deep bidirectional LSTM network architecture were adapted for joint estimation of the target (RV and HR) signals.

Currently, external recordings of cardiac and respiratory activity during the scan are needed to model and/or correct for low-frequency RV and HR effects in fMRI data. However, it is not always possible to acquire clean external physiological measures during fMRI, and a large number of existing datasets lack such measures altogether. Data-driven approaches have been developed for removing low-frequency effects, such as global signal regression and temporal ICA, yet these methods cannot unambiguously separate RV and HR influences from neurally driven fMRI signals without the use of recorded physiological waveforms for reference [4,5,10,15]. Therefore, it is crucial to have methods for inferring physiological fluctuations directly from the fMRI data itself.

Toward this goal, two recent studies have shown that the RV time course can be directly reconstructed from fMRI data [1,21]. However, these studies focused only on predicting RV while neglecting HR. HR has been associated with functional circuits for emotion regulation [20] and found to correlate with dynamic variation in rs-fMRI networks [7]; thus, the ability to reconstruct the HR signal would contribute important information to fMRI studies. In addition, previous work focuses only on resting-state data (with no other task conditions), and utilizes a small dataset, which may limit the generalizability.

In this paper, we propose a novel approach for inferring prominent sources of fMRI signal variation: slow changes in breathing (respiratory volume; RV) and heart rate (HR). We build upon previous work and address two key questions: (1) whether HR can also be inferred from fMRI itself, which may be more challenging because measures of HR tend to be more noisy than those of RV; (2) whether the proposed approach is transferable to fMRI scans of varying durations and different (task) fMRI conditions. We address the first question by constructing joint learning (multi-task learning; MTL) architectures, and train our models on resting-state fMRI data from the Human Connectome Project (HCP). To answer the second question, we evaluate our best model on three separate task fMRI datasets from HCP.

The remainder of the paper is organized as follows. Section 2 briefly explains the data, pre-processing steps, and the preferred joint learning architecture. Experimental results are demonstrated in Sect. 3. In Sect. 4, we discuss the implications of this work and considerations for performance.

2   Data and Preprocessing

2.1   HCP Data

Resting-State fMRI. The scans used for training comprised a subsample of resting-state fMRI (rs-fMRI) data from the HCP 1200-subject release [28]. Subjects were scanned up to 4 times, twice on one day and twice on a second day. We included only those subjects who had all 4 runs, and whose physiological signals were reported to have passed a quality check in both Power et al. [17] and Xifra-Porcas et al. [29]. This resulted in a dataset with 375 subjects (n = 1500 scans). Rs-fMRI data were acquired with the following parameters: temporal resolution (TR) of 0.72 s, duration of 1200 frames per run (14.4 min), and spatial resolution of 2 mm isotropic.

Task fMRI. The scans used for external validation comprised a subsample of task fMRI data from the HCP 1200 Subject Release. The three task-fMRI datasets considered here are *emotion processing (EP)*, *social cognition (SC)* and *working memory (WM)*. We employed automated selection criteria to remove scans with poor-quality physiological data, which checked for (i) clipping of waveforms (values clamped at 0 and 4095), (ii) unrealistic heart rates (below 30 bpm or above 97 bpm, or constant at 48 bpm [29]), (iii) missing waveform. As each subject was scanned at most two times for each task condition, we included only those subjects for which both runs were available, were not included in the resting state study and whose physiological data were labeled as 'clean' for all three task conditions by the above criteria. The selection process resulted in a total of

214 scans (107 subjects). Task fMRI data were acquired with the same acquisition parameters as the rs-fMRI data, except for the duration, which was shorter for the tasks (EP = 176 frames, SC = 274 frames, and WM = 405 frames per run (https://protocols.humanconnectome.org/HCP/3T/).

Physiological Data. Cardiac and respiratory data were continuously monitored during the scans at 400 samples/s, using a pulse oximeter on the right index finger and a pneumatic respiratory belt strapped around the abdomen, respectively.

2.2 Preprocessing

Both task and resting-state fMRI scans had undergone the HCP Minimal Preprocessing pipeline [11]. Beyond this, we applied linear and quadratic detrending to remove slow scanner drifts, followed by band-pass filtering (0.01–0.15 Hz) and temporal downsampling by a factor of 2. We did not regress out head motion, as it has been shown that respiration can create (pseudo) head motions [18]; therefore, retaining head motion may provide information relevant to predicting RV and HR.

For the physiological data of each scan, we extracted HR as the inverse of the mean inter-beat-interval in sliding windows of 6 s centered at each fMRI TR (every 0.72 s). Likewise, an RV signal was calculated as the temporal standard deviation of the raw respiration waveform in a window of 6 s centered at each TR [8]. Both RV and HR were then band-pass filtered and down-sampled by a factor of 2, in a manner identical to the fMRI data.

In this work, an initial dimensionality reduction step was applied to the fMRI data by parcellating the brain into regions of interest based on four published atlases: Schaefer Cerebral Cortex Atlas (400 regions embedded into 17 networks) [22], Pandora TractSeg White Matter Atlas (72 regions) [13], Melbourne Subcortex Atlas (16 regions) [24] and Ascending Arousal Network (AAN) Atlas (9 regions) [9] (Fig. 1). For each atlas, the mean fMRI time series was extracted from all voxels within each ROI. Because the units of BOLD signal are arbitrary and vary in amplitude across scans and subjects, all time-series signals are temporally normalized to zero mean and unit variance.

All atlases were available in MNI space, and when necessary, we resampled to 2mm isotropic voxels to match the resolution of the preprocessed fMRI scans. From the Pandora tractography-based white matter atlases [13], we used the HCP probabilistic atlas that was created using TractSeg method [26]. A threshold of 95% was used to exclude voxels of lower confidence to minimize the overlap between white matter and gray matter regions. From the Melbourne hierarchical subcortex atlases, we used the 7 Tesla HCP Scale I.

3 Experiments

3.1 Joint Training

The network architectures were selected based on their potential for learning time-series dynamics and ability to operate on fMRI scans of varying lengths.

Given that RV and HR are often closely coupled, we hypothesized that learning both of these signals jointly - in a manner that leverages correlations between them - may act as data augmentation and enhance model accuracy.

Baseline Models. In previous work, three network architectures were proposed for reconstructing RV signals directly from fMRI data: U-Net and bi-LSTM [1] as well as a 5-layer separable convolutional (separable CONV1d) network [21]. These published models were used as baselines.

Transformer. Many successful multivariate time-series prediction approaches are adapted from natural language processing (NLP) domain. Hence, in this work, we also investigated a Transformer (attention) network adapted for time-series prediction [8]. In contrast to [8], the original positional encoding is used [25]. We employed an attention window of size 8 and h = 8 parallel attention heads. The query, keys, values are vectors of size 8 and the output vector is size of dmodel = 2048. All sub-layers in the model, as well as the embedding layers, produce outputs of dimension dmodel = 2048. As a crucial step to accommodate learning of both RV and HR, another decoder block is added to the original architecture. Lastly, we added linear layers at the end of each decoder block, without activation functions, to allow the network to output predictions of arbitrary amplitude. Both encoder and decoder blocks are composed of a stack of N = 2 identical layers.

The aforementioned parameters, as well as the hyperparameters (learning rate = 1.0e−3, dropout = 0.3, and batchsize = 4), were chosen empirically. The models were allowed to train up to 200 epochs.

Deep Bidirectional LSTM. A bidirectional LSTM (bi-LSTM) architecture [12] was adapted. The models were trained with a hidden state (h) size (depth) 2000, learning rate *(lr)* of 1.0e−3, a batch size of 16, and were allowed to train up to 20 epochs saving only the best models using validation accuracy. The aforementioned hyperparameters were chosen empirically.

Implementation Details. We adapted the all of aforementioned models to solve our joint prediction problem. For the baseline and bi-LSTM models, two linear layers were added to the end of each architecture to force the network models to simultaneously learn both physiological signals of interest. Input to the networks consisted of ROI time series, which were provided to the network as different channels. The two outputs of our networks are the estimated RV and HR time-series. All models were trained using ADAM optimizer with default parameters. The experiments were implemented using the PyTorch deep learning library.

Evaluation. The goal of our study was to accurately reconstruct RV and HR time courses from fMRI data alone. Pearson correlation was used when evaluating model performance (i.e., to compare the predicted versus actual RV and HR time series) as well as for the loss function, with the exception of the baseline model *SepCONV* (which was trained using MSE as its loss function, consistent with its original implementation [21]). All models were trained in a 5-fold cross-validation paradigm using rs-fMRI data. Within each fold, the training partition was further split into training (80%) and validation (15%) sets). The best models were saved based on the validation loss to avoid overfitting. Additionally, three task-fMRI datasets were used to evaluate the generalization of our models to scans with different cognitive (task) conditions. These task datasets and subjects were not seen during the model training.

3.2   Investigating Relevant Regions of Interest

Individual Regions of Interest. We hypothesized that the fMRI signals of certain brain regions, especially those which are most susceptible to RV and HR modulation, may contribute to the learning of RV and HR more than others. To investigate the relative importance of the regions across our four atlases, we conducted a heuristic experiment. Using our deep bi-LSTM network architecture and the rs-fMRI dataset, we trained models using each ROI time-series signal separately. This single-ROI analysis resulted in a total of 497 unique models.

Exclusion of Brainstem and Cerebellum. Given that brainstem and cerebellar regions are often absent from clinical fMRI acquisitions, we also assess the performance of our technique using all regions except those that overlap with brainstem and cerebellum (total of 477 ROIs).

4   Results

Table 1 shows the performance comparison between the new network models, along with recently published models as baselines. Here, validation performance on the rs-fMRI data (using all 497 ROIs) is shown. For a fair comparison, we modified the baseline networks for joint learning by adding an additional fully connected (linear) layer at the end of each architecture.

Table 1. Model Performance on rs-fMRI data. For each tested model, median Pearson correlation score (with IQR) is shown together with the loss function that was used in training, and the time it takes to predict RV and HR for a single scan.

Model	RV (median $r \pm$ IQR)	HR (median $r \pm$ IQR)	Loss	Time (secs)
SepCONV [21]	0.683 $\pm$ 0.23	0.097 $\pm$ 0.13	MSE	0.06
U-Net [1]	0.668 $\pm$ 0.22	0.584 $\pm$ 0.17	Pearson	0.025
Bi-LSTM [1]	0.642 $\pm$ 0.24	0.582 $\pm$ 0.16	Pearson	0.087
Transformer	0.665 $\pm$ 0.22	0.586 $\pm$ 0.16	Pearson	0.08
Deep Bi-LSTM	**0.689 $\pm$ 0.22**	**0.627 $\pm$ 0.16**	Pearson	0.173

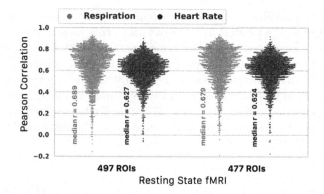

Fig. 2. Physiological signal prediction accuracy on resting state fMRI data.
Hyperparameter selection and design choices were made using 497 ROIs. An additional
analysis was conducted using 477 ROIs (497 ROIs minus brainstem and cerebellar
regions). Each marker represents the Pearson correlation score between the observed
and predicted signal for all scans in the unseen test data in 5-fold cross-validation.
(Color figure online)

The validation performance of our models that was trained solely on long-
duration rs-fMRI scans is shown in Fig. 2. The similarity between observed and
predicted time courses for both RV (green) and HR (red) is shown for cross
validated resting state (1500 scans from 375 subjects). Figure 3 shows the model
performance on each of the three external validation task datasets (comprising
214 scans each, from 107 unseen subjects).

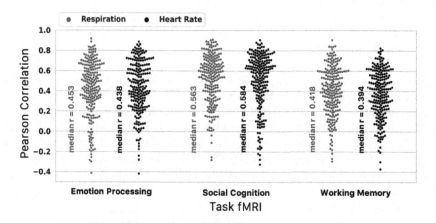

Fig. 3. Model generalizability assesment on task fMRI data. Three task
datasets were used to assess the across-condition generalization. From the models that
had been trained on resting-state data under 5-fold CV paradigm, a model was randomly selected for this assesment. Pearson correlation scores were calculated between
the observed and predicted RV and HR signals.

In Fig. 4, we visualize the predictiveness of each ROI from our heuristic experiment by mapping the mean Pearson correlation score (across 1500 scans) of single ROI models onto their respective atlas location. Many white matter regions were observed to be more predictive of both physiological signals than gray matter ROIs, particularly within areas of the brainstem and cerebellum.

Nonetheless, in a separate experiment, retraining the model without brainstem and cerebellar regions (i.e., using 477 total ROIs) resulted in comparable performance on the resting-state data compared to the full model (which used all 497 ROIs; Fig. 2).

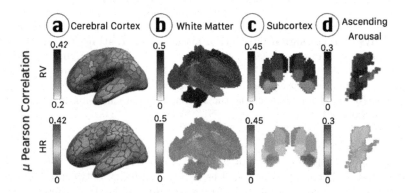

Fig. 4. Heuristic experiment on individual ROIs. Predictiveness of a given ROI for RV or HR was assessed using Pearson correlation on rs-fMRI validation data. Results were visualized by projecting the Pearson correlation onto (a) cerebral cortex (400 ROIs) (b) white matter bundle regions (72 ROIs), (c) subcortical regions (16 ROIs) and (d) ascending arousal network (9 ROIs).

5   Discussion

This work presents a novel approach to infer heart rate fluctuations jointly with respiration, and directly from fMRI dynamics. To the best of our knowledge this is the only method to reconstruct both of these physiological time series from fMRI, which can provide valuable knowledge about brain-body interactions and autonomic function - and help clarify the interpretation of fMRI signals - when physiological recordings are missing. Our proposed joint learning model was found to reconstruct HR as well as RV signals with very high accuracy (median r ∼ 0.69 for RV, and median r ∼ 0.63 for HR), and was transferable to scans with different fMRI task conditions.

In this study, we have considerably expanded, relative to previous work, the amount of HCP rs-fMRI data that is utilized for physiological signal estimation (N = 1500). Although an initial quality-check was performed on respiration and heart rate recordings, we visually observed that a number of scans still contained artifacts such as clipping and imperfect heart-beat detection. However, the marked improvement in performance beyond the published work [1,21] suggests that a larger, albeit less strictly vetted dataset allowed the networks to harness the power of big data. Other factors that may have positively impacted the current performance include the expanded set of ROIs (including non-gray matter tissue types) and not regressing out head motion from the fMRI data, as suggested in [21].

In investigating relevant regions of interest, our heuristic, individual ROI analysis hinted that some white matter regions in the brainstem and cerebellum were very predictive of respiration volume and heart rate. Nonetheless, we also verified that our physiological signals can be successfully reconstructed even without brainstem and cerebellar regions. This finding is important, since brainstem and cerebellum are not always included in fMRI acquisitions. While there is still much to explore in terms of regional relevance of physiological signals, we also found that the good prediction from tractography-based ROIs [13] suggest that incorporating parcellation boundaries derived from multiple imaging modalities (such as those based on diffusion-weighted MRI) and non-gray-matter regions could be beneficial to fMRI prediction tasks.

Perhaps surprisingly, while Transformer architecture was comparable in performance, it did not outperform the other models examined here. One possible reason stems from the fact that it does not draw upon information about the temporal ordering in the input, unlike the Bi-LSTM. We might speculate that further tailoring aspects of the network to fMRI data, such as the positional encoding, could improve its accuracy on the task at hand, and will be an important avenue of future work.

In summary, we present a novel joint learning framework for reconstructing key physiological signals directly from fMRI data. The models generalized to unseen datasets and fMRI paradigms. The choices were made for this study significantly improved predictions compared to previously published results. An interesting extension to this work could explore other cohorts (such as aging or development) that exhibit greater variability in brain anatomy and physiology.

Acknowledgements. This work was supported by NIH grant K22 ES028048 (C.C.).

References

1. Bayrak, R.G., Salas, J.A., Huo, Y., Chang, C.: A deep pattern recognition approach for inferring respiratory volume fluctuations from fMRI data. In: Martel, A.L., et al. (eds.) MICCAI 2020. LNCS, vol. 12267, pp. 428–436. Springer, Cham (2020). https://doi.org/10.1007/978-3-030-59728-3_42
2. Birn, R.M., Diamond, J.B., Smith, M.A., Bandettini, P.A.: Separating respiratory-variation-related fluctuations from neuronal-activity-related fluctuations in fMRI. Neuroimage **31**(4), 1536–1548 (2006)
3. Birn, R.M., Smith, M.A., Jones, T.B., Bandettini, P.A.: The respiration response function: the temporal dynamics of fMRI signal fluctuations related to changes in respiration. Neuroimage **40**(2), 644–654 (2008)
4. Bright, M.G., Whittaker, J.R., Driver, I.D., Murphy, K.: Vascular physiology drives functional brain networks. NeuroImage **217**, 116907 (2020)

5. Caballero-Gaudes, C., Reynolds, R.C.: Methods for cleaning the bold fMRI signal. Neuroimage **154**, 128–149 (2017)
6. Chang, C., Cunningham, J.P., Glover, G.H.: Influence of heart rate on the bold signal: the cardiac response function. Neuroimage **44**(3), 857–869 (2009)
7. Chang, C., Metzger, C.D., Glover, G.H., Duyn, J.H., Heinze, H.J., Walter, M.: Association between heart rate variability and fluctuations in resting-state functional connectivity. Neuroimage **68**, 93–104 (2013)
8. Cohen, M., Charbit, M., Corff, S.L., Preda, M., Nozière, G.: End-to-end deep meta-modeling to calibrate and optimize energy loads. arXiv preprint arXiv:2006.12390 (2020)
9. Edlow, B.L., et al.: Neuroanatomic connectivity of the human ascending arousal system critical to consciousness and its disorders. J. Neuropathol. Exp. Neurol. **71**(6), 531–546 (2012)
10. Glasser, M.F., et al.: Using temporal ICA to selectively remove global noise while preserving global signal in functional MRI data. NeuroImage **181**, 692–717 (2018)
11. Glasser, M.F., et al.: The minimal preprocessing pipelines for the human connectome project. Neuroimage **80**, 105–124 (2013)
12. Graves, A., Schmidhuber, J.: Framewise phoneme classification with bidirectional LSTM and other neural network architectures. Neural Netw. **18**(5–6), 602–610 (2005)
13. Hansen, C.B., et al.: Pandora: 4-D white matter bundle population-based atlases derived from diffusion MRI fiber tractography. bioRxiv (2020)
14. Kassinopoulos, M., Mitsis, G.D.: Identification of physiological response functions to correct for fluctuations in resting-state fMRI related to heart rate and respiration. Neuroimage **202**, 116150 (2019)
15. Kundu, P., Inati, S.J., Evans, J.W., Luh, W.M., Bandettini, P.A.: Differentiating bold and non-bold signals in fMRI time series using multi-echo EPI. Neuroimage **60**(3), 1759–1770 (2012)
16. Murphy, K., Birn, R.M., Bandettini, P.A.: Resting-state fMRI confounds and cleanup. Neuroimage **80**, 349–359 (2013)
17. Power, J.D., Lynch, C.J., Dubin, M.J., Silver, B.M., Martin, A., Jones, R.M.: Characteristics of respiratory measures in young adults scanned at rest, including systematic changes and "missed" deep breaths. Neuroimage **204**, 116234 (2020)
18. Power, J.D., Lynch, C.J., Silver, B.M., Dubin, M.J., Martin, A., Jones, R.M.: Distinctions among real and apparent respiratory motions in human fMRI data. NeuroImage **201**, 116041 (2019)
19. Power, J.D., Plitt, M., Laumann, T.O., Martin, A.: Sources and implications of whole-brain fMRI signals in humans. Neuroimage **146**, 609–625 (2017)
20. Sakaki, M., Yoo, H.J., Nga, L., Lee, T.H., Thayer, J.F., Mather, M.: Heart rate variability is associated with amygdala functional connectivity with mPFC across younger and older adults. Neuroimage **139**, 44–52 (2016)
21. Salas, J.A., Bayrak, R.G., Huo, Y., Chang, C.: Reconstruction of respiratory variation signals from fMRI data. NeuroImage **225**, 117459 (2021)
22. Schaefer, A., et al.: Local-global parcellation of the human cerebral cortex from intrinsic functional connectivity MRI. Cerebral Cortex **28**(9), 3095–3114 (2018)
23. Shmueli, K., et al.: Low-frequency fluctuations in the cardiac rate as a source of variance in the resting-state fMRI bold signal. Neuroimage **38**(2), 306–320 (2007)
24. Tian, Y., Margulies, D.S., Breakspear, M., Zalesky, A.: Hierarchical organization of the human subcortex unveiled with functional connectivity gradients. bioRxiv (2020)

25. Vaswani, A., et al.: Attention is all you need. In: Advances in Neural Information Processing Systems, vol. 30, pp. 5998–6008 (2017)
26. Wasserthal, J., Neher, P., Maier-Hein, K.H.: TractSeg-fast and accurate white matter tract segmentation. NeuroImage **183**, 239–253 (2018)
27. Wise, R.G., Ide, K., Poulin, M.J., Tracey, I.: Resting fluctuations in arterial carbon dioxide induce significant low frequency variations in bold signal. Neuroimage **21**(4), 1652–1664 (2004)
28. WU-Minn, H.: 1200 subjects data release reference manual (2017). https://www.humanconnectome.org
29. Xifra-Porxas, A., Kassinopoulos, M., Mitsis, G.D.: Physiological and head motion signatures in static and time-varying functional connectivity and their subject discriminability. bioRxiv (2020)

Multi-head GAGNN: A Multi-head Guided Attention Graph Neural Network for Modeling Spatio-temporal Patterns of Holistic Brain Functional Networks

Jiadong Yan[1], Yuzhong Chen[1], Shimin Yang[1], Shu Zhang[2], Mingxin Jiang[1],
Zhongbo Zhao[1], Tuo Zhang[3], Yu Zhao[4], Benjamin Becker[1], Tianming Liu[5],
Keith Kendrick[1], and Xi Jiang[1(✉)]

[1] School of Life Science and Technology, MOE Key Lab for Neuroinformation,
University of Electronic Science and Technology of China, Chengdu, China
xijiang@uestc.edu.cn
[2] Center for Brain and Brain-Inspired Computing Research, School of Computer Science,
Northwestern Polytechnical University, Xi'an, China
[3] School of Automation, Northwestern Polytechnical University, Xi'an, China
[4] Syngo Innovation, Siemens Healthineers, Malvern, USA
[5] Computer Science Department, The University of Georgia, Athens, USA

Abstract. It has been widely demonstrated that complex brain function is mediated by the interaction of multiple concurrent brain functional networks, each of which is spatially distributed across specific brain regions in a temporally dynamic fashion. Therefore, modeling spatio-temporal patterns of those holistic brain functional networks provides a foundation for understanding the brain. Compared to conventional modeling approaches such as correlation, general linear model, and matrix decomposition methods, recent deep learning methodologies have shown a superior performance. However, the existing deep learning models either underutilized both spatial and temporal characteristics of fMRI during model training, or merely focused on modeling only one targeted brain functional network at a time while ignoring holistic ones, resulting in a significant gap in our current understanding of how the brain functions. To bridge this gap, we propose a novel Multi-Head Guided Attention Graph Neural Network (Multi-Head GAGNN) to simultaneously model spatio-temporal patterns of multiple brain functional networks. In Multi-Head GAGNN, the spatial patterns of multiple brain networks are firstly modeled in a multi-head attention graph U-net, and then adopted as guidance for modeling the corresponding temporal patterns of multiple brain networks in a temporal multi-head guided attention network model. Results based on two task fMRI datasets from the public Human Connectome Project demonstrate superior ability and generalizability of Multi-Head GAGNN in simultaneously modeling spatio-temporal patterns of holistic brain functional networks compared to other state-of-the-art models. This study offers a new and powerful tool for helping understand complex brain function.

Keywords: Brain functional network · Functional MRI · Spatio-temporal patterns · Multi-head guided attention · Graph Neural Network

© Springer Nature Switzerland AG 2021
M. de Bruijne et al. (Eds.): MICCAI 2021, LNCS 12907, pp. 564–573, 2021.
https://doi.org/10.1007/978-3-030-87234-2_53

1   Introduction

Mounting evidence based on advanced neuroimaging techniques such as functional MRI (fMRI) has demonstrated that complex brain function processes emerge from and are realized by the interaction of multiple brain functional networks, each of which is spatially distributed across specific brain regions in a temporally dynamic fashion [1, 2]. Therefore, simultaneously modeling both spatial and temporal patterns of holistic brain networks based on fMRI data facilitates the understanding of their functional mechanisms [3, 4]. Conventional modeling approaches including correlation [5], general linear model (GLM) [6], as well as matrix decomposition methods such as principal component analysis (PCA) [7], independent component analysis (ICA) [8], and sparse representation (SR) [9–11], are however limited to modeling the spatially overlapping brain networks [9] or optimizing model parameters [11].

Recently, deep learning-based methods have been developed and applied to modeling spatial and/or temporal patterns of brain functional networks with superior performance [12–15]. Initially, a recurrent neural network (RNN) model [12] was adopted to model the dynamical temporal pattern of targeted brain functional networks using recurrent connections, while the spatial pattern of brain network was further obtained by a simple correlation operation. Subsequently, a convolutional autoencoder (CAE) [13] was proposed to model the brain network's temporal pattern, and to obtain its spatial pattern using regression. Later on, a restricted Boltzmann machine (RBM) model [14] was adopted to model the temporal pattern first and next the spatial pattern using LASSO. Recently, a spatio-temporal convolutional neural network (ST-CNN) [15] was proposed to model the spatial and temporal patterns of a targeted brain functional network using 3D U-Net and 1D CAE, respectively. Although achieving promising results [12–15], existing deep learning methods only focused on one pattern (either spatial or temporal) of brain networks, leading to the underutilization of both spatial and temporal characteristics of fMRI. Moreover, the previous deep learning methods merely focused on modeling one targeted brain functional network at a time and ignored holistic ones.

In order to solve the above two limitations, we propose a novel Multi-Head Guided Attention Graph Neural Network (Multi-Head GAGNN) to simultaneously model both spatial and temporal patterns of multiple brain networks based on individual fMRI data. The proposed model consists of two parts: spatial part and temporal part. In the spatial part, we aim to simultaneously model the spatial patterns of multiple brain functional networks using a novel multi-head attention graph U-net. In the temporal part, we aim to simultaneously model the temporal patterns of the multiple brain functional networks using a multi-head guided attention network model [16] guided by the modeled spatial patterns of multiple brain functional networks. Based on data from two task-based fMRI (t-fMRI) experiments from the same 200 subjects in the Human Connectome Project (HCP), the proposed Multi-Head GAGNN shows superior ability and generalizability in simultaneously modeling both spatial and temporal patterns of holistic brain functional networks compared to other state-of-the-art (SOTA) models.

2 Methods

2.1 Data Acquisition, Preprocessing and Identification of Training Labels

The t-fMRI data of healthy subjects from the public Human Connectome Project (HCP) S900 release [17] were adopted. We randomly selected emotion and motor t-fMRI data from the same 200 subjects in this study. The major t-fMRI acquisition parameters are referred to [17]. The t-fMRI data was preprocessed based on FSL FEAT and normalized to 0–1 distribution the same as in [15]. The data was further down-sampled to a spatial size of 48*56*48 and temporal size of 176 before feeding into the model to fit the computational capacity. We used the ten most representative resting state networks (RSNs) [18] as the example targeted brain functional networks to evaluate the proposed model. The training labels of both spatial and temporal patterns of the ten RSNs in each individual brain were obtained the same way as in [15].

2.2 Model Architecture of Multi-head GAGNN

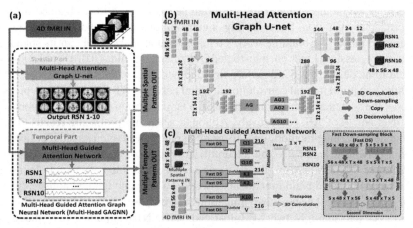

Fig. 1. (a) An overview of the Multi-Head GAGNN framework. Detailed model architecture of Multi-Head GAGNN consists of (b) the spatial part and (c) the temporal part.

An overview of the proposed framework is in Fig. 1(a). The whole-brain fMRI data are the input, and both spatial and temporal patterns of the ten targeted RSNs are the outputs. The model consists of two core parts: the spatial part and the temporal part.

Attention Mechanism and 'Attention Graph' Block. We first briefly introduced the attention mechanism [19]. We adopted Q, K, and V to represent the Query, Key, and Value matrices obtained from the extracted features of input 4D fMRI data. We calculated the attention matrix $Softmax\left(\frac{QK^T}{\sqrt{M}}\right)$ (M is the feature number of K, T is the transpose operation), and obtained the attention output as defined in Eq. (1):

$$attention\ output = Softmax\left(\frac{QK^T}{\sqrt{M}}\right) \times V \tag{1}$$

We then introduced the 'Attention Graph' (AG) block in Fig. 1(b). We used A, X, and W to represent the graph adjacency matrix, input and weight, respectively. The output of the basic graph convolution [20] was defined in Eq. (2):

$$graph\ convolution\ output = AXW \tag{2}$$

Different from Eq. (2), the proposed AG block replaced A with QK^T in Eq. (1) since QK^T is not only an adjacency matrix but also includes attention. Therefore, the output of the AG block was able to integrate both graph convolution and an attention mechanism as defined in Eq. (3):

$$AG\ block\ output = fold\left(norm\left(QK^T\right)XW\right) \tag{3}$$

where $Q = unfold\left(conv_Q(X)\right)$, $K = unfold(conv_k(X))$, $conv_Q(\cdot)$ and $conv_k(\cdot)$ are two different 3D convolutions; $unfold(\cdot)$ is the operation to change the 4D matrix with size D*H*W*T to 2D one with size (D*H*W)*T; $fold(\cdot)$ is the operation to change the 2D matrix with size (D*H*W)*T to the 4D one with size D*H*W*T; $norm(\cdot)$ is the min-max normalization (0–1 distribution) operation.

Spatial Part. The detailed model architecture of the spatial part is called a 'Multi-Head Attention Graph U-net' as illustrated in Fig. 1(b). The spatial part was based on the 3D U-net adopting 3D convolution with stride 2 for down-sampling and 3D deconvolution for up-sampling. The spatial pattern training models of the ten RSNs shared the same network layers until the first single AG block was involved. Ten different AG blocks (AG1, AG2, etc.) were then introduced and followed the single AG block in order to branch spatial pattern modeling of the ten RSNs. We took AG1 as an example. AG1 was performed the up-sampling part of the 3D U-net first and then a 3D CNN with three layers to reduce the temporal dimension, and finally output the modeled spatial pattern of RSN1. The modeling processes for the other nine RSNs were the same. The size of all 3D filters was 3*3*3 and each 3D convolution layer was followed by a batch norm layer and a rectified linear unit (ReLU). We adopted the 'overlap rate' [15] to evaluate the spatial similarity between the modeled pattern P_m and the training label P_l of the same RSN in Eq. (4). The loss function of the spatial part was defined as the averaged negative overlap rate of the ten RSNs.

$$overlap\ rate = \frac{sum(\min(P_m, P_l))}{(sum(P_m) + sum(P_l))/2} \tag{4}$$

Temporal Part. The detailed model architecture of the temporal part is called a 'Multi-Head Guided Attention Network' as illustrated in Fig. 1(c). We first introduced a 'Fast Down-sampling' block (Fast DS) to help quickly down-sample the three spatial dimensions of the 4D input data (D*H*W*T) in order to prevent model overfitting as well as to fit in the GPU memory. The down-sampled 4D data was with size S*S*S*T (S was set to 6 in this study). The size of all 3D filters was set to 3*3*3 and each 3D convolution layer was followed by a batch norm layer and a ReLU. We then introduced the complete architecture of the temporal part. The modeled spatial patterns of the ten RSNs together with the 4D fMRI data were adopted as the input. We multiplied the modeled spatial pattern with the 3D spatial block of each time point of the 4D fMRI in order to extract the corresponding temporal features within the 4D fMRI data for guidance. Then after the Fast DS and *unfold*($\cdot$) operations, we branched the temporal pattern modeling of the ten RSNs. Note that the ten branches shared the same V, but with different Q and K. We took the first branch as an example (Fig. 1(c)). We performed the attention mechanism on V, $Q1$ and $K1$ in Eq. (1) to obtain the 2D attention output with size (S*S*S)*T, and then an average operation to output the modeled temporal pattern of RSN1 with size 1*T. The modeling process for the other nine RSNs was the same. We adopted the Pearson correlation coefficient (PCC) to evaluate the temporal similarity between the modeled pattern and the training label of the same RSN [15]. The loss function of the temporal part was then defined as the averaged negative PCC of the ten RSNs.

2.3   Training Scheme of Multi-head GAGNN Model

The 200 subjects for the emotion t-fMRI were randomly split into 160 for training and 40 for testing. The motor t-fMRI from the same 200 subjects was used as independent testing data. There were two stages of the entire model training corresponding to the spatial part and the temporal part, respectively. We set 0.00001 and 0.001 as the learning rate for the first and second stage, respectively and Adam [21] was adopted as the optimizer. After 200 epochs, the spatial part loss converged and reached to -0.66 (Fig. 2(a)), and the temporal part loss converged and reached to -0.80 (Fig. 2(b)).

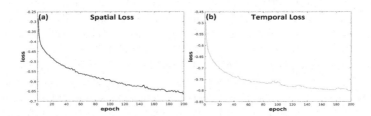

Fig. 2. Training loss curves of (a) the spatial part and (b) the temporal part.

2.4   Evaluation and Validation of Multi-head GAGNN Model

To evaluate the performance of the proposed Multi-Head GAGNN model, we calculated both spatial (Eq. (4)) and temporal similarity (Sect. 2.2) between the modeled pattern and the training label pattern across the ten RSNs. We also calculated spatial similarity of the modeled pattern with the RSN spatial template [18]. Furthermore, we compared the proposed model with other SOTA models including ST-CNN [15] and SR [9–11]. We further adopted the emotion t-fMRI data from the 40 testing subjects as well as that from the motor t-fMRI of all 200 subjects which was not used in the training to evaluate the generalizability of the proposed model.

3 Results

3.1 Spatio-temporal Pattern Modeling of Ten RSNs in Emotion T-fMRI

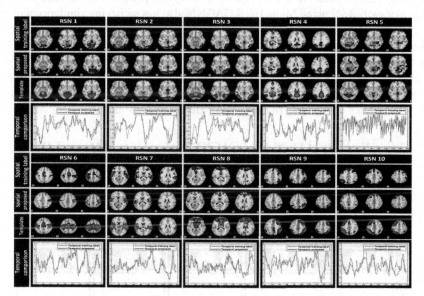

Fig. 3. Spatio-temporal pattern modeling of the ten RSNs in emotion t-fMRI data.

Table 1. Spatio-temporal pattern modeling of all testing subjects in emotion t-fMRI.

RSN ID	Spatial similarity between modeled ones and training labels	Temporal similarity between modeled ones and training labels	Spatial similarity with RSN spatial templates	
			Proposed	SR
RSN1	0.409 ± 0.084	0.679 ± 0.215	**0.247 ± 0.027**	0.138 ± 0.037
RSN2	0.442 ± 0.058	0.753 ± 0.125	**0.163 ± 0.014**	0.110 ± 0.018
RSN3	0.422 ± 0.066	0.700 ± 0.126	**0.284 ± 0.026**	0.196 ± 0.026
RSN4	0.366 ± 0.050	0.669 ± 0.159	**0.210 ± 0.014**	0.127 ± 0.023
RSN5	0.373 ± 0.060	0.361 ± 0.242	**0.125 ± 0.020**	0.107 ± 0.022
RSN6	0.356 ± 0.059	0.546 ± 0.205	**0.204 ± 0.032**	0.149 ± 0.019
RSN7	0.358 ± 0.046	0.549 ± 0.192	**0.246 ± 0.022**	0.174 ± 0.033
RSN8	0.349 ± 0.045	0.536 ± 0.298	**0.204 ± 0.022**	0.154 ± 0.028
RSN9	0.326 ± 0.059	0.626 ± 0.194	**0.241 ± 0.020**	0.151 ± 0.028
RSN10	0.354 ± 0.040	0.629 ± 0.200	**0.229 ± 0.022**	0.177 ± 0.027

We randomly selected one subject's emotion t-fMRI data as an example and visualized the modeled spatial and temporal patterns of the ten RSNs in Fig. 3. It can be seen that for each RSN, the modeled spatial pattern using Multi-Head GAGNN has reasonable similarity with the training label as well as with the spatial template. Moreover, Multi-Head GAGNN has a superior performance to the training label in terms of modeling more clustered brain areas and less noise points in the spatial pattern. The modeled temporal pattern of each RSN using Multi-Head GAGNN also has satisfying similarity with the training label. We further calculated the mean spatial and temporal similarity values of the modeled patterns across all 40 testing subjects in the emotion t-fMRI in Table 1. It can be seen that both spatial and temporal patterns of all ten RSNs achieve satisfying similarity with the training labels across all testing subjects. Moreover, the RSN spatial template has a higher mean and lower standard deviation of spatial similarity compared with the modeled spatial pattern using Multi-Head GAGNN than that using SR.

3.2   Generalizability of Multi-head GAGNN Performance in Motor T-fMRI

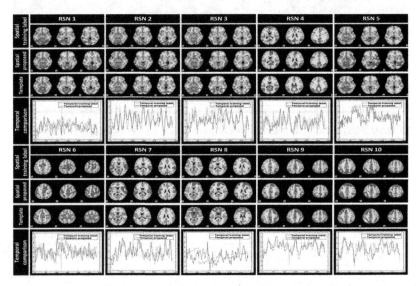

Fig. 4. Spatio-temporal pattern modeling of the ten RSNs in motor t-fMRI data.

We directly applied the trained model to motor t-fMRI data to evaluate the generalizability of Multi-Head GAGNN. Similarly, we randomly selected one subject as an example and the result is visualized in Fig. 4. We see that both spatial and temporal patterns of the modeled RSNs have satisfying similarity with the training labels. Moreover, Multi-Head GAGNN has superior performance in modeling the spatial pattern of RSN with more clustered brain areas and less noise points compared to the training label. Table 2 reports the mean spatial and temporal similarity values across all 200 testing subjects in the motor t-fMRI. Similarly, the proposed Multi-Head GAGNN achieves satisfying spatial and temporal similarity with the training label. Moreover, Multi-Head GAGNN has a higher mean and lower standard deviation of spatial similarity with the RSN spatial templates compared to that of SR.

Table 2. Spatio-temporal pattern modeling of all testing subjects in motor t-fMRI.

RSN ID	Spatial similarity between modeled ones and training labels	Temporal similarity between modeled ones and training labels	Spatial similarity with RSN spatial templates	
			Proposed	SR
RSN1	0.349 ± 0.054	0.550 ± 0.218	**0.136 ± 0.029**	0.133 ± 0.029
RSN2	0.347 ± 0.035	0.682 ± 0.151	0.089 ± 0.016	0.103 ± 0.022
RSN3	0.330 ± 0.047	0.469 ± 0.183	**0.182 ± 0.031**	0.157 ± 0.032
RSN4	0.336 ± 0.052	0.593 ± 0.178	**0.140 ± 0.021**	0.108 ± 0.025
RSN5	0.320 ± 0.053	0.251 ± 0.262	0.062 ± 0.019	0.074 ± 0.023
RSN6	0.333 ± 0.053	0.482 ± 0.192	**0.140 ± 0.032**	0.126 ± 0.026
RSN7	0.344 ± 0.048	0.502 ± 0.192	**0.200 ± 0.026**	0.160 ± 0.034
RSN8	0.338 ± 0.056	0.466 ± 0.266	**0.163 ± 0.022**	0.124 ± 0.038
RSN9	0.319 ± 0.046	0.536 ± 0.236	**0.168 ± 0.023**	0.132 ± 0.026
RSN10	0.341 ± 0.047	0.453 ± 0.198	**0.192 ± 0.024**	0.143 ± 0.027
Averaged	0.336 ± 0.049	0.498 ± 0.208	**0.147 ± 0.024**	0.126 ± 0.028

3.3   Comparison of Different Model Structures

Table 3. Comparison of spatio-temporal pattern modeling between proposed and ST-CNN.

Datasets	Spatial similarity with training label		Temporal similarity with training label	
	Proposed	ST-CNN	Proposed	ST-CNN
Emotion	**0.210 ± 0.014**	0.167 ± 0.044	**0.669 ± 0.159**	0.601 ± 0.184
Motor	0.140 ± 0.021	0.160 ± 0.025	**0.593 ± 0.178**	0.414 ± 0.156

In addition to comparisons with a conventional data decomposition method (i.e., SR) in Sects. 3.1 and 3.2, we also compared Multi-Head GAGNN with the SOTA deep learning model ST-CNN [15]. Since ST-CNN can only model one targeted brain network at one time, we randomly selected RSN4 as an example. Table 3 shows that Multi-Head GAGNN has a superior or at least comparable spatial and temporal similarity compared to ST-CNN, suggesting that our proposed model has a satisfactory ability to simultaneously model both spatial and temporal patterns of multiple functional networks while still achieving competing spatial/temporal similarity with other SOTA models which merely focus on modeling only one targeted brain network.

Moreover, we replaced the AG block in Fig. 1(b) with another three methods, and compared the modeled spatial similarity in Table 4 based on emotion t-fMRI. The averaged spatial similarity of the ten RSNs shows that the proposed AG block achieves the best modeling ability, indicating the effectiveness of integrating both GCN and attention in the proposed AG block for brain functional network modeling.

Table 4. Comparison of different model structures for spatial modeling.

	Proposed (Attention Graph)	Attention	GCN	CNN
RSN1	**0.247 ± 0.027**	0.210 ± 0.022	0.181 ± 0.026	0.206 ± 0.014
RSN2	0.163 ± 0.014	0.120 ± 0.010	**0.171 ± 0.024**	0.145 ± 0.011
RSN3	**0.284 ± 0.026**	0.268 ± 0.021	0.251 ± 0.031	0.273 ± 0.016
RSN4	**0.210 ± 0.014**	0.196 ± 0.018	0.160 ± 0.015	0.147 ± 0.016
RSN5	0.125 ± 0.020	0.129 ± 0.011	0.132 ± 0.020	**0.132 ± 0.012**
RSN6	0.204 ± 0.032	**0.237 ± 0.021**	0.191 ± 0.023	0.210 ± 0.016
RSN7	**0.246 ± 0.022**	**0.246 ± 0.016**	0.232 ± 0.033	0.196 ± 0.019
RSN8	0.204 ± 0.022	**0.228 ± 0.017**	0.176 ± 0.029	0.220 ± 0.017
RSN9	0.241 ± 0.020	0.214 ± 0.012	**0.255 ± 0.024**	0.216 ± 0.013
RSN10	0.229 ± 0.022	0.232 ± 0.018	**0.241 ± 0.022**	0.210 ± 0.018
Averaged	**0.215 ± 0.022**	0.208 ± 0.017	0.199 ± 0.025	0.196 ± 0.015

4 Conclusion

In this work, a novel Multi-Head GAGNN was proposed to model spatio-temporal patterns of holistic brain functional networks. As far as we know, it is one of the earliest studies to model both spatial and temporal patterns of multiple brain networks in a deep learning fashion. Based on two t-fMRI datasets from the same 200 subjects, experimental evaluations demonstrated that the proposed model had a better ability and generalizability in simultaneously modeling spatio-temporal patterns of multiple brain functional networks compared to other SOTA models. In the future, we plan to improve the current model structure to an unsupervised one. We also plan to apply the proposed model to identify brain network abnormalities in mental disorders such as autism.

Acknowledgements. This work was partly supported by the National Natural Science Foundation of China (61976045), Sichuan Science and Technology Program (2021YJ0247), Key Scientific and Technological Projects of Guangdong Province Government (2018B030335001), National Natural Science Foundation of China (62006194), the Fundamental Research Funds for the Central Universities (3102019QD005), High-level researcher start-up projects (06100-20GH020161), and National Natural Science Foundation of China (31971288 and U1801265).

References

1. Fox, M.D., Snyder, A.Z., Vincent, J.L., Corbetta, M., Van Essen, D.C., Raichle, M.E.: The human brain is intrinsically organized into dynamic, anticorrelated functional networks. Proc. Natl. Acad. Sci. U.S.A. **102**(27), 9673–9678 (2005)
2. Fedorenko, E., Duncan, J., Kanwisher, N.: Broad domain generality in focal regions of frontal and parietal cortex. Proc. Natl. Acad. Sci. U.S.A. **110**(41), 16616–16621 (2013)

3. Naselaris, T., Kay, K.N., Nishimoto, S., Gallant, J.L.: Encoding and decoding in fMRI. Neuroimage **56**(2), 400–410 (2011)
4. Logothetis, N.K.: What we can do and what we cannot do with fMRI. Nature **453**(7197), 869–878 (2008)
5. Bandettini, P.A., Jesmanowicz, A., Wong, E.C., Hyde, J.S.: Processing strategies for time-course data sets in functional MRI of the human brain. Magn. Reson. Med. **30**(2), 161–173 (1993)
6. Friston, K.J., Holmes, A.P., Worsley, K.J., Poline, J.P., Frith, C.D., Frackowiak, R.S.: Statistical parametric maps in functional imaging: a general linear approach. Hum. Brain Mapp. **2**(4), 189–210 (1994)
7. Andersen, A.H., Gash, D.M., Avison, M.J.: Principal component analysis of the dynamic response measured by fMRI: a generalized linear systems framework. Magn. Reson. Imag. **17**(6), 795–815 (1999)
8. McKeown, M.J., Hansen, L.K., Sejnowski, T.J.: Independent component analysis of functional MRI: what is signal and what is noise? Curr. Opin. Neurobiol. **13**(5), 620–629 (2003)
9. Lv, J.L., et al.: Sparse representation of whole-brain fMRI signals for identification of functional networks. Med. Image Anal. **20**(1), 112–134 (2015)
10. Jiang, X., et al.: Sparse representation of HCP grayordinate data reveals novel functional architecture of cerebral cortex. Hum. Brain Mapp. **36**(12), 5301–5319 (2015)
11. Zhang, W., et al.: Experimental comparisons of sparse dictionary learning and independent component analysis for brain network inference from fMRI data. IEEE Trans. Biomed. Eng. **66**(1), 289–299 (2019)
12. Hjelm, R.D., Plis, S.M., Calhoun, V.: Recurrent neural networks for spatiotemporal dynamics of intrinsic networks from fMRI data. In: NIPS: Brains and Bits (2016)
13. Huang, H., et al.: Modeling task fMRI data via deep convolutional autoencoder. IEEE Trans. Med. Imag. **37**(7), 1551–1561 (2018)
14. Zhang, W., et al.: Hierarchical organization of functional brain networks revealed by hybrid spatiotemporal deep learning. Brain Connect. **10**(2), 72–82 (2020)
15. Zhao, Y., et al.: Four-dimensional modeling of fMRI data via spatio-temporal convolutional neural networks (ST-CNNs). IEEE Trans. Cognit. Developm. Syst. **12**(3), 451–460 (2020)
16. Wang, Z.Y., Zou, N., Shen, D.G., Ji, S.W.: Non-local U-nets for biomedical image segmentation. Proc. AAAI Conf. Artif. Intell. **34**(4), 6315–6322 (2020)
17. Van Essen, D.C., Smith, Barch, D.M., Behrens, T.E.J., Yacoub, E., Ugurbil, K.: The WU-Minn human connectome project: an overview. Neuroimage **80**, 62–79 (2013)
18. Smith, S.M., et al.: Correspondence of the brain's functional architecture during activation and rest. Proc. Natl. Acad. Sci. **106**(31), 13040–13045 (2009)
19. Vaswani, A., et al.: Attention is all you need. In: 31st Annual Conference on Neural Information Processing Systems, pp. 6000–6010 (2017)
20. Thomas, N.K., Max, W.: Semi-supervised classification with graph convolutional networks. In: International Conference on Learning Representations 2017 (2017)
21. Kingma, D., Ba, J.: Adam: a method for stochastic optimization. In: International Conference on Learning Representations 2015 (2015)

Building Dynamic Hierarchical Brain Networks and Capturing Transient Meta-states for Early Mild Cognitive Impairment Diagnosis

Mianxin Liu[1] , Han Zhang[2], Feng Shi[3], and Dinggang Shen[1(✉)]

[1] School of Biomedical Engineering, ShanghaiTech University, Shanghai 201210, China
dgshen@shanghaitech.edu.cn
[2] Institute of Brain-Intelligence Technology, Zhangjiang Lab, Shanghai 201210, China
[3] Department of Research and Development, Shanghai United Imaging Intelligence Co., Ltd., Shanghai 200232, China

Abstract. Latest diagnostic studies at the preclinical stage of Alzheimer's disease focus on dynamic functional connectivity network (dFCN) from resting-state fMRI. However, the existing methods fall short in at least two aspects: 1) Single-scale atlas is generally used for building the dFCN, while functional interactions at and cross multiple spatial scales are largely neglected; 2) Features extracted from dFCN at each time segment are often simply pooled together, whereas the disease related meta-states, i.e., dFCN configurations, appear transiently and may not be sensitively captured. In the presented study, we designed multiscale atlas-based graph convolutional network, and utilized a multiple-instance-learning pooling to tackle these issues. First, we leveraged those previously established multiscale atlases to build hierarchical brain networks, represented by multiscale graphs, which were also applied to different time segments to form dFCNs. At each time segment, we processed these multiscale graphs by our specially designed multiscale graph convolutional networks that were connected based on the inter-scale hierarchy. A long short-term memory (LSTM) architecture was then implemented to process temporal information of the dFCN. The output from the LSTM was pooled with attention-based multiple instance learning to dynamically assign larger weights to disease related (more diagnostic) transient states. Experiments on 481 subjects show that our method achieved 77.78% accuracy (with 75.00% sensitivity and 78.57% specificity) in healthy control *vs.* early mild cognitive impairment (eMCI) classification, which outperformed the state-of-the-art methods. Our study *not only* fits the practical needs of eMCI diagnosis with resting-state fMRI *but also* highlights that the pathological of eMCI could manifest as abnormal transient meta-states of multiscale functional interactions.

Keywords: Dynamic functional connectivity · Early mild cognitive impairment · Graph convolutional network · Multiple instance learning · Multiscale atlases

© Springer Nature Switzerland AG 2021
M. de Bruijne et al. (Eds.): MICCAI 2021, LNCS 12907, pp. 574–583, 2021.
https://doi.org/10.1007/978-3-030-87234-2_54

1 Introduction

The early diagnosis of Alzheimer's disease (AD) at its preclinical stage, i.e., mild cognitive impairment (MCI), would allow on-time intervention to delay or prevent progression. However, at an even earlier stage than MCI, namely early MCI (eMCI), mesoscale brain structural alterations or behavioral changes can hardly be detected and used for reliable diagnosis. A recent endeavor is to seek possible functional abnormality of the brain under neurodegeneration with resting-state functional magnetic resonance imaging (fMRI), which probes neural activity by the fluctuations in the blood-oxygen-level-dependent (BOLD) signals. Functional connectivity (FC) can be calculated between the BOLD signals from each pair of brain regions to form a static graph, namely FC network (FCN), to approximate interactions among neural populations. However, the static FCN assumes temporal stationarity of such interactions and thus disregards the moment-to-moment variability in the brain communicating activities [1]. A dynamic FCN (dFCN) from the fMRI time segments constructed by a sliding window on the BOLD signals has been developed to capture the temporal details of the brain dynamics, which is more capable of the assessment of the early neurodegenerative conditions than the static FCN-based methods [1].

In the conventional disease diagnosis studies, dFCN was reshaped as a vector that was directly sent to classifiers such as support vector machine (SVM) [2, 3]. More recent studies processed dFCNs by graph convolutional network (GCN) [4–7] to distill high-level features, and some studies employed deep time series learning such as the long short-term memory (LSTM) network in the temporal domain to detect diseases. Despite the novelty and success of these methods, we argue that their performances are still largely hindered by two major aspects as described below.

First, the brain is known as a multiscale hierarchical system and thus disease-induced dysfunctions could happen in the FC *within* and *across* multiple spatial scales. GCN can learn certain topological hierarchy in the brain network built on a single scale atlas in a purely data-driven manner. However, it inevitably ignores the biologically validated network hierarchy that can be defined by certain dedicated multiscale brain functional parcellation atlases. Prior knowledge from the multiscale atlases could guide GCN to capture multiscale brain functional interactions more effectively, leading to more successful disease detection.

Second, most of the current studies blindly pool the learned high-level information from the entire dFCN in the whole time span [7, 8]. However, increasing evidence has shown that brain disease-related information is more likely to be contained in a few specific "meta-state", defined as reproducible and transient patterns of the dFCN configurations in time segments [3, 9, 10]. Identifying and focusing on these informative meta-states could improve the sensitivity of diagnostic models. In the machine learning field, the task of capturing the disease related meta-states for a more accurate diagnosis can be addressed by the methods of multiple instance learning (MIL) [11], which effectively identifies the informative instance among multiple instances during the learning and prediction.

To this end, we proposed a multiscale atlas-based GCN, with LSTM and MIL pooling, to provide better early brain disease diagnosis. Specifically, 1) we apply pre-defined multiscale atlases to construct multiscale dFCN and guide the GCN to integrate the multiscale information based on the inter-scale hierarchy; 2) After using a LSTM network to process temporal information, we pool the outputted high-order dFCN features of all time segments from the LSTM by attention-based multiple instance learning to focus on disease-related meta-states.

2 Methods

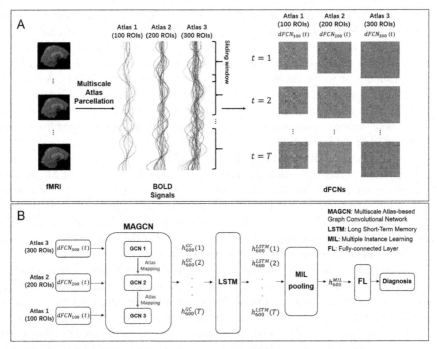

Fig. 1. Schematic illustration of our proposed Multiscale Atlas-based GCN (MAGCN) with LSTM and MIL pooling. (A) Multiscale atlases were applied to fMRI to extract regional averaged BOLD signals and build dFCNs at different scales. (B) These dFCNs were fed into the MAGCN, LSTM, and MIL pooling for eMCI diagnosis. The details of MAGCN are depicted in Fig. 3.

In this work, we presented a Multiscale Atlas-based GCN (MAGCN) with LSTM and MIL pooling for better utilizing the information in multiple spatiotemporal scale brain dynamics to detect eMCI-related abnormality in an individualized manner (see flowchart in Fig. 1). Briefly, a set of pre-defined multiscale atlases (Fig. 2) were applied to fMRI to obtain regional averaged BOLD signals and build dFCNs at different scales (Fig. 1A). Then, MAGCN integrated the information contained in the multiple spatial-scale dFCNs, based on the predefined multiscale atlases and the relationship between neighboring scales (Fig. 3). We further characterized temporal evolution in the dFCN by LSTM, whose outputs were then pooled by an attention-based MIL pooling layer, followed by a fully-connected layer to classify the disease status of the subject (Fig. 1B).

2.1 Multiscale Atlases for Parcellation

The multiscale atlases (Schaefer's atlases) from [12] were applied to the fMRI data to capture neural activities at multiple spatial scales. Such atlases provide a set of brain functional parcellations at multiple scales ranging from 100 regions of interest (ROIs) to 1000 ROIs. These ROIs belong to seven resting-state functional networks (RSNs) defined in [13], namely salience network (SAL), somatomotor network (SM), limbic network (LIM), visual network (VIS), default mode network (DMN), frontoparietal network (FP), and attention network (ATT). In this study, we used the first three atlases, with 100, 200, and 300 ROIs, respectively (Fig. 2).

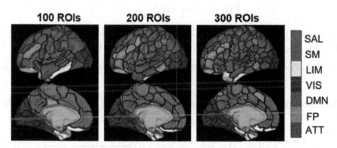

Fig. 2. The implemented multiscale atlases naturally capture brain functional hierarchy.

2.2 Dynamic Functional Connectivity Network

The dFCNs were constructed to characterize spatiotemporal fluctuations of the brain functional interactions (Fig. 1A). Based on the parcellations at different spatial scales, for each subject, we extracted the BOLD signals within each ROIs and computed the ROI-averaged signals. A sliding window with a length of 180 s and a stride of 15 s was applied on the ROI signals to separate them into a number of time segments. Within each segment, a graph was constructed with ROIs as nodes and pairwise Pearson correlations among the ROI signals as edges. When using atlas with R ROIs ($R = 100$, 200 or 300 in this paper) and a total of T segments, the computation leads to a dFCN with dimension $R \times R \times T$ for each subject. The dFCN at the segment t is denoted as $dFCN_R(t)$, $t = 1, 2, \ldots, T$.

2.3 Multiscale Atlas-Based GCN (MAGCN)

We designed MAGCN to distill information from the brain multiscale hierarchical functional interactions (Fig. 3). We used the spectral graph convolution [4] to build the GCNs, each of which was with a ReLU activation function and a dropout (rate = 0.3).

Atlas Mapping. We first defined the atlas mapping to introduce inter-scale connections among the GCNs at different scales. This is because, though the dFCNs at different scales are expected to capture scale-specific functional communications, a ROI from the coarse parcellation can be sub-divided into several small ROIs in the next finer parcellation and thus their features should be roughly similar. This constraint of inter-scale hierarchy could lead to more accurate characterizations of the brain states, when compared to independent operation of graph convolution at each scale.

We used spatial overlapping of ROIs defined at different scales to characterize the inter-scale hierarchy. We constructed the mapping matrix between atlases by checking spatial overlapping between ROIs of different atlases. Formally, from the atlas $\mathcal{R}$ (with R ROIs/nodes) to atlas $\wp$ (P ROIs/nodes), the atlas mapping matrix $M_{\mathcal{R} \to \wp}$ is defined as below.

$$M_{\mathcal{R} \to \wp}(i, j) = \begin{cases} 1 & (\textit{if ROI i in R overlaps ROI j in } \wp) \\ 0 & (\textit{otherwise}) \end{cases}. \tag{1}$$

The atlas mapping is used to map the node features defined in different atlases. The GCN outputted feature vector h_R^{GC} defined in atlas $\mathcal{R}$ can be mapped to h_P^{AM} for the atlas $\wp$ by a simple matrix multiplication with the mapping matrix, $h_R^{GC} M_{\mathcal{R} \to \wp} = h_P^{AM}$.

Multiscale Atlas-based Graph Convolutional Network (MAGCN)

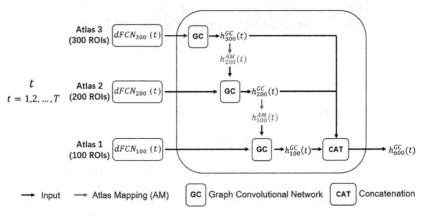

Fig. 3. The proposed hierarchical learning and fusing architecture of MAGCN.

MAGCN. After defining the atlas mapping, we then proposed the MAGCN. In the presented work, we obtained three dFCNs, as $dFCN_{100}, dFCN_{200}$, and $dFCN_{300}$ based on the three atlases. For each time segment t, the $dFCN_{100}(t)$, $dFCN_{200}(t)$, and $dFCN_{300}(t)$ were first fed into the MAGCN. MAGCN started with a graph convolution between the node features (identity matrix used here) and the dFCN from the finest (300-ROI) parcellation. Using the atlas mapping, the outputted feature $h_{300}^{GC}(t)$ was then converted to $h_{200}^{AM}(t)$ with the mapping matrix, which was further used in the graph convolution with the $dFCN_{200}(t)$. This step could introduce constraint on the graph convolution at the coarser scale. The hierarchical graph convolution continued till a graph convolution with dFCN from the coarsest atlas (the 100-ROI atlas here). The outputs from the three steps were concatenated to form a fused representation $h_{600}^{GC}(t)$ for the state of the multiscale brain system.

The concatenated features $h_{600}^{GC}(t)$ at each t were then forwarded to a classical LSTM (with all-zero matrix as initial hidden state) to capture temporal variations in the time segments. The output of the LSTM was then denoted as $h_{600}^{LSTM}(t)$.

2.4 Attention Based Multiple Instance Learning Pooling

The disease related meta-state may transiently appear in the temporal domain without a fixed temporal order. Therefore, we formulated the problem as multiple instance learning (MIL) problem and pooled the output temporal features from the LSTM layer using attention-based MIL pooling [14]. The temporal feature from the LSTM at each subseries was thus viewed as one instance that may contain disease related information. The attention-based MIL pooling performed weighted average on the temporal features, where the weights were decided by the features themselves and the relationship between weight and feature was learned by a neural network. Formally, the temporal features $h_{600}^{LSTM}(t)$ were assigned with a weight $\alpha(t)$ and averaged to obtain the pooled feature h_{600}^{MIL},

$$h_{600}^{MIL} = \sum_{t=1}^{T} \alpha(t) h_{600}^{LSTM}(t) \qquad (2)$$

and $\alpha(t)$ is computed by

$$\alpha(t) = \frac{\exp\left\{\mathbf{W}\tanh\left(\mathbf{V}h_{600}^{LSTM}(t)\right)\right\}}{\sum_{t=1}^{T} \exp\left\{\mathbf{W}\tanh\left(\mathbf{V}h_{600}^{LSTM}(t)\right)\right\}}, \qquad (3)$$

where $\mathbf{W}$ and $\mathbf{V}$ are parameters to be learned in the neural network. Also note that $\alpha(t)$ is yielded by using a SoftMax normalization from $\mathbf{W}\tanh\left(\mathbf{V}h_{600}^{LSTM}(t)\right)$.

The pooled feature h_{600}^{MIL} is finally fed into one fully-connected layer with Softmax function for generating diagnostic assessment.

When using the attention mechanism to reveal eMCI related and unrelated meta-states, we checked the original amplitude of the unnormalized attention weight $\mathbf{W}\tanh\left(\mathbf{V}h_{600}^{LSTM}(t)\right)$ to set the separating threshold. With a proper separation of eMCI related and unrelated meta-states, we could further explore mesoscale neural mechanisms of eMCI.

3 Experiments

3.1 Implementation

The proposed model was implemented using Pytorch and trained with epoch = 100, learning rate = 0.001 and batch size = 30. Adam was used as optimizer with weight decay = 0.01 to avoid overfitting. The training of the network was accelerated by one Nvidia GTX 3080 GPU, and it took around 30 min in the current dataset. To address the sample imbalance problem, we applied a weighted cross-entropy as the loss function, where the weights were the inverse of the sample ratio in the training set (ranging from 1/3 to 1/4).

3.2 Dataset

We used the fMRI data from the Alzheimer's Disease Neuroimaging Initiative (ADNI2 and ADNIgo) [15, 16]. Each fMRI data was acquired with TR = 3 s and 420 s duration (140 volumes). The first 10 volumes of each image were discarded to ensure magnetization equilibrium. The remaining volumes were further performed with a series of standard preprocessing steps, including slice timing correction, motion correction, covariates removal, normalization to the standard space, temporal filtering and spatial smoothing.

The dataset employed in this study contains 481 fMRI data (one fMRI for each subject), which includes normal control (NC, $N_{NC} = 364$) and eMCI ($N_{eMCI} = 117$) subjects. The data were shuffled and randomly split into a training set ($N_{NC} = 225$, $N_{eMCI} = 76$), a validation set ($N_{NC} = 69$, $N_{eMCI} = 21$), and a testing set ($N_{NC} = 70$, $N_{eMCI} = 20$). The model performing best in the validation set was applied to the testing set to evaluate the performance. The dataset split scheme was the same when comparing different algorithms.

3.3 Comparison of Different Methods

We demonstrated the effectiveness of our method by two parts of comparisons.

MIL Pooling. First, the network architecture from [7] was used as the baseline analysis. They proposed the graph convolutional LSTM (GC-LSTM), which performed spectral graph convolution operation on single-scale dFCNs and forwarded the feature from graph convolution to a LSTM architecture. Three fully-connected layers (FLs) following GC-LSTM was used to generate prediction of the status of subjects. We compared both *the GC-LSTM with FL* and *the GC-LSTM with MIL pooling* at the corresponding single scale to demonstrate the effectiveness of MIL pooling.

Multiscale Atlases Fusion. Second, we demonstrated the effectiveness of multiscale atlases fusion and our proposed scheme. A simple fusion of information was first done by a majority voting among the predictions from independently trained classifiers (GC-LSTMs with MIL pooling established in the first part) at different scales. The prediction from each classifier is regarded as one vote, and the final predicted label is the most voted label among classifiers. In addition, we compared the prediction using MAGCN without atlas mapping to highlight the importance of adding the constraint of inter-scale hierarchy among GCNs at different scales.

3.4 Results

In Table 1, we observed that GC-LSTM with FLs showed low accuracy and sensitivity, because FLs estimated fixed weights for different time segments and blindly pooled the information from the dFCN, where some meta-states irrelevant to the eMCI condition were assigned with high weights. Using MIL pooling can avoid this issue and improve performances.

Table 1. The HC-*vs*-eMCI classification performance by different methods on the dFCs, based on multiscale atlases. The best performances in the test set were reported.

	Method	Accuracy	Sensitivity	Specificity
Single atlas with FLs	100 ROIs GC-LSTM	66.66%	35.00%	75.71%
	200 ROIs GC-LSTM	67.78%	45.00%	74.29%
	300 ROIs GC-LSTM	56.67%	40.00%	61.43%
Single atlas with MIL pooling	100 ROIs GC-LSTM with MIL	71.11%	55.00%	75.71%
	200 ROIs GC-LSTM with MIL	70.00%	60.00%	72.86%
	300 ROIs GC-LSTM with MIL	73.33%	55.00%	78.57%
Multi-scale atlases fusion	Voting	73.33%	60.00%	75.71%
	MAGCN (without atlas mapping) with LSTM and MIL	75.55%	55.00%	**81.43%**
	MAGCN with LSTM and MIL (**proposed**)	**77.78%**	**75.00%**	78.57%

Besides, it can be observed that trivial voting among the single-scale based classifiers cannot achieve comparable accuracy compared to our MAGCN based methods, because the classifiers are trained independently and the improvement can be limited and not guaranteed.

In addition, the MAGCN without atlas mapping also exhibited lower performance than the MAGCN with atlas mapping. This validated the inter-scale hierarchy among different atlases introduced by atlas mapping can lead to a more proper description of the internal state of the multiscale brain system.

To detect the eMCI related meta-states, we further checked the attention weights in MIL using correctly predicted subjects in the testing set (55 healthy and 15 eMCI subjects, including 1050 meta-states). The distributions in Fig. 4A suggested a threshold as a clear separation for disease and normal meta-states. Using that threshold, we identified 168 and 882 meta-states as eMCI-related and normal, respectively. About 74.67% of the meta-states from eMCI subjects are suggested to be eMCI-related. Figure 4B provided an observation of the inter-state transition in the brain dynamics within a examplified subject. In Fig. 4C, we computed and visualized the averaged pattern for eMCI and normal meta-states. The eMCI meta-states show lower connectivity and less pronounced module structures (involving intra- and inter-hemisphere VIS, SM, and ATT networks) than that of the normal meta-states.

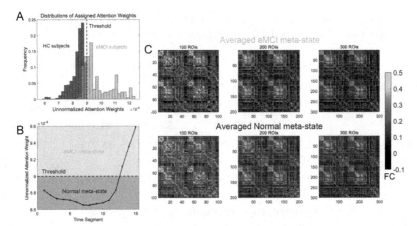

Fig. 4. Detection of eMCI related meta-state in the testing set. (A) The distribution of unnormalized attention weight in these correctly predicted HC and eMCI subjects. (B) Changing trajectory of unnormalized attention weight from one exemplified eMCI subject. The y-axis ranges of NC and eMCI referred to the distributions in (A). (C) The visualization of respectively averaged patterns of all detected eMCI and normal meta-states. The diagonal elements were set to zero to emphasize non-diagonal structures.

4 Conclusion

We proposed a novel method for eMCI diagnosis based on fused information from multiple scale dFCNs defined with multiscale atlases. Using datasets from ADNI, we demonstrated that our method can differentiate eMCI from NC with 77.78% accuracy. We illustrated that the improvement of diagnosis accuracy originates from the use of prior knowledge on the brain parcellation and a proper design to fuse the multiscale spatial information stepwise in a stacked GCN framework. In addition, the introduction of MIL pooling can improve the sensitivity of the model predictions compared to the FL pooling, which is consistent with our understanding that the disease-related information is exhibited in certain dFCN meta-states.

References

1. Filippi, M., Spinelli, E.G., Cividini, C., Agosta, F.: Resting state dynamic functional connectivity in neurodegenerative conditions: a review of magnetic resonance imaging findings. Front. Neurosci. **13**, 657 (2019)
2. Yan, B., et al.: Quantitative identification of major depression based on resting-state dynamic functional connectivity: a machine learning approach. Front. Neurosci. **14**, 191 (2020)
3. Vergara, V.M., Mayer, A.R., Kiehl, K.A., Calhoun, V.D.: Dynamic functional network connectivity discriminates mild traumatic brain injury through machine learning. NeuroImage. Clin. **19**, 30–37 (2018)
4. Kipf, T.N., Welling, M.: Semi-supervised classification with graph convolutional networks (2016). https://arxiv.org/abs/1609.02907

5. Defferrard, M., Bresson, X., Vandergheynst, P.: Convolutional neural networks on graphs with fast localized spectral filtering. In: Advances in Neural Information Processing Systems, pp. 3844–3852 (2016)
6. Meszlényi, R.J., Buza, K., Vidnyánszky, Z.: Resting state fMRI functional connectivity-based classification using a convolutional neural network architecture. Front. Neuroinform. **11**, 61 (2017)
7. Xing, X., et al.: DS-GCNs: connectome classification using dynamic spectral graph convolution networks with assistant task training. Cereb. Cortex **31**(2), 1259–1269 (2021)
8. Chen, X., Zhang, H., Gao, Y., Wee, C.Y., Li, G., Shen, D.: The Alzheimer's disease neuroimaging initiative: high-order resting-state functional connectivity network for MCI classification. Hum. Brain Mapp. **37**(9), 3282–3296 (2016)
9. Jones, D.T., et al.: Non-stationarity in the "resting brain's" modular architecture. PLoS One **7**(6), e39731 (2012)
10. Kim, J., et al.: Abnormal intrinsic brain functional network dynamics in Parkinson's disease. Brain **140**(11), 2955–2967 (2017)
11. Wang, X., Yan, Y., Tang, P., Bai, X., Liu, W.: Revisiting multiple instance neural networks. Pattern Recogn. **74**, 15–24 (2018)
12. Schaefer, A., et al.: Local-global parcellation of the human cerebral cortex from intrinsic functional connectivity MRI. Cereb. Cortex **28**(9), 3095–3114 (2018)
13. Yeo, B.T., et al.: The organization of the human cerebral cortex estimated by intrinsic functional connectivity. J. Neurophysiol. **106**(3), 1125–1165 (2011)
14. Ilse, M., Tomczak, J., Welling, M.: Attention-based deep multiple instance learning. In: the 35th International Conference on Machine Learning, pp. 2127–2136. PMLR (2018)
15. Jack, C.R., Jr., et al.: Magnetic resonance imaging in Alzheimer's disease neuroimaging initiative 2. Alzheimer's Dement. **11**(7), 740–756 (2015)
16. Aisen, P.S., Petersen, R.C., Donohue, M., Weiner, M.W.: Alzheimer's disease neuroimaging initiative: Alzheimer's disease neuroimaging initiative 2 clinical core: progress and plans. Alzheimer's Dement. **11**(7), 734–739 (2015)

Recurrent Multigraph Integrator Network for Predicting the Evolution of Population-Driven Brain Connectivity Templates

Oytun Demirbilek and Islem Rekik$^{(\boxtimes)}$ (iD)

BASIRA Lab, Faculty of Computer and Informatics, Istanbul Technical University,
Istanbul, Turkey
irekik@itu.edu.tr
http://basira-lab.com

Abstract. Learning how to estimate a connectional brain template (CBT) from a population of brain multigraphs, where each graph (e.g., functional) quantifies a particular relationship between pairs of brain regions of interest (ROIs), allows to pin down the unique connectivity patterns shared across individuals. Specifically, a CBT is viewed as an integral representation of a set of highly heterogeneous graphs and ideally meeting the centeredness (i.e., minimum distance to all graphs in the population) and discriminativeness (i.e., distinguishes the healthy from the disordered population) criteria. So far, existing works have been limited to only integrating and fusing a population of brain multigraphs acquired at a single timepoint. In this paper, we unprecedentedly tackle the question: *"Given a baseline multigraph population, can we learn how to integrate and forecast its CBT representations at follow-up timepoints?"* Addressing such question is of paramount in predicting common alternations across healthy and disordered populations. To fill this gap, we propose Recurrent Multigraph Integrator Network (ReMI-Net), the *first graph recurrent neural network* which infers the baseline CBT of an input population t_1 and predicts its longitudinal evolution over time $(t_i > t_1)$. Our ReMI-Net is composed of recurrent neural blocks with graph convolutional layers using a cross-node message passing to first learn hidden-states embeddings of each CBT node (i.e., brain region of interest) and then predict its evolution at the consecutive timepoint. Moreover, we design a novel time-dependent loss to regularize the CBT evolution trajectory over time and further introduce a cyclic recursion and learnable normalization layer to generate well-centered CBTs from time-dependent hidden-state embeddings. Finally, we derive the CBT adjacency matrix from the learned hidden state graph representation. ReMI-Net significantly outperformed benchmark methods in both centeredness and discriminative connectional biomarker discovery criteria in demented patients. Our ReMI-Net GitHub code is available at https://github.com/basiralab/ReMI-Net.

Electronic supplementary material The online version of this chapter (https://doi.org/10.1007/978-3-030-87234-2_55) contains supplementary material, which is available to authorized users.

M. de Bruijne et al. (Eds.): MICCAI 2021, LNCS 12907, pp. 584–594, 2021.
https://doi.org/10.1007/978-3-030-87234-2_55

Keywords: Longitudinal multigraphs · Recurrent graph convolution · Graph population template evolution forecasting · Connectional brain template

1 Introduction

The development of network neuroscience [1] aims to present a holistic representation of the brain graph (also called network or connectome), a universal map of heterogeneous pairwise brain region relationships (e.g., correlation in neural activity or dissimilarity in morphology). Due to its multi-fold complexity, the underlying causes of neurological and psychiatric disorders, such as Alzheimer's disease, autism, and depression remain largely unknown and difficult to pin down [2,3]. How these brain disorders unfold at the individual and population scales remains one of the most challenging obstacles to understanding how the *brain graph* gets altered by disorders, let alone a *brain multigraph*. Conventionally, a brain multigraph is composed of a set of graphs, each capturing a unique 'view' of the brain wiring network (such as morphology or function) [1,4]. A single view of the brain graph is represented as a symmetric adjacency matrix where each element stores the connectivity weight between a pair of anatomical regions of interest (ROIs), encapsulating a particular type of interaction between them. Learning how to integrate a population of brain multigraphs remains a formidable challenge to identify the most representative and shared brain alterations caused by a specific disorder. Very recent works addressed this challenge by learning a *connectional brain template* –in short CBT– from a heterogeneous population of brain multigraphs. Such integral and compact encoding of a connectomic population of brain graphs into a single connectivity matrix (i.e., CBT) presents a powerful and easy tool for comparing connectomic populations of brain multigraphs [5–7] in different states (e.g., healthy and disordered) [8,9]. A well-representative CBT is well-centered (i.e., minimum distance to all graphs in the population) and discriminative (i.e., distinguishes the healthy from the disordered population) [8].

So far, existing CBT learning methods [8–10] have been limited to only integrating and fusing a population of brain multigraphs acquired at a single timepoint, limiting their generalizability to *longitudinal* (i.e., time-series) multigraph populations. Ideally, one would design a model to not only integrate a multigraph population at a single baseline timepoint t_1, but also *forecast* the time-dependent evolution of the learned baseline CBT at follow-up timepoints $t_i > t_1$. Adding to the difficulty of integrating a set of heterogeneous brain connectomes, predicting the future of a population via the compact CBT representation, to eventually map out and discover disorder-specific alternations, presents a big jump in the field of network neuroscience, which we set out to take in this paper. **To the best of our knowledge, no existing works attempted to solve the challenging problem of CBT evolution prediction from a baseline multigraph population observed at t_1.** Recently, [8] proposed the netNorm framework that leverages similarity network fusion [11] for integrating multi-view brain graphs by building a high-order population graph capturing the most centered brain connectivity weights across individuals. Although pioneering, such method resorted to using Euclidean distance for graph comparison which overlooks the non-Euclidean nature of a graph as well as its node-specific topological properties. To address this limitation, recently [10] proposed a supervised multi-topology network cross-diffusion (SM-Net Fusion) to learn a population CBT using a weighted combination of the multi-topological matrices that encapsulate the various topologies in a heterogeneous graph population. Nonetheless, SM-Net Fusion can only handle a population of graphs derived from a single view or neuroimaging modality, failing to generalize to multigraphs. To better model the complex interactions at the *individual* brain multigraph level (i.e., between ROIs) and at the *population* level (i.e., between multigraphs), one can leverage the power of the emerging graph neural networks (GNNs) [12–14] in learning end-to-end mapping for our CBT integration and evolution forecasting tasks. Very recently, [9] proposed deep graph normalizer (DGN), the state-of-the-art method to integrate a population of brain multigraphs into a representative CBT using GNNs. Although compelling, DGN remains a fully integrational –and not predictive– GNN architecture, that is not customized to *dynamic* multigraph populations.

To address all these limitations, we propose our **Recurrent Multi-graph Integrator Network (ReMI-Net)**, the first recurrent graph convolutional neural network architecture for *integrating* a baseline multigraph population into a CBT and *predicting* its time-dependent evolution in a progressive manner. *First*, our ReMI-Net inputs a baseline population of multigraphs and learns how to integrate them into a baseline CBT representation which evolves with time through our recurrent graph convolutional block. Specifically, our model includes a learnable view-normalization block since each view of the individual multigraph can differ in scale and distribution from other views. This learns how to integrate the multigraph population into a baseline CBT at initial timepoint t_1. *Second*, we propose a novel *graph-based* recurrent block that predicts the baseline CBT evolution by learning a low-dimensional hidden-state vector for each node (ROI) at each timepoint. Each time-specific cell of our recurrent block is a message passing network that captures the non-linearities between brain regions (i.e., multigraph nodes) in a graph convolutional end-to-end fashion. *Third*, for each subject in the population and at each timepoint $t_i \geq t_1$, we pass the hidden state embeddings of each node (ROI) to next cell to learn its embedding at the consecutive timepoint t_{i+1}. *Fourth*, we design a novel time-dependent loss to regularize the CBT evolution trajectory over time. *Finally*, we derive the CBT connectivity matrix at timepoint t_i by computing the pairwise distance between the hidden-state node embeddings. We present several major contributions to the state-of-the-art at three different levels. *(1) Methodological level.* Our ReMI-Net is the first GNN architecture for brain CBT learning and time-dependent trajectory prediction from a baseline timepoint. We design a novel message passing network with recurrent multigraph inputs and derive our CBT from the learned time-dependent hidden-state node embeddings. We also propose a novel time-dependent loss to regularize the forecasted CBTs over time. *(2) Conceptual level.* Our framework introduced the concept of a longitudinal network template (i.e., 'atlas') evolution prediction from a baseline multigraph population. *(3) Clinical level.* We demonstrate that ReMI-Net generates reliable and biologically sound brain templates that fingerprint the temporal evolution of demented brains and reveal time-dependent biomarkers.

2 Proposed Method

In the following, we explain the main blocks of the proposed ReMI-Net architecture[1] for forecasting representative, centered and discriminative CBTs from a given multigraph population at baseline timepoint. Figure 1 displays the five key blocks of our ReMI-Net: **A)** Tensor representation of longitudinal multigraph data, **B)** Prediction of the *subject-specific* CBT from learned recurrent hidden-state node embeddings, **C)** Novel message passing network to perform recurrent convolutional operations on multigraphs and generate hidden-state node embeddings, **D)** Proposed CBT centeredness loss $\mathcal{L}_c$ and time regularization loss $\mathcal{L}_t$, and **E)** Generation of the *population-specific* CBT evolution trajectory.

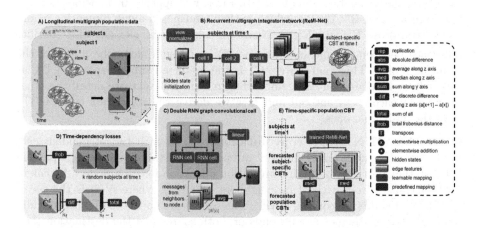

Fig. 1. Overview of our ReMI-Net architecture for integrating a population of multigraphs into a representative CBT and forecasting its evolution over time. Each block of our architecture is detailed in the Method section with reference to components in this figure.

A) Longitudinal multigraph population data. Let $\mathbb{G}_c = \{\mathcal{S}_1^t, \ldots, \mathcal{S}_{n_s}^t\}_{t=1}^{n_t}$ be a set of longitudinal tensors $\{\mathcal{S}^t\}_{t=1}^{n_t}$, where each tensor $\mathcal{S}^t \in \mathbb{R}^{n_r \times n_r \times n_v}$ represents a brain multigraph with n_v views and n_r nodes (Fig. 1–A). Particularly, each edge is a vector $\mathbf{e}_{i,j}^t \in \mathbb{R}^{n_v}$ that connects the node i to j storing connectivity weights across the n_v views.

---

[1] https://github.com/basiralab/ReMI-Net.

B) Recurrent multigraph integrator network. Next, we give an overview of our ReMI-Net architecture and detail each of its fundamental operations and components. *1) View normalization layer.* Since the value range of connectivities in a brain multigraph might largely vary across views, the CBT learning task might be biased towards particular views. To address this issue, we first design a view-normalizer that normalizes the views between 0-1 for each training subject. Specifically, our normalizer includes a normalization layer from [15] that adds learnable parameters –weights and biases– to the z-score normalization, then an additional sigmoid layer shifts the scale to 0-1. *2) Cyclic recursion.* Next, we input the *normalized* multigraph at baseline timepoint t_1 to each of the time-dependent ReMI-Net cells, each learning the hidden-state embeddings $\mathbf{h}^t \in \mathbb{R}^{n_r \times n_h}$ of multigraph nodes across views at baseline and follow-up timepoints $t_i \geq t_1$. n_h denotes the dimensionality of the embedding space. For initialization, we introduce a prior hidden-state vector $\mathbf{h}^0$ composed of zero elements for each node. To reinforce the learning of the baseline cell in our recurring architecture, we feed the hidden-state node embeddings from the last cell at n_t back to the first one –noted as a single learning cycle. *3) Subject-specific CBT prediction over time.* After learning the hidden-state embeddings $\mathbf{h}^t$ at each timepoint t, we derive the subject-specific CBT $\hat{\mathbf{C}}_s^t$ for subject s at timepoint t by several tensor operations (Fig. 1–B). First, we replicate $\mathbf{h}^t$ n_r times producing a tensor $\mathcal{R}^t \in \mathbb{R}^{n_r \times n_h \times n_r}$, which is then transposed along dimension x and z (i.e., transpose $\mathcal{R}^t_{xyz}$ to $\mathcal{R}^t_{zyx}$). Next, we compute the element-wise absolute difference between $\mathcal{R}^t$ and its transpose $\mathcal{R}^{t^T}$ and sum along the y axis of the resulting tensor. This final matrix $\hat{\mathbf{C}}_s^t$ is the predicted subject-specific CBT at time t from hidden-state node embedding matrix $\mathbf{h}^t$. Next, we detail the inner working of each recurrent cell and how we derive the *population-based* CBTs from the individual ones.

C) Double RNN graph convolutional cell. In each time-dependent cell, the hidden-state node embeddings $\mathbf{h}^t$ are propagated through a message passing network that performs a graph convolution operation for each node i, by calculating messages from each of its neighbors $j \in \mathcal{N}(i)$, where $\mathcal{N}(i)$ is the set of neighbors of node i. According to [16], a message from node j to i is theoretically more effective if the message depends on both source and the destination (i.e., pair messages). Therefore, a pair message function can be defined in the spirit of [17]: $\mathbf{m}_{i,j}^{t+1} = \mathcal{M}(\mathbf{h}_i^t, \mathbf{h}_j^t, \mathbf{e}_{i,j}^t)$, where $\mathbf{m}_{i,j}^{t+1} \in \mathbb{R}^{n_h}$ is the message vector from $j \in \mathcal{N}(i)$ to node i computed at the next timepoint. In order to compute the next hidden-state node embeddings in a graph convolutional fashion that captures nonlinear dependencies between nodes, we further propose a novel message function as: $\mathcal{M}(\mathbf{h}_i^t, \mathbf{h}_j^t, \mathbf{e}_{i,j}^t) = f_\Theta(\mathbf{e}_{i,j}^t, \mathbf{h}_i^t) \odot f_\Theta(\mathbf{e}_{i,j}^t, \mathbf{h}_j^t)$ (Fig. 1–C). This is operated via each ReMI-Net cell composed of two RNN cells that share and pass messages to generate the hidden-state node embeddings at the next timepoint $t+1$. Both RNN cells input the baseline edge features $\mathbf{e}_{i,j}^{t=1}$ in addition to taking the hidden embeddings $\mathbf{h}_i^t$ for node i and $\mathbf{h}_j^t$ for node j, respectively (Fig. 1–C). Next, we perform an element-wise multiplication between the outputs of both RNN cells as a message to node i from j. We then average the messages across neighbors to eventually generate the hidden-state embedding vector $\mathbf{h}_i^{t+1} \in \mathbb{R}^{n_h}$ for node i at follow-up timepoint:

$$\mathbf{h}_i^{t+1} = \Theta \cdot \mathbf{h}_i^t + \frac{1}{|\mathcal{N}(i)|}\left[\sum_{j \in \mathcal{N}(i)} f_\Theta(\mathbf{e}_{i,j}^t, \mathbf{h}_i^t) \odot f_\Theta(\mathbf{e}_{i,j}^t, \mathbf{h}_j^t)\right];$$

f_Θ is the proposed RNN cell function to learn, which maps for each node a pair of edge features and corresponding hidden state embeddings to an n_h-dimensional vector. A separate set of learnable parameters –weights and biases– ($\Theta \in \mathbb{R}^{n_h \times n_h}$) is defined to generalize the convolution and consider the case of the node level effects [17]. Our designed f_Θ function is also generalizable to long-short term memory (LSTM) [18] and gated recurrent unit (GRU) [19] networks.

D) Time-dependent loss. To learn a centered and well-representative CBT, we randomly sample k subjects from the input population and take the Frobenius distance ($d_F(A, B) = \sqrt{\sum_i \sum_j |A_{ij} - B_{ij}|^2}$) between each subject-specific CBT $\hat{\mathbf{C}}_s^t$ and each random training subject (Fig. 1–D). Next, we sum the resulting distances across random subjects and average across timepoints. The random training sampling has a regularization effect as it reduces the risk of overfitting [9]. We formulate our *subject-specific* CBT centeredness loss for subject s as follows:

$$\mathcal{L}_c(s) = \frac{1}{n_t}\sum_{t=1}^{n_t}\sum_{v=1}^{n_v}\sum_{\mathcal{S}_k^t \in \mathbb{K}}^{n_k} \lambda_v \|\hat{\mathbf{C}}_s^t - \mathcal{S}_k^t\|_F; \quad \lambda_v = \frac{max\{\mu_v\}_{v=1}^{n_v}}{\mu_v}$$

where λ_v is a view-normalizer calculated for each subject and each timepoint separately, $\mathbb{K}$ is a randomly selected subset and $\mathcal{S}_k^t$ is a random training sample. μ_v is the mean value of each view. To stabilize our training over time, we further propose a time regularization loss where we constrain two consecutive CBTs to be similar as we hypothesize that connectivity brain changes are local and sparse:

$$\mathcal{L}_t(s) = \frac{1}{n_t - 1}\sum_{t=1}^{n_t - 1}\sqrt{\sum_i \sum_j (\hat{\mathbf{C}}_{s_{i,j}}^{t+1} - \hat{\mathbf{C}}_{s_{i,j}}^t)^2}$$

Lastly, time regularizer loss $\mathcal{L}_t$ has also a weight α as a hyperparameter in total loss: $\mathcal{L}_{total} = \frac{1}{n_s}\sum_{s=1}^{n_s}[\mathcal{L}_c(s) + \alpha\mathcal{L}_t(s)]$.

E) Generation of the *population-specific* CBT evolution trajectory. So far, each of the estimated subject-specific CBTs is somewhat biased towards a specific subject. Hence, we perform an additional post-training operation by passing all training samples through the trained ReMI-Net to generate subject-specific CBTs (Fig. 1–E). Since we aim to obtain population representative CBTs at all timepoints t, we use the element-wise median to select the most centered connectivities across all subject specific CBTs. This results in population center $\hat{\mathbf{P}}^t$ at time t.

3   Results and Discussion

Longitudinal Brain Multigraph Dataset. We evaluated our architecture on 67 subjects (32 subjects diagnosed with Alzheimer's Diseases (AD) and 35 with Late Mild Cognitive Impairment (LMCI)) from the Alzheimer's Disease Neuroimaging Initiative (ADNI) dataset [7], each represented by four cortical brain networks derived from T1-weighted magnetic resonance imaging for each cortical hemisphere, respectively. Each network comprises $n_r = 35$ nodes and is measured at $n_t = 2$ timepoints (baseline and a 6-month follow up). The multigraph tensor views are derived from the following measures: maximum principal curvature, the mean cortical thickness, the mean sulcal depth, and the average curvature. Each edge weight encodes the average dissimilarity in morphology between two brain cortical ROIs using Desikan-Killiany cortical surface parcellation atlas.

Evaluation and Parameter Setting. We evaluated our model against its variants and benchmark methods using 5-fold cross-validation where we learned the CBT from all samples but for each hemisphere independently. The recurrent block is composed of 3 layers of ReMI-Net cells, each with hidden-state node embeddings size 12, 36 and 24, respectively. All models are trained using gradient descent with Adam optimization, a learning rate of 0.0008 and 100 epochs. We empirically set the hyperparameter $\alpha = 0.3$ in our loss function and the number of random training samples in the centeredness loss to $k = 10$. As for the recursive cycling, we only cycle once.

Benchmarks and ReMI-Net Variations. Since our ReMI-Net is the first method aiming to forecast CBT evolution from a single baseline population, we benchmarked our model against DGN [9] at each timepoint independently. We further evaluated 3 variants of our model: **(1) Standard vanilla:** 3 layers of graph convolutional cells without normalization and cyclic recursion to the baseline timepoint; **(2) Standard cyclic:** 3 layers of graph convolutional cells with cyclic recursion operation and without normalization; and **(3) Cyclic min-max norm:** 3 layers of graph convolutional cells with cyclic recursion and min-max view normalization.

CBT Representativeness Test. To evaluate our ReMI-Net at different timepoints, we calculated the average Frobenius distance between the predicted training CBT (i.e., generated from the training set) and all samples in the testing set. We inspect the results at each timepoint independently using two different evaluation strategies. **(1) Last model:** We train each model for a complete 100

Table 1. Reproducibility rate between ROIs selected by CBT-based methods and an independent learner [20].

Overlap rate	netNorm [8]	DGN [9]	**ReMI-Net**
AD-LMCI Left Hem.	0.60	**0.73**	**0.73**
AD-LMCI Right Hem.	0.33	0.40	**0.80**

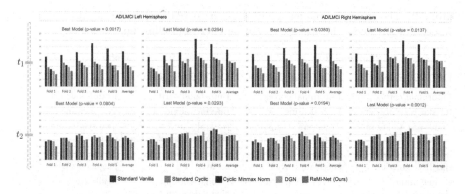

Fig. 2. Evaluation of the centeredness of the learned CBTs by DGN [9] and ReMI-Net variants at baseline and follow-up timepoints t_1 and t_2, respectively.

epochs for easy comparison across methods and **(2) Best model:** We pick the best model that achieves the best results on the testing samples by adopting an early stop training strategy (i.e., number of epochs < 100). Figure 2 displays the average Frobenius distance quantifying the centeredness and representativeness of the CBT forecasted by ReMI-Net and its 3 variants at each timepoint. **Table 1-supp** displays the MAE in node strength where our ReMI-Net outperforms benchmarks. **Fig. 1-supp** also shows that our model achieves the best performance on a simulated connectomic dataset with 6 timepoints. We also report the CBT results by training DGN [9] at each timepoint, *independently*. Lastly, we report the p-values using a two-tailed paired t-test between DGN (the best performing method among benchmarks) and our ReMI-Net. Clearly, our model significantly outperforms DGN and its variants in terms of CBT representativeness across all test folds, evaluation datasets (left and right hemispheres) and at all timepoints.

CBT Discriminativeness and Biomarker Reproducibility Test. By acting as a connectional brain fingerprint, a well-centered CBT can capture the most discriminative ROIs of a multigraph population between two brain states (e.g., LMCI *vs.* AD) and identify their alternations over time caused by a progressive disease such as dementia. To test this hypothesis, we train our ReMI-Net on each class-specific population –LMCI and AD independently, to predict the CBT at baseline and follow-up timepoints. Next, we compute the residual matrix between the CBT^{LMCI} and CBT^{AD} via absolute element-wise difference. Then we assign a discriminability score for each brain ROI by summing the elements of its row in the residual matrix. The ROIs with the highest discriminability scores better disentangle both brain states. To further evaluate the reproducibility of the discovered biomarker regions, we train an independent learner, specifically the generalized multiple kernel learning (GMKL) [20] which classifies samples and learns a weight for each ROI using 5-fold cross-validation. Next, we inspect the reproducibility overlap rate of the top $K = 15$ most discriminative ROIs between GMKL and each of CBT learning models. Table 1 shows that ReMI-Net has a

Table 2. The most discriminative ROIs (top to bottom) between AD and LMCI patients for left and right hemispheres at baseline and follow-up timepoints.

Timepoint t_1		Timepoint t_2	
Left Hemisphere	Right Hemisphere	Left Hemisphere	Right Hemisphere
Entorhinal cortex	Entorhinal cortex	Entorhinal cortex	Entorhinal cortex
Inferior temporal gyrus	Pericalcarine cortex	Inferior temporal gyrus	Pericalcarine cortex
Transverse temporal cortex	Transverse temporal cortex	Rostral anterior cingulate cortex	Transverse temporal cortex
Rostral anterior cingulate cortex	Rostral anterior cingulate cortex	Transverse temporal cortex	Rostral anterior cingulate cortex
Caudal anterior cingulate cortex	Inferior temporal gyrus	Caudal anterior cingulate cortex	Insula cortex

Table 3. The top brain ROIs marking the difference between the baseline and follow-up CBTs generated for AD patients by our ReMI-NET.

Left Hemisphere	Right Hemisphere
Pericalcarine cortex	Pericalcarine cortex
Caudal anterior cingulate cortex	Caudal anterior cingulate cortex
Medial orbital frontal cortex	Medial orbital frontal cortex
Postcentral gyrus	Postcentral gyrus
Precuneus cortex	Precuneus cortex

significant increase in the overlap reproducibility in the right hemisphere and a similar performance to state-of-the-art DGN [9] in the left hemisphere.

Clinical Discoveries. 1) *Discriminability approach.* Table 2 displays the top 5 most discriminative between AD and LMCI at baseline and follow-up timepoints t_1 and t_2. The entorhinal cortex is found as the most discriminative ROI between both groups, which replicates the findings of independent clinical studies on LMCI and AD [21–23]. **2)** *Progressive approach.* Table 3 shows the most affected ROIs by AD with the disorder progression over time. The results are significantly similar between left and right hemispheres. As reported in [21], we also identify the pericalcarine cortex, precuneus cortex and medial orbital frontal cortex as the most dynamically changing regions in AD patients.

4   Conclusion

In this paper, we introduced the *first* Recurrent Multi-graph Integrator Network to forecast the connectional brain template evolution over time using a single baseline population. The proposed method outperformed both state-of-the-art and variant benchmarks in terms of all evaluation measures. Moreover, we introduced our own recurrent graph convolution method and demonstrated the high reproducibility of biomarkers revealed by our learned time-dependent CBTs in a demented population. In our future work, we will evaluate the generalizability of ReMI-Net to large-scale brain multigraph populations derived from different neuroimaging modalities. We will also explore other operations to derive the CBT from the learned embeddings including graph-based similarity measures.

Acknowledgments. I. Rekik is supported by the European Union's Horizon 2020 research and innovation programme under the Marie Sklodowska-Curie Individual Fellowship grant agreement No 101003403 (http://basira-lab.com/normnets/) and the Scientific and Technological Research Council of Turkey under the TUBITAK 2232 Fellowship for Outstanding Researchers (no. 118C288, http://basira-lab.com/reprime/). However, all scientific contributions made in this project are owned and approved solely by the authors.

References

1. Bassett, D.S., Sporns, O.: Network neuroscience. Nat. Neurosci. **20**, 353 (2017)
2. Fornito, A., Zalesky, A., Breakspear, M.: The connectomics of brain disorders. Nat. Rev. Neurosci. **16**, 159–172 (2015)
3. Van den Heuvel, M.P., Sporns, O.: A cross-disorder connectome landscape of brain dysconnectivity. Nat. Rev. Neurosci. **20**, 435–446 (2019)
4. Mheich, A., Wendling, F., Hassan, M.: Brain network similarity: methods and applications. Netw. Neurosci. **4**, 507–527 (2020)
5. Essen, D., et al.: The human connectome project: a data acquisition perspective. Neuroimage **62**, 2222–31 (2012)
6. Van Essen, D., Glasser, M.: The human connectome project: progress and prospects. In: Cerebrum: The Dana Forum on Brain Science 2016 (2016)
7. Mueller, S., et al.: The Alzheimer's disease neuroimaging initiative. Neuroimaging Clin. N. Am. **15**, 869 (2005)
8. Dhifallah, S., Rekik, I., Initiative, A.D.N., et al.: Estimation of connectional brain templates using selective multi-view network normalization. Med. Image Anal. **59**, 101567 (2020)
9. Gurbuz, M.B., Rekik, I.: Deep graph normalizer: a geometric deep learning approach for estimating connectional brain templates. In: Martel, A.L., et al. (eds.) MICCAI 2020. LNCS, vol. 12267, pp. 155–165. Springer, Cham (2020). https://doi.org/10.1007/978-3-030-59728-3_16
10. Mhiri, I., Mahjoub, M.A., Rekik, I.: Supervised multi-topology network cross-diffusion for population-driven brain network atlas estimation. In: Martel, A.L., et al. (eds.) MICCAI 2020. LNCS, vol. 12267, pp. 166–176. Springer, Cham (2020). https://doi.org/10.1007/978-3-030-59728-3_17
11. Wang, B., et al.: Similarity network fusion for aggregating data types on a genomic scale. Nat. Methods **11**, 333–337 (2014)
12. Zhou, J., et al.: Graph neural networks: a review of methods and applications. arXiv preprint arXiv:1812.08434 (2018)
13. Zhang, S., Tong, H., Xu, J., Maciejewski, R.: Graph convolutional networks: a comprehensive review. Comput. Soc. Netw. **6**(1), 1–23 (2019). https://doi.org/10.1186/s40649-019-0069-y
14. Bessadok, A., Mahjoub, M.A., Rekik, I.: Graph neural networks in network neuroscience. arXiv preprint arXiv:2106.03535 (2021)

15. Ba, J.L., Kiros, J.R., Hinton, G.E.: Layer normalization (2016)
16. Gilmer, J., Schoenholz, S.S., Riley, P.F., Vinyals, O., Dahl, G.E.: Neural message passing for quantum chemistry. In: International Conference on Machine Learning, PMLR, pp. 1263–1272 (2017)
17. Battaglia, P.W., Pascanu, R., Lai, M., Rezende, D., Kavukcuoglu, K.: Interaction networks for learning about objects, relations and physics (2016)
18. Yu, Y., Si, X., Hu, C., Zhang, J.: A review of recurrent neural networks: LSTM cells and network architectures. Neural Comput. **31**, 1235–1270 (2019)
19. Chung, J., Gulcehre, C., Cho, K., Bengio, Y.: Empirical evaluation of gated recurrent neural networks on sequence modeling. arXiv preprint arXiv:1412.3555 (2014)
20. Varma, M., Babu, B.: More generality in efficient multiple kernel learning, p. 134 (2009)
21. Yang, H., et al.: Study of brain morphology change in Alzheimer's disease and amnestic mild cognitive impairment compared with normal controls. Gen. Psychiatry **32**, e100005 (2019)
22. Zhou, M., Zhang, F., Zhao, L., Qian, J., Dong, C.: Entorhinal cortex: a good biomarker of mild cognitive impairment and mild Alzheimer's disease. Rev. Neurosci. **27**, 185–195 (2016)
23. Howett, D., et al.: Differentiation of mild cognitive impairment using an entorhinal cortex-based test of virtual reality navigation. Brain **142**, 1751–1766 (2019)

Efficient Neural Network Approximation of Robust PCA for Automated Analysis of Calcium Imaging Data

Seungjae Han[1], Eun-Seo Cho[1], Inkyu Park[2], Kijung Shin[1,2],
and Young-Gyu Yoon[1(✉)]

[1] School of Electrical Engineering, KAIST, Daejeon, South Korea
ygyoon@kaist.ac.kr
[2] Graduate School of AI, KAIST, Daejeon, South Korea

Abstract. Calcium imaging is an essential tool to study the activity of neuronal populations. However, the high level of background fluorescence in images hinders the accurate identification of neurons and the extraction of neuronal activities. While robust principal component analysis (RPCA) is a promising method that can decompose the foreground and background in such images, its computational complexity and memory requirement are prohibitively high to process large-scale calcium imaging data. Here, we propose BEAR, a simple bilinear neural network for the efficient approximation of RPCA which achieves an order of magnitude speed improvement with GPU acceleration compared to the conventional RPCA algorithms. In addition, we show that BEAR can perform foreground-background separation of calcium imaging data as large as tens of gigabytes. We also demonstrate that two BEARs can be cascaded to perform simultaneous RPCA and non-negative matrix factorization for the automated extraction of spatial and temporal footprints from calcium imaging data. The source code used in the paper is available at https://github.com/NICALab/BEAR.

Keywords: Calcium imaging · Robust principal component analysis · Neural network · Non-negative matrix factorization

1 Introduction

Recent advances in calcium imaging techniques have enabled imaging population neuronal activity across a large volume [21]. State-of-the-art functional imaging methods can generate more than a gigabyte of data per second [2, 6, 23, 25] that is prohibitively large to be analyzed by humans, which necessitates the development of automated analysis algorithms. Furthermore, calcium imaging data suffers from a high level of background fluorescence, which stems from the intrinsic property of calcium indicator molecules. Neurons with a calcium indicator

Electronic supplementary material The online version of this chapter (https://doi.org/10.1007/978-3-030-87234-2_56) contains supplementary material, which is available to authorized users.

have a certain baseline fluorescence level at rest (i.e., no activity), which becomes brighter by 20 to 40% when there is a single spike. Because most of the neurons are silent at a given time, the images consist of mostly background and a very small portion of foreground, which makes it challenging to identify neurons in calcium imaging data. Therefore, it is desirable to first separate the foreground (i.e., activity) from the backgrounds [13,17] before applying downstream analysis algorithms, such as image-based neuron detection algorithms [7,12], NMF-based signal extraction methods [15,18], and even volume reconstruction [25].

Robust principal component analysis (RPCA) [4] is a promising method for the separation of the foreground in calcium imaging data. Neuronal activities can be separated from the background fluorescence by exploiting that the background is nearly stationary, while the activity is spatiotemporally sparse [25]. The separation can be formulated as the following optimization problem:

$$\min_{L,S}(rank(L) + \lambda ||S||_0) \text{ subject to } Y = L + S, \tag{1}$$

where Y, L, S, $||S||_0$, and λ are a data matrix, a low-rank matrix, a sparse matrix, the L_0 norm of S, and a hyperparameter, respectively. While this problem is known to be computationally intractable, the exact solution can be obtained under weak assumptions through the surrogate optimization as follows [4,5]:

$$\min_{L,S}(||L||_* + \lambda ||S||_1) \text{ subject to } Y = L + S, \tag{2}$$

where $||L||_*$ is the nuclear norm of L, and $||S||_1$ is the L_1 norm of S.

However, conventional RPCA based on the principal component pursuit method (PCP) [4] for solving the optimization problem has limited capabilities when it comes to handling extremely large data because it involves singular value decomposition (SVD) of the entire data matrix, which is computationally expensive. In addition, solving the exact optimization problem requires storage of the entire data in the main memory, which is not feasible for large-scale calcium imaging data sets. Thus, many variants of RPCA have been developed to increase the speed [10,27,28] or to process large data [8,19,24], but it remains a challenge to simultaneously achieve high speed and scalability.

Here, we introduce a Bilinear neural network for Efficient Approximation of RPCA (BEAR) which is a computationally efficient implementation of RPCA as a neural network. BEAR has a number of advantages over conventional RPCA algorithms. It (a) has significantly lower time complexity of $O(nmr)$ than conventional RPCA with time complexity of $O(nm^2 + n^2m)$ for single iteration, where n and m are the sizes of the data along two dimensions and r is the rank of L which is typically a small integer, (b) can decompose extremely large data, and (c) can be naturally combined with other neural networks for end-to-end training.

2   Methods

2.1   BEAR as a Robust Principal Component Analysis

BEAR is based on the following surrogate optimization:

$$\min_{L,S} ||S||_1 \text{ subject to } Y = L + S \text{ and } rank(L) \le r, \tag{3}$$

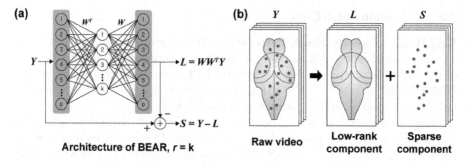

Fig. 1. Bilinear neural network for efficient approximation of RPCA (BEAR). (a) The low rank component L is obtained by passing the input data Y through the bilinear network. The sparse component S is obtained as $S = Y - L$. (b) BEAR can decompose calcium imaging data into background and neural activity.

which is attained by replacing the minimization of $||L||_*$ in (2) by the maximum rank constraint on L. A low-rank representation of the data is obtained by choosing a small integer r. The constraint $rank(L) \leq r$ is enforced by setting $L = WMY$, where $W \in \mathbb{R}^{n \times r}$ and $M \in \mathbb{R}^{r \times n}$. We note that $L = WMY$ is a stronger condition than $rank(L) \leq r$, and this additional condition allows us to implement the optimization process as training the network shown in Fig. 1a. Furthermore, we imposed an additional constraint on W and M by setting $M = W^T$ to reduce the number of trainable parameters and make the training more stable. Thus, the final form of the optimization problem is expressed as

$$\min_{W} ||S||_1 \text{ subject to } Y = L + S \text{ and } L = WW^T Y. \tag{4}$$

As the three constraints, (a) $rank(L) \leq r$, (b) $Y = L + S$, and (c) $M = W^T$, are set by the network architecture, the optimization problem can be solved by simply training the network to minimize the loss function $\mathcal{L} = ||S||_1$ (Algorithm S1 in the supplementary material), using gradient-based optimization algorithms, such as stochastic gradient descent, RMSprop [22], and Adam [11]. This allows us to use mini-batches for training the network; hence, it can be applied to an extremely large data matrix Y. This property is ideal for separating foregrounds in large-scale calcium imaging data (Fig. 1b). It should be noted that both forward propagation and backpropagation through the network require only two matrix multiplications thereby achieving the computational complexity of $O(nmr)$, whereas other methods that employ SVD have the computational complexity of $O(nm^2 + n^2m)$. Furthermore, the network can be used to infer the low-rank components without updating W, assuming that the network is trained with the earlier data points and the low-rank components remain stationary. This inference-only mode can significantly improve the computation time.

2.2   Extensions of BEAR

BEAR with Greedy Rank Estimation. BEAR requires the integer number r to be explicitly set which has to come from prior knowledge regarding the rank of the low-rank matrix L, whereas general RPCA methods do not. To employ BEAR in the absence of such prior knowledge, we introduce an extension of BEAR with greedy rank estimation (Greedy BEAR).

Rank estimation is done by adding an outer loop to BEAR in which the target rank is gradually increased. As soon as $rank(L) + \lambda ||S||_1$ increases, where $rank(L)$ is the target rank and $||S||_1$ is the loss after training the BEAR with the target rank, the algorithm terminates, and the network from the earlier iteration of the outer loop is chosen as the final model. Note that this algorithm (Algorithm S2 in the supplementary material) attempts to solve $\min_{L,S}(rank(L) + \lambda ||S||_1)$, which is similar to the original optimization problem expressed in (1).

BEAR as Non-negative Matrix Factorization. We note that BEAR can be modified to solve non-negative matrix factorization (NMF) by changing the loss function and introducing a non-negativity constraint on W as follows:

$$\min_{W} ||R||_F \text{ subject to } Y = L + R \text{ and } L = WW^T Y \text{ and } W \geq 0. \tag{5}$$

This optimization problem is equivalent to projective NMF [26]. It is known to produce localized sparse footprints, which is ideal for neuron segmentation.

BEAR Cascaded with Downstream Networks. Because BEAR is a neural network, it can be combined with other networks for downstream tasks and trained end-to-end. This can be implemented by cascading BEAR with another

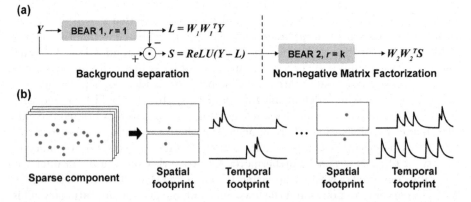

Fig. 2. (a) By cascading two BEARs, RPCA and NMF can be simultaneously performed. (b) The second BEAR in **a** takes sparse component S from the first BEAR and performs NMF. The coefficient matrix W_2 and $W_2^T S$ correspond to the extracted spatial and temporal footprints, respectively.

network (Cascaded BEAR). The loss function can be set as $\mathcal{L} = \mathcal{L}_1 + \mu \mathcal{L}_2$, where $\mathcal{L}_1$ and $\mathcal{L}_2$ are the loss functions of the first and second networks, respectively, and μ is a hyperparameter. As an example, two BEARs can be cascaded to simultaneously perform RPCA and NMF as shown in Fig. 2a. The sparse component from the first BEAR is fed to a ReLU layer to ensure non-negativity and then fed to the second BEAR for NMF (Algorithm S3 in the supplementary material) to extract the spatial and temporal footprints as illustrated in Fig. 2b.

3 Experiments

To validate and evaluate our algorithm, we compared Greedy BEAR with existing RPCA algorithms, PCP [4], IALM [14], GreGoDec [28] and OMWRPCA [24]. Next, we applied the cascaded BEAR to calcium imaging datasets to demonstrate its capability to perform RPCA and NMF simultaneously. We ran these tests on a PC with Intel i7-9700K CPU, NVIDIA GeForce RTX 2080 Ti GPU, and 128GB of RAM. The source code of PCP, IALM, and GreGoDec was from LRSLibrary [3,20], a publicly available repository of multiple RPCA algorithms. The source code of OMWRPCA was from [24]. PCP, IALM, GreGoDec, and OMWRPCA were run on the CPU. BEAR was implemented using Pytorch, and its performance on the GPU and the CPU was tested separately. The computation time of BEAR includes the time taken for both training and inference. The default maximum number of iterations of PCP was reduced by a factor of 10. The animal experiments conducted for this study were approved by the Institutional Animal Care and Use Committee (IACUC) of KAIST (KA2019-13).

3.1 Phase Diagram in Rank and Sparsity

We evaluated the ability of the algorithms to recover varying rank matrices from data matrices that have error with varying levels of sparsity. We randomly generated each data matrix $Y \in \mathbb{R}^{n \times n}$ by adding a low-rank matrix L and a sparse matrix S, where n was set as 1,000. Each low rank matrix was generated as $L = XY^T$, where the entities of $X \in \mathbb{R}^{n \times r}$ and $Y \in \mathbb{R}^{n \times r}$ were drawn from a normal distribution with the mean value of zero and the variance of $1/n$. Each sparse matrix S was generated by drawing each entity from a probability mass function as follows: $P(x = 0.1) = \rho/2, P(x = -0.1) = \rho/2, P(x = 0) = 1 - \rho$.

For each (r, ρ) pair, we generated 5 random matrices, each of which was solved with the four algorithms. The evaluation metrics were the relative error, defined as $||L - \hat{L}||_F / ||L||_F$ where $\hat{L}$ was the recovered low-rank matrix, and the computation time. We averaged the relative error and computation time from 5 trials for each (r, ρ) pair. We used the Adam optimizer with the batch size of 1,000. The learning rate was 0.003, and the number of training epochs was 50.

Figure 3 shows the obtained phase diagrams. Greedy BEAR achieved a relatively uniform level of error across a wide range of r and ρ, while other algorithms failed in the high-rank or low-sparsity phase (Fig. 3a). Greedy BEAR was the fastest under all conditions (Fig. 3b). In the rightmost plots in Fig. 3, each grid,

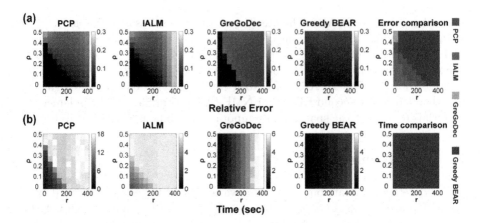

Fig. 3. Phase diagram for PCP, IALM, GreGoDec, and Greedy BEAR on 1000×1000 matrices. (a) Relative errors. (b) Computation times.

which represents each (r, ρ) pair, is assigned to the best algorithm in terms of each measure. The percentages of (r, ρ) pairs which Greedy BEAR achieved the best relative error and the computation time were 69% and 100%, respectively.

3.2 Performance on Large Calcium Imaging Data

We measured the computation times of the algorithms using 3-D calcium imaging data. The data was acquired by performing calcium imaging of larval zebrafish brains 4 Hz using a custom-designed epi-fluorescence microscope. The larvae expressing GCaMP7a pan-neuronally were imaged at 4 days post fertilization. The larvae were paralyzed in standard fish water containing 0.25 mg/ml of pancuronium bromide for 2 min prior to imaging and then embedded in agarose for immobilization.

The sizes of the videos were $480(x) \times 270(y) \times 41(z) \times 150(t)$ and $480(x) \times 270(y) \times 41(z) \times 1000(t)$, respectively, and they were reshaped to 5313600×150 (3.2GB) and 5313600×1000 (21.3GB), respectively. For BEAR with inference-only mode, the network was trained with the first one third of the data, and inference was performed on the entire data. We used the Adam optimizer with

Table 1. Computation times of four algorithms (in seconds). *Out of Memory. **Predicted based on small number of iterations. †Inference-only mode.

Algorithms data size	PCP	IALM	GreGoDec	OMWRPCA	Greedy BEAR		BEAR†
	CPU	CPU	CPU	CPU	CPU	GPU	GPU
5313600×150	13814	1211	429	37883**	371	134	**42**
5313600×1000	OOM*	OOM*	OOM*	80883**	1345	537	**377**

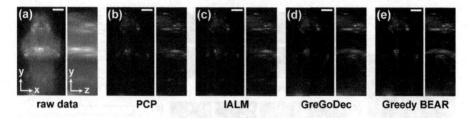

Fig. 4. Decomposition of the zebrafish calcium imaging data. Maximum-intensity projections along x and z axes are shown. (a) Input image. (b–e) Sparse components from PCP, IALM, GreGoDec, and BEAR. Scale bars, 100 μm.

the batch size of 64. The learning rate was 0.00005, and the number of training epochs was 45.

The computation times of the four algorithms are summarized in Table 1. Greedy BEAR with GPU acceleration was faster than the existing algorithms and was capable of processing the largest data. The decomposition results obtained by the four algorithms were visually nearly indistinguishable as shown in Fig. 4 and Supplementary Video 1.

3.3 Simultaneous RPCA and NMF Using Cascaded BEAR

We applied cascaded BEAR illustrated in Fig. 2a to the publicly available mouse brain calcium imaging data obtained by two-photon microscopy [9] for simultaneous RPCA and NMF. The dimensions of the video were $80(x) \times 60(y) \times 2000(t)$, which were reshaped to 4800×2000. The ranks for the first and second BEARs were set as 1 and 8, respectively. We used the Adam optimizer with the batch size of 512. The learning rate was 0.0002 and the number of training epochs was 5,000. As shown in Fig. 5, the spatial components from cascaded BEAR were confined and sparse, whereas those from conventional NMF [1,16] were not. In addition, the temporal signals from cascaded BEAR showed lower baseline fluctuation due to less signal mixing.

Next, we applied cascaded BEAR to a zebrafish calcium imaging video (Fig. 6a). It was acquired by imaging a larval zebrafish brain, expressing nuclear localized GCaMP6s pan-neuronally, 10 Hz using a spinning disk confocal microscope with a 25x 1.1NA objective lens. The dimensions of the video were $768(x) \times 768(y) \times 593(t)$, which were reshaped to 589824×593. The ranks for the first BEAR and second BEAR were set as 1 and 50, respectively. We used the Adam optimizer with the batch size of 64. The learning rate was 0.0001, and the number of training epochs was 1,000. As shown in Fig. 6, cascaded BEAR was able to process the calcium imaging data with a very large number of neurons. The obtained spatial footprints were confined and they corresponded well with the neurons (Fig. 6b), whereas the results obtained by conventional NMF were not confined, which manifested as low color contrast (Fig. 6c).

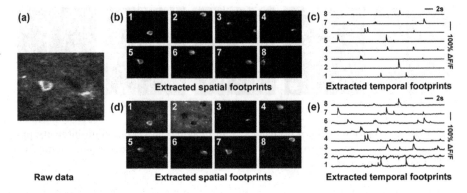

Fig. 5. Cascaded BEAR for analysis of neuronal activity of a mouse. (a) Single frame from the data. (b) Spatial footprints extracted using cascaded BEAR. (c) Temporal signals extracted using cascaded BEAR. (d) As in **b**, but using conventional NMF. (e) As in **c**, but using conventional NMF.

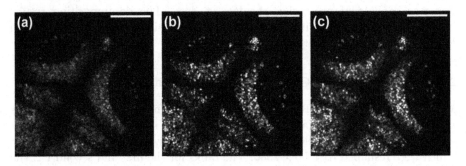

Fig. 6. Cascaded BEAR for analysis of neuronal activity of a larval zebrafish. (a) A confocal image of a larval zebrafish brain expressing nuclear localized GCaMP6s. (b) Spatial footprints from cascaded BEAR are colored and overlaid. (c) As in **b**, but from conventional non-negative matrix factorization. Same random color palette as in **b** is assigned to each spatial footprint. Scale bars, 100 μm.

4   Conclusion

In summary, we proposed BEAR, a bilinear neural network for analyzing large calcium imaging data. Because BEAR can be trained using the gradient from mini-batches, it can decompose arbitrary sized data while exploiting GPU acceleration for speed improvement. Furthermore, BEAR can be cascaded with other networks for end-to-end training. For example, two BEARs can be cascaded for simultaneous RPCA and NMF to identify neurons in a large calcium imaging data. Although our demonstration was focused on processing calcium imaging data, BEAR could be used for other general RPCA applications.

Acknowledgements. This research was supported by National Research Foundation of Korea (2020R1C1C1009869), the Korea Medical Device Development Fund grant funded by the Korea government (202011B21-05), Institute of Information & communications Technology Planning & Evaluation (IITP) grant funded by the Korea government (MSIT) (No.2019-0-00075, Artificial Intelligence Graduate School Program (KAIST)), and 2020 KAIST-funded AI Research Program. The zebrafish lines used for calcium imaging were provided by the Zebrafish Center for Disease Modeling (ZCDM), Korea.

References

1. Berry, M.W., Browne, M., Langville, A.N., Pauca, V.P., Plemmons, R.J.: Algorithms and applications for approximate nonnegative matrix factorization. Comput. Stat. Data Anal. **52**(1), 155–173 (2007)
2. Bouchard, M.B., et al.: Swept confocally-aligned planar excitation (scape) microscopy for high-speed volumetric imaging of behaving organisms. Nat. Photonics **9**(2), 113–119 (2015)
3. Bouwmans, T., Sobral, A., Javed, S., Jung, S.K., Zahzah, E.H.: Decomposition into low-rank plus additive matrices for background/foreground separation: a review for a comparative evaluation with a large-scale dataset. Comput. Sci. Rev. **23**, 1–71 (2017)
4. Candès, E.J., Li, X., Ma, Y., Wright, J.: Robust principal component analysis? J. ACM (JACM) **58**(3), 1–37 (2011)
5. Chandrasekaran, V., Sanghavi, S., Parrilo, P.A., Willsky, A.S.: Rank-sparsity incoherence for matrix decomposition. SIAM J. Optim. **21**(2), 572–596 (2011)
6. Cong, L., et al.: Rapid whole brain imaging of neural activity in freely behaving larval zebrafish (danio rerio). Elife **6**, e28158 (2017)
7. Dong, M., et al.: Towards neuron segmentation from macaque brain images: a weakly supervised approach. In: Martel, A.L., et al. (eds.) MICCAI 2020. LNCS, vol. 12265, pp. 194–203. Springer, Cham (2020). https://doi.org/10.1007/978-3-030-59722-1_19
8. Feng, J., Xu, H., Yan, S.: Online robust PCA via stochastic optimization. Adv. Neural. Inf. Process. Syst. **26**, 404–412 (2013)
9. Giovannucci, A., et al.: Caiman an open source tool for scalable calcium imaging data analysis. Elife **8**, e38173 (2019)
10. Hovhannisyan, V., Panagakis, Y., Parpas, P., Zafeiriou, S.: Multilevel approximate robust principal component analysis. In: Proceedings of the IEEE International Conference on Computer Vision Workshops, pp. 536–544 (2017)
11. Kingma, D.P., Ba, J.: Adam: a method for stochastic optimization. arXiv preprint arXiv:1412.6980 (2014)
12. Kirschbaum, E., Bailoni, A., Hamprecht, F.A.: DISCo: deep learning, instance segmentation, and correlations for cell segmentation in calcium imaging. In: Martel, A.L., et al. (eds.) MICCAI 2020. LNCS, vol. 12265, pp. 151–162. Springer, Cham (2020). https://doi.org/10.1007/978-3-030-59722-1_15
13. Li, C., et al.: Fast background removal method for 3D multi-channel deep tissue fluorescence imaging. In: Descoteaux, M., Maier-Hein, L., Franz, A., Jannin, P., Collins, D.L., Duchesne, S. (eds.) MICCAI 2017. LNCS, vol. 10434, pp. 92–99. Springer, Cham (2017). https://doi.org/10.1007/978-3-319-66185-8_11
14. Lin, Z., Chen, M., Ma, Y.: The augmented lagrange multiplier method for exact recovery of corrupted low-rank matrices. arXiv preprint arXiv:1009.5055 (2010)

15. Nejatbakhsh, A., et al.: Demixing calcium imaging data in *C. elegans* via deformable non-negative matrix factorization. In: Martel, A.L., et al. (eds.) MICCAI 2020. LNCS, vol. 12265, pp. 14–24. Springer, Cham (2020). https://doi.org/10.1007/978-3-030-59722-1_2

16. Paatero, P., Tapper, U.: Positive matrix factorization: a non-negative factor model with optimal utilization of error estimates of data values. Environmetrics **5**(2), 111–126 (1994)

17. Peng, T., Wang, L., Bayer, C., Conjeti, S., Baust, M., Navab, N.: Shading correction for whole slide image using low rank and sparse decomposition. In: Golland, P., Hata, N., Barillot, C., Hornegger, J., Howe, R. (eds.) MICCAI 2014. LNCS, vol. 8673, pp. 33–40. Springer, Cham (2014). https://doi.org/10.1007/978-3-319-10404-1_5

18. Pnevmatikakis, E.A., et al.: Simultaneous denoising, deconvolution, and demixing of calcium imaging data. Neuron **89**(2), 285–299 (2016)

19. Rahmani, M., Atia, G.K.: High dimensional low rank plus sparse matrix decomposition. IEEE Trans. Signal Process. **65**(8), 2004–2019 (2017)

20. Sobral, A., Bouwmans, T., Zahzah, E.H.: Lrslibrary: low-rank and sparse tools for background modeling and subtraction in videos. Robust Low-Rank and Sparse Matrix Decomposition: Applications in Image and Video Processing (2016)

21. Stevenson, I.H., Kording, K.P.: How advances in neural recording affect data analysis. Nat. Neurosci. **14**(2), 139–142 (2011)

22. Tieleman, T., Hinton, G.: Lecture 6.5-rmsprop: divide the gradient by a running average of its recent magnitude. COURSERA: Neural Netw. Mach. Learn. **4**(2), 26–31 (2012)

23. Voleti, V., et al.: Real-time volumetric microscopy of in vivo dynamics and large-scale samples with scape 2.0. Nat. Methods **16**(10), 1054–1062 (2019)

24. Xiao, W., Huang, X., He, F., Silva, J., Emrani, S., Chaudhuri, A.: Online robust principal component analysis with change point detection. IEEE Trans. Multimedia **22**(1), 59–68 (2019)

25. Yoon, Y.G., et al.: Sparse decomposition light-field microscopy for high speed imaging of neuronal activity. Optica **7**(10), 1457–1468 (2020)

26. Yuan, Z., Yang, Z., Oja, E.: Projective nonnegative matrix factorization: sparseness, orthogonality, and clustering. Neural Process. Lett. 11–13 (2009)

27. Zhou, T., Tao, D.: Godec: randomized low-rank & sparse matrix decomposition in noisy case. In: Proceedings of the 28th International Conference on Machine Learning, ICML 2011 (2011)

28. Zhou, T., Tao, D.: Greedy bilateral sketch, completion & smoothing. In: Carvalho, C.M., Ravikumar, P. (eds.) Proceedings of the Sixteenth International Conference on Artificial Intelligence and Statistics, vol. 31, 650–658. PMLR (2013)

Text2Brain: Synthesis of Brain Activation Maps from Free-Form Text Query

Gia H. Ngo[1(✉)], Minh Nguyen[1,2], Nancy F. Chen[3], and Mert R. Sabuncu[1,4]

[1] School of Electrical and Computer Engineering, Cornell University, Ithaca, USA
ghn8@cornell.edu
[2] Computer Science Department, University of California, Davis, USA
[3] Institute of Infocomm Research (I2R), A*STAR, Singapore, Singapore
[4] Radiology, Weill Cornell Medicine, New York City, USA

Abstract. Most neuroimaging experiments are under-powered, limited by the number of subjects and cognitive processes that an individual study can investigate. Nonetheless, over decades of research, neuroscience has accumulated an extensive wealth of results. It remains a challenge to digest this growing knowledge base and obtain new insights since existing meta-analytic tools are limited to keyword queries. In this work, we propose Text2Brain, a neural network approach for coordinate-based meta-analysis of neuroimaging studies to synthesize brain activation maps from open-ended text queries. Combining a transformer-based text encoder and a 3D image generator, Text2Brain was trained on variable-length text snippets and their corresponding activation maps sampled from 13,000 published neuroimaging studies. We demonstrate that Text2Brain can synthesize anatomically-plausible neural activation patterns from free-form textual descriptions of cognitive concepts. Text2Brain is available at https://braininterpreter.com as a web-based tool for retrieving established priors and generating new hypotheses for neuroscience research.

Keywords: Coordinate-based meta-analysis · Transformers · Information retrieval · Image generation

1 Introduction

Decades of neuroimaging research have yielded an impressive repertoire of findings and greatly enriched our understanding of the cognitive processes governing the mind. However, individual brain imaging experiments are often under-powered [1,2], constrained by the number of subjects and psychological processes that each experiment can probe [3]. To synthesize reliable trends across

G. H. Ngo and M. Nguyen—Equal contribution.
N. F. Chen and M. R. Sabuncu—Equal contribution.

Electronic supplementary material The online version of this chapter (https://doi.org/10.1007/978-3-030-87234-2_57) contains supplementary material, which is available to authorized users.

© Springer Nature Switzerland AG 2021
M. de Bruijne et al. (Eds.): MICCAI 2021, LNCS 12907, pp. 605–614, 2021.
https://doi.org/10.1007/978-3-030-87234-2_57

such experiments, researchers often perform meta-analysis on the coordinates of the most significant effect (such as 3D location of peak brain activation in response to a task). Most meta-analyses require the expert selection of relevant experiments (e.g. [4–6]). One key challenge with conducting meta-analysis on neuroimaging experiments is the consolidation of synonymous terms. As neuroscientific research constantly evolves, different denominations might be used in different contexts or invented to refine existing ideas. For instance, "self-generated thought", one of the most highly studied functional domains of the human brain [7], can be referred to by different terms such as "task-unrelated thought" [8].

Manual selection of experiments for meta-analysis can be replaced by automated keyword search through data automatically scraped from the neuroimaging literature [9–11]. For example, Neurosynth [9] and more recently Neuroquery [10] both use automated keyword search to retrieve relevant studies to synthesize brain activation maps from text queries. However, Neurosynth and Neuroquery only allow for rigid queries formed out of predefined keywords and rely on superficial lexical similarity via co-occurrences of keywords for inference of longer or rarer queries. We propose an alternative approach named Text2Brain, which permits more flexible free-form text queries. Text2Brain also characterizes more fine-grained and implicit semantic similarity via vector representations from neural modeling in order to retrieve more relevant studies. Moreover, existing approaches estimate voxel-wise activations using either univariate statistical testing or regularized linear regression. In contrast, Text2Brain generates whole-brain activation maps using a 3D convolutional neural network (CNN) for more accurate construction of both coarse and fine details.

We compare Text2Brain's predictions with those from established baselines where we used article titles as free-form queries. Furthermore, we assess model predictions on independent test datasets, including reliable task contrasts and meta-analytic activation maps of well-studied cognitive domains predicted from their descriptions. Our analysis shows that Text2Brain generates activation maps that better match the target images than the baselines do. Given its flexibility in taking input queries, Text2Brain can be used as an educational aid as well as a tool for synthesizing prior maps for future research.

2 Materials and Methods

2.1 Overview

Figure 1 shows the overview of our approach. For each research article, full text and activation coordinates are extracted to create training samples (Sect. 2.2). Text2Brain model consists of a transformer-based text encoder and a 3D CNN (Sect. 2.3). The transformer uses attention to encode the input text into vector representation [12, 13]. Thus, over many text-brain activation map pairs, the model automatically learns the association between activation at a spatial location with the most relevant words in the input text. Unlike classical keyword search that mainly exploits co-occurrence of keywords regardless of context, a transformer refines the vector representation depending on the specific phrasing of the text inputs (i.e. context) [14]. This allows Text2Brain to map synonymous text to a similar activation map. Instead of explicitly searching through articles, Text2Brain stores the articles' content in its parameters [15] and outputs a relevant vector representation when presented with an input query. Thus, we use an augmented data sampling strategy to encourage the model to construct and store rich many-to-one mappings between textual description and activation maps (Sect. 2.4).

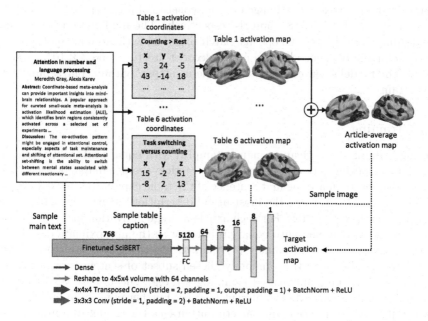

Fig. 1. Overview of data preprocessing, the Text2Brain model, and training procedure. All activation maps are 3D volumes, but projected to the surface for visualization.

2.2 Data Preprocessing

Coordinates of peak activation were scraped from tables of results reported in more than 13,000 neuroimaging articles and previously released in [10]. Each table has the corresponding article's PubMed ID, the table's ID as originally numbered in the article, and coordinates of peak activation converted to MNI152 coordinate system [16]. Following the preprocessing procedure of [10], a Gaussian sphere with full width at half maximum (FWHM) of 9 mm is placed at each of the coordinates of peak activation. Thus, an activation map for each table in an article is generated from the set of activation foci associated with the table. An article-average activation map is also generated by averaging the activation maps of all the tables in the article. The articles' full text are scraped using their PubMedID via NCBI API[1] and Elsevier E-utilities API[2].

2.3 Model

Figure 1 shows the basic schematic of Text2Brain, which consists of a text encoder based on SciBERT [17] and a 3D CNN as the image decoder. Output embedding of the text encoder is of dimension 768 and projected via a fully-connected layer, then reshaped to a 3D volume of dimension $4 \times 5 \times 4$ and 64 channels at each voxel. The image decoder consists of 3 transposed 3D convolutional layers with 32, 16, 8 channels respectively. The model was trained using mean-squared error for 2000 epochs, batch size of 24 with Adam [18]. The learning rate for the text encoder and image decoder are 10^{-5} and 3×10^{-2}, respectively. The model's source code is available at https://github.com/sabunculab/text2brain.

2.4 Training

During training, an activation map is sampled with equal probability from the set of table-specific activation maps and the article-average map. For each table-specific activation map, the first sentence of the table caption (as our data exploration suggested this to be the most useful description) is also extracted as the image's corresponding text. For each article-average activation map, one of four types of text is sampled with equal probability as the approximate description of the activation pattern, namely (1) the article's title (2) one of the article's keywords (2) abstract (3) a randomly chosen subset of sentences from the discussion section. This augmented sampling strategy encourages Text2Brain to generalize over input texts of different lengths. Furthermore, sampling multiple text snippets for an activation pattern encourages the model to automatically infer keywords present across queries and implicitly learn the association between different but synonymous words with an activation map. Supplemental Fig. S2 shows an ablation study on the sampling strategy.

3   Experimental Setup

3.1   Predict Activation Maps from Article Titles

From the dataset of 13000 articles, 1000 articles are randomly sampled as the test set such that the keywords (defined by the articles' authors) are not included in the training and validation articles. Of the remaining articles, 1000 are randomly held out as a validation set for parameters tuning. For each article, the article-average activation map is predicted from its title using Text2Brain and the two baselines of Neurosynth and Neuroquery.

---

[1] https://www.ncbi.nlm.nih.gov/books/NBK25501/.
[2] https://dev.elsevier.com/.

3.2   Predict Activation Maps from Contrast Descriptions

The Human Connectome Project (HCP) offers neuroimaging data from over 1200 subjects, including task fMRI (tfMRI) of 86 task contrasts from 7 domains [19]. While detailed descriptions of task contrasts are provided by HCP, we instead use the more concise contrast descriptions provided by the Individual Brain Charting (IBC) project [20], which includes fMRI data from 12 subjects and 180 task contrasts, 43 of which are also studied in the HCP. The reason for using the IBC contrast descriptions is because they are more succinct and thus more favorable to the baselines. The target (ground-truth) activation maps are the group-average contrast maps provided by the HCP, as the large number of subjects provides more reliable estimates of the contrast maps. In our analyses, we use the agreement between the IBC and HCP maps as a measure of reliability. Note that despite using similar experimental protocols, there are subtle differences between the IBC and HCP experiments. For example, while the original HCP language task was conducted in English, the corresponding language task in the IBC project was conducted in French. Overall, Text2Brain and the two baselines were evaluated on the 43 HCP task contrasts.

3.3   Baselines

The first baseline, Neurosynth [9], collected all peak activation coordinates across neuroimaging articles that mention a given keyword and performed a statistical test at every voxel to determine a significant association. For longer query, we performed statistical test using activation coordinates reported in all articles that contain at least one of the keywords in the input text.

The second baseline, Neuroquery [10], builds upon Neurosynth by extending the vocabulary of keywords via manual selection from other sources. The keyword encoding is obtained after performing non-negative matrix factorization of the articles' abstract (as a bag of keywords) represented with term frequency - inverse document frequency (TF-IDF) features [21]. A ridge regression model was trained to learn the mapping from the text encoding to the activation at each voxel. The inference of a keyword is smoothed by a weighed summation with most related keywords (in the TF-IDF space). For longer queries, the predicted activation is the average of maps predicted from all keywords in the input.

3.4   Evaluation Metrics

To measure the similarity of predicted and target activation maps at different levels of detail, we compute Dice scores [22] at various thresholds. This evaluation procedure is similar to that used in [10] for a thresholded target map, but we apply the same thresholding to both the target and predicted map. For example, at a lower threshold (e.g., considering the 5% most activated voxels), the Dice score measures the correspondence of the fine-grained details between the target and predicted activation maps. At higher thresholds (e.g. 25% most activated voxels), this metric captures gross agreement of activation clusters. We also compute an approximated integration of Dice scores across all thresholds (from 5% up to 30%), i.e. the area under the Dice curve (AUC), as a summary measure. Supplemental Fig. S1 shows the Dice curve for an example pair of target-predicted activation maps. We only consider up to 30% to be fair to the baselines, as the portion of activated voxels predicted by Neuroquery only extends up to 30% of the gray matter mask.

4   Results

4.1   Validate Activation Maps Predicted from Article Title

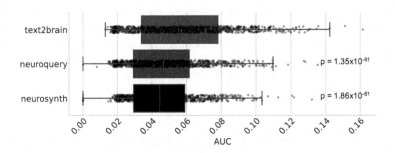

Fig. 2. Evaluation of article-average activation maps predicted from their titles measured in area under the Dice curve (AUC) score. The p-values are computed from paired-sample t-tests between Text2Brain and each of the 2 baselines.

Figure 2 compares the quality of activation maps predicted from the titles of 1000 articles. Text2Brain model (mean AUC = 0.0576) outperforms Neuroquery (mean AUC = 0.0478) and Neurosynth (mean AUC = 0.0464). Paired-sample t-tests show that this performance gap is statistically very significant. The p-value for the comparison between Neuroquery and Neurosynth is $p = 0.015$. While Text2Brain can make a prediction for all samples, Neurosynth and Neuroquery fail to make prediction for some article titles, resulting in zero AUCs values.

4.2   Prediction of Task Contrast Maps from Description

Figure 3 shows the AUC scores for the prediction of the three models and the IBC average contrasts, against the HCP target maps. The 22 contrasts with the HCP-IBC's AUC score above the average, considered to be the reliable contrasts, are shown. Across all 43 HCP contrasts, Text2Brain (mean AUC = 0.082) performs better than the baselines, i.e. Neuroquery (mean AUC = 0.0755, $p = 0.08$), Neurosynth (mean AUC = 0.047, $p = 1.5 \times 10^{-5}$), where p-values are computed from the paired t-test between Text2Brain's and the baselines' prediction. As reference, IBC contrasts yield mean AUC = 0.094 ($p = 0.077$).

Figure 4 shows the prediction from three most reliable task contrasts (having the highest HCP-IBC AUC), thresholded at the top 20% most activated voxels. The three contrasts correspond to different HCP task groups, namely "WORKING MEMORY", "SOCIAL", and "MOTOR". Text2Brain's prediction improves over the baselines for the three contrasts. Neurosynth was not able to generate activation maps for two of the contrast descriptions ("2-back vs 0-back" and "Move tongue"). On the other hand, for the "Move tongue" contrast, Neuroquery predicts activation in the primary cortex, but the peak is in the wrong

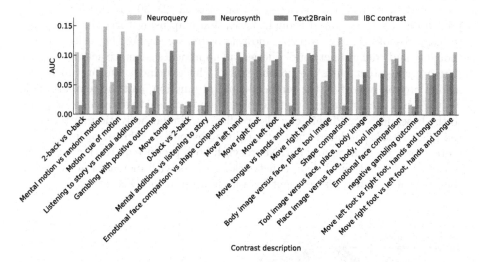

Fig. 3. AUC of predicted HCP task activation maps from contrasts' description.

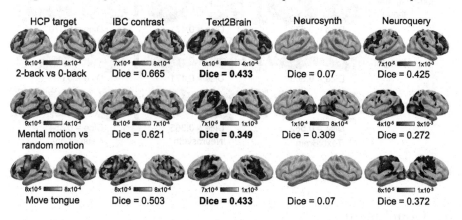

Fig. 4. Task activation maps predicted from contrasts' description. The Dice scores are computed between the binarized map of 20% most activated voxels in the predicted and target brain maps.

location, shifted more toward the hand region of the homunculus. Additionally, there is a false positive prediction in the occipital cortex.

Finally, we are interested in examining the prediction for "Self-generated thought", which is one of the most commonly studied functional domains, due to its engagement in a wide range of cognitive processes that do not require external stimuli [8], and is associated with the default network [23]. The ground-truth map for self-generated thought, taken from [24], is estimated using activation likelihood estimation (ALE) [25–27], a well established tool for coordinate-based meta-analysis, applied on 1812 activation foci across 167 imaging studies over 7 tasks based on strict selection criteria [28–30]. Figure 5 shows the prediction of self-generated thought activation map using three different query terms, thresholded at the top 20% most activated voxels. For the "self-generated thought" and "default network" queries, all approaches generate activation maps that are consistent with the ground-truth, which includes the precuneus, the medial prefrontal cortex, the temporo-parietal junction, and the temporal pole. Text2Brain's prediction best matches the ground-truth activation map compared to the baselines. Text2Brain can also replicate a similar activation pattern from the query "task-unrelated thought", evident by only a slight drop in the Dice score. However, Neuroquery and Neurosynth both produce activation maps that deviate from the typical default network's regions with increased activation in the prefrontal cortex, also evident by a large drop in the Dice scores.

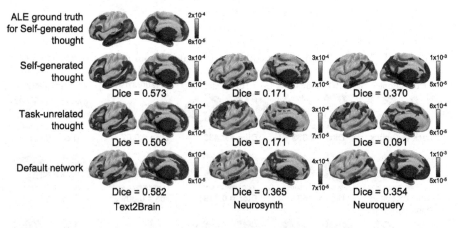

Fig. 5. Prediction of self-generated thought activation map using synonymous queries

5   Conclusion

In this work, we present a model named Text2Brain for generating activation maps from free-form text query. By finetuning a high-capacity SciBERT-based text encoder to predict coordinate-based meta-analytic maps, Text2Brain captures the rich relationship in the language representational space, allowing the model to generalize its prediction for synonymous queries. This is evident in the better performance of Text2Brain in predicting the self-generated thought activation map using different descriptions of the functional domain. Text2Brain's capability to implicitly learn relationships between terms and images will help the model stays relevant and useful even as neuroimaging literature continues to evolve with new information and rephrasing of existing concepts. We also show that Text2Brain accurately predicts most of the task contrasts included in the HCP dataset based on their description, validating its capability to make prediction for longer, arbitrary queries. Text2Brain also avoids the failure cases suffered by Neurosynth and Neuroquery in which they cannot predict if the input words are not defined in their vocabularies, even though the queries are relevant to neuroscience research such as the title of an article or a contrast description. In the future, we will work on the interpretability of the approach, such as to attribute regions of activation in the generated map to specific word in the input query, as well as to efficiently match activation maps and research text most relevant to the synthesized images.

Acknowledgment. This work was supported by NIH grants R01LM012719, R01AG053949, the NSF NeuroNex grant 1707312, the NSF CAREER 1748377 grant and Jacobs Scholar Fellowship.

References

1. Carp, J.: The secret lives of experiments: methods reporting in the fMRI literature. Neuroimage **63**(1), 289–300 (2012)
2. Button, K.S., et al.: Power failure: why small sample size undermines the reliability of neuroscience. Nat. Rev. Neurosci. **14**(5), 365–376 (2013)
3. Church, J.A., Petersen, S.E., Schlaggar, B.L.: The "Task B problem" and other considerations in developmental functional neuroimaging. Hum. Brain Mapp. **31**(6), 852–862 (2010)
4. Costafreda, S.G., Brammer, M.J., David, A.S., Fu, C.H.Y.: Predictors of amygdala activation during the processing of emotional stimuli: a meta-analysis of 385 pet and FMRI studies. Brain Res. Rev. **58**(1), 57–70 (2008)
5. Minzenberg, M.J., Laird, A.R., Thelen, S., Carter, C.S., Glahn, D.C.: Meta-analysis of 41 functional neuroimaging studies of executive function in schizophrenia. Arch. Gen. Psychiatry **66**(8), 811–822 (2009)
6. Shackman, A.J., Salomons, T.V., Slagter, H.A., Fox, A.S., Winter, J.J., Davidson, R.J.: The integration of negative affect, pain and cognitive control in the cingulate cortex. Nat. Rev. Neurosci. **12**(3), 154–167 (2011)
7. Smallwood, J.: Distinguishing how from why the mind wanders: a process-occurrence framework for self-generated mental activity. Psychol. Bull. **139**(3), 519 (2013)
8. Andrews-Hanna, J.R., Smallwood, J., Spreng, R.N.: The default network and self-generated thought: component processes, dynamic control, and clinical relevance. Ann. N. Y. Acad. Sci. **1316**(1), 29 (2014)
9. Yarkoni, T., Poldrack, R.A., Nichols, T.E., Van Essen, D.C., Wager, T.D.: Large-scale automated synthesis of human functional neuroimaging data. Nat. Methods **8**(8), 665–670 (2011)

10. Dockès, J., et al.: NeuroQuery, comprehensive meta-analysis of human brain mapping. Elife **9**, e53385 (2020)
11. Rubin, T.N., Koyejo, O., Gorgolewski, K.J., Jones, M.N., Poldrack, R.A., Yarkoni, T.: Decoding brain activity using a large-scale probabilistic functional-anatomical atlas of human cognition. PLoS Comput. Biol. **13**(10), e1005649 (2017)
12. Vaswani, A., et al.: Attention is all you need. arXiv preprint arXiv:1706.03762 (2017)
13. Devlin, J., Chang, M.W., Lee, K., Toutanova, K.: Bert: pre-training of deep bidirectional transformers for language understanding. arXiv preprint arXiv:1810.04805 (2018)
14. Tenney, I., et al.: What do you learn from context? Probing for sentence structure in contextualized word representations. arXiv preprint arXiv:1905.06316 (2019)
15. Petroni, F., et al.: Language models as knowledge bases? In: Proceedings of the 2019 Conference on Empirical Methods in Natural Language Processing and the 9th International Joint Conference on Natural Language Processing (EMNLP-IJCNLP), pp. 2463–2473 (2019)
16. Lancaster, J.L., et al.: Bias between MNI and Talairach coordinates analyzed using the ICBM-152 brain template. Hum. Brain Mapp. **28**(11), 1194–1205 (2007)
17. Beltagy, I., Lo, K., Cohan, A.: SciBERT: a pretrained language model for scientific text. arXiv preprint arXiv:1903.10676 (2019)
18. Loshchilov, I., Hutter, F.: Decoupled weight decay regularization. In: Proceedings of ICLR (2018)
19. Barch, D.M., et al.: Function in the human connectome: task-fMRI and individual differences in behavior. Neuroimage **80**, 169–189 (2013)
20. Pinho, A.L., et al.: Individual brain charting dataset extension, second release of high-resolution fMRI data for cognitive mapping. Sci. Data **7**(1), 1–16 (2020)
21. Salton, G., Buckley, C.: Term-weighting approaches in automatic text retrieval. Inf. Process. Manag. **24**(5), 513–523 (1988)
22. Dice, L.R.: Measures of the amount of ecologic association between species. Ecology **26**(3), 297–302 (1945)
23. Buckner, R.L., Andrews-Hanna, J.R., Schacter, D.L.: The brain's default network: anatomy, function, and relevance to disease (2008)
24. Ngo, G.H., et al.: Beyond consensus: embracing heterogeneity in curated neuroimaging meta-analysis. Neuroimage **200**, 142–158 (2019)
25. Turkeltaub, P.E., Eden, G.F., Jones, K.M., Zeffiro, T.A.: Meta-analysis of the functional neuroanatomy of single-word reading: method and validation. Neuroimage **16**(3), 765–780 (2002)
26. Laird, A.R., et al.: ALE meta-analysis: controlling the false discovery rate and performing statistical contrasts. Hum. Brain Mapp. **25**(1), 155–164 (2005)
27. Eickhoff, S.B., Laird, A.R., Grefkes, C., Wang, L.E., Zilles, K., Fox, P.T.: Coordinate-based activation likelihood estimation meta-analysis of neuroimaging data: a random-effects approach based on empirical estimates of spatial uncertainty. Hum. Brain Mapp. **30**(9), 2907–2926 (2009)
28. Spreng, R.N., Mar, R.A., Kim, A.S.N.: The common neural basis of autobiographical memory, prospection, navigation, theory of mind, and the default mode: a quantitative meta-analysis. J. Cogn. Neurosci. **21**(3), 489–510 (2009)
29. Mar, R.A.: The neural bases of social cognition and story comprehension. Annu. Rev. Psychol. **62**, 103–134 (2011)
30. Sevinc, G., Spreng, R.N.: Contextual and perceptual brain processes underlying moral cognition: a quantitative meta-analysis of moral reasoning and moral emotions. PLoS ONE **9**(2), e87427 (2014)

Estimation of Spontaneous Neuronal Activity Using Homomorphic Filtering

Sukesh Kumar Das[1]([✉])(iD), Anil K. Sao[1](iD), and Bharat Biswal[2](iD)

[1] Indian Institute of Technology Mandi, Mandi 175005, HP, India
d17025@students.iitmandi.ac.in, anil@iitmandi.ac.in
[2] New Jersey Institute of Technology, Newark, NJ 07102, USA
bharat.biswal@njit.edu

Abstract. Blood Oxygen Level-Dependent (BOLD) signal changes in functional magnetic resonance imaging (fMRI) measures neuronal activities blurred by hemodynamic response function (HRF) and hence may not be reliable to estimate functional connectivity (FC). Several methods have been attempted to estimate the neuronal activity signal (NAS) from the observed BOLD signal. Using this as a blind source separation problem, these methods assume a parametric model of HRF. But it is not clear if these models accurately reflect the biophysical process. In this paper, we have proposed an approach based on a homomorphic filter (HMF) to deconvolve NAS from resting state fMRI (rs-fMRI) time course. It exploits the hypothesis that the HRF has predominantly low frequency energy in comparison to the NAS. Hence, by choosing an appropriate value of cutoff quefrency with the help of thresholded BOLD signal after HMF, HRF can be suppressed from observed BOLD signal to get an estimate of NAS. The estimated NAS, in the framework of dictionary learning (DL), is able to produce subtle resting state networks (RSNs) in comparison with the existing blind deconvolution (BD) method. The Jaccard similarity distance between RSNs by taking random samples and the entire subjects underpins the robustness of the estimated RSNs. Another quantitative comparison has also been drawn to show the efficacy of the HMF in the estimation of NAS using the maximum normalized cross-correlation coefficient (MNCC) distribution for different RSNs.

Keywords: Deconvolution · Homomorphic filtering · rs-fMRI · RSNs

1 Introduction

Blood oxygen level-dependent (BOLD) signal acquired during functional magnetic resonance imaging (fMRI), attempts to measure the neuronal activity in the human brain. Studying spontaneous brain activity without any stimulus has emerged in an alternative way to study the brain function [1]. Functional connectivities (FC), computed using the observed BOLD time course, over the different spatial regions, without any explicit stimulus are called resting state networks

© Springer Nature Switzerland AG 2021
M. de Bruijne et al. (Eds.): MICCAI 2021, LNCS 12907, pp. 615–624, 2021.
https://doi.org/10.1007/978-3-030-87234-2_58

(RSNs). The neuronal activity, observed in the BOLD signal, is modulated by hemodynamic response function (HRF) and mathematically expressed as [2]

$$m_o[n] = s[n] * h[n] + \epsilon[n], \qquad (1)$$

where, $s[n]$ is the underlying neuronal activity signal (NAS) that carries timing information of neuronal events, $h[n]$ is the HRF which models the dynamic changes in the blood flow and $\epsilon[n]$ is the noise evoked in measurement. As the HRF blurs the NAS and varies across the voxels, its effect should be removed from the BOLD signal to estimate the FC resulted due to NAS. Several methods have been proposed to estimate $s[n]$ from the observed $m_o[n]$ [3–5], so that RSNs can be estimated efficiently. The estimation of $s[n]$ from $m_o[n]$ is challenging in rs-fMRI because it is void of the experimental paradigm [5]. The NAS can be estimated from rs-fMRI using a parametric blind deconvolution approach with the help of spontaneous pseudo-events [5]. Here, HRF is estimated as a weighted combination of three bases at minimum noise variance and then Wiener filter deconvolves the spontaneous neuronal activity. Karahanouglu and colleagues introduced total activation to estimate activity related signals by imposing data fitting and spatio-temporal regularizations [4]. Later, the activity related signal is used in inverse differential hemodynamic operator [6] and activity signal is obtained for the rest and task conditions. A joint estimation of the HRF and NAS is carried out by solving a semi-blind deconvolution problem with the constraint that the derivative of NAS is sparse [3]. All of the deconvolution methods assume either a parametric model of the HRF or uniformity of HRF across the brain regions and also limited due to the presence of noise (physiological and head movement). The ambiguity in the case of parametric model of HRF is that the optimal number of parameters is still unclear to represent the actual bio-physical process [7] and intra and inter-subject variability of HRF is also a serious concern in the deconvolution of NAS from rs-fMRI [8].

In this paper, we have proposed an approach, to estimate NAS from m_o, which does not assume any parametric model of HRF. It is based on the observation that the HRF is relatively slow varying in nature and can be associated with low frequency range of spectrum of BOLD signal. This observation may suggest to use a high pass filter (HPF) to suppress the effect of HRF in observed BOLD signal and the resultant BOLD signal will have mostly the information of NAS. However, a simple HPF cannot be used to separate both HRF and NAS, as they are multiplicative in nature in the frequency domain. The proposed method is based on homomorphic filtering (HMF) to estimate NAS from the BOLD rs-fMRI data. In this filtering, the observed BOLD time course is transformed into cepstrum domain, where HRF and NAS are additive in nature. The separation of them in cepstrum domain is done by high pass liftering (HPL) with an optimal value of cutoff quefrency (Q_c). The Q_c is obtained by maximizing the value of a metric, normalized cross correlation coefficients (NCC) between estimated NAS and thresholded rs-fMRI BOLD time course (we call it probable neuronal variable (PNV)). The resultant NAS is employed in the framework of online dictionary learning (ODL) to estimate RSNs [9,10]. It should be noted that the

HMF was studied for task fMRI [7], where the knowledge of experimental design aids in the estimation of NAS. For task related fMRI, it is assumed that the NAS is resulted due to the similar external stimuli. So, the NAS is not much variant across the subjects, especially for a particular brain region associated with the task. On the other hand, the rs-fMRI hasn't such direct advantage and hence, NAS is assumed to be more variant as activations occur due to intrinsic stimuli. Therefore, there is a need to estimate voxel-wise NAS for the subtle RSNs.

The organization of the paper is as follows: Sect. 2 explains the proposed method including estimation of NAS and RSNs using DL method. Section 3 describes the data and tools used in the method. Experimental results are explained in Sect. 4 and the discussion is provided in Sect. 5.

2   Method

Schematic diagram for the estimation of the RSNs using the proposed method, has been illustrated in Fig. 1. Here, the preprocessed BOLD signal is empirically decomposed (EMD) and the first three intrinsic mode functions (IMFs) are considered in HMF to suppress the drift [11,12]. The lower oscillatory modes (forth and onwards) can be associated with drift component present in BOLD signal. The NAS is then estimated for every individual voxel using homomorphic deconvolution of the obtained signal and is used to form a time-voxel matrix. Finally, the matrix is decomposed using ODL to estimate RSNs. The Homomorphic deconvolution is performed on the noise suppress signal with the idea that the HMF converts convolution of two time domain signals into a linear sum of them in the cepstrum domain [13]. Following that, if two signals have characteristics of energy concentrated in different ranges of frequencies, the separation of the same can be relatively easier in the cepstrum domain. As the HRF has energy

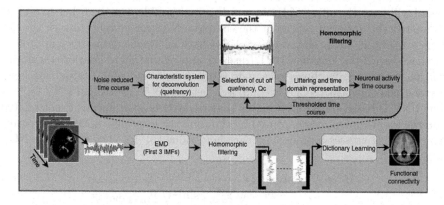

Fig. 1. Block diagram for estimation of RSNs using NAS obtained by HMF

concentrated mostly in low frequency (in comparison to NAS) because of its slowly varying nature. Thus, separation of the NAS and HRF can be relatively easier using HMF. The noise reduced BOLD signal can be written as

$$m_d[n] = s[n] * h[n], \tag{2}$$

In the frequency domain, it can be written as

$$M_d(e^{j\omega}) = S(e^{j\omega})H(e^{j\omega}), \tag{3}$$

where, $M_d(e^{j\omega})$, $S(e^{j\omega})$ and $H(e^{j\omega})$ are the Fourier transform (FT) of $m_d[n]$, $s[n]$ and $h[n]$ respectively. In Cepstrum domain, m_d can lead to be

$$\tilde{m}_d[n] = F^{-1}\{\log M_d(e^{j\omega})\} = F^{-1}\{logS(e^{j\omega})\} + F^{-1}\{logH(e^{j\omega})\} = \tilde{s}[n] + \tilde{h}[n], \tag{4}$$

Now, the $\tilde{m}_d[n]$ is represented as a linear sum of NAS and HRF in the cepstrum domain. Thus, after HPL using a suitable Q_c, NAS can be separated in cepstrum domain. Now, estimated NAS in cepstrum domain,

$$\hat{\tilde{s}}[n] = HPL\{\tilde{m}_d[n]\}, \tag{5}$$

Time domain representation of the estimated NAS is obtained by inverse FT of exponential of the FT of the liftered cepstrum, therefore

$$\hat{s}[n] = F^{-1}\{e^{F\{\hat{\tilde{s}}[n]\}}\} \tag{6}$$

2.1   Selection of the Cut Off Quefrency

Selection of a suitable Q_c to estimate $\hat{s}[n]$ from the observed $m_o[n]$ in rs-fMRI is challenging, because if we choose a very small value of Q_c then, the resultant signal will have the effect of HRF, which is undesirable. On the other hand, the estimated $\hat{s}[n]$ may not represent the complete NAS for a large value of the Q_c. Hence, to obtain an optimal value of Q_c, we have followed a data driven approach. We have varied the length of DFT (discrete approximation of Fourier transform) and Q_c and for each choice of a pair (DFT length and Qc), normalized cross correlation coefficient (NCC) is computed between NAS ($\hat{s}[n]$) and PNV ($s_p[n]$). The PNV is based on the hypothesis that rs-fMRI is spontaneous event related and the events are mostly captured by choosing the locations, in the BOLD time course, which crosses a predefined threshold value [14–16]. The maximum NCC is given as

$$MNCC = \underset{\hat{s}[n]}{\text{argmax}} \frac{max\{|r_{\hat{s}s_p}|\}}{\sqrt{\sum_{n=1}^{a} \hat{s}[n]^2}\sqrt{\sum_{n=1}^{b} s_p[n]^2}} \tag{7}$$

where a and b are the signal lengths of $\hat{s}[n]$ and $s_p[n]$ respectively and $r_{\hat{s}s_p}$ is the cross correlation of the signals, $\hat{s}[n]$ and $s_p[n]$.

2.2   Connectivity Maps Using Online Dictionary Learning (ODL)

The DL exploits the sparse behavior of the RSNs over an over-complete dictionary [10,17–19]. It decomposes multi-subject rs-fMRI data (time-voxel matrix) to produce spatial maps (RSNs). The NAS, $\mathbf{S} \in \mathbb{R}^{nm \times v}$ is decomposed as

$$\mathbf{S} \approx \mathbf{DA} \ \text{with} \ \mathbf{D} \in \mathbb{R}^{nm \times k} \ \text{and} \ \mathbf{A} \in \mathbb{R}^{k \times v}. \tag{8}$$

The $\mathbf{S}$ is obtained by temporally concatenating n number of consecutive NAS volumes obtained from m number of subjects each with v number of voxels. $\mathbf{D}$ consists k dense temporal atoms (every column) and the coefficient matrix, $\mathbf{A}$ consists k sparse spatial maps (every row). The atoms or basis vectors in $\mathbf{D}$ learned such that the rows ($\mathbf{a}_i$) of the matrix $\mathbf{A}$ corresponds to k number of RSNs. For the decomposition, sparsity inducing penalty is combined with the data fitting term and it leads to the following optimization problem:

$$\operatorname*{argmin}_{\mathbf{D},\mathbf{A}} \|\mathbf{S} - \mathbf{DA}\|_F^2 + \lambda \| \mathbf{A}^T \|_1 \mathbf{s.t.} \forall j, \|\mathbf{D}_j\|_2 \leq 1. \tag{9}$$

This equation is solved using ODL [9]. Further, all time points, in a time course at subject level, are mapped to low rank (p) subspace and final rank k decomposition is performed over concatenated data. The decomposition of the data with suitable initialization of atoms extracts reliable RSNs [10,20]. The optimization problem (Eq. 9) with respect to the $\mathbf{D}$ and the $\mathbf{A}$ is not jointly convex and is solved by minimizing over one while keeping the other one fixed.

3   Data Description and Tools

The experiments for the proposed method are carried out using rs-fMRI data taken from UCLA consortium for Neuropsychiatric phenomics data set[1]. A total of 40 healthy participants (age = 21–50, 18F, 22M) are considered in this study. The data were acquired on a 3T Siemens Trio scanner. The fMRI data were acquired using the following scan parameters: TR (repetition time) = 2,000 ms; TE (echo time) = 29 ms; field of view = 192 mm; flip angle = 72°; # slices (axial) = 34; slice thickness = 4 mm with gap = 0; matrix size = 64 × 64; #images = 152. The data were pre-processed using SPM12[2] with MATLAB R2019a. The initial 4 time points across subjects were discarded to get a steady magnetization. Anatomical images were reoriented w.r.t. the Montreal Neurological Institute (MNI) space and functional images were reoriented w.r.t. the anatomical images. Functional images were realigned followed by slice time correction. Anatomical images were co-registered to functional images and it was segmented into gray matter (GM), white matter (WM) and cerebro-spinal fluid (CSF). Functional images were then normalized to MNI space using the deformation field obtained in segmentation. The normalized functional images were smoothed with a Gaussian kernel of FWHM 5 mm × 5 mm × 5 mm. The time courses are band passed to reduce the high frequency physiological noise.

---

[1] https://openneuro.org/datasets/ds000030/versions/00016.

[2] http://www.fil.ion.ucl.ac.uk/spm/.

4 Experimental Results

The output of the steps involved in estimation of NAS using HMF has been demonstrated in Fig. 2. A voxel time course $(m[n])$, taken from the left lateral parietal cortex (lLPC; MNI coordinate: $-48, 62, 34$) of default mode network (DMN) is represented in Fig. 2(a). The drift is then subtracted and transformed into cepstrum domain using HMF. The Q_c, for the time course is 3 at DFT length of 158. In cepstrum, the right part of the Q_c is considered as the NAS and the left part of the Q_c can be associated with the HRF. Time domain reconstruction of HRF and NAS are shown in Figs. 2(b) and (c) respectively. The PNV (blue) is obtained by thresholding (threshold value = 1) the standardized noise suppressed time course, $m_d[n]$. For the given signal, computed MNCC is 0.544 ($p < 0.05$).

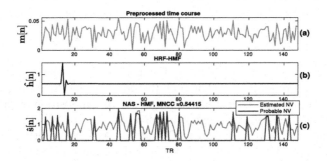

Fig. 2. Deconvolution using HMF: (a) preprocessed time course from the lLPC, (b) reconstructed signal related to HRF and (c) estimated NAS superimposed with PNV (Color figure online)

4.1 Connectivity Maps During Rest Using Estimated NAS

The RSNs are estimated for the healthy control in rest and obtained using pre-processed BOLD time course and estimated NAS by BD and HMF methods with the ODL optimization Eq. 9. The reduced dimensions (p) used in the experiment is same as the value of the $k(= 28)$ and the values of the sparsity controlling parameter (λ) is 10. The six exemplary RSNs (anterior DMN - aDMN, medial sensory motor network - MSMN, auditory network - Aud, posterior DMN - pDMN, dorsal attention network - DAN and superior visual network - SVN) have been illustrated in Fig. 3. We could identify corresponding regions associated with each of the RSNs using NAS obtained by HMF. For example, in aDMN, we could identify the medial pre-frontal cortex, anterior cingulate and inferior parietal region. Similarly, the auditory network includes the Heschl gyrus, primary and associative auditory cortices, superior temporal, and insular cortices. Here, all six networks appear well by preserving the symmetricity in the case of NAS estimated by HMF. The RSNs show good localization with less spurious than the maps obtained by pre-processed and blind deconvolved (BD) time course.

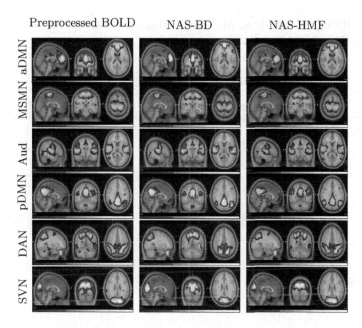

Fig. 3. Sagital, coronal and axial views of RSNs (from top: aDMN, MSMN, Aud, pDMN, DAN and SVN) obtained with ODL using conventionally preprocessed BOLD, NAS using BD and HMF respectively. All RSNs are presented in same scale (0–0.3)

4.2 Robustness of the RSNs

For the lack of the ground truth, the robustness of the estimated RSNs is demonstrated using two quantitative metric namely Jaccard similarity distances (JSD) and MNCC. We have used two sub-sampling schemes while using ODL and the decomposition was repeated for 25 times to obtain RSNs. In each repetition, out of 40 subjects, 20 subjects and 30 subjects are taken as a random subset and voxel-wise NAS is estimated using BD and HMF. The six RSNs (aDMN, MSMN, Aud, pDMN, DAN and SVN) are identified manually for every repetition of the two methods. A total of 150 RSNs is obtained using each of the two methods with one of the sub-sampling schemes. Each of these RSNs obtained with random subsample is compared with the respective RSNs obtained with 40 subjects (Fig. 3) using JSD [21]. The similarity distances with the two subsample schemes (row wise) for the two methods have been presented in Figs. 4(a) and 4(b) respectively. The average of the similarity scores of 150 RSNs with 20 random subjects for the BD method and the HMF methods are 0.57 ± 0.16 and 0.66 ± 0.09 respectively. For the 30 random subjects, the average similarity scores are 0.61 ± 0.17 and 0.74 ± 0.09 for the two methods respectively. We have also computed the MNCC between PNV and estimated NAS. The time courses were extracted from the six RSNs obtained using the proposed approach and MNCCs were computed for both proposed and BD approaches for 40 healthy control subjects. The mean of the MNCC for BD and HMF methods for the given six RSNs are (0.201, 0.261), (0.198, 0.258), (0.201, 0.266), (0.198, 0.259), (0.202, 0.274) and (0.203, 0.256) respectively.

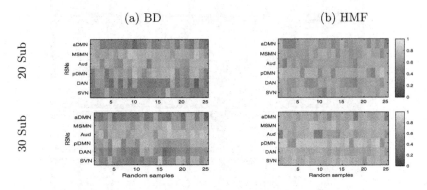

Fig. 4. JSD of RSNs between random 20 subject FCs (first row) and 30 subject FCs (second column) for BD (Fig. 4(a)) and HMF (Fig. 4(b)) deconvolution.

5   Discussion

In HMF, the BOLD time courses can be represented as a linear sum of HRF and NAS in the cepstrum domain. By choosing a suitable Q_c, NAS and HRF are separated. The separation was relatively easier because of the low frequency characteristic of the HRF and it is achieved with a hypothesis driven PNV which is used to maximize the NCC. The mean of the MNCC demonstrates the statistical significance of NAS in comparison with the BD for the predominant RSNs ($\rho < 0.05$).

It has been demonstrated that the estimated NAS can produce RSNs with fair spatial localization and symmetricity in the ODL framework. The method is more effective in removing spurious voxels than the traditional BOLD or BD method (see Fig. 3). The RSNs estimated using HMF demonstrates that the average similarity with the entire subjects increases with the increased number of random subjects which is expected but there is not much difference in variance for every case (see Fig. 4). The mean and the standard deviation of the JSD underpins that the reliability of the RSNs, obtained using the proposed method is more than that of BD approach. The reliability for the aDMN and DAN are very less in the case of BD method even if the size of the random sample is increased. Moreover, the work does not require the optimal number of parameters which can accurately model the HRF. As HRF is variant across the brain and subject, parametric model may not be a reliable choice in deconvolution whereas the proposed method estimates voxel wise NAS across the brain without necessitating any parametric assumption of HRF. We have kept selection of the Q_c as data driven procedure with the help of point process analysis [15,16,22,23]. In the future, we will demonstrate our work in different pathological conditions.

References

1. Bijsterbosch, J., Smith, S.M., Beckmann, C.F.: Introduction to Resting State fMRI Functional Connectivity. Oxford University Press, Oxford (2017)
2. Glover, G.H.: Deconvolution of impulse response in event-related bold fMRI. Neuroimage **9**(4), 416–429 (1999)
3. Cherkaoui, H., Moreau, T., Halimi, A., Ciuciu, P.: Sparsity-based blind deconvolution of neural activation signal in fMRI. In: IEEE International Conference on Acoustics, Speech and Signal Processing (ICASSP), pp. 1323–1327. IEEE (2019)
4. Karahanoğlu, F.I., Caballero-Gaudes, C., Lazeyras, F., Van De Ville, D.: Total activation: fMRI deconvolution through spatio-temporal regularization. Neuroimage **73**, 121–134 (2013)
5. Wu, G.R., Liao, W., Stramaglia, S., Ding, J.R., Chen, H., Marinazzo, D.: A blind deconvolution approach to recover effective connectivity brain networks from resting state fMRI data. Med. Image Anal. **17**(3), 365–374 (2013)
6. Friston, K.J., Mechelli, A., Turner, R., Price, C.J.: Nonlinear responses in fMRI: the balloon model, volterra kernels, and other hemodynamics. Neuroimage **12**(4), 466–477 (2000)
7. Sreenivasan, K.R., Havlicek, M., Deshpande, G.: Nonparametric hemodynamic deconvolution of fMRI using homomorphic filtering. IEEE Trans. Med. Imaging **34**(5), 1155–1163 (2014)
8. Deshpande, G., Sathian, K., Hu, X.: Effect of hemodynamic variability on granger causality analysis of fMRI. Neuroimage **52**(3), 884–896 (2010)
9. Mairal, J., Bach, F., Ponce, J., Sapiro, G.: Online learning for matrix factorization and sparse coding. J. Mach. Learn. Res. **11**(1) (2010)
10. Mensch, A., Varoquaux, G., Thirion, B.: Compressed online dictionary learning for fast resting-state fMRI decomposition. In: Proceedings of 13th International Symposium on Biomedical Imaging (ISBI), pp. 1282–1285. IEEE (2016)
11. Aggarwal, P., Gupta, A., Garg, A.: Joint estimation of hemodynamic response function and voxel activation in functional MRI data. In: Navab, N., Hornegger, J., Wells, W.M., Frangi, A.F. (eds.) MICCAI 2015. LNCS, vol. 9349, pp. 142–149. Springer, Cham (2015). https://doi.org/10.1007/978-3-319-24553-9_18
12. Huang, N.E., et al.: The empirical mode decomposition and the hilbert spectrum for nonlinear and non-stationary time series analysis. Proc. R. Soc. Lond. Ser. A **454**(1971), 903–995 (1998)
13. Oppenheim, A., Schafer, R.: Homomorphic analysis of speech. IEEE Trans. Audio Electroacoust. **16**(2), 221–226 (1968)
14. Esfahlani, F.Z., et al.: High-amplitude co-fluctuations in cortical activity drive functional connectivity. bioRxiv p. 800045 (2020)
15. Liu, X., Duyn, J.H.: Time-varying functional network information extracted from brief instances of spontaneous brain activity. Proc. Natl. Acad. Sci. **110**(11), 4392–4397 (2013)
16. Tagliazucchi, E., Balenzuela, P., Fraiman, D., Chialvo, D.R.: Criticality in large-scale brain fMRI dynamics unveiled by a novel point process analysis. Front. Physiol. **3**, 15 (2012)
17. Das, S., Sao, A.K., Biswal, B.: Precise estimation of resting state functional connectivity using empirical mode decomposition. In: Mahmud, M., Vassanelli, S., Kaiser, M.S., Zhong, N. (eds.) BI 2020. LNCS (LNAI), vol. 12241, pp. 75–84. Springer, Cham (2020). https://doi.org/10.1007/978-3-030-59277-6_7

18. Iqbal, A., Seghouane, A.K.: Dictionary learning algorithm for multi-subject fMRI analysis via temporal and spatial concatenation. In: Proceedings of International Conference on Acoustics, Speech and Signal Processing (ICASSP), pp. 2751–2755. IEEE (2018)
19. Lee, K., Tak, S., Ye, J.C.: A data-driven sparse GLM for fMRI analysis using sparse dictionary learning with MDL criterion. IEEE Trans. Med. Imaging **30**(5), 1076–1089 (2010)
20. Halko, N., Martinsson, P.G., Tropp, J.A.: Finding structure with randomness: probabilistic algorithms for constructing approximate matrix decompositions. SIAM Rev. **53**(2), 217–288 (2011)
21. Jaccard, P.: The distribution of the flora in the alpine zone. 1. New Phytologist **11**(2), 37–50 (1912)
22. Li, W., Li, Y., Hu, C., Chen, X., Dai, H.: Point process analysis in brain networks of patients with diabetes. Neurocomputing **145**, 182–189 (2014)
23. Tagliazucchi, E., Siniatchkin, M., Laufs, H., Chialvo, D.R.: The voxel-wise functional connectome can be efficiently derived from co-activations in a sparse spatio-temporal point-process. Front. Neurosci. **10**, 381 (2016)

A Matrix Autoencoder Framework to Align the Functional and Structural Connectivity Manifolds as Guided by Behavioral Phenotypes

Niharika Shimona D'Souza[1]([⊠]), Mary Beth Nebel[2,3], Deana Crocetti[2], Joshua Robinson[2], Stewart Mostofsky[2,3,4], and Archana Venkataraman[1]

[1] Department of Electrical and Computer Engineering, Johns Hopkins University, Baltimore, USA
Shimona.Niharika.Dsouza@jhu.edu
[2] Center for Neurodevelopmental and Imaging Research, Kennedy Krieger Institute, Baltimore, USA
[3] Department of Neurology, Johns Hopkins School of Medicine, Baltimore, USA
[4] Department of Psychiatry and Behavioral Science, Johns Hopkins School of Medicine, Baltimore, USA

Abstract. We propose a novel matrix autoencoder to map functional connectomes from resting state fMRI (rs-fMRI) to structural connectomes from Diffusion Tensor Imaging (DTI), as guided by subject-level phenotypic measures. Our specialized autoencoder infers a low dimensional manifold embedding for the rs-fMRI correlation matrices that mimics a canonical outer-product decomposition. The embedding is simultaneously used to reconstruct DTI tractography matrices via a second manifold alignment decoder and to predict inter-subject phenotypic variability via an artificial neural network. We validate our framework on a dataset of 275 healthy individuals from the Human Connectome Project database and on a second clinical dataset consisting of 57 subjects with Autism Spectrum Disorder. We demonstrate that the model reliably recovers structural connectivity patterns across individuals, while robustly extracting predictive and interpretable brain biomarkers in a cross-validated setting. Finally, our framework outperforms several baselines at predicting behavioral phenotypes in both real-world datasets.

Keywords: Matrix autoencoder · Manifold alignment · Functional connectivity · Structural connectivity · Phenotypic prediction

Electronic supplementary material The online version of this chapter (https://doi.org/10.1007/978-3-030-87234-2_59) contains supplementary material, which is available to authorized users.

M. de Bruijne et al. (Eds.): MICCAI 2021, LNCS 12907, pp. 625–636, 2021.
https://doi.org/10.1007/978-3-030-87234-2_59

1   Introduction

The brain is increasingly viewed as an interconnected network. Two key elements of this network are the structural pathways between brain regions and the functional signaling that rides on top. This structural connectivity information can be measured via Diffusion Tensor Imaging (DTI) tractography [1,23]. Likewise, resting-state fMRI (rs-fMRI) captures inter-regional co-activation, which can be used to infer functional connectivity [13,25]. Several studies have found both direct and indirect correspondences between structural and functional connectivity [14,16]. Going a step further, structural and functional connectivity have been shown to be predictive of each other at varying scales [8,31,39]. Hence, multimodal integration of these viewpoints has become a key area of focus for characterizing neuropsychiatric disorders such as autism and ADHD [28,40].

In the clinical neuroscience realm, techniques for integrating structural and functional connectivity focus on group-wise discrimination. These works include statistical tests on edge-based features to identify significant differences in Alzheimer's disease [15], parallel ICA using structure and function to identify discriminative biomarkers of schizophrenia [36], and classical machine learning techniques to predict diagnosis [6]. While highly informative at a group level, these methods do not directly address inter-individual variability, for example by predicting finer grained patient characteristics. This divide has been partially bridged by end-to-end deep learning models. Examples include MLPs [26] for age prediction from functional connectomes and convolutional neural networks [20] for predicting cognitive and motor measures from structural connectomes. The work of [11] takes the alternate approach of combining a dictionary learning model on the functional connectomes coupled with an Artificial Neural Network (ANN) that predicts multiple clinical measures, while also preserving interpretability. Even so, these models focus exclusively on a single neuroimaging modality and do not exploit the interplay between function and structure.

Geometric learning frameworks have recently shown great promise in multimodal connectomics studies, both for conventional manifold learning [38] and in the context of Graph Convolutional Networks (GCN) [27,40]. Their primary advantage is the ability to directly incorporate and exploit the underlying data geometry. Beyond associative analyses, the work of [4,39] employ multi-GCNs combined with a Generative Adversarial Network (GAN) for the alignment problem. Particularly, [39] examines the problem of recovering structural connectomes from patient functional connectomes While this paper marks a seminal contribution to multimodal integration, the representations learned by end-to-end GCNs can be hard to interpret. It can also be difficult to train GANs on modest-sized datasets [19].

We propose an end-to-end matrix autoencoder that maps rs-fMRI correlation matrices to structural connectomes obtained from DTI tractography. Inspired by recent work in Riemannian deep learning [9,17], our matrix autoencoder, estimates a low dimensional embedding from rs-fMRI correlation matrices while taking into account the geometry of the functional connectivity (FC) manifold. Our second matrix decoder uses this embedding to reconstruct patient structural

connectivity (SC) matrices akin to a manifold alignment [37] between the FC
and SC data spaces. For regularization, the FC embedding is also used to predict
behavioral phenotypes. We demonstrate that our framework reliably traverses
from function to structure and extracts meaningful brain biomarkers.

2   A Matrix Auto-encoder for Connectome Manifolds

Figure 1 illustrates our matrix autoencoder framework consisting of an encoder-
decoder for functional connectivity (gray box), manifold alignment for estimating
structural connectivity (blue box), and ANN for prediction of behavioral phenotypes (green box). Let N be the number of patients and P be the number of ROIs
in our brain parcellation. We denote the rs-fMRI correlation matrix for patient
n by $\mathbf{\Gamma}_n \in \mathcal{R}^{P \times P}$. $\mathbf{A}_n \in \mathcal{R}^{P \times P}$ is the corresponding structural connectivity
profile, and $\mathbf{y}_n \in \mathcal{R}^{M \times 1}$ is a vector of M concatenated phenotypic measures.

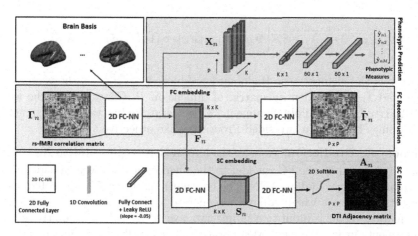

Fig. 1. A Matrix Autoencoder for aligning the FC and SC manifolds **Gray
Box:** Matrix encoder-decoder for functional connectomes. **Blue Box:** Alignment
Decoder for estimating DTI connectomes **Green Box:** ANN for predicting behavioral
phenotypes (Color figure online)

Functional Connectivity Reconstruction: By construction, the correlation
matrices $\mathbf{\Gamma}_n$ belong to the manifold of symmetric positive semi-definite matrices
$\mathcal{P}_P^+$. Our matrix autoencoder estimates a latent functional embedding $\mathbf{F}_n \in \mathcal{P}_K^+$
using a 2D fully connected (2D FC-NN) layer [9,17]. Formally, this mapping
$\mathbf{\Phi}_{\mathrm{ec}}(\cdot) : \mathcal{P}_P^+ \to \mathcal{P}_K^+$ is parametrized by weights $\mathbf{W} \in \mathcal{R}^{P \times K}$ and is computed as a
cascade of two linear layers with tied weights: $\mathbf{F}_n = \mathbf{\Phi}_{\mathrm{ec}}(\mathbf{\Gamma}_n) = \mathbf{W}^T \mathbf{\Gamma}_n \mathbf{W}$. Our
decoder is another 2D FC-NN that estimates $\tilde{\mathbf{\Gamma}}_n$ from $\mathbf{F}_n$ via a similar transfor-
mation $\mathbf{\Phi}_{\mathrm{dc}}(\cdot) : \mathcal{P}_K^+ \to \mathcal{P}_P^+$ that shares weights with the encoder. Mathematically,
our FC reconstruction loss is represented as follows:

$$\mathcal{L}_{\mathrm{FC}} = \frac{1}{N} \sum_n ||\boldsymbol{\Phi}_{\mathrm{dc}}(\boldsymbol{\Phi}_{\mathrm{ec}}(\boldsymbol{\Gamma}_n)) - \boldsymbol{\Gamma}_n||_F^2 = \frac{1}{N} \sum_n ||\mathbf{W}\mathbf{W}^T \boldsymbol{\Gamma}_n \mathbf{W}\mathbf{W}^T - \boldsymbol{\Gamma}_n||_F^2 \quad (1)$$

The second term of Eq. (1) encourages the columns of the brain basis $\mathbf{W}$ to be orthonormal. Conceptually, this specialized loss helps us learn uncorrelated patterns that explain the rs-fMRI data well while acting as an implicit regularizer.

Structural Connectivity Estimation: The structural connectivity matrices $\mathbf{A}_n$ are derived from DTI tractography and belong to the manifold of symmetric (non PSD) matrices $\mathcal{S}_P$. Our alignment decoder first generates an SC embedding $\mathbf{S}_n \in \mathcal{R}^{K \times K}$ from $\mathbf{F}_n$ via a 2D FC-NN layer $\boldsymbol{\Phi}_{\mathrm{align}}(\cdot) : \mathcal{P}_K^+ \to \mathcal{P}_K^+$, followed by a second 2D FC-NN layer $\boldsymbol{\Phi}_{\mathrm{est}}(\cdot) : \mathcal{P}_K^+ \to \mathcal{P}_P^+$ which maps to the structural connectivity matrices. For stability our SC matrices do not have self-connections and are normalized to $||\mathbf{A}_n||_1 = 1$. Accordingly, at the output layer, we suppress the diagonal elements and apply a 2D softmax $\mathcal{SF}(\cdot)$ to generate the final output $\tilde{\mathbf{A}}_n \in \mathcal{R}^{P \times P}$. Our SC estimation objective is represented as follows:

$$\mathcal{L}_{\mathrm{SC}} = \frac{1}{N} \sum_n \left|\left| \mathcal{SF}\left[\boldsymbol{\Phi}_{\mathrm{est}}(\boldsymbol{\Phi}_{\mathrm{align}}(\mathbf{F}_n)) \circ [\mathbf{1}\mathbf{1}^T - \mathcal{I}_P] \right] - \mathbf{A}_n \right|\right|_F^2 \quad (2)$$

where $\circ$ is the element-wise Hadamard product. $\mathbf{1} \in \mathcal{R}^{P \times 1}$ is the vector of all ones, and $\mathcal{I}_P$ is the identity matrix of dimension P. Conceptually, the loss in Eq. (2) is akin to manifold alignment [37] between the functional and structural embeddings based on a two sided Procrustes-like objective.

Phenotypic Prediction: We map the intermediate representation $\mathbf{X}_n = \boldsymbol{\Gamma}_n \mathbf{W} \in \mathcal{R}^{P \times K}$ learned by the FC encoder to the phenotypes $\mathbf{y}_n$ via a cascade of a 1D convolutional layer and an ANN. The convolutional layer $\mathcal{F}_{\mathrm{conv}}(\cdot)$ collapses $\mathbf{X}_n$ along its rows via a weighted sum to generate a K dimensional feature vector. This feature vector is input to a simple two layered ANN $\mathcal{G}(\cdot)$ to jointly estimate the elements in $\hat{\mathbf{y}}_n$. We use a Mean Squared Error (MSE) loss function:

$$\mathcal{L}_{\mathrm{phen}} = \frac{1}{NM} \sum_n ||\hat{\mathbf{y}}_n - \mathbf{y}_n||_F^2 = \frac{1}{NM} \sum_n ||\mathcal{G}(\mathcal{F}_{\mathrm{conv}}(\mathbf{X}_n)) - \mathbf{y}_n||_2^2 \quad (3)$$

This prediction task is a secondary regularizer that encourages our matrix autoencoder to learn representations predictive of inter-subject variability.

Implementation Details: We train our framework on a joint objective that combines Eqs. (1), (2) and (3) as follows:

$$\mathcal{L} = \mathcal{L}_{\mathrm{FC}} + \gamma_1 \mathcal{L}_{\mathrm{SC}} + \gamma_2 \mathcal{L}_{\mathrm{phen}} \quad (4)$$

where γ_1 and γ_2 balance the tradeoff for the SC estimation and phenotypic prediction relative to the FC reconstruction objective. We employ an ADAM optimizer [22] with learning rate 0.005 and weight decay regularization [29] ($\delta = 0.0005$) run for a maximum of 400 epochs. Optimization parameters were fixed based on a validation set consisting of 30 additional patients from the HCP database. We use this strategy to set the dimensionality of our autoencoder embedding at $K = 15$ and loss penalities to $\{\gamma_1, \gamma_2\} = 10^3, 3$. Finally, we utilize a spectral initialization for the encoder-decoder weights $\mathbf{W}$ in Eq. (1) based on the top K eigenvectors of the average patient correlation matrix $\bar{\mathbf{\Gamma}}_n$ for the training set. We use a similar initialization based on $\bar{\mathbf{A}}_n$ for $\mathbf{\Phi}_{\text{est}}(\cdot)$, and default initialization [24] for the remaining layers. Our model has a runtime of 10–12 min on an 8 core machine with 32GB RAM implemented in PyTorch (v1.5.1).

2.1 Baseline Methods for Phenotypic Prediction

Matrix AE without rs-fMRI Decoder: We start with the architecture in Fig. 1 but omit the rs-fMRI decoder loss ($\mathcal{L}_{\text{FC}}$) in Eq. (4). This helps us evaluate the benefit of a tied encoder-decoder model for the rs-fMRI matrices.

Matrix AE without DTI Decoder: We start with the architecture in Fig. 1 but remove the DTI decoder loss ($\mathcal{L}_{\text{SC}}$) in Eq. (4). This helps us evaluate the benefit of manifold alignment to constrain the functional embedding.

Decoupled Matrix AE + ANN: We start with the architecture in Fig. 1 but decouple the representation learning on the connectomics data from the prediction of phenotypic measures by training the models separately.

BrainNetCNN: This baseline integrates multimodal connectivity data via the BrainNetCNN [20]. We modify the original architecture, which is designed for a single modality, to have two branches, one for the rs-fMRI correlation matrices $\mathbf{\Gamma}_n$, and another for the DTI connectomes $\mathbf{A}_n$. The ANN is modified to output M measures of clinical severity. We set the hyperparameters according to [20]

rs-fMRI Dictionary Learning + ANN: The framework in [11] uses rs-fMRI correlation matrices for the prediction of multiple clinical measures. The model combines a dictionary learning with a neural network predictor, with these two blocks optimized in an end-to-end fashion via a coupled optimization objective.

3   Experimental Evaluation and Results

HCP Dataset and Preprocessing: We download rs-fMRI and DTI scans for of 275 healthy individuals from the Human Connectome Project (HCP) S1200 database [12]. Rs-fMRI data was pre-processed according to the standard HCP pipeline [35], which accounts for motion and physiological confounds. We selected a 15-minute interval from the rs-fMRI scans to remain commensurate with standard rs-fMRI protocols. We used the Neurodata MR Graphs package [21] to pre-process the DTI scans and estimate fiber bundles via streamline tractography. Our phenotypic measure was the Cognitive Fluid Intelligence Score (CFIS)

[5,10] adjusted for age, which is obtained via a battery of tests measuring cognitive reasoning (dynamic range: 70–150). We used the Automatic Anatomical Labelling (AAL) atlas with 116 ROIs as our parcellation. From here, we compute the Pearson correlation between the mean rs-fMRI time course in the two regions; we subtract the contribution of the top (roughly constant) eigenvector to obtain the input Γ_n. The input $\mathbf{A}_n$ are obtained by dividing the number of tracts between ROIs by the total fiber tracts across all ROI pairs.

Predicting Behavioral Phenotypes: Table 1 (and Fig. 1 in Supplementary) compares the model against the baselines when predicting CFIS in a five-fold cross validated setting. Lower Median Absolute Error (MAE) and higher Normalized Mutual Information (NMI) and correlation coefficient (R) signify improved performance. Our framework outperforms the baselines during testing, though the model of [11] comes in a close second. This suggests that the Matrix Autoencoder faithfully models subject-specific variation even in unseen patients.

Functional to Structural Association: We evaluate three aspects of our functional to structural manifold alignment. First is our ability to recover structural connectivity matrices during testing. Here, we compare two distance metrics: (1) F_{self} is the Frobenius norm between a test example $\mathbf{A}_n$ and the model prediction for the same example $\hat{\mathbf{A}}_n$, and (2) F_{other} is $\hat{\mathbf{A}}_n$ and other SC matrices $\mathbf{A}_m, (m \neq n)$. As shown in the left of Fig. 2(a), F_{self} is consistently smaller than F_{other}, with statistical significance determined using the Wilcoxon rank sum test. This indicates that individual differences in SC are preserved by our framework. In the same plot, we also benchmark the recovery performance of our framework against a baseline Matrix encoder-decoder (gray box in Fig. 1) with the SC matrices as *input and output*. We also compare against a linear regression between the vectorized upper diagonal FC features (input) and SC features (output) to help evaluate the benefit of our matrix decomposition. As seen, our function $\rightarrow$ structure decoding achieves similar performance as directly encoding/decoding the structural connectivity. At the same time, the linear regression baseline performs worse than both of these techniques. This suggests that the

Table 1. CFIS prediction on the HCP dataset against the baselines using Median Absolute Error (MAE), Normalized Mutual Information (NMI) for training and testing, and Correlation Coefficient (R) for the test set. Best performance is highlighted in bold.

Method	MAE train	MAE test	NMI train	NMI test	R
No rs-fMRI dec.	6.31 ± 5.61	16.42 ± 12.41	0.85	0.61	0.07
No DTI dec.	6.30 ± 5.80	15.44 ± 13.00	0.86	0.61	0.11
Decoupled.	2.53 ± 2.41	14.90 ± 13.60	0.87	0.59	0.10
BrainNetCNN	6.80 ± 6.25	14.95 ± 12.74	0.88	0.59	0.12
Dict. Learn. + ANN	**3.19 ± 2.19**	15.26 ± 13.99	**0.89**	0.66	0.29
Our framework	3.19 ± 2.47	**14.08 ± 11.85**	0.86	**0.69**	**0.30**

ability to directly leverage the low rank matrix structure is key to preserving individual differences during reconstruction.

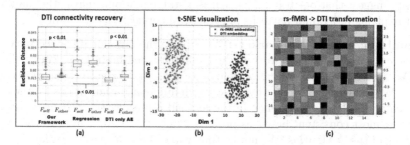

Fig. 2. (a) Recovery of SC for **(L)**: Our Framework **(M)**: Linear Regression **(R)**: DTI only Autoencoder **(b)** t-SNE visualization for FC and SC embeddings **(c)** Coefficient of Variation (C_v) (log scale) for the weights of $\mathbf{\Phi}_{\text{align}}(\cdot)$. Cold colors imply small deviations, i.e. better stability (Color figure online)

Second, we use t-SNE to visualize the symmetric FC and SC embeddings, $\mathbf{F}_n$ and $\mathbf{S}_n$, respectively. Figure 2(b) displays the 2D t-SNE representation computed from the upper-triangle entries of the embedding. As seen, the FC and SC are clustered in two different locations within this space. Interestingly, the learned representations are non-overlapping without explicit enforcement. This suggests that the alignment decoder $\mathbf{\Phi}_{\text{align}}(\cdot)$ is learning a conversion between manifolds.

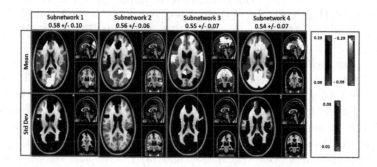

Fig. 3. Top four bases learned by the Matrix Autoencoder measured by the absolute correlation coefficient across cross validation folds and initializationa.

Third, we examine the stability of the transformation learned by the alignment decoder, i.e. the weights $\mathbf{W}_{\text{align}} \in \mathcal{R}^{K \times K}$ of $\mathbf{\Phi}_{\text{align}}(\cdot)$. We first match the columns of $\mathbf{W}_{\text{align}}$ across cross validation folds according to correspondences between the functional brain basis. For each entry of $\mathbf{W}_{\text{align}}$, we compute the coefficient of variation (C_v), i.e. the ratio of the standard deviation to the mean (in absolute value). Lower values of C_v indicate smaller deviations from the mean values, i.e. better stability. Figure 2(c) displays the log coefficient of variation $\log(C_v)$, where the cool colors indicate smaller C_v. As seen, a majority of the entries of $\mathbf{W}_{\text{align}}$ have low variation over the mean pattern value. Overall, our results suggest that our framework learns a stable mapping across the manifolds that explains individual patterns of structural connectivity faithfully.

Examining the FC Biomarkers: We explore the functional connectivity patterns learned by our framework by first matching the brain bases (i.e., columns of $\mathbf{W}$) across the cross validation folds based on the absolute correlation coefficient. We run this experiment five times with different initializations for the ANN branch to check for consistency in the learned representation. Figure 3 displays the four most consistent bases, as projected onto the brain using the region definitions of the AAL atlas. We notice that while there is spatial overlap between the bases, the standard deviations are small, which indicates that our framework is learning stable patters in the data. Subnetwork 1 highlights regions from the default mode network, which is widely inferred within the resting state literature, and known to play a critical role in consolidating working memory [34]. Subnetworks 1, 3 and 4 highlight regions from the somatomotor network and visual cortex, together believed to be important functional biomarkers of cognitive intelligence [7]. Finally, Subnetwork 2 and 4 displays contributions from the frontoparietal network and the medial prefrontal network. These areas are believed to play a role in working memory, attention, and decision making, all of which are associated with cognitive intelligence [30].

Application to a Secondary Dataset: We evaluate our framework on a second clinical cohort of 57 children with high-functioning Autism Spectrum Disorder (ASD). Rs-fMRI and DTI scans were acquired on a Philips $3T$ Achieva scanner (**rs-fMRI:** EPI, TR/TE = 2500/30 ms, res = 3.05 × 3.15 × 3 mm, duration = 128 or 156 time samples; **DTI:** EPI, TR/TE = 6356/75 ms, res = 0.8 × 0.8 × 2.2 mm, b-value = 700 s/mm^2, 32 gradient directions). Rs-fMRI preprocessing includes motion correction, normalization to MNI, spatial and temporal filtering, and nuisance regression with CompCorr [3]. DTI data was preprocessed using the FDT pipeline in FSL [18], with tractography performed using the BEDPOSTx and PROBTRACKx functions in FSL [2]. We define $\mathbf{y}_n$ using three phenotypes that characterize various impairments associated with ASD: (1) the Autism Diagnostic Observation Schedule (ADOS) [33], (2) the Social Responsiveness Scale (SRS) [33], and (3) Praxis [32]. We carry forward the same parcellation scheme and model parameters as used for the HCP dataset. Table 2 compares the *multi-score* prediction testing performance of ADOS, SRS, and Praxis in a five fold cross validation setting. We observe that only our framework and the model of [11] can *simultaneously predict all three measures*. In contrast, the other baselines achieve good testing performance on one or two of the measures (for example, No DTI decoder baseline for ADOS and SRS) but cannot generalize all three. Our supplementary document includes additional results for SC recovery and FC biomarker extraction, and scatter plots for phenotypic prediction on the ASD datatset. Overall, our experiments on both healthy (HCP) and clinical (ASD) populations suggest that our model is robust across cohorts and generalizes effectively even with modest dataset sizes.

Table 2. Multi-score performance on the ASD dataset using Median Absolute Error (MAE), Normalized Mutual Information (NMI), and Correlation Coefficient (R) for testing. Best performance is highlighted in bold. Near misses are underlined

Measure	Method	MAE test	NMI test	R
ADOS	No rs-fMRI dec.	3.11 ± 2.74	<u>0.46</u>	0.089
	No DTI dec.	**2.61 ± 2.59**	0.41	0.14
	Decoupled	**2.64 ± 2.30**	**0.49**	0.35
	BrainNetCNN	3.89 ± 2.80	0.35	0.05
	Dict. Learn.+ANN	<u>2.71 ± 2.40</u>	0.43	<u>0.50</u>
	Our framework	<u>2.71 ± 1.84</u>	**0.49**	**0.51**
SRS	No rs-fMRI dec.	16.84 ± 16.01	0.77	0.039
	No DTI dec.	**15.65 ± 12.69**	0.81	0.24
	Decoupled	17.40 ± 14.16	0.74	0.02
	BrainNetCNN	17.50 ± 15.18	0.73	0.15
	Dict. Learn.+ANN	<u>16.79 ± 13.83</u>	**0.89**	**0.37**
	Our framework	<u>16.04 ± 13.40</u>	<u>0.83</u>	<u>0.34</u>
Praxis	No rs-fMRI dec.	14.03 ± 10.80	0.74	0.02
	No DTI dec.	19.65 ± 13.18	0.81	0.23
	Decoupled	17.08 ± 12.23	0.76	0.09
	BrainNetCNN	19.35 ± 12.56	0.74	0.20
	Dict. Learn.+ANN	**13.19 ± 10.75**	<u>0.82</u>	**0.37**
	Our framework	**13.14 ± 10.78**	**0.86**	<u>0.32</u>

4   Conclusion

We have introduced a novel matrix autoencoder to map the manifold of rs-fMRI functional connectivity to the manifold of DTI structural connectivity. Our framework is strategically designed to leverage the underlying geometry of the data spaces and robustly recover brain biomarkers that are simultaneously explanative of behavioral phenotypes. We demonstrate that our framework offers both interpretability and generalizability, even for multi-score prediction on modest sized datasets. Finally, our framework makes minimal assumptions, and can potentially find application both within and outside the medical realm.

Acknowledgements. This work was supported by the National Science Foundation CRCNS award 1822575, National Science Foundation CAREER award 1845430, National Institute of Mental Health (R01 MH085328-09, R01 MH078160-07, K01 MH109766 and R01 MH106564), National Institute of Neurological Disorders and Stroke (R01NS048527-08), and the Autism Speaks foundation.

References

1. Assaf, Y., Pasternak, O.: Diffusion tensor imaging (DTI)-based white matter mapping in brain research: a review. J. Mol. Neurosci. **34**(1), 51–61 (2008)
2. Behrens, T.E., et al.: Probabilistic diffusion tractography with multiple fibre orientations: what can we gain? NeuroImage **34**(1), 144–155 (2007)
3. Behzadi, Y., et al.: A component based noise correction method (CompCor) for bold and perfusion based FMRI. NeuroImage **37**(1), 90–101 (2017)
4. Bessadok, A., Mahjoub, M.A., Rekik, I.: Topology-aware generative adversarial network for joint prediction of multiple brain graphs from a single brain graph. In: Martel, A.L., et al. (eds.) MICCAI 2020. LNCS, vol. 12267, pp. 551–561. Springer, Cham (2020). https://doi.org/10.1007/978-3-030-59728-3_54
5. Bilker, W.B., et al.: Development of abbreviated nine-item forms of the Raven's standard progressive matrices test. Assessment **19**(3), 354–369 (2012)
6. Castellazzi, G., et al.: A machine learning approach for the differential diagnosis of Alzheimer and vascular dementia fed by MRI selected features. Front. Neuroinformatics **14**, 25 (2020)
7. Chén, O.Y., et al.: Resting-state brain information flow predicts cognitive flexibility in humans. Sci. Rep. **9**(1), 1–16 (2019)
8. Chu, S.H., et al.: Function-specific and enhanced brain structural connectivity mapping via joint modeling of diffusion and functional MRI. Sci. Rep. **8**(1), 4741 (2018)
9. Dong, Z., et al.: Deep manifold learning of symmetric positive definite matrices with application to face recognition. In: Proceedings of the AAAI Conference on Artificial Intelligence, vol. 31 (2017)
10. Duncan, J.: Frontal lobe function and general intelligence: why it matters. Cortex J. Devoted Study Nervous Syst. Behav. **41**(2), 215–217 (2005)
11. D'Souza, N.S., Nebel, M.B., Wymbs, N., Mostofsky, S., Venkataraman, A.: Integrating neural networks and dictionary learning for multidimensional clinical characterizations from functional connectomics data. In: Shen, D., et al. (eds.) MICCAI 2019. LNCS, vol. 11766, pp. 709–717. Springer, Cham (2019). https://doi.org/10.1007/978-3-030-32248-9_79
12. Essen, V., et al.: The WU-MINN human connectome project: an overview. Neuroimage **80**, 62–79 (2013)
13. Fox, M.D., Raichle, M.E.: Spontaneous fluctuations in brain activity observed with functional magnetic resonance imaging. Nat. Rev. Neuro. **8**(9), 700–711 (2007)
14. Fukushima, M., et al.: Structure-function relationships during segregated and integrated network states of human brain functional connectivity. Brain Struct. Funct. **223**(3), 1091–1106 (2018)
15. Hahn, K., et al.: Selectively and progressively disrupted structural connectivity of functional brain networks in Alzheimer's disease—revealed by a novel framework to analyze edge distributions of networks detecting disruptions with strong statistical evidence. Neuroimage **81**, 96–109 (2013)
16. Honey, C.J., et al.: Predicting human resting-state functional connectivity from structural connectivity. Proc. of the Nat. Acad. Sci. **106**(6), 2035–2040 (2009)
17. Huang, Z., Van Gool, L.: A Riemannian network for SPD matrix learning. In: Proceedings of the AAAI Conference on Artificial Intelligence, vol. 31 (2017)
18. Jenkinson, M., et al.: FSL. NeuroImage **62**(2), 782–790 (2012)
19. Karras, T., et al.: Training generative adversarial networks with limited data. arXiv preprint arXiv:2006.06676 (2020)

20. Kawahara, J., et al.: BrainNetCNN: convolutional neural networks for brain networks; towards predicting neurodevelopment. NeuroImage **146**, 1038–1049 (2017)
21. Kiar, G., et al.: ndmg: Neurodata's MRI graphs pipeline. Zenodo (2016)
22. Kingma, D.P., Ba, J.: Adam: a method for stochastic optimization. arXiv preprint arXiv:1412.6980 (2014)
23. Le Bihan, D., et al.: Diffusion tensor imaging: concepts and applications. J. Magn. Reson. Imaging Official J. Int. Soc. Magn. Reson. Med. **13**(4), 534–546 (2001)
24. LeCun, Y.A., Bottou, L., Orr, G.B., Müller, K.-R.: Efficient BackProp. In: Montavon, G., Orr, G.B., Müller, K.-R. (eds.) Neural Networks: Tricks of the Trade. LNCS, vol. 7700, pp. 9–48. Springer, Heidelberg (2012). https://doi.org/10.1007/978-3-642-35289-8_3
25. Lee, M.H., et al.: Resting-state FMRI: a review of methods and clinical applications. Am. J. Neuroradiol. **34**(10), 1866–1872 (2013)
26. Lin, L., et al.: Predicting healthy older adult's brain age based on structural connectivity networks using artificial neural networks. Comput. Meth. Prog. Biomed. **125**, 8–17 (2016)
27. Liu, J., et al.: Community-preserving graph convolutions for structural and functional joint embedding of brain networks. In: 2019 IEEE International Conference on Big Data (Big Data), pp. 1163–1168. IEEE (2019)
28. Liu, S., et al.: Multimodal neuroimaging computing: a review of the applications in neuropsychiatric disorders. Brain Inform. **2**(3), 167–180 (2015)
29. Loshchilov, I., Hutter, F.: Decoupled weight decay regularization. arXiv preprint arXiv:1711.05101 (2017)
30. Menon, V.: Large-scale brain networks and psychopathology: a unifying triple network model. Trends Cogn. Sci. **15**(10), 483–506 (2011)
31. Messé, A., et al.: Predicting functional connectivity from structural connectivity via computational models using MRI: an extensive comparison study. NeuroImage **111**, 65–75 (2015)
32. Mostofsky, S.H., et al.: Developmental dyspraxia is not limited to imitation in children with autism spectrum disorders. J. Int. Neuropsychol. Soc. JINS **12**(3), 314 (2006)
33. Payakachat, N., et al.: Autism spectrum disorders: a review of measures for clinical, health services and cost-effectiveness applications. Expert Rev. Pharmacoecon. Outcomes Res. **12**(4), 485–503 (2012)
34. Sestieri, C., et al.: Episodic memory retrieval, parietal cortex, and the default mode network: functional and topographic analyses. J. Neurosci. **31**(12), 4407–4420 (2011)
35. Smith, S.M., et al.: Resting-state FMRI in the human connectome project. Neuroimage **80**, 144–168 (2013)
36. Sui, J., et al.: Combination of resting state fMRI, DTI, and sMRI data to discriminate schizophrenia by N-way MCCA+ JICA. Front. Human Neurosci. **7**, 235 (2013)
37. Wang, C., et al.: Manifold alignment using procrustes analysis. In: Proceedings of the 25th International Conference on Machine Learning, pp. 1120–1127 (2008)
38. Wong, E., Anderson, J.S., Zielinski, B.A., Fletcher, P.T.: Riemannian regression and classification models of brain networks applied to autism. In: Wu, G., Rekik, I., Schirmer, M.D., Chung, A.W., Munsell, B. (eds.) CNI 2018. LNCS, vol. 11083, pp. 78–87. Springer, Cham (2018). https://doi.org/10.1007/978-3-030-00755-3_9

39. Zhang, L., Wang, L., Zhu, D.: Recovering brain structural connectivity from functional connectivity via Multi-GCN based generative adversarial network. In: Martel, A.L., et al. (eds.) MICCAI 2020. LNCS, vol. 12267, pp. 53–61. Springer, Cham (2020). https://doi.org/10.1007/978-3-030-59728-3_6

40. Zhang, W., Zhan, L., Thompson, P., Wang, Y.: Deep representation learning for multimodal brain networks. In: Martel, A.L., et al. (eds.) MICCAI 2020. LNCS, vol. 12267, pp. 613–624. Springer, Cham (2020). https://doi.org/10.1007/978-3-030-59728-3_60

Clinical Applications - Neuroimaging - Others

Topological Receptive Field Model for Human Retinotopic Mapping

Yanshuai Tu[1] 🆔, Duyan Ta[1], Zhong-Lin Lu[2,3] 🆔, and Yalin Wang[1(✉)] 🆔

[1] Arizona State University, Tempe, AZ 85201, USA
ylwang@asu.edu
[2] New York University, New York, NY, USA
[3] NYU Shanghai, Shanghai, China

Abstract. The mapping between visual inputs on the retina and neuronal activations in the visual cortex, i.e., retinotopic map, is an essential topic in vision science and neuroscience. Human retinotopic maps can be revealed by analyzing the functional magnetic resonance imaging (fMRI) signal responses to designed visual stimuli *in vivo*. Neurophysiology studies summarized that visual areas are topological (i.e., nearby neurons have receptive fields at nearby locations in the image). However, conventional fMRI-based analyses frequently generate non-topological results because they process fMRI signals on a voxel-wise basis, without considering the neighbor relations on the surface. Here we propose a topological receptive field (tRF) model which imposes the topological condition when decoding retinotopic fMRI signals. More specifically, we parametrized the cortical surface to a unit disk, characterized the topological condition by tRF, and employed an efficient scheme to solve the tRF model. We tested our framework on both synthetic and human fMRI data. Experimental results showed that the tRF model could remove the topological violations, improve model explaining power, and generate biologically plausible retinotopic maps. The proposed framework is general and can be applied to other sensory maps.

Keywords: Retinotopic map · Population receptive field · Topological

1 Introduction

It is of great interest to quantify, simulate, and understand the relation between the visual inputs and the neuronal response with the retinotopic maps [1, 2]. A precise retinotopic map provides opportunities to understand or even simulate various aspects of the visual system. For instance, the complex-log [3] model, which depicted a rough position map between the retina and the primal visual cortex (V1), explained the dynamics of spiral

The work was supported in part by NIH (R21AG065942, RF1AG051710 and R01EB025032) and Arizona Alzheimer Consortium.

Electronic supplementary material The online version of this chapter (https://doi.org/10.1007/978-3-030-87234-2_60) contains supplementary material, which is available to authorized users.

M. de Bruijne et al. (Eds.): MICCAI 2021, LNCS 12907, pp. 639–649, 2021.
https://doi.org/10.1007/978-3-030-87234-2_60

visual illusions [4]. The retinotopic maps discovered some stable visual regions [5]. Additionally, retinotopic map research holds great promise in understanding brain plasticity and improving rehabilitation's efficacy from visual impairments. Further, retinotopic maps have been clinically adopted to monitor the progress and recovery under amblyopia treatment [6], a disorder that affects about 2% of children and may cause significant visual impairment if untreated.

Typically, human retinotopic maps are obtained *in vivo* by analyzing cortical functional magnetic resonance imaging (fMRI) response signals to visual stimuli [7, 8]. Since the first fMRI work on retinotopic mapping, several analysis models were proposed [9–11] to decode perception parameters from the noisy fMRI signals. Among them, the population receptive field (pRF) model [10] generates state-of-the-art results and becomes a cornerstone in fMRI signal analysis of retinotopic maps [12].

Although the pRF model achieved great success, the pRF results are usually not topological. The topological condition is an essential requirement of retinotopic maps since neurophysiology studies have revealed nearby neurons have receptive fields at nearby locations in the image [13, 14] (the topological condition). The topological condition is also the requirement of the vision system's hierarchical organization [1]: each visual area represents a unique map of a portion of retina. If there are duplicated representations in a visual area, this visual area should be further divided into more areas. Besides, it is challenging to infer accurate visual-related quantification without a topological retinotopic map, e.g., visual boundaries, cortical magnification factor, visual anisomery. Therefore, topological retinotopic map results are vital for neuroscience research.

A variety of methods were developed to reduce topological violations in the post-processing of pRF results [15–20]. For instance, the model fitting [15] is widely used to fit the decoded parameters with an algebraic model for V1–V3. Smoothing methods, e.g., Laplacian smoothing [16], process visual parameters one by one but cannot ensure the topological condition. Surface registration of retinotopic mesh is also developed for post-processing of pRF results [18, 21]. To our knowledge, however, *none* of the existing methods have considered the topological condition when decoding fMRI signals.

It is advantageous but challenging to impose the topological condition when decoding fMRI. First, the topological condition is different across visual areas, while the precise visual areas are delineated upon the decoding results. One may think the segmentation is available through the anatomical surface. Unfortunately, all anatomical-based segmentation is not accurate [21], limited by its modality. Second, since the fMRI signal is noisy, it is challenging to segment visual areas from the noisy decoded results.

We propose a topological receptive field (tRF) framework, which imposes the topological conditions by integrating the topology-preserving segmentation and topological fMRI decoding into a coherent system. More specifically, we first segmented the visual areas by preserving the prior connectivity pattern of visual areas based on initial decoding and then modeled the fMRI decoding with the topological condition within each visual area. Since the new fMRI decoding results can refine the segmentation, we repeated the segmentation and decoding until theoretically guaranteed convergence.

We validated the proposed framework on both synthetic data and human retinotopy data. Our experiments showed the superiority of tRF over other state-of-the-art methods, including the pRF model, post-processing by model fitting, smoothing, or registration methods. To our knowledge, (1) it is the first work that enforces the topological condition in decoding retinotopic fMRI signals; (2) it is also the first automatic visual area segmentation with the topology organization preserved. Our framework is general and can be extended to other sensory maps since most of the human perception maps are spatially organized, e.g., auditory map [22] is spatially related to sound frequency.

2 Method

2.1 Retinotopic Mapping Experiment

We begin with a brief introduction to the retinotopic mapping experiment. Such an experiment acquires both structural MRI and fMRI signals, as illustrated in Fig. 1. The structural MRI (Fig. 1g) is used to reconstruct the cortical surface. The fMRI (Fig. 1b) is acquired to monitor neuron activities during the visual stimuli (Fig. 1a). The visual stimulation is carefully designed to encode the visual space (Fig. 1f) uniquely. Before the analysis, the fMRI volumes are preprocessed (Fig. 1c) and projected onto the cortical surface (Fig. 1h). Eventually, for each subject, there is a high-quality anatomical/structural cortical surface in the form of discrete manifold $S = (V, E, F)$ (where V is the surface vertex set, E is the edge set, and F is the face set), together with fMRI time series $y_i(t)$ for each vertex $V_i \in V$ on the cortical surface.

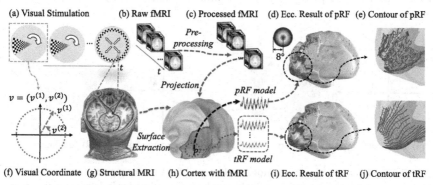

(a) Visual Stimulation   (b) Raw fMRI   (c) Processed fMRI   (d) Ecc. Result of pRF   (e) Contour of pRF

(f) Visual Coordinate   (g) Structural MRI   (h) Cortex with fMRI   (i) Ecc. Result of tRF   (j) Contour of tRF

Fig. 1. The retinotopic map experiment and the pipelines of pRF/tRF models.

2.2 pRF Model

The population receptive field (pRF) model decodes fMRI signals vertex by vertex. Namely, on the measurement $y_i(t)$, the pRF predicts perception parameters, including the population perception center $v_i = \left(v_i^{(1)}, v_i^{(2)}\right)$ and population receptive field size σ_i. We take the polar coordinates for the visual fields, as shown in (Fig. 1f). Namely, $v_i^{(1)}$ and $v_i^{(2)}$ are the eccentricity and polar angle, respectively. Given the receptive model $r\left(v'; v_i, \sigma\right)$ [10], and the hemodynamic model $h(t)$ [23], the predicted fMRI is, $\hat{y}(t; v_i, \sigma) = \beta\left(\int r\left(v'; v_i, \sigma\right)s\left(t, v'\right)dv'\right)*h(t)$, where β is the activation level (a time-independent scalar). The perception center v_i and population receptive field size σ_i are estimated by minimizing the error between the measured and predicted fMRI activation signal,

$$(v_i, \sigma_i) = \arg \min_{(v_i, \sigma_i)} E_p = \int \left(\hat{y}(t; v_i, \sigma_i) - y_i(t) \right)^2 dt. \qquad (1)$$

Solving Eq. 1 for all points generates the pRF retinotopic map. We show a typical pRF retinotopic map on the visual cortex in Fig. 1d. Unfortunately, a large portion of the result is not topological (the visual coordinates' contour curves are messed up in Fig. 1e). We are motivated to propose the tRF (Fig. 1i, j) to improve the retinotopic maps.

2.3 Parametrization

To simplify the discussion, we first establish a parametrization between the 3D visual cortex and the 2D planar disk. We cut a geodesic patch containing the visual areas. In specific, we picked a point p_0 on the cortex, then enclosed a patch P consist of cortical points within a certain geodesic to P_0, as illustrated in Fig. 2b. Then we computed the conformal parametrization for patch P to planar disk D. If $c : P \rightarrow D$ is the conformal parameterization, then $u_i = c(V_i)$ where $u_i \in \mathbb{R}^2$ is the parametric coordinate for V_i. The conformal mapping was done via spherical conformal mapping of the double coverings of surface patches, followed by stereographic projection [24]. With the parametrization c, we can discuss the topology condition within the planar domains.

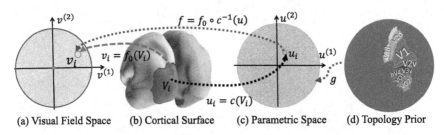

(a) Visual Field Space   (b) Cortical Surface   (c) Parametric Space   (d) Topology Prior

Fig. 2. Parametrization and segmentation: (a) the visual field, (b) the region of interest (ROI), (c) conformal parametrization of ROI, and (d) the topology-prior (the label is defined for each point).

2.4 Topology Conditions

The topology conditions have two inferences. First, the visual coordinates are topological with respect to the surface parametrization within each visual area (we call this condition the *topological condition* within each visual area). Second, since the human visual cortex is organized into several visual areas, and the visual areas hold the same organization (for instance, V1 is adjacent to V2d and V2v) across subjects (to distinguish, we call this condition *topology-preserving condition* across visual areas).

Both conditions can be quantified by the Beltrami coefficient of quasiconformal maps [25]. Namely, we will use the Beltrami coefficient to enforce a topology-preserving surface segmentation and use the concept again to ensure the topological condition for the retinotopic map within each visual area. Next, we introduce the Beltrami coefficient's definition and then explain how it is adopted into these two conditions.

An orientation-preserving homeomorphism $\varphi(z) : \mathbb{C} \to \mathbb{C}$ is *quasiconformal* if it satisfies the Beltrami equation $\frac{\partial \varphi(z)}{\partial \bar{z}} = \mu_\varphi \frac{\partial \varphi(z)}{\partial z}$, where μ_φ is the *Beltrami coefficient* satisfying $|\mu_\varphi|_\infty < 1$. In particular, if $\mu_\varphi = 0$ then φ is conformal.

The topological condition requires the retinotopic map from each visual area to the retina to be orientation consistent. The orientation (biologically the visual field sign, *VFS*) for the visual area can be either positive or negative, depending on the parametrization and visual area. Once fixing a parametrization, the map (from parametric space to retina) orientation for a specific visual area is then fixed. Of course, one can treat the negative visual field area to be positive by flipping the parametrization vertically ($u'^{(2)} = -u^{(2)}$). Without losing generality, considering a positive visual area, we require the VFS is consistently positive. Mathematically, it requires the Beltrami coefficient μ_f, associated with the retinotopic map from parametric space (within the unit disk) to visual field space $f : u_i \mapsto v_i$, satisfies $|\mu_f|_\infty < 1$.

We used a registration-akin method for topology-preserving segmentation. The method diffeomorphically morphs the topology-prior (which has the right topology) to the subject's parametrization disk and then uses prior's label information for the subject. The segmentation is topology-preserving if the morphing is diffeomorphic. Let $g : D \to D$ be the *morph function*. If the Beltrami coefficient μ_g satisfies $|\mu_g|_\infty < 1$, g is diffeomorphic [18]. Compared to retinotopic registration [18], where visual coordinates are required when defining a template, we only require the prior label, which is easier to give and reduces the model visual coordinates hypothesis.

2.5 tRF Model

We introduce our tRF model, which includes fMRI decoding $T = \{(v_i, \sigma_i)|V_i \in V\}$ and visual areas segmentation $L = \{L_i \in \mathbb{N}|V_i \in V\}$, (one value for an area) altogether.

Topology-Preserving Segmentation
We search a wrapping function g that maximizes the label alignment between subject and wrapped-prior. However, there are no labels before segmentation. As an alternative, we maximize the alignment between VFS of topology prior and VFS of a subject.

Let $J = (F_J, E_J, u_J, L_J)$ be the topology-prior mesh, where F_J is face set, E_J is edge set, $u_J = \{u_j\}$ is vertex set and $L_J = \{L_j \in \mathbb{N}|u_j \in u_J\}$ is label set, $\hat{s}_g : D \to \{-1, 1\}$ be the VFS for topology-prior after wrapping, and $s := \text{sign}(1 - |\mu_f|)$ be the VFS function of the subject, where $\mu_f = \frac{\partial v}{\partial \bar{u}}/\frac{\partial v}{\partial u}$ ($v = v^{(1)} + iv^{(2)}$ and $u = u^{(1)} + iu^{(2)}$). We model the registration by $E = \int_D |\hat{s}_g - s| + \lambda_g \int_D |\nabla g|^2 du$, such that $|\mu_g| < 1$, where $|\hat{s}_g - s|$ is the *VFS mismatch cost,* and λ_g is a positive constant controls the smoothness of g. The mismatch cost is formulated to encourage the subject, and the wrapped topology-prior shares the same VFS at each position.

Topological fMRI Decoding

Without losing generality, for a positive orientated visual area V^+, the topological fMRI decoding model is to solve within the V^+ area, by $T = \arg\min_{T} \sum_{V_j \in V^+} \int |\hat{y}(t; v_j, \sigma_j) - y_j(t)|^2 dt + \lambda_v |\nabla v|_i^2 du$, s.t. $|\mu_f| < 1$, where $|\nabla v|^2 = |\nabla v^{(1)}|^2 + |\nabla v^{(2)}|^2$. Similarly, the tRF model can be applied to the negative area by flipping the parametrization vertically. To avoid over smoothing, we choose the values of λ_v based on the Generalized Cross-Validation [26, 27].

tRF Model

The tRF finds the visual parameters for the visual cortex via two modules,

$$\hat{L} = L_J(g), \text{ where } g = \arg\min_{g} \int_D |\hat{s}_g - s| + \lambda_g \int_D |\nabla g|^2 du, \text{ s.t.} |\mu_g| < 1 \quad (2)$$

$$\hat{T} = \arg\min_{\hat{T}} \sum_{u_i \in D} \int (\hat{y} - y_i)^2 dt + \lambda_v |\nabla v|_i^2, \quad \text{ s.t. } s(u_i) = \hat{s}(u_i | \hat{L}). \quad (3)$$

2.6 Numerical Method

A direct search for $(\hat{L}, \hat{T})$ is computationally infeasible since the number of parameters is typically high and $\hat{y}$ is also computationally heavy. Naturedly, we divide the problem into two subproblems, segmentation and fMRI decoding, and update them iteratively. Since both subproblems constrain the norm of Beltrami, we introduce the technique in advance, which is called *topology-projection*.

Topology-Projection with Smoothing

Given a map f, the topology-projection computes a smooth and topological map f' with small changes of f. According to Measurable Riemannian Mapping Theorem [25], the Beltrami coefficient uniquely encodes a map upon suitable normalization. Therefore, we can manipulate a map by its Beltrami coefficient. If a maps whose $|\mu_f| > 1$ for some points, we adjust its norm for those points by $\mu'_f = a\mu_f/|\mu_f|$ ($a = 0.95$ in this work). Then we use μ'_f to recovery the topological map f'. In specific, one can find f' by solving $\nabla \cdot A \nabla f' = 0$, with Dirichlet boundary condition. Here A is a 2-by-2 matrix upon μ'_f [28], ∇ and $\nabla\cdot$ are the gradient and divergence operators, respectively. f' can be solved efficiently by writing $\nabla \cdot A \nabla$ as a matrix (see Supplementary Materials - *SM*). After the topology-projection, we then apply *Laplacian smoothing* [27] to f' to make it smooth.

Topology-Preserving Segmentation

To solve Eq. 2, we divide it into two subproblems: naïve g searching and topology-projection. Specifically, (1) for $u \in u_J$, we update the target of $g(u)$ to nearest point g' such that $s(u_i) = \hat{s}(g')$, since it minimizes the VFS mismatch; (2) used the topological projection to fix the topological condition for g' (see *SI* for segmentation results).

Topological fMRI Decoding

We also split Eq. 3 into two subproblems: parameter searching and topology-projection. Specifically, we (1) used the topology-projection for each visual area respectively to get topological visual parameters T'; (2) used the gradient descent to update T' parameters, $T' \leftarrow T' - \eta \left\{ \left(\frac{\partial E_p}{\partial v^{(1)}}, \frac{\partial E_p}{\partial v^{(2)}}, \frac{\partial E_p}{\partial \sigma} \right) \right\}$, where η is a constant (*updating step*). The updated result T' is further used to improve the segmentation, described in Eq. 2.

2.7 Algorithm

We now summarize the overall procedures of the tRF framework in Algorithm 1. Note that we decay the updating step by $\eta^{(t+1)} = k_\beta \eta^{(t)}$, $k_\beta < 1$, to ensure tRF converges. Its convergence proof is provided in SM.

Algorithm 1: The tRF framework

Input: Discrete region of interest $S = (V, E, F)$; fMRI signal y_i for each point; a tolerance $\epsilon \in \mathbb{R}^+$; and a decay factor $k_\beta < 1$.

Results: The topology-preserving segmentation L and topological retinotopic map $T = \{(v_i, \sigma_i)\}$ that sufficiently explains the fMRI signal.

Solve the pointwise pRF parameter (v_i, σ_i) by **Eq. 1**
Compute the conformal disk parametrization $u_i = c(V_i), V_i \in V$
Initialize $T = \{(v_i, \sigma_i)\}$, and $\delta \leftarrow \infty$
while $\delta > \epsilon$ **do**
    Compute the segmentation L, by solving **Eq. 2**
    Update T' on given segmentation L by solving **Eq. 3**
    $\delta = \max(T' - T)$, $T \leftarrow T'$, and $\eta^{(t+1)} \leftarrow k_\beta \eta^{(t)}$
end

---

3 Dataset

3.1 Synthetic Data

We first evaluated our method on synthetic data. We used double-sech model [15] as ground truth and assigned a receptive field size $\sigma = k_2 v^{(1)}$ ($k_2 = 0.01, 0.02, 0.03$ for V1, V2, and V3, respectively). Then we generated the normalized fMRI signal $y_0(t)$ with the specific stimulation pattern in [29] and added two levels of noise to the fMRI signal $y(t) = y_0(t) + r(t)$, where $r \sim N(0, \gamma)$ ($\gamma = 0.1$ and $\gamma = 0.5$ respectively).

3.2 Real Data with 7T MRI System

We also tested the framework on the Human Connectome Project (HCP) dataset [30]. The HCP dataset is a publicly available retinotopy dataset on 7T fMRI scanners with high spatial and temporal resolutions.

4 Results

4.1 Results on Synthetic Data

We compared the performance of the proposed method with several methods, including the pRF model, the model fitting result [18], the Laplacian smoothing, and the registration method (which registers pRF to another parameterized Double-Sech template and uses the template's value) [31]. More specifically, given the noisy functional signal $y(t)$ for each vertex, we compared the errors between methods' output and ground truth in Table 1, including the perception center error $|\Delta v|$, receptive field size error $|\Delta\sigma|$, and the numbers of triangles that violates the required VFS within its segmentation.

One can see, (1) only the proposed method can make topological results ($T_n = 0$), and the smoothing achieved the worst results, which means the smoothing does not contribute to our topological results in topology-projection; (2) On the other hand, the smoothing method also achieved decent precision ($|\Delta v| \sim 2.8$); (3) The model-fitting method is promising in topological condition ($T_n < 5$). However, it is only applicable to the V1–V3 complex but not to other visual areas (see Table 2 the flipping number is significant due to regions beyond V3); (4) The TPS registration method can ensure the topology in theory. However, since the noisy measure of retinotopic coordinates, the anchors/landmarks are noisy and destroyed the topological condition. Those results suggest the proposed method achieve the best accuracy under topological condition. We now provide an illustrative comparison in Fig. 3 for the mentioned methods.

Table 1. The performance compare (the smaller, the better) on synthetic data. Results for $\gamma = 0.1$ (small noise) and $\gamma = 0.5$ (big noise) are separated by the "/" symbol.

Method	$|\Delta v|$	$|\Delta\sigma|$	T_n
pRF Method	2.924/2.313	0.902/1.100	393/483
Smoothing	2.827/2.288	0.863/1.021	469/528
Registration	3.549/3.518	0.974/0.971	440/444
Model-fitting	3.559/3.590	0.902/1.100	3/5
tRF (proposed)	**2.485/1.801**	**0.768/0.857**	**0/0**

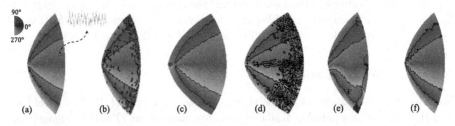

Fig. 3. Comparison of various methods for the V1–V3 model: (a) Ground-truth, (b) pRF, (c) model-fitting, (d) smoothing, (e) registration, and (f) the proposed tRF.

Table 2. The $\bar{e}/T_n$ (fitness/topology) of different methods for the first three HCP observers.

Observers	pRF	Model-fitting	Smoothing	Registration	tRF(proposed)
S1	0.300/870	0.335/489	0.301/842	0.305/953	**0.296/0**
S2	0.252/1119	0.288/934	0.253/1126	0.287/1141	**0.250/0**
S3	0.276/1194	0.296/808	0.276/1197	0.298/1082	**0.273/0**

4.2 Results on the Real Data

We applied the framework to subjects in the HCP dataset. Since no ground-truth is available for the real dataset, we compared the methods based on the fMRI fitting goodness (the RMSE of fitting error $\bar{e}$) and the number of non-topology triangles (T_n). We list the fitting error $\bar{e}$ and topology violations T_n for the first three observers in Table 2. The proposed method achieved the best fMRI fitting and zero topology violations. We also showed the intuitive comparison between the methods for the first observer in Fig. 4a–e. The proposed tRF (Fig. 4e) fixed the topology violations ($T_n = 0$) compared to the pRF model (Fig. 4a, $T_n = 870$), model-fitting (Fig. 4b, $T_n = 489$), smoothing (Fig. 4c, $T_n = 842$), or registration (Fig. 4d, $T_n = 953$). Besides, our method achieved the smallest fMRI fitting error (RMSE is **0.296**), compared to pRF (0.300), model-fitting (0.335), smoothing (0.301), or registration method (0.305). We further report the average results for the first twenty observers in the dataset. Our method reduced the mean fitting error from the second-best method's 0.376 (pRF method) to 0.372, and the topology violations were reduced from 1234 (median value) to **0**. Those results showed that the proposed tRF not only reduced the topology violation (i.e., compatible with neurophysiology conclusions) but also improved/kept the fitting power.

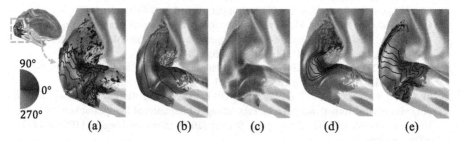

90°

0°

270°

(a)   (b)   (c)   (d)   (e)

Fig. 4. Results of (a) pRF, (b) model-fitting, (c) smoothing, (d) registration, and (e) tRF.

References

1. Hubel, D.H., Wiesel, T.N.: Receptive fields and functional architecture of monkey striate cortex. J. Physiol. **160**, 106–154 (1962). https://doi.org/10.1113/jphysiol.1968.sp008455
2. Wang, Y., Gaborski, R.S.: Computational modeling of topographic arrangements in human visual cortex (2012). https://doi.org/10.1016/j.tics.2014.03.008

3. Schwartz, E.L.: Spatial mapping in the primate sensory projection: analytic structure and relevance to perception. Biol. Cybern. **25**, 181–194 (1977). https://doi.org/10.1007/BF0188 5636
4. Schwartz, E.L.: Computational anatomy and functional architecture of striate cortex: a spatial mapping approach to perceptual coding. Vision Res. **20**, 645–669 (1980). https://doi.org/10. 1016/0042-6989(80)90090-5
5. Glasser, M.F., et al.: A multi-modal parcellation of human cerebral cortex. Nature **536**, 171–178 (2016). https://doi.org/10.1038/nature18933
6. Li, X., Dumoulin, S.O., Mansouri, B., Hess, R.F.: The fidelity of the cortical retinotopic map in human amblyopia. Eur. J. Neurosci. **25**, 1265–1277 (2007). https://doi.org/10.1111/j.1460-9568.2007.05356.x
7. Olman, C.A., Van de Moortele, P.F., Schumacher, J.F., Guy, J.R., Uğurbil, K., Yacoub, E.: Retinotopic mapping with spin echo BOLD at 7T. Magn. Reson. Imag. **28**, 1258–1269 (2010). https://doi.org/10.1016/j.mri.2010.06.001
8. Ogawa, S., et al.: Functional brain mapping by blood oxygenation level-dependent contrast magnetic resonance imaging. A comparison of signal characteristics with a biophysical model. Biophys. J. **64**, 803–812 (1993). https://doi.org/10.1016/S0006-3495(93)81441-3
9. Sato, T.K., Nauhaus, I., Carandini, M.: Traveling waves in visual cortex. Neuron **75**(2), 218–229 (2012)
10. Dumoulin, S.O., Wandell, B.A.: Population receptive field estimates in human visual cortex. Neuroimage **39**, 647–660 (2008). https://doi.org/10.1016/j.neuroimage.2007.09.034
11. Kay, K.N., Winawer, J., Mezer, A., Wandell, B.A.: Compressive spatial summation in human visual cortex. J. Neurophysiol. **110**, 481–494 (2013). https://doi.org/10.1152/jn.00105.2013
12. Van Essen, D.C., Smith, S.M., Barch, D.M., Behrens, T.E.J., Yacoub, E., Ugurbil, K.: The WU-Minn human connectome project: an overview. Neuroimage **80**, 62–79 (2013). https://doi.org/10.1016/j.neuroimage.2013.05.041
13. Wandell, B.A., Dumoulin, S.O., Brewer, A.A.: Visual field maps in human cortex. Neuron **56**, 366–383 (2007). https://doi.org/10.1016/j.neuron.2007.10.012
14. Warnking, J., et al.: fMRI retinotopic mapping—step by step. Neuroimage **17**, 1665–1683 (2002). https://doi.org/10.1006/NIMG.2002.1304
15. Schira, M.M., Tyler, C.W., Spehar, B., Breakspear, M.: Modeling magnification and anisotropy in the primate foveal confluence. PLoS Comput. Biol. **6**, e1000651 (2010)
16. Qiu, A., Bitouk, D., Miller, M.I.: Smooth functional and structural maps on the neocortex via orthonormal bases of the Laplace-Beltrami operator. IEEE Trans Med Imaging. **25**, 1296–1306 (2006). https://doi.org/10.1109/TMI.2006.882143
17. Benson, N.C., et al.:: The HCP 7T retinotopy dataset: description and pRF analysis. bioRxiv. 308247 (2018). https://doi.org/10.1101/308247
18. Tu, Y., Ta, D., Gu, X., Lu, Z.L., Wang, Y.: Diffeomorphic registration for retinotopic mapping via quasiconformal mapping. In: Proceedings - International Symposium on Biomedical Imaging, pp. 687–691. IEEE Computer Society (2020). https://doi.org/10.1109/ISBI45749. 2020.9098386
19. Tu, Y., Tal, D., Lu, Z.L., Wang, Y.: Diffeomorphic smoothing for retinotopic mapping. In: Proceedings - International Symposium on Biomedical Imaging, pp. 534–538. IEEE Computer Society (2020). https://doi.org/10.1109/ISBI45749.2020.9098316
20. Sereno, M.I., Mcdonald, C.T., Allman, J.M.: Analysis of retinotopic maps in extrastriate cortex. Cereb. Cortex. **4**, 601–620 (1994). https://doi.org/10.1093/cercor/4.6.601
21. Benson, N.C., Winawer, J.: Bayesian analysis of retinotopic maps. Elife. 7, (2018). https://doi.org/10.7554/eLife.40224
22. Berlot, E., Formisano, E., De Martino, F.: Mapping frequency-specific tone predictions in the human auditory cortex at high spatial resolution. J. Neurosci. **38**, 4934–4942 (2018). https://doi.org/10.1523/JNEUROSCI.2205-17.2018

23. Lindquist, M.A., Meng Loh, J., Atlas, L.Y., Wager, T.D.: Modeling the hemodynamic response function in fMRI: efficiency, bias and mis-modeling. Neuroimage **45**, S187–S198 (2009). https://doi.org/10.1016/j.neuroimage.2008.10.065

24. Ta, D., Shi, J., Barton, B., Brewer, A., Lu, Z.-L., Wang, Y.: Characterizing human retinotopic mapping with conformal geometry: a preliminary study. In: Ourselin, S., Styner, M.A. (eds.) Medical Imaging 2014: Image Processing, p. 90342A (2014). https://doi.org/10.1117/12.2043570

25. Ahlfors, L.V., Earle, C.J.: Lectures on quasiconformal mappings. Van Nostrand (1966). https://doi.org/10.1090/ulect/038

26. Shahraray, B., Anderson, D.J.: Optimal estimation of contour properties by cross-validated regularization. IEEE Trans. Pattern Anal. Mach. Intell. **11**, 600–610 (1989). https://doi.org/10.1109/34.24794

27. Eilers, P.H.C.: A perfect smoother. Anal. Chem. **75**, 3631–3636 (2003). https://doi.org/10.1021/ac034173t

28. Lui, L.M., Lam, K.C., Wong, T.W., Gu, X.: Texture map and video compression using Beltrami representation. SIAM J. Imag. Sci. **6**, 1880–1902 (2013). https://doi.org/10.1137/120866129

29. Kay, N., et al.: The HCP 7T Retinotopy Dataset. https://osf.io/bw9ec/

30. Benson, N.C., et al.: The human connectome project 7 tesla retinotopy dataset: description and population receptive field analysis. J. Vis. **18**, 1–22 (2018). https://doi.org/10.1167/18.13.23

31. Sprengel, R., Rohr, K., Stiehl, H.S.: Thin-plate spline approximation for image registration. In: Annual International Conference of the IEEE Engineering in Medicine and Biology – Proceedings, pp. 1190–1191. IEEE (1996). https://doi.org/10.1109/IEMBS.1996.652767

SegRecon: Learning Joint Brain Surface Reconstruction and Segmentation from Images

Karthik Gopinath[✉], Christian Desrosiers, and Herve Lombaert

ETS Montreal, Montreal, Canada
karthik.gopinath.1@etsmtl.net

Abstract. Commonly-used tools for cortical reconstruction and parcellation, such as FreeSurfer, are central to brain surface analysis but require extensive computation times. This paper proposes SegRecon, a fast learning approach where an integrated end-to-end deep learning method does *simultaneously* reconstruct and segment cortical surfaces directly from an MRI volume, all in a single step. We train a volume-based neural network to predict, for each voxel, the signed distance to the white-to-grey-matter interface along with its corresponding spherical representation in the registered atlas space. The continuous representation of the spherical coordinates enables our approach to naturally extract an implicit isolevel surface for its reconstruction and obtain the parcel labels from the spherical atlas. We illustrate the advantages of our method with thorough experiments on the MindBoggle dataset. Our parcellation results show more than 4% improvements in average Dice accuracy with respect to FreeSurfer and a drastic speed-up from hours to seconds of computation.

Keywords: Brain surface reconstruction · Surface segmentation · Parcellation

1 Introduction

The accurate reconstruction and segmentation of cortical surfaces from MRI are essential to a variety of brain analyses [8,23]. Standard pipelines for surface reconstruction [4,6,13,17,24] follow a sequence of costly operations that often include: white matter segmentation, surface mesh generation from the segmentation masks, mesh smoothing and projection to a sphere, topological correction of the projected mesh, and fine-tuning of re-projected mesh on the segmented volume. The segmentation of the cortical surface into neuroanatomical parcels is then performed in a subsequent and typically more expensive step of up-to 4 h, which involves the re-projection of each surface to a sphere via a metric-preserving inflation process, registration to a spherical atlas [7,16] and cortical parcellation using atlas labels [5].

© Springer Nature Switzerland AG 2021
M. de Bruijne et al. (Eds.): MICCAI 2021, LNCS 12907, pp. 650–659, 2021.
https://doi.org/10.1007/978-3-030-87234-2_61

Recently, Henschel et al. [12] developed a pipeline called FastSurfer that accelerates processing times using deep learning for brain segmentation and spectral embedding for registration to a spherical atlas. Despite considerably reducing processing times compared to traditional approaches, this pipeline still requires the processing of volume segmentation and surface reconstruction in two consecutive steps, via a combination of different techniques. To overcome this limitation, Cruz et al. [3] has proposed a deep learning model for cortical surface reconstruction, called DeepCSR. Inspired by [21], DeepCSR reconstructs a surface without the need for an explicit segmentation, by sampling points on a reference grid of arbitrary resolution. However, this process is highly expensive for surfaces with hundreds of thousands of points, in terms of both computation and memory. Additionally, DeepCSR *only* enables surface reconstruction, and *not* parcellation which is one of the most time-costly operations in standard neuroimaging pipelines. Approaches in the lines of [9,10,18,19,25] have proposed algorithms to operate directly on surface data for cortical parcellation. Spectral embeddings of surface meshes in a low-dimensional space [18] were also used to predict cortical parcellation labels, but their main limitation was that mesh nodes are considered separately instead of jointly. To better exploit the connectivity information of a mesh graph, recent work has been proposed with graph convolutional networks (GCN) [9–11,25]. While this strategy provides a faster and more accurate parcellation of the cortical surface, it is, however, sensitive to any error from a separate surface mesh reconstruction.

This paper proposes a novel deep learning model, SegRecon, for the joint reconstruction and parcellation of cortical surfaces. Our end-to-end model works directly on MRI volumes and predicts a dense set of surface points along with their corresponding parcellation labels. The proposed architecture, built upon the 3D-UNet [2], is trained to predict for each voxel of an input volume a vector encoding the brain hemisphere of the voxel, its signed distance to the white-to-grey-matter interface, and its spherical coordinates in the registered atlas space. By learning this multi-task problem, the network can thereafter be used to reconstruct and segment implicit brain surfaces both efficiently and topologically accurate [1].

The main contributions of our work are the following: (1) To our knowledge, we propose the first deep learning model to take an MRI volume as input and parcelate and jointly reconstruct the brain surfaces. This contrasts with existing approaches, which either perform surface reconstruction and segmentation in separate steps [12], are limited to reconstruction [3], or require a pregenerated mesh as input [9,18,25]; (2) Compared to DeepCSR, the proposed network implements a fully-convolutional architecture that densely predicts the location relative to cortical surfaces for each voxel of the input image, in a single feed-forward pass; (3) For parcellation, with respect to the widely-used FreeSurfer software, our method achieves a 4.3% improvement in Dice, while being several orders of magnitude faster as well as generating cortical surface.

In the next section, we present the proposed deep learning model for these tasks, describing in detail the network architecture, training losses, and inference steps. The performance of our method is evaluated on the MindBoggle dataset [15]. The ablation study and comparison to the state-of-art in our experiments demonstrate the benefits of our method.

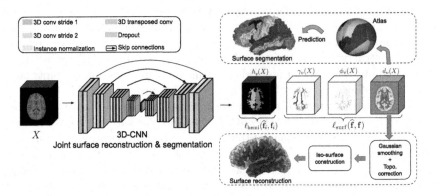

Fig. 1. Overview of SegRecon: The 3D-CNN model takes as input MRI volume X for joint learning of the signed distance to white-to-grey matter interface and its corresponding spherical coordinates in the registered atlas space. (*Red*) The cortical surface is reconstructed by applying Gaussian smoothing and topological correction on the predicted signed distance map prediction $d_v(X)$, followed by iso-surface reconstruction via the Marching Cubes algorithm. (*Blue*) In parallel, the predicted spherical atlas coordinates $(\gamma_v(X), \phi_v(X))$ and hemisphere label $(h_v(X))$ are used to propagate atlas parcellation labels to near-surface voxels v. An illustration for left hemisphere is shown here. (Color figure online)

2   Method

An overview of SEGRECON is shown in Fig. 1 with the end-to-end surface construction and segmentation steps illustrated. Let $\mathcal{D} = \{(X_i, S_i, Y_i)\}_{i=1}^n$ be a training set composed of 3D volumes, $X \in \mathbb{R}^{|\Omega|}$ with voxel set $\Omega \subset \mathbb{R}^3$; surfaces defined by m points, $S_i \in \mathbb{R}^{m \times 3}$; and segmentation labels, $Y_i \in \mathbb{R}^{m \times c}$ where c is the number of segmentation classes. The goal is to learn a function f parameterized by θ which maps an input 3D volume X to a surface with points S and corresponding labels Y.

One of the main challenges in this task comes from the disparity between the well-defined grid space of images X and the domain of surfaces S where the number of points can vary from one surface to another and points can lie anywhere in 3D space. In [3], this problem is solved by giving as input to model f both the image X and a query point $p \in \mathbb{R}^3$ in the template space. The model then predicts if p belongs to the surface in X, or alternatively its distance to this surface. To reconstruct a surface at inference time, the model is queried over a fixed reference grid. While this strategy allows reconstructing a surface at arbitrary resolution, it suffers from two important drawbacks. First, since the template points which can be in the hundreds of thousands are queried independently, reconstructing a surface requires significant time and computation. Moreover, unlike dense prediction approaches, this strategy does not exploit the spatial relationship between points. Last, because feature maps need to be computed for the whole 3D volume X, it also needs a large amount of memory.

To overcome these drawbacks, we instead learn a model that densely projects voxels of the input volume X to a spherical atlas space. Specifically, f maps each voxel $v \in \Omega$ to a vector

$$f_v(X) = \left[d_v(X), \phi_v(X), \gamma_v(X), h_v^{lh}(X), h_v^{rh}(X), h_v^{bg}(X) \right], \tag{1}$$

where $d_v(X)$ is the signed distance from v to its nearest surface point, such that $d_v(X) \leq 0$ if v is inside the surface else $d_v(X) > 0$, $\phi_v(X)$, $\gamma_v(X)$ are the polar angle and azimuthal angle of $v \in \Omega$ defining its position in the spherical atlas, and $h_v^{lh}(X)$, $h_v^{rh}(X)$, $h_v^{bg}(X) \in [0, 1]$ are the probabilities that v is in the left hemisphere, right hemisphere and background, respectively. Here, polar and azimuthal angles are normalized so to lie in the $[-1, 1]$ range. A further topological correction step [1] over the predicted surface points prevents the extraction of critical points yielding topological defects. The resulting surface is defined implicitly as the 0-levelset of the distance map and can be efficiently reconstructed using an iso-surface extraction algorithm such as the Marching Cubes [20].

2.1 Training the Model

Denote $\widehat{\mathbf{f}}_i = f(X_i)$ as the predicted vector for an image X_i and let $\mathbf{f}_i$ be the corresponding ground-truth. To train the model, we use the following loss function

$$\mathcal{L}(\theta; \mathcal{D}) = \sum_{i=1}^{n} \ell_{\text{surf}}(\widehat{\mathbf{f}}_i, \mathbf{f}_i) + \lambda \, \ell_{\text{hemi}}(\widehat{\mathbf{f}}_i, \mathbf{f}_i), \tag{2}$$

The first loss term, ℓ_{surf}, ensures that the signed distance of voxels to the surface, as well as their position in the spherical atlas space, are well predicted. Dropping index i for simplicity, it is defined as:

$$\ell_{\text{surf}}(\widehat{\mathbf{f}}, \mathbf{f}) = \sum_{v \in \Omega} \mathbb{1}_{|d_v| \leq \epsilon} \cdot \left[(\widehat{d}_v - d_v)^2 + \min\left\{ (\widehat{\phi}_v - \phi_v)^2, (1 + \widehat{\phi}_v - \phi_v)^2 \right\} \right.$$
$$\left. + \min\left\{ (\widehat{\gamma}_v - \gamma_v)^2, (1 + \widehat{\gamma}_v - \gamma_v)^2 \right\} \right]. \tag{3}$$

where $\mathbb{1}_P$ is the indicator function, equal to 1 if predicate P is true else, 0 otherwise. We only consider voxels within a distance of ϵ to the nearest surface point in order to focus learning on relevant points close to our surface. This is achieved with function $\mathbb{1}_{|d_v| \leq \epsilon}$ in Eq. (3). Additionally, we consider the non-uniqueness of spherical coordinates (e.g., $-\pi \equiv \pi$) by computing, for each angle, the minimum L_2 distance from the predicted angle or this angle plus 1 to the ground-truth. The distance d_v is, therefore, defined between the center of the voxel v in image space and the nearest point on surface S. In this work, we use the surface mesh generated by FreeSurfer for training. The sign of d_v is determined using the white-matter segmentation mask, with voxels inside the white matter having a negative distance. Likewise, the ground-truth spherical coordinates ϕ_v and γ_v are obtained using FreeSurfer [6] with the Desikan-Killiany-Tourville (DKT) atlas [16].

The second term, ℓ_{hemi} enables the network to predict in which hemisphere lies a voxel v. This prediction is necessary since the surface atlas is defined separately for each hemisphere. Here, we use cross-entropy as loss function:

$$\ell_{\text{hemi}}(\widehat{\mathbf{f}}, \mathbf{f}) = -\sum_{v \in \Omega} \sum_{c \in \{lh, rh, bg\}} h_v^c \log \widehat{h}_v^c. \tag{4}$$

The ground-truth hemisphere masks are once again obtained from FreeSurfer.

2.2   Surface Reconstruction and Segmentation

Once the network is trained, it can be used to reconstruct and segment surfaces directly from a test volume X. First, we feed the volume to the network to obtain a prediction vector for all voxels. Since the network is fully-convolutional, this can be done efficiently in a single feed-forward pass. Next, we apply a small-width Gaussian filter on the predicted 3D distance map $\widehat{d}$ using a single convolution operation and employ a topological correction step [1] to overcome any defects in the surface.

To segment the surface, we first compute the near-surface voxels in each hemisphere as follows:

$$S^c = \left\{ v \in \Omega \mid |d_v| \le \epsilon \wedge c = \arg\max_{c'} \widehat{h}_v^{c'} \right\}, \quad c \in \{lh, rh\}. \tag{5}$$

We then find the nearest-neighbor to a given reference atlas R^c for all the near-surface voxels $v \in S^c$ using their predicted angles $\widehat{\phi}_v$ and $\widehat{\gamma}_v$. The segmentation labels from this reference atlas R^c are then projected back to the near-surface voxels S^c.

2.3   Implementation Details

The overall architecture of SEGRECON is shown in Fig. 1. As an input, we provide the skull-stripped, intensity normalized 3D T1-MRI volume. We use a 3D-UNet architecture similar to [2] in order to map the input voxel to a point in the spherical atlas space.

We apply a softmax activation in the first three output channels to predict the probability of a voxel belonging to the background, left hemisphere, or right hemisphere. The polar and azimuthal angles, $\widehat{\phi}_v$ and $\widehat{\gamma}_v$, are predicted with a tanh activation. The last output channel produces the signed distance map $\widehat{d}_v$ for each voxel. The network parameters, θ, are optimized using a stochastic gradient descent with the Adam optimizer [14]. During training, we pick ϵ in Eq. 3 to be 2.5. The surface is reconstructed using the Marching Cubes [20] on the 0-levelset of a distance map of the predicted signed distance, smoothed with a Gaussian kernel of sigma $= 0.5$ and topologically correct [1]. We use an i7 desktop machine with 16Gb RAM and Nvidia RTX 2080 GPU for our work.

3   Experiments and Results

To benchmark the performance of our method, we use one of the largest publicly available dataset containing manual surface parcellation, MindBoggle [15]. This dataset contains 101 subjects with MRI volumes, FreeSurfer processed meshes, and 32 manually-labeled cortical parcels. We split the dataset randomly into training, validation, and testing using a ratio of 70–10–20%. The qualitative results of the reconstructed surface is shown in Fig. 2. The reconstruction error in terms of Hausdorff distances between the predicted and FreeSurfer meshes is found to be 1.313 voxels only. In a first experiment, we evaluate the effect of varying the reference atlas template for predicting parcellation labels, and show that a robust parcellation can be achieved by combining the predictions from multiple atlases. Finally, we highlight the advantages of joint reconstruction and parcellation methods against state-of-the-art methods.

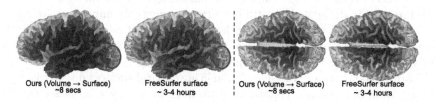

Ours (Volume → Surface)   FreeSurfer surface   Ours (Volume → Surface)   FreeSurfer surface
~8 secs   ~ 3-4 hours   ~8 secs   ~ 3-4 hours

Fig. 2. Surface visualization: Comparison of a cortical surface predicted by our method (Gaussian smoothing $\sigma = 0.001$) and FreeSurfer [7]. Our SegRecon method in comparison with FreeSurfer meshes yields a reconstruction error of 1.313 voxels (avg. Hausdorff distance) with visually similar results while being orders of magnitude faster.

3.1   Effect of Reference Atlas on Parcellation

Instead of predicting class probabilities for each voxel, as in standard 3D segmentation networks, the proposed network predicts spherical atlas coordinates (i.e., angles $\widehat{\phi}_v$ and $\widehat{\gamma}_v$). This has two important advantages: i) considerably reducing the number of outputs for the number of classes to only two, and ii) providing information on the precise location of a voxel inside a parcel instead of simply measuring if a voxel is inside a parcel or not. As we will show in the next section, this continuous prediction strategy leads to a higher accuracy compared to a standard segmentation approach. However, the final predicted labels depend on the reference atlas.

For assessing the impact of the reference atlas on segmentation performance, we randomly select five subjects from the training set and use the spherical coordinates and parcellation labels of their surface mesh nodes as different atlases $Ref_1, \ldots, Ref_5$. Table 1 reports the mean Dice score obtained for test subjects using each of the five atlases. While a high accuracy is obtained in all cases, the performance also varies significantly from 84.60% to 87.33%.

To provide a greater robustness to the choice of atlas, we apply a simple multi-atlas strategy in which a separate prediction is obtained for each atlas, and individual predictions are then combined using majority voting. As shown in Table 1 (last column) this strategy leads to an important boost in Dice score to 88.69% compared to the average of 85.63% computed across all atlases.

3.2 Comparison with the State-of-the-Art

We next compare our joint reconstruction and parcellation method against several baselines and recent approaches. Table 2 reports the performance of tested methods in terms of average Dice scores, mean Hausdorff distances and runtime. To evaluate the benefit of predicting cortical parcels using spherical atlas coordinates, we first train a 3D-UNet to predict the parcellation label probabilities directly at the voxel level as in standard 3D segmentation networks. This baseline, called DirectSeg in Table 2, gives a low Dice score of 79.95%.

We also evaluate the FreeSurfer parcellation against the manual labels provided in the MindBoggle dataset. FreeSurfer considerably improves parcellation accuracy compared to DirectSeg with a Dice score of 84.39%. However, this comes at the price of a significant increase in computation times, from 300 ms per volume for DirectSeg to a few hours for FreeSurfer.

Third, we show the advantage of predicting cortical surfaces directly from 3D images, as in our method, compared to working with surface meshes computed previously. Toward this goal, we test two mesh-based models, named FS + SRF and FS + GCN in the results. The first one, Spectral Random Forest (SRF) [18], performs a spectral embedding of nodes in the FreeSurfer mesh graph using the main eigen-components of its Laplace matrix. The labels of embedded nodes are then predicted separately using a Random Forest classifier. In the latter, the connectivity of nodes in the mesh graph is also exploited in the prediction using a graph convolutional network (GCN) [9]. As can be seen, predicting labels for all nodes simultaneously in FS + GCN, instead of individually in FS + RF, largely improves Dice score by 6.72%. However, as both approaches require generating surface meshes in a former step, which can take around 2 h for FreeSurfer, their total run time remains substantial. In comparison, our method achieves a mean Dice score of 88.69% with an average total run time of 8 s per volume. That is a 4.30% improvement over the Dice score of FreeSurfer, at a fraction of its computational cost.

Next, we evaluate the performance of our method SEGRECON in two different settings. First, we show the importance of the hemisphere prediction loss of Eq. (4) on performance. To do so, we reduce the weight $\lambda = 0.0001$ of the loss term ℓ_{hemi} during training in Eq. (2). This ablation baseline is denoted as

Table 1. Effect of reference atlas: Column 1–5: The average Dice overlap (in %) obtained after using five different references as an atlas for label propagation. The last column shows the results when we vote across five different atlas references.

Ref$_1$	Ref$_2$	Ref$_3$	Ref$_4$	Ref$_5$	Voting
84.60 ±1.90	85.85 ±1.79	85.29 ±1.93	85.08 ±1.54	87.33±1.90	**88.69±1.84**

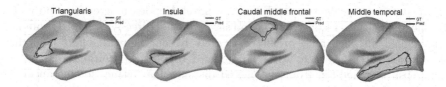

Fig. 3. Parcellation performance: The manual parcellation boundaries are shown in red, with our predicted parcellation boundaries in black. Our model segments 32 parcels in total on the brain surface. We show four parcels, namely, Triangularis, Insula, Caudal middle frontal and middle temporal of the left hemisphere for qualitative analysis. The cortical mesh is inflated here for visualization. (Color figure online)

Table 2. Evaluation of SegRecon: Comparison against approaches in terms of Dice scores (in %), Hausdorff Distances (in mm), and computational time. The first row shows the performance of a DirectSeg a 3D-CNN network on surface parcellation. The second row illustrates the results of the traditional FreeSurfer algorithm for parcellation. In the third and fourth row, we show the ability of a Spectral Random Forest (SRF) and GCN learning based approach to segment the cortical surface. The fifth row shows the importance of learning hemisphere segmentation in our work. Finally, in the last row, we show the performance of our model.

Methods	Dice overlap (%)	Hausdorff (mm)	Time
DirectSeg	79.95 ±2.58	–	∼ 300 ms
FreeSurfer	84.39 ±1.91	2.11 ±0.29	∼ 4 h
FS + SRF	79.89 ±2.62	1.97 ±0.40	∼ 2 h + 18 s
FS + GCN	86.61 ±2.45	1.66 ±0.44	∼ 2 h + 3 s
w/o hemisphere	59.28 ±12.20	3.94 ±3.14	∼ 8 s
SegRecon (Ours)	**88.69 ±1.84**	**1.20 ±1.36**	∼ 8 s

w/o hemisphere in Table 2. As can be observed, the lack of accurate hemisphere prediction results in ambiguous label prediction for surface voxels in both hemispheres which results in a low Dice score of 59.28%. Finally, we present the setting of our model for predicting a distance $\widehat{d}_v$ for each voxel. In this way, our model predicts the iso-surface for surface reconstruction. The accurate prediction of polar and azimuthal angles (i.e., $\widehat{\phi}_v$ and $\widehat{\gamma}_v$) for obtaining parcel labels from the atlas yields an average Dice score of 88.69%. Similar improvements of our method compared to other approaches are also found for the Hausdorff distance metric. Qualitative results obtained by our surface segmentation method are shown in Fig. 3, where we illustrate the differences between the predicted and manual label boundaries for four different parcels or regions.

4 Conclusion

In this work, we presented SEGRECON, a novel deep learning end-to-end model for the joint surface reconstruction and segmentation directly from MRI volumes. The signed distance map predicted densely by our network offers a implicit description of the surface. After enforcing a topological guarantee [1], a surface mesh is generated from the signed distance map using, for instance, the Marching Cubes [20]. Our experiments first analyzed the impact of the reference atlas used for transferring cortical parcellation labels to the surface. A robust performance with Dice score of 88.69% can be achieved for cortical parcellation via a multi-atlas strategy where the predictions for different atlases are combined using majority voting. We also compared our method against several baselines and state-of-the-art approaches for cortical parcellation. Our approach has higher Dice score over 3D-UNet (79.9%), without offering a reconstructed cortical surface, higher Dice score (84.3%) and lower computation time over FreeSurfer (hours vs. seconds) and can perform a joint surface reconstruction and parcellation with higher Dice score compared to GCN (86.6%) that require precomputed surfaces as input. Furthermore, while state-of-art methods obtained a mean Hausdorff distance close to 2 mm, our model achieved a lowest mean distance of 1.20 mm. While Gaussian smoothing and topological correction can alleviate reconstruction artifacts, this technique are performed as a post processing step. Additionally, imposing topological constraints during network training and local smoothing based on anisotropic diffusion [22] could help regularize the mesh while better preserving cortical folding patterns.

References

1. Bazin, P.L., Pham, D.L.: Topology correction of segmented medical images using a fast marching algorithm. Comput. Meth. Prog. Biomed. **88**(2), 182–190 (2007)
2. Çiçek, Ö., Abdulkadir, A., Lienkamp, S.S., Brox, T., Ronneberger, O.: 3D U-Net: learning dense volumetric segmentation from sparse annotation. In: MICCAI (2016)
3. Cruz, R.S., Lebrat, L., Bourgeat, P., Fookes, C., Fripp, J., Salvado, O.: DeepCSR: a 3D deep learning approach for cortical surface reconstruction. arXiv preprint arXiv:2010.11423 (2020)
4. Dahnke, R., Yotter, R.A., Gaser, C.: Cortical thickness and central surface estimation. Neuroimage **65**, 336–348 (2013)
5. Desikan, R.S., et al.: An automated labeling system for subdividing the human cerebral cortex on MRI scans into gyral based regions of interest. Neuroimage **31**(3), 968–980 (2006)
6. Fischl, B., et al.: Automatically parcellating the cortex. Cereb. Cortex **14**(1), 11–22 (2004)

7. Fischl, B., Sereno, M.I., Dale, A.M.: Cortical surface-based analysis: Ii: inflation, flattening, and a surface-based coordinate system. Neuroimage **9**(2), 195–207 (1999)
8. Glasser, M.F., et al.: A multi-modal parcellation of human cerebral cortex. Nature **536**(7615), 171–178 (2016)
9. Gopinath, K., Desrosiers, C., Lombaert, H.: Graph convolutions on spectral embeddings for cortical surface parcellation. Med. Image Anal. **54**, 297–305 (2019)
10. Gopinath, K., Desrosiers, C., Lombaert, H.: Graph domain adaptation for alignment-invariant brain surface segmentation. arXiv preprint arXiv:2004.00074 (2020)
11. He, R., Gopinath, K., Desrosiers, C., Lombaert, H.: Spectral graph transformer networks for brain surface parcellation. In: ISBI (2020)
12. Henschel, L., Conjeti, S., Estrada, S., Diers, K., Fischl, B., Reuter, M.: Fastsurfer-a fast and accurate deep learning based neuroimaging pipeline. NeuroImage (2020)
13. Kim, J.S., et al.: Automated 3-D extraction and evaluation of the inner and outer cortical surfaces using a Laplacian map and partial volume effect classification. Neuroimage **27**(1), 210–221 (2005)
14. Kingma, D.P., Ba, J.: Adam: stochastic optimization. In: ICLR (2014)
15. Klein, A., et al.: Mindboggling morphometry of human brains. PLOS Comput. Biol. **13**(2), e1005350 (2017)
16. Klein, A., Tourville, J.: 101 labeled brain images and a consistent human cortical labeling protocol. Front. Neurosci. **6**, 171 (2012)
17. Kriegeskorte, N., Goebel, R.: An efficient algorithm for topologically correct segmentation of the cortical sheet in anatomical MR volumes. NeuroImage **14**(2), 329–346 (2001)
18. Lombaert, H., Criminisi, A., Ayache, N.: Spectral forests: learning of surface data, application to cortical parcellation. In: MICCAI (2015)
19. López-López, N., Vázquez, A., Poupon, C., Mangin, J.F., Ladra, S., Guevara, P.: GeoSP: a parallel method for a cortical surface parcellation based on geodesic distance. In: EMBC (2020)
20. Lorensen, W.E., Cline, H.E.: Marching cubes: a high resolution 3D surface construction algorithm. ACM SIGGRAPH Comput. Graph. **21**(4), 163–169 (1987)
21. Park, J.J., Florence, P., Straub, J., Newcombe, R., Lovegrove, S.: DeepSDF: Learning continuous signed distance functions for shape representation. In: CVPR (2019)
22. Perona, P., Malik, J.: Scale-space and edge detection using anisotropic diffusion. IEEE Trans. Pattern Anal. Mach. Intell. **12**(7), 629–639 (1990)
23. Querbes, O., et al.: Early diagnosis of Alzheimer's disease using cortical thickness: impact of cognitive reserve. Brain **132**(Pt 8), 2036–2047 (2009)
24. Shattuck, D.W., Leahy, R.M.: Brainsuite: an automated cortical surface identification tool. Med. Image Anal. **6**(2), 129–142 (2002)
25. Wu, Z., et al.: Intrinsic patch-based cortical anatomical parcellation using graph convolutional neural network on surface manifold. In: MICCAI (2019)

LG-Net: Lesion Gate Network for Multiple Sclerosis Lesion Inpainting

Zihao Tang[1,2(✉)], Mariano Cabezas[2], Dongnan Liu[1,2], Michael Barnett[2,3], Weidong Cai[1], and Chenyu Wang[2,3]

[1] School of Computer Science, University of Sydney, Sydney, NSW 2008, Australia
ztan1463@uni.sydney.edu.au
[2] Brain and Mind Centre, University of Sydney, Sydney, NSW 2050, Australia
[3] Sydney Neuroimaging Analysis Centre, Sydney, NSW 2050, Australia

Abstract. Multiple sclerosis (MS) is an immune-mediated neurodegenerative disease that results in progressive damage to the brain and spinal cord. Volumetric analysis of the brain tissues with Magnetic Resonance Imaging (MRI) is essential to monitor the progression of the disease. However, the presence of focal brain pathology leads to tissue misclassifications, and has been traditionally addressed by "inpainting" MS lesions with voxel intensities sampled from surrounding normal-appearing white matter. Based on the characteristics of brain MRIs and MS lesions, we propose a Lesion Gate Network (LG-Net) for MS lesion inpainting with a learnable dynamic gate mask integrated with the convolution blocks to dynamically select the features for a lesion area defined by a noisy lesion mask. We also introduce a lesion gate consistency loss to support the training of the gated lesion convolution by minimizing the differences between the features selected from the brain with and without lesions. We evaluated the proposed model on both public and in-house data and our method demonstrated a faster and superior performance than the state-of-the-art inpainting techniques developed for MS lesion and general image inpainting tasks.

Keywords: Multiple sclerosis · Lesion inpainting · Dynamic gate mask · Gate consistency

1 Introduction

Multiple sclerosis (MS) is a progressive, inflammatory demyelinating disease that affects the central nerve system. Most clinical manifestations are driven by focal inflammatory lesions located in the white matter (WM). The measurement of brain volumetrics and its longitudinal change is a sensitive and reliable surrogate for the severity of MS due to the correlation between irreversible brain tissue loss

Electronic supplementary material The online version of this chapter (https://doi.org/10.1007/978-3-030-87234-2_62) contains supplementary material, which is available to authorized users.

and cognitive and physical disability [15]. Brain Magnetic Resonance Imaging (MRI) has been routinely used in MS clinics for diagnosis and disease progression monitoring, and now is also widely used for quantitative brain volume analysis through tissue segmentation in the research setting [14]. However, this analysis is influenced by the presence of focal brain lesions [3,10]. Hence a precise tissue analysis is crucial to monitor and study the MS disease progression.

Lesion inpainting methods based on traditional statistical models [11,12] follow the assumption that the lesion area should be filled with patches or intensities that are similar to normal-appearing WM. In practice, this assumption is incorrect since the lesions may be located at the grey-white matter boundary and, in some cases, the manually or automatically generated MS lesion masks may also contain errors and overlap with adjacent grey matter (GM) or cerebrospinal fluid (CSF) structures. In these situations, traditional approaches tend to inaccurately fill these regions and lead to tissue segmentation and brain volumetric errors.

From a computer vision perspective, the objective for a general inpainting model is to fill the contents in the masked regions such that new areas are visually consistent with the remaining image [2]. Recently deep learning generative models that learn semantic and hidden representations in an end-to-end fashion have emerged for image inpainting tasks [8,9,18,19]. However, the ultimate goal of the MS inpainting framework is not only to fill the lesion area with intensities similar to normal-appearing tissue, but also for filled zones to retain morphological consistency with other regional brain structures.

In this study, we introduce our Lesion Gate Network (LG-Net), a novel, light-weight, and efficient deep learning based MS lesion inpainting framework, which incorporates a dynamic gating mechanism to represent the lesion areas by learning features from only valid regions of interest (ROIs) during the convolution process. Furthermore, we designed a gate consistency loss to improve the performance of the gating mechanism by capturing the normal-appearing tissue representations for lesion regions. Finally, we evaluated our inpainting framework against several inpainting algorithms on the public IXI dataset and an in-house healthy control (HC) dataset. Our evaluation results show that the proposed LG-Net achieves the highest performance from a clinical and computer vision perspective with a reduced number of parameters, and time and memory cost when compared to the other state-of-the-art approaches. A qualitative analysis of the inpainted images also illustrates the capability of the network to tackle noisy MS lesion masks in the real-world clinical practice.

2 Methods

2.1 Data Preparation

We applied the standard preprocessing procedure for raw MR brain images. Skull-stripping was first applied to define the ROI (all non-brain tissues are excluded) using a pre-trained 3D U-Net [4]. N4 bias correction [16] was then performed on the skull-stripped images.

Since we are interested on the effect of different lesion distributions in a controlled environment, we generated a synthesized MS lesion bank by collecting 12,143 lesions from 524 patients labelled by different experts using FLAIR images. Afterwards, we registered them to the MNI space [5] to create a lesion map. We then randomly selected a set of the lesions and linearly registered them to the skull-stripped healthy cases. Using these registered lesions, we created a synthetic lesion mask that covered less than 10% of the subject's brain volume. To simulate the scenario of noisy labelling, we kept synthetic lesions with a small intersection with GM or CSF (less than 10% of the lesion volume). This synthesis pipeline was applied to all datasets for evaluation.

2.2 MS Lesion Inpainting Framework

Our proposed MS lesion inpainting framework, LG-Net, follows the network architecture from Yu et al. [19]. The details of the framework are shown in Fig. 1. LG-Net takes three consecutive axial brain slices (the target slice and its adjacent slices) as a pseudo 3D input together with the binary lesion masks; and outputs the corresponding inpainted results for all three slices (we pick the target slice only for inference). The reason we used axial slices over coronal or sagittal slices for training is that axial slices contain symmetric information. Even though we did not use a 3D model, we used pseudo 3D slices to give the network more spatially continuous information on tissue structure, while keeping a fast and light-weight model.

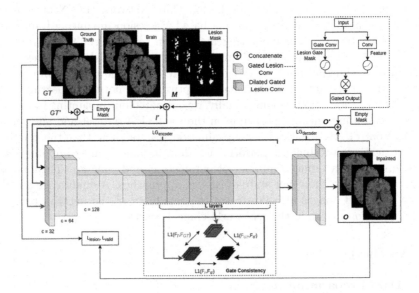

Fig. 1. The training procedure of the proposed LG-Net, where c is the channel number of the convolution layers and **L** is the layers we chose to extract the features for gate consistency. $O\prime$ and $GT\prime$ only go through $LG_{encoder}$ for gate consistency.

2.3   Learnable Dynamic Gate Mask

Convolutional blocks apply the same kernel around each pixel (x, y) of the original image. These blocks extract local features for all the valid spatial locations in the original image. However, lesion voxels usually present features that are different to those of normal-appearing tissues. Therefore, vanilla convolution does not provide a plausible solution for lesion inpainting, as all the voxels are treated in the same manner. For this reason, partial convolutions were proposed [8] to re-weight the convolution output by using the valid mask in each convolution block. Despite the fact that partial convolution shows a superior performance on inpainting tasks when compared with vanilla convolution, it assumes that the valid mask is perfect which only contains relevant voxels. However, MRI scans usually also include a proportion of background with no useful information for inpainting MS lesions.

To adapt the idea of partial convolution to inpainting with noisy MS lesion masks, we replaced the masks with learnable dynamic gate masks [19] for each convolution layer in the learning phase. The lesion gate mask and the corresponding gated lesion convolution are defined as:

$$O'(x) = activation(Conv_{feature}(x)) \odot sigmoid(Conv_{mask}(x)), \qquad (1)$$

where $Conv_{mask}$ is the lesion mask convolution used for dynamic re-weighting process for all the spatial locations across all channels. $activation()$ can be any activation function (here we use LeakyReLU(0.2)). Compared to masks used by partial convolution, the proposed learnable gate mask dynamically selects the features for inpainting in each convolution layer and disregards the background region.

2.4   Lesion Gate Consistency

We designed a triple lesion gate consistency (LGC) loss specifically for lesion inpainting to avoid using adversarial techniques and contextual attention which provide limited performance gain and a more complex structure in [19]. According to the lesion gate mechanism, lesion voxel features are represented by features from only the valid tissue regions. We thus would like to have these lesion features (inside the lesion gate mask) to be similar to those corresponding normal-appearing tissues (outside the lesion gate mask). Based on this assumption, the LGC loss is computed between all the possible pairs between the input (unhealthy brain concatenated with lesion mask, noted as I'), inpainted (synthesized healthy brain concatenated with empty mask, noted as O') and ground truth images (healthy brain concatenated with empty mask, noted as GT'). Finally, L_1 distances are calculated between each pair of the corresponding features F_x^l (extracted from l-th chosen layer from $LG_{encoder}(x)$) as follows:

$$GC_n(I', O', GT') = ||F_{I'}^l - F_{O'}^l||_1 + ||F_{I'}^l - F_{GT'}^l||_1 + ||F_{O'}^l - F_{GT'}^l||_1, \qquad (2)$$

where l is one of the L chosen layers located between the first dilated gated lesion convolution layer and the last convolution layer before up-sampling and the final LGC loss is:

$$L_{LGC}(I', O', GT') = \sum_{l=1}^{L} GC_l(I', O', GT').$$ (3)

2.5 Loss Function

The loss function used for training consists of a reconstruction loss and the proposed gate consistency loss. For the reconstruction loss we used the L_1 distance between the inpainted results and the ground truth, which is computed as $L_{lesion} = ||M \odot (GT - O))||_1$ and $L_{valid} = ||((1 - M) \odot B \odot (GT - O))||_1$. Here, M and B represent the lesion mask and brain mask, respectively. Since the L_1 loss for lesion is relatively small (lesions usually represent a small portion of the whole brain) and the gate consistency is designed to support the training but not dominate the main synthesis task, we use a relatively small weight for L_{LGC}. The final loss used for the proposed LG-Net model is defined as:

$$L_{total} = 10 \times L_{lesion} + L_{valid} + 0.1 \times L_{LGC}.$$ (4)

2.6 Implementation Details

We used the Adam optimizer with an initial learning rate of 0.0001 and the batch size was set to 8 in training phase. The size of the input pseudo 3D brain image slice was $3 \times 256 \times 256$ and no data augmentation was applied since the input brain MR images usually present a similar orientation. Models were trained for 50 epochs with early stop in all experiments on a single NVIDIA GeForce Tesla V100-SXM2 GPU with PyTorch (v1.1.0) and Intel Xeon E2698-v4 CPU.

3   Datasets

Public Dataset. IXI public dataset contains 581 MR brain images of healthy controls [6]. The acquisition protocol for each subject in the IXI dataset includes T1, T2, proton density (PD), and 15-direction diffusion-weighted images. We used only T1-weighted MR brain images in our experiment, however, our framework can be extended to fit other imaging sequences. In order to perform volumetric evaluation, we removed all the cases that failed quality control after automatic tissue segmentation, focusing our analysis on 529 cases. 449 images were randomly selected for the training (including 9 images for evaluation), totalling 42,480 pseudo 3D slices. The remaining 80 3D T1 brain images were used for testing. The intensities of all brain images were scaled to the [0, 1] range before feeding into the inpainting network.

In-house HC Dataset. 12 healthy control T1 brain MRIs were acquired on a 3T GE MR750 using fast spoiled gradient echo (FSPGR) with magnetization-prepared inversion recovery (IR) pulse (TE/TI/TR = 2.8/900/5.9 ms, FA = 10). This HC dataset was not involved in any of the training phases of the models.

4   Experimental Results

For comparison, we included the FSL lesion filling (LF) [3], a U-Net with vanilla convolutions [13], a partial convolution (PConv) approach [8], and a partial convolution approach with non-lesion attention (NLPC) [17] in our evaluation. All deep learning approaches including the proposed method follow the same settings stated in Sect. 2.6. U-Net, PConv and NLPC follow the same architecture design with skip connections.

4.1   Inpainting Quality Analysis

We evaluated all the methods with common image similarity metrics as illustrated in Table 1 (evaluation on in-house dataset is listed in **Supplementary Material** Table 1). MAE and norm MSE stand for mean absolute error and mean squared error with the images scaled to [0, 1] range, respectively. As shown in Table 1, our method outperforms the other compared methods with the minimum MAE and normalized MSE, and also the highest PSNR on public IXI dataset. LG-Net also achieved the best overall results on the in-house HC dataset (MAE = 159.1792 ± 35.9891; nMSE = 0.0025 ± 0.0015; and 26.6632 ± 2.4188 for MAE, normalized MSE, and PSNR, respectively). Moreover, LG-Net has the least trainable parameters in the network and fastest inference speed on each case in the testing dataset. Furthermore, no extra network and parameters (e.g., perceptual loss [7] uses an extra VGG network to compute the features extracted by the main network) are required for LG-Net to compute lesion gate consistency loss.

Table 1. Comparison of the image quality of inpainted brain images generated by different methods on IXI testing dataset. Best performance measures (the lowest MAE and norm MSE; and the highest PSNR) are marked in bold.

Method	MAE ↓	norm MSE ↓	PSNR ↑	#params	Sec/case
LF [3]	29.74 (24.27)	0.0036 (0.0018)	25.07 (2.54)	N/A	67.06
U-Net [13]	24.57 (20.16)	0.0018 (0.0010)	28.06 (2.59)	32.865 M	6.30
PConv [8]	22.17 (17.25)	0.0016 (0.0008)	28.64 (2.49)	32.865 M	10.96
NLPC [17]	21.28 (16.96)	0.0014 (0.0007)	29.01 (2.39)	65.884 M	18.39
w/o LGC	18.40 (14.10)	0.0011 (0.0006)	30.03 (2.28)	3.044 M	2.62
LG-Net	**17.22 (13.64)**	**0.0009 (0.0005)**	**30.79 (2.29)**	3.044 M	2.68

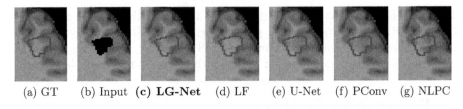

(a) GT   (b) Input **(c) LG-Net**   (d) LF   (e) U-Net   (f) PConv   (g) NLPC

Fig. 2. Inpainted results by different compared methods with (b) as the input of the network. (a) is the ground truth image, (c)–(g) are the corresponding inpainted brain images generated by the proposed LG-Net, Lesion Fill (LF), U-Net, PConv, and NLPC, respectively. The synthetic lesion area is enclosed with red contour (Color figure online).

Noisy Labelling. MS lesions are normally expected to be seen only in WM from routine clinical scans, but they can usually appear at the GM/WM and CSF/WM boundary (e.g. juxtacortical and periventricular lesions). Hence error-free manual labelling is challenging for MS lesion annotators and automated lesion segmentation algorithms. Therefore, noisy masks are to be expected. Lesion inpainting methods based on the assumption that masks are perfect, will fill the lesion area with all WM intensity and that will result in an inaccurate tissue classification. Thus, the ability of a lesion inpainting method to perform compatibly on noisy labelling is crucial. As stated in Sect. 2.1, the synthetic lesion masks were generated from real-world MS lesion analysis to simulate noisy labelling. The visual inpainted results of a representative synthetic lesion which slightly overlaps with GM are shown in Fig. 2. The proposed method generates a smoother and more closely matched intensity distribution with the ground truth compared to other methods.

4.2   Brain Tissue Volumetric Analysis

The ultimate goal for lesion inpainting is to reduce the tissue misclassification errors due to the presence of MS lesions. To validate the generated images in terms of segmentation, we have included two publicly available tissue segmentation methods: tissue segmentation methods from SPM12 (v7487) [1] and FAST [20] from the FSL package (v5.0.10), to cross-validate the performance of our inpainting framework. These methods have been widely used in clinical application for brain tissue volumetric analysis.

All the tissue segmentation methods were run separately on each original brain image in the testing dataset and its corresponding inpainted image. The results of the brain volumetric analysis using SPM12 on IXI dataset are shown in Table 2 (see Table 2 in **Supplementary Material** for results generated using FAST; and Table 3 and Table 4 for corresponding results on HC dataset). BNATD and LNATD denote the Brain-wise and Lesion-wise Normal Appearing Tissue Difference. In brief, BNATD was calculated as the absolute tissue difference between ground truth and inpainted images within the lesion masks divided by the whole brain volume. Similarly, tissue differences were normalised by the whole lesion volume for LNATD. Despite the fact that NLPC

Table 2. Comparison of the brain tissue volumetric analysis (using SPM12) of inpainted results generated by different methods on IXI testing data. Best performance measures (the lowest BNATD and LNATD; and the highest precision, recall and Dice coefficient) are marked in bold.

Method	BNATD ↓	LNATD ↓	Precision ↑	Recall ↑	Dice ↑
Grey matter					
LF	0.1920 (0.0729)	2.6053 (0.8167)	61.23 (4.95)	67.29 (5.45)	0.6398 (0.0441)
U-Net	0.1526 (0.1004)	2.0970 (1.2507)	71.35 (6.45)	75.47 (5.54)	0.7315 (0.0498)
PConv	0.1334 (0.0618)	1.8515 (0.8164)	72.91 (5.55)	73.79 (5.07)	0.7322 (0.0445)
NLPC	0.1130 (0.0671)	1.5483 (0.8506)	74.36 (5.28)	72.64 (6.07)	0.7339 (0.0512)
w/o LGC	0.1187 (0.0768)	1.5992 (0.9097)	79.06 (3.68)	**76.17 (5.49)**	0.7749 (0.0401)
LG-Net	**0.1023 (0.0630)**	**1.3712 (0.7172)**	**86.06 (3.51)**	76.14 (4.49)	**0.8074 (0.0356)**
White matter					
LF	0.4130 (0.1757)	3.8831 (1.2514)	95.03 (1.52)	91.41 (2.03)	0.9318 (0.0157)
U-Net	0.2211 (0.1591)	2.1120 (1.3435)	95.84 (1.47)	95.26 (1.74)	0.9554 (0.0119)
PConv	0.2079 (0.0974)	2.0213 (0.9318)	95.67 (1.48)	95.59 (1.39)	0.9562 (0.0116)
NLPC	0.2293 (0.1127)	2.2419 (1.0740)	**96.21 (1.50)**	95.15 (1.67)	0.9567 (0.0124)
w/o LGC	0.2005 (0.1337)	1.8879 (1.1388)	95.92 (1.59)	97.21 (0.64)	0.9656 (0.0096)
LG-Net	**0.1742 (0.1077)**	**1.6453 (0.9079)**	95.89 (1.38)	**98.56 (0.50)**	**0.9720 (0.0083)**

had the highest precision for white matter segmentation, LG-Net achieved the best performance on all the other tissue metrics. That indicates the proposed method generated fewer tissue discrepancies in the lesion regions.

4.3 Effectiveness of the Lesion Gate Consistency

To explore the effectiveness of the gate mechanism and lesion gate consistency, ablation studies were conducted on both datasets for all metrics. The results shown in Table 1 and 2 reveal that the use of the proposed lesion gate consistency increases the performance of inpainted brain in most of the metrics.

5 Conclusions

We have proposed Lesion Gate Network (LG-Net) for MS lesion inpainting via dynamic learnable gate masks. The dynamic mask updating mechanism is integrated into the architecture to fill the lesion area with morphologically and texturally consistent information with respect to the normal-appearing tissues. The integration of the lesion gate consistency improves the performance of the gated lesion convolution. We evaluated our method on the public and in-house datasets with synthetic lesions and the results show that our proposed framework outperformed other inpainting methods from both a clinical and computer vision perspective. LG-Net could be deployed as an automatic system to support any tissue classification or segmentation method to reduce the effect of MS lesions on brain volumetric analysis in an efficient manner.

Acknowledgments. The authors acknowledge the support of Australian Government Cooperative Research Centres Project grant (CRCPFIVE000141) and Research Training Program (RTP) Scholarship.

Compliance with Ethical Standards. The study was approved by the University of Sydney Human Research and Ethics Committee and all procedures adhered the tenets of the Declaration of Helsinki; we also appreciate the efforts devoted to collect and share the IXI brain dataset for open access [6].

References

1. Ashburner, J., Friston, K.J.: Unified segmentation. Neuroimage **26**(3), 839–851 (2005)
2. Barnes, C., Goldman, D.B., Shechtman, E., Finkelstein, A.: The PatchMatch randomized matching algorithm for image manipulation. Commun. ACM **54**(11), 103–110 (2011)
3. Battaglini, M., Jenkinson, M., Stefano, N.D.: Evaluating and reducing the impact of white matter lesions on brain volume measurements. Hum. Brain Mapp. **33**(9), 2062–2071 (2012)
4. Çiçek, Ö., Abdulkadir, A., Lienkamp, S.S., Brox, T., Ronneberger, O.: 3D U-Net: learning dense volumetric segmentation from sparse annotation. In: Ourselin, S., Joskowicz, L., Sabuncu, M.R., Unal, G., Wells, W. (eds.) MICCAI 2016. LNCS, vol. 9901, pp. 424–432. Springer, Cham (2016). https://doi.org/10.1007/978-3-319-46723-8_49
5. Chau, W., McIntosh, A.R.: The Talairach coordinate of a point in the MNI space: how to interpret it. Neuroimage **25**(2), 408–416 (2005)
6. IXI dataset. https://brain-development.org/ixi-dataset/. Accessed 15 Sept 2021
7. Johnson, J., Alahi, A., Fei-Fei, L.: Perceptual losses for real-time style transfer and super-resolution. In: Leibe, B., Matas, J., Sebe, N., Welling, M. (eds.) ECCV 2016. LNCS, vol. 9906, pp. 694–711. Springer, Cham (2016). https://doi.org/10.1007/978-3-319-46475-6_43
8. Liu, G., Reda, F.A., Shih, K.J., Wang, T.-C., Tao, A., Catanzaro, B.: Image inpainting for irregular holes using partial convolutions. In: Ferrari, V., Hebert, M., Sminchisescu, C., Weiss, Y. (eds.) ECCV 2018. LNCS, vol. 11215, pp. 89–105. Springer, Cham (2018). https://doi.org/10.1007/978-3-030-01252-6_6
9. Liu, H., Jiang, B., Song, Y., Huang, W., Yang, C.: Rethinking image inpainting via a mutual encoder-decoder with feature equalizations. In: Vedaldi, A., Bischof, H., Brox, T., Frahm, J.-M. (eds.) ECCV 2020. LNCS, vol. 12347, pp. 725–741. Springer, Cham (2020). https://doi.org/10.1007/978-3-030-58536-5_43
10. Ma, Y., et al.: Multiple sclerosis lesion analysis in brain magnetic resonance images: techniques and clinical applications. arXiv preprint arXiv:2104.10029 (2021)
11. Magon, S., et al.: White matter lesion filling improves the accuracy of cortical thickness measurements in multiple sclerosis patients: a longitudinal study. BMC Neurosci. **15**, 106 (2014). https://doi.org/10.1186/1471-2202-15-106
12. Prados, F., et al.: A multi-time-point modality-agnostic patch-based method for lesion filling in multiple sclerosis. Neuroimage **139**, 376–384 (2016)
13. Ronneberger, O., Fischer, P., Brox, T.: U-Net: convolutional networks for biomedical image segmentation. In: Navab, N., Hornegger, J., Wells, W.M., Frangi, A.F. (eds.) MICCAI 2015. LNCS, vol. 9351, pp. 234–241. Springer, Cham (2015). https://doi.org/10.1007/978-3-319-24574-4_28

14. Sastre-Garriga, J., et al.: MAGNIMS consensus recommendations on the use of brain and spinal cord atrophy measures in clinical practice. Nat. Rev. Neurol. **16**, 171–182 (2020)
15. De Stefano, N.: Clinical relevance of brain volume measures in multiple sclerosis. CNS Drugs **28**(2), 147–156 (2014). https://doi.org/10.1007/s40263-014-0140-z
16. Tustison, N.J., et al.: N4ITK: improved N3 bias correction. IEEE Trans. Med. Imaging **29**(6), 1310–1320 (2010)
17. Xiong, H., Wang, C., Barnett, M., Wang, C.: Multiple sclerosis lesion filling using a non-lesion attention based convolutional network. In: Yang, H., Pasupa, K., Leung, A.C.-S., Kwok, J.T., Chan, J.H., King, I. (eds.) ICONIP 2020. LNCS, vol. 12532, pp. 448–460. Springer, Cham (2020). https://doi.org/10.1007/978-3-030-63830-6_38
18. Yu, J., Lin, Z., Yang, J., Shen, X., Lu, X., Huang, T.S.: Generative image inpainting with contextual attention. In: CVPR, pp. 5505–5514 (2018)
19. Yu, J., Lin, Z., Yang, J., Shen, X., Lu, X., Huang, T.S.: Free-form image inpainting with gated convolution. In: ICCV, pp. 4471–4480 (2019)
20. Zhang, Y., Brady, J.M., Smith, S.: Hidden Markov random field model for segmentation of brain MR image. In: Medical Imaging 2000: Image Processing, vol. 3979, pp. 1126–1137 (2000)

Self-Supervised Lesion Change Detection and Localisation in Longitudinal Multiple Sclerosis Brain Imaging

Minh-Son To[1,2(✉)], Ian G. Sarno[2], Chee Chong[2,3], Mark Jenkinson[4,5], and Gustavo Carneiro[5]

[1] FHMRI, Flinders University, Bedford Park, Australia
minhson.to@flinders.edu.au
[2] SAMI, Flinders Medical Centre, Bedford Park, Australia
[3] Dr Jones & Partners Medical Imaging, Eastwood, Australia
[4] FMRIB Centre, University of Oxford, Oxford, UK
[5] AIML, University of Adelaide, Adelaide, Australia

Abstract. Longitudinal imaging forms an essential component in the management and follow-up of many medical conditions. The presence of lesion changes on serial imaging can have significant impact on clinical decision making, highlighting the important role for automated change detection. Lesion changes can represent anomalies in serial imaging, which implies a limited availability of annotations and a wide variety of possible changes that need to be considered. Hence, we introduce a new unsupervised anomaly detection and localisation method trained exclusively with serial images that do not contain any lesion changes. Our training automatically synthesises lesion changes in serial images, introducing detection and localisation pseudo-labels that are used to self-supervise the training of our model. Given the rarity of these lesion changes in the synthesised images, we train the model with the imbalance robust focal Tversky loss. When compared to supervised models trained on different datasets, our method shows competitive performance in the detection and localisation of new demyelinating lesions on longitudinal magnetic resonance imaging in multiple sclerosis patients. Code for the models will be made available at https://github.com/toson87/MSChangeDetection.

Keywords: Change detection · Siamese networks · Multiple sclerosis

This paper was partially supported by an Avant Doctor in Training Research Scholarship and the Australian Research Council through grants DP180103232 and FT190100525.

Electronic supplementary material The online version of this chapter (https://doi.org/10.1007/978-3-030-87234-2_63) contains supplementary material, which is available to authorized users.

M. de Bruijne et al. (Eds.): MICCAI 2021, LNCS 12907, pp. 670–680, 2021.
https://doi.org/10.1007/978-3-030-87234-2_63

1 Introduction

Diagnostic imaging interpretation routinely involves comparisons with prior imaging to identify new lesions or detect changes to existing structures and lesions. Change detection can be a tedious and difficult task for the radiologist [3] and importantly, the presence of change can alter clinical management. For example, magnetic resonance imaging (MRI) plays a central role in monitoring disease progression in multiple sclerosis [30] and new demyelinating lesions on follow-up imaging may require modification to immunotherapy [25]. Change detection requires a balance between highlighting relevant differences, and suppressing trivial perceptual differences. The latter may be related to imperfect image registration or technical differences in acquisition, such as different scanners, different magnetic field strength (e.g. 1.5T versus 3T MRI), and different imaging spatial resolution.

Deep learning models have gained significant attention in recent years due to their unprecedented performance in a variety of computer vision tasks [31]. Many of these have been translated to medical imaging applications [23], including supervised lesion detection and classification [14], organ and structural segmentation [16,36], and image enhancement [27]. A drawback of supervised deep learning models is the requirement for a sufficiently large and balanced annotated training set [13]. Unfortunately, certain types of lesion appearance and change detection, such as MS in brain MRI, are anomalous events, where a large variety of changes can occur. For such problems, it is prohibitively inefficient to annotate every type of change to collect a large and balanced training set. Thus, supervised models for change detection may not generalise well for broad application in medical imaging.

In this paper, we propose a solution to mitigate the lack of annotated lesion change samples with a new unsupervised anomaly detection and localisation method trained only with serial images that do not contain any change. Our training simulates changes in the serial images, enabling the production of detection and localisation pseudo-labels that are used to self-supervise the training of our model. The images are synthesised by mixing super-pixels [2] from the original image and its reconstructed image generated by a variational autoencoder [18]. To tackle the class imbalance problem, we employ a focal Tversky loss [1,32]. Experiments on the problem of detection and localisation of new demyelinating lesions on longitudinal MRI in multiple sclerosis patients show that our method produces competitive detection and localisation results compared with other fully-supervised methods [9,19] trained on different datasets.

2 Related Work

Automated change detection is a long-standing problem in medical imaging [6] and other fields [17,28]. Previous work on detecting change in longitudinal multiple sclerosis imaging include subtraction techniques to visualise areas of change [26], statistical modelling [34], and deep learning [5,19,24,35].

Approaches that jointly solve image registration and change detection have also been devised [7,11].

We propose a data augmentation strategy based on mixing super-pixels [2]. Unlike CutOut [10] and CutMix [38], mixing based on super-pixels utilises a segmentation that is not restricted to horizontally- or vertically-oriented edges. Furthermore, as super-pixels group pixels with perceptually similar characteristics, mixing super-pixels is more likely to preserve the integrity of anatomical and lesional boundaries and maintain the saliency of transposed portions.

3   Materials and Methods

3.1   Data Acquisition and Processing

Research ethics for this study was granted by Bellberry Limited. Longitudinal volumetric T2 FLAIR examinations performed by a private imaging provider (Dr Jones & Partners Medical Imaging) between August 2018 and October 2020 across multiple sites were extracted. Based on findings in the radiologist report, pairs of scans were separated into two sets, namely: pairs containing lesion changes (Change), and pairs without lesion changes (NoChange). The training set contains pairs of scans in the NoChange set and is denoted by $\mathcal{D} = \{\mathbf{x}_i, \hat{\mathbf{x}}_i\}_{i=1}^{|\mathcal{D}|}$, where $\mathbf{x}_i, \hat{\mathbf{x}}_i \in \mathcal{X} \subset \mathbb{R}^{H \times W \times D}$ represent a pair of the MRI scans from the same patient. The testing set, with pairs of scans from the Change set, is represented by $\mathcal{T} = \{(\mathbf{x}_i, \hat{\mathbf{x}}_i, \hat{\mathbf{y}}_i)\}_{i=1}^{|\mathcal{T}|}$, where $\mathbf{x}_i, \hat{\mathbf{x}}_i \in \mathcal{X}$ denote a pair of scans from the same patient with a lesion change indicated in the binary map $\hat{\mathbf{y}}_i \in \mathcal{Y} \subset \{0,1\}^{H \times W \times D}$.

The images in both sets were pre-processed with a resampling to 1.0 mm × 1.0 mm × 1.0 mm isometric voxels, greyscale intensity normalisation to $[0,1]$ (0th and 99th percentile), skull stripping [15], rigid body registration [12], and N4 bias correction [22]. For pairs of scans in the Change set, the ground truth masks for regions of change were manually annotated by a trainee radiologist (IGS) using ITK-SNAP [39] based on the findings in the report as a guide. In total, 237 pairs of NoChange scans were used for model training, while 94 pairs of Change scans were used for model testing.

3.2   Methods

The training of our model follows a two-stage process that uses only the NoChange dataset $\mathcal{D}$. The first stage consists of a variational auto-encoder (VAE) [18] combined with a new data augmentation strategy related to Cut-Mix [38], that automatically generates images and annotation maps with synthetic lesion changes. The second step comprises training of a siamese 3D U-net [29] that takes two serial images of the same patient and returns a segmentation map of the lesion change.

Generating Synthetic Lesion Changes. The generation of synthetic lesion changes starts with a VAE [18] that relies on an encoder $f_\theta : \mathcal{X} \to \mathcal{M} \times \mathcal{S}$, a sampling from the Gaussian model to generate the embedding $\mathbf{z} \sim \mathcal{N}(\mu, \Sigma)$, where $\mathbf{z} \in \mathcal{Z}$, $\mu \in \mathcal{M}$ and $\Sigma \in \mathcal{S}$, and a reconstruction from the decoder $g_\gamma : \mathcal{Z} \to \mathcal{X}$. This VAE is trained with the images $\mathbf{x} \in \mathcal{D}$ using an voxel-wise L1 reconstruction loss and a regularisation loss to minimise the Kulback-Leibler divergence between the embedding distribution and a zero-mean identity-covariance Gaussian. To increase the diversity of the reconstructions, a perturbation is also applied to the embedding, $\tilde{\mathbf{z}} = \mathbf{z} \times \Delta$, where $\Delta \sim \mathcal{U}(-\delta, \delta)$ ($\mathcal{U}(.)$ denoting a uniform distribution). Hence, for each image $\mathbf{x}_i \in \mathcal{D}$, we sample new reconstructions, denoted by $\tilde{\mathbf{x}}_i = g_\gamma(\mathbf{z}_i \times \Delta)$, with $\mathbf{z}_i \sim \mathcal{N}(\mu_i, \Sigma_i)$, and $\mu_i, \Sigma_i = f_\theta(\mathbf{x}_i)$.

To synthesise lesion changes, we propose a super-pixel mixing data augmentation strategy, referred to as SuperMix. Using the scan $\mathbf{x}_i$ and its sampled reconstruction $\tilde{\mathbf{x}}_i$, we first produce a super-pixel segmentation [2] that tesselates $\mathbf{x}_i$ to produce n_{seg} binary maps, each represented by $\mathbf{b}_{i,t} \in \{0,1\}^{H \times W \times D}$ for $t \in \{0, ..., n_{\text{seg}} - 1\}$ (in this map, a voxel is labelled with 1, if it belongs to the tth super-pixel). The synthesised image and lesion change are defined by

$$\mathbf{x}_i' = \sum_{t=0}^{n_{\text{seg}}-1} (\lambda_t)(\mathbf{b}_{i,t} \odot \mathbf{x}_i) + (1 - \lambda_t)(\mathbf{b}_{i,t} \odot \tilde{\mathbf{x}}_i),$$

$$\hat{\mathbf{y}}_i = \sum_{t=0}^{n_{\text{seg}}-1} (\lambda_t)(\mathbf{0}_{H \times W \times D}) + (1 - \lambda_t)(\mathbf{b}_{i,t}),$$

$$(1)$$

where $\lambda_t = 1$ if $u < \tau$, with $u \sim \mathcal{U}(0,1)$, and $\mathbf{0}_{H \times W \times D}$ denotes a volume of size $H \times W \times D$ containing only zeros. This allows us to build a synthesised set of Change, represented by $\hat{\mathcal{D}} = \{(\mathbf{x}_i', \hat{\mathbf{x}}_i, \hat{\mathbf{y}}_i)\}_{i=1}^{|\mathcal{D}|}$. Figure 1 shows examples of the synthesised lesion changes for different values of n_{seg}.

Siamese 3D U-Net. Our proposed siamese 3D U-net takes a sample $(\mathbf{x}_i', \hat{\mathbf{x}}_i, \hat{\mathbf{y}}_i) \in \hat{\mathcal{D}}$ to form two inputs and one output. The inputs are $\mathbf{x}_i^{(1)} = (\mathbf{x}_i', |\mathbf{x}_i' - \hat{\mathbf{x}}_i|) \in \mathcal{X}^2$ and $\mathbf{x}_i^{(2)} = (\hat{\mathbf{x}}_i, |\mathbf{x}_i' - \hat{\mathbf{x}}_i|) \in \mathcal{X}^2$, and the output is the segmentation map $\hat{\mathbf{y}}_i$. Siamese network configurations enable independent processing of image inputs that may improve feature extraction prior to fusion and comparison of features [8]. Our siamese 3D U-net consists of a down-sampling network, represented by $e_\phi : \mathcal{X}^2 \times \mathcal{X}^2 \to \mathcal{K}$, followed by an up-sampling network, denoted by $d_\psi : \mathcal{K} \to [0,1]^{H \times W \times D}$. As shown in Fig. 2 the model contains skip connections that propagate data directly from the encoder to decoder at the same spatial resolution, providing local contextual information to the up-sampling process. Each block of the model is formed by inception modules [37] that incorporate convolutional filters of different sizes with dimensionality reduction achieved by $1 \times 1 \times 1$ convolutional layers to enable efficient processing of features at multiple scales. Since areas of change typically occupy small regions in the entire image, the ability to focus the network to only relevant areas improves utilisation of the network resources. To that end, attention gates were built into

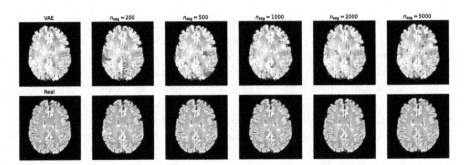

Fig. 1. SuperMix examples showing the effect of n_{seg} on the size of super-pixels. The SuperMix mixing parameter τ in (1) is fixed at 0.98. The perturbed VAE reconstruction ($\delta = 5$) and overlaid SuperMix masks (red) are shown in the top row. The resultant synthesised lesion changes are shown in the bottom row (Color figure online).

the network to filter information being passed through the skip connections [33]. Down-sampling operations in the U-net encoder can result in loss of information flow in the deeper layers. To mitigate this, we provided down-sampled inputs to the deep encoding layers. We also encouraged the intermediate layers of the U-net decoder to learn meaningful segmentations by incorporating deep supervision [20]. This was implemented by passing the outputs of the intermediate layers through a $1 \times 1 \times 1$ convolutional layer with sigmoid activation.

The training of this siamese 3-D model is based on the minimisation of the focal Tversky loss [1,32] that can handle the class imbalance problem associated with segmenting small and localised areas of change. This loss is defined by

$$\ell(\hat{\mathcal{D}}, \phi, \psi) = \frac{1}{|\mathcal{D}|} \sum_{i=1}^{|\mathcal{D}|} (1 - \ell_{tv}(\hat{\mathbf{y}}_i, \tilde{\mathbf{y}}_i))^{\gamma}, \tag{2}$$

where $\gamma \in \mathbb{R}^+$, $\tilde{\mathbf{y}}_i = d_\psi(e_\phi(\mathbf{x}_i^{(1)}, \mathbf{x}_i^{(2)})))$, and $\ell_{tv}(.)$, the Tversky index, generalises the Dice coefficient, with

$$\ell_{tv}(\hat{\mathbf{y}}_i, \tilde{\mathbf{y}}_i) = \frac{TP}{TP + \alpha FN + \beta FP}, \tag{3}$$

where TP, FN and FP are the numbers of true positive, false negative and false positive between the annotation $\hat{\mathbf{y}}_i$ and the model output $\tilde{\mathbf{y}}_i$, respectively. In (3), the parameters α and β can be tuned to emphasise recall over precision ($\alpha > \beta$), or vice versa ($\alpha < \beta$), such that $\alpha + \beta = 1$ (Supplemental Fig. 3).

After training the model, inference is performed by running it on a test sample composed of a pair of serial images $(\mathbf{x}, \hat{\mathbf{x}})$ from $\mathcal{T}$. The model produces $\tilde{\mathbf{y}} = d_\psi(e_\phi(\mathbf{x}^{(1)}, \mathbf{x}^{(2)}))$ with $\mathbf{x}^{(1)} = (\mathbf{x}, |\mathbf{x} - \hat{\mathbf{x}}|)$ and $\mathbf{x}^{(2)} = (\hat{\mathbf{x}}, |\mathbf{x} - \hat{\mathbf{x}}|)$. Then, we threshold this output to produce a binary output with $\tilde{\mathbf{y}} > \kappa$, which is then used by connected component analysis (CCA) to produce segmentation blobs of size $\geq \{20, 100\}$. This binarised result can then be compared with the ground truth annotation in $\hat{\mathbf{y}}$ from $\mathcal{T}$.

Siamese U-Net

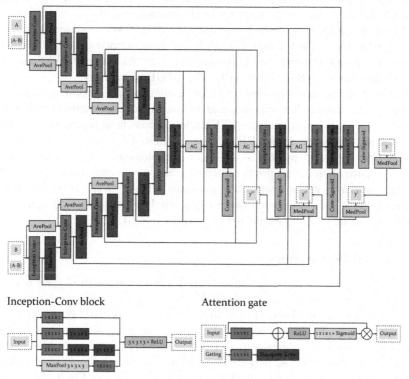

Fig. 2. Schematic of proposed siamese U-net model.

4   Experiments and Evaluation

Training for the VAE consisted of 200 iterations in batches of 4096 volume crops randomly sampled from the set of NoChange scans, with mini batch size of 2. Each crop measured $192 \times 192 \times 16$ voxels. The effect of δ to parameterise the uniform distribution that influences the VAE generation is linked to the VAE encoder architecture, such that replacing max-pooling with strided convolutions results in greater mottling in the generator output (Supplemental Fig. 1). For all experiments we utilised the VAE (MaxPool) architecture and δ was fixed at 5.0. The training for the lesion change detector consisted of 60 iterations in batches of 100 NoChange pairs augmented by SuperMix, with mini batch size of 2. The SuperMix mixing parameter τ in (1) was fixed at 0.98, with $\log n_{\text{seg}} \sim \mathcal{U}(\log 200, \log 5000)$. The recall/precision balance of the model is tunable by the α, β, and γ parameters of the focal loss in (2) and (3) (Supplemental Fig. 3). For all outputs α was set to 0.75, while $\gamma = 0.75$ for intermediate layers, and we tested $\gamma = \{0.75, 1.0\}$ for the final layer. In a series of ablation experiments, we explored the contributions of network architecture (non-siamese or siamese U-net), loss function (binary cross-entropy or focal Tversky loss), multi-scale inputs, deep supervision, attention gates, and Inception modules. We included L2 weight regularisation in the siamese U-net.

Adam optimiser was used. A learning rate of 0.00005 was used to train the VAE. For the lesion change detector, an initial learning rate of 0.0002 and learning decay rate of 0.001 were used. Models were written in Keras/Tensorflow and trained on a GeForce RTX 3090 GPU with 24 GB of memory.

Model performance was evaluated by generating prediction masks on the Change pairs, thresholding $\kappa = 0.1$, and comparing against ground truth lesions. The lesion-wise true positive rate (LTPR), lesion-wise false positive rate (LFPR) and positive predictive value (PPV) were calculated for lesion changes in each Change pair and averaged across all pairs (Supplemental D). Overlap between a ground truth change and predicted change lesion was significant (i.e. a lesion-wise true positive) if the intersection over union (IoU) of the blobs was 0.01 or greater. Others have considered overlap of a single voxel to be sufficient [4], but we decided to be more conservative with an IoU $\geq .01$.

4.1   Model Performance

Ablation experiments show the incremental improvements in change detection provided by incorporation of the focal Tversky loss, attention gates, multiscale inputs and deep supervision (Fig. 3 and Table 1). Connected components postprocessing with a larger blob size (100 compared to 20) generally improve PPV, with a corresponding reduction in LTPR. Siamese architectures yielded the best performance. The focal loss parameter γ did not strongly influence segmentation performance. Example change mask predictions are shown in Fig. 4, demonstrating the model's ability to detect changes in different parts of the brain, as well as

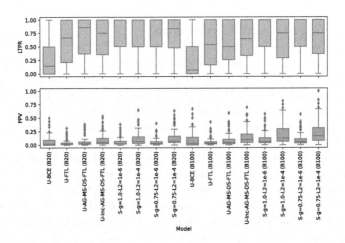

Fig. 3. Performance of different model configurations. U, U-net; BCE, binary cross-entropy; FTL, focal Tversky loss; AG, attention gate; MS, multiscale input; DS, deep supervision; B, minimum blob size; g, γ parameter of final layer FTL; L2, L2 weight regularisation factor.

Table 1. Performance of selected model configurations. AG, attention gate; MS, multiscale input; DS, deep supervision; BCE, binary cross-entropy; FTL, focal Tversky loss. See also Fig. 3.

Model	Loss	γ	L2 reg	Blob	LTPR	LFPR	PPV
U-net	BCE	–	–	20	0.330	0.932	0.0684
U-net	FTL	1.0	–	20	0.618	0.960	0.0404
U-net-AG-MS-DS	FTL	1.0	–	20	0.664	0.945	0.0546
U-net-Inc-AG-MS-DS	FTL	1.0	–	20	0.645	0.909	0.0914
Siamese U-net	FTL	1.0	10^{-6}	20	**0.758**	0.942	0.0579
Siamese U-net	FTL	1.0	10^{-4}	20	0.708	0.884	0.116
Siamese U-net	FTL	0.75	10^{-6}	20	0.747	0.946	0.0541
Siamese U-net	FTL	0.75	10^{-4}	20	0.634	**0.874**	**0.126**
Li et al. 2018 [21]	–	–	–	–	0.133	0.997	0.0031

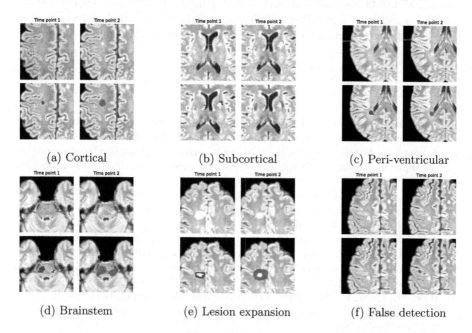

(a) Cortical   (b) Subcortical   (c) Peri-ventricular

(d) Brainstem   (e) Lesion expansion   (f) False detection

Fig. 4. Example predicted change masks. Ground truth (blue) and predicted (red) change masks are shown overlaid the scan at time point 2 (Color figure online).

changes to existing lesions. High LFPR values may correspond to other regions of change (Fig. 4(f)), distinct from lesion changes. Some may be changes unrelated to multiple sclerosis, but their relevance is not explored here.

The lesion change recall (i.e. LTPR) of the siamese models is comparable to published supervised models [9,19]. We also test a publicly-available deep fully convolutional network model for white matter hyperintensities (WMH) segmentation that was trained on different datasets [21]. In this case, the difference between the predicted WMH segmentations of each input image is taken to be the predicted change segmentation. Despite reporting an average recall of 0.84 on the original data, that model [21] performs poorly on our data. We observed that lesions tended to be under-segmented and lesion changes were frequently missed, resulting in a low LTPR the model in [21] (Table 1).

5   Conclusions

The generation of images and annotation maps with synthetic lesion changes and the siamese U-net are the main contributions of this paper. These contributions mitigate the lack of annotated lesion change datasets with a self-supervised training of the anomaly detection and localisation methods. Change detection is optimised by the incorporation of the focal Tversky loss, attention gates, multiscale inputs and deep supervision. Our approach, which does not require detailed, voxel-level annotated training sets, demonstrates high detection rates for lesion changes in multiple sclerosis imaging, comparable to fully-supervised models.

References

1. Abraham, N., Khan, N.M.: A novel focal Tversky loss function with improved attention U-Net for lesion segmentation. arXiv:1810.07842 (2018)
2. Achanta, R., Shaji, A., Smith, K., et al.: SLIC superpixels compared to state-of-the-art superpixel methods. IEEE Trans. Pattern Anal. Mach. Intell. **34**, 2274–2282 (2012)
3. Altay, E.E., Fisher, E., Jones, S.E., et al.: Reliability of classifying multiple sclerosis disease activity using magnetic resonance imaging in a multiple sclerosis clinic. JAMA Neurol. **70**, 338 (2013)
4. Aslani, S., Dayan, M., Storelli, L., et al.: Multi-branch convolutional neural network for multiple sclerosis lesion segmentation. NeuroImage **196**, 1–15 (2019)
5. Birenbaum, A., Greenspan, H.: Multi-view longitudinal CNN for multiple sclerosis lesion segmentation. Eng. Appl. Artif. Intell. **65**, 111–118 (2017)
6. Bosc, M., Heitz, F., Armspach, J.P., et al.: Automatic change detection in multimodal serial MRI: application to multiple sclerosis lesion evolution. NeuroImage **20**(2), 643–656 (2003)
7. Bu, S., Li, Q., Han, P., et al.: Mask-CDNet: a mask based pixel change detection network. Neurocomputing **378**, 166–178 (2020)
8. Daudt, R.C., Saux, B.L., Boulch, A.: Fully convolutional Siamese networks for change detection. arXiv:1810.08462 (2018)
9. Denner, S., Khakzar, A., Sajid, M., et al.: Spatio-temporal learning from longitudinal data for multiple sclerosis lesion segmentation. arXiv:2004.03675 (2020)
10. DeVries, T., Taylor, G.W.: Improved regularization of convolutional neural networks with cutout. arXiv:1708.04552 (2017)
11. Dufresne, E., Fortun, D., Kumar, B., et al.: Joint registration and change detection in longitudinal brain MRI. In: 2020 IEEE 17th International Symposium on Biomedical Imaging (ISBI), pp. 104–108. IEEE (2020)

12. Garyfallidis, E., Brett, M., Amirbekian, B., et al.: Dipy, a library for the analysis of diffusion MRI data. Front. Neuroinform. **8**, 8 (2014)
13. Hofmanninger, J., Prayer, F., Pan, J., Röhrich, S., Prosch, H., Langs, G.: Automatic lung segmentation in routine imaging is primarily a data diversity problem, not a methodology problem. Eur. Radiol. Exp. **4**(1), 1–13 (2020). https://doi.org/10.1186/s41747-020-00173-2
14. Huang, X., Shan, J., Vaidya, V.: Lung nodule detection in CT using 3D convolutional neural networks. In: 2017 IEEE 14th International Symposium on Biomedical Imaging (ISBI 2017) (2017)
15. Isensee, F., Schell, M., Pflueger, I., et al.: Automated brain extraction of multisequence MRI using artificial neural networks. Hum. Brain Mapp. **40**(17), 4952–4964 (2019)
16. Kayalibay, B., Jensen, G., van der Smagt, P.: CNN-based segmentation of medical imaging data. arXiv:1701.03056 (2017)
17. Khelifi, L., Mignotte, M.: Deep learning for change detection in remote sensing images: comprehensive review and meta-analysis. IEEE Access **8**, 126385–126400 (2020)
18. Kingma, D.P., Welling, M.: Auto-encoding variational Bayes. arXiv:1312.6114 (2013)
19. Krüger, J., Opfer, R., Gessert, N., et al.: Fully automated longitudinal segmentation of new or enlarged multiple sclerosis lesions using 3D convolutional neural networks. NeuroImage: Clin. **28**, 102445 (2020)
20. Lee, C.Y., Xie, S., Gallagher, P., et al.: Deeply-supervised nets. arXiv:1409.5185 (2014)
21. Li, H., Jiang, G., Zhang, J., et al.: Fully convolutional network ensembles for white matter hyperintensities segmentation in MR images. NeuroImage **183**, 650–665 (2018)
22. Lowekamp, B.C., Chen, D.T., Ibáñez, L., Blezek, D.: The design of SimpleITK. Front. Neuroinform. **7**, 45 (2013)
23. Lundervold, A.S., Lundervold, A.: An overview of deep learning in medical imaging focusing on MRI. Zeitschrift für Medizinische Physik **29**, 102–127 (2019)
24. McKinley, R., Wepfer, R., Grunder, L., et al.: Automatic detection of lesion load change in multiple sclerosis using convolutional neural networks with segmentation confidence. NeuroImage: Clin. **25**, 102104 (2020)
25. McNamara, C., Sugrue, G., Murray, B., MacMahon, P.J.: Current and emerging therapies in multiple sclerosis: implications for the radiologist, part 1–mechanisms, efficacy, and safety. AJNR **38**, 1664–1671 (2017)
26. Patel, N., Horsfield, M.A., Banahan, C., et al.: Detection of focal longitudinal changes in the brain by subtraction of MR images. AJNR **38**, 923–927 (2017)
27. Plassard, A.J., Davis, L.T., Newton, A.T., et al.: Learning implicit brain MRI manifolds with deep learning. arXiv:1801.01847 (2018)
28. Radke, R.J., Andra, S., Al-Kofahi, O., Roysam, B.: Image change detection algorithms: a systematic survey. IEEE Trans. Image Process. **14**(3), 294–307 (2005)
29. Ronneberger, O., Fischer, P., Brox, T.: U-Net: convolutional networks for biomedical image segmentation. arXiv:1505.04597 (2015)
30. Àlex, R., Wattjes, M.P., Tintoré, M., et al.: MAGNIMS consensus guidelines on the use of MRI in multiple sclerosis–clinical implementation in the diagnostic process. Nat. Rev. Neurol. **11**, 471–482 (2015)
31. Russakovsky, O., Deng, J., Su, H., et al.: ImageNet large scale visual recognition challenge. Int. J. Comput. Vis. **115**, 211–252 (2015)

32. Salehi, S.S.M., Erdogmus, D., Gholipour, A.: Tversky loss function for image segmentation using 3D fully convolutional deep networks. arXiv:1706.05721 (2017)
33. Schlemper, J., Oktay, O., Schaap, M., et al.: Attention gated networks: learning to leverage salient regions in medical images. Med. Image Anal. **53**, 197–207 (2019)
34. Schmidt, P., Pongratz, V., Küster, P., et al.: Automated segmentation of changes in FLAIR-hyperintense white matter lesions in multiple sclerosis on serial magnetic resonance imaging. NeuroImage: Clin. **23**, 101849 (2019)
35. Sepahvand, N.M., Arnold, D.L., Arbel, T.: CNN detection of new and enlarging multiple sclerosis lesions from longitudinal MRI using subtraction images. In: 2020 IEEE 17th International Symposium on Biomedical Imaging (ISBI), pp. 127–130 (2020)
36. Snaauw, G., Gong, D., Maicas, G., et al.: End-to-end diagnosis and segmentation learning from cardiac magnetic resonance imaging. In: 2019 IEEE 16th International Symposium on Biomedical Imaging (ISBI 2019), pp. 802–805. IEEE (2019)
37. Szegedy, C., Liu, W., Jia, Y., et al.: Going deeper with convolutions. In: 2015 IEEE Conference on Computer Vision and Pattern Recognition (CVPR), pp. 1–9 (2015)
38. Yun, S., Han, D., Oh, S.J., et al.: CutMix: regularization strategy to train strong classifiers with localizable features. arXiv:1905.04899 (2019)
39. Yushkevich, P.A., Piven, J., Hazlett, H.C., et al.: User-guided 3D active contour segmentation of anatomical structures: significantly improved efficiency and reliability. Neuroimage **31**(3), 1116–1128 (2006)

SyNCCT: Synthetic Non-contrast Images of the Brain from Single-Energy Computed Tomography Angiography

Florian Thamm[1,2]($\boxtimes$), Oliver Taubmann[2]($\boxtimes$), Felix Denzinger[1,2]($\boxtimes$),
Markus Jürgens[2]($\boxtimes$), Hendrik Ditt[2]($\boxtimes$), and Andreas Maier[1]($\boxtimes$)

[1] Friedrich-Alexander University Erlangen-Nuremberg, Erlangen, Germany
{florian.thamm,felix.denziger,andreas.maier}@fau.de
[2] Siemens Healthcare GmbH, Forchheim, Germany
{oliver.taubmann,markus.juergens,hendrik.ditt}@siemens-healthineers.com

Abstract. By injecting contrast agent during a CT acquisition, the vascular system can be enhanced. This acquisition type is known as CT Angiography (CTA). However, due to typically lower dose levels of CTA scans compared to non-contrast CT acquisitions (NCCT) and the employed reconstruction designed specifically for vessel reconstruction, soft tissue contrast in the brain parenchyma is usually subpar. Hence, an NCCT scan is preferred for the visualization of such tissue. We propose SyNCCT, an approach which synthesizes NCCT images from the CTA domain by removing enhanced vessel structures and improving soft tissue contrast. Contrary to virtual non-contrast (VNC) images based on dual energy scans, which target the physically accurate removal of iodine rather than generating a realistic NCCT with improved gray/white matter separation, our approach only requires a conventional single-energy acquisition. By design, our method integrates prior domain knowledge and employs residual learning as well as a discriminator to achieve perceptual realism. In our data set of patients with ischemic stroke, the absolute differences in automatic ASPECT scoring, which rates early signs of an occlusion in the anterior circulation on a scale from 0 (most severe) to 10 (no signs), was 0.78 ± 0.75 (median of 1) when comparing our SyNCCT to the real NCCT images. Qualitatively, realistic appearance of the images was confirmed by means of a Turing test with a radiologist, who classified 64% of 64 (32 real, 32 generated) images correctly. Two other physicians classified 65% correctly, on average.

Keywords: Computed tomography angiography · Domain transfer · Single energy CT

Electronic supplementary material The online version of this chapter (https://doi.org/10.1007/978-3-030-87234-2_64) contains supplementary material, which is available to authorized users.

1   Introduction

Computed Tomography Angiography (CTA) is a commonly used modality for the visualization of the blood vessels, for instance the cerebral arteries. Compared to regular non-contrast (NCCT) acquisitions of the head, a CTA scan is typically acquired with a lower radiation dose as this is sufficient for distinguishing the vasculature from surrounding tissue. However, the low dose combined with a reconstruction tailored for sparse, high-density objects leads to subpar image contrast in the remaining brain parenchyma – particularly between gray and white brain matter. Hence, while a transfer from the CTA into the NCCT domain is far from trivial as it involves more than merely "masking out" contrasted vessels, it would be highly beneficial precisely because of the improved soft tissue contrast – provided it is both believable and accurate to the patient's overall anatomy and potential pathologies. The reversed way, the prediction of contrast agent in non-contrasted issue has been done for CT in [8] and MRI in [3].

Related work to the domain transfer from contrasted images to non-contrasted images, is found in the field of Dual Energy CT (DECT). Using the two energy spectra inherent in DECT, a CTA image can be decomposed into its iodine and non-iodine components creating the Virtual Contrast image (VC) and Virtual Non-Contrast (VNC) image, respectively [18]. Motivated primarily by considerations of physical correctness, material decomposition algorithms have several practical limitations, resulting in low signal-to-noise ratios and overall poor high frequency contents [15]. Next to the demand for the physically correct removal of iodine, related work improved the visual appearance of the DE VNC and reduced its distance to the real NCCT images. Poirot et al. [15] proposed to apply a ResNet [5] in a supervised setting. It receives dual-energy CTA scans alongside a VNC image approximated using a look-up table that is generated by DECT and NCCT voxel correspondences. In this setup, Poirot et al. were able to improve the predicted VNC quantitatively and qualitatively in relation to real NCCT scans. However, DECT scanners are not as widely available as conventional ones and the results presented in [15] qualitatively still exhibit room for improvement, especially regarding high frequency information. Another problem related to this, is the domain transfer from single energy to dual energy. This has been shown in [9,12]. Such a domain transfer, given its physically correct, would consequently enable the prediction of a VNC image in a second step. Another problem related to this, us the domain transfer from single energy to dual energy. This has been shown in [9,12]. Such a domain transfer, given the physical correctness, would consequently enable the prediction of a VNC image in a second step.

In this work, we want to investigate the feasibility of a domain transfer from single energy CTA to NCCT. With this transfer, our main target is to improve the information gain from a single CTA scan by using it to predict a synthetic NCCT (SyNCCT) image that offers soft tissue contrast comparable to a real NCCT. Although physically correct removal of iodine contrast cannot be guaranteed in this setup, our work still aims to achieve a high degree of realism in the appearance of the SyNCCT images. To evaluate this, we conducted a Turing test with several physicians. Secondly, we want to ensure that the domain transfer neither introduces new pathologies nor removes existing disease patterns. For patients with ischemic stroke, the ASPECTS (Alberta Stroke Program Early CT Score, [1]) is a clinically established rating scheme to assess the severity of the infarction. Therefore, we performed an automated ASPECT scoring on the SyNCCT images and compared the results to the outcomes determined on the corresponding real NCCT images for such patients. We also compared our approach quantitatively to baseline methods with respect to both the ASPECT score and distance based measurements. For the latter, we distinguished between the vessels and brain tissue.

2   Methods

2.1   Data

In total, 166 data sets of ischemic stroke patients were available, each consisting of a CTA scan covering the entire head/neck region together with a co-registered NCCT scan of the brain. Of these, 16 data sets were removed due to inaccurate registration or strong metal artifacts. The remaining 150 data sets consist of 39223 individual slices. All data was acquired from one site with a Somatom Definition AS+ (Siemens Healthineers, Forchheim, Germany). The dose length product of NCCT scans was three times higher than of CTA scans. In each data set, automatic segmentation of the brain was performed and background was removed. All (axial) CTA slices had a size of 512×512 voxels. Thin slices were used to ensure a precise distinction of dense nearby vessel structures in the CTA scan, with 15 data sets reconstructed with a slice thickness of 1 mm and 135 with a thickness of 0.5 mm. For the following experiments, we split the data on patient level into 107 data sets (28302 slices) for training, 20 data sets (5237 slices) for validation and 23 data sets (5684 slices) for testing purposes.

2.2   Preprocessing

We normalized the images based on their relative cumulative histograms computed inside the brain tissue. To this end, we determined the 0.1% and 99.9% values for each volume in the training set and subsequently averaged them. The images are then linearly transformed such that the lower and upper values are mapped to -1 and 1, respectively, while clipping values outside the range. The boundary values are reused for the validation and testing phase.

2.3   SyNCCT

The proposed method – outlined in Fig. 1 – is driven by the idea to train a neural network to synthesize NCCT images by providing it with additional input channels derived from the CTA scan that we assume to be beneficial for learning the task at hand. In addition to the single-energy CTA input, the network receives

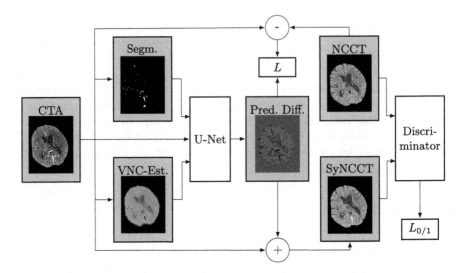

Fig. 1. Proposed pipeline including the optional discriminator network. External data is marked in blue: the CTA as input to the pipeline and the NCCT as the training target. Gray boxes highlight computed results: the vessel segmentation, and the statistically estimated VNC image (VNC-Est.), the raw prediction of the U-Net and the final synthesized NCCT (SyNCCT). (Color figure online)

a binary segmentation mask of the cerebral arteries and a rough estimation of a contrast-adjusted, VNC-like image (VNC-Est.) modeled by the statistical relation between the NCCT and CTA voxels. A U-Net [16] receives these three components and predicts the difference to the NCCT scan in a residual setup. This architecture is further extended by a discriminator learning mechanism.

Segmentation. In a CTA scan, vessel tree enhancement is the result of injected iodine. The idea is to extend the network input with the vessel segmentation, thus providing hints on what to remove. Since NCCT images are not available during inference, the segmentation must be performed based on the CTA images only. Hence, the use of Digital Subtraction Angiography is ruled out. The segmentation was performed with an image processing pipeline by Thamm et al. designed to work on CTA data [17].

Estimation. The VNC-Est. image is modeled based on the statistical relation between the voxels of the CTA and the NCCT data set. To this end, a 2D histogram of all CTA-NCCT voxel correspondences is computed. To decrease noise, all volumes were slice-wise median filtered with a kernel size of 5. Vessels were removed by the segmentation and excluded from the histogram computation. As most voxel correspondences are between 30 HU and 70 HU, the graphical representation has been normalized for each CTA HU level (column-wise in Fig. 2). The result is shown in Fig. 2 averaged over all training data sets and

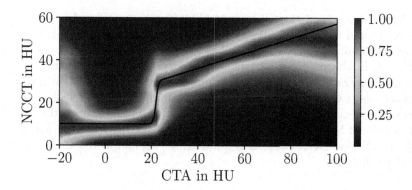

Fig. 2. Averaged column-wise normalized 2D histogram. Colors indicate the relative occurrence of NCCT values w.r.t. to the occurrence of the most common NCCT value for each CTA value. The black line (defined in Eq. 1) models the distribution along the most likely occurring NCCT value for each CTA value. (Color figure online)

smoothed using a Gaussian kernel. We fitted the distribution with the function $f(x_{CTA}) \approx x_{NCCT}$ which approximately maps voxel values x_{CTA} from the CTA domain to the distribution of x_{NCCT} over the NCCT domain. f is defined as

$$f(x_{CTA}) = \alpha \tanh(x_{CTA} + \beta) + \gamma \frac{x_{CTA} + \beta}{1 - e^{-(x_{CTA} + \beta)}} + \delta \qquad (1)$$

and marked as a black line in Fig. 2. For the training set, described in Sect. 2.1, the best fitting parameters, determined using a least-squares fit, were $\alpha = 10$, $\beta = -22$, $\gamma = 0.35$ and $\delta = 20$.

Model and Training. The three input channels, consisting of the CTA data, the vessel segmentation and the VNC-Est. are processed by a 2D U-Net [16]. Due to memory constraints and to increase the sample variance, the images were processed patch-wise by tiles of 64 × 64 pixels. During training these tiles do not overlap. All U-Nets presented in the following sections had the same dimensions. In total, 4 pooling layers/transposed convolutions, each with a stride of 2, were applied. On each stage, after every pooling/transposed convolutional layer, three convolutional layers with a kernel size of 3 and a reflection padding of 1 were applied. ReLU activations were used in between the convolutional layers as non-linearities. With every downsampling/upsampling step, the number of feature maps doubles/halves beginning with 32 feature maps in the first and 256 feature maps in the fourth, deepest stage. Skip connections between each encoding and decoding stage were established by concatenation. In all configurations, the U-Net had 1.9 million parameters and was trained with a batch size of 8 for 200 epochs and a learning rate of 10^{-5} using the Adam optimizer [7]. We used the PyTorch (1.6.0) library [14] and Python (3.8). The first loss denoted as L in Fig. 1 supervises the network to predict the difference between the NCCT image and the CTA input. This can be considered a residual connection around the network. We tested the Mean Squared Error (MSE) and the Structural Similarity Index SSIM [19] as loss functions for L and compared both in Sect. 3. In the proposed variant, the supervised U-Net is treated as generator in a generator-discriminator learning mechanism (Generative Adversarial Network, short GAN [4]). A PatchGAN discriminator [6] with 400k parameters detects fake/real images on a per-patch basis, pairwise between the SyNCCT and the actual NCCT image, and is optimized by the discriminator loss denoted as $L_{0/1}$. The SyNCCT image is computed by the sum of the CTA image and U-Net prediction.

Postprocessing. Patch-wise processing may produce mosaicing artifacts in the final restitched image. Therefore, during inference, the 64 × 64 tiles are processed with an overlap of 32 pixels in each dimensions (stride of 32) and cropped to their center area of size 36 × 36, with overlapping margin averaged across tiles.

3   Results

We used either the MSE loss or the SSIM loss for the supervised part as well as both of these variants extended with the discriminator mechanism. As one baseline, we used a CycleGAN [20] consisting of the same U-Nets and discriminators used for SyNCCT, where each prediction (the fake CTA and the fake NCCT) is additionally supervised. Without this additional pairwise supervision, no meaningful results could be achieved. No additional channels next to the NCCT and CTA scans were incorporated for the CycleGAN setup. As a second baseline, we modified the approach of Poirot et al. [15] to be compatible to the single energy problem at hand. We made two changes to its setup. Instead of computing the look-up table based on the different energy levels, we instead used the estimated VNC as the look-up reference. Secondly, we fed the network with just one energy level. A comparison to both baselines can be considered as a quasi-ablation study since both baselines are simplifications of the proposed method in the most relevant aspects. Mod. Poirot uses a different architecture and has no additional segmentation channel. CycleGANs do not use (1) and neither a segmentation nor a residual connection. We trained the modified network on the before-mentioned patches after brain segmentation to enable a direct comparison between this approach and the others. We compare all 6 setups qualitatively and quantitatively.

Qualitative Comparison. Example images from the test set are displayed in Fig. 3. More examples are presented in the supplementary material. All setups with no discriminator (MSE, SSIM, Modified Poirot, upper row in Fig. 3) exhibit an over-smoothed appearance. In terms of perceptual realism, these setups fall short compared to the desired NCCT image. Additionally, the modified Poirot method failed to remove the majority of the blood vessels which are still visible as bright dots. Adding a discriminator forced the network to generate high frequency information as this is present in the real NCCT images as well. The discriminator mechanism also led to a significantly better gray/white matter contrast. Thus, the SSIM+Dis, MSE+Dis and CycleGAN approaches produced more realistic images overall (bottom row in Fig. 3). However, the CycleGAN setup failed to remove contrast-enhanced vessel structures (red arrow in (h)) in contrast to the SSIM+Dis and MSE+Dis setup. MSE+Dis on the other hand generated some repetitive patterns (red arrow in (g)), especially in homogeneous regions. The setup we deemed best, also in a direct comparison to the NCCT data, is the SSIM+Dis version. Thus, for further investigation we only considered the SSIM+Dis variant.

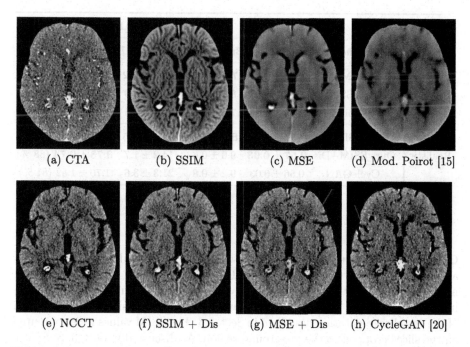

(a) CTA   (b) SSIM   (c) MSE   (d) Mod. Poirot [15]

(e) NCCT   (f) SSIM + Dis   (g) MSE + Dis   (h) CycleGAN [20]

Fig. 3. Example image on 4 (b,c,f,g) variants of SyNCCT and 2 baselines (d,h) alongside with the CTA (input) and the NCCT (target output). (Color figure online)

Turing Test. To measure the quality of the SSIM+Dis variant and its proximity to the NCCT images, a Turing test with one radiologist and two physicians has been conducted. In phase one of the test, 64 individual axial slices from the test data were shown to the subjects with the task to blindly classify each image as stemming from a real NCCT or a generated SyNCCT. In phase two, 32 pairs of one real and one generated image of the same slice index were shown next to each other and the subject had to decide which one of both is synthetic. To avoid bias, the subjects were not informed how the images were created. The data sets were randomly selected and windowing was chosen to be in a typical range for soft tissue contrast (C = 30 HU, W = 50 HU) [2, 10]. In the first phase, the radiologist classified 64% of the cases correctly while both physicians were correct in 65% the cases on average. In the second phase, the radiologist achieved an accuracy of 88% and the two other physicians an average accuracy of 47%, respectively. This suggests that distinguishing the generated SyNCCT images from real scans is far from trivial even for experts, in particular without direct comparison. Detailed results are shown in the supplementary material.

Table 1. Quantitative comparison of all approaches on the test data.

	Method	SSIM	MAE Brain (HU)	MAE Vessels (HU)	ASPECTS
Smooth	Ours MSE	$\mathbf{0.60 \pm 0.03}$	$\mathbf{7.2 \pm 0.5}$	$\mathbf{11.6 \pm 2.2}$	1.22 ± 1.17 (83 %)
	Ours SSIM	0.59 ± 0.03	8.9 ± 0.6	13.1 ± 1.9	$\mathbf{1.04 \pm 1.14}$ (83 %)
	Mod. Poirot [1]	0.59 ± 0.03	7.5 ± 0.5	11.7 ± 1.5	1.61 ± 1.64 (77 %)
Realistic	Ours MSE+Dis.	$\mathbf{0.57 \pm 0.03}$	$\mathbf{8.7 \pm 0.6}$	$\mathbf{13.7 \pm 2.5}$	1.00 ± 1.45 (85 %)
	Ours SSIM+Dis.	$\mathbf{0.57 \pm 0.03}$	9.0 ± 0.6	14.2 ± 1.7	$\mathbf{0.78 \pm 0.74}$ (88 %)
	CycleGAN	0.56 ± 0.03	9.2 ± 0.8	21.3 ± 3.6	1.70 ± 1.52 (81 %)

Quantitative Evaluation. In Table 1 we compared all models in terms of their average performance on patient-level. The respective best epochs were determined on the validation data. We subdivided the evaluation of the Mean Absolute Error (MAE) on brain parenchyma and vessel sections using a dilated version of the vessel segmentation described in Sect. 2.3. The SSIM values were computed on the slices cropped to the foreground region. As discussed in Sect. 3, SyNCCT variants without the discriminator component as well as the modified Poirot method led to overly *smooth* results. In this category, the best performance was achieved with the MSE variant of SyNCCT. In contrast, the generative approaches were perceptually more *realistic*. Here, the MSE+Dis. configuration achieved the best SSIM and MAE on both brain tissue and vessels. Due to the noisy nature of the images and slight registration inaccuracies, distance-based metrics do not properly represent the perception. Therefore, direct supervision led to smooth results, which is a common problem using MSE. As the GAN methods introduce high frequencies, the MAE becomes substantially different.

We also evaluated SyNCCT w.r.t. its consistency in an automated evaluation of the ASPECT score [1]. The purpose of ASPECTS is to rate the severity of ischemic strokes based on NCCT and ranges from 0 (most severe) to 10 (no signs). It is calculated by deducting one point from a maximum score of 10 for each region that appears affected by infarction out of a standardized set of 10 vascular territories. It is desired that pathologies, such as early signs of ischemic stroke, are properly preserved and translated during domain transfer. Thus, we postulate that ASPECT scores determined on SyNCCT images be consistent with those determined in the real NCCT images. Numerous clinical publications, like [11,13], have shown that the performance of such tools is comparable to that of a radiologist. The MAE of the ASPECT scores between the real and the predicted images is listed in Table 1 with the coverage between the regions resulting from the various configurations to the NCCT regions in percent. The scores were automatically determined using commercial software (syngo.CT ASPECTS, version VB40, Siemens Healthineers, Forchheim, Germany). The SSIM+Dis configuration outperformed all other approaches with an MAE of 0.78 ± 0.74.

4 Conclusion

We present the—to the best of our knowledge—first method enabling the domain transfer from CTA to NCCT using a single energy level. At its core, the method combines a supervised U-Net and a generator-discriminator learning mechanism. We provide the U-Net with additional input channels derived from the CTA that foster robust learning of the task at hand, such as a segmentation of the vessels and a VNC-like, contrast-adjusted image. We were able to confirm by means of a Turing test that the generated SyNCCT images could not be reliably distinguished from real NCCT scans. We determined the ASPECT scores on the NCCT data sets and compared the results to different variations of the network and baseline approaches. The best performance according to the ASPECT criterion was achieved using the SSIM loss together with a discriminator. In terms of ASPECT scores, we achieved an MAE of 0.78 and a median error of 1, which indicates that no artificial stroke patterns were added and existing signs of stroke were correctly carried over to the NCCT domain.

References

1. Barber, P.A., Demchuk, A.M., Zhang, J., Buchan, A.M., Group, A.S., et al.: Validity and reliability of a quantitative computed tomography score in predicting outcome of hyperacute stroke before thrombolytic therapy. The Lancet **355**(9216), 1670–1674 (2000)
2. Bibb, R., Eggbeer, D., Paterson, A.: Medical Modelling: The Application of Advanced Design and Rapid Prototyping Techniques in Medicine. Woodhead Publishing (2014)
3. Bône, A., et al.: Contrast-enhanced brain MRI synthesis with deep learning: key input modalities and asymptotic performance. In: 2021 IEEE 18th International Symposium on Biomedical Imaging (ISBI), pp. 1159–1163. IEEE (2021)
4. Goodfellow, I.J., et al.: Generative adversarial networks. arXiv preprint arXiv:1406.2661 (2014)

5. He, K., Zhang, X., Ren, S., Sun, J.: Deep residual learning for image recognition. In: Proceedings of the IEEE Conference on Computer Vision and Pattern Recognition, pp. 770–778 (2016)
6. Isola, P., Zhu, J.Y., Zhou, T., Efros, A.A.: Image-to-image translation with conditional adversarial networks. In: Proceedings of the IEEE Conference on Computer Vision and Pattern Recognition, pp. 1125–1134 (2017)
7. Kingma, D.P., Ba, J.: Adam: a method for stochastic optimization. arXiv preprint arXiv:1412.6980 (2014)
8. Klimont, M., et al.: Deep learning for cerebral angiography segmentation from non-contrast computed tomography. PLoS ONE 15(7), 1–15 (2020). https://doi.org/10.1371/journal.pone.0237092
9. Lee, D., Kim, H., Choi, B., Kim, H.J.: Development of a deep neural network for generating synthetic dual-energy chest x-ray images with single x-ray exposure. Phys. Med. Biol. 64(11), 115017 (2019)
10. Lev, M., Gonzalez, R.: 17 - CT angiography and CT perfusion imaging. In: Toga, A.W., Mazziotta, J.C. (eds.) Brain Mapping: The Methods, 2nd edn, pp. 427–484. Academic Press, San Diego (2002). https://doi.org/10.1016/B978-012693019-1/50019-8, https://www.sciencedirect.com/science/article/pii/B9780126930191500198
11. Li, L., et al.: Comparison of the performance between frontier aspects software and different levels of radiologists on assessing CT examinations of acute ischaemic stroke patients. Clin. Radiol. 75(5), 358–365 (2020)
12. Lyu, T., et al.: Estimating dual-energy CT imaging from single-energy CT data with material decomposition convolutional neural network. Med. Image Anal. 70, 102001 (2021)
13. Maegerlein, C., et al.: Automated calculation of the alberta stroke program early CT score: feasibility and reliability. Radiology 291(1), 141–148 (2019)
14. Paszke, A., et al.: Pytorch: an imperative style, high-performance deep learning library. arXiv preprint arXiv:1912.01703 (2019)
15. Poirot, M.G., et al.: Physics-informed deep learning for dual-energy computed tomography image processing. Sci. Rep. 9(1), 1–9 (2019)
16. Ronneberger, O., Fischer, P., Brox, T.: U-Net: convolutional networks for biomedical image segmentation. In: Navab, N., Hornegger, J., Wells, W.M., Frangi, A.F. (eds.) MICCAI 2015. LNCS, vol. 9351, pp. 234–241. Springer, Cham (2015). https://doi.org/10.1007/978-3-319-24574-4_28
17. Thamm, F., Jürgens, M., Ditt, H., Maier, A.: VirtualDSA++: automated segmentation, vessel labeling, occlusion detection and graph search on CT-angiography data. In: Kozlíková, B., Krone, M., Smit, N., Nieselt, K., Raidou, R.G. (eds.) Eurographics Workshop on Visual Computing for Biology and Medicine. The Eurographics Association (2020). https://doi.org/10.2312/vcbm.20201181
18. Toepker, M., et al.: Virtual non-contrast in second-generation, dual-energy computed tomography: reliability of attenuation values. Eur. J. Radiol. 81(3), e398–e405 (2012)
19. Wang, Z., Bovik, A.C.: A universal image quality index. IEEE Signal Process. Lett. 9(3), 81–84 (2002)
20. Zhu, J.Y., Park, T., Isola, P., Efros, A.A.: Unpaired image-to-image translation using cycle-consistent adversarial networks. In: Proceedings of the IEEE International Conference on Computer Vision, pp. 2223–2232 (2017)

Local Morphological Measures Confirm that Folding Within Small Partitions of the Human Cortex Follows Universal Scaling Law

Karoline Leiberg$^{(\boxtimes)}$, Christoforos Papasavvas, and Yujiang Wang

Newcastle University, Newcastle upon Tyne NE1 7RU, UK
K.Leiberg2@Newcastle.ac.uk

Abstract. The universal scaling law of cortical morphology describes cortical folding as a tight relationship between average grey matter thickness, pial surface area, and exposed surface area. It applies for mammalian species, humans, and across lobes, however it remains to be shown that local cortical folding obeys the same rules. Here, we develop a method to obtain morphological measures for small regions across the cortex and correct surface areas by curvature to account for differences in patch size, resulting in a map of local morphology. It enables a near-pointwise analysis of morphological variables and their regional changes due to processes such as healthy ageing. We confirm empirically that the theorised relation of morphological measures still holds at this level of local partition sizes as predicted, justifying the use of independent variables derived from the scaling law to identify regional differences in folding, subject-specific abnormalities, and local effects of ageing.

1   Introduction

Cortical morphology is a useful imaging-based biomarker for a range of applications, including aging and disease. Measures that describe the shape and folding of a brain, such as gyrification and cortical thickness, are being used both over the entire cortex and locally to identify differences in individuals and cohorts of subjects [2–5]. Such measures are often studied separately, without taking their interaction into account.

Only recently, a universal scaling law describing the interaction of cortical morphology measures has been proposed [6]. It is theoretically derived from minimising effective free energy of growing surfaces under compressing forces e.g. at cortical development [6]. This scaling law captures the folding of the cortex as the dependence between the measure of average cortical thickness T, total pial surface area A_t and exposed surface area A_e in the equation

$$A_t\sqrt{T} = kA_e^{1.25}, \tag{1}$$

© Springer Nature Switzerland AG 2021
M. de Bruijne et al. (Eds.): MICCAI 2021, LNCS 12907, pp. 691–700, 2021.
https://doi.org/10.1007/978-3-030-87234-2_65

where k is a constant. Empirically, this relation has been shown to hold for different mammalian species [6], in human hemispheres [10], and within human cortices across the lobes [11]. It says that the three variables are interdependent. For example, if two cortices of a similar age have the same total area, but one is thicker, it will also have more exposed area - and hence less gyrification. The scaling law can be visualised by the analogy of crumpling a piece of paper into a ball: the thickness of the paper, its total area and the force used to crush it determine the exposed surface area of the ball [6].

The scaling law additionally allows us to derive three linearly independent, interpretable morphological variables as linear combinations of $\log T^2$, $\log A_t$ and $\log A_e$ [9]:

$$K = \log A_t + \frac{1}{4} \log T^2 - \frac{5}{4} \log A_e, \tag{2}$$

$$I = \log A_t + \log T^2 + \log A_e, \tag{3}$$

$$S = \frac{3}{2} \log A_t - \frac{9}{4} \log T^2 + \frac{3}{4} \log A_e. \tag{4}$$

K is derived directly from Eq. (1) by taking logarithms and arranging for $\log k$. It is a measure of tension/pressure acting on the cortex, or, in the analogy, pressure applied to the paper ball. I describes the overall size of the brain or paper ball. Changes in I correspond to an isometric scaling of the cortex. Lastly, S is the inner product of K and I that contains the remaining information about shape; it can be thought of as the "folding technique" of the paper ball. An increase in S indicates a cortex being thinner and more gyrified, but overall flatter. Using K, I, and S avoids covarying variables, whilst keeping them interpretable. Otherwise, analysing cortical thinning without accounting for surface area changes could miss signs of atrophy that can be discovered when accounting for the covariance [9]. For example, if in a longitudinal analysis of a subject the thickness of its cortex does not reduce (or only a little), but the total surface area decreases, i.e. the cortex is flattening, this would be reflected in a reduction of K, but would be missed in typical studies that only consider changes in thickness. Even comparing two cohorts in just one variable can suffer from the same problem [9].

However, one distinct challenge remains in regionalising these independent measures of cortical morphology, K, I, and S. It is clear that the scaling law cannot hold for arbitrarily small regions. The goal of this paper is to show that the scaling law still holds for areas smaller than lobes, defined independently of them. We first develop a method to find the raw variables T, A_t, and A_e in small patches across the cortex, and correct the surface areas so they are independent of the size chosen for patches. This allows us to compare the folding within the regions to that of other partitions of the cortex or even the full hemisphere [11].

If local cortical folding follows the scaling law, it justifies the use of K, I, and S for further morphological analyses. For example, we will show that they can be used to get a better understanding of regional age-related cortical atrophy.

2   Methods

2.1   MRI Data and Processing

To asses the rules of local cortical folding in a cohort of healthy subjects in a large age range, we used the NKI Rockland Sample data [7].

The MRIs were preprocessed with the FreeSurfer 6.0 recon-all pipeline to obtain the fine, triangular mesh representing the grey matter surface, as well as the grey matter thickness on each point of the mesh. We then ran the local gyrification index processing stream (https://surfer.nmr.mgh.harvard.edu/fswiki/ LGI) to obtain the outer smooth pial surface, which can be thought of as a surface which tightly wraps around the pial surface, computed by closing the sulci of the pial surface with a sphere of 15 mm in diameter [8]. We compute this smooth surface to later use the area on it as the exposed surface area A_e.

Out of the 929 subjects for whom data was available, 95 subjects were rejected due to inadequate image quality, motion artifacts or parts of their grey matter missing. We performed visual quality inspections on a random sample of 50 subjects and applied manual corrections on the 50 subjects where needed. We deemed overall image quality and FreeSurfer processing to be adequate. The remaining sample consists of 834 subjects, 508 female and 326 male, in an age range from 6 to 85 years. A full list of subjects used can be found on github (https://github.com/KarolineLeiberg/folding_pointwise).

2.2   Extraction of Local Morphological Measures

The following describes the method in which morphological measures were computed for each point on the pial surface. The full code is available on github: https://github.com/KarolineLeiberg/folding_pointwise.

The pial surface and smooth pial surface are reduced to 5% and 10% of their original resolution respectively, using the downsampling function of the MATLAB iso2mesh toolbox. This retains the main features of cortical folds, but reduces the number of vertices in the pial surface to about 7000 per hemisphere.

The thickness map is converted from the original pial surface to the down-sampled pial surface by assigning the value of the nearest pial vertex for each downsampled pial point. The pial thickness is subsequently transformed from being a pointwise measure to a measure for each face in the mesh (https:// github.com/cnnp-lab/CorticalFoldingAnalysisTools).

Then, for each point p on the downsampled pial surface, a local patch around p is defined as all vertices within a radius of 25 mm, measured as Euclidean distance, that are directly connected to p through neighbouring points which are also in the patch - this is to avoid e.g. disconnected patches across two gyri that are very close to each other, but the connecting sulcus is not included. We then close any potential holes in the patch. A visualisation of the process can be found in Fig. 1 a–c. A corresponding smooth patch is defined by first finding the nearest downsampled pial point for each downsampled smooth pial point,

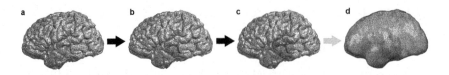

Fig. 1. Defining a patch on the pial surface around a point. **a–c** Extracts of the process of adding neighbouring points to the patch. **c** The final pial surface patch around this point. **d** The corresponding patch on the smooth pial surface.

and then including smooth pial points if their neighbouring pial point is in the patch. (Fig. 1 d).

The total surface area A_t^p for the patch around p is then computed as the sum of all triangular faces of the downsampled pial surface which are contained in p's patch. Similarly, the average cortical thickness T^p is the average thickness across those faces. The exposed surface area A_e^p is computed as the sum of face areas within the patch on the downsampled smooth pial surface.

A convex hull, the smallest convex surface that contains all points, is fitted over p's downsampled smooth pial patch, and the integrated Gaussian Curvature of the patch around p is approximated as the sum of Gaussian curvatures of all points on the convex hull which do not lie on the edge of the smooth patch (https://github.com/cnnp-lab/CorticalFoldingAnalysisTools). We use the convex hull to estimate the integrated curvature rather than the smooth surface since its convex property ensures we get a positive value of curvature, which can be used to compute the proportion of curvature of the patch in relation to the full hemisphere's curvature.

Having computed the morphological measures for regions around each point on the downsampled pial surface, the data is converted back to the original, full pial surface, again using values of the nearest neighbour. This is done so the data can later be projected to the FreeSurfer average subject to compare across subjects.

2.3 Surface Area Reconstruction

As shown in Wang *et al.* [11], the size of regions into which a cortex is partitioned fully determines its surface areas and thus its location in the space of A_t, A_e and T, making it impossible to compare patches to each other and infer whether they align on the plane predicted by the scaling law. Each patch's surface areas are thus reconstructed to what they would be if the patch was a full hemisphere of the same gyrification index (the ratio of A_t to A_e), using the proportion of curvature it contains, approximated via its convex hull. This correction is done using the formulas [11]

$$A_t'^p = A_t^p * \frac{4\pi}{I_G^p}, \tag{5}$$

$$A_e'^p = A_e^p * \frac{4\pi}{I_G^p}, \tag{6}$$

where A_t^p and A_e^p are the total and exposed surface areas of patch p before correction, and $A_t'^p$ and $A_e'^p$ are the values after correction using the integrated Gaussian curvature I_G^p over p. This correction of a patch's surface areas by its integrated Gaussian curvature preserves its average cortical thickness, gyrification index, and the average Gaussian curvatures of its total and exposed areas [11].

Points that lie on particularly flat parts of the cortex or have very small surface areas tend to have convex hulls with integrated curvatures close to zero. In such cases, the correction of the surface areas leads to an overcorrection, inflating the surface area to unreasonably large values. For simplicity, we will exclude these points from our analysis at this stage, choosing not to investigate their compliance with the scaling law. All points within 10 mm to the left and right of the midline of the brain and all points with a curvature below 0.16 are excluded. We will discuss later on how future work may be able to investigate these more challenging parts of the cortex, where the Gaussian curvature of the convex hull is not a good representation of the proportion of the patch of interest.

2.4 Fitting the Scaling Law Within Subjects

We assess if local folding follows the scaling law by regressing within each subject over all patches in the two-dimensional projection of the scaling law, where

$$X = \log A_e'^p, \tag{7}$$

$$Y = \log(A_t'^p \sqrt{T^p}). \tag{8}$$

Here, $A_t'^p$ and $A_e'^p$ are the surface areas after correcting by curvature, T^p is the observed thickness. The slope of the regression is a subject-specific estimation of the exponent of A_e in Eq. (1). We verify if each subject's local folding follows the scaling law by seeing how close the slope is to 1.25.

2.5 Local Age Effects

To quantify the effect ageing has on morphology in the raw measures and in the independent variables, we used the FreeSurfer function mri_surf2surf to convert all subjects' morphology maps (i.e. pial surfaces with morphological measures computed at each point) to the same surface space, where they can be compared and analysed as a group. We compute the independent variables K, I, and S on each point of the pial surface. We then fit linear regression models pointwise across all subjects, including both sex and age as covariates. Points represented in fewer than 10 subjects of either sex were excluded to ensure the regression was representative of the population. We then use the coefficient of the ageing covariate at each point and for each variable as an indicator of the local effect of age-related atrophy on that measure.

3   Results

3.1   Surface Area Reconstruction by Gaussian Curvature Breaks down in Insula and on Midline

We apply Gaussian curvature corrections (Eq. (5), (6)) to all points to reconstruct surface areas of patches to what they would be for a full hemisphere, but find that some points have approximately zero integrated Gaussian curvature (Fig. 2a shows the convex hull over one example patch). When we look across all points of the cortex, we observe a distinct subset of points that display zero curvature (Fig. 2b). These points are usually located in the insula or particularly deep sulci, where the patch has a small, overall convex, exposed surface, or on the midline, where the exposed surface might be large, but very flat. The midline additionally contains points with very large curvature, where the convex hull over a patch can contain an angle of 90° or more.

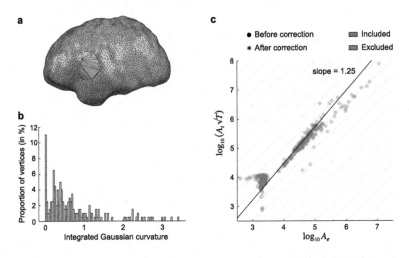

Fig. 2. Gaussian curvature and its effect on reconstructing surface areas in one example subject. **a** Example of patch with zero curvature. **b** Distribution of curvature across patches. **c** Sample of 150 patches plotted in the X-Y-plane, both before (round marker) and after (asterisk) surface area corrections. Yellow lines with slope 1 indicate the shift of points caused by the curvature corrections. Green points are included in further analysis, purple points are excluded due to their low curvature or position relative to the midline. The black line is a regression line through the green points after correction. (Color figure online)

Correcting points by extremely small curvatures leads to overcorrections (Fig. 2c): Whilst most points naturally align in the X-Y-plane (Eq. (7), (8)) with a slope of 1.25 after the surface areas are reconstructed, points with curvatures near zero are being overcorrected to the top-right of the plot. For simplicity, we therefore exclude these points from our further analysis, acknowledging that Gaussian curvature of the convex hull is not a good representation of the proportion of those patches. This affects around 20% of all points on the pial surface.

3.2   Local Folding Follows Scaling Law

To verify whether the scaling law still applies in small patches of the cortex, meaning if the measures of T, A_t, and A_e covary locally as predicted, we fit the scaling law within subjects, estimating the slope between points in X and Y as described in Eq. (7) and (8). We find that the slope for each subject, i.e. the subject-specific estimates of the coefficient of A_e in Eq. (1), are distributed around a mean of 1.23 (Fig. 3), only slightly lower than the observed coefficient of 1.25 for hemispheres and lobes. This shows that the scaling law still holds at this level of patch sizes, meaning that when looking at small, local regions of the cortex with spherical radii of 25 mm, the average thickness, total area and exposed area covary predictably according to Eq. (1). Note that the subjects follow the scaling law without any age or sex corrections, because age differences only affect the scaling law intercept (K), but not the slope [10].

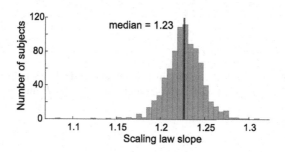

Fig. 3. Distribution of slopes observed in subjects when fitting a regression through points across the cortex in the X-Y-plane. The black line marks the median of observed slopes.

3.3 K, I and S Have Additional Value When Observing Ageing Effects

As we have seen, the folding follows the scaling law even locally, which implies that K, I, and S are theoretically independent. We convert to these variables to see if they add insight to the differences in local morphology.

When we look at the effect of healthy ageing on local morphology, we see a decrease in the raw variables A_t, A_e, and T at varying rates in most areas of the brain (Fig. 4 a–c). A decrease in surface areas can be interpreted as a flattening of the cortex, since less surface area within a constant radius indicates less folding.

The independent variables show an isometric shrinking of most areas, with an overall loss of tension (Fig. 4 d–e).

In the raw variables, we do not see systematic differences in how the areas around the upper and lower precentral gyrus are affected by atrophy. We do however see such a difference in the shape term S (Fig. 4f): S decreases in the upper precentral gyrus with ageing, but the lower part is relatively unaffected. This is one example of the added value from switching to independent variables; we gain information otherwise hidden in the covariance of A_t, A_e, and T.

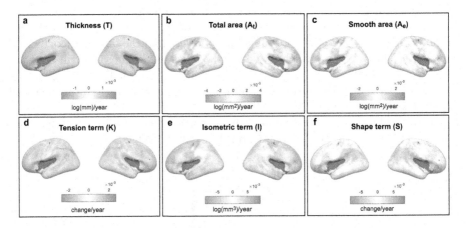

Fig. 4. Local effects of healthy ageing in the left and right hemisphere. **a–c** Raw variables T, A_t, and A_e (logged). **d-f** Independent variables K, I, and S.

4   Discussion

On a big data set covering a large age range, we have demonstrated that the local folding of the brain follows the universal scaling law, meaning the grey matter thickness, total surface area and exposed surface area in small regions covary according to the same rule as the whole cortex. This result extends previous work, which had shown that morphology in full hemispheres and lobes adheres to the scaling law. We empirically confirmed that the minimum size of regions for which the folding rule applies has a radius of under 25 mm. This result confirms our hypothesis, allowing us to use the independent variables K, I and S on much smaller scale than before.

Our method works well for most of the cortex. However, we had to exclude around 20% of points on the pial surface from our analysis, predominantly located on the midline and in the insula. For these points, we founds that the Gaussian curvature was not an adequate representation of the proportion of the patch. For future iterations of the method, we plan to improve on this. An alternative way to approximate patch curvature would be to compute the integrated curvature of the patch's smooth pial surface, rather than its convex hull. However, this surface may already contain too much local information of folding for our purpose. Another idea is to inflate the pial surface to a sphere and use the proportion of surface area or curvature of the patch's representation on the sphere for corrections.

Another improvement of the method could be achieved by making the radius defining the patch around each point on the pial surface adaptive depending on the thickness at that point. This flexible way of patch definitions uses the smallest radius required to find enough exposed surface area at each point, making the morphological map as close to pointwise as possible. We expect the ideal patch radii to be around 20–30 mm for human brains (spanning at least one sulcus/gyrus), which is why we chose a value in this range for the fixed-size method.

The spatial normalisation (to compare subjects to each other or perform group analyses) as we implemented is susceptible to misregistration issues. Such problems could be improved upon, e.g. by converting the variables K, I and S to neighbourhood-based histogram features [1].

The local applicability of the scaling law allows us to use it in terms of independent components derived from it, to further analyse local morphology. In our results we have shown local differences in how the cortex is affected by the process of ageing in the independent variables, which were not clear from the raw variables alone. Our method could have other practical applications, such as finding local abnormalities in patient groups compared to a control cohort, indicating regional effects of dysfunction. It might also be used to detect subject-specific abnormalities, indicating areas of the brain that fold atypically.

References

1. Awate, S.P., Leahy, R.M., Joshi, A.A.: Kernel methods for Riemannian analysis of robust descriptors of the cerebral cortex. In: Niethammer, M., et al. (eds.) IPMI 2017. LNCS, vol. 10265, pp. 28–40. Springer, Cham (2017). https://doi.org/10.1007/978-3-319-59050-9_3

2. Chaudhary, S., et al.: Cortical thickness and gyrification index measuring cognition in Parkinson's disease. Int. J. Neurosci. 1–10 (2020)

3. Frangou, S., Modabbernia, A., Williams, S.C.R., Fuentes-claramonte, P., Glahn, D.C.: Cortical thickness across the lifespan: data from 17, 075 healthy individuals aged 3–90 years. Hum. Brain Mapp. 1–21 (2021)

4. Galovic, M., et al.: Progressive cortical thinning in patients with focal epilepsy. JAMA Neurol. **76**(10), 1230–1239 (2019)

5. Libero, L.E., Schaer, M., Li, D.D., Amaral, D.G., Nordahl, C.W.: A longitudinal study of local gyrification index in young boys with autism spectrum disorder. Cerebral Cortex **29**(6), 2575–2587 (2019)

6. Mota, B., Herculano-Houzel, S.: Cortical folding scales universally with surface area and thickness, not number of neurons. Science **349**(6243), 74–77 (2015)

7. Nooner, K.B., et al.: The NKI-rockland sample: a model for accelerating the pace of discovery science in psychiatry. Front. Neurosci. **6**, 152 (2012)

8. Schaer, M., Bach Cuadra, M., Tamarit, L., Lazeyras, F., Eliez, S., Thiran, J.P.: A Surface-based approach to quantify local cortical gyrification. IEEE Trans. Med. Imaging **27**(2), 161–170 (2008)
9. Wang, Y., et al.: Independent components of human brain morphology. NeuroImage **226**, 117546 (2021)
10. Wang, Y., Necus, J., Kaiser, M., Mota, B.: Universality in human cortical folding in health and disease. Proc. Nat. Acad. Sci. U.S.A. **113**(45), 12820–12825 (2016)
11. Wang, Y., Necus, J., Rodriguez, L.P., Taylor, P.N., Mota, B.: Human cortical folding across regions within individual brains follows universal scaling law. Commun. Biol. **2**(1), 1–8 (2019)

Exploring the Functional Difference of Gyri/Sulci via Hierarchical Interpretable Autoencoder

Lin Zhao[1], Haixing Dai[1], Xi Jiang[2], Tuo Zhang[3], Dajiang Zhu[4], and Tianming Liu[1(✉)]

[1] Cortical Architecture Imaging and Discovery Lab, Department of Computer Science and Bioimaging Research Center, The University of Georgia, Athens, GA, USA
tliu@cs.uga.edu
[2] School of Life Science and Technology, University of Electronic Science and Technology of China, Chengdu, China
[3] School of Automation, Northwestern Polytechnical University, Xi'an, China
[4] Department of Computer Science and Engineering, The University of Texas at Arlington, Arlington, TX, USA

Abstract. Understanding the functional mechanism of human brain has been of intense interest in the brain mapping field. Recent studies suggested that cortical gyri and sulci, the two basic cortical folding patterns, play different functional roles based on various data-driven methods from local time scale to global perspective. However, given the evidence that the brain's neuronal organization follows a hierarchical principle both spatially and temporally, it is unclear whether there exists temporal and spatial hierarchical functional differences between gyri and sulci due to the lack of suitable analytical tools. To answer this question, in this paper, we proposed a novel Hierarchical Interpretable Autoencoder (HIAE) to explore the hierarchical functional difference between gyri and sulci. The core idea is that hierarchical features learned by autoencoder can be embedded into a one-dimensional vector which interprets the features as spatial-temporal patterns, with which the region-based analysis in gyri and sulci can be further performed. We evaluated our framework using the Human Connectome Project (HCP) fMRI dataset, and the experiments showed that our framework is effective in terms of revealing meaningful hierarchical spatial-temporal features. Analysis based on Activation Ratio (AR) metric suggested that gyri have more low-frequency/global features while sulci have more high-frequency/local features. Our study provided novel insights to understand the brain's folding-function relationship.

Keywords: Gyri/Sulci · Cortical folding · Hierarchical Interpretable Autoencoder · Functional difference · fMRI

1 Introduction

Exploring the functional mechanism of human brain and its relationship with the brain's structural substrate has been of intense interest for neuroscience for centuries [1].

© Springer Nature Switzerland AG 2021
M. de Bruijne et al. (Eds.): MICCAI 2021, LNCS 12907, pp. 701–709, 2021.
https://doi.org/10.1007/978-3-030-87234-2_66

Recently, gyri and sulci, two basic structural folding patterns of cerebral cortex, have been demonstrated to play different functional roles based on various data-driven methods [2–5]. For example, in the spatial domain, sparse dictionary learning (SDL) was employed to decompose the whole-brain fMRI data into global networks, and it was found that gyri had stronger spatial overlap patterns of such global networks than sulci [2, 3]. In the temporal domain, a recent study leveraged convolutional neural networks (CNN) to investigate local temporal characters of fMRI signals from gyri and sulci [5]. The results suggested that gyral signals are of lower frequency with less diversity compared with sulcal signals, and it was suggested that global networks are synchronized at a lower frequency [5]. Despite these studies which significantly advanced our understanding of functional differences between gyri and sulci, both the data analytical tools and result interpretation approaches in previous works are limited to a single scale functional architecture. In fact, the real-world stimuli unfold at different time scales and over extended regions, such that the neuronal organization and its activities patterns were suggested to follow a hierarchical principle, both spatially and temporally [6], in which a higher level receives inputs from lower ones with larger receptive fields. Therefore, whether there exists such a temporal and spatial hierarchical functional difference between gyri and sulci remains to be elucidated, and a new analytical tool to answer this question is much needed.

A recent deep convolutional autoencoder (DCAE) [6] model, which extracted dozens of features in different layers hierarchically, seems suitable to the above objective. Interestingly, the data-driven features derived from DCAE have certain neuroscientific meanings, demonstrating the great promise of using deep learning methods for exploring and modeling the hierarchical organization of gyri and sulci. However, current deep learning models such as the DCAE are still limited in the sense that features from deep layers are much more abstract and complex without a straightforward neuroscientific meaning, e.g., the CNN model adopted in [5] was only composed of one convolutional layer for the ease of interpretation. Also, features at each layer are usually multi-dimensional and consequently, it is difficult to derive the spatial distributions of those learned features for region-based analysis. For example, the DCAE model in [6] just demonstrated the representative features from one dimension at each layer, while other dimensions remained unexplored.

To overcome these limitations and to develop a new method to explore the hierarchical functional difference between gyri and sulci, in this paper, we proposed a novel Hierarchical Interpretable Autoencoder (HIAE). The key idea is that hierarchical features learned by the autoencoder in different hierarchical layers can be abstracted and embedded into a one-dimensional vector via a carefully designed Feature Interpreter (FI) which is jointly optimized with the autoencoder. The embedded vector can be interpreted as the spatial patterns of the learned features and can be further regressed with the fMRI signals to obtain the temporal patterns. Thus, with those spatial-temporal patterns at each layer, we are then able to perform region-based analysis in gyri and sulci. We evaluated our framework on publicly available Human Connectome Project (HCP) fMRI dataset, and extensive experiments showed our framework's effectiveness in terms of both meaningful spatial-temporal features and gyral/sulcal functional differences.

2 Materials and Method

2.1 Overview

The proposed framework is shown in Fig. 1. We firstly introduce the data and pre-processing steps in Sect. 2.2. Then the HIAE model, the approach to obtaining the spatial-temporal patterns, and the optimization of HIAE model are detailed in Sect. 2.3. In Sect. 2.4, we propose the Activation Ratio (AR) metric to investigate the hierarchical functional difference between gyri and sulci.

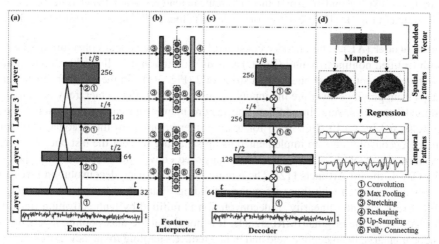

Fig. 1. Illustration of the HIAE framework. (a) The encoder extracts features hierarchically. The numbers above/beside the rectangular are the lengths/numbers of channels of feature maps, respectively. (b) The Feature Interpreter abstracts feature maps into an embedded vector, the digits of which are represented by colored circles corresponding to the colored squares in (d). (d) Interpreting the hierarchical features. The activation value of each digit can be mapped back to cortical surface mesh to reveal the spatial patterns. Then, the spatial patterns are regressed with original signals to obtain the corresponding temporal patterns.

2.2 Data Acquisition and Pre-processing

We adopted HCP grayordinate-based task fMRI (tfMRI) data from Human Connectcome Project of Q3 release (https://db.humanconnectome.org/) [9, 10] to evaluate our framework. HCP provides the high-resolution surface mesh with accurate mapping of fMRI time series from volume space to surface vertices, which greatly facilitates our analyses. Among all the seven task paradigms, Emotion (176 volume frames) and Motor (284 volume frames) tfMRI data were used as the testing bed. More details of data acquisition and task designs are referred to [9].

The HCP tfMRI data were preprocessed by HCP minimal preprocessing pipelines [11]. In addition, we normalized all the fMRI signals with zero mean and standard deviation one as in [5, 8]. From all participants in HCP Q3 release, we randomly selected

200 subjects for each task and extracted fMRI signals only from the cortical surface vertices. The training/validation dataset was composed of 8,000,000/2,000,000 tfMRI time series randomly sampled from all the extracted signals, respectively. In experimental stage, all subjects were included with whole-brain fMRI signals for analysis. We use the convexity ("sulc" in FreeSurfer) to segment gyri (>0.1) from sulci (<-0.1).

2.3   Hierarchical Interpretable Autoencoder (HIAE)

In this section, we introduce a novel Hierarchical Interpretable Autoencoder (HIAE) which can extract and interpret the hierarchical features from fMRI time series. As illustrated in Fig. 1, HIAE consists of a 4-layer autoencoder and 4 corresponding FIs.

Autoencoder (AE). The encoder consists of 4 one-dimensional convolutional layers with rectified nonlinearity unit (ReLU) as activation function, followed by a max pooling layer (except the last one) with a stride of 2 (Fig. 1(a)). In the encoder, the feature maps from previous layer are the inputs for the next layer, thus a hierarchy is naturally formed between layers. In addition, the feature maps from each convolutional layer are also fed into a FI (Fig. 1(b)) for feature interpretation. In the decoder, the symmetric deconvolutions are implemented as up-sampling (except the last one) followed by normal convolutions rather than the transposed convolution because of the potential Checkerboard Artifacts [12]. The feature maps from each deconvolution layer are firstly concatenated with those from corresponding FI and then fed into the next layer (Fig. 1(c)). The non-linearity of deconvolution is fulfilled by tanh activation function for recovering the original fMRI signals. It is noted that the max pooling layers in the encoder enlarge the receptive field and in fact reduce sampling frequency of time series. In this sense, features from subsequent convolutional layer are of lower frequency and more global in time scale.

Feature Interpreter (FI). Feature Interpreter consists of two fully connected (FC) layers: one for embedding the feature maps into a one-dimensional vector as latent representation; another for mapping the embedded vector back to reconstruct the input feature maps (Fig. 1(b)). Specifically, given a feature map $f \in \mathbb{R}^{k \times t'}$ from a single time series where k is the number of channels and t' is the length of the features in time, it is firstly stretched into a vector $f' \in \mathbb{R}^{1 \times kt'}$ and then fed into the first FC layer, the output of which is the embedded vector $v \in \mathbb{R}^{1 \times d}$ where d is the number of digits in embedded vector. For the whole brain fMRI signals $X \in \mathbb{R}^{n \times t}$, where n is the number of vertices and t is the number of fMRI time points, we can concatenate the embedded vector of each signal along the first dimension and get the spatial distribution patterns $S \in \mathbb{R}^{n \times d}$, based on which the temporal patterns $T \in \mathbb{R}^{d \times t}$ can be obtained by regression as in [2, 3]:

$$min\|X - ST\|_F^2 \tag{1}$$

In this way, the FI can interpret the hierarchical features learned by AE both temporarily and spatially and facilitate our region-based analysis in Sect. 2.4.

Optimization. The parameters in HIAE are optimized by minimizing the Mean Square Error (MSE) between the original fMRI signals $X \in \mathbb{R}^{n \times t}$ and their final reconstructions $X' \in \mathbb{R}^{n \times t}$:

$$min \frac{1}{2} \|X - X'\|_F^2 \qquad (2)$$

The setting of hyperparameters for our framework is summarized in Table 1. The numbers in Encoder and Decoder are referred to kernel size/input channels/output channels. For these hyperparameters we followed the setting in [8]. The number in FI is the number of digits in embedded vector at each layer, which impacts on the interpretability of our framework. To our best knowledge, there has been no published studies adopting the same structure. Thus, we empirically investigated its effects in Sect. 3.3.

Table 1. Hyperparameters setting of HIAE.

	Layer1	Layer2	Layer3	Layer4
Encoder	21/1/32	9/32/64	9/64/128	9/128/256
Decoder	9/256/128	9/256/64	9/128/32	21/64/1
FI	16	16	16	16

The proposed framework was implemented by PyTorch v1.7.1 (https://pytorch.org/). We used the Adam optimizer [13] to minimize the loss function in Eq. (2). Training was performed for 100 epochs with a batch size of 128 on a single GTX 1080Ti GPU.

2.4 Activation Ratio

As discussed in Sect. 2.3, features from deeper convolutional layer of AE are of lower frequency and more global compared with those from shallower layers. To investigate the potential difference between gyri and sulci in terms of represented features of FI, we defined an Activation Ratio for each subject:

$$AR_h = r \times \frac{1}{d} \sum_{i=1}^{d} \frac{g_{hi}}{s_{hi}} \qquad (3)$$

where g_{hi}/s_{hi} denotes the number of gyral/sulcal vertices with activation value larger than threshold h regarding the ith temporal feature or digits (d digits in total). r is the ratio of the total number of sulcal vertices to the total number of gyral vertices. If AR_h is greater than one, it means gyri are more activated in this layer and tend to have more corresponding characteristics than sulci, and vice versa. In this study, we empirically set different threshold h and computed the mean/variance activation value of all subjects in each layer. The experimental results are reported in Sect. 3.2.

3   Experimental Results

3.1   Interpretability of HIAE

To evaluate the interpretability of HIAE in terms of hierarchical spatial-temporal patterns, for each digit in the FI of the last layer (Layer 4), we identified the corresponding digits in preceding FIs based on maximum cosine similarity to form a hierarchy among different layers. In Fig. 2, we demonstrate the group-averaged spatial distribution pattern as well as its preceding counterparts of a randomly selected digit in the FI of Layer 4.

In Fig. 2, we found that the spatial patterns in the first layer are randomly distributed in cortical surface. From the second layer to the fourth layer, meaningful patterns become more observable in a hierarchical manner. The spatial patterns of this randomly selected digit for Emotion task are relevant to Visual Network in [15] and the ones for Motor task are correlated with the Sensorimotor Network in [15]. It is noted that these observations are well reproduced across all digits in the Layer 4.

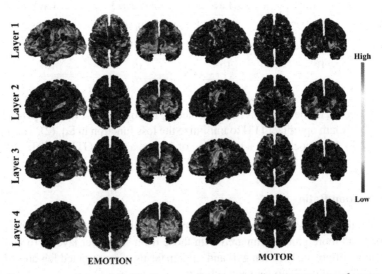

Fig. 2. Group-averaged spatial distribution as well as its preceding counterparts of one randomly selected digit in the Layer 4. The regions with red color have larger activation values than blue regions. (Color figure online)

We further regressed the spatial patterns in Fig. 2 with the original fMRI signals and obtained the corresponding temporal patterns, which are demonstrated in Fig. 3.

In Fig. 3, the temporal patterns in lower layers (e.g., Layer 1) are in faster or high-frequency oscillation than those in higher layers (e.g., Layer 4), which is greatly consistent with our rationale that features from deep layers are of relatively low frequency. In addition, the temporal patterns in higher layer are much more correlated with task design compared with those in lower layers.

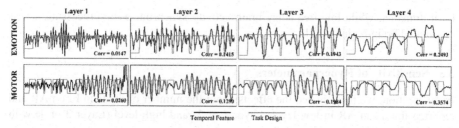

Fig. 3. The corresponding temporal patterns derived from the spatial patterns in Fig. 2. The orange curve represents the task design and the blue curve is the derived temporal pattern. (Color figure online)

Overall, the spatial and temporal patterns extracted and interpreted by our HIAE model have straightforward neuroscientific meaning, indicating the effectiveness and interpretability of HIAE model.

3.2 Gyral/Sulcal Functional Differences

Based on the activation value of each digit in FI across all surface vertices, we calculated the AR values defined in Sect. 2.4 and report them in Table 2.

Table 2. Mean ($\pm$ standard deviation) AR (Activation Ratio) value over all subjects with different threshold for two tasks. Abbreviations: EMO: Emotion; MOT: Motor; L: Layer. More gyral vertices are activated if $AR_h > 1$, and vice versa.

		$h = 0$	$h = 0.1$	$h = 0.2$	$h = 0.3$
L1	EMO	$100.04 \pm 0.98 \times 10^{-2}$	$99.39 \pm 1.29 \times 10^{-2}$	$98.54 \pm 2.78 \times 10^{-2}$	$\mathbf{97.53 \pm 3.08 \times 10^{-2}}$
	MOT	$99.47 \pm 1.41 \times 10^{-2}$	$98.79 \pm 1.60 \times 10^{-2}$	$97.89 \pm 2.11 \times 10^{-2}$	$\mathbf{96.69 \pm 2.98 \times 10^{-2}}$
L2	EMO	$100.26 \pm 1.33 \times 10^{-2}$	$99.72 \pm 1.69 \times 10^{-2}$	$99.30 \pm 2.27 \times 10^{-2}$	$99.86 \pm 3.13 \times 10^{-2}$
	MOT	$100.19 \pm 1.48 \times 10^{-2}$	$99.18 \pm 2.05 \times 10^{-2}$	$97.71 \pm 3.09 \times 10^{-2}$	$96.25 \pm 4.62 \times 10^{-2}$
L3	EMO	$00.10 \pm 1.64 \times 10^{-2}$	$99.95 \pm 2.47 \times 10^{-2}$	$100.48 \pm 3.64 \times 10^{-2}$	$103.11 \pm 5.24 \times 10^{-2}$
	MOT	$98.49 \pm 1.86 \times 10^{-2}$	$99.03 \pm 2.53 \times 10^{-2}$	$100.67 \pm 3.59 \times 10^{-2}$	$104.80 \pm 5.10 \times 10^{-2}$
L4	EMO	$99.87 \pm 1.40 \times 10^{-2}$	$110.79 \pm 4.23 \times 10^{-2}$	$143.07 \pm 13.32 \times 10^{-2}$	$\mathbf{247.92 \pm 69.05 \times 10^{-2}}$
	MOT	$99.25 \pm 1.49 \times 10^{-2}$	$104.54 \pm 2.68 \times 10^{-2}$	$114.58 \pm 5.45 \times 10^{-2}$	$\mathbf{131.06 \pm 11.37 \times 10^{-2}}$

In general, from lower layers to higher layers, the ARs gradually increase from less than 1 to larger than 1, especially with a larger threshold (e.g., $h = 0.3$, highlighted in bold font). This indicates that the sulci activate more in lower layers and gyri activate more in higher layers. Given that the features in lower layers are of relatively higher frequency/more local scale, we can conclude that gyri have more low-frequency/global features and sulci have more high-frequency/local features. This observation is consistent with the studies in [5] and is in line with previous studies arguing that gyri are global functional center and sulci are local functional unit [16]. It also agrees with neuroscience study suggesting that high-frequency brain activity reflects the cortical processing in local domains, while low-frequency brain activity synchronize the processing across distributed brain regions [17].

3.3 Sensitivity of Hyperparameters

In this section, we investigated the effects about the number of digits in FI on AR and reported the mean AR in low level (Layer 1 or 2) and high level (Layer 3 or 4) with threshold $h = 0.3$ under different settings.

In Fig. 4, the mean AR is larger than 1 in high level but smaller than 1 in low level even though it varies among different settings, suggesting the robustness and insensitivity of our HIAE model to hyperparameters in gyro-sulcal analysis.

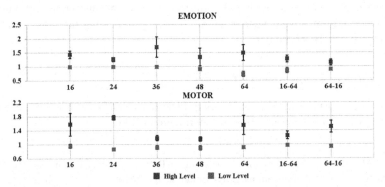

Fig. 4. Mean ($\pm$ standard deviation) AR (Activation Ratio) from low level and high level with different number of digits settings on x-axis. The single number in x-axis represents that the number of digits are the same for each FI. 16–64 denotes that the number of digits from 1st FI to the 4th FI varies from 16 to 64 with an interval of 16 and vice versa.

3.4 Validation via Wavelet Entropy

We computed the wavelet entropy of gyral/sulcal fMRI signals in terms of high/low frequency components to further validate our results. Generally, the entropy for low-frequency components of gyri (EMO: 2.61 ± 0.11; MOT: 3.11 ± 0.16) is larger than that of sulci (EMO: 2.49 ± 0.09; MOT: 2.94 ± 0.15) for both tasks. Conversely, the entropy for high-frequency components of gyri (EMO: 1.77 ± 0.15; MOT: 1.69 ± 0.22) is smaller than that of sulci (EMO: 1.92 ± 0.12; MOT: 1.91 ± 0.20). These differences are confirmed by two-sample one-tailed pair-wise t-tests. This implies that gyral signals are more complex in low-frequency band while sulcal signals are more complex in the high-frequency band. It is consistent with our findings in Sect. 3.2, indicating the effectiveness our HIAE model to extract and interpret hierarchical features and the reliability of our results.

4 Discussion and Conclusion

We proposed a novel Hierarchical Interpretable Autoencoder to explore the gyral/sulcal hierarchical functional difference. Our method is able to extract and interpret meaningful spatial and temporal features hierarchically. Moreover, we found that gyri have more low-frequency and global features compared with sulci when the receptive fields increase from local scale to global scale in both emotion and motor processing. These findings were also validated by wavelet entropy, suggesting the effectiveness and validity of our HIAE model to explore the hierarchical difference between gyri and sulci. The setting of optimal hyperparameters in HIAE model, e.g., the number of digits, can be fully explored in the future. Another future work would be applying the HIAE model on resting state fMRI data to investigate the intrinsic hierarchical functional difference between gyri and sulci.

References

1. Park, H.J., Friston, K.: Structural and functional brain networks: from connections to cognition. Science **342**(6158), 1238411 (2013)
2. Jiang, X., et al.: Sparse representation of HCP grayordinate data reveals novel functional architecture of cerebral cortex. Hum. Brain Mapp. **36**(12), 5301–5319 (2015)
3. Jiang, X., et al.: Temporal dynamics assessment of spatial overlap pattern of functional brain networks reveals novel functional architecture of cerebral cortex. IEEE Trans. Biomed. Eng. **65**(6), 1183–1192 (2016)
4. Liu, H., et al.: Elucidating functional differences between cortical gyri and sulci via sparse representation HCP grayordinate fMRI data. Brain Res. **1672**, 81–90 (2017)
5. Liu, H., et al.: The cerebral cortex is bisectionally segregated into two fundamentally different functional units of gyri and sulci. Cereb. Cortex **29**(10), 4238–4252 (2019)
6. Betzel, R.F., Bassett, D.S.: Multi-scale brain networks. Neuroimage **160**, 73–83 (2017)
7. Hasson, U., Yang, E., Vallines, I., Heeger, D.J., Rubin, N.: A hierarchy of temporal receptive windows in human cortex. J. Neurosci. **28**(10), 2539–2550 (2008)
8. Huang, H., et al.: Modeling task fMRI data via deep convolutional autoencoder. IEEE Trans. Med. Imaging **37**(7), 1551–1561 (2017)
9. Barch, D.M., et al.: Function in the human connectome: task-fMRI and individual differences in behavior. Neuroimage **80**, 169–189 (2013)
10. Van Essen, D.C., et al.: The WU-Minn human connectome project: an overview. Neuroimage **80**, 62–79 (2013)
11. Glasser, M.F., et al.: WU-Minn HCP Consortium: the minimal preprocessing pipelines for the human connectome project. Neuroimage **80**, 105–124 (2013)
12. Odena, A., Dumoulin, V., Olah, C.: Deconvolution and checkerboard artifacts. Distill **1**(10), e3 (2016)
13. Kingma, D.P., Ba, J.: Adam: a method for stochastic optimization. arXiv preprint arXiv:1412. 6980 (2014)
14. Sang, Y.F., Wang, D., Wu, J.C., Zhu, Q.P., Wang, L.: Wavelet-based analysis on the complexity of hydrologic series data under multi-temporal scales. Entropy **13**(1), 195–210 (2011)
15. Smith, S.M., et al.: Correspondence of the brain's functional architecture during activation and rest. Proc. Natl. Acad. Sci. **106**(31), 13040–13045 (2009)
16. Deng, F., et al.: A functional model of cortical gyri and sulci. Brain Struct. Funct. **219**(4), 1473–1491 (2013)
17. Buzsáki, G., Logothetis, N., Singer, W.: Scaling brain size, keeping timing: evolutionary preservation of brain rhythms. Neuron **80**(3), 751–764 (2013)

Personalized Matching and Analysis of Cortical Folding Patterns via Patch-Based Intrinsic Brain Mapping

Jiong Zhang and Yonggang Shi[✉]

USC Stevens Neuroimaging and Informatics Institute, University of Southern
California (USC), Los Angeles, CA 90033, USA
yshi@loni.usc.edu

Abstract. The temporal cortex is one of the earliest regions with tau
pathology and associated gray matter atrophy in Alzheimer's disease
(AD). Surface mapping has conventionally been widely used to provide
one-to-one correspondences and hence the detection of thickness changes
in the cortical ribbon. The presence of very different topography of the
sulcal and gyral folds across subjects, however, makes it challenging to
have meaningful sulcus-to-sulcus and gyrus-to-gyrus matching. This is
critical for the quantification of thickness changes because sulcal and
gyral areas have different thickness profiles. In this paper, we propose
a novel framework for personalized and localized cortical folding pattern analysis to address this challenge. Given a pair of source and target
patches, intrinsic surface mapping based on Riemannian metric optimization on surfaces (RMOS) is first employed to compute the fine-grained
maps. Afterwards, we design an edge-distortion based pattern matching
method to detect locally well-matched folding patterns between temporal cortical patches. A patch-based similarity measure is then defined
to establish a personalized atlas set for each individual source patch.
Finally, a personalized z-score map is computed for normality assessment in disease groups and the detection of atrophy with respect to the
normal controls. The proposed framework is validated on a large-scale
dataset from the Alzheimer's Disease Neuroimaging Initiative (ADNI)
to demonstrate the effectiveness of the proposed framework for personalized analysis and increased power in the detection of atrophy in AD
and mild cognitive impairment (MCI).

Keywords: Cortical folding patterns · Personalized matching ·
Surface mapping · Brain atrophy

1 Introduction

Alzheimer's disease (AD) is the most prevalent form of dementia that causes
cognitive impairment with memory deficits and earlier neuropathological alter-

This work was supported by the National Institute of Health (NIH) under
grants RF1AG056573, RF1AG064584, R01EB022744, R21AG064776, R01AG062007,
P41EB015922, P30AG066530.

M. de Bruijne et al. (Eds.): MICCAI 2021, LNCS 12907, pp. 710–720, 2021.
https://doi.org/10.1007/978-3-030-87234-2_67

ations [9]. Many studies [8,11] have shown that the temporal region is one of the sub-regions that is particularly damaged in AD and may lead to a predictable pattern of brain atrophy. To quantify brain changes, surface mapping techniques have been widely applied in different cortical regions to establish detailed one-to-one correspondence for tracking geometric variations [10,13,15]. This requires precise local correspondences of the cortical folding patterns in order to achieve accurate detection of sensitive changes in respective cortical regions such as the temporal cortex.

However, limited by the large cortical folding pattern variations across subjects, surface mapping techniques have difficulty in finding anatomically meaningful (*sulcus-to-sulcus* and *gyrus-to-gyrus*) correspondences for many local patterns. This will be especially problematic for the analysis and detection of pathological variations during disease progression. Many studies have thus focused on computing thickness measures over a large cortical area, e.g., FreeSurfer parcellations, in group comparisons due to this difficulty in obtaining local thickness measures with comparable sulcal/gyral anatomy across subjects. As shown in Fig. 1, the large variations of the temporal folding patterns between different subjects make it challenging to find perfect *sulcus-to-sulcus* and *gyrus-to-gyrus* pattern matching. Many of the gyral patterns on one surface might be matched to the sulcal patterns of another surface after standard mapping. In particular, we can observe from Fig. 1(a) that there exist very obvious cortical thickness differences between sulcal areas (2.65 ± 0.52 mm) and gyral areas (3.30 ± 0.74 mm) in the temporal region of a normal control subject. Thus, the intrinsic thickness differences could be amplified between mismatched sulcal and gyral areas, which can obscure the minor brain atrophy changes at the early stage of AD.

To explore representative cortical folding patterns, Li *et al.* [2] proposed a method to group folding patterns into different clusters for more precise analysis and applied it to different cortical regions such as the superior temporal gyrus and precuneus. Gahm *et al.* [7] set up a framework to develop a distributed atlas of the transentorhinal cortex to reduce anatomical misalignment by mapping only between similar patches. Despite several research attempts to study subtypes of cortical folding patterns, there is still a strong need for establishing a general framework by considering more personalized and localized cortical folding pattern analysis of different brain regions.

In this work, we propose a general surface mapping and pattern matching framework for personalized analysis of local cortical folding patterns. A dedicated patch-based surface mapping technique is employed to provide detailed one-to-one correspondences between source and target temporal patches, and a distortion based pattern matching method is designed to areas with similar folding patterns. A similarity degree is defined to establish a personalized atlas set for each individual source patch by selecting its most similar temporal patches. This will more accurately describe the local changes by comparing with the similar patterns within the atlas set. Finally, a map of $\mathcal{Z}$-score based normality assessment is defined to quantify the personalized deviations of pathological cases to the normally distributed atlas set. By using the large-scale Alzheimers

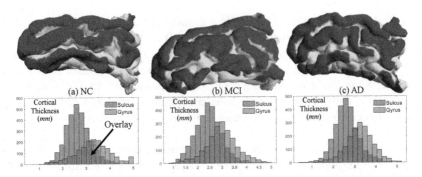

Fig. 1. Cortical thickness differences between temporal sulcus and gyrus respectively in (a) NC, (b) AD and (c) mild cognitive impairment (MCI) subjects. Top row: sulcal (yellow) and gyral (red) areas on the temporal patches. Bottom row: histogram of thickness measures in sulcal and gyral areas. Dark brown area indicates the overlapping transparency. (Color figure online)

Disease Neuroimaging Initiative (ADNI) [12] dataset, we demonstrate the effectiveness of the proposed framework for describing the personalized and localized variation of AD and MCI individuals with respect to their atlas set of NC subjects. We show that our method can significantly improve the statistical power in characterizing the cortical atrophy of temporal regions at the early stage of AD.

2   Methodology

In this section, we provide the technical details of our novel framework for localized matching and shape analysis of cortical folding patterns in the temporal cortex. We will firstly explain the automatic patch generation of the temporal cortex followed by the intrinsic RMOS mapping of patches. Afterwards, we will define the folding pattern similarity measure between patches by considering their edge-based distortions. Then the locally matched patterns between patches are obtained based on the similarity ratio. Finally, we will perform personalized normality assessment for both AD and mild cognitive impairment (MCI) groups by comparing to similar patterns in the NC group.

2.1   3D Surface Reconstruction and Automatic Patch Extraction

To automatically extract patches of the temporal cortex, a continuous cortical surface representation $\mathcal{M} = \{\mathcal{V}, \mathcal{T}\}$ is firstly constructed by applying FreeSurfer reconstruction on T1-weighted MRI [1], where $\mathcal{V} = \{\mathcal{V}_i | i = 1, \cdots, N_\mathcal{V}\}$ and $\mathcal{T} = \{\mathcal{T}_j | j = 1, \cdots, N_\mathcal{T}\}$ denote the set of vertices and triangles, respectively. Meanwhile, cortical surface parcellation [4] are also employed to divide cerebral cortex into different patch regions. Based on FreeSurfer's cortical parcellation results ($aparc+aseg$), we aim to extract the primary temporal cortex by combing neighboring regions such as the superior temporal gyrus, inferior temporal gyrus, middle temporal gyrus, temporal pole, banks of superior temporal sulcus, transverse temporal, fusiform gyrus, entorhinal cortex, and parahippocampal gyrus to ensure a continuous patch representation. As a result, the temporal patch for each subject is generated and then it is decimated to around 5000 vertices for computational efficiency. In Fig. 1, we have shown that large cortical folding variations across subjects are existed in the temporal region. As such, precise local pattern matchings and reliable feature comparisons between different groups become challenging. To better understand the local variations across subjects, the dedicated RMOS mapping technique [6] will be specifically applied as an initial step on temporal patches for personalized folding pattern analysis.

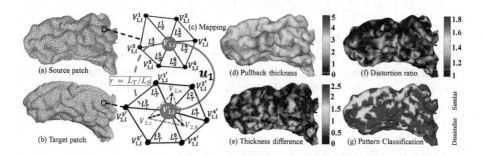

Fig. 2. Illustration of patch-based RMOS mapping and pattern matching between (a) source and (b) target patches; The mapping from a source vertex to a projected target vertex and edge distortions are illustrated in (c); (d) gives the pullback thickness on the source surface and (e) shows the thickness difference map between source and target patches; (f) represents the distortion ratios after surface mapping and (g) provides the classification map of similar and dissimilar temporal folding patterns. (Color figure online)

2.2   Patch-Based RMOS Mapping and Edge Distortion Metric

Riemannian metric optimization on surfaces (RMOS) mapping [5,6] has been designed to establish detailed one-to-one correspondences in the Laplace–Beltrami embedding space across different cortical surfaces. It is intrinsic to surface geometry and generally applicable to both cortical and subcortical structures. Based on the extracted temporal patches from the whole cortical surfaces, we take advantage of the RMOS framework to achieve a fine mapping between patches. The basic procedure of the patch-based RMOS mapping on temporal patches is given as follows. Given the triangular mesh representation of two temporal patches $\mathcal{M}_1$ and $\mathcal{M}_2$, as well as their corresponding edge weights, the Laplace–Beltrami eigen-system of the mesh surfaces $\mathcal{M}$ with the set of eigenfunctions $\Phi = \{f_0, f_1, f_2, \dots\}$ defines the LB embeddings $I_{\mathcal{M}}^{\Phi} : \mathcal{M} \mapsto \mathbb{R}^{\infty}$, which can be written as

$$I_{\mathcal{M}}^{\Phi}(x) = (\frac{f_1(x)}{\sqrt{\lambda_1}}, \frac{f_2(x)}{\sqrt{\lambda_2}} \cdots, \frac{f_n(x)}{\sqrt{\lambda_n}}, \cdots) \quad \forall x \in \mathcal{M} , \tag{1}$$

where λ_n is the n-th eigenvalue of the eigenfunction f_n.

The energy function for feature-driven surface mapping with patch-based RMOS is defined as $E = E_F + \gamma E_R$, where E_F is the data fidelity term for matching the provided features and E_R represents the regularization term. γ provides the weight between them. Given $\xi_1^j : \mathcal{M}_1 \mapsto \mathbb{R}$ and $\xi_2^j : \mathcal{M}_2 \mapsto \mathbb{R}$ representing the L feature functions $(j = 1, \cdots, L)$ respectively defined on the two temporal patch surfaces, the data term E_F defined with the patch map from $\mathcal{M}_1$ to $\mathcal{M}_2$, i.e., $u_1 : \mathcal{M}_1 \to \mathcal{M}_2$, and the inverse map from $\mathcal{M}_2$ to $\mathcal{M}_1$, i.e., $u_2 : \mathcal{M}_2 \to \mathcal{M}_1$ is given by:

$$E_F(\mathcal{M}_1, \mathcal{M}_2) = \sum_{j=1}^{L} \left[\int_{\mathcal{M}_1} (\xi_1^j - \xi_2^j \circ u_1)^2 d\mathcal{M}_1 + \int_{\mathcal{M}_2} (\xi_2^j - \xi_1^j \circ u_2)^2 d\mathcal{M}_2 \right] . \tag{2}$$

The matching of the LB embeddings is iteratively updated to achieve the energy minimization of E in a gradient descent way. For more details, refer to [6]. In this work, we propose to use the geometric thickness feature $\hat{f}$ generated from FreeSurfer as the data term for the iterative matching. In Fig. 2, we provide a typical example of the patch-based RMOS mapping on two temporal surfaces (Fig. 2(a) and (b)). The one-to-one vertex mapping from the source to target surface is shown in Fig. 2(c).

In the RMOS maps u_1 and u_2, each point in one patch is discretized as a linear combination of vertex positions in the other patch. For a given vertex $\mathcal{V}_{1,i}$ on the source patch $\mathcal{M}_1$, its map onto the target patch $\mathcal{M}_2$ is within a triangle $\mathcal{T}_{2,j} \in \mathcal{T}_2$ with three vertices $(\mathcal{V}_{2,a}, \mathcal{V}_{2,b}, \mathcal{V}_{2,c})$, as shown in Fig. 2(c). Thus, the mapped vertex position $\mathcal{V}'_{1,i}$ of $\mathcal{V}_{1,i}$ on $\mathcal{M}_2$ can be written as $\mathcal{V}'_{1,i} = \alpha \mathcal{V}_{2,a} + \beta \mathcal{V}_{2,b} + \gamma \mathcal{V}_{2,c}$, and the pull-back feature an be represented as $\hat{f}_2(u_1(\mathcal{V}_{1,i})) = \alpha \hat{f}_2(\mathcal{V}_{2,a}) + \beta \hat{f}_2(\mathcal{V}_{2,b}) + \gamma \hat{f}_2(\mathcal{V}_{2,c})$, where the triplet (α, β, γ) is the linear operator denoted by the RMOS map u_1. Both the vertices and edges have their one-to-one corresponding deformations after RMOS mapping. Such changes can be used to describe the local folding similarity between two patch surfaces. Therefore, we employ the edge distortions after RMOS mapping to represent the degree of local expansion or shrinkage from the source patch to the target patch and vice versa. As shown in Fig. 2(c), the local edge lengths at the surrounding of vertex $\mathcal{V}_{1,i}$ on the source patch $\mathcal{M}_1$ can be computed as the ℓ_2-norm of the vertex position differences, i.e., $L_S^h = |\mathcal{V}_{1,i} - \mathcal{V}_{1,i}^h|_{\ell_2}$, with $h = \{1, 2, \ldots, N_h\}$, where N_h denotes the number of neighboring vertices. Similarly, we compute the corresponding local edge lengths in the neighborhood of the vertex $\mathcal{V}'_{1,i}$ on the target $\mathcal{M}_2$ as $L_T^h = |\mathcal{V}'_{1,i} - \mathcal{V}_{1,i}^{h'}|_{\ell_2}$. The edge distortion ratio of local surface expansion after RMOS mapping can be defined by $r_e = L_T/L_S > 1$. Note that both surface expansion (with $L_S > L_T$) and shrinkage (with $L_S < L_T$) should be treated equally since we only focus on the degrees of local variations after RMOS mapping. Thus, the edge distortion ratio of local surface shrinkage is defined by $r_s = 1/r_e > 1$. The most similar folding patterns will have $r_e = L_T/L_S \approx 1$

with almost no local surface changes. The edge distortion ratios from the source patch $\mathcal{M}_1$ to the target patch $\mathcal{M}_2$ are given by

$$r_i = \frac{1}{N_h} \sum_{j=1}^{N_h} r_i^j, \text{ with } \begin{cases} r = r_e, & \text{if } r_e > 1, \\ r = r_s, & \text{if } r_e < 1. \end{cases} \tag{3}$$

2.3   Localized Cortical Folding Pattern Matching

In this subsection, we aim to localize the similar cortical folding patterns based on the distortion ratios r_i obtained from the mapping between source and target vertices. A distortion ratio of 1 indicates the most similar folding patterns. The larger the distortion ratio, the lower the shape similarity of the one-to-one correspondences. In our work, we will develop personalized analysis of brain atrophy by focusing on highly similar folding patterns as measured by the distortion ratios. We will use a relatively strict criterion in this work by setting a threshold value of $T_r = 1.1$ to separate the similar and dissimilar patterns. The degree of the similarity θ between the source and target is defined as the fraction of vertices satisfying this criterion.

To demonstrate the matching process, in Fig. 2 we show an example between a source patch and a target patch for similarity localization. In Fig. 2(a)–(b) we can observe the large folding pattern variations for both gyrus and sulcus. Figure 2(f)–(g) show the edge distortion ratio and the binarized patch surfaces with similar pattern (in yellow color) and dissimilar pattern (in red color). From Fig. 2(e) and (g) the thickness differences, we can roughly observe that most of the similar patterns present relatively smaller thickness differences, which are useful for the detection of sensitive pathological changes. In contrast, those dissimilar patch regions with large thickness differences may obscure the minor brain atrophy and bring confounding factors to the analysis of neurodegeneration.

2.4   Personalized Normality Assessment of Abnormal Subjects

To explore the potential of using the proposed local pattern matching framework for more precise AD study, we develop a new approach for characterizing the personalized variation of a MCI/AD individual by mapping it to an atlas set of NC subjects. More specifically, for each MCI/AD subject, we will treat it as a source patch and all NC subjects as target patches. Afterwards, our pattern matching is employed to find the most similar NC subjects as a personalized atlas set for the source patch. Hence, each individual pathological case will have an independent atlas set. To describe how likely an individual MCI/AD subject deviates from its normally distributed atlas, the $\mathcal{Z}$-score, also known as the standard deviation score is used to measure the thickness distance between the individual and its atlas population. The absolute $\mathcal{Z}$-score value of a source vertex is given by $\mathcal{Z}_i = |(F_{\mathcal{S}}^i - \mu_i)/\sigma_i|$ with $i = \{1, \cdots, N_\mathcal{V}\}$, where $F_{\mathcal{S}}^i$ represents the thickness of vertex $\mathcal{V}_{1,i}$ of the source patch, μ_i and σ_i are the mean and standard deviation of the thickness measures of corresponding vertices from the target patches belonging to the personalized atlas set from the NC group. In Fig. 3, we show the $\mathcal{Z}$-score distributions of both MCI and AD individuals based on the standard mapping, which uses the corresponding vertices of all subjects, and the proposed pattern matching based on vertices from a personalized atlases set. From Fig. 3(d) and (h) we can observe higher $\mathcal{Z}$-score deviations from the personalized NC atlas compared to the standard mapping shown in Fig. 3(b) and (f). In our case, larger absolute $\mathcal{Z}$-score values indicate higher deviation of the

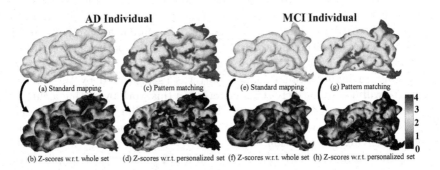

AD Individual   **MCI Individual**

(a) Standard mapping   (c) Pattern matching   (e) Standard mapping   (g) Pattern matching

(b) Z-scores w.r.t. whole set   (d) Z-scores w.r.t. personalized set   (f) Z-scores w.r.t. whole set   (h) Z-scores w.r.t. personalized set

Fig. 3. Comparison of $\mathcal{Z}$-score distribution based on standard mapping and the proposed pattern matching method. In (c) and (g), only vertices labeled with yellow color were considered into the personalized normality assessment. (Color figure online)

data from the NC group and hence more powerful for detecting changes due to disease. However, the NC altas sets having higher thickness standard deviations (without pattern matching) will produce smaller $\mathcal{Z}$-score values (closer to 0) for the MCI/AD individuals. This makes it hard to distinguish the individuals from the altas distributions, and thus it increases the challenges for classification. In the experimental section, we will use the $\mathcal{Z}$-score based normality assessment to compare the localized pattern matching with conventional analysis based on standard surface mapping to demonstrate the benefits of developing personalized and localized analysis.

3   Experimental Results

In this section, we will evaluate the proposed framework using the T1-weighted MRI scans of 200 NC, 50 MCI and 46 AD subjects from the clinical ADNI dataset. All the temporal patches and their corresponding cortical thickness are initially generated using Freesurfer [3]. The diagnostic criteria in ADNI was previously described and informed written consent was obtained from all participants at each site [14].

3.1 Patch-Based and Region-Based Analysis for Temporal Pattern Matching

To validate if the pattern matching is useful to the thickness comparisons across subjects, we first perform an exploratory analysis over the 200 NC subjects at surface patch level. By randomly selecting one subject as source and the remaining 199 subjects as targets, the pattern matching approach is applied to detect similar temporal folding patterns between the source and each of the target patches. Afterwards, 50 of the most similar and dissimilar temporal patches are separately selected based on the similarity degree θ defined in Sect. 2.3. Vertex-wise thickness difference F^i are calculated by taking the absolute value of the thickness difference between the source patch F_S^i and each of the projected target patches $F_T^{m,i}$, i.e., $F^{m,i} = |F_S^i - F_T^{m,i}|$, where $F_T^{m,i}$ represents the thickness feature of vertex $\mathcal{V}_{1,i}'$ on the m_{th} projected target patch with $m = \{1, \cdots, 50\}$. Afterwards, the mean thickness difference F_μ^m of each source and target pair is obtained by averaging $F^{m,i}$ over all vertices $\mathcal{V}_{1,i}$ on the source patch.

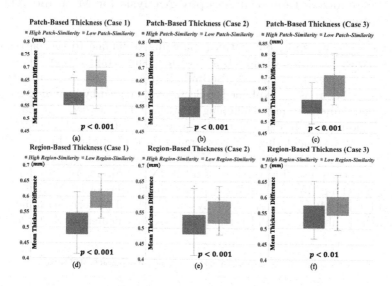

Fig. 4. Statistical analysis of mean thickness difference at patch and region levels.

Statistical analysis is performed using Wilcoxon rank sum to compare the mean patch thickness differences F_μ^m of the similar and dissimilar groups. In Fig. 4, we show the statistical results on three random cases with their respectively selected patch as source and the remained 199 patches as targets for pattern matching. In Fig. 4(a)–(c), we can observe that F_μ^m present very significant patch-based thickness differences between the similar and dissimilar group of each subject (with $p < 0.001$). The group with high temporal folding pattern similarity presents much smaller thickness differences compared with the low similarity group. This suggests that without considering the high local folding pattern variations and the limitations in surface mapping, large thickness differences due to mismatched comparisons may obscure minor changes caused by pathological conditions. In addition, we perform a localized region-based analysis to compare the similar and dissimilar patterns within each temporal patch of the 50 most similar subjects. Therefore, the mean thickness difference F_μ^m of each patch will be separated into two parts, i.e., the mean thickness difference of similar regions $F_{\mu 1}^m$ and dissimilar regions $F_{\mu 2}^m$. Then, statistical analysis is performed on these two thickness measures across the 50 most similar subjects. In Fig. 4(d)–(f), strong group differences and statistical significance ($p < 0.001$) are obtained between the similar and dissimilar folding patterns. Regions with similar patterns present much smaller thickness differences compared to dissimilar patterns. These findings indicate that the proposed pattern matching framework is potentially useful for the reduction of confounding factors caused by large variations of temporal folding patterns across subjects.

3.2 Personalized Temporal Atrophy Analysis for MCI and AD Groups

In this section, we apply our pattern matching technique to the pathological MCI set (with 50 subjects) and AD set (with 46 subjects) to study its potential in achieving better personalized discrimination against the NC subjects. The personalized NC atlas set explained in Sect. 2.4 is defined by including 50 of the most similar subjects from the mapping between 200 NC subjects and each of the MCI/AD individuals. In Fig. 5, normality assessment of both pathological groups are performed by calculating a $\mathcal{Z}$-score maps for each subject. In the standard mapping method for comparison, the $\mathcal{Z}$-score calculation at each vertex uses the thickness at corresponding vertices of all 200NC subjects. With our localized pattern matching (by establishing a personalized atlas of 50 NC subjects for each individual MCI/AD case), we only consider local patterns with more than five matched subjects as valid vertices for $\mathcal{Z}$-score calculation. Statistical tests

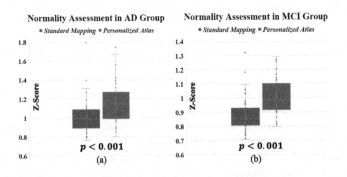

Fig. 5. Statistical analysis of Z-score based normality analysis by automatically generating a personalized atlas from NC subjects for each MCI/AD individual.

are performed on the mean value of each $\mathcal{Z}$-score distribution in the MCI/AD group. We can observe from Fig. 5 that consistently higher normality scores are obtained by our localized pattern matching with significant differences to results from the standard mapping method. This implies the possibility of extending the proposed framework for early pathological detection and classification of AD.

4   Conclusion

In this work, we have proposed a personalized and localized mapping and pattern matching framework to solve the practical challenges of large cortical folding pattern variations in atrophy detection. The proposed framework is applied to the temporal cortex and shows higher statistical power in detecting cortical thickness changes in pathological MCI/AD conditions. The validation also indicates the high potential of extending our method to study cortical folding patterns in other brain regions.

References

1. Dale, A.M., Fischl, B., Sereno, M.I.: Cortical surface-based analysis: I. segmentation and surface reconstruction. Neuroimage **9**(2), 179–194 (1999)
2. Duan, D., et al.: Exploring folding patterns of infant cerebral cortex based on multi-view curvature features: methods and applications. Neuroimage **185**, 575–592 (2019)
3. Fischl, B., Sereno, M.I., Dale, A.M.: Cortical surface-based analysis: Ii: inflation, flattening, and a surface-based coordinate system. Neuroimage **9**(2), 195–207 (1999)
4. Fischl, B., et al.: Automatically parcellating the human cerebral cortex. Cerebral Cortex **14**(1), 11–22 (2004)
5. Gahm, J.K., Shi, Y.: Riemannian metric optimization for connectivity-driven surface mapping. In: Ourselin, S., Joskowicz, L., Sabuncu, M.R., Unal, G., Wells, W. (eds.) MICCAI 2016. LNCS, vol. 9900, pp. 228–236. Springer, Cham (2016). https://doi.org/10.1007/978-3-319-46720-7_27
6. Gahm, J.K., Shi, Y., Initiative, A.D.N., et al.: Riemannian metric optimization on surfaces (RMOS) for intrinsic brain mapping in the Laplace-Beltrami embedding space. Med. Image Anal. **46**, 189–201 (2018)
7. Gahm, J.K., Tang, Y., Shi, Y.: Patch-based mapping of transentorhinal cortex with a distributed atlas. In: Frangi, A.F., Schnabel, J.A., Davatzikos, C., Alberola-López, C., Fichtinger, G. (eds.) MICCAI 2018. LNCS, vol. 11072, pp. 689–697. Springer, Cham (2018). https://doi.org/10.1007/978-3-030-00931-1_79
8. Holland, D., Brewer, J.B., Hagler, D.J., Fennema-Notestine, C., Dale, A.M., Initiative, A.D.N., et al.: Subregional neuroanatomical change as a biomarker for alzheimer's disease. Proc. Nat. Acad. Sci. **106**(49), 20954–20959 (2009)
9. Jack, C.R., Jr., et al.: NIA-AA research framework: toward a biological definition of alzheimer's disease. Alzheimer's Dementia **14**(4), 535–562 (2018)
10. Li, G., et al.: Mapping region-specific longitudinal cortical surface expansion from birth to 2 years of age. Cerebral Cortex **23**(11), 2724–2733 (2013)
11. Miller, M.I., et al.: The diffeomorphometry of temporal lobe structures in preclinical alzheimer's disease. NeuroImage Clin. **3**, 352–360 (2013)

12. Mueller, S.G., et al.: The alzheimer's disease neuroimaging initiative. Neuroimaging Clin. **15**(4), 869–877 (2005)
13. Nie, J., Guo, L., Li, G., Faraco, C., Miller, L.S., Liu, T.: A computational model of cerebral cortex folding. J. Theor. Biol. **264**(2), 467–478 (2010)
14. Petersen, R.C., et al.: Alzheimer's disease neuroimaging initiative (ADNI): clinical characterization. Neurology **74**(3), 201–209 (2010)
15. Thompson, P.M., et al.: Mapping hippocampal and ventricular change in alzheimer disease. Neuroimage **22**(4), 1754–1766 (2004)

Clinical Applications - Oncology

Clinical Applications: Oncology

A Location Constrained Dual-Branch Network for Reliable Diagnosis of Jaw Tumors and Cysts

Jiacong Hu[1], Zunlei Feng[1,4(✉)], Yining Mao[1], Jie Lei[2], Dan Yu[3], and Mingli Song[1,4]

[1] Zhejiang University, Hangzhou, China
zunleifeng@zju.edu.cn
[2] Zhejiang University of Technology, Hangzhou, China
[3] The First Affiliated Hospital, Zhejiang University School of Medicine, Hangzhou, China
[4] Zhejiang Provincial Key Laboratory of Service Robot, College of Computer Science, Zhejiang University, Hangzhou, China

Abstract. The jaw tumors and cysts are usually painless and asymptomatic, which poses a serious threat to patient life quality. Proper and accurate detection at the early stage will effectively relieve patients' pain and avoid radical segmentation surgery. However, similar radiological characteristics of some tumors and cysts bring challenges for accurate and reliable diagnosis of tumors and cysts. What's more, existing transfer learning based classification and detection methods for diagnosis of tumors and cysts have two drawbacks: a) diagnosis performance of the model is highly reliant on the number of lesion samples; b) the diagnosis results lack reliability. In this paper, we proposed a Location Constrained Dual-branch Network (LCD-Net) for reliable diagnosis of jaw tumors and cysts. To overcome the dependence on a large number of lesion samples, the features extractor of LCD-Net is pretrained with self-supervised learning on massive healthy samples, which are easier to collect. For similar radiological characteristics, the auxiliary segmentation branch is devised for extracting more distinguishable features. What's more, the dual-branch network combined with the patch-covering data augmentation strategy and localization consistency loss is proposed to improve the model's reliability. In the experiment, we collect 872 lesion panoramic radiographs and $10,000$ healthy panoramic radiographs. Exhaustive experiments on the collected dataset show that LCD-Net achieves SOTA and reliable performance, which provides an effective tool for diagnosing jaw tumors and cysts.

Keywords: Jaw cysts and tumors · Panoramic radiography · Deep learning · Dual-branch framework · Self-supervised learning · Reliability

1 Introduction

Odontogenic tumors and cysts of the jaw are the second most common disease after tooth impaction in the oral and maxillofacial areas. Jaw tumors and cysts

© Springer Nature Switzerland AG 2021
M. de Bruijne et al. (Eds.): MICCAI 2021, LNCS 12907, pp. 723–732, 2021.
https://doi.org/10.1007/978-3-030-87234-2_68

are usually painless and asymptomatic unless they grow so large as to involve the entire jawbone, causing noticeable swelling or weakening it to cause pathologic fractures [9,14]. Those manifested symptoms pose a severe threat to patient life quality. The majority of these cyst and tumor lesions can be identified at an earlier stage through a routine radiographic exam called the panoramic radiograph or orthopantomogram [6]. The treatment modalities for different types of tumors and cysts are different. Precise preoperative diagnosis of these tumors and cysts can help oral and maxillofacial surgeons plan appropriate treatment. What's more, early diagnosis also can reduce morbidity and mortality through long-term follow-up and early intervention.

Accurate diagnosis of different types of tumors and cysts is a challenging task. On the one hand, panoramic radiographs are captured by a sensor/plate that rotates around the patients head, causing the superimposition of all the bony structures of the facial skeleton. The superimposition characteristic brings difficulty to accurate diagnosis, even for experienced professionals. On the other hand, some tumors and cysts have very similar radiological characteristics. For example, ameloblastomas and other cystic lesions in the orofacial areas have similar radiological characteristics [2]. Those similar radiological characteristics may lead to misdiagnose occasionally. So, an accurate and reliable auxiliary diagnostic method for jaw tumors and cysts is significant.

Recently, deep learning approaches have achieved promising results in the medical image analysis area [1,8,10]. Inspired by the successful application of deep learning, several works [3,11,12,15,23] adopted Convolutional Neural Network (CNN) to diagnose radiolucent lesions of the jaw. Poedjiastoeti et al. [15] first applied CNN to detect jaw tumors based on the pretrained VGG-16 [21] with ImageNet [7]. Lee et al. [12] also adopted GoogLeNet Inception-v3 [22] to diagnose cystic lesions using panoramic and cone beam computed tomographic images. Unlike CNN-based classification methods, Ariji et al. [3] adopted deep learning object detection technique to detect and classify radiolucent lesions in the mandible. Kwon et al. [11] adopted a deep CNN modified from YOLOv3 [17] for detecting and classifying odontogenic jaw tumors and cysts. Yang et al. [23] also adopted the deep convolutional neural network YOLOv2 [16] for detecting tumors and cysts of the jaw on panoramic radiographs.

However, the above deep learning-based methods' performance heavily relies on a large number of labeled datasets. Existing CNN-based methods [12,15] first pre-train the whole classification network on ImageNet [7], and then finetune the network on several hundreds of lesion samples. The domain difference between normal image dataset ImageNet [7] and medical radiographs is huge, which heavily reduces the robustness and performance of the above transfer learning based methods [12,15]. On the other hand, detection based methods [3,11,23] adopted YOLO [16,17] to detect lesion areas and classify the type of tumors and cysts, which achieves good detection results but poor classification performance.

On the other hand, the reliability and interpretability of auxiliary diagnostic methods is an important and essential factor for diagnosing tumors and cysts. Poedjiastoeti et al. [15] adopted gradient based class activation mappings

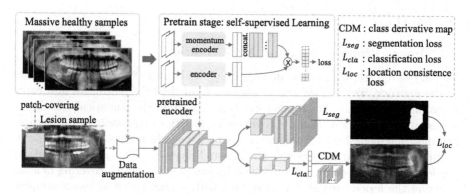

Fig. 1. The flow diagram of LCD-Net. Self-supervised learning is firstly adopted to pretrain an encoder on massive healthy samples. Then, the two branches are trained with segmentation loss L_{seg} and classification loss L_{cla} on the data augmented samples. The two branches share the same pretrained encoder. The location consistency loss L_{loc} is devised for constraining the consistency of lesion locations predicted by segmentation and classification branch, which enhances the reliability of predicted results.

(Grad-CAM) [20] to generate heatmap visualizations of the areas of each digital panoramic radiographic image. However, Grad-CAM usually show some unrelated areas, which indicates that the interpretability of CAMs is not that reliable. The poor classification results of detection based methods [3,11,23] are unpractical for the diagnosis of jaw cysts and tumors.

In this study, we proposed a Location Constrained Dual-branch Network (LCD-Net) for reliable diagnosis of jaw tumors and cysts. As shown in Fig. 1, LCD-Net is comprised of a classification branch and a segmentation branch that share the same encoder. To relieve the requirement of massive lesion samples, the self-supervised learning framework is adopted to pretrain the feature extractor with massive healthy samples. Then, the patch-covering data augmentation strategy, which randomly covers the lesion and healthy area with a patch, is designed to increase the diversity of lesion samples. Furthermore, we devise the location consistency loss function for constraining the area associated with the predicated category inside or near the lesion mask, which can improve the reliability of classified results. In the inference stage, LCD can synchronously predict the category and lesion mask of samples.

Our main contribution can be summarized as follows: 1) We propose a dual-branch network for synchronously predicting the category and lesion mask of samples, which can provide reliable evidence of diagnosis results. 2) The patch-covering data augmentation strategy and the localization consistency loss are devised to improve the diversity of lesion samples and constrain the location consistency between the classification and segmentation branches, which both can effectively improve the reliability of predicted results. 3) LCD-Net achieves SOTA performance with limited lesion training samples through pretraining on massive healthy samples.

2   Method

2.1   Image Preprocessing and Augmentation

The size of the original panoramic radiograph is about 3000×1500. To fit the size of the network, we crop the center patches from the original images and resize them into 512×256. The cropped size is set based on the statistical calculation of lesion area position. Based on the characteristics of panoramic radiograph, the data augmentation strategies we adopted include horizontal flipping, cut-and-pasting and patch-covering. The cut-and-pasting strategy denotes cutting the lesion area and pasting it on a healthy sample. The patch-covering denotes covering the lesion area and healthy area with a gray patch of lesion samples, which can augment diversity of lesion samples. We do a survey of the lesion area size on all lesion samples, which gives the minimal and maximal size of lesion area. For the lesion area of a lesion sample, the cover-patch size is randomly generated between the size of the lesion area and the maximal size. For the healthy area of a lesion sample, the cover-patch size is randomly generated between the minimal size and the maximal size. In our experiment, each lesion sample will be covered with 20 patches, where half the patches are generated for covering the lesion area and the rest half patches are used for covering the healthy area of the lesion sample. it's worth noting that for the same lesion sample, the patch-covering can generate very similar samples, which is an advantage for enhancing the reliability of predicted results.

2.2   Location Constrained Dual-Branch Network

Figure 1 gives the flow diagram of LCD-Net. The training of LCD contains two stages: the self-supervised pretrain stage and the two-branch network training stage. In the self-supervised pretrain stage, the encoder is trained using self-supervised learning with massive healthy samples. Based on the pretrained encoder, the dual-branch framework is trained with the data augmented samples.

Self-supervised Learning. To relieve the requirement on the massive lesion samples, we adopted self-supervised MoCoV2 [5] for training the feature extractor. With the pretrain on massive healthy samples, the encoder has the basic capability for extracting features from the panoramic radiograph of the jaw, which avoids the domain gap caused by transfer learning based methods.

Dual-Branch Network. As shown in Fig. 1, LCD-Net contains segmentation and classification branches. The two branches share the same encoder, which has been pretrained with self-supervised learning. In the training stage, the standard cross-entropy loss is adopted to supervise the training of the classification branch. The multi-label classification loss L_{cla} is defined as follows:

$$\mathcal{L}_{cla} = \frac{1}{K} \sum_{k=1}^{K} y_k \log(p_k), p_k = \mathcal{F}_\theta(I_k), \tag{1}$$

where K is the total training samples, y_k is the ground truth of k-th sample and p_k is the predicted probability of k-th sample I_k with classification branch $\mathcal{F}_\theta$.

In order to assist the classification sub-branch in extracting more distinguishable features, we introduce the segmentation branch for segmenting the lesion area. What is noteworthy is that the segmentation branch predicts the lesion area, which increases the interpretability of the diagnosis. It's a benefit for the dentist to confirm the diagnosis result further. The segmentation loss $\mathcal{L}_{seg}$ of segmentation branch is defined as follows:

$$\mathcal{L}_{seg} = \frac{1}{K} \sum_{k=1}^{K} ||\widetilde{I_k^m} - I_k^m||_2^2, \widetilde{I_k^m} = \mathcal{F}_\phi(I_k), \tag{2}$$

where, I_k^m is the ground truth mask of k-th sample and $\widetilde{I_k^m}$ is the predicted mask of k-th sample I_k with segmentation branch $\mathcal{F}_\phi$.

Location Consistency Constraint. For enhancing the reliability of classified results, we devise the location consistency loss $\mathcal{L}_{loc}$. Intuitively, the lesion area predicted by the segmentation sub-network should be consistent with the area that classification branch has a high response. So, we calculate the Class Derivative Map (CDM) F' of the last convolution layer as follows:

$$F' = \sum_{t=1}^{T} sigmoid(f'_t), [f'_1, f'_2, ..., f'_T] = \mathcal{F}'_f(\mathcal{L}_{cla}), \tag{3}$$

where T is the feature map number of the last convolution layer, f'_t is the differential map of $\mathcal{L}_{cla}$ with respect to t-th feature map f_t. For the k-th input image I_k, the corresponding CDM is F'_k. So, the high response of CDM F'_k should be inside or around the lesion area. For the healthy sample, the high response of CDM F'_k has no restriction. So, the location consistency loss $\mathcal{L}_{loc}$ is formulated as follows:

$$\mathcal{L}_{loc} = \frac{1}{K} \sum_{k=1}^{K} \left\{ \frac{\sum_{n=1}^{N} \mathbf{1}(\overline{I_k^m}[n] = 0) * F'_k[n]}{\sum_{n=1}^{N} \mathbf{1}(\overline{I_k^m}[n] = 0)} - \frac{\sum_{n=1}^{N} \mathbf{1}(\overline{I_k^m}[n] = 1) * F'_k[n]}{\sum_{n=1}^{N} \mathbf{1}(\overline{I_k^m}[n] = 1)} \right\},$$
$$\overline{I_k^m} = \mathfrak{D}_r(\widehat{I_k^m}), I_k \in S, \tag{4}$$

where, N is the multiplication of width and height of the last layer feature map, where $\mathbf{1}(*)$ is the function that returns 1 if its input is true and 0 otherwise, $F'_k[n]$ denotes the n-th value of CDM F'_k, $\widehat{I_k^m}$ denotes the resized GT mask of k-th image I_k, the width and height of $\widehat{I_k^m}$ are the same with the last layer feature map, $\mathfrak{D}_r$ denote the dilation operation with disk strel of radius r, $\overline{I_k^m}$ denotes the dilated mask, S is the set of lesion samples. The location consistency loss $\mathcal{L}_{loc}$ is only used to update the parameters after the last convolution layer in the classification branch.

Based on the pretrained encoder, the dual-branch framework is trained with $\mathcal{L} = \mathcal{L}_{cla} + \alpha \mathcal{L}_{seg}$, where α is the balance parameter. The parameters after the last convolution layer of the classification branch are trained with $\mathcal{L}_{loc}$. The training with $\mathcal{L}$ and $\mathcal{L}_{loc}$ is iterative. In the testing stage, the two-branch network predicts the category of lesion samples and segments the lesion area. The segmented results can be used as the diagnosis reference for the dentist to further diagnose the jaw tumors and cysts.

3   Experiments

Dataset. In this study, we collect $10,000$ panoramic radiographs of healthy peoples and 872 lesion samples. The dentigerous cysts (DCs) and periapical cysts (PCs) occupy most odontogenic cyst lesions, which are benign and non-invasive. Since 2005, WHO has labeled odontogenic keratocysts (OKCs) as keratocystic odontogenic tumors (KCOTs) and has classified OKCs as tumors according to their behavior. Except for KCOTs, ameloblastomas (ABs) is another primary odontogenic tumor of the jaw. The collected 872 lesion samples contain 648 cyst samples (DCs: 356, PCs: 292) and 224 tumor samples (ABs: 94, KCOTs: 130). For each lesion sample, the experienced dentist annotates the lesion area mask and lesion category. In our experiment, healthy panoramic radiographs are split into 9500 samples for pretraining and training, 300 samples for validation and 200 samples for testing. For lesion samples, 70%, 20% and 10% samples are used for training, validation and testing, respectively.

Experiment Setting. The segmentation network architecture is the same as the Unet [19]. The classification branch shares the encoder with the segmentation network, and has an average pooling layer, 2048 neurons in the fully connection layer after the encoder. In the experiment, the balance parameter α is set to 1.0, and the radius r is a random value between 6 and 12. The learning rate for the classification and segmentation network are all set to be $1e^{-3}$ and $1e^{-2}$, respectively.

Metric. The classification metric contains the Accuracy (Acc.), Precious (Pre.), Sensitivity (Sen.), Specificity (Spe.) and F1-score (F1). The metric contains Pixel Accuracy (PA), Sen., Spe., and IoU for the evaluation of segmentation. The metric for detection contains Average Precision (AP), Pre., Sen. and IoU.

3.1   Quantitative Evaluation

To verify the effectiveness of the proposed method, we compare LCD-Net with all existing CNN based classification methods [12,15] and detection based methods [3,11,23] for diagnosing jaw tumors and cysts. Tables 1 and 2 give the quantitative evaluation of classification and detection. The CNN based classification methods [12,15] are firstly pretrained on ImageNet then are trained on the collected training datasets. From Table 1, we can see that the detection based methods [3,11,23] achieve the worst classification results. The CNN based classification methods [12,15] achieve better performance than those CNN based

Table 1. The classification performance. 'Acc.','Pre.','Sen.' ,'Spe.', and 'F1' denote Accuracy, Precious, Sensitivity (Recall), Specificity and F1-score, respectively.

	Ariji *et al.* [3]					Kwon *et al.* [11]					Yang *et al.* [23]				
Category	Acc.	Pre.	Sen.	Spe.	F1	Acc.	Pre.	Sen.	Spe.	F1	Acc.	Pre.	Sen.	Spe.	F1
DCs	72.03	53.85	49.30	81.82	51.47	69.79	49.09	38.57	83.03	43.20	70.09	50.00	34.29	85.37	40.68
PCs	71.19	22.92	26.19	80.93	24.44	74.89	33.33	40.48	82.38	36.56	74.36	32.00	38.10	82.29	34.78
ABs	85.17	18.52	27.78	89.91	22.22	85.11	16.00	22.22	90.32	18.60	84.62	17.86	27.78	89.35	21.74
KCOTs	83.05	22.22	24.00	90.05	23.08	82.98	24.14	28.00	89.52	25.93	80.77	13.79	16.67	88.10	15.09
Healthy	65.68	49.28	42.50	77.56	45.64	66.38	50.67	47.50	76.13	49.03	63.68	46.84	46.25	72.73	46.54
Means	75.42	33.36	33.95	84.05	33.37	75.83	34.65	35.35	84.28	34.66	74.70	32.10	32.62	83.57	31.77
	Poedjiastoeti *et al.* [15]					Lee *et al.* [12]					LCD-Net				
Category	Acc.	Pre.	Sen.	Spe.	F1	Acc.	Pre.	Sen.	Spe.	F1	Acc.	Pre.	Sen.	Spe.	F1
DCs	82.91	73.44	67.14	89.63	70.15	81.09	70.31	63.38	88.62	66.67	86.32	78.79	74.29	91.46	76.47
PCs	85.04	57.78	61.90	90.10	59.77	85.29	57.45	64.29	89.80	60.67	88.89	68.18	71.43	92.71	69.77
ABs	90.17	36.36	47.06	93.55	41.03	88.66	33.33	42.11	92.69	37.21	91.45	45.00	50.00	94.91	47.37
KCOTs	89.32	48.00	50.00	93.81	48.98	88.66	44.00	45.83	93.46	44.90	91.45	58.33	58.33	95.24	58.33
Healthy	80.77	73.08	70.37	86.27	71.70	79.83	71.79	68.29	85.90	70.00	85.47	78.75	78.75	88.96	78.75
Means	85.64	57.73	59.29	90.67	58.33	84.71	55.38	56.78	90.09	55.89	88.72	65.81	66.56	92.66	66.14

Table 2. The detection performance.

	Ariji *et al.* [3]				Kwon *et al.* [11]				Yang *et al.* [23]				LCD-Net			
Category	AP	Pre.	Sen.	IoU	AP	Pre.	Sen.	IoU	AP	Pre.	Sen.	IoU	AP	Pre.	Sen.	IoU
DCs	67.78	54.23	57.38	52.83	71.73	50.21	52.14	65.72	69.66	48.51	59.63	53.67	72.02	61.32	72.36	71.26
PCs	62.17	39.86	42.24	57.22	65.29	42.83	61.97	64.38	60.52	52.33	42.17	55.42	69.54	57.50	63.49	72.34
ABs	54.91	34.22	42.80	43.61	57.61	44.70	58.01	51.02	52.31	36.75	40.28	58.93	65.43	49.88	51.12	68.58
KCOTs	53.26	45.87	54.66	49.25	58.44	50.92	56.79	54.21	49.52	42.36	50.25	57.66	64.32	51.19	63.37	69.77
Means	59.53	43.55	49.27	50.72	63.27	47.17	57.23	58.83	58.00	44.99	48.08	56.42	67.83	54.97	62.59	70.49

classification methods, which benefits from the pretrain on ImageNet. LCD-Net achieves the highest classification scores than all existing methods, which verifies the effectiveness of LCD-Net for the diagnosis of jaw tumors and cysts. What's more, LCD-Net achieves comparative detection performance on par with those detection based methods [3,11,23], which is shown in Table 2. The detection performance of LCD-Net is calculated based on the bounding of the predicted segmentation mask. The segmentation performance of LCD-Net is given in Table 3, where we can see that cysts (DCs and PCs) have more accurate results.

3.2 Ablation Study

To verify the effectiveness of each component, we conduct an ablation study on self-supervision learning (denoted by '-pretrain'), path-covering data augmentation ('-covering'), segmentation branch ('-segment'), and location consistency loss ('-location'). LCD-Net is the integration of those components. Table 4 shows the quantitative results of the above methods. LCD-Net gains more than 20% performance boost than '-pretrain' on most metrics, which verifies that self-supervised pretrain with massive healthy samples has greatly benefit for

a) Input   b) Segmentation   c) Result without $\mathcal{L}_{loc}$ d) Result with $\mathcal{L}_{loc}$

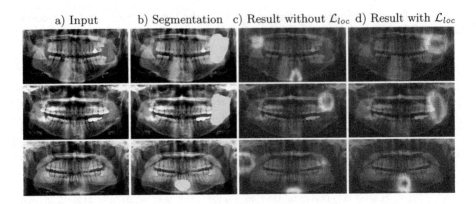

Fig. 2. The segmentation results and visual results using Grad-CAM [20].

the diagnosis of jaw tumors and crysts. What's more, '-covering', '-segment', and '-location' drop by about 6%, 7%, and 6% compared with LCD-Net, which demonstrate the effectiveness of those components.

Table 3. The segmentation performance.

Category	PA	Sen.	Spe.	mIoU
Unet[18]	67.43	63.18	68.21	69.02
PSPNet [24]	64.21	58.86	65.62	67.93
PAN [13]	65.84	59.59	67.21	68.04
DeeplabV3+ [4]	67.92	64.01	69.27	69.34
LCD-Net	68.11	64.09	70.64	70.84

Table 4. The Ablation study.

Method	Acc.	Pre.	Sen.	Spe.	F1
-pretrain	75.33	34.45	35.67	84.48	34.22
-covering	85.63	56.32	58.44	90.71	57.84
-segment	84.35	54.23	59.01	89.78	53.32
-location	86.24	57.83	60.22	89.99	54.34
LCD-Net	88.72	65.81	66.56	92.66	66.14

3.3   Explainability and Reliability

Due to the specificity of medical diagnosis, the explainability and reliability are important and essential factors of the diagnostic methods. In this study, patch-covering, segmentation branch and location consistency loss are devised for strengthening the explainability and reliability. Figure 2 shows the segmentation results, the visual results of LCD-Net without and with location consistency loss using GRAD-CAM [20]. The predicted masks indicate the lesion areas, which are important gist for doctors to make further diagnosis. Grad-CAM [20] can generate heatmap visualizations of areas, which are positively related to the predicted class. Figure 2(c) shows that some areas unrelated to lesions are highlighted, which means that the reliability of method without location consistency loss are relatively weak. On the contrary, LCD-Net with location consistency loss decreases the response of healthy area for lesion category, which effectively enhances the model's reliability.

4   Conclusion

In this paper, we propose a Location Constrained Dual-branch Network (LCD-Net) for diagnosing jaw tumors and cysts. Unlike existing transfer learning based methods, the proposed LCD-Net can intensify model's performance through pre-train on massive healthy samples with self-supervised learning. In LCD-Net, the segmentation branch is introduced to segment lesion areas, improving the classification branch's ability to extract more distinguishable features. What's more, patch-covering strategy and localization consistency loss are devised to improve the model's reliability. Extensive experiments demonstrate that LCD-Net achieves SOTA performance and more reliable results, which provides a useful tool for diagnosing jaw tumors and cysts. In the future, we will focus on the extension of localization consistency on more general classification tasks.

Acknowledgments. This work is supported by National Natural Science Foundation of China (No.62002318), Key Research and Development Program of Zhejiang Province (2020C01023), Zhejiang Provincial Science and Technology Project for Public Welfare (LGF21F020020), Ningbo Natural Science Foundation 202003N4318), the Fundamental Research Funds for the Central Universities (2021FZZX001-23), and the Major Scientific Research Project of Zhejiang Lab (No. 2019KD0AC01).

References

1. Altaf, F., Islam, S., Akhtar, N., Janjua, N.K.: Going deep in medical image analysis: Concepts, methods, challenges, and future directions. IEEE Access **7**, 99540–99572 (2019)
2. Apajalahti, S., Kelppe, J., Kontio, R., Hagström, J.: Imaging characteristics of ameloblastomas and diagnostic value of computed tomography and magnetic resonance imaging in a series of 26 patients. Oral Surg. Oral Med. Oral Pathol. Oral Radiol. **120**(2), e118–e130 (2015)
3. Ariji, Y., et al.: Automatic detection and classification of radiolucent lesions in the mandible on panoramic radiographs using a deep learning object detection technique. Oral Surg. Oral Med. Oral Pathol. Oral Radiol. **128**(4), 424–430 (2019)
4. Chen, L., Papandreou, G., Adam, H.: Rethinking atrous convolution for semantic image segmentation. arXiv (2017)
5. Chen, X., Fan, H., Girshick, R., He, K.: Improved baselines with momentum contrastive learning. arXiv preprint arXiv:2003.04297 (2020)
6. Choi, J.W.: Assessment of panoramic radiography as a national oral examination tool: review of the literature. Imaging Sci. Dent. **41**(1), 1 (2011)
7. Deng, J., Dong, W., Socher, R., Li, L.J., Li, K., Fei-Fei, L.: ImageNet: a large-scale hierarchical image database. In: 2009 IEEE Conference on Computer Vision and Pattern Recognition, pp. 248–255. IEEE (2009)
8. Feng, Z., et al.: Edge-competing pathological liver vessel segmentation with limited labels. In: AAAI Conference on Artificial Intelligence (2021)
9. González-Alva, P., et al.: Keratocystic odontogenic tumor: a retrospective study of 183 cases. J. Oral Sci. **50**(2), 205–212 (2008)
10. Ker, J., Wang, L., Rao, J., Lim, T.: Deep learning applications in medical image analysis. IEEE Access **6**, 9375–9389 (2018)

11. Kwon, O., et al.: Automatic diagnosis for cysts and tumors of both jaws on panoramic radiographs using a deep convolution neural network. Dentomaxillofacial Radiol. **49**(8), 20200185 (2020)

12. Lee, J.H., Kim, D.H., Jeong, S.N.: Diagnosis of cystic lesions using panoramic and cone beam computed tomographic images based on deep learning neural network. Oral Dis. **26**(1), 152–158 (2020)

13. Li, H., Xiong, P., An, J., Wang, L.: Pyramid attention network for semantic segmentation. In: BMVC, p. 285 (2018)

14. Meara, J.G., Shah, S., Li, K.K., Cunningham, M.J.: The odontogenic keratocyst: a 20-year clinicopathologic review. Laryngoscope **108**(2), 280–283 (1998)

15. Poedjiastoeti, W., Suebnukarn, S.: Application of convolutional neural network in the diagnosis of jaw tumors. Healthcare Inform. Res. **24**(3), 236 (2018)

16. Redmon, J., Farhadi, A.: Yolo9000: better, faster, stronger. In: Proceedings of the IEEE Conference on Computer Vision and Pattern Recognition, pp. 7263–7271 (2017)

17. Redmon, J., Farhadi, A.: Yolov3: an incremental improvement. arXiv preprint arXiv:1804.02767 (2018)

18. Ronneberger, O., Fischer, P., Brox, T.: U-Net: convolutional networks for biomedical image segmentation. In: Navab, N., Hornegger, J., Wells, W.M., Frangi, A.F. (eds.) MICCAI 2015. LNCS, vol. 9351, pp. 234–241. Springer, Cham (2015). https://doi.org/10.1007/978-3-319-24574-4_28

19. Ronneberger, O., Fischer, P., Brox, T.: U-Net: convolutional networks for biomedical image segmentation. In: International Conference on Medical Image Computing and Computer-Assisted Intervention (2015)

20. Selvaraju, R.R., Cogswell, M., Das, A., Vedantam, R., Parikh, D., Batra, D.: Grad-CAM: visual explanations from deep networks via gradient-based localization. In: Proceedings of the IEEE International Conference on Computer Vision, pp. 618–626 (2017)

21. Simonyan, K., Zisserman, A.: Very deep convolutional networks for large-scale image recognition. arXiv preprint arXiv:1409.1556 (2014)

22. Szegedy, C., Vanhoucke, V., Ioffe, S., Shlens, J., Wojna, Z.: Rethinking the inception architecture for computer vision. In: Proceedings of the IEEE Conference on Computer Vision and Pattern Recognition, pp. 2818–2826 (2016)

23. Yang, H., et al.: Deep learning for automated detection of cyst and tumors of the jaw in panoramic radiographs. J. Clin. Med. **9**(6), 1839 (2020)

24. Zhao, H., Shi, J., Qi, X., Wang, X., Jia, J.: Pyramid scene parsing network. In: CVPR, pp. 6230–6239 (2017)

Motion Correction for Liver DCE-MRI with Time-Intensity Curve Constraint

Yuhang Sun[1,2], Dongming Wei[3], Zhiming Cui[2,4], Yujia Zhou[1], Caiwen Jiang[2], Jiameng Liu[2], Qianjin Feng[1(✉)], and Dinggang Shen[2,5(✉)]

[1] School of Biomedical Engineering, Southern Medical University, Guangzhou, China
[2] School of Biomedical Engineering, ShanghaiTech University, Shanghai, China
dgshen@shanghaitech.edu.cn
[3] School of Biomedical Engineering, Shanghai Jiao Tong University, Shanghai, China
[4] Department of Computer Science, The University of Hong Kong, Pok Fu Lam, Hong Kong
[5] Shanghai United Imaging Intelligence Co., Ltd, Shanghai, China

Abstract. Motion correction is a fundamental preprocessing step for liver dynamic contrast-enhanced magnetic resonance imaging (DCE-MRI), which can help accurate diagnosis of benign and malignant tumors. Previous studies have difficulty in aligning small structures, *e.g.*, tumors and vessels, due to the remarkable intensity changes over different images. Except for measuring physiologic parameters in DCE-MRI, the time-intensity curves (TICs) can also be used to constrain the alignment of small anatomical structures such as small tumors and vessels. In this work, we propose a coarse-to-fine motion correction scheme with smoothness constraint of TICs to correct the motion in liver DCE-MRI. Specifically, the proposed motion correction scheme consists of two major stages. First, different time point images are registered to the selected fixed image pairwisely via a fully convolutional network (FCN), which outputs their corresponding coarse displacement vector fields (DVFs). Second, all of the coarse DVFs are further refined simultaneously under the group similarity of the warped time points and the fixed image, together with the smoothness constraint of TICs at a fine level. To our knowledge, our work is the first in constraining the motion correction using TICs for better alignment of small structures. Experimental results on liver DCE-MRI demonstrate that our proposed method can obtain a more accurate alignment of small structures (*e.g.*, tumors and vessels) than state-of-the-art methods.

Keywords: Motion correction · DCE-MRI · Smoothness of time-intensity curves

1 Introduction

Dynamic contrast-enhanced magnetic resonance image (DCE-MRI) provides the quantitative perfusion information for different tissues. The benign and malignant tumors can be distinguished by monitoring the dynamic change of the

© Springer Nature Switzerland AG 2021
M. de Bruijne et al. (Eds.): MICCAI 2021, LNCS 12907, pp. 733–742, 2021.
https://doi.org/10.1007/978-3-030-87234-2_69

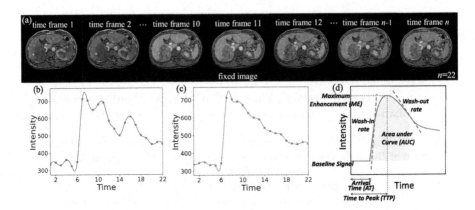

Fig. 1. (a) Illustration the original DCE-MRI time series; (b) the TIC of one voxel (*i.e.*, green point in time series); (c) the TIC after motion correction of the corresponding voxel; and (d) some parameters calculated from the TIC. (Color figure online)

injected contrast agent in the tumor region [1]. Generally, the time-intensity curves (TICs) are used to measure several physiologic parameters (*e.g.*, wash-in and wash-out rates) in the region of interest (ROI) (see Fig. 1(d)) [2]. The process of DCE-MRI acquisition usually lasts several minutes. During this period, the patient's motion, including heartbeat, respiration, and bowel movement, potentially makes the TICs inaccurate to reflect the perfusion (Fig. 1(b)). Therefore, the motion correction of DCE-MRI time series is a necessary preprocessing step for the diagnosis of cancer.

Existing DCE-MRI motion correction methods can be summarized into three categories. (1) Pharmacokinetic (PK) model based methods [3–5] quantitatively model the wash-in and wash-out process of the injected contrast agents, which heavily relies on the pre-defined tissue-specific physiological characteristic; (2) Adaptive loss function based methods [6–8] define the similarity between different time points, which is quite challenging when coping with dynamic intensity change in DCE-MRI; (3) De-enhanced based methods [9–11] have been proposed to remove the contrast agent components from the original DCE-MRI time series. Although the similarity of de-enhanced DCE-MRI can be well measured, it requires a robust de-enhancing method. Overall, the existing DCE-MRI motion correction methods neglect the fact that the smooth intensity change along with the sequence time points, which is also evaluated to improve the motion correction performance of small structures (*e.g.*, tumors and vessels) in our proposed method. In addition, the existing methods perform the DCE-MRI motion correction in a pairwise manner. Instead, the motion correction performance of multiple time points in DCE-MR can be further improved under the similarity among all of the time points.

To tackle the challenges mentioned above, we propose a method that combines pairwise and groupwise motion correction for DCE-MRI. As illustrated in Fig. 2, this motion correction method consists of two sequential stages: 1) pairwise motion correction stage, and 2) groupwise motion correction stage. In the pairwise motion correction stage, one time point is selected as the fixed image (*i.e.*, the 11th time point in our study). All the other time points are registered with this fixed image via a pairwise manner to obtain coarse displacement vector fields (DVFs), respectively. For groupwise motion correction stage, the similarities between different moving images are also taken into account to optimize the refined DVFs. Besides, we propose to utilize the smoothness of the TICs as a constraint to guide the groupwise motion correction.

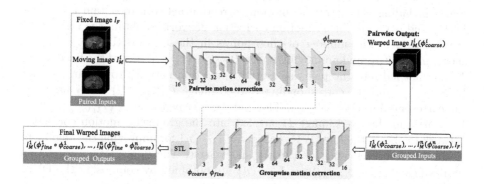

Fig. 2. Illustration of our proposed motion correction framework, including a pairwise motion correction network to obtain the global DVF (top panel), and a groupwise motion correction network to obtain the refined DVFs (bottom panel). The input images of the pairwise network consist of a random moving image I_M^i of the moving image set and the fixed image I_F. The input images of the groupwise network are the warped moving image set $\{I_M^1(\phi_{coarse}^1), ..., I_M^n(\phi_{coarse}^n)\}$ and the fixed image I_F.

2   Materials and Method

2.1   Clinical Dataset

Twenty subjects with liver cancer are collected in this study, and each subject was scanned by 1.5T T1-weighted gradient echo pulse sequences combined with TR of 11 ms and TE of 3 ms at 22 different time points. The contrast agent injection happens from the 4th to the 8th time point. A 3D volume was obtained at each time point with the slice number varying from 22 to 26, and the image size of each slice varying from 288×288 to 320×320. The voxel spacing ranges from $0.9375 \times 0.9375 \times 4 \, mm^3$ to $1.0938 \times 1.0938 \times 4.15 \, mm^3$. Before the acquisition of each time point, patients are required to hold their breath after expiration, and the whole acquisition lasts around 500s.

Time points are mainly classified into three periods based on the intensity changes during DCE-MRI, *i.e.*, (1) the baseline period before the injection of contrast agent; (2) the wash-in period after the injection of contrast agent, where the intensity increases rapidly as the contrast agent washes in; and (3) the wash-out period, where the intensity decreases as the contrast agent washes out. Considering the intensity distribution varies in DCE-MRI, we choose the MRI scan at the 11th time point in the wash-in period as the fixed image, since it has the smaller intensity changes between the fixed image and all other moving images.

2.2   Proposed Method

Our proposed motion correction framework for liver DCE-MRI is illustrated in Fig. 2, including (1) a pairwise motion correction network to obtain the coarse DVFs, and (2) a groupwise motion correction network to obtain the fine DVFs.

Denote n as the number of time points in the DCE-MRI time series. As shown in Fig. 2, a random moving image $I_M^i, (i \in \{1, \cdots, n\})$, and the fixed image I_F are first fed into the pairwise motion correction network to obtain a coarse DVF ϕ_{coarse}^i and the corresponding warped image $I_M^i(\phi_{coarse}^i)$ of I_M^i based on the corresponding DVF ϕ_{coarse}^i. Then, all the warped images $\{I_M^i(\phi_{coarse}^i)|i = 1, \cdots, n\}$ and the fixed image I_F are fed into the groupwise motion correction network to obtain the fine DVFs $\{\phi_{fine}^i|i = 1, \cdots, n\}$ for improving the alignment of subtle structures, such as tumors and vessels. The final motion correction results $\{I_M^i(\phi_{fine}^i \circ \phi_{coarse}^i)|i = 1, \cdots, n\}$ are obtained by composing the DVFs of the first stage and the second stage.

Pairwise Motion Correction Stage. Pairwise motion correction network consists of a series of DVF prediction layers and a spatial transformer layer (STL) [12], akin to VoxelMorph [13] (see top panel in Fig. 2). We adopt a U-Net-like CNN as our network backbone, which consists of an encoder part, a decoder part, and several skip connections bridging these two parts. Each layer of the encoder and the decoder consists of a convolution, where a $3 \times 3 \times 3$ kernel is used to generate the feature map with a stride of 1, which is followed by a Leaky ReLU activation. The loss function of the pairwise motion correction could be formulated as follows:

$$\mathcal{L}_{pair} = -\mathcal{L}_{sim}(I_M(\phi_{coarse}^i), I_F) + \lambda_1 \mathcal{L}_{smooth}(\phi_{coarse}^i), i \in \{1, \cdots, n\}. \tag{1}$$

where $\mathcal{L}_{sim}(\cdot, \cdot)$ measures the approximate mutual information [14] between $I_M(\phi_{coarse}^i)$ and I_F, $\mathcal{L}_{smooth}(\cdot)$ is a regularization term that constrains the DVF to be smooth, and λ_1 is the regularization parameter.

Groupwise Motion Correction Stage. We consider further aligning all the images in a groupwise manner, as more than one images need to be aligned to the fixed image in DCE-MRI time series.

As shown in the bottom panel of Fig. 2, the network structure is similar with the pairwise network, but with fewer decoders to reduce the training parameters. The inputs are all the images in the warped moving set, *i.e.*, $\{I_M^i(\phi_{coarse}^i)|i = 1, \cdots, n\}$, and the fixed image I_F of the pairwise stage, and the outputs are the fine DVFs $\{\phi_{fine}^i|i = 1, \cdots, n\}$. After composing ϕ_{coarse} of the pairwise stage and ϕ_{fine} of the groupwise stage, we obtain the final motion correction results $\{I_M^i(\phi_{fine}^i \circ \phi_{coarse}^i)|i = 1, \cdots, n\}$ by STL [12] in a coarse-to-fine manner.

Instead of computing the similarity between each pair of images in the group [15], we simplify this process by computing the similarity between each image and the corresponding groupmean image I_{Mean}^i, $(i \in \{1, \cdots, n\})$, of the rest images in the group at each iteration. The loss function of the groupwise motion correction network is defined as follows:

$$\mathcal{L}_{group} = \sum_{i=1}^{n} -\mathcal{L}_{sim}(I_M^i(\phi_{fine}^i \circ \phi_{coarse}^i), I_F)$$

$$+ \sum_{i=1}^{n} -\mathcal{L}_{sim}(I_M^i(\phi_{fine}^i \circ \phi_{coarse}^i), I_{Mean}^i), \quad (2)$$

Considering that some subtle structures could not obtain satisfactory alignment, due to the remarkable intensity changes over different time points, we propose to utilize the smoothness of the TICs to regularize the motion correction process in our work. Ideally, during wash-in and wash-out periods, the TICs should be smooth and free of fluctuations and unaffected by patient movement (see Fig. 1(d)). For achieving such a goal, a voxelwise loss is introduced to measure the consistency of voxel intensity along time points. We constrain L1-norm of the second derivative of the time-intensity curve, as defined in the following:

$$\mathcal{L}_{TICs} = \left\| \frac{\Delta^2 S}{\Delta^2 t} \right\|_1, \quad (3)$$

where S denotes the TICs for all the voxels, and t represents different time points. Specially, $t \neq t_c$, where t_c is the starting time point to inject contrast agents (with remarkable intensity changes at this time point). This loss constrains TICs to be smooth during each period, besides the time point of starting contrast agent injection. Combining with Eq. 2, the overall objective function of groupwise motion correction network is defined as:

$$\mathcal{L}_{fine} = \mathcal{L}_{group} + \lambda_2 \mathcal{L}_{TICs} + \lambda_3 \sum_{i=1}^{n} \mathcal{L}_{smooth}(\phi_{fine}^i \circ \phi_{coarse}^i), \quad (4)$$

where λ_2 and λ_3 are the trade-off parameters. And the final warped images are obtained by the combined DVFs of the first stage and the second stage:

$$\phi_{final} = \phi_{fine} \circ \phi_{coarse}. \quad (5)$$

Implementation. Both networks are implemented based on PyTorch and trained on a workstation equipped with two NVIDIA Tesla V100 GPU (32G). Data augmentation is conducted through random rotating and flipping. We first trained the pairwise motion correction network for $15,000$ iterations with a batch size of 6. Then, we trained the groupwise motion correction network for $1,500$ iterations with a batch size of 21. Adam is adopted as the optimizer to train both models with a learning rate of 10^{-4}. And the hyperparameters λ_1, λ_2, λ_3 in the Eqs. 1 and 4 are set as 1, 5×10^{-4} and 10^4, respectively.

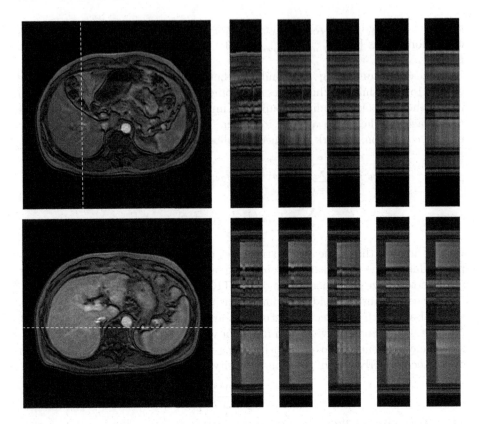

Fig. 3. Two examples of comparisons of different motion correction methods. The images in each row are a transverse view of one subject with a white dashed line indicating the location of time-cut images; time-cut image before motion correction; time-cut image after ANTs; time-cut image after pairwise VoxelMorph; time-cut image after the variant of our method (without $\mathcal{L}_{TICs}$); and time-cut image after our proposed method.

3   Experiments

3.1   Experimental Setup

There are two preprocessing steps for the DCE-MRI time series. First, the intensity is normalized to $[0, 1]$. Second, all the images are cropped and resampled to be $320 \times 320 \times 32$ for having $1 \times 1 \times 1$ mm^3 resolution. Our clinical dataset includes 20 subjects and each subject contains 22 images of different time points. We partition our DCE-MRI dataset into 90% for training and 10% for testing. Three-fold cross-validation is performed.

In our method, we propose a two-stage learning strategy (pairwise and groupwise motion correction) and introduce the TICs smooth constraint to network learning process. To validate the effectiveness of our method on motion correction, we compare our method with traditional registration toolbox ANTs [16]

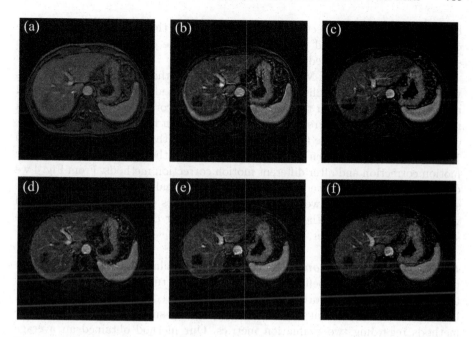

Fig. 4. Illustration of subtraction images of different motion correction methods. (a) Original image of one subject at the 7th time point. (b) Subtraction image between the images of the 7th time point and the 1st time point before motion correction. (c–f): Subtraction images after ANTs, pairwise VoxelMorph, the variant of our proposed method (without $\mathcal{L}_{TICs}$), and our proposed method, respectively.

with mutual information similarity, pairwise VoxelMorph [13], and the variant of our method (with two-stage learning strategy and without the constraint of TICs smoothness). For the pairwise VoxelMorph, the hyperparameter and network structure are the same as the pairwise stage of our proposed method. And for the variant of our method, the only difference from our proposed method is that it does not contain the voxelwise constraint, $i.e.$, $\mathcal{L}_{TICs}$, which is a comparison method to investigate the contribution of $\mathcal{L}_{TICs}$.

3.2   Experimental Results

Qualitative Evaluation. The time-cut images are generated from a livelong time series combined with a pixel-wide line, which represents the visualization of the temporal motion correction performance. The time-cut images should be orderly with an accurate motion correction method. Figure 3 illustrates two examples of time-cut images before motion correction and after motion correction of different methods, which are generated from the location of the white dash line in the transverse view images. Note that our proposed method utilizes the voxelwise constraint of time-intensity curves to guide the motion correction process, while other methods do not. As shown in Fig. 3, our proposed method has the best motion correction performance, where the structures in the white dash line are well aligned visually. This implies the effectiveness of the TICs smoothness constraint used in our work.

One challenge of DCE-MRI motion correction is that small structure such as vessels could not be well aligned by traditional methods. In our method, we add a voxelwise regularization to constrain the motion correction process to obtain more accurate motion correction results. As shown in Fig. 4, subtraction images of the post-contrast image (the 7th time point) and the pre-contrast image (the 1st time point) have been generated to illustrate the vessel structure before motion correction and after different motion correction methods. From Fig. 4 we can see that not only the edge of organs but also small structures in the organs such as vessels could be well aligned (fewer artifacts on the edges of liver and vessels) by our method. These results imply that our $\mathcal{L}_{TICs}$ could handle subtle structure alignment better.

Quantitative Evaluation. Quantitatively, we evaluate our method through two metrics: (1) Dice coefficients of the liver and tumor, (2) target registration error (TRE) of important anatomies, such as edges of vessels. As shown in Table 1, we can see that our method is generally superior to the other three methods regarding two evaluation metrics. Our method obtained an average Dice score of 92.96% for liver and 78.94% for tumors. And the Dice score of the variant of our proposed method (without $\mathcal{L}_{TICs}$) is 0.82% higher than the pairwise VoxelMorph for liver, which demonstrates the effectiveness of groupwise motion correction. As for TRE, our method achieved 2.11 mm, which is 0.27 mm lower than the second-best result. These results indicate that our method could handle not only the alignment of the large organs but also the subtle structures, such as vessels.

Table 1. Quantitative comparison of four different motion correction methods on the DCE-MRI dataset.

Method	ANTs	VoxelMorph (pairwise)	Our Method (w/o $\mathcal{L}_{TICs}$)	Our method
Dice (liver) (%)	91.58 ± 1.37	91.52 ± 0.76	92.34 ± 0.53	**92.96±0.58**
Dice (tumor) (%)	77.84 ± 3.86	78.13 ± 3.45	78.37 ± 2.98	**78.94±2.57**
TRE (vessel) (mm)	2.46 ± 1.04	2.44 ± 0.88	2.38 ± 0.68	**2.11±0.43**

4 Conclusion and Future Work

In this paper, we propose a liver DCE-MRI time series motion correction method, which has potential to assist clinicians to distinguish benign and malignant tumors. Particularly, we developed a coarse-to-fine motion correction network (pairwise and groupwise motion correction) by a two-stage learning. In addition, to obtain a more accurate alignment of small structures, we introduced a voxelwise constraint of smoothness of the time-intensity curve at each voxel. The experimental results showed that our method could obtain more accurate alignment of vessels than other motion correction methods. Future work could include developing other specific losses for other significant organs and the strategy of groupwise motion correction for the DCE-MRI time series.

References

1. Jackson, A., O'Connor, J.P., Parker, G.J., Jayson, G.C.: Imaging tumor vascular heterogeneity and angiogenesis using dynamic contrast-enhanced magnetic resonance imaging. Clin. Cancer Res. **13**(12), 3449–3459 (2007)
2. Schnall, M.D., et al.: Diagnostic architectural and dynamic features at breast MR imaging: multicenter study. Radiology **238**(1), 42–53 (2006)
3. Tofts, P.S.: T1-weighted DCE imaging concepts: modelling, acquisition and analysis. Signal **500**(450), 400 (2010)
4. Xiaohua, C., Brady, M., Lo, J.L.-C., Moore, N.: Simultaneous segmentation and registration of contrast-enhanced breast MRI. In: Christensen, G.E., Sonka, M. (eds.) IPMI 2005. LNCS, vol. 3565, pp. 126–137. Springer, Heidelberg (2005). https://doi.org/10.1007/11505730_11
5. Buonaccorsi, G.A., et al.: Tracer kinetic model-driven registration for dynamic contrast-enhanced MRI time-series data. Magnetic resonance in medicine: an official journal of the international society for magnetic resonance in medicine **58**(5), 1010–1019 (2007)
6. Rohlfing, T., Maurer, C.R., Bluemke, D.A., Jacobs, M.A.: Volume-preserving non-rigid registration of MR breast images using free-form deformation with an incompressibility constraint. IEEE Trans. Med. Imaging **22**(6), 730–741 (2003)
7. Ghaffari, A., Fatemizadeh, E.: Sparse-induced similarity measure: mono-modal image registration via sparse-induced similarity measure. IET Image Process. **8**(12), 728–741 (2014)
8. Ghaffari, A., Fatemizadeh, E.: RISM: single-modal image registration via rank-induced similarity measure. IEEE Trans. Image Process. **24**(12), 5567–5580 (2015)
9. Hamy, V., et al.: Respiratory motion correction in dynamic MRI using robust data decomposition registration-application to DCE-MRI. Med. Image Anal. **18**(2), 301–313 (2014)
10. Zhou, Y., et al.: Correlation-weighted sparse representation for robust liver DCE-MRI decomposition registration. IEEE Trans. Med. Imaging **38**(10), 2352–2363 (2019)
11. Sun, Y., Feng, Q.: Liver DCE-MRI registration based on sparse recovery de-enhanced curves. In: IEEE 17th International Symposium on Biomedical Imaging (ISBI). IEEE 2020, pp. 705–708 (2020)
12. Jaderberg, M., Simonyan, K., Zisserman, A., Kavukcuoglu, K.: Spatial transformer networks. arXiv preprint arXiv:1506.02025 (2015)
13. Balakrishnan, G., Zhao, A., Sabuncu, M.R., Guttag, J., Dalca, A.V.: An unsupervised learning model for deformable medical image registration. In: Proceedings of the IEEE Conference on Computer Vision and Pattern Recognition, pp. 9252–9260 (2018)

14. Belghazi, M.I., et al.: Mutual information neural estimation. In: International Conference on Machine Learning, PMLR, pp. 531–540 (2018)
15. Liu, Q., Wang, Q.: Groupwise registration of brain magnetic resonance images: a review. J. Shanghai Jiaotong Univ. (Sci.) **19**(6), 755–762 (2014)
16. Avants, B.B., Tustison, N., Song, G.: Advanced normalization tools (ants). Insight J. **2**(365), 1–35 (2009)

Parallel Capsule Networks for Classification of White Blood Cells

Juan P. Vigueras-Guillén[1]([📧]) [iD], Arijit Patra[2] [iD], Ola Engkvist[3,4] [iD],
and Frank Seeliger[1] [iD]

[1] CVRM Safety, Clinical Pharmacology and Safety Science, R&D, AstraZeneca,
Gothenburg, Sweden
JuanPedro.ViguerasGuillen@astrazeneca.com
[2] Digitisation and AI, Clinical Pharmacology and Safety Science, R&D, AstraZeneca,
Cambridge, UK
[3] Molecular AI, Discovery Science, R&D, AstraZeneca, Gothenburg, Sweden
[4] Department of Computer Science & Engineering, Chalmers University of
Technology, Gothenburg, Sweden

Abstract. Capsule Networks (CapsNets) is a machine learning architecture proposed to overcome some of the shortcomings of convolutional
neural networks (CNNs). However, CapsNets have mainly outperformed
CNNs in datasets where images are small and/or the objects to identify
have minimal background noise. In this work, we present a new architecture, parallel CapsNets, which exploits the concept of branching the network to isolate certain capsules, allowing each branch to identify different entities. We applied our concept to the two current types of CapsNet
architectures, studying the performance for networks with different layers
of capsules. We tested our design in a public, highly unbalanced dataset
of acute myeloid leukaemia images (15 classes). Our experiments showed
that conventional CapsNets show similar performance than our baseline
CNN (ResNeXt-50) but depict instability problems. In contrast, parallel
CapsNets can outperform ResNeXt-50, is more stable, and shows better
rotational invariance than both, conventional CapsNets and ResNeXt-50.

Keywords: CapsNets · Dynamic routing · ResNeXt · Leukocytes

1 Introduction

In the last decade, Convolutional Neural Networks (CNNs) have shown remarkable performance for a wide range of computer vision tasks [5,8,11]. However,
CNNs have many drawbacks, such as the inability to learn viewpoint invariant
representations and the need for large amount of training data. Capsule Networks (CapsNets) [6,19] is a new neural network architecture that tackles those
shortcomings by using capsules. A capsule is a group of neurons (depicted as a
vector) whose output represents the various perspectives of an entity, such as
pose, texture, scale, or the relative relationship between the entity and its parts.

This technique has immense potential in the medical field, such as in cell
classification, where (i) different types of cells are classified depending on the

© Springer Nature Switzerland AG 2021
M. de Bruijne et al. (Eds.): MICCAI 2021, LNCS 12907, pp. 743–752, 2021.
https://doi.org/10.1007/978-3-030-87234-2_70

hierarchical relationship of the cell and its parts (shape of the nucleus, texture of the cytoplasm, presence of subcellular organelles), and where (ii) rotational invariance is crucial. However, CapsNets require a large number of parameters when the network is enlarged, and they have mainly shown promising performance for small images and/or with barely any background noise.

In this work, we present the concept of CapsNets parallelization, where parts of the network are subdivided in branches to isolate capsules, helping the network to (i) identify different entities in different branches, and (ii) avoid instability problems when capsule layers are enlarged. This concept is applied in both types of current CapsNets [6,19]. We also propose a variation to the Sabour et al.'s CapsNets [19]; our proposal entails to lose the spatial information in the first layer of capsules, forcing the middle layer of capsules to encode whole entities. We also show how, against general assumption, conventional CapsNets do not seem to perform proficiently as more capsule layers are added, and they are not more robust than CNNs for small datasets.

1.1   Capsule Networks

DR-CapsNets. Sabour et al. [19] proposed the first CapsNet based on dynamic routing (DR), with one CNN and two capsule layers, to solve the MNIST dataset (images of 28×28 px). Their first layer of capsules, Primary-Caps, took the output of a convolution ($6 \times 6 \times 256$) and considered that every 8 elements along the feature axis would represent a capsule instantiation, thereby creating 32 capsules, each one evaluated in a grid of 6×6. The latter simply entails that –in the subsequent steps– the weights ($\mathbf{W}_{ij}$) that multiply a lower-layer capsule i to produce the next-layer capsule j are shared between the capsules of the grid. Furthermore, they proposed that the module of a capsule vector should represent a probability (with range 0–1) and thus they defined a squashing function

$$\mathbf{v}_j = \frac{\|\mathbf{s}_j\|^2}{1+\|\mathbf{s}_j\|^2} \frac{\mathbf{s}_j}{\|\mathbf{s}_j\|}, \qquad \mathbf{s}_j = \sum_i c_{ij}\hat{\mathbf{u}}_{j|i}, \qquad \hat{\mathbf{u}}_{j|i} = \mathbf{W}_{ij}\mathbf{u}_i \qquad (1)$$

where $\mathbf{v}_j$ and $\mathbf{s}_j$ are the squashed and non-squashed capsule j, respectively, and the capsule $\mathbf{s}_j$ is computed as indicated above for all (except the first) layer of capsules, where c_{ij} are the coefficients obtained by the dynamic routing and $\mathbf{u}_i$ are the squashed capsules from the lower layer. Briefly, the dynamic routing aims to determine how close the predicted vectors $\hat{\mathbf{u}}_{j|i}$ are to the mean predicted vector $\mathbf{v}_j$ (by using the scalar product), giving a higher c_{ij} to those closer. They also defined a specific loss function, named 'margin loss'. Further details in [19].

EM-CapsNets. Hinton et al. [6] proposed a different network (with one CNN and four capsule layers) to solve the smallNORB dataset (images of 96×96 px). They extended the concept of CapsNets in the following ways: (i) capsules were depicted as matrices instead of vectors; (ii) a routing based on the Expectation-Maximization (EM) algorithm was proposed, where the matching between capsules is done by considering that a higher-layer capsule represents a Gaussian and the lower-layer capsules are data-points (further details in [6]); (iii) convolutional capsules were presented, where a higher-layer capsule is computed only based on the neighbouring lower-layer capsules (K in Fig. 1); (iv) a technique called 'Coordinate Addition' was proposed (applied only in the last convolutional capsule), which adds the position of the capsule to its vote matrix in order to keep the spatial information; (v) a new loss function, spread loss, was defined.

1.2 Related Work

Many publications have used CapsNets to perform tasks in medical images. The majority simply applied Sabour et al.'s network [19] to small patches to perform tasks such as detection of diabetic retinopathy (fundus images) and mitosis (histology images, H&E) [10], classification of breast cancer [2,9] and colorectal tissue [18] (H&E), detection of liver lesions (CT, adding attention gates to the network) [7], and blood vessel segmentation (fundus, with an inception block [21] as CNN) [12]. Others used Sabour et al.'s network in larger images by adding more CNN blocks (glaucoma detection, OCT [4]) or heavily downsampling the images (brain tumour classification, MRI [1]). Only a few explored the use of a 3-capsule-layer DR-CapsNet for the classification of white cells [15] and detection of cell apoptosis [17], and an interesting use of DR within a CNN architecture has been used for classification of thoracic disease (CT) [20]. Special attention goes to LaLonde et al., who has developed CapsNets with several capsule layers to perform polyp classification [14] and expanded the concept of CapsNets to perform segmentation (CT images of the lungs) [13]. Hinton et al.'s architecture [6] has seldom been applied in the literature.

2 Methods

2.1 Parallel Capsule Networks

Parallelization can be applied in several ways, depending where to start and merge the branches. In this work, we focus in studying networks with 3 capsule layers (3-CapsLayer). These contain Primary-, Mid-, and Class-Caps, and the parallelization is performed by creating a unique set of CNNs and Primary-Caps for each Mid-Cap, which are then concatenated to be routed to the Class-Caps (Fig. 1, top). Several CNN-blocks precede the capsule section, allowing the image to be reduced to a suitable size for the capsules. The first CNNs, which generate basic features, are common to all branches. Our chosen CNN is ResNeXt [22].

Since there is only one Mid-Cap per branch, there is no routing between Primary-Caps and Mid-Caps, which means that the algorithms only need to find the most appropriate transformation matrix $\mathbf{W}_{i|j=1}$ per Primary-Cap. This allows each branch to focus in generating specialized features (either in the CNNs or Primary-Caps) that are suitable for the Mid-Cap.

For comparative purposes, networks with 2 and 4 capsule layers were also tested. The 2-CapsLayer does not include Mid-Caps, and thus the different sets of Primary-Caps are concatenated before routing to the Class-Caps. The 4-CapsLayer performs the merging before the Class-Caps.

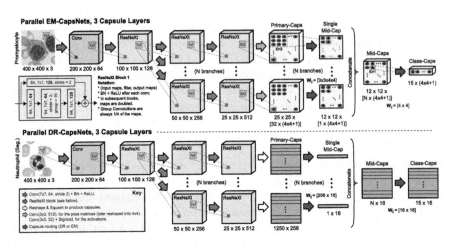

Fig. 1. The proposed Parallel-CapsNets in two types of capsule architectures: EM (top and middle), and DR (bottom). An arrow represents an action, whereas a block represents the output of the action. *Zoom for details.*

Parallel EM-CapsNets, 3 Capsule Layers. Our proposed network (Fig. 1, top) applies parallelization to Hinton et al.'s architecture [6]. All capsules have a pose matrix of 4×4 and an activation value. In a branch, there are 32 Primary-Caps and 1 Mid-Cap. The window (K) used in the convolutional capsules is 3×3 for the Mid-Caps and 1×1 for the Class-Caps, with a stride of 2 for the Mid-Caps. Coordinate addition is not employed.

Parallel DR-CapsNets, 3 Capsule Layers. Our DR network (Fig. 1, bottom) introduces two changes to Sabour et al.'s network [19]: the weights $\mathbf{W}_{ij}$ are not shared among the capsules of the grid, and the Primary-Caps are 256 elements long (two capsules per grid-point). The former entails that the spatial information is lost in the first layer, forcing the Mid-Caps to encode the whole existence of an entity. Furthermore, since all Primary-Caps are directed to a single Mid-Cap, the algorithm can only discard a useless Primary-Cap by setting its $\mathbf{W}_{i|j=1}$ to zero and, thus, not sharing weights can help to such discard. Both, the Mid-Caps and Class-Caps, have a size of 16 elements.

Parallel CapsNets, 2 Capsule Layers. In the EM network, the number of Primary-Caps are evenly distributed among the different branches, and the last CNN-block is duplicated to allow another stride. In the DR network, the length of the Primary-Caps is reduced based on the number of branches.

2.2 Data, Baseline Networks, Implementation Details, and Metrics

Data Description. We evaluated the proposed networks on a public dataset of white blood cells (leukocytes), which were present in patients with acute myeloid leukaemia (AML), a blood-type cancer that leads to the overproduction of abnormal leukocytes, published by Matek et al. [16] (Laboratory of Leukemia Diagnostics at Munich University Hospital, Germany). The dataset contains 18,365 single-cell images of 400×400 pixels from 15 highly-unbalanced classes (see Table 2). This dataset is particularly interesting because (i) leukocytes show hierarchical structures (the nucleus might depict a prominent nucleolus or be formed by different segments, and cytoplasm might show different textures), (ii) it is highly unbalanced, and (ii) many red cells appear in the background (noise).

Baseline. The network ResNeXt-50 [22] was chosen as baseline. This network was also used by the dataset's authors [16], but we achieved a higher performance with the same network and setup: +6% in PRE and SEN. DenseNets [8] and ResNets [5] were also tested, but they provided significantly inferior performance. An adapted single block of ResNeXt was chosen for the CNN-blocks in CapsNets, which also gave better results than other alternatives. We also tested the non-parallel versions of both architectures (EM and DR).

Implementation Details. All networks were implemented in Tensorflow 2.2 on a single NVIDIA V100 GPU with 32 GB of memory. In order to determine the most appropriate number of branches in the networks, each class in the dataset was subdivided into 5 folds, using 4 for training and 1 for validation. Once established the best branching, a 5-fold cross-validation (CV) was performed on the whole dataset for a final comparison (specifically, we use the same aforementioned 5 folds in the CV setup). The batches contained one example of each class (15 images), randomly shuffling the order within the batch to avoid bias. Data augmentation was performed by flipping the images up-down and left-right and by rotating the images ($0°$–$180°$). For the CNN baseline, the loss function was categorical cross-entropy. For the CapsNets, we used the loss functions suggested by their original authors (margin loss and spread loss). Nadam optimizer [3] was used, with a learning rate of 0.001. We defined an epoch as 500 iterations, and we trained for 700 epochs (no early stop), which required from 6 to 20 days to train, depending on the number of layers.

Evaluation Metrics. To quantify the performance, we used the weighted (*WAcc*) and non-weighted categorical accuracy (*Acc*),

$$\text{WAcc} = \sum_i \frac{\text{TP}^i + \text{TN}^i}{\text{TP}^i + \text{TN}^i + \text{FP}^i + \text{FN}^i}, \quad \text{Acc} = \frac{\text{TP} + \text{TN}}{\text{TP} + \text{TN} + \text{FP} + \text{FN}}, \quad (2)$$

reported as a percentage, where i is the class, TP is true positives, TN is true negatives, FP is false positives, and FN is false negatives. We also defined an agreement metric (*Agr*) that measures the percentage of images that were classified as the same type in all eight basic orientations (flipped up-down and left-right), regardless of whether the classification was correct. A higher agreement should suggest better rotational invariance. For the final architectures, we estimate the sensitivity, SEN = TP/(TP+FN), and precision, PRE = TP/(TP+FP).

3   Results

Non-parallel Architectures. In the DR-CapsNets, the 2-layer type performed rather poorly (Table 1), the 3-layer increased performance but diverged with more than two Mid-Caps (Fig. 2-B), and the 4-layer was unable to converge. Indeed, our proposed modification in DR networks (losing spatial information in the Primary-Caps) is not suitable for networks with only two layers. Regarding the EM network, it allowed for deeper architectures, but performance decreased with the addition of layers and/or Mid-Caps (Fig. 2). Building a 4-layer EM network was feasible, but performance was poor in all possible setups (*WAcc* in the range of 60–65). This suggests that, contrary to popular belief, the original EM-CapsNet does not perform as expected when the capsule section is increased in layers. EM-CapsNets were also highly sensitive to the position where capsule strides were placed, not being advised to place it in the Primary-Caps. Other instability problems were observed: EM networks were unable to converge when either (i) the image was only reduced to 50×50 before entering the capsule section (regardless if further striding was performed in the capsule section), or (ii) the Primary-Caps were increased from 32 to 64 (regardless the number of capsule layers). In summary, these findings suggest that both routing algorithms fail when a large number of capsules take part. Overall, only the 2-layer EM network (better *WAcc*) and the 3-layer DR network (better *Acc* and *Agr*) were slightly better than ResNeXt-50 (Table 1), and their number of parameters were notably lower (Table 1).

Parallel Architectures. As expected, the parallelization of 2-layer networks was rather detrimental: even though DR-CapsNets provided slightly better *Acc* and *Agr* in all cases, *WAcc* was always lower (Fig. 2-A), and EM-CapsNet provided lower results in all three metrics for all cases. In contrast, parallelizing 3-layer CapsNets was highly beneficial, particularly for our DR network (Fig. 2-B), which did not depict convergence problems with the addition of branches. Regarding EM-CapsNets, the right number of branches yielded a peak in accuracy (Fig. 2-B) (*Acc* and *Agr* were always higher than their non-parallel

Table 1. Categorical accuracy, weighted (*WAcc*) and non-weighted (*Acc*), and the agreement metric (*Agr*) in the validation set for the best setups in each case (N is the number of branches), along with the number of parameters in the network (NPar).

Network	Layers	Non-parallel					Parallel				
		N	WAcc	Acc	Agr	NPar	N	WAcc	Acc	Agr	NPar
DR-CapsNet	2	-	65.9	95.2	94.5%	77.3M	4	59.4	95.4	95.1%	77.6M
	3	2	70.6	95.5	96.7%	11.3M	12	75.4	**96.1**	**97.4%**	73.2M
EM-CapsNet	2	-	**76.0**	94.7	96.1%	3.6M	2	75.4	93.9	94.6%	4.6M
	3	2	69.4	93.0	94.9%	3.1M	10	72.3	93.9	93.8%	30.1M
ResNeXt-50	–	–	73.8	94.9	96.2%	23.3M	–	–	–	–	–

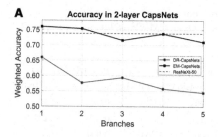
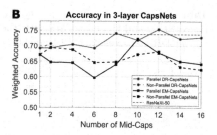

Fig. 2. Weighted accuracy in the validation set for the (A) 2-CapsLayer and (B) 3-CapsLayer networks, based on the number of branches (N) or Mid-Caps. For the 2-layer CapsNets, a branch 1 indicates the non-parallel setup.

counterparts). In general, this agrees with our hypothesis that parallelization allows each branch to detect an entity independently (in its Mid-Cap) without affecting the remaining Mid-Caps. The drawback is the high increase in parameters. Specifically, it increased 3M parameters per each new branch in the EM network (only 0.01M per each Mid-Cap in the non-parallel version) and 6M in the DR network (5M in the non-parallel version). However, the increase in computational time was not excessive (a 0–50% more). Overall, DR-CapsNet outperformed ResNeXt-50 with the right selection of branches (much higher *Acc* and *Agr*, Table 1), but EM-CapsNet was still most proficient with only 2-layers and no branching (none of the EM nets –parallel or not– outperformed ResNeXt-50 in *Acc* and *Agr*).

Size of the Capsules. The original CapsNets [6,19] employed 16 elements to encode a capsule, but it was not discussed the relevance of that size or whether capsules from different layers should have different sizes. For this, we experimented with several sizes (9, 16, and 25, and their equivalent matrices 3×3, 4×4, 5×5) in all possible combinations, and we found that (i) a size of 16 is the most optimal among our options, (ii) it is preferable –but not crucial– that all layers have the same capsule size, (iii) a smaller capsule (9) worked surprisingly well as long as all layers employ it, and (iv) a larger size (25) started overfitting.

Performance for Small Training Data. To evaluate this, we split the dataset in the following way: up to x images per class were used to create 5 folds, using 4 folds for training and 1 for testing, placing the remaining images above x in the test set (being $x = [10, 25, 50, 100, 250, 500, 1000]$). For the lowest x, classes were balanced, becoming more unbalanced as x increases. This was tested in the baseline ResNeXt-50 and the best CapsNets. Interestingly, our experiments depicted a similar behavior as x increased for all cases, which contradicts the assumption that CapsNets outperform CNNs in smaller datasets.

Overall Benefits of Our Proposed Networks. The non-parallel CapsNets showed convergence problems when a capsule layer became slightly large, which

Table 2. Precision and sensitivity for the different classes in a 5-fold CV, for the ResNeXt-50 and the best CapsNets: the default 2-layer EM-CapsNet, and the parallel 3-layer DR-CapsNet with 12 branches.

Type of cell	Images	ResNeXt-50		EM-CapsNet		Par. DR-CapsNet	
		Precision	Sensitivity	Precision	Sensitivity	Precision	Sensitivity
Neutrophil (segm.)	8,484	0.987	0.977	0.992	0.958	0.988	0.980
Neutrophil (band)	109	0.330	0.523	0.190	0.596	0.350	0.520
Lymphocyte (typ.)	3,937	0.988	0.924	0.968	0.964	0.991	0.930
Lymphocyte (atyp.)	11	0.125	0.091	0.667	0.182	0.750	0.273
Monocyte	1,789	0.913	0.908	0.918	0.911	0.930	0.917
Eosinophil	424	0.964	0.943	0.983	0.932	0.955	0.962
Basophil	79	0.732	0.658	0.630	0.798	0.658	0.705
Myeloblast	3,268	0.904	0.975	0.940	0.956	0.934	0.988
Promyelocyte	70	0.597	0.571	0.662	0.643	0.630	0.612
Promyelocyte (bil.)	18	0.393	0.611	0.282	0.611	0.465	0.611
Myelocyte	42	0.579	0.524	0.525	0.500	0.611	0.524
Metamyelocyte	15	0.316	0.400	0.333	0.267	0.380	0.466
Monoblast	26	0.436	0.923	0.414	0.923	0.418	0.820
Erythroblast	78	0.819	0.872	0.892	0.846	0.852	0.873
Smudge cell	15	0.563	0.600	0.643	0.600	0.633	0.600
Total	18,365	0.643	0.700	0.669	0.712	0.703	0.727

limits the number of entities to detect in the layer. However, there is no supportive evidence to believe that capsule layers should be fully connected between them. Indeed, a capsule (object) would only need to have connections to a reduced number of lower capsules (parts of the object), and different objects might be formed by completely different parts. Thus, branching would help to overcome the aforementioned limitation, although at the expense of a higher number of parameters. Moreover, the 3-layer CapsNets showed an interesting behaviour: the case with one single Mid-Cap was able to encode a generic white cell in just a capsule of 16 element that, subsequently, was transformed to 15 different classes. This highlights the power of capsules to encode entities and transformations. It also suggests that branching should be considered with heterogeneous datasets (groups of objects with morphological dissimilarities among the groups).

Final Results. Our proposed 3-layer parallel DR-CapsNet provided slightly better sensitivity and precision (Table 2), but there was not a clear pattern (the detection of low-represented classes was not highly improved by CapsNets). Overall, we believe that three layers of capsules is appropriate for this dataset because of the morphological structure of white cells: the CNNs might denoise the image from background red cells while retaining the important features from the white cells, Primary-Caps might encode those basic features into capsules, Mid-Caps might then encode whole entities (nucleus, cytoplasm, or even generic whole cells), and Class-Caps might simply be the connection (and transformation) of different Mid-Caps entities. Our experiments also seemed to suggest that losing the spatial information in the layer previous to the merging is the most appropriate approach to exploit branching, but we could not test that hypothesis in the EM network due to lack of time. Many other experiments could also be tested to further improve the performance: branches with different sizes, merging some branches at different layers, etc.

4   Conclusions

Our work suggests that, for the classification of white cells, original CapsNets (i) do not generally outperform a well-established CNN (ResNeXt-50) unless it is a simple 2-layer network, therefore (ii) adding more capsule layers is usually detrimental, (iii) they are not more robust for small training data, (iv) they tend to be very sensitive to the tuning parameters, (v) they are unable to converge if a layer contains too many capsules, and (vi) their rotational encoding does not seem to be outstanding. In contrast, our proposed parallel DR-CapsNet seems to better learn the viewpoint invariant representations (highest Agr), provides better accuracy (highest Acc), and does not suffer from convergence problems.

References

1. Afshary, P., Mohammadiy, A., Plataniotis, K.N.: Brain tumor type classification via Capsule Networks. In: 25th IEEE International Conference on Image Processing (ICIP), pp. 3129–3133, Athens, Greece (2018)
2. Anupama, M.A., Sowmya, V., Soman, K.P.: Breast cancer classification using Capsule Network with preprocessed histology images. In: 2019 International Conference on Communication and Signal Processing (ICCSP), Chennai, India (2019)
3. Dozat, T.: Incorporating Nesterov momento into Adam. In: International Conference on Learning Representations Workshop (ICLRW), San Juan, Puerto Rico (2016)
4. Gaddipati, D.J., Desai, A., Sivaswamy, J., Vermeer, K.A.: Glaucoma assessment from OCT images using Capsule Network. In: 41st Annual International Conference of the IEEE Engineering in Medicine and Biology Society (EMBC), pp. 5581–5584, Berlin, Germany (2019)
5. He, K., Zhang, X., Ren, S., Sun, J.: Deep residual learning for image recognition. In: 2016 IEEE Conference on Computer Vision and Pattern Recognition (CVPR), pp. 770–778, Las Vegas, NV, USA (2016)
6. Hinton, G.E., Sabour, S., Frosst, N.: Matrix capsules with EM routing. In: International Conference on Learning Representations (ICLR) (2018)
7. Hoogi, A., Wilcox, B., Gupta, Y., Rubin, D.L.: Self-attention Capsule Networks for object classification. arXiv 1904.12483 (2019)
8. Huang, G., Liu, Z., van der Maaten, L., Weinberger, K.Q.: Densely connected convolutional networks. In: 2017 IEEE Conference on Computer Vision and Pattern Recognition (CVPR), pp. 2261–2269, Honolulu, HI, USA (2017)
9. Iesmantas, T., Alzbutas, R.: Convolutional Capsule Network for classification of breast cancer histology images. In: 15th International Conference on Image Analysis and Recognition (ICIAR), Póvoa de Varzim, Portugal (2018)

752   J. P. Vigueras-Guillén et al.

10. Jiménez-Sánchez, A., Albarqouni, S., Mateus, D.: Capsule networks against medical imaging data challenges. In: Stoyanov, D., et al. (eds.) LABELS/CVII/STENT -2018. LNCS, vol. 11043, pp. 150–160. Springer, Cham (2018). https://doi.org/10. 1007/978-3-030-01364-6_17

11. Krizhevsky, A., Sutskever, I., Hinton, G.E.: ImageNet classification with deep convolutional neural networks. In: Advances in Neural Information Processing Systems (NIPS), pp. 1097–1105 (2012)

12. Kromm, C., Rohr, K.: Inception Capsule Network for retinal blood vessel segmentation and centerline extraction. In: IEEE 17th International Symposium on Biomedical Imaging (ISBI), pp. 1223–1226, Iowa City, IA, USA (2020)

13. LaLonde, R., Bagci, U.: Capsules for object segmentation. In: Medical Imaging with Deep Learning (MIDL) Conference. Amsterdam, The Netherlands (2018)

14. LaLonde, R., Kandely, P., Spampinatox, C., Wallacey, M.B., Bagci, U.: Diagnosing colorectal polyps in the wild with Capsule Networks. In: 17th IEEE International Symposium on Biomedical Imaging (ISBI), pp. 1086–1090, Iowa City, IA, USA (2020)

15. Liu, Y., Fu, Y., Chen, P.: WBCaps: a capsule architecture-based classification model designed for white blood cells identification. In: 41st Annual International Conference of the IEEE Engineering in Medicine and Biology Society (EMBC), pp. 7027–7030, Berlin, Germany (2019)

16. Matek, C., Schwarz, S., Spiekermann, K., Marr, C.: Human-level recognition of blast cells in acute myeloid Leukaemia with convolutional neural networks. Nat. Mach. Intell. 1, 538–544 (2019)

17. Mobiny, A., Lu, H., Nguyen, H.V., Roysam, B., Varadarajan, N.: Automated classification of apoptosis in phase contrast microscopy using Capsule Network. IEEE Trans. Med. Imaging 31(1), 1–10 (2019)

18. Nguyen, H., Blank, A., Dawson, H.E., Lugli, A., Zlobec, I.: Classification of colorectal tissue images from high throughput tissue microarrays by ensemble deep learning methods. Nat. Sci. Rep. 11, 2371 (2021)

19. Sabour, S., Frosst, N., Hinton, G.E.: Dynamic routing between capsules. In: Proceedings of the 31st International Conference on Neural Information Processing Systems (NIPS), pp. 3859–3869 (2017)

20. Shen, Y., Gao, M.: Dynamic routing on deep neural network for thoracic disease classification and sensitive area localization. In: Shi, Y., Suk, H.-I., Liu, M. (eds.) MLMI 2018. LNCS, vol. 11046, pp. 389–397. Springer, Cham (2018). https://doi. org/10.1007/978-3-030-00919-9_45

21. Szegedy, C., et al.: Going deeper with convolutions. In: 2015 IEEE Conference on Computer Vision and Pattern Recognition (CVPR), pp. 1–9, Boston, MA, USA (2015)

22. Xie, S., Girshick, R., Dollár, P., Tu, Z., He, K.: Aggregated residual transformations for deep neural networks. In: IEEE Conference on Computer Vision and Pattern Recognition (CVPR), pp. 5987–5995, Honolulu, HI, USA (2017)

Incorporating Isodose Lines and Gradient Information via Multi-task Learning for Dose Prediction in Radiotherapy

Shuai Tan[1], Pin Tang[1], Xingchen Peng[2], Jianghong Xiao[3], Chen Zu[4], Xi Wu[5], Jiliu Zhou[1,5], and Yan Wang[1(✉)]

[1] School of Computer Science, Sichuan University, Chengdu, China
[2] Department of Radiation Oncology, Cancer Center, West China Hospital, Sichuan University, Chengdu, China
[3] Department of Biotherapy, Cancer Center, West China Hospital, Sichuan University, Chengdu, China
[4] Department of Risk Controlling Research, JD.com, Beijing, China
[5] School of Computer Science, Chengdu University of Information Technology, Chengdu, China

Abstract. Radiation therapy has been widely used in the treatment of cancer. However, a high-quality radiotherapy plan often requires dosimetrists to tweak repeatedly in a trial-and-error manner based on experience, causing it quite time-consuming and subjective. In this paper, we present a multi-task dose prediction (MTDP) network to automatically predict the dose distribution from computer tomography (CT) image. Specifically, the MTDP network consists of three highly-related tasks: a main dose prediction task for generating fine-grained dose value for each pixel, an auxiliary isodose lines prediction task for providing coarse-grained dose range for each pixel, and an auxiliary gradient prediction task for capturing subtle gradient information such as radiation patterns and edges of the dose distribution map, to obtain a more accurate and robust dose distribution map. The three related tasks are integrated via a shared encoder, following the multi-task learning strategy. To strengthen the correlations of different tasks, we also introduce two additional constraints, i.e., isodose consistency loss and gradient consistency loss, to enforce the match between the dose distribution features produced by the two auxiliary tasks and the main task. The experiments conducted on an in-house dataset with 110 rectum cancer patients have demonstrated the effectiveness and superiority of our method compared with the state-of-the-art methods. Code is available at https://github.com/DeepMedLab/MTDP-network.

Keywords: Radiotherapy treatment · Dose prediction · Multi-task learning · Isodose lines · Gradient information

1 Introduction

As one of the essential treatments for cancer patients, radiotherapy has achieved enormous advancement in recent decades. When making the radiotherapy plan, it is of great

© Springer Nature Switzerland AG 2021
M. de Bruijne et al. (Eds.): MICCAI 2021, LNCS 12907, pp. 753–763, 2021.
https://doi.org/10.1007/978-3-030-87234-2_71

importance to distribute the clinically acceptable dose on the planning target volume (PTV) and minimize the dose distributed on the organs at risk (OARs). To achieve this goal, current radiotherapy plan is always tuned manually in a trial-and-error manner by dosimetrists, which is quite boring and time-consuming [1]. Besides, the quality of the plan might be highly variable among dosimetrists due to their different expertise and experience [2]. Therefore, it is particularly crucial to work out an efficient and robust method which can automatically predict the dose distribution for cancer patients to lighten the burden on dosimetrists.

In recent years, deep learning architectures have achieved significant progress, especially in the radiotherapy dose prediction tasks [4–13]. Particularly, inspired by the idea of U-net [3], Dan et al. [4] constructed a seven-level hierarchy U-net architecture to automatically predict dose distribution of prostate cancer. To ensure the information extracted from the earlier layers in the network can be effectively propagated to the deeper layers, Kearney et al. [6] proposed the DoseNet for the radiotherapy of prostate cancer patients by utilizing 3D U-net and residual blocks [14]. Song et al. [7] applied the DeepLab v3 + [15] for dose prediction of rectum cancer. Besides, inspired by the tremendous success of generative adversarial networks (GANs), Mahmood et al. [8] introduced the GAN to obtain 3D dosimetric results of oropharyngeal cancer, which outperforms the U-net based prediction model in terms of satisfying the clinical criteria. Cao et al. [9] proposed an adaptive multi-organ loss based GAN to generate the dose map for cervical cancer patients.

Although the deep learning methods have achieved promising performance, they always dedicate to predicting the dose map through a single task, neglecting some influential details of the dose distribution, such as isodose lines and gradient information. Such details are actually highly related to the dose prediction task. Without loss of generality, some examples are presented in Fig. 1, where (a) and (b) represent the computer tomography (CT) image and PTV segmentation of a rectum cancer patient, and (c) is the dose map planned by an experienced dosimetrist. In radiotherapy, it requires that the PTV reaches the prescribed dose while the dose deposition on the OARs is minimized, so the dose distribution tends to decrease rapidly from the PTV to its surroundings, showing the decline trend and the gradient properties in the dose map. Concretely, by discretizing the continuously distributed dose values in Fig. 1(c), the dose map can be divided into several different areas, resulting in an isodose lines map which reflects the coarse-grained dose range for each pixel, as shown in Fig. 1(d). In addition, we visually observe the gradient properties, i.e., radiation patterns and the obvious edges of the dose map, from Fig. 1(c). To capture such gradient properties of dose map, we employ a Sobel operator to obtain the gradient map, as shown in Fig. 1(e).

In this paper, to capture these influential but neglected information, we harness the spirit of multi-task learning (MTL) to integrate the highly related tasks and propose a novel end-to-end framework, namely multi-task dose prediction (MTDP) network, to automatically predict the clinically acceptable dose distribution of rectum cancer. The contributions of this paper can be summarized as follows: (1) Compared with the previous single-task dose prediction methods, our method consists of three highly-related tasks: a main dose prediction task, an auxiliary isodose lines prediction task and an auxiliary gradient prediction task. The main dose prediction task aims at generating the fine-grained dose value for each pixel while the auxiliary isodose lines prediction task, in contrast, is designed to provide the coarse-grained dose range for each pixel. In this manner, multi-granularity information of the dose distribution can be obtained to produce a more robust dose distribution map. Meanwhile, the auxiliary gradient prediction task tries to capture the subtle gradient information such as radiation patterns and edges of the dose distribution. By integrating these three related tasks into an end-to-end model, we argue that each task could benefit from the relevant features jointly learnt by other tasks. (2) Considering that traditional MTL methods always integrate different tasks by sharing the shallow layers, neglecting the correlations of the output layers for different tasks, we introduce two additional constraints, i.e., isodose consistency loss and gradient consistency loss, to enforce the dose distribution features produced by the auxiliary tasks matching well with those generated by the main task, thus leading to a more accurate dose prediction without incurring additional parameters. (3) Our model in this study can be employed as a clinical guidance tool for radiotherapy treatment planning. Concretely, the dose map generated by our model can be converted into the appropriate objective functions and provides dosimetrists with an initial point close to the ideal plan. In this manner, the trial-and-error steps and planning time can be reduced. In addition, oncologists can also have an accurate expectation of the treatment plan through the predicted dose maps. Experiments conducted on an in-house dataset of 110 rectum cancer patients show that our method has obvious superiority compared with the state-of-the-art methods.

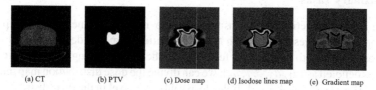

(a) CT   (b) PTV   (c) Dose map   (d) Isodose lines map   (e) Gradient map

Fig. 1. Examples of a rectum cancer patient.

2 Methodology

The proposed network is illustrated in Fig. 2, which jointly implements three relevant tasks, i.e., a main dose prediction task, an auxiliary isodose lines prediction task and an auxiliary gradient prediction task. Compared with the single-task model, the auxiliary tasks of our multi-task model provide richer supervision, yielding stronger features which help to increase the dose prediction accuracy. The whole network takes the original CT images together with PTV segmentation results of the rectal tumor as inputs and outputs the corresponding dose distribution maps, isodose lines maps and gradient maps, respectively. Note that the encoder is shared by the three tasks and has the same parameters while the parameters in the task-specific decoders are different for generating specific target images. In this fashion, the encoder can be forced to learn features related to the isodose lines and gradient properties of the dose distribution map, leading to a low disparity between the output and ground-truth. It should be pointed out that we employ two extra constraints termed isodose consistency loss and gradient consistency loss for decoder features refinement, which is different from vanilla MTL methods. Moreover, to process the tremendous non-local features in the dose prediction task, we embed a self-attention (SA) [16] module in our framework to capture the long-range dependencies. The details of the network are described in the following sub-sections.

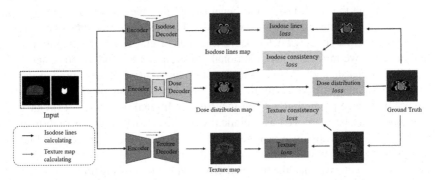

Fig. 2. Schematic view of the proposed MTDP network.

2.1 Auxiliary Tasks

Isodose Lines Prediction Task. Inspired by the iso-height lines in geography, we assign an isodose lines map to each dose distribution map and design a novel isodose lines prediction task to capture this feature. Specifically, assuming that there are N isodose lines denoted as $\{d_1, d_2, \ldots, d_n\}$, we can divide the dose distribution map into $N - 1$ areas as $\{A_{1,2}, A_{2,3}, \ldots, A_{n-1,n}\}$. If the dose value of a pixel in the dose distribution map falls between d_i and d_{i+1}, it is labeled by $A_{i,i+1}$. In this way, the isodose lines maps can be easily obtained from the dose distribution maps and are further used as the labels of this task. Different from the dose prediction task which is a regression task, the isodose lines prediction task is an $(N - 1)$-label classification task actually.

Gradient Prediction Task. We employ the commonly used Sobel operator to extract the gradient map from the manually optimized dose map. As shown in Fig. 2, the obtained gradient maps are further utilized as the labels to train the gradient prediction network. Through this manner, the shared encoder is forced to extract the features related to gradient properties, leading to a more delicate dose distribution map.

2.2 Architecture

We employ three encoder-decoder subnetworks for the three tasks. Specifically, the shared encoder is composed of 5 down-sampling blocks and each block consists of two convolutional layers with a kernel size of 3×3, a batch normalization (BN) layer, and a rectified linear unit (ReLU) activation function. Moreover, each block is followed by a max-pooling layer except the last one. The number of feature channels is doubled and the feature size is halved after each block. Correspondingly, all the three task-specific decoders have 8 3×3 convolutional layers and every two convolutional layers accompanying BN and ReLU form an up-sampling block. Contrary to the down-sampling blocks in the encoder, each up-sampling block is followed by a deconvolutional layer to double the size of the feature maps and halve the channel numbers. Similar to [3], skip connections are also applied for multi-level feature aggregation and feature reuse. The outputs of the three subnetworks are different due to the different tasks. To be specific, given the input image $I^{H \times W \times 2}$, the dose prediction task and gradient prediction task respectively generate a dose distribution map $D^{H \times W \times 1}$ and a gradient map $G^{H \times W \times 1}$, while isodose lines prediction task outputs $N-1$ probability maps $P^{H \times W \times (N-1)}$. Hence, we append a 1×1 convolutional layer with a sigmoid function at the end of each decoder to generate the final outputs with corresponding size. Moreover, we apply a self-attention module to the feature maps passed by the deepest two skip connections between the shared encoder and the decoder of dose prediction task, which can extract non-local features significantly.

2.3 Objective Functions

Multi-task Learning Loss Functions. In our MTDP model, each task has its own loss function. For the dose prediction task, we propose to use an L1 loss as follows:

$$Loss_{dose} = \frac{1}{W \times H} \sum_{i=1}^{W \times H} \left\| Dec_{dose}(F)^i - y^i \right\|_1, \tag{1}$$

where F is the feature maps obtained by the shared encoder from the input CT slices and the PTV segmentation results. y denotes the manually optimized dose map (ground-truth). The superscript i denotes pixel in the slice. $Dec_{dose}(\cdot)$ represents the decoder designed for the dose prediction task.

As for the isodose lines prediction task, we apply a cross entropy (CE) loss since it is a $(N-1)$-label classification task, which is calculated as follows.

$$Loss_{iso} = -\frac{1}{W \times H} \sum_{i=1}^{W \times H} \sum_{c}^{N-1} A^{i,c} \cdot log\left(Dec_{iso}(F)^{i,c}\right), \tag{2}$$

where $Dec_{iso}(\cdot)$ is the decoder of the isodose lines prediction task. $A^{i,c}$ is the classification label for pixel i and category c, and it is generated from the ground-truth y by isodose lines calculating as mentioned in Sect. 2.1. $Dec_{iso}(F)^{i,c}$ represents the probability that pixel i belongs to category c.

The loss function for the gradient prediction task is an L2 loss defined as:

$$Loss_{gra} = \frac{1}{W \times H} \sum_{i=1}^{W \times H} \left\| Dec_{gra}(F)^i - S(y)^i \right\|_2, \tag{3}$$

where $Dec_{gra}(\cdot)$ is the decoder of the gradient prediction task and $S(\cdot)$ is the Sobel operator.

Consistency Loss Functions. We also impose two consistency constraints, i.e., isodose consistency loss $Loss_{IC}$ and gradient consistency loss $Loss_{GC}$, to enforce the match between the dose distribution features produced by the two auxiliary tasks and the main task. For isodose lines information, considering the pixel i belonging to the area $\{d_m^i, d_{m+1}^i\}$, the dose value of each pixel is constrained to be in the correct area by using Eq. 4.

$$Loss_{IC} = \frac{1}{W \times H} \sum_{i=1}^{W \times H} \begin{cases} \left\| Dec_{dose}(F)^i - d_{m+1}^i \right\|_2, & Dec_{dose}(F)^i > d_{m+1}^i \\ 0, & d_m^i \le Dec_{dose}(F)^i > d_{m+1}^i \\ \left\| Dec_{dose}(F)^i - d_m^i \right\|_2, & Dec_{dose}(F)^i > d_m^i \end{cases}$$
$$(4)$$

For gradient related features, we derive a pseudo gradient map from the predicted dose distribution map with $S(\cdot)$. Then, we use an L2 loss to minimize the gap between the pseudo gradient map and the gradient map generated from ground-truth.

$$Loss_{GC} = \frac{1}{W \times H} \sum_{i=1}^{W \times H} S \left\| (Dec_{dose}(F))^i - S(y)^i \right\|_2 \tag{5}$$

The total loss function in the training phase is formulated by the weighted sum of the above losses.

$$Loss_{total} = \lambda_1 Loss_{dose} + \lambda_2 Loss_{iso} + \lambda_3 Loss_{gra} + \lambda_4 Loss_{IC} + \lambda_5 Loss_{GC}, \tag{6}$$

where λ s are hyper-parameters to balance these terms.

2.4 Training Details

Our network is implemented with PyTorch framework with the default initial weights and is trained for 200 epochs totally with Adaptive moment estimation (Adam) optimizer. Initial learning rates are different for different model components. Specifically, we set 1e-5 for the shared encoder, and 1e-5, 3e-6, 4e-4 for the decoders of dose prediction task, gradient prediction task and isodose lines prediction task, respectively. Then the learning rates gradually decrease by 1% after training for 100 epochs. Moreover, we set the number of areas, i.e., $N - 1$, in the isodose lines prediction task as 7. $\lambda_1, \lambda_2, \lambda_4$ and λ_5 in Eq. 6 equal to 10, 5, 5 and 5, respectively. Note that, λ_3 is set to 20 for the initial 30 epochs since the gradient features are harder to learn. Then, it gradually decreases to 5 in the next 15 epochs and keeps unchanged in the remaining 155 epochs.

3 Experiment and Analysis

3.1 Dataset and Evaluation

We evaulate our method on an in-house dataset which consists of 110 rectal cancer patients. Concretely, the images of 88 patients are randomly selected as the training set and the remaining images of the other 22 patients are treated as test samples. Then, the 3D CT images, with the size of $512 \times 512 \times 172$, are split into 2D slices and the slices without dose value are abandoned for the sake of computational efficiency. After that, totally 7228 slices are obtained for experiments. We assess the performance of the proposed algorithm using evaluation metrics including conformity index (CI) [17], heterogeneity index (HI) [18], and the absolute differences (AD) at patient level by stacking the predicted 2D slices successively into a complete 3D dose map. To intuitively illustrate the disparity between CI and HI metrics of the different methods and the ground-truth, we also calculate the difference $|\Delta|$. Moreover, we introduce the dose volume histogram (DVH) [19] as another important indicator of the dose prediction accuracy. For intuitive comparison, the DVHs of the ground-truth and the predicted plan are plotted together. If the DVH curves of the ground-truth and the predicted plan are closer, we can affirm that the predicted plan is more accurate.

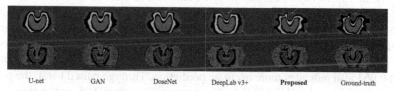

U-net   GAN   DoseNet   DeepLab v3+   **Proposed**   Ground-truth

Fig. 3. Visualization of the isodose maps (first row) and the gradient maps (second row) of the SOTAs and the proposed method. Obvious improvement achieved by our method is pointed by the red arrows and circles. (Color figure online)

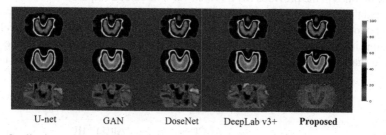

U-net   GAN   DoseNet   DeepLab v3+   **Proposed**

Fig. 4. Qualitative comparison between the SOTAs and the proposed. From top to bottom: the ground-truth, predicted dose distribution map and the difference map.

Table 1. Quantitative comparison with state-of-the-art methods.

Methods	CI	$|\Delta|(\%)$	p-value	HI	$|\Delta|(\%)$	p-value	AD
U-net [5]	0.835	2.68	0.0023	0.068	18.07	0.0326	0.078
GAN [8]	0.841	1.98	0.0425	0.071	14.46	0.0470	0.071
DoseNet [6]	0.839	2.21	0.0355	0.073	12.05	0.0387	0.074
DeepLab v3 + [7]	0.847	1.28	0.0455	0.070	15.66	0.0505	0.070
Proposed	**0.854**	**0.47**	–	**0.081**	**2.41**	–	**0.063**
Ground-truth	0.858	–	–	0.083	–	–	–

3.2 Comparison with State-Of-The-Art Methods

To verify the superiority of our network, we take several state-of-the-art (SOTA) methods along the dose prediction direction for comparison, including U-net [5], GAN [8], DoseNet [6] and Deeplab v3 + [7]. The quantitative comparison is given in Table 1. As can be seen, compared with SOTAs, the CI and HI of the proposed method are much closer to the ground-truth, achieving a difference of 0.47% and 2.41%, and the AD of 0.063 also indicates more accurate prediction. Through paired t-test, the p-value between the proposed and the SOTAs are almost all less than 0.05, suggesting that the improvements over SOTAs are statistically meaningful.

To prove that our method can learn the isodose lines and gradient information effectively, visualizations of these two types of maps are displayed in Fig. 3. As observed, the isodose lines in the previous SOTAs are over-smoothed compared with the proposed model and the ground-truth, as indicated by the red arrows. Furthermore, the gradients of the previous SOTAs are relatively ambiguous but the proposed model can excellently restore this information. Besides, Fig. 4 gives the difference maps of these methods and it is clear that the proposed MTDP achieves the least disparity from the ground truth.

Table 2. Quantitative comparison with the ablation models.

Methods	CI	$|\Delta|(\%)$	HI	$|\Delta|(\%)$
(1) DPT	0.835 ± 0.019	2.68	0.068 ± 0.020	18.07
(2) DPT + GPT	0.836 ± 0.022	2.56	0.096 ± 0.023	12.04
(3) DPT + GPT + $Loss_{GC}$	0.838 ± 0.023	2.39	0.077 ± 0.010	7.23
(4) DPT + GPT + IPT + $Loss_{GC}$	0.842 ± 0.023	1.86	0.074 ± 0.015	10.84
(5) DPT + GPT + IPT + $Loss_{GC}+Loss_{IC}$	0.849 ± 0.025	1.04	0.074 ± 0.011	10.84
(6) Proposed	**0.854 ± 0.018**	**0.47**	**0.081 ± 0.024**	**2.41**
Ground-truth	0.858 ± 0.035	–	0.083 ± 0.017	–

3.3 Ablation Study

To investigate the effectiveness of the components of the proposed method, we conduct the ablation study in a progressive way. The experiment arrangement can be concluded

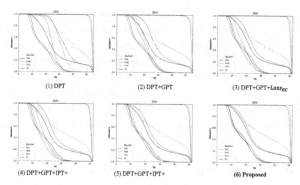

Fig. 5. The DVHs of the manually optimized plan (solid line) and the predicted results (dashed line) in terms of PTV and OARs.

as: (1) U-net as the main dose prediction task (DPT), (2) dose prediction task + gradient prediction task (DPT + GPT), (3) dose prediction task + gradient prediction task + gradient consistency loss (DPT + GPT + $Loss_{GC}$), (4) all three tasks + gradient consistency loss (DPT + GPT + IPT + $Loss_{GC}$), (5) all three tasks + two consistency losses (DPT + GPT + IPT + $Loss_{GC}$+$Loss_{IC}$), (6) DPT + GPT + IPT + $Loss_{GC}$+$Loss_{IC}$+SA (proposed).

Effectiveness of the Auxiliary Tasks: To verify the contributions of the two auxiliary tasks, we compare the prediction accuracy using (1) and (2) for gradient prediction task, and using (3) and (4) for isodose lines prediction task. As can be seen in Table 2, after adding the gradient prediction task, the $|\Delta|$ s of CI and HI drop by 0.12% and 6.03%, respectively. As for the isodose lines prediction task, the CI value of (4) improves 0.04 absolutely over (3), leading to a decrease of 0.53% on $|\Delta|$. For qualitative comparison, we plot DVHs in Fig. 5, from which we can find that the dashed lines (predicted plan) of (2) and (4) are closer to the solid lines (ground truth), compared with (1) and (3). Both qualitative and quantitative results demonstrate the contributions of the two auxiliary tasks.

Effectiveness of Consistency Constraints: To study the effectiveness of the proposed two consistency constraints, i.e., $Loss_{GC}$ and $Loss_{IC}$, we compare the experimental results between (2) and (3), (4) and (5), respectively. As shown in Table 2, the gradient consistency loss and the isodose consistency loss respectively achieve 0.17% and 0.82% improvements on $|\Delta|$. Moreover, the solid lines and dashed lines of (3) and (5) in the DVHs are much closer than (2) and (4) as shown in Fig. 5.

Besides, the experimental results of (5) and (6) in both Table 2 and Fig. 5 clarify the contribution of the self-attention module employed in this work.

4 Conclusion

In this paper, we propose a multi-task dose prediction network (MTDP) to automatically predict the dose distribution for the radiotherapy of rectal cancer patients. Assuming that the isodose lines and gradient information are beneficial to the accuracy of the dose prediction, we explore an isodose lines prediction task and a gradient prediction task as two auxiliary tasks to help improving the performance of the main dose prediction task. The proposed model is trained in an end-to-end manner which integrates the three tasks via a shared encoder. To strengthen the correlations of the outputs for different tasks, we devise two additional constraints, i.e., isodose consistency loss and gradient consistency loss, for further performance improvement. The experimental results demonstrate that our method has obvious superiority compared with the state-of-the-art methods.

Acknowledgments. This work is supported by National Natural Science Foundation of China (NSFC 62071314) and Sichuan Science and Technology Program (2021YFG0326, 2020YFG0079).

References

1. Murakami, Y., et al.: Possibility of chest wall dose reduction using volumetric-modulated arc therapy (VMAT) in radiation-induced rib fracture cases: comparison with stereotactic body radiation therapy (SBRT). J. Radiat. Res. **59**(3), 327–332 (2018)
2. Nelms, B.E., et al.: Variation in external beam treatment plan quality: an inter-institutional study of planners and planning systems. Pract. Radiat. Oncol. **2**(4), 296–305 (2012)
3. Ronneberger, O., Fischer, P., Brox, T.: U-net: convolutional networks for biomedical image segmentation. In: Navab, N., Hornegger, J., Wells, W.M., Frangi, A.F. (eds.) Medical Image Computing and Computer-Assisted Intervention – MICCAI 2015: 18th International Conference, Munich, Germany, October 5-9, 2015, Proceedings, Part III, pp. 234–241. Springer International Publishing, Cham (2015). https://doi.org/10.1007/978-3-319-24574-4_28
4. Nguyen, D., et al.: Dose prediction with U-net: a feasibility study for predicting dose distributions from contours using deep learning on prostate IMRT patients. arXiv preprint arXiv: 1709.09233 (2017)
5. Nguyen, D., et al.: A feasibility study for predicting optimal radiation therapy dose distributions of prostate cancer patients from patient anatomy using deep learning. Sci. Rep. **9**(1), 1–10 (2019)
6. Kearney, V., et al.: DoseNet: a volumetric dose prediction algorithm using 3D fully-convolutional neural networks. Phys. Med. Biol. **63**(23), 235022 (2018)
7. Song, Y., et al.: Dose prediction using a deep neural network for accelerated planning of rectal cancer radiotherapy. Radiother. Oncol. **149**, 111–116 (2020)
8. Mahmood, R., et al.: Automated treatment planning in radiation therapy using generative adversarial networks. In: Machine Learning for Healthcare Conference. PMLR (2018)
9. Cao, C., et al.: Adaptive multi-organ loss based generative adversarial network for automatic dose prediction in radiotherapy. In: IEEE 18th International Symposium on Biomedical Imaging. IEEE (2021)
10. Nguyen, D., et al.: 3D radiotherapy dose prediction on head and neck cancer patients with a hierarchically densely connected U-net deep learning architecture. Phys. Med. Biol. **64**(6), 065020 (2019)
11. Murakami, Y., et al.: Fully automated dose prediction using generative adversarial networks in prostate cancer patients. PLoS ONE **15**(5), e0232697 (2020)
12. Babier, A., et al.: Knowledge-based automated planning with three-dimensional generative adversarial networks. Med. Phys. **47**(2), 297–306 (2020)

13. Barragán-Montero, A.M., et al.: Three-dimensional dose prediction for lung IMRT patients with deep neural networks: robust learning from heterogeneous beam configurations. Med. Phys. **46**(8), 3679–3691 (2019)
14. He, K., et al.: Deep residual learning for image recognition. In: Proceedings of the IEEE Conference on Computer Vision and Pattern Recognition (2016)
15. Chen, L.-C., et al.: Encoder-decoder with atrous separable convolution for semantic image segmentation. In: Ferrari, V., Hebert, M., Sminchisescu, C., Weiss, Y. (eds.) ECCV 2018. LNCS, vol. 11211, pp. 833–851. Springer, Cham (2018). https://doi.org/10.1007/978-3-030-01234-2_49
16. Zhang, H., et al.: Self-attention generative adversarial networks. In: International Conference on Machine Learning. PMLR (2019)
17. Paddick, I.: A simple scoring ratio to index the conformity of radiosurgical treatment plans: technical note. J. Neurosur. **93**(supplement_3), 219–222 (2000)
18. Helal, A., Abbas, O.: Homogeneity index: effective tool for evaluation of 3DCRT. Pan Arab J. Oncol. **8**(2), 20–24 (2015)
19. Graham, M.V., et al.: Clinical dose–volume histogram analysis for pneumonitis after 3D treatment for non-small cell lung cancer (NSCLC). Int. J. Radiat. Oncol. Biol. Phys. **45**(2), 323–329 (1999)

Sequential Learning on Liver Tumor Boundary Semantics and Prognostic Biomarker Mining

Jie-Neng Chen[1(✉)], Ke Yan[2], Yu-Dong Zhang[3], Youbao Tang[2], Xun Xu[3], Shuwen Sun[3], Qiuping Liu[3], Lingyun Huang[4], Jing Xiao[4], Alan L. Yuille[1], Ya Zhang[5], and Le Lu[2]

[1] Johns Hopkins University, Baltimore, USA
[2] PAII Inc, Bethesda, USA
[3] The First Affiliated Hospital of Nanjing Medical University, Nanjing, China
[4] Ping An Technology, Shenzhen, China
[5] Shanghai Jiao Tong University, Shanghai, China

Abstract. The boundary of tumors (hepatocellular carcinoma, or HCC) contains rich semantics: capsular invasion, visibility, smoothness, folding and protuberance, etc. Capsular invasion on tumor boundary has proven to be clinically correlated with the prognostic indicator, microvascular invasion (MVI). Investigating tumor boundary semantics has tremendous clinical values. In this paper, we propose the first and novel computational framework that disentangles the task into two components: spatial vertex localization and sequential semantic classification. (1) A HCC tumor segmentor is built for tumor mask boundary extraction, followed by polar transform representing the boundary with radius and angle. Vertex generator is used to produce fixed-length boundary vertices where vertex features are sampled on the corresponding spatial locations. (2) The sampled deep vertex features with positional embedding are mapped into a sequential space and decoded by a multilayer perceptron (MLP) for semantic classification. Extensive experiments on tumor capsule semantics demonstrate the effectiveness of our framework. Mining the correlation between the boundary semantics and MVI status proves the feasibility to integrate this boundary semantics as a valid HCC prognostic biomarker.

1 Introduction

Microvascular invasion (MVI) has been clinically identified as a prognostic factor of hepatocellular carcinoma (HCC) after surgical treatment, whereas it is undetectable preoperatively on diagnostic imaging [3,19]. Microscopic features of HCC such as tumor size, capsule and margin are hypothesized as important predictors of MVI [1,27]. Tumor capsule, specific for hepatocarcinogenesis, was observed in 70% of progressed HCC [12]. Histologically, tumor capsule contains two layers: the inner layer is composed of tight fibrous tissue containing thin,

© Springer Nature Switzerland AG 2021
M. de Bruijne et al. (Eds.): MICCAI 2021, LNCS 12907, pp. 764–774, 2021.
https://doi.org/10.1007/978-3-030-87234-2_72

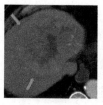

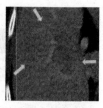
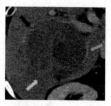

(a) Class 1: complete capsule   (b) Class 2: incomplete capsule with indistinct disruption   (c) Class 3: incomplete capsule with significant disruption   (d) Combination of 3 classes: Green-Class1; Blue-Class2; Red-Class3

Fig. 1. Categories of tumor capsule invasion as a type of boundary semantics.

slit-like vascular channels, and the outer layer is composed of looser fibrovascular tissue [6]. Tumor capsule invasion increases the risk of vascular invasion and intrahepatic metastasis, generally indicating poor cancer patient prognosis. However, despite its strong potential, assessing the presence and integrity of radiological HCC capsule is an expert-based subjective evaluation. Inter- and intra-observer variations and lack of reproducibility are the major roadblocks limiting its wide adoption [8]. There is a critical unmet need to develop new yet effective computational methods and means to objectively and quantitatively examine the capsular invasion and focal extensional nodule to predict the precision prognosis of patients with HCC.

The capsule in radiology can be interpreted as the semantics on tumor boundary, and this problem falls into learning the tumor boundary semantics, i.e., dense classification on the boundary pixels. We propose a novel framework that disentangles the task into two pillars: the spatial localization and the sequential classification. (1) The spatial localization aims to precisely identify the vertices on tumor boundary that is extracted from predicted tumor mask. A polar coordinate transform is used to represent the boundary with radius and angle so that the boundary can be divided to N grids with equidistant angle. Then we perform an efficient vertex generator to produce the localized vertices' coordinates. (2) Our sequential learning tackles the semantic classification on the localized vertices' coordinates via sampled deep vertex features (that are concatenated with positional embedding). Finally the formed sequence features are decoded by a multilayer perceptron (MLP) for semantic classification.

To the best of our knowledge, this is the first work to solve the dense tumor boundary semantics mining problem and formulate it in a sequential learning manner. The polar coordinate transform enables us to obtain spatially uniform boundary coordinates. The vertex sequential features are sampled from multiscale pyramid features, permitting to naturally integrate low-/mid-/high-level cues on boundary semantics. We demonstrate the effectiveness of our approach on two tumor boundary semantic datasets: capsular invasion (CAP) and focal extensional nodule (FEN). Our method improves the baseline of entangled optimization by 23.26% F1 score on CAP dataset and 10.32% F1 score on FEN dataset. Moreover, we conduct a study of prognostic biomarker mining to validate the clinical correlation between boundary semantics and MVI status.

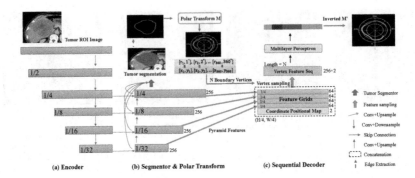

Fig. 2. Overview of our tumor boundary semantics modeling framework.

Previous Work. UPI-Net is proposed in [17] to detect semantic contour in Placental Ultrasound by binary boundary segmentation which is also studied in nature images by [2,9,20]. In contrast, we detect multi-classes boundary semantics independent to object semantic and/or instance category. Polar representation is used [18] to localize cell via star-convex polygons and to model foreground instances for instance segmentation [24,25]. Sequential learning is a common task in natural language processing such as machine translation [21,23], and various vision tasks [4,5,7,16]. Moreover, to tackle with the class imbalance problem in sequential learning, Li *et al.* [13] introduced a sequential dice loss into NLP tasks as the training objective, which is adopted in our work.

2   Method

Given a tumor RoI image $\mathbf{x} \in \mathbb{R}^{H \times W \times C}$, our goal is to predict the corresponding pixel-wise label map along tumor boundary. Unlike existing approaches of directly training a segmentation network (*e.g.*, U-Net), our method converts the problem into conducting the sequential prediction on a 1D band label map $\mathbf{y} \in \mathbb{R}^{N_{angle}}$. Our overall framework is depicted in Fig. 2.

2.1   Tumor Boundary Spatial Localization

Our encoder is initialized with a ResNet-50 network [10]. ResNet features in scale $\mathbf{S} \in \{\frac{1}{4}, \frac{1}{8}, \frac{1}{16}, \frac{1}{32}\}$ will pass through the U-shape blocks, where each block consists of a 2×upsampling operator, a skip-connection, a 3 × 3 convolution layer, and a ReLU layer successively. Multi-scale pyramid features are generated with each scale feature $\mathbf{x} \in \mathbb{R}^{HS \times WS \times 256}$. We first build a tumor segmentation model (See Fig. 2.b) on top of the multi-scale pyramid features that upsample every scale feature into scale 1/4 and merge them with an add operator, followed by a 3 × 3 convolution layer, 4× upsampling operator and a softmax activation layer to predict the binary tumor mask $\mathbf{y_t} \in \mathbb{R}^{H \times W}$. This tumor segmentor is trained with Dice loss and cross-entropy loss. After obtaining the tumor mask, we employ the residual of the dilation and the erosion of predicted tumor mask as in [22] to generate the tumor boundary. Gaussian blurring with a 5×5 kernel is utilized to make the boundary thickness closer to the human annotation and avoid the discontinuity in the polar coordinate.

Polar coordinates is a two-dimensional coordinate system where each point on a plane is determined by a distance from a reference point and an angle from a reference direction. The reference point (analogous to the origin of a Cartesian coordinate system) is called the pole, and the ray from the pole in the reference direction is the polar axis. The distance from the pole is called the radial coordinate, and the angle is called the angular coordinate. Hence, (x_i, y_i) in Cartesian coordinate system is denoted as (r_i, θ_i) in polar coordinate system M; the pole $(0, 0)$ in polar system is exactly the centroid of tumor mask (x_c, y_c) in Cartesian system. Cartesian position of the vertex can be recovered from the inverted transform M', where $x_i = r_i \cos \theta_i + x_c$ and $y_i = r_i \sin \theta_i + y_c$.

We propose an efficient vertex generator to produce N boundary vertices $(x_1, y_1), \ldots, (x_N, y_N)$. N rectangle grids in polar coordinate system standing for N rays with equidistant angle $\Delta\theta = \frac{360°}{N}$ are generated. Specifically, grid $k \in \{1, 2, \ldots, N\}$ is filled with a set of candidate vertices $(r_g, \theta_g) \in \mathbb{G}^{R \times 3}$ in polar representation, where $\theta_g \in \{(k-1) * \Delta\theta, k * \Delta\theta, (k+1) * \Delta\theta\}$, $r_g \in \{r | \theta = \theta_g\}$, and R approximates to the $\max(r | \theta = k * \Delta\theta) - \min(r | \theta = k * \Delta\theta)$. For robustness, we randomly sample a vertex point each time in N grids iteratively as a data augmentation strategy.

2.2 Deep Sequential Learning on Tumor Semantics

There are four scales of features from the feature pyramid (in Sect. 2.1) being processed to the same feature size $\frac{H}{4} \times \frac{W}{4} \times 64$ with bilinear upsampling and 3×3 convolution, and concatenated channel-wise to generate the feature $\mathbf{x_p} \in \mathbb{R}^{\frac{H}{4} \times \frac{W}{4} \times 256}$. To retain the positional information of each tumor boundary vertex and to ensure the structural relationship to be learned, we make use of the coordinate positional map [14,15] $\mathbf{x_{coor}} \in \mathbb{R}^{\frac{H}{4} \times \frac{W}{4} \times 2}$, where the channels representing the x and y Cartesian coordinates are normalized to $[-1, 1]$. Last we concatenate $\mathbf{x_p}$ and $\mathbf{x_{coor}}$ channel-wise to form the feature grids $\mathbf{x_g} \in \mathbb{R}^{\frac{H}{4} \times \frac{W}{4} \times 256+2}$. Given an arbitrary point (x_i, y_i), a corresponding grid of feature $\in \mathbb{R}^{258}$ can be sampled from $\mathbf{x_g}$. Recall that each time the efficient vertex generator will generate a set of boundary vertices $(x_1, y_1), (x_2, y_2), \ldots, (x_N, y_N)$, where total N grids of feature will be sampled from $\mathbf{x_g}$ and further sequentialized to $\mathbf{x_{seq}} \in \mathbb{R}^{N \times 258}$. A multilayer perceptron (MLP) is adopted to decode $\mathbf{x_{seq}} \in \mathbb{R}^{N \times 258}$ into 1D sequential classification prediction of $f(\mathbf{x}) \in \mathbb{R}^N$. The MLP contains two layers with a GELU non-linearity and a softmax activation. To alleviate the class imbalance where statistically the first class makes accounts for 70%, we exploit the 1D sequential dice loss [13] with cross entropy loss as the training objective for the sequential decoding. Let $\mathcal{D}$ be the dataset and the labeled data pair

$\mathbf{x}_i, \mathbf{y}_i \in \mathcal{D}$, the sequential loss $\mathcal{L}_{seq}$ is formulated as:

$$\mathcal{L}_{seq} = 1 - \underbrace{\sum_{\mathbf{x}_i, \mathbf{y}_i \in \mathcal{D}} 2 \frac{\sum f(\mathbf{x}_i) \mathbf{y}_i}{\sum f(\mathbf{x}_i) + \sum \mathbf{y}_i}}_{1D \ dice \ loss} - \underbrace{\sum_{\mathbf{x}_i, \mathbf{y}_i \in \mathcal{D}} \mathbf{y}_i \cdot \log\left(f(\mathbf{x}_i)\right)}_{cross \ entropy \ loss} \quad (1)$$

For clarification, $X_p, X_{coor}, X_g, X_{seq}$ correspond to "Feature Grids", "Coordinate Positional Map", "the combination of Feature Grids and Coordinate Positional Map", "Vertex Feature Sequence" in Fig. 2 respectively.

2.3 Prognostic Tumor Biomarker Mining

Microvascular invasion (MVI) score is a significant prognostic indicator of HCC in pathological imaging findings. We aim to fully exploit the correlation between MVI and our radiological imaging measurements. For each patient with multiple slices of images, we inference the data using the algorithm described above in a slice-by-slice fashion and stack the results as the prediction. The number of pixels of each class in the 3D tumor boundary is counted and divided by the total number of pixels to obtain a feature vector with three variables (e.g. [0.6, 0.1, 0.3]), where two of them are independent variables making up the patient-specific capsular biomarker. A logistic regression (LR) classifier is employed to analyze the correlation between the capsular biomarker and MVI.

3   Experiments and Discussion

Dataset Collection. A total of 364 unique patients (4049 axial slices) with pathologically confirmed liver tumor (HCC) were included in our study, which are split to 193: 58: 113 for training, validation and testing, respectively. All patients underwent standard multi-phase contrast-enhanced abdominal CT imaging within 2 weeks before surgery. *Capsular invasion* (**CAP**) was categorized using the following three-point scale: (1) complete capsule or invisible capsule with smooth tumor margin; (2) incomplete capsule with indistinct disruption; or (3) incomplete capsule with significant disruption. We also collect a type of tumor boundary semantics named *focal extensive nodule* (**FEN**) to measure the degree of protruding into the non-tumor parenchyma. FEN was assessed via a three-point criterion: (1) smooth margin without FEN; (2) slight FEN (the number of FEN is less than 3); or (3) significant FEN (the number of FEN is 3 or more). CAP and FEN classes were labeled manually using the referring three-point criteria by three board-certified radiologists on 5mm portal-venous CT images slice-by-slice. For the prognostic biomarker mining, we collected the patient-level labels of **microvascular invasion (MVI)** from their associated histopathologic examinations, where 0 stands for MVI negative; 1 for MVI positive. For inter-reader variability analysis, a subset from test-set consisting of 62 patients with 591 axial slices are labelled repeatedly by 3 radiologists Y2, Y4 and Y10, whose year-of-practice are 2, 4 and 10 respectively.

Evaluation Metrics. To evaluate the sequential classification, we perform five quantitative measures including F1 score, accuracy, AUC score, precision and recall. We adopted Dice Similarity coefficient (DSC) to evaluate the tumor segmentation accuracy that is associated with boundary extraction quality. Following [26], we perform quantitative measures including sensitivity, specificity, AUC, and Youden's J index (J) to evaluate the logistic regression model of MVI prediction. The class weights in all LR models are adjusted to maximize Youden's J index of (sensitivity+specificity-1).

Implementation Details. For all experiments, we apply simple data augmentations, e.g., random rotation, random resize and flipping. Models are trained with SGD optimizer with learning rate 0.01, momentum 0.9 and weight decay 1e-4. The default batch size is 64 and default training epoch is 300. The tumor segmentation branch and the sequential classification branch are optimized jointly in training. In addition, during training, we use the proposed train-time efficient vertex generator while during testing we directly generate the vertices from predicted tumor boundary.

Quantitative Comparisons. The baseline mentioned in Sect. 2 is a multi-classes segmentation network consisting of the same ResNet-50 encoder as ours and an UNet segmentation decoder. The baseline directly produces 2D semantic boundary mask. To make it comparable with the proposed method that is evaluated in sequence space, the baseline prediction mask is converted to 1D sequential band prediction by the vertex-based mask sampling, following our test-time vertex generating and sampling in Sect. 2.1. There are two branches in our framework: tumor segmentation and sequential prediction. Table 1 shows our framework can achieve 87.74% tumor DSC on the capsular invasion (CAP) dataset and 88.29% tumor DSC on the focal extensional nodule (FEN) dataset, facilitating high-quality boundary extraction and vertices generation. For sequential prediction, our framework outperforms the baseline on all measures, where the F1 score is superior by 23.25% on CAP and by 10.24% on FEN dataset. The average tumor segmentation accuracy achieves 88% DSC, indicating there is a margin of error between predicted and ground-truth tumor boundary. An upper-bound performance of our overall framework may be expected by replacing the predicted boundary with ground-truth in vertex localization. However, the quantitative results in Table 3 (*GT_Vertices*) indicate otherwise. This observation can be explained from [11] that claims the reliability of human-labelled boundary ground-truth is questionable due to ill-posed nature of boundary detection and uncertainty caused by human-annotated error. This result suggests that our framework is robust against slight shifts of boundaries due to its built-in efficient vertex generator and pyramid features. We also provide **qualitative comparison** results on the Capsular invasion prediction, as shown in Fig. 3. From there, our model is observed as the better solution to recover the boundary shape while the baseline fails sometimes (in the second row); and has stronger representation power to encode the context and distinguish the boundary semantics.

Table 1. Comparisons on CAP dataset and FEN dataset. The evaluation on sequential classification are based on F1, accuracy, AUC, precision and recall(%). We use DSC(%) to evaluate the tumor segmentation in our method.

Dataset	Model	F1	Accuracy	AUC	Precision	Recall	DSC-Tumor
CAP	Baseline	25.28	47.57	51.48	33.09	23.99	−
	Ours	**48.42**	**64.19**	**64.85**	**54.15**	**48.23**	87.74
FEN	Baseline	26.26	44.68	49.36	34.58	25.62	−
	Ours	**36.58**	**52.37**	**59.59**	**42.52**	**37.46**	88.29

Comparison with Inter-reader Variability. To quantify the inter-reader variability issue and how our approach measures against it, we compare the three radiologists' annotations (Y2, Y4, Y10) with each other and our prediction against them. The relatively poor inter-reader consistency in Table 2 show that the task is intrinsically challenging for human readers and the boundary semantics annotations lack of objectivity. Taking the ten-year practicing radiologist (Y10) as standard, our automated performance is closely equivalent to the four-year experienced radiologist (Y4), clearly proving the usability of our algorithm.

Table 2. Inter-reader variability analysis. R1 and R2 are annotation providers. Taking the ten-year practiced radiologist (Y10) as the standard, our algorithm's performance is comparable to the four-year practiced radiologist's (Y4).

R1	R2	F1	Accuracy	AUC	Precision	Recall
Y2	Y10	51.40	59.18	66.05	55.98	51.78
Y2	Y4	54.86	62.23	65.78	57.04	55.80
Y4	Y10	49.87	57.92	63.74	50.87	53.86
Average		52.04	59.78	65.19	54.63	53.81
Ours	Y2	38.78	54.16	62.52	38.00	46.28
Ours	Y4	42.49	56.67	63.15	41.70	50.20
Ours	**Y10**	48.60	67.65	65.24	48.72	53.26
Average		43.29	59.49	63.64	42.81	49.91

Ablation Study. We conduct the ablation study to analyze the effectiveness of different algorithm modules. The results are summarized in Table 3, where *PyraFeat* for pyramid feature, *CoordPos* for coordinate positional map, and *N_Vertices* for number of vertices/rays. (1) We investigate the impact of pyramid feature. In ablation, we only keep the $\frac{1}{4}$ scale of feature for vertex feature sampling, instead of four scales pyramid feature. From the second row of Table 3 without the pyramid feature, the model performance drops significantly on all

Table 3. Verification of the upper bound with *GT_Vertices* and the ablation study on pyramid feature, coordinate positional map, and number of vertices. All experiments are run on CAP dataset.

PyraFeat	CoordPos	N_Vertices	GT_Vertices	F1	Accuracy	AUC	Precision	Recall
✓		90		44.99	63.28	64.2	51.15	44.47
	✓	90		41.44	61.76	63.69	48.12	40.52
✓	✓	30		47.84	62.71	63.94	54.19	47.44
✓	✓	90	✓	47.5	**64.23**	64.79	53.63	47.23
✓	✓	90		**48.42**	64.19	**64.85**	**54.15**	**48.23**

measures. This is because the boundary semantics in our work are affected by several factors including tumor size, tumor type and the slice position, and consequently the pyramid feature in multi-scale facilitates the integrated use of low-/mid-/high level cues naturally. (2) Adding the coordinate positional map demonstrates performance robustness improvement, as the sequential feature encodes the positional dependencies. (3) Due to the vital role of boundary vertices in our framework, we analyze the impact of different numbers of vertices (i.e., the number of rays). By comparing the third row versus the fifth row in Table 3, the setting of 30 vertices yields inferior results than the default 90 vertices, implying that denser vertices are favored by our task.

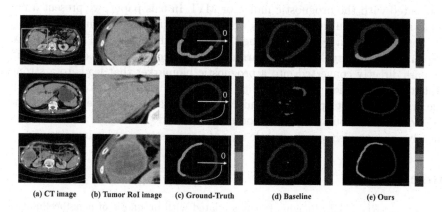

| (a) CT image | (b) Tumor RoI image | (c) Ground-Truth | (d) Baseline | (e) Ours |

Fig. 3. Qualitative comparison. (a) Whole CT image, (b) Tumor RoI image, (c) Ground-Truth, (d) Baseline prediction, (e) Our prediction. The sequence is flatten clockwise with the starting ray of 0 angle shown in (c).

Prognostic Biomarker Mining. The inter-reader variability and the ambiguous annotation of capsular invasion in CT scans may undermine the performance since the reported results in Table 1 are far less than perfect. This motivates us to further examine the effectiveness of our system on histopathology, which are widely considered as objective observation rather than subjective interpretation.

The capsular biomarkers in the UpperBound setting uses features (pixel ratio of three classes, see Sect. 2.3) calculated from the ground-truth CAP annotations while those biomarkers in "our prediction" uses features calculated from our model's prediction. Table 4 reports the correlation of the patient-specific capsular biomarkers and the MVI analyzed by logistic regression. It is surprising that our prediction slightly outperforms the upper bound in AUC, suggesting our predicted capsular biomarker performs comparable to radiologists' manual annotations for the clinically important task of prognostic MVI prediction.

Table 4. The evaluated results (%) of logistic regression for predicting MVI from the capsular biomarker.

	Sensitivity	Specificity	AUC	J
UpperBound (Ground-truth)	55.88	82.85	69.36	38.73
Our prediction	67.64	72.85	70.25	40.50

4   Conclusion

Capsular invasion on tumor boundary has been clinically hypothesized of being correlated with the prognostic indicator MVI. In this paper, we present a novel quantitative computing framework on modeling the tumor boundary semantics of capsular invasion by disentangling this task to efficient spatial localization and sequential boundary semantics learning. The detected tumor boundary semantics are directly converted into a prognostic biomarker that leads to a stronger statistical correlation with MVI than the version using human annotation. For the first time we interpret the boundary semantics as an effective tumor prognostic biomarker through objective computation, and provide an alternative non-invasive way to discover the subtle sign of prognostic vascular invasion.

References

1. An, C., Kim, M.J.: Imaging features related with prognosis of hepatocellular carcinoma. Abdominal Radiol. **44**(2), 509–516 (2019)
2. Bertasius, G., Shi, J., Torresani, L.: High-for-low and low-for-high: efficient boundary detection from deep object features and its applications to high-level vision. In: Proceedings of the IEEE International Conference on Computer Vision, pp. 504–512 (2015)
3. Chan, A.W., et al.: Development of pre and post-operative models to predict early recurrence of hepatocellular carcinoma after surgical resection. J. Hepatol. **69**(6), 1284–1293 (2018)
4. Chen, G., Chen, J., Lienen, M., Conradt, J., Röhrbein, F., Knoll, A.C.: Flgr: fixed length gists representation learning for rnn-hmm hybrid-based neuromorphic continuous gesture recognition. Front. Neuroscience **13**, 73 (2019)

5. Chen, J., et al.: Transunet: Transformers make strong encoders for medical image segmentation. arXiv preprint arXiv:2102.04306 (2021)
6. Choi, J.Y., Lee, J.M., Sirlin, C.B.: Ct and mr imaging diagnosis and staging of hepatocellular carcinoma: part i. development, growth, and spread: key pathologic and imaging aspects. Radiology **272**(3), 635–654 (2014)
7. Dosovitskiy, A., et al.: An image is worth 16 × 16 words: transformers for image recognition at scale. In: ICLR (2021)
8. Ehman, E.C., et al.: Rate of observation and inter-observer agreement for li-rads major features at ct and mri in 184 pathology proven hepatocellular carcinomas. Abdominal Radiol. **41**(5), 963–969 (2016)
9. Hariharan, B., Arbeláez, P., Bourdev, L., Maji, S., Malik, J.: Semantic contours from inverse detectors. In: 2011 International Conference on Computer Vision, pp. 991–998. IEEE (2011)
10. He, K., Zhang, X., Ren, S., Sun, J.: Deep residual learning for image recognition. In: Proceedings of the IEEE Conference on Computer Vision and Pattern Recognition, pp. 770–778 (2016)
11. Hou, X., Yuille, A., Koch, C.: Boundary detection benchmarking: beyond f-measures. In: Proceedings of the IEEE Conference on Computer Vision and Pattern Recognition, pp. 2123–2130 (2013)
12. Kojiro, M.: Histopathology of liver cancers. Best Pract. Res. Clin. Gastroenterol. **19**(1), 39–62 (2005)
13. Li, X., Sun, X., Meng, Y., Liang, J., Wu, F., Li, J.: Dice loss for data-imbalanced nlp tasks. arXiv preprint arXiv:1911.02855 (2019)
14. Liang, J., Homayounfar, N., Ma, W.-C., Xiong, Y., Hu, R., Urtasun, R.: Poly-transform: deep polygon transformer for instance segmentation. In: Proceedings of the IEEE/CVF Conference on Computer Vision and Pattern Recognition, pp. 9131–9140 (2020)
15. Liu, R., et al.: An intriguing failing of convolutional neural networks and the coord-conv solution. arXiv preprint arXiv:1807.03247 (2018)
16. Mao, J., Xu, W., Yang, Y., Wang, J., Huang, Z., Yuille, A.: Deep captioning with multimodal recurrent neural networks (m-rnn). arXiv preprint arXiv:1412.6632 (2014)
17. Qi, H., Collins, S., Alison Noble, J.: Upi-net: semantic contour detection in placental ultrasound. In: Proceedings of the IEEE/CVF International Conference on Computer Vision Workshops, p. 0 (2019)
18. Schmidt, U., Weigert, M., Broaddus, C., Myers, G.: Cell detection with star-convex polygons. In: Frangi, A.F., Schnabel, J.A., Davatzikos, C., Alberola-López, C., Fichtinger, G. (eds.) MICCAI 2018. LNCS, vol. 11071, pp. 265–273. Springer, Cham (2018). https://doi.org/10.1007/978-3-030-00934-2_30
19. Shah, S.A., et al.: Recurrence after liver resection for hepatocellular carcinoma: risk factors, treatment, and outcomes. Surgery **141**(3), 330–339 (2007)
20. Shen, W., Wang, X., Wang, Y., Bai, X., Zhang, Z.: Deepcontour: a deep convolutional feature learned by positive-sharing loss for contour detection. In: Proceedings of the IEEE Conference on Computer Vision and Pattern Recognition, pp. 3982–3991 (2015)
21. Sutskever, I., Vinyals, O., Le, Q.V.: Sequence to sequence learning with neural networks. arXiv preprint arXiv:1409.3215 (2014)
22. Tang, Y., Tang, Y., Zhu, Y., Xiao, J., Summers, R.M.: E2net: an edge enhanced network for accurate liver and tumor segmentation on ct scans. In: International Conference on Medical Image Computing and Computer-Assisted Intervention, pp. 512–522. Springer (2020)

23. Vaswani, A., et al.: Attention is all you need. In: Advances in Neural Information Processing Systems, pp. 5998–6008 (2017)
24. Xie, E., et al.: Polarmask: single shot instance segmentation with polar representation. In: Proceedings of the IEEE/CVF Conference on Computer Vision and Pattern Recognition, pp. 12193–12202 (2020)
25. Xu, W., Wang, H., Qi, F., Lu, C.: Explicit shape encoding for real-time instance segmentation. In: Proceedings of the IEEE/CVF International Conference on Computer Vision, pp. 5168–5177 (2019)
26. Yao, J., Shi, Y., Lu, L., Xiao, J., Zhang, L.: DeepPrognosis: preoperative prediction of pancreatic cancer survival and surgical margin via contrast-enhanced CT imaging. In: Martel, A.L., et al. (eds.) MICCAI 2020. LNCS, vol. 12262, pp. 272–282. Springer, Cham (2020). https://doi.org/10.1007/978-3-030-59713-9_27
27. Zhu, F., Yang, F., Li, J., Chen, W., Yang, W.: Incomplete tumor capsule on preoperative imaging reveals microvascular invasion in hepatocellular carcinoma: a systematic review and meta-analysis. Abdominal Radiol. 44(9), 3049–3057 (2019)

Do We Need Complex Image Features to Personalize Treatment of Patients with Locally Advanced Rectal Cancer?

Iram Shahzadi[1,2,3], Annika Lattermann[1,2,3,4], Annett Linge[1,2,3,4,5],
Alexander Zwanenburg[1,2,3,5], Christian Baldus[6], Jan C. Peeken[3,7,8,9],
Stephanie E. Combs[3,7,8,9], Michael Baumann[1,3,4], Mechthild Krause[1,2,3,4,5,10],
Esther G. C. Troost[1,2,3,4,5,10], and Steffen Löck[1,2,3,4(✉)]

[1] OncoRay – National Center for Radiation Research in Oncology, Faculty of Medicine and University Hospital Carl Gustav Carus, Technische Universität Dresden, Helmholtz-Zentrum Dresden – Rossendorf, Dresden, Germany
Steffen.Loeck@OncoRay.de
[2] German Cancer Consortium (DKTK) Partner Site Dresden, Dresden, Germany
[3] German Cancer Research Center (DKFZ), Heidelberg, Germany
[4] Department of Radiotherapy and Radiation Oncology, Faculty of Medicine and University Hospital Carl Gustav Carus, Technische Universität Dresden, Dresden, Germany
[5] National Center for Tumor Diseases (NCT), Partner Site Dresden, Dresden, Germany
[6] Department of Radiology, Faculty of Medicine and University Hospital Carl Gustav Carus, Technische Universität Dresden, Dresden, Germany
[7] German Cancer Consortium (DKTK) Partner Site Munich, Munich, Germany
[8] Department of Radiation Oncology, Klinikum rechts der Isar, Technische Universität München, Munich, Germany
[9] Department of Radiation Sciences (DRS), Institute of Radiation Medicine (IRM), Helmholtz Zentrum München, Neuherberg, Germany
[10] Helmholtz-Zentrum Dresden-Rossendorf, Institute of Radiooncology-OncoRay, Dresden, Germany

Abstract. Radiomics has shown great potential for outcome prognosis and presents a promising approach for improving personalized cancer treatment. In radiomic analyses, features of different complexity are extracted from clinical imaging datasets, which are correlated to the endpoints of interest using machine-learning approaches. However, it is generally unclear if more complex features have a higher prognostic value and show a robust performance in external validation. Therefore, in this study, we developed and validated radiomic signatures for outcome prognosis after neoadjuvant radiochemotherapy in locally advanced rectal cancer (LARC) using computed tomography (CT) and T2-weighted magnetic resonance imaging (MRI) of two independent institutions (training/validation: 94/28 patients). For the prognosis of tumor response and freedom from distant metastases (FFDM), we used different imaging features extracted from the gross

E. G. C. Troost and S. Löck—Shared last authorship.

Electronic supplementary material The online version of this chapter (https://doi.org/10.1007/978-3-030-87234-2_73) contains supplementary material, which is available to authorized users.

tumor volume: less complex morphological and first-order (MFO) features, more complex second-order texture (SOT) features, and both feature classes combined. Analyses were performed for both imaging modalities separately and combined. Performance was assessed by the area under the curve (AUC) and the concordance index (CI) for tumor response and FFDM, respectively. Overall, radiomic features showed prognostic value for both endpoints. Combining MFO and SOT features led to equal or higher performance in external validation compared to MFO and SOT features alone. The best results were observed after combining MRI and CT features (AUC = 0.76, CI = 0.65). In conclusion, promising biomarker signatures combining MRI and CT were developed for outcome prognosis in LARC. Further external validation is pending before potential clinical application.

Keywords: Rectal cancer · Tumor response · Distant metastases · Biomarkers

1   Introduction

The personalization of treatment is a central aim in cancer therapy to improve the outcome of patient populations with heterogeneous treatment response. In particular, for patients with locally advanced rectal cancer (LARC) there is increased interest in the adaptation of organ-preserving and low-morbidity surgeries or watch-and-wait strategies in case of pathologically complete response (pCR) after neoadjuvant radiochemotherapy (nRCT) [1]. To identify these well responding patients, validated biomarkers related to tumor radiochemosensitivity have to be available.

Several studies have analyzed molecular data as potential biomarkers of LARC response to nRCT [2–5]. The inclusion of biomarkers from clinical imaging may further increase the robustness and accuracy of corresponding prognostic models since they depict and possibly characterize the entire tumor instead of the biopsy site. Radiomic analyses employ modern machine learning algorithms to identify such biomarkers based on different imaging modalities [6–8]. For outcome prognosis in LARC, magnetic resonance imaging (MRI) based radiomic models were developed for tumor response to nRCT and freedom from distant metastases (FFDM) [9–14]. So far, however, few groups have considered radiomic features extracted from computed tomography (CT) imaging [15, 16] or a combination of CT and MRI features [17, 18]. Although the results of these analyses are encouraging, important methods, such as assessing feature robustness, were not always considered and external validation was rarely performed.

One key challenge in radiomics is the selection of features that correlate well with the endpoint of interest. Feature classes of different complexity are commonly extracted, either directly from the images or after applying different filters: (i) morphological features that describe the shape of the region of interest (ROI), (ii) first-order features (FO) that describe the voxel intensity distribution, and (iii) second-order texture features (SOT) that describe statistical inter-relationships between neighboring voxels [19]. However, it is generally unclear if the addition of the more complex and more difficult to interpret SOT features is necessary for creating translatable models.

In the present study, we therefore aimed to identify and validate different radiomic signatures for the prognosis of tumor response to nRCT and FFDM in patients with

LARC based on pre-treatment CT and T2-weighted (T2-w) MRI datasets. We investigated the prognostic value of (i) morphological and first-order features (MFO), (ii) second-order texture features (SOT), and (iii) the combination of both feature classes. In addition, we combined the most promising features from CT and MRI to build a multimodal signature, and validated the results with an independent, external dataset.

2 Materials and Methods

2.1 Patient Data

In this retrospective study, radiomics signatures were developed and validated within the German Cancer Consortium-Radiation Oncology Group (DKTK-ROG) based on data from two partner sites with a total of 122 patients. All patients were diagnosed with histopathologically confirmed LARC and underwent neoadjuvant radiochemotherapy (nRCT) followed by surgery between 2006 and 2014. Patients of the DKTK-ROG partner site Dresden were allocated to the training data (N = 94) and patients of the partner site Munich were allocated to the validation data (N = 28). Inclusion criteria for our study were: histologically confirmed LARC and nRCT followed by surgery, availability of pre-treatment MRI, treatment planning CT-scans with sufficient image quality, and availability of endpoint information. Patient characteristics are summarized in Table S1 and the study design is presented in Fig. 1a. The study was approved by the local Ethics Committee of both institutions.

2.2 Endpoints, Image Preprocessing, and Feature Extraction

In our study, we developed radiomic signatures for the prognosis of tumor response and FFDM in patients with LARC using pre-treatment CT and T2-w MRI. Tumor response was determined by expert pathologists from the work-up of the surgical specimens. The patients were stratified into two groups based on the tumor regression grade (TRG): (i) responders (corresponding to TRG 3 and 4) and (ii) non-responders (corresponding to TRG 0–2) following Dworak et al. [20]. The survival endpoint FFDM was calculated from the first day of nRCT to the day of event or censoring.

Figure 1b presents the process of image preprocessing and feature extraction. MR images were first corrected for background phase variation by multiplying the image with an image mask created by Canny edge detection [21]. MR images were then corrected for bias fields using N4ITK [22] followed by intensity normalization using the 95th percentile of image intensities. Both MR and CT images were resampled to an isotropic voxel size of $1.0 \times 1.0 \times 1.0$ mm^3 using trilinear interpolation. Laplacian of Gaussian (LoG) filters with 5 different kernel widths (1 mm, 2 mm, 3 mm, 4 mm, 5 mm) were applied individually to the base MR and CT images. The five response maps were averaged to a single image. Table S2 summarizes the image acquisition parameters for MRI and CT data. After image preprocessing, imaging features were computed using the publicly available Python module MIRP [23]. A set of 18 statistical, 2 local-intensity-based, 29 morphological, 38 intensity-histogram-based, and 95 second-order texture (SOT) features. SOT features were extracted from the 3D volume based on the grey level co-occurrence matrix (GLCM), grey level run length matrix (GLRLM), grey level size zone matrix (GLSZM), grey level distance zone matrix (GLDZM), neighbourhood grey tone dependence matrix (NGTDM), and neighbouring grey level dependence matrix (NGLDM) from the 3D gross tumor volume (GTV) on the treatment-planning CT and on the pre-treatment T2-w MRI using baseline and LoG transformed images.

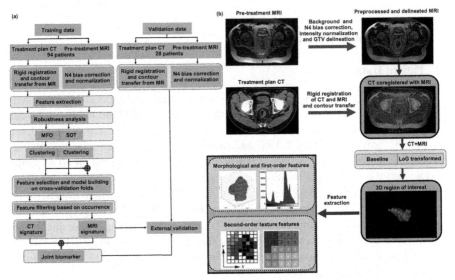

Fig. 1. (a) Study design. After image preprocessing, radiomic features were extracted from each imaging modality, analyzed for robustness, separated into morphological and first-order (MFO) features and second-order texture (SOT) features, and clustered. Radiomic signatures for both CT and MRI were developed in a cross-validation approach and externally validated using (i) MFO, (ii) SOT, and (iii) MFO + SOT features for the endpoints tumor response and freedom from distant metastases. Finally, a joint MRI and CT signature was created and externally validated for each endpoint. (b) Image preprocessing and feature extraction pipeline. MR images were preprocessed and the gross tumor volume (GTV) was delineated centrally by one experienced radiation oncologist and one radiologist. GTV contours were then transferred to CT after rigid registration. All features were extracted from the GTV on the original and the Laplacian of Gaussian (LoG) transformed CT and MR images using a 3D approach.

Image preprocessing and feature extraction in MIRP was implemented according to the recommendations of the Image Biomarker Standardisation Initiative (IBSI) [19]. Feature definitions and extraction algorithms are stated in the IBSI reference manual. Image processing parameters used for feature extraction are summarized in Table S3. Features were filtered for stability under small image perturbations (Gaussian noise, translation, and volume growth/shrinkage) by removing features with the lower boundary of the 95% confidence interval (CI) of the intra-class correlation coefficient (ICC) below 0.8 [24]. Afterwards, features were separated into two feature classes: (i) MFO features and (ii) SOT features. Features with high mutual correlation in each class were clustered by hierarchical clustering with average linkage, using a mutual Spearman correlation of 0.8. The feature with the highest mutual information with the endpoint was selected as the representative for each cluster.

2.3 Radiomic Modeling Workflow

The three feature classes containing (i) MFO, (ii) SOT, and (iii) MFO + SOT features obtained after clustering were used to create radiomic signatures for tumor response and FFDM prognosis in MRI and CT data separately. Finally, a joint signature was created for each endpoint combining MRI and CT features. For analysis of each feature class, a workflow of four major processing steps was applied: (i) feature preprocessing, (ii) feature-selection and signature development, (iii) model building, and (iv) external validation. Steps (i)-(iii) were first performed within 33 repetitions of 3-fold cross-validation (CV) on the training dataset to identify an optimal signature, i.e., the steps were repeatedly performed on the internal training part and validated on the internal validation part of the cross-validation folds. After identifying the final signature in each feature class, a final model was developed on the entire training data, which was validated on the external validation data.

The following procedure was performed for each of the 99 cross-validation runs: (i) Features were transformed using the Yeo-Johnson transformation to align their distribution to a normal distribution. Afterwards, features were z-transformed to mean zero and standard deviation one. Both transformations were performed on the internal training part and the resulting parameters were applied to the features of the internal validation part. (ii) Features in the internal training part were ranked according to their relevance using mutual information feature selection (MIFS) for tumor response and maximum relevance minimum redundancy (MRMR) feature selection for FFDM. To avoid potential overfitting, only the five most relevant features were selected. (iii) These features were used to build a prognostic model on the internal training part, which was validated on the internal validation part. Multivariable logistic regression was applied for the prognosis of tumor response and Cox regression for FFDM.

The average model performance was assessed by the median CV area under the curve (AUC) and concordance index (CI) for tumor response and FFDM prognosis, respectively. The occurrence of every feature in the 99 modeling steps was counted and features were ordered with increasing occurrence. Features with > 40% occurrence were selected. If a subset of these features showed a mutual Spearman correlation > 0.5 on the entire training data, only the feature with the highest occurrence was considered. The resulting final radiomic signature was then used to build prognostic models on 100 bootstraps of the entire training data. (iv) The ensemble of 100 trained models was then applied to the external validation data.

Finally, a joint signature based on MRI and CT data was created by combining those features from the MFO and SOT signatures of both modalities that were also part of the respective MFO + SOT signature. The same procedure as described in the last paragraph was then performed: clusters with Spearman correlation > 0.5 were reduced to one feature, models were trained on 100 bootstraps and validated.

2.4 Statistical Analysis

The following baseline clinical parameters were available: gender, age, tumor localization, UICC stage, grading, cT stage, cN stage, type of chemotherapy, and radiotherapy dose. The following categorical variables were binarised for univariable analysis: tumor localization (0 for localization 3–6 cm, 1 for localization $\geq$ 6 cm from anal verge), cT stage (0 for cT < 4 and 1 for cT = 4), cN stage (0 for cN = 0 and 1 cN $\geq$ 1), grading (0 for grading $\leq$ 2 and 1 for grading = 3), chemotherapy (0 for 5FU-infusion, 5FU-infusion + oxaliplatine, 5FU-infusion + capecitabine and 1 for capecitabine, capecitabine + other). All clinical features were associated to tumor response by univariable logistic regression and to FFDM by univariable Cox regression. Categorical variables were compared between the training and validation data by the χ^2 test, whereas continuous variables were compared using the Mann-Whitney-U test.

Associations between the final model predictions and the endpoints were evaluated by the AUC for tumor response and by the CI for FFDM prognosis. The estimated value and the 95% confidence interval of these metrics were computed. For association with FFDM, patients were stratified into a low and a high-risk group using the median of the averaged linear predictors of the Cox regression over the 100 bootstraps on the training data. The cutoff was transferred to the validation data. The difference in FFDM between the stratified patient groups was assessed using the log-rank test. Features used to build a Cox model for FFDM prognosis were also tested for proportional hazard assumption using Schoenfeld residual test [25]. All tests were two-sided with a significance level of 0.05. Statistical and FFDM analyses were performed in R 3.6.0. Python 3.8 was used for tumor response analyses.

3   Results

The endpoints and clinical characteristics were similar for training and validation data (p > 0.05) except for radiotherapy dose (p < 0.001) (Table S1). None of the clinical features was significantly associated to the endpoints.

364 radiomic features were extracted from the GTV of each baseline and LoG transformed T2-w MR and CT imaging dataset. Stability analysis reduced these features to 323 and 358 in MRI and CT data, respectively. Clustering of correlated features further reduced their number (MFO: MRI = 28, CT = 41; SOT: MRI = 20, CT = 28). Based on these features, radiomic signatures were first developed and validated individually for both imaging modalities using 3 feature classes: (i) MFO features only, (ii) SOT features only, and (iii) MFO + SOT features for both endpoints. Finally, a joint signature was developed by combining T2-w MRI and CT signatures for each endpoint.

Table 1 presents the results for the prognosis of tumor response, including the names of the finally selected features. In internal cross-validation, SOT features showed a higher prognostic value compared to MFO features on both imaging modalities (MRI: AUC_{MFO} = 0.56, AUC_{SOT} = 0.66; CT: AUC_{MFO} = 0.60, AUC_{SOT} = 0.63). In external validation, the MFO signature performed slightly better for MRI, while SOT features achieved a higher AUC for the CT (MRI: AUC_{MFO} = 0.59, AUC_{SOT} = 0.55; CT: AUC_{MFO} = 0.56, AUC_{SOT} = 0.69). Higher external validation performance was achieved by using MFO + SOT features (MRI: $AUC_{MFO+SOT}$ = 0.63, CT: $AUC_{MFO+SOT}$ = 0.75). Integrating MRI and CT-based features to one signature led to slightly better AUC in external validation results (MRI + CT: AUC = 0.76, Fig. 2a).

Table S4 presents the results for the prognosis of FFDM. The proportional hazards assumption was not violated for any feature or model presented in this table. In internal cross-validation, MFO and SOT features showed a comparable prognostic value on both imaging modalities (MRI: $CI_{MFO} = 0.63$, $CI_{SOT} = 0.64$; CT: $CI_{MFO} = 0.62$, $CI_{SOT} = 0.63$). In external validation, only the MFO signature based on MRI features achieved a reasonable result (MRI: $CI_{MFO} = 0.57$). An improved performance in external validation was again achieved by combining MFO + SOT features (MRI: $CI_{MFO+SOT} = 0.59$, CT: $CI_{MFO+SOT} = 0.54$). Integrating MRI and CT-based features to one signature led to the best external validation results (MRI + CT: CI = 0.65) for FFDM. Due to the low number of events in the validation data, confidence intervals of the CI were wide. The stratification of the training data and the independent validation data into patient groups at low and high risk of distant metastases was performed based on the MFO + SOT models for MRI, CT, and MRI + CT. While the stratification on the training data was significant in all cases ($p < 0.05$), the individual MRI and CT-based signatures did not achieve a significant stratification on the external validation data ($p = 0.97$ and $p = 0.75$, respectively). The joint MRI + CT signature, however, revealed statistical significance ($p = 0.049$ Fig. 2b).

Table S5 contains exemplary model and transformation parameters for the developed multimodal signatures for both endpoints. Training was performed on the entire training data. The given data allow for independent external validation of the models.

Table 1. Median area under the curve (AUC) (95% confidence interval) for tumor response prognosis for each feature class based on CT, MRI, and joint MRI + CT using cross-validation (CV) on the training data, the entire training data (final training), and external validation data. For feature definitions, see e.g. [19].

Modality	Feature level	CV train	CV valid	Signature	Final training	External validation
MRI	**MFO**	0.73	0.56	log_stat_min log_morph_com	0.69 (0.56–0.82)	0.59 (0.39–0.79)
	SOT	0.77	0.66	dzm_zdnu_norm_3d_fbn_n32 log_rlm_lrhge_3d_v_mrg_fbn_n32	0.76 (0.67–0.85)	0.55 (0.35–0.80)
	MFO + SOT	0.77	0.63	dzm_glnu_3d_fbn_n32 log_stat_min log_morph_com	0.71 (0.59–0.82)	0.63 (0.40–0.89)
CT	**MFO**	0.77	0.60	ivh_v50 log_morph_comp_1	0.71 (0.59–0.83)	0.56 (0.38–0.77)
	SOT	0.77	0.63	dzm_ldhge_3d_fbn_n32	0.73 (0.64–0.83)	0.69 (0.46–0.93)
	MFO + SOT	0.77	0.63	ivh_v50 dzm_ldhge_3d_fbn_n32	0.75 (0.65–0.85)	0.75 (0.55–0.94)
MRI + CT				CT_ivh_v50 CT_dzm_ldhge_3d_fbn_n32 MRI_log_stat_min MRI_log_morph_com	0.78 (0.70–0.85)	0.76 (0.54–0.93)

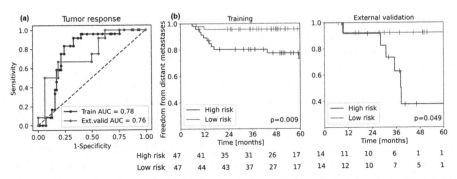

Fig. 2. (a) Receiver operating characteristics (ROC) curves for tumor response prognosis and (b) Kaplan-Meier estimates for risk-group stratification in LARC in training and validation based on the respective joint MRI + CT signature.

4 Discussion

In the present study, we developed and validated radiomics signatures including pre-treatment T2-w MRI and treatment planning CT imaging features for the prognosis of tumor response to nRCT and FFDM in patients with LARC. The discriminative performance of MFO, SOT, and MFO + SOT features was externally validated. MFO + SOT features were associated with a higher performance in external validation compared to MFO and SOT features alone. Furthermore, the joint MRI + CT signatures showed the best performance for both endpoints (tumor response: AUC = 0.76, FFDM: CI = 0.65).

One major issue in radiomic analyses is feature reproducibility and the lack of consensual guidelines on which features have to be extracted from clinical imaging data. Furthermore, numerous features of different complexity can be extracted, and frequently their number is larger than the study population, which can lead to substantial model overfitting. In this study, we observed that more complex SOT features showed a similar or higher performance in internal cross-validation compared to less complex MFO features. However, in external validation, the performance of SOT features decreased. In the joint MRI + CT signatures, MFO features showed a high occurrence (3 out of 4 features for both endpoints), reflecting their relevance in these best performing signatures. One may thus conclude that complex SOT features can be less reproducible in external datasets, e.g., due to different image acquisition parameters or protocols. Still, they have the potential to improve performance as was observed for the signatures based on combining MFO + SOT features in our study, i.e., both MFO and SOT features may contain relevant and independent prognostic information.

Overall, the performance of our best performing MRI signature for tumor response (AUC = 0.63) was somewhat lower than other validated results, e.g., Antunes et al. [9] (AUC = 0.71) and Dinapoli et al. [10] (AUC = 0.75). Only a few studies exist on CT-based or CT + MRI-based radiomics in LARC [14–17]. While these studies showed a high prognostic value, most were not externally validated. In several studies, first-order radiomics features extracted from T2-w MRI were associated with pathological complete response [26, 27], while other authors evaluated more complex radiomic features and demonstrated promising results [28, 29]. External validation of these results is pending.

Limitations of this study are its retrospective nature and the relatively low number of patients, leading to a small number of events for both endpoints. In particular, for the endpoint FFDM, only 6 distant metastases were observed in the external validation data. Due to this low number of events, wide confidence intervals were reported in Table S4, i.e., the external validation results have a large uncertainty. We aimed to mitigate this problem by internal cross-validation on the training data for feature selection and by bootstrapped model validation. It is planned to further validate our results on additional datasets of the DKTK-ROG, once available.

In conclusion, we developed and externally validated radiomic signatures for the prognosis of tumor response and FFDM in patients with LARC after nRCT based on T2-w MRI and CT imaging. We studied feature classes of differing complexity and observed that a combination of MFO and SOT features from MRI and CT led to the highest prognostic value. After further external validation, the developed radiomic signatures may be applied for interventional trials on personalized treatment strategies.

References

1. Dossa, F., Chesney, T.R., Acuna, S.A., Baxter, N.N.: A watch-and-wait approach for locally advanced rectal cancer after a clinical complete response following neoadjuvant chemoradiation: a systematic review and meta-analysis. Lancet Gastroenterol. Hepatol. 2(7), 501–513 (2017)

2. Das, P., Skibber, J.M., Rodriguez-Bigas, M.A., Feig, B.W., Chang, G.J., Wolff, R.A., et al.: Predictors of tumor response and downstaging in patients who receive preoperative chemoradiation for rectal cancer. Cancer 109(9), 1750–1755 (2007)

3. Ryan, J.E., Warrier, S.K., Lynch, A.C., Ramsay, R.G., Phillips, W.A., Heriot, A.G.: Predicting pathological complete response to neoadjuvant chemoradiotherapy in locally advanced rectal cancer: a systematic review. Colorectal Dis. 18(3), 234–246 (2016)

4. Ojima, E., Inoue, Y., Miki, C., Mori, M., Kusunoki, M.: Effectiveness of gene expression profiling for response prediction of rectal cancer to preoperative radiotherapy. J. Gastroenterol. 42(9), 730–736 (2007)

5. Watanabe, T., Komuro, Y., Kiyomatsu, T., Kanazawa, T., Kazama, Y., Tanaka, J., et al.: Prediction of sensitivity of rectal cancer cells in response to preoperative radiotherapy by DNA microarray analysis of gene expression profiles. Can. Res. 66(7), 3370–3374 (2006)

6. Parmar, C., Grossmann, P., Bussink, J., Lambin, P., Aerts, H.J.: Machine learning methods for quantitative radiomic biomarkers. Sci. Rep. 5(1), 1–11 (2015)

7. Gillies, R.J., Kinahan, P.E., Hricak, H.: Radiomics: images are more than pictures, they are data. Radiology 278(2), 563–577 (2016)

8. Song, J., Yin, Y., Wang, H., Chang, Z., Liu, Z., Cui, L.: A review of original articles published in the emerging field of radiomics. Eur. J. Radiol. 127, 108991 (2020)

9. Antunes, J.T., Ofshteyn, A., Bera, K., Wang, E.Y., Brady, J.T., et al.: Radiomic features of primary rectal cancers on baseline T2-weighted MRI are associated with pathologic complete response to neoadjuvant chemoradiation: a multisite study. J. Magn. Reson. Imaging 52(5), 1531–1541 (2020)

10. Dinapoli, N., et al.: Magnetic resonance, vendor-independent, intensity histogram analysis predicting pathologic complete response after radiochemotherapy of rectal cancer. Int. J. Radiat. Oncol. Biol. Phys. **102**(4), 765–774 (2018)

11. Yi, X., Pei, Q., Zhang, Y., Zhu, H., Wang, Z., Chen, C., et al.: MRI-based radiomics predicts tumor response to neoadjuvant chemoradiotherapy in locally advanced rectal cancer. Front. Oncol. **9**, 552 (2019)

12. Horvat, N., Veeraraghavan, H., Khan, M., Blazic, I., Zheng, J., Capanu, M., et al.: MR imaging of rectal cancer: radiomics analysis to assess treatment response after neoadjuvant therapy. Radiology **287**(3), 833–843 (2018)

13. Nie, K., Shi, L., Chen, Q., Hu, X., Jabbour, S.K., Yue, N., et al.: Rectal cancer: assessment of neoadjuvant chemoradiation outcome based on radiomics of multiparametric MRI. Clin. Cancer Res. **22**(21), 5256–5264 (2016)

14. Jeon, S.H., Song, C., Chie, E.K., Kim, B., Kim, Y.H., Chang, W., et al.: Delta-radiomics signature predicts treatment outcomes after preoperative chemoradiotherapy and surgery in rectal cancer. Radiat. Oncol. **14**(1), 1–10 (2019)

15. Chee, C.G., Kim, Y.H., Lee, K.H., Lee, Y.J., Park, J.H., Lee, H.S., et al.: CT texture analysis in patients with locally advanced rectal cancer treated with neoadjuvant chemoradiotherapy: a potential imaging biomarker for treatment response and prognosis. PLoS ONE **12**(8), e0182883 (2017)

16. Bibault, J.E., Giraud, P., Housset, M., Durdux, C., Taieb, J., Berger, A., et al.: Deep learning and radiomics predict complete response after neo-adjuvant chemoradiation for locally advanced rectal cancer. Sci. Rep. **8**(1), 1–8 (2018)

17. Li, Z.Y., Wang, X.D., Li, M., Liu, X.J., Ye, Z., Song, B., et al.: Multi-modal radiomics model to predict treatment response to neoadjuvant chemotherapy for locally advanced rectal cancer. World J. Gastroenterol. **26**(19), 2388 (2020)

18. Zhang, Y., He, K., Guo, Y., Liu, X., Yang, Q., Zhang, C., et al.: A novel multimodal radiomics model for preoperative prediction of lymphovascular invasion in rectal cancer. Front. Oncol. **10**, 457 (2020)

19. Zwanenburg, A., Vallières, M., Abdalah, M.A., Aerts, H.J., Andrearczyk, V., et al.: The image biomarker standardization initiative: standardized quantitative radiomics for high-throughput image-based phenotyping. Radiology **295**(2), 328–338 (2020)

20. Dworak, O., Keilholz, L., Hoffmann, A.: Pathological features of rectal cancer after preoperative radiochemotherapy. Int. J. Colorectal Dis. **12**(1), 19–23 (1997)

21. Canny, J.: A computational approach to edge detection. IEEE Trans. Pattern Anal. Mach. Intell. **6**, 679–698 (1986)

22. Tustison, N.J., et al.: N4ITK: improved N3 bias correction. IEEE Trans. Med. Imaging **29**(6), 1310–1320 (2010)

23. Zwanenburg, A., Leger, S., Starke, S.: GitHub-oncoray/mirp: medical image radiomics processor. https://github.com/oncoray/mirp. Accessed January 1 2021

24. Zwanenburg, A., et al.: Assessing robustness of radiomic features by image perturbation. Sci. Rep. **9**(1), 1–10 (2020)

25. Schoenfeld, D.: Partial residuals for the proportional hazards regression model. Biometrika **69**(1), 239–241 (1982)

26. De Cecco, C.N., et al.: Performance of diffusion-weighted imaging, perfusion imaging, and texture analysis in predicting tumoral response to neoadjuvant chemoradiotherapy in rectal cancer patients studied with 3T MR: initial experience. Abdom. Radiol. **41**(9), 1728–1735 (2016)

27. Meng, Y., et al.: MRI texture analysis in predicting treatment response to neoadjuvant chemoradiotherapy in rectal cancer. Oncotarget **9**(15), 11999 (2018)

28. Aker, M., Ganeshan, B., Afaq, A., Wan, S., Groves, A.M., Arulampalam, T.: Magnetic resonance texture analysis in identifying complete pathological response to neoadjuvant treatment in locally advanced rectal cancer. Dis. Colon Rectum **62**(2), 163–170 (2019)
29. Liu, Z., Zhang, X.Y., Shi, Y.J., Wang, L., Zhu, H.T., Tang, Z., et al.: Radiomics analysis for evaluation of pathological complete response to neoadjuvant chemoradiotherapy in locally advanced rectal cancer. Clin. Cancer Res. **23**(23), 7253–7262 (2017)

Multiple Instance Learning with Auxiliary Task Weighting for Multiple Myeloma Classification

Talha Qaiser[1], Stefan Winzeck[1], Theodore Barfoot[1,2], Tara Barwick[3,4], Simon J. Doran[5], Martin F. Kaiser[5,6], Linda Wedlake[6], Nina Tunariu[6], Dow-Mu Koh[5,6], Christina Messiou[5,6], Andrea Rockall[3,4], and Ben Glocker[1(✉)]

[1] BioMedIA Group, Department of Computing, Imperial College London, London, UK
{t.qaiser,b.glocker}@imperial.ac.uk
[2] Biomedical Engineering and Imaging Sciences, King's College London, London, UK
[3] Department of Surgery and Cancer, Imperial College London, London, UK
[4] Department of Imaging, Imperial College Healthcare NHS Trust, London, UK
[5] The Institute of Cancer Research, London, UK
[6] The Royal Marsden NHS Foundation Trust, London, UK

Abstract. Whole body magnetic resonance imaging (WB-MRI) is the recommended modality for diagnosis of multiple myeloma (MM). WB-MRI is used to detect sites of disease across the entire skeletal system, but it requires significant expertise and is time-consuming to report due to the great number of images. To aid radiological reading, we propose an auxiliary task-based multiple instance learning approach (ATMIL) for MM classification with the ability to localize sites of disease. This approach is appealing as it only requires patient-level annotations where an attention mechanism is used to identify local regions with active disease. We borrow ideas from multi-task learning and define an auxiliary task with adaptive reweighting to support and improve learning efficiency in the presence of data scarcity. We validate our approach on both synthetic and real multi-center clinical data. We show that the MIL attention module provides a mechanism to localize bone regions while the adaptive reweighting of the auxiliary task considerably improves the performance.

1 Introduction

Caused by the aberrant proliferation of plasma cells within the bone marrow, multiple myeloma (MM) is one of the most common hematologic malignancies. While MM is incurable, it can be treated when diagnosed appropriately [12]. Diffusion-weighted (DW) WB-MRI offers high sensitivity to diagnose MM in an

Electronic supplementary material The online version of this chapter (https://doi.org/10.1007/978-3-030-87234-2_74) contains supplementary material, which is available to authorized users.

M. de Bruijne et al. (Eds.): MICCAI 2021, LNCS 12907, pp. 786–796, 2021.
https://doi.org/10.1007/978-3-030-87234-2_74

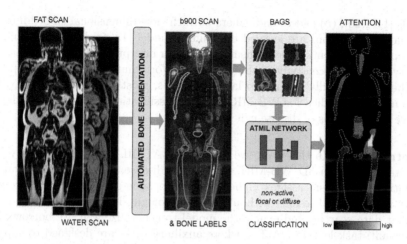

Fig. 1. Schematic overview of the proposed framework. Fat and water WB-MRI scans are used to automatically segment bone structures. For each b900 scan, 3D instances from segmented bone regions are sampled and combined into a bag. From these, the ATMIL network derives a disease class and an attention map.

early stage [18]. DW WB-MRI is a relatively new diagnostic tool, only recently being recommended for patients with myeloma but not yet widely used. To our knowledge, there are currently no computational approaches for MM classification using DW WB-MRI. Prior work in this domain mostly focused on PET/CT MM imaging [15,23]. Broadly, bone marrow appearances of patients imaged for suspected myeloma can be classified as: healthy, focal lesions, diffuse infiltration, or focal lesions on a background of diffuse infiltration. The presence of more than one focal lesion with diameter > 5 mm indicates the need for treatment, smaller lesions will require monitoring [16]. In clinical practice, myeloma patterns are identified by visual examination of WB-MRI scans. However, this compares to the search of a needle in a haystack as WB-MRI consists of hundreds of 2D images and lesions can be scattered across the skeletal system and make up only a small fraction ($\approx\leq$ 10%). Automated MM classification with the ability to localize disease could improve diagnostic accuracy and reduce the reading time.

Convolutional neural networks (CNNs) are successfully applied for whole-image disease classification, but lack the ability to localize disease. To train a CNN for local disease classification one would require region-wise or voxel-wise annotations, which are difficult to obtain in large quantities. Weakly supervised approaches, such as multiple instance learning (MIL), are appealing as these can be trained on global, patient-level labels (e.g. disease categories) while having the ability to localize the contributing image regions. Instead of predicting labels for individual samples (e.g. patches), MIL frameworks aim to derive a single label from a set of instances, known as a *bag*. Broadly, deep MIL can be categorized into two groups (a) multi-stage, where an encoder maps bag instances into a low-dimensional feature space and an aggregation model is then used to predict bag labels [1,14,20], (b) end-to-end, where both the low-dimensional embedding and bag label predictions are learned by a single model [4,6,9]. End-to-end models equipped with attention mechanisms offer better interpretability and localisation ability and have recently been shown to outperform multi-stage models [4]. MIL has been successfully applied to large data sets, mostly in histopathology, but have not been studied for WB-MRI disease classification where training data is scarce and patterns of disease can be subtle.

Contributions. In this work, we propose an ATMIL network to classify MM disease patterns in 3D WB-MRI. To localize disease, a permutation-invariant attention mechanism assigns higher weights to bone structures that are more likely to carry MM lesions without the need for any localized annotations. To combat data scarcity and enable the model to generalize well, we borrow ideas from multi-task learning (MTL) where auxiliary tasks are designed to support learning of the main task [13,22]. We propose adaptive reweighting by minimizing the divergence between our main and auxiliary task. We have compared the performance of our approach in combination with different auxiliary losses. At first, we evaluated our framework's performance on Morpho-MNIST [2], a synthetic data set we use to mimic a 4-class classification problem. We then validate the performance on real, multi-center clinical data with diffusion weighted WB-MRI. Our results suggest that the auxiliary task improves the generalization and predictive performance of the attention-based MIL framework with the ability to identify anatomical regions containing local disease patterns.

2   Methodology

Our proposed framework contains three main components (Fig. 1): an automated bone segmentation to identify regions of interests for sampling bag instances, an ATMIL network for MM classification with an attention mechanism to localize disease, and an integrated adaptive weighting scheme to quantify the usefulness of an auxiliary task to support learning of the main task.

2.1   Multiple Instance Learning

In a fully-supervised learning problem the objective is to predict from images $I \in \mathbb{R}^D$ a target class label $Y \in \{1, ..., y\}$. In a MIL setting, each I is represented by a bag of instances, $\delta = \{i_1, ..., i_N\}$ which is associated with only a single label Y. A classifier $g_\theta(\cdot)$ processes the entire bag of instances to predict the bag label. In our case, each scan represents one bag (δ) and we sample 3D instances only from segmented bone regions. Figure 2 illustrates the architecture of the proposed framework. First, a group of convolution and fully connected layers transform i_N into a non-linear feature representation $\boldsymbol{h}_N$. Further, we need a symmetric function $S(\cdot)$ to aggregate instance-level embeddings $\boldsymbol{h}_N$ and to enable permutation invariance within the model:

$$g_\theta(I) = S(f(i_1), ..., f(i_N))$$

(1)

where $h_N = f(i_N)$. For that, pooling operators (e.g. mean, max *etc.*) could be applied, but standard methods are non-trainable and may hamper identifying key instances within a bag.

2.2 Attention-Based Multiple Instance Learning

A soft-attention mechanism offers an adaptive and reliable alternative to pooling operators and which can also be trained in an end-to-end manner. The objective here is to use instance embeddings $H = \{h_1, ..., h_N\}$ and learn a set of attention weights $A = \{a_1, ..., a_N\}$. Further, we compute the weighted sum of instance embeddings and attention weights to obtain bag-level predictions:

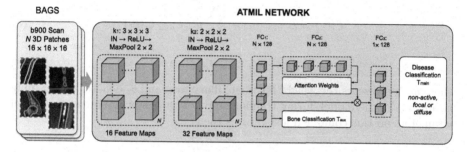

Fig. 2. Illustration of the proposed ATMIL framework architecture. Bags of 3D instances (from segmented bone regions) are passed through convolutional (k1 & k2) and fully connected layers (fc1, fc2 & fc3). The network provides a disease classification (T_{main}), an auxiliary bone classification (T_{aux}) and attention maps.

$$z = \sum_{n=1}^{N} a_n h_n \qquad \text{with} \qquad a_n = \frac{\exp\{u^T \tanh(V h_n^T)\}}{\sum_{j=1}^{N} \exp\{u^T \tanh(V h_j^T)\}} \qquad (2)$$

where u and V are model parameters. The attention weights also represent the importance of each instance in a bag, which we later use to compute attention maps on the WB-MR image level (as further discussed in the Results section).

2.3 Adaptive Reweighting of Auxiliary Task

Transferring knowledge from a related auxiliary task can provide additional supervision for the main task using the same data. It has been previously observed that multi-task learning with shared feature representation offers improved generalization [7] and overcome data inefficiency in both supervised and reinforcement learning tasks [13]. The most straightforward way for optimizing MTL parameters is to combine loss functions uniformly or use a predefined hyperparameter coupled with some annealing factor (e.g. based on epochs) to reduce the influence of auxiliary task as the training proceeds [21]. However, while optimizing the weights for MTL, it is not obvious at what stage of the training process and how much an auxiliary task may assist the main task. Hence, quantifying the benefit of an auxiliary task is highly relevant for MIL.

Our main task (T_{main}) is to predict disease labels on image-level and we employ an auxiliary task (T_{aux}) to predict anatomical labels for bone regions (obtained from the predicted segmentation maps). The corresponding loss terms for both tasks are $\mathcal{L}_{main}$ and $\mathcal{L}_{aux}$, respectively. The proposed framework has a shared backbone with learning parameters θ and a separate head for each task. The objective is to jointly optimize model weights for the main task while leveraging the auxiliary task for additional supervision to learn more generalized features. Our framework's total loss is formalized as:

$$\mathcal{L}(\theta_t) = \mathcal{L}_{main}(\theta_t) + \sum_{m=1}^{M} w_m \mathcal{L}_{aux,m}(\theta_t) \tag{3}$$

$$\theta_{t+1} = \theta_t - \alpha_t \nabla_{\theta_t} \mathcal{L}(\theta_t) \tag{4}$$

where w_m represents the weight for m auxiliary task and θ_t denotes the learning parameters of the model during training step t. The weight update of the model parameters on this combined loss term is then:

$$\theta_t \leftarrow \theta_{t-1} - \alpha_t(-\nabla \log p(T_{main}|\theta_{t-1}) - \sum_{m=1}^{M} w_m \nabla \log p(T_{aux,m}|\theta_{t-1})) \tag{5}$$

One way of estimating w_m is to minimize the Fisher divergence between gradients of our T_{main} and T_{aux}. This has been reported to be beneficial for semi- and unsupervised learning [22] and offers better divergence than Kullback-Leibler [24]. Higher values of w_m correspond to T_{aux} which are more advantageous for T_{main} or where parameter distributions of both tasks are similar.

$$w \leftarrow w - \beta \nabla_w ||\nabla \log p(T_{main}|\theta_t) - \sum_{m=1}^{M} w_m \nabla \log p(T_{aux,m}|\theta_t)||_2^2 \tag{6}$$

with learning rate β for the adaptive task weight. For MM disease classification, we are using a single auxiliary task (bone labels) by setting $M = 1$.

3   Data and Implementation Details

Morpho-MNIST (proof-of-concept). The Morpho-MNIST [2] data set contains morphologically perturbed digits. With randomly varying intensity and thickness transformation we aim to mimic MM disease patterns (Fig. 3). These include plain digits with no transformations (healthy), fragmented digits that are split into several components (inactive), local thickness (focal) as well as combined global and local thickness (diffuse).

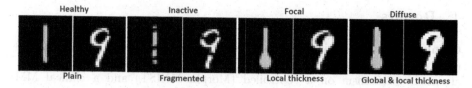

Fig. 3. Examples from the Morpho-MNIST Data Set. Digits were transformed to fragments, local thickness and global & local thickness to imitate inactive, focal and diffuse MM classes, respectively.

WB-MRI (real Clinical Data). Images were collected at The Royal Marsden Hospital and Imperial College Healthcare NHS Trust either on a Siemens MAGNETOM Aera or Avanto scanner. Structural (Dixon: fat & water images) and diffusion (b $= 50, 900$ s/mm^2) WB-MRI data were acquired for 176 MM patients at multiple time points. The total data set of 304 scans comprises 25 healthy, 52 inactive, 110 focal as well as 117 diffuse lesion scans. The healthy and inactive classes were coined 'non-active', and focal and diffuse classes refer to 'active' disease patterns. Our analysis made use of 75 manually annotated fat and water scans for bone segmentation and focused on b900 WB-MRI for disease classification, as higher b-values highlight MM [17,18]. All images were resampled to 2 mm isotropic voxel space and standardized via Nyul's histogram matching [19] to account for scanner differences.

Bone Segmentation. DeepMedic [10] was trained and validated on 55 and 20 subjects, respectively, to segment 18 bone structures from fat and water scans (three pathways, RMSprop, learning rate $= 0.001$, momentum $= 0.6$, patch size $= 37 \times 37 \times 37$ voxels, L1 and L2 regularization, standard scaling of input images). To remove spurious predictions, segmentations were corrected via comparison to an atlas based segmentation, positioning to the midsagittal plane and morphological operators. Final segmentations on validation data showed average Dice scores per bone class between 0.690 (sacrum) and 0.919 (left femur). A new model was trained on all 75 annotated scans and applied to all 304 scans. Segmentations were again corrected as described above and dilated to be more inclusive.

Disease Classification. We sampled 3D patches of size $16 \times 16 \times 16$ that overlapped with the bone segmentation by 50%, yielding on average $\approx$755 instances per WB MR image. The ATMIL network was trained with Adam optimizer [11] with learning rate $\alpha = 0.0005$, a decaying factor of 10 (applied after 100 epochs), and adaptive task weight $\beta = 0.05$. We trained the ATMIL network for 200 and 1000 epochs for Morpho-MNIST and WB-MRI, respectively. Random data augmentations (cropping, rotating, horizontally flipping, elastically deformation and Gaussian smoothing ($0.0 < \sigma \leq 1.0$)) was applied to WB-MRI. Further details can be found online: https://github.com/biomedia-mira/atmil.

4   Results

Comparative Analysis. The main objectives for reporting this analysis is (a) to investigate how the proposed framework performs with and without attention mechanism on a controlled (Morpho-MNIST) and a clinical MM (WB-MRI) dataset, (b) how different task reweighting approaches could provide gradient directions that yield additional supervision to the main task. For comparative analysis, we used the following auxiliary task reweighting methods (I) uniform (same weights for both losses), (II) weighted loss (WL) [21], $(1 - \gamma^\eta)\nabla \log p(T_{main}|\theta) + \gamma^\eta \nabla \log p(T_{aux}|\theta)$ where γ represents a predefined weights ($\gamma = 0.5$) and η denotes number of epochs, (III) GradNorm (GN) [3] to balance both tasks' gradient norm, (IV) AdaLoss (AL) [8] performs log operator around the auxiliary task, (V) CosineSim (CS) [5], the auxiliary task only contribute if having positive cosine similarity, $\cos(\nabla \log p(T_{main}|\theta), \nabla \log p(T_{aux}|\theta))$ (VI) OL-AUX (OA) [13] assigns high weight based on inner product of both tasks, $\nabla \log p(T_{main}|\theta)\nabla \log p(T_{aux}|\theta)^T$. We perform three repetitions per experiments for the clinical MM data and report the average performance and standard error of the mean. Table 1 reports the disease classification accuracy on Morpho-MNIST data (the auxiliary task was to learn if a given image contains more than one connected component or not) and, Table 2 presents results for 3D WB-MRI when comparing with different task reweighting approaches. For a fair analysis, we randomly split our data on patient-level by selecting around 75% for training, 5% for validation, and 20% for the final testing. The test data contains equal number of samples from each class (20 WB-MR scans/class). In the ATMIL setting, we are only concerned with the performance of our main task. Evidently, auxiliary task reweighting offers better generalization as compared to the uniform summation of loss terms (uniform) and other weighting methods.

Table 1. Classification accuracy on Morpho-MNIST data. Each bag contains 100 instances and the test data consists of 1000 bags with equal representation of positive (focal, diffuse, healthy, inactive) and negative (healthy, inactive) bags.

Number of training bags	100	150	200	300	500
MIL + max	0.778	0.782	0.826	0.91	0.947
MIL + att	0.805	0.786	0.857	0.975	0.981
MIL + att + uniform	0.863	0.923	0.936	**0.985**	0.983
MIL + att + WL [21]	0.817	0.940	0.943	0.978	0.974
MIL + att + GN [3]	0.844	0.916	0.968	0.969	0.985
MIL + att + CS [5]	0.824	0.939	0.964	0.983	0.985
MIL + att + AL [8]	0.973	0.923	**0.981**	0.975	0.986
MIL + att + OA [13]	0.931	0.953	0.963	0.982	0.987
ATMIL (ours)	**0.975**	**0.960**	0.951	0.976	**0.989**

Table 2. Classification results on 3D WB-MRI of MM patients. Experiments were performed 3 times and average ± standard error of the mean for specificity, sensitivity, and F1-score (macro-average) are reported.

Method	Specificity	Sensitivity	F1-score
MIL + max	0.753 ± 0.023	0.527 ± 0.023	0.539 ± 0.012
MIL + att	0.844 ± 0.013	0.689 ± 0.027	0.702 ± 0.027
MIL + att + uniform	0.836 ± 0.010	0.672 ± 0.020	0.689 ± 0.021
MIL + att + WL [21]	0.839 ± 0.010	0.678 ± 0.019	0.686 ± 0.018
MIL + att + GN [3]	0.861 ± 0.010	0.722 ± 0.014	0.727 ± 0.019
MIL + att + CS [5]	0.842 ± 0.010	0.683 ± 0.095	0.712 ± 0.010
MIL + att + AL [8]	0.833 ± 0.010	0.667 ± 0.018	0.671 ± 0.017
MIL + att + OA [13]	0.850 ± 0.010	0.701 ± 0.019	0.702 ± 0.018
ATMIL (ours)	**0.869 ± 0.010**	**0.737 ± 0.016**	**0.744 ± 0.019**

CosineSim [5] acts as a filter that only combines both loss terms when they follow the same direction and AdaLoss [8] simply wraps the auxiliary loss term with a log operator. Performance is comparable on Morpho-MNIST, but less good on clinical data. GradNorm [3] is a gradient norm balancer and shows relatively stable performance. The proposed ATMIL approach outperforms other methods on both data sets, and most importantly on the real clinical application.

Qualitative Assessment. Figure 4 shows qualitative, visual results for the attention maps generated by our approach. Anatomical regions that are relevant for triggering the bag label prediction are assigned high weights. We show four example patients for which different disease patterns were reported during clinical assessment. The attention maps seem to reflect well how much a scan was affected by MM. It is important to highlight that our framework was trained entirely with patient-level disease labels, and yet, is capable of highlighting burden of disease. The clinical utility of attention maps remains to be evaluated, but they may provide an interpretable output for whole-image disease classification, and could potentially even directly support radiological reading when used as a visual guide. This could be particularly useful for less experienced readers, and may generally reduce reading time of WB-MRI.

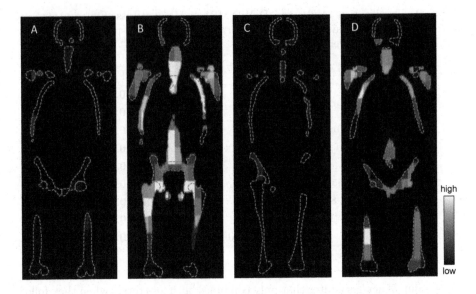

Fig. 4. Attention maps generated by our ATMIL network. Examples of patients with different MM reports: low (A) and high (B) occurrence of diffuse lesions & low (C) and high (D) count of bones with focal lesions. Grey outlines show segmented bones.

5   Conclusion

In this work, we proposed an MIL framework that leverages additional supervision from an auxiliary task using the same data. The presented method provides an effective way for adjusting the weight (to measure the usefulness) of the auxiliary task, minimizing the divergence between auxiliary and main tasks in an MIL setting. In future work it may be worth considering multiple related auxiliary tasks. As a real-world use case, we report results on WB-MRI of MM patients for which only patient-level disease labels were used during training. Attention weights enable the model to identify and localize instances corresponding to bone regions that likely show signs of disease. We believe the proposed approach carries the potential to solve other problems where only image-level labels are available and where data scarcity is a major challenge for effective training.

Acknowledgements. This work is supported by the National Institute of Health Research (EME Project: 16/68/34). SW is funded by the UKRI London Medical Imaging & Artificial Intelligence Centre for Value Based Healthcare. TB and AR are supported by the Imperial NIHR BRC and the Imperial CRUK Centre.

References

1. Campanella, G., et al.: Clinical-grade computational pathology using weakly supervised deep learning on whole slide images. Nat. Med. **25**(8), 1301–1309 (2019)
2. Castro, D.C., Tan, J., Kainz, B., Konukoglu, E., Glocker, B.: Morpho-mnist: quantitative assessment and diagnostics for representation learning. J. Mach. Learn. Res. **20**(178), 1–29 (2019)
3. Chen, Z., Badrinarayanan, V., Lee, C.Y., Rabinovich, A.: Gradnorm: gradient normalization for adaptive loss balancing in deep multitask networks. In: International Conference on Machine Learning, pp. 794–803. PMLR (2018)
4. Chikontwe, P., Kim, M., Nam, S.J., Go, H., Park, S.H.: Multiple instance learning with center embeddings for histopathology classification. In: Martel, A.L., et al. (eds.) MICCAI 2020. LNCS, vol. 12265, pp. 519–528. Springer, Cham (2020). https://doi.org/10.1007/978-3-030-59722-1_50
5. Du, Y., Czarnecki, W.M., Jayakumar, S.M., Farajtabar, M., Pascanu, R., Lakshminarayanan, B.: Adapting auxiliary losses using gradient similarity. arXiv preprint arXiv:1812.02224 (2018)
6. Fernando, B., Bilen, H.: Deep multiple instance learning with gaussian weighting (2019)
7. Girshick, R.: Fast r-cnn. In: Proceedings of the IEEE International Conference on Computer Vision, pp. 1440–1448 (2015)
8. Hu, H., Dey, D., Hebert, M., Bagnell, J.A.: Learning anytime predictions in neural networks via adaptive loss balancing. Proc. AAAI Conf. Artif. Intell. **33**, 3812–3821 (2019)
9. Ilse, M., Tomczak, J., Welling, M.: Attention-based deep multiple instance learning. In: International conference on machine learning, pp. 2127–2136. PMLR (2018)
10. Kamnitsas, K., et al.: Efficient multi-scale 3d cnn with fully connected crf for accurate brain lesion segmentation. Med. Image Anal. **36**, 61–78 (2017)
11. Kingma, D.P., Ba, J.: Adam: a method for stochastic optimization. arXiv preprint arXiv:1412.6980 (2014)
12. Kumar, S.K., et al.: Multiple myeloma. Nat. Rev. Dis. Primers **3**(1), 17046 (2017)
13. Lin, X., Baweja, H.S., Kantor, G., Held, D.: Adaptive auxiliary task weighting for reinforcement learning. Advances in Neural Information Processing Systems, vol. 32 (2019)
14. Lu, M.Y., Williamson, D.F., Chen, T.Y., Chen, R.J., Barbieri, M., Mahmood, F.: Data efficient and weakly supervised computational pathology on whole slide images. arXiv preprint arXiv:2004.09666 (2020)
15. Mesguich, C., et al.: Improved 18-fdg pet/ct diagnosis of multiple myeloma diffuse disease by radiomics analysis. Nuclear Medicine Communications (2021)
16. Messiou, C., et al.: Guidelines for acquisition, interpretation, and reporting of whole-body mri in myeloma: myeloma response assessment and diagnosis system (my-rads). Radiology **291**(1), 5–13 (2019)
17. Messiou, C., Kaiser, M.: Whole body diffusion weighted mri-a new view of myeloma. Br. J. Haematol. **171**(1), 29–37 (2015)
18. Messiou, C., Kaiser, M.: Whole-body imaging in multiple myeloma. Magn. Reson. Imaging Clin. **26**(4), 509–525 (2018)
19. Nyúl, L.G., Udupa, J.K.: On standardizing the mr image intensity scale. Magn. Reson. Med. Official J. Int. Soc. Magn. Reson. Med. **42**(6), 1072–1081 (1999)
20. Ozdemir, O., Russell, R.L., Berlin, A.A.: A 3d probabilistic deep learning system for detection and diagnosis of lung cancer using low-dose ct scans. IEEE Trans. Med. Imaging **39**(5), 1419–1429 (2019)

21. Sadafi, A., et al.: Attention based multiple instance learning for classification of blood cell disorders. In: Martel, A.L., et al. (eds.) MICCAI 2020. LNCS, vol. 12265, pp. 246–256. Springer, Cham (2020). https://doi.org/10.1007/978-3-030-59722-1_24

22. Shi, B., Hoffman, J., Saenko, K., Darrell, T., Xu, H.: Auxiliary task reweighting for minimum-data learning. arXiv preprint arXiv:2010.08244 (2020)

23. Xu, L., et al.: Automated whole-body bone lesion detection for multiple myeloma on 68ga-pentixafor pet/ct imaging using deep learning methods. Contrast Media Molecular Imaging 2018 (2018)

24. Yang, Y., Martin, R., Bondell, H.: Variational approximations using fisher divergence. arXiv preprint arXiv:1905.05284 (2019)

Correction to: Highly Reproducible Whole Brain Parcellation in Individuals via Voxel Annotation with Fiber Clusters

Ye Wu, Sahar Ahmad, and Pew-Thian Yap

Correction to:
Chapter "Highly Reproducible Whole Brain Parcellation
in Individuals via Voxel Annotation with Fiber Clusters"
in: M. de Bruijne et al. (Eds.): *Medical Image Computing*
and Computer Assisted Intervention – MICCAI 2021,
LNCS 12907, https://doi.org/10.1007/978-3-030-87234-2_45

In the published paper (https://link.springer.com/chapter/10.1007/978-3-030-87234-2_45) the authors failed to acknowledge one of the supporting grants.

The relevant note should be changed from "This work was supported in part by United States National Institutes of Health (NIH) grants MH125479 and EB006733." to "This work was supported in part by United States National Institutes of Health (NIH) grants EB008374, MH125479, and EB006733."

The updated version of this chapter can be found at
https://doi.org/10.1007/978-3-030-87234-2_45

Author Index

Printed in the United States
by Baker & Taylor Publisher Services